Fourth Edition

MAP USE & ANALYSIS

John Campbell

JL

Boston Burr Ridge, IL Dubuque, IA Madison, WI New York San Francisco St. Louis
Bangkok Bogotá Caracas Lisbon London Madrid
Mexico City Milan New Delhi Seoul Singapore Sydney Taipei Toronto

McGraw-Hill Higher Education

A Division of The ***McGraw-Hill*** *Companies*

MAP USE & ANALYSIS, FOURTH EDITION

Published by McGraw-Hill, an imprint of The McGraw-Hill Companies, Inc., 1221 Avenue of the Americas, New York, NY 10020.

 This book is printed on recycled, acid-free paper containing 10% postconsumer waste.

10 11 12 13 14 QPD/QPD 13 12 11 10 09

ISBN: 978-0-07-303748-6
MHID: 0-07-303748-6

Vice president and editor-in-chief: *Kevin T. Kane*
Publisher: *JP Lenney*
Sponsoring editor: *Robert Smith*
Developmental editor: *Kassi Radomski*
Editorial assistant: *Jenni Lang*
Senior marketing manager: *Lisa L. Gottschalk*
Editing associate: *Joyce Watters*
Senior production supervisor: *Sandra Hahn*
Designer: *K. Wayne Harms*
Cover designer: *Anis Leung*
Cover images: *Wides & Holl/FPG International*
Photo research coordinator: *John C. Leland*
Compositor: *Shepherd, Inc.*
Typeface: *10/12 Adobe Garamond*
Printer: *Quebecor Printing Book Group/Dubuque, IA*

Library of Congress Cataloging-in-Publication Data

Campbell, John, 1932–
Map use & analysis / John Campbell. — 4th ed.
p. cm.
Includes bibliographical references and index.
ISBN 0–07–303748–6
1. Maps. I. Title: Map use & analysis. II. Title.

GA151 .C25 2001
912—dc21

00–020989
CIP

www.mhhe.com

For Hazel, as always

Contents

Preface

This book is designed to serve as an introduction to the fascinating world of maps. It explains how to use maps to obtain information about a wide variety of topics. Throughout the book, maps are viewed in a broad framework. Thus, the discussion includes mental maps, aerial photographs, remotely sensed images, computer-assisted cartography, and geographical information systems, in addition to traditional printed maps. The writing style is neither formalistic nor casual, with an emphasis on clarity of explanation. The discussions assume that the reader has no specific prior knowledge of the topic, so that even novice map users can understand and use the information and techniques presented.

Chapter 1 explains why maps are useful and lists the many types of maps available. Then, mapping processes are discussed in Chapter 2 because an understanding of how maps are produced provides a necessary basis for their proper use. For similar reasons, important aspects of the size and shape of the earth are reviewed. Chapter 3 surveys the characteristics of map projections, especially from the standpoint of the appropriateness of certain projections for specific purposes.

A variety of locational and land-ownership systems, which are essential for specifying location and for describing land ownerships, are explained in Chapter 4. These include latitude and longitude, the Universal Transverse Mercator and State Plane Coordinate systems, metes-and-bounds surveys, the U.S. Public Land Survey System, and the similar Canada Land Survey System. The link between scale and generalization is discussed in Chapter 5, as are methods of scale determination.

Because maps are used for measuring distances and areas, Chapter 6 discusses methods for carrying out these measurements. Techniques used in route selection and navigation on land and water and in the air are examined in Chapter 7.

Chapter 8 examines how terrain is represented on maps. Contour interpretation techniques and techniques for producing profiles and determining slopes are provided in Chapter 9. Chapter 10 follows with extensive illustrations of a variety of landform types. The examples shown are, in large part, selected from the definitive U.S. Geological Survey collection *A Set of 100 Topographic Maps Illustrating Specified Physiographic Features*.

More abstract topics are introduced in the following three chapters. Chapter 11 discusses methods by which thematic maps are used to convey a variety of qualitative and quantitative information. Chapter 12 surveys the characteristics of mapped distributions, such as the shape and pattern of point distributions. The analysis of stream patterns and transportation systems through the abstract concepts of networks and trees is covered in Chapter 13.

Cartograms are surveyed in Chapter 14. This chapter also discusses a variety of special-purpose maps, including highway and street maps, weather and climate maps, geologic maps, maps of the past, maps of the moon and the planets, and maps in journalism and on television. Chapter 15 is devoted to the interpretation of graphs. Graphs are important in their own right, but they are also extensively used as map symbols and as map supplements. Map misuse and the use of maps as powerful propaganda tools are examined in Chapter 16.

Remote-sensing techniques are becoming increasingly important and accessible to map users. Therefore, aerial photography as well as remote sensing from space are explained in Chapters 17 and 18.

Chapter 19 describes modern techniques of computer-assisted cartography, and Chapter 20 discusses applications of digital maps, including electronic atlases, electronic road maps, and automated automobile navigation. Geographic information systems are examined in Chapter 21. Both computer-assisted cartography and geographic information systems are increasingly used to produce unique products and analyses of interest to map users.

At appropriate points in the text, supplementary treatments of a variety of topics are provided, including discussions of the analemma, dates and time zones, units of measurement and map scales, the National Map Accuracy Standards, and levels of measurement.

A survey of U.S. and Canadian map producers and map sources is provided in Appendix A. This appendix also includes comments on major map collections that map users will want to visit when their travels permit. Next, Appendix B briefly discusses the special problems often encountered when dealing with foreign maps. Other appendixes discuss background topics, including copyright protection, the use of the magnetic compass in the field, map storage and cataloging systems, and the British National Grid.

New to This Edition

A number of improvements were made to this fourth edition. The more significant ones are listed here for your convenience.

- All of the chapters were carefully reviewed for accuracy and to make their content as up-to-date as possible. Similarly, the addresses and telephone numbers of mapping companies and agencies were verified and updated. (Some companies and agencies were removed because they no longer emphasize mapping activities.) In addition, recent changes in the organization of mapping activities in Canada were incorporated into Appendix A.
- E-mail addresses and World Wide Web URLs were removed from the text due to the frequency with which they change. Instead, they will be placed on the book's World Wide Web site, (www.mhhe.com/earthsci/geography/campbell/), and will be regularly updated there.
- Overall, there are 17 new or revised illustrations in the text. These include a new radar image and a 1-meter resolution remote-sensing image, which only recently became available.
- Chapter 1 includes the addition of a discussion of common map elements, including typography, neatline, scale, orientation, and types of insets.
- Chapter 2 now includes additional coverage of methods of longitude determination, including the importance of John Harrison and his development of the chronometer.
- Chapter 3 was extensively rewritten with the aim of reinforcing an appreciation for the wide range of distortions associated with the map projection process and the importance of projection selection in map use. Four new illustrations were created to support this presentation.
- A brief discussion of the use of maps as tools for analyzing the culture landscape was added to Chapter 4.
- In Chapter 5, the origin of the acre as a measure of land area in the U.S. Public Land Survey is explored.
- A discussion of the importance of map projection and scale on the accuracy of map-based measurements was added to Chapter 6.
- A description of direction determination methods, including cardinal compass directions, points of the compass, and degrees, was added to Chapter 7. Also included in this chapter is an update regarding the various ways of distributing the *Notice to Mariners* publications.
- In Chapter 9, a discussion of measures of surface inclination, including slope and gradient, was added.
- A description of fire insurance maps, especially Sanborn Maps, was added to Chapter 14.
- Chapter 18 now includes updated information on the status of Landsat, including Landsat 7 and the Enhanced Thematic Mapper Plus system. Also in Chapter 18, the description of SPOT was updated to include SPOT 4. In addition, a new color plate illustrating remote-sensing image classification was obtained for this chapter.
- Chapter 19 contains a revised discussion of the Geographic Names Information System (GNIS) and of TIGER®.
- Two changes were made specifically to assist students. (1) The information about map producers, map sources, and foreign maps (formerly Chapters 22 and 23) were changed to appendices. This information was always designed to serve simply as a reference source, and not to be learned in the same manner as the subject-matter chapters. At the same time, the agency and company names were changed from boldface to regular-face type and removed from the glossary. (2) Similarly, the names of landform types in Chapter 10 were removed from the glossary and are no longer shown in boldface type. Students are not expected to memorize the many landform types illustrated in the chapter. The goal of the chapter is simply to illustrate the usefulness of topographic maps in revealing information about the configuration of the earth's surface.

Acknowledgments

As has always been the case, many people have helped in various ways during the process of preparing the fourth edition. I thank all of them for their general support. I would especially like to thank the reviewers who offered many helpful suggestions for revisions to this edition: Andrew J. Bach, Western Washington University; Kent B. Barnes, Towson State University; Michael Camille, Northeast Louisiana University; Richard A. Crooker, Kutztown University; Charles Ehlschlaeger, Hunter College; Robert M. Hordon, Rutgers University; Thomas W. Paradis, Northern Arizona University; and Charles Roberts, Florida Atlantic University. These reviewers' recommendations were instrumental in my revision decisions. I followed their advice as fully as I could, although some useful ideas could not realistically be implemented.

Daryl Bruflodt and Bob Smith, both editors at McGraw-Hill, generously supported the preparation of a new edition, and Kassi Radomski, freelance editor, provided helpful advice during the revision process. Joyce Watters and the rest of the production team at McGraw-Hill also contributed their special skills to the publication of this book. Without the expertise of all these people, the project would have been impossible.

Finally, I must thank my wife Hazel for her assistance in important parts of the project, and for her continuing encouragement and support.

John Campbell
Racine, Wisconsin

CHAPTER

1 INTRODUCTION

WHY USE A MAP?

Maps are an ancient invention, probably having existed for over four thousand years. Remnants of clay-tablet maps dating to about 2200 B.C. have been found (Figure 1.1). In addition, primitive, maplike scratchings in the sand, as well as maps in other forms, likely existed for many millennia before that.

These early maps performed useful functions for their creators, functions that maps still perform for us today. For example, maps are especially effective devices for recording and communicating information about the environment. Most importantly, they clearly preserve the locational attributes of that information; that is, they show the relationships between one feature and another. Not only do they show that there is a forest outside of town, but they also indicate the extent and limits of both forest and town. This is something that other forms of recording and communicating information, such as written descriptions, tables, and graphs, generally do not do as effectively or efficiently. Furthermore, maps can also indicate distances and directions between locations, or the areas occupied by different types of land uses or features.

In more recent times, maps have offered other benefits. For example, they are useful for determining the patterns formed by many types of distributions on the earth's surface. The arrangements of cultural features, such as the spacing between towns or the patterns of roads and other transportation facilities, can be analyzed using maps. Similarly, the alignments of physical

Figure 1.1 Clay tablet map, c. 2200 B.C. This is the oldest known map, discovered in the ruins of GA.SUR (modern Yorghan Tepe) in northern Mesopotamia by the American School in Baghdad and Harvard University. The map is oriented with east at the top. Two rivers join at the left of the map (north) and flow toward the upper right corner (northeast). Three hundred acres of arable land are shown just below the confluence of the rivers. Localities of special interest are designated by circles within which the names are written. (Size 7.6 × 6.5 cm.)

Source: From the collections of the Semitic Museum, Harvard University. Used with permission.

features, such as streams and lakes and the crests of hills and the floors of valleys, can be determined. The arrangements of such features can be explored, organized, and analyzed by visual and statistical methods, using maps as the database. These techniques have helped investigators to generate hypotheses about why the patterns are as they are. These investigations have suggested, for example, that certain economic relationships underlie the locational pattern of settlement and that the pattern of some physical features affects the location of specific human activities. Similarly, regular physical relationships underlie the patterns of stream erosion, the arrangement of drumlins in glaciated regions, and the alignment of river tributaries, to name just a few of the vast multitude of possible examples.

Maps are not limited only to showing information about physical and cultural features on the earth's surface. They are also used to show distributions of more abstract features, such as the flow of trade, the use of communications, the extent of political influence, or the areas occupied by peoples of various races, languages, or religions. They provide a major source of historical documentation and are used for regional planning and property-assessment purposes. In addition, features found on the surface of extraterrestrial bodies, such as the moon and the planets, have been mapped, as have imaginary environments used as the settings for works of fiction.

Regardless of the topic or the locale, maps frequently play a role in providing information and explanations regarding topics of interest. The rest of this introductory chapter examines just what constitutes a map.

WHAT IS A MAP?

The answer to the question, What is a map? may seem rather obvious to someone who is interested enough in maps to read this book. Maps, after all, are neatly drawn, bird's-eye views of the earth's surface. They show where places and things are located and help us find our way from one place to another. We are all familiar with many kinds of maps—from those confusingly folded, multicolored road maps that are stuffed into the glove compartments of our automobiles; to the beautifully drawn illustrations in books and atlases that provide information about the distribution of climates, vegetation types, languages, income, political patterns, and myriad other topics; to the topographic sheets whose intricate contours and drainage patterns serve as guides for our hiking or fishing expeditions. A visit to the map collection of your community or university library will convince you that a tremendous variety of maps is available. The International Cartographic Association has defined **cartography** as the art, science, and technology of making such maps, together with their study as scientific documents and works of art. This book deals with both aspects of the definition, to some extent, but does not deal with the specifics of traditional map production processes.

In recent years, new methods of gathering, analyzing, and presenting information about the earth have been introduced, including aerial photography, satellite-based remote sensing, and computer methods. These developments have made it necessary to recognize that the conventional maps mentioned previously are only part of the contemporary map picture and that the definition of maps must include an extremely broad range of "products." Some of these products (such as road maps), show topics that are physical, some (such as language maps), show more social or cultural topics, and some (such as maps of income levels), show even more abstract subjects. Many products, such as aerial photographs or satellite images, serve purposes similar to those served by maps, even though they differ from conventional maps. It also has been suggested that some "maps" are not even visible. Invisible maps may exist only as bits of electronic information stored in the memory of a computer or, even more abstractly, only in our minds. Finally, maps are not limited to representing information about the earth. They are equally useful, for example, for showing features found on the moon or the other planets, and they can present patterns that exist on the ground (topographic maps), under the ground (geologic maps), or above the ground (weather maps).

Given the variety of possible maps, this text uses a broad, flexible definition: A **map** is any concrete or abstract representation of the features that occur on or near the surface of the earth or other celestial bodies. The great variety of map concepts that fall within this definition may be classified as either (1) real maps or (2) virtual maps.[1]

A **real map (cartographic map)** is any tangible map product that has a permanent form and that can be directly viewed (often referred to as *hard copy*). Conventional drawn or printed products (traditionally called *maps*) fit into this category, as do maplike aerial photographic products or the end product of some other type of remote sensing, maps produced using devices controlled by computers, block diagrams and similar drawings, and relief models and globes constructed to represent some part or all of the earth's surface.

Virtual maps are related to real maps in one way or another and have qualities that allow them to be converted into real maps. They are divided into three types.

[1]The classes used here are simplified from those proposed by Harold Moellering in "Real Maps, Virtual Maps, and Interactive Cartography," in *Spatial Statistics and Models*, ed. Gary L. Gaile and Cort J. Willmott (Dordrecht, Holland: D. Reidel Publishing, 1984), 109–32.

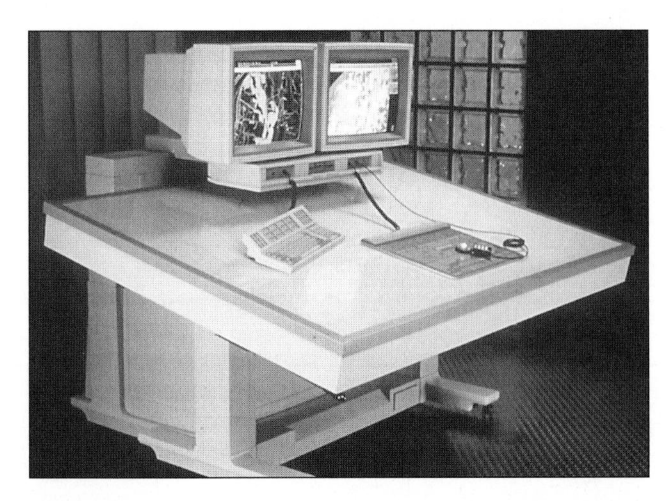

Figure 1.2 Computer workstation, with map images displayed on the screens of two monitors.
Source: Courtesy of Intergraph Corporation.

One type of virtual map consists of images that can be directly viewed but are not permanent. A map image projected onto the screen of a computer monitor is an example (Figure 1.2). Such an image is real while the device is turned on, and the information that it presents is similar or identical to that of a real map of the same topic. The difference is that as soon as the monitor is turned off, the image vanishes.

A second type of virtual map consists simply of mental images that are in many ways the conceptual equivalent of a conventional printed map.

The third type of virtual map consists of information gathered by researchers in the field or obtained by remote-sensing methods about such topics as surface elevations, rock types, soils, ethnic types, income levels of inhabitants, types of crops, and names and locations of geographic features. Such geographic data are traditionally stored in written notes, books, or computer printouts, all of which can be directly viewed. Increasingly, however, they are stored in digital form in a computer memory or storage device, such as a magnetic disk. This form of map is common for remotely sensed information, which is often gathered and stored directly in digital form, but also frequently includes information converted from printed or written sources. Computer software and hardware systems permit this type of information to be seen in tabular form or in the form of a visual image.

All three types of virtual maps can be converted into visible, "real" map products.

While there is not complete agreement on the use of the term *virtual map,* the label does convey the idea that a map is simply a special format for the storage of geographic data. Whether the information is called *map data* or *virtual maps,* there is no doubt that it is an integral element of the mapping process. Virtual maps are discussed frequently in this book, beginning with the next section.

VIRTUAL MAPS

Mental Maps

As already noted, the current concept of maps must encompass much more than the artifacts traditionally called *maps.* Indeed, the most familiar but, at the same time, most unusual forms of maps are the virtual maps called *cognitive maps,* or *mental maps.*

Mental maps are images we have in our minds. These images provide us with an awareness of the location of places in the world, the relationships between places in terms of direction and distance, the size and characteristics of regions, and so on. Mental maps have been called "the environmental image, the generalized mental picture of the exterior physical world that is held by an individual."[2] Some mental maps are a kind of miniature map in the mind, sometimes rather vague and ill-formed but sometimes complete with accurate details. Other mental maps are more subtle and difficult to define because they are somewhat more abstract and are tailored to one's individual conceptions, experiences, and needs, even to the point of distortion. Indeed, it has been suggested that mental maps "are quite unlike [real] maps . . . because they are personal, fragmentary, incomplete, and presumably, frequently erroneous."[3]

One aspect of the mental map concept is easily illustrated. Simply visualize the route that you usually follow from your home to the shopping center, or recall the general outline of the United States and the locations of its cities and states. Consider how easily you move about in different environments with the help of your mental images. The framework that mental maps provide allows you to tell others about the route that you follow, and they may be able to visualize it as well. Furthermore, such a mental image provides a frame of reference to which you can add information on the basis of new experiences. A mental map can be converted into a more conventional real map by sketching its image on a piece of paper (Figure 1.3). We all do this when we want to guide our friends to our new house or to indicate to a stranger the best route to follow to see the local tourist attractions.

Another aspect of mental maps is illustrated by considering where in the country you would prefer to live if you were given a choice. You are almost sure to have a ready answer to the question. Certain regions,

[2]Kevin Lynch, *The Image of the City* (Cambridge, Mass.: M.I.T. Press, 1960), 4. Much of the discussion that follows is based on this book.
[3]John S. Keates, *Understanding Maps* (New York: John Wiley & Sons, Inc., A Halsted Press Book, 1982), 53.

Figure 1.3 Mental map images of the same neighborhood, drawn by three different boys: Dave, Ernest, and Ralph.
Source: From Peter Gould and Rodney White, *Mental Maps* (Pelican Books, 1974). Reprinted by permission.

states, or cities undoubtedly appeal to you, whereas others have absolutely no allure, and still others fall somewhere in between. Almost everyone, it seems, has somehow endowed different locations with images of relative attractiveness or repulsiveness. Such images are often shared with others who have been exposed to the same physical and cultural environments.

A number of considerations undoubtedly affect your feelings about where you would like to live. For example, what is the climate like? Is there a year-round monotony about it, or is there a variation between summer and winter weather, with the spring and fall transitions that many people find attractive and invigorating? If you are a sports buff, you may prefer a location that promises a long snow season to facilitate skiing, or a year-round warm climate suited to swimming and other water sports.

Other factors that you may consider include scenery: Is the view that meets the eye an unvarying, flat plain, or are there mountains or seashores to bring variety? How about pollution? Are the air and water clear, or is there likely to be contamination of either or both? Even the reputation of a state for having a good or bad government can enter into your mental rating.

Social contacts and preferences may also affect your mental image of the place you would like to live. Regional reputations differ with regard to the sociability and openness of the residents toward newcomers, and one may appeal to you more than another, depending upon your own social style. Economic conditions are also important. If an area has a reputation for economic growth and activity, it may appeal to you because it provides the potential for prosperity.

In the end, your mental image of a region is a composite of a whole range of considerations. Certainly, not all considerations are equally important. If one element has greater importance, you will give it greater weight than another less important element. Such weighting is not necessarily systematic, but the relative importance of the factors is taken into account in forming your mental map. Obviously, not everyone's mental map will be the same. All of the preferences are subjective, but they are nevertheless real.

How are these subjective images of desirability or repulsiveness formed? Partly by direct experience. You may have traveled in certain areas, for example, and formed impressions based on your experiences and observations. Even such direct experience must be recognized as subject to error. This is especially true if the period of observation is too short, or if it occurs under conditions that are different from those of your ordinary, day-to-day existence.

Other, less direct, factors may also enter into the formation of opinions about the relative desirability of different locales. You may be influenced by geography courses, news broadcasts, maps, television programs or movies, or conversations. If the information you obtain is inadequate, inaccurate, or otherwise misleading, your conclusions will be faulty. Although opinions based on indirect sources may not be as solid as those drawn from direct experience, they are no less influential.

Experimentation has shown that there are certain similarities among people's mental maps of where they would like to live, even though they find themselves at different locations. In Figure 1.4, some mental map concepts have been converted to more familiar map form. These examples summarize the spatial preferences of groups of students enrolled in universities located in different regions of the United States. The students were asked to rank the forty-eight contiguous states in terms of desirability (the higher the score, the greater the desirability).

The rating maps have general tendencies in common. For example, some parts of the country are consistently viewed as more attractive. Western California, central Colorado, and, to a lesser extent, northern New England fall into this category. Other areas, such as the South (especially Alabama and Mississippi) and the Dakotas, are generally seen as less attractive. On the other hand, the maps also show that people like areas with which they are familiar. In each case, the home territory of the raters tends to be viewed as relatively attractive. Alabamians, for example, have high regard for their home state and adjoining parts of Florida and Georgia. Minnesotans and Pennsylvanians have similar views of their home territories. Each of these ratings contrasts with the views of the other three groups. Californians also view western California as attractive, thus reinforcing the general trend. In addition, however, they extend their preference area farther north and south along the coast. You may find it interesting to compare these maps with your own preferences.

Digital Data

Increasingly, computer methods are being used in mapping and related fields. Computers are being applied to the processing and analysis of remote-sensing data, to the production of maps (called **computer-assisted cartography**), and to the operation of **geographic information systems.** These applications are discussed in Chapters 19 and 21.

Computer techniques can introduce a number of potential advantages, including speedier, easier, and more accurate map design and production. Other manipulations, including scale and projection changes and the statistical analysis of data, can also be more easily accomplished using computer techniques. Increased timeliness is also anticipated because updated information can be made immediately available to all users from a centralized database. Overall, computer methods should provide the cartographer with the flexibility to produce more innovative and effective maps.

Although the potential advantages of computer applications in cartography are not always achieved, progress is being made, and more frequent computer applications are inevitable in the future. Some of the results produced by computer methods will be indistinguishable from conventional maps. Other results, however, may be quite different, and some knowledge of the methods by which they were produced will help you to understand and evaluate them. For all of these reasons, portions of this book are devoted to computer applications in mapping.

REAL MAPS

Virtual maps have limitations that frequently make the use of more conventional real maps a necessity. It has been suggested that "one of the main reasons for making maps must be that 'mental maps' are inadequate as useful stores of locational information."[4] For this reason, much of this book deals with actual, physical map products and with concepts and measures related to them. The sections that follow briefly summarize the many types of maps and maplike products.[5] The topics introduced here are discussed in greater detail throughout the rest of the text.

Common Map Elements

Most real maps have a number of elements in common. These elements, some of which are discussed in detail in other parts of the book, are summarized here in an introductory fashion.

Typography

Typographical information (including titles, legends, names, and notes) is part of almost all maps. This type of information is crucial to the understanding of individual maps.

The title, and any subtitles it may contain, indicates the purpose for which the map was prepared. A properly worded **title** helps the map user by stating the subject of the map, the time period to which it applies, and other aspects of its content. The title of a map, therefore, should suggest whether the map is likely to be suitable for the purpose at hand.

Most maps also contain legends, sometimes in a special legend box. **Legends** show map symbols, along with an explanation of their meaning. Legends

[4]Keates, *Understanding Maps*, 53.
[5]Modified from Morris M. Thompson, *Maps for America*, 3d ed. (Washington, D.C.: U.S. Department of the Interior, U.S. Geological Survey, 1987), 15–17.

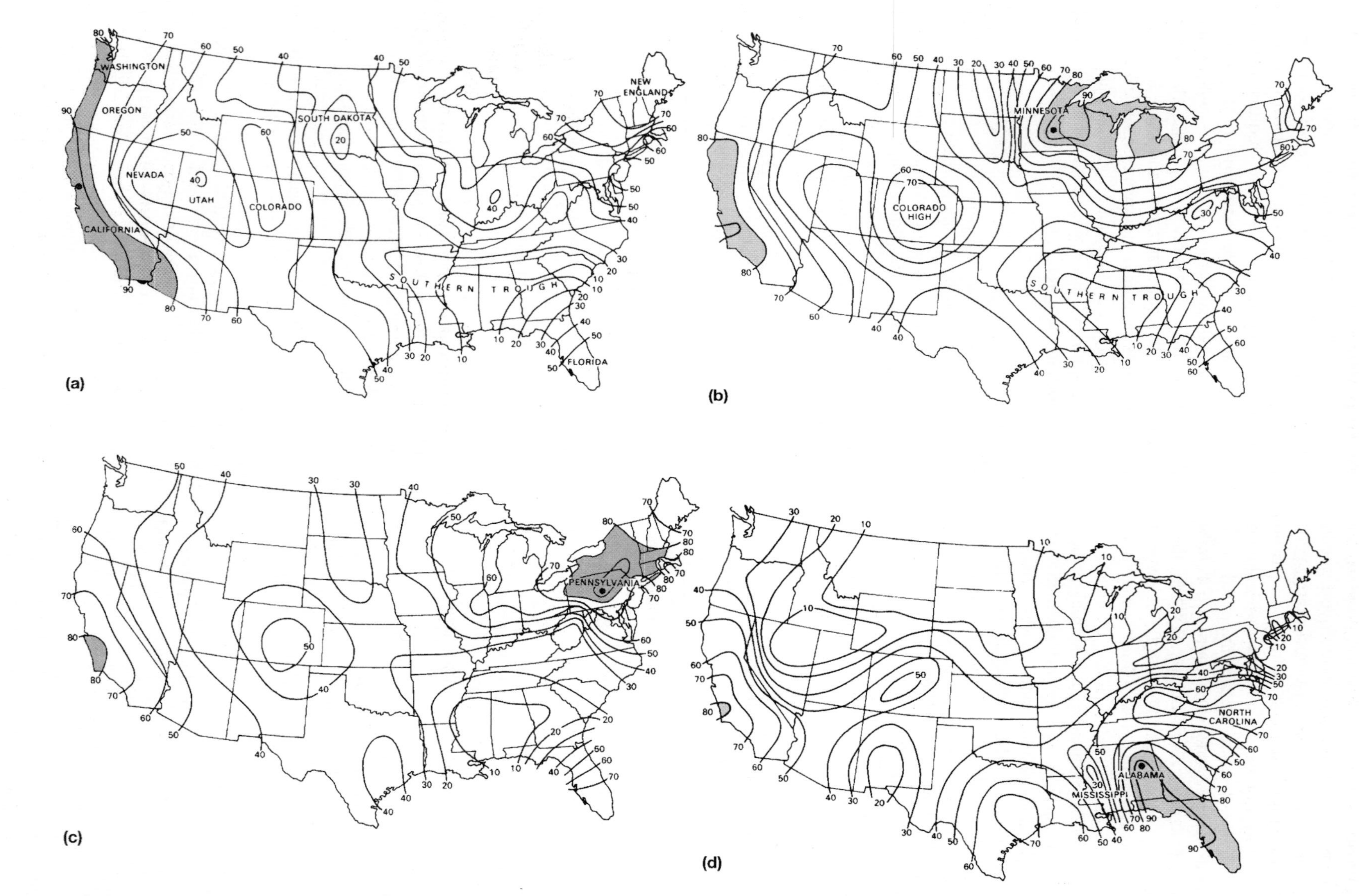

Figure 1.4 Mental maps representing students' views of the desirability of the forty-eight contiguous states. Views are from: (*a*) California, (*b*) Minnesota, (*c*) Pennsylvania, and (*d*) Alabama.

Source: Adapted from Peter Gould and Rodney White, *Mental Maps* (Pelican Books, 1974). Reprinted by permission.

are required because maps use symbols to convey their messages. These symbols may be simple lines, geometric shapes, more complex drawings such as proportional circles, or even graphs. They may also be in black and white or in different colors. Some common conventions for symbolization, such as the use of a fine black or blue line to represent a coastline or a river, are usually easily recognized and are unlikely to appear in a legend. Legends are often necessary, however, because there is no prescribed set of universally used map symbols. Therefore, the symbols used on one map may be very different from those used on others. Thus, a railroad line may be symbolized by a narrow black line with widely spaced cross-ticks on one map and by a dashed red line without cross-ticks on another. There is even wide variation in the way that common features, such as roads and boundaries, are presented on different maps. Because less conventional symbols often require explanation, it is important to consult the legend of a map before trying to use and interpret it. Any unique symbols used on the map should be shown in the legend. Methods of symbolization are discussed and illustrated in many parts of this book, but particularly in Chapters 7, 8, 11, 14, 15, and 16.

Other typographical information tells the map user the names of the feature types shown on the map. This includes the names of political or physical features and any other elements requiring identification. In addition, an explanation is often given for the source of the information shown on the map. The explanation usually takes the form of a note near the neatline or in the legend box. The absence of such information should make us question the validity of the map—is it simply based on the imagination of the cartographer, or is it based on reliable data, such as surveys or census? This type of concern applies whether the map shows physical features, political information, or statistical data.

Neatline

It is common to frame a map with a neatline or border (the two terms are often used interchangeably, but there is a slight difference between them). A **neatline** is simply a narrow line that frames the mapped area. In some cases, such as USGS topographic quadrangle maps, the neatline follows the lines of latitude and longitude. (On these maps, the area outside the neatline, called the map collar, contains the title, scale, declination diagram, and other supporting information.) In other cases, the neatline is simply a box. A border is very similar to a neatline but is often a design element and may consist of single or double lines or designs that are even more elaborate. Not all maps have neatlines or borders; some simply run off the edge of the page, in a style called a bleed. The presence or absence of a neatline or border is usually not crucial to the function of a map.

Scale

Scale is another feature that maps have in common. **Scale** is the ratio between the size of features on the map and the size of the same features on the ground. In some cases, the scale of the map is not crucial to understanding the message of the map, so an indication of scale may be omitted. In most cases, however, an indication of scale is helpful to the map user. The indication may take the form of a representative fraction, a word statement, or a graphic scale, as discussed in Chapter 5.

Orientation

The **orientation** of a map consists of the way that it is aligned, relative to the earth's surface. Although most maps are oriented with north at the top, this is not universally true. Therefore, the orientation of each map must be communicated to the map user. This is commonly done by showing a north arrow, as discussed in Chapter 7. In some cases, however, it may be misleading to include a north arrow, because of the characteristics of a particular map projection. Direction can still be indicated in such a case by including the latitude and longitude graticule or a locational grid, as discussed in Chapter 4.

Insets

Sometimes small additional maps are included as one or more **insets** in the main map. Insets take four general forms. The first type is an enlargement of a portion of the mapped area. For example, if the main map shows a country, the inset may show a detail, such as the street layout of the major city in the country (Figure 1.5a). The second type of inset is a locator map, which shows where the mapped region lies, in relation to a larger, better known region. For example, if the main map shows a country, the locator map might show where the country lies within a broader region (Figure 1.5b). Insets of a third type show areas related to the main map, such as remote islands that fall under the political control of the country shown in the main map (Figure 1.5c). Finally, a fourth type of inset provides additional information (Figure 1.5d). For example, the main map might be a political map of a country, showing its boundaries and other information. Insets on such a map might show the distribution of climatic regions, the distribution of farms, the distribution of population, the occurrence of industrial development, or any of myriad other important aspects of the country. Most supplemental insets are treated as individual maps, each with its own legend, scale, directional indication, and other features, if needed.

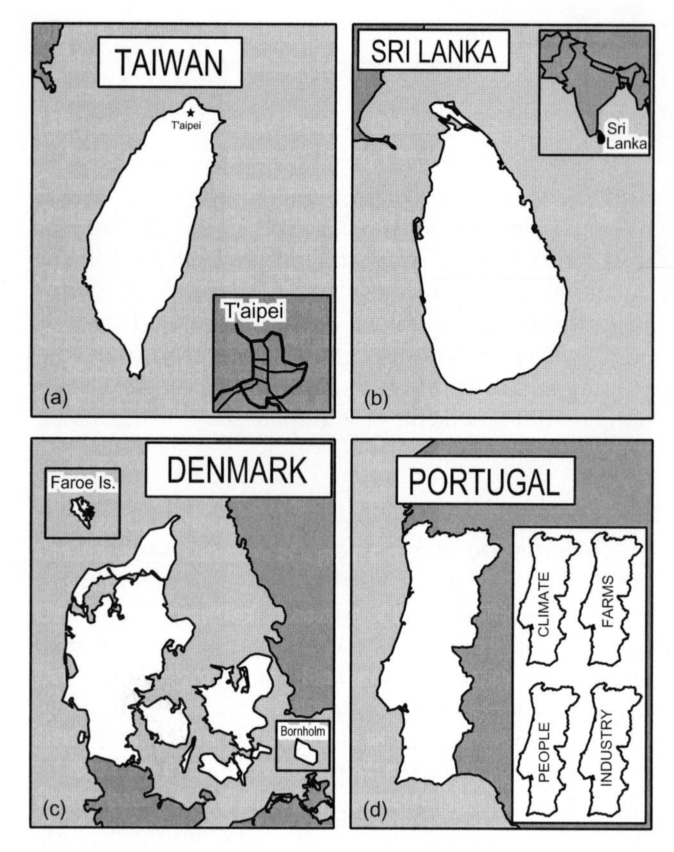

Figure 1.5 Types of insets. (*a*) Enlargement of important region. (*b*) Area locator. (*c*) Related areas. (*d*) Additional information.

Planimetric Maps

A variety of information about the many types of features found on the earth's surface is shown on maps. Maps that do not attempt to show the relief features in measurable form are perhaps the simplest type and are called **planimetric maps.**

Planimetric **base maps** are often used to provide the framework for thematic maps, which present information about some special subject (see Figure 1.13). Depending on the topic being presented, base maps range from very simple to very complex. Simple base maps are often adapted from existing maps and may contain only the shorelines and boundaries that outline a country or region. Other base maps may be more complex and may include rivers, political boundaries, transportation routes, and other information. The amount of detail shown on a base map depends on the purposes of the particular project. Base maps are often available in digital form.

Outline maps are outlines that provide a framework for plotting information (Figure 1.6). They facilitate the preparation of quick sketch maps for study purposes or classroom use.

The information shown on **cadastral maps** usually includes the location of property-ownership lines, with their bearings and lengths, the ownership and size of land parcels, and similar information (Figure 1.7). When

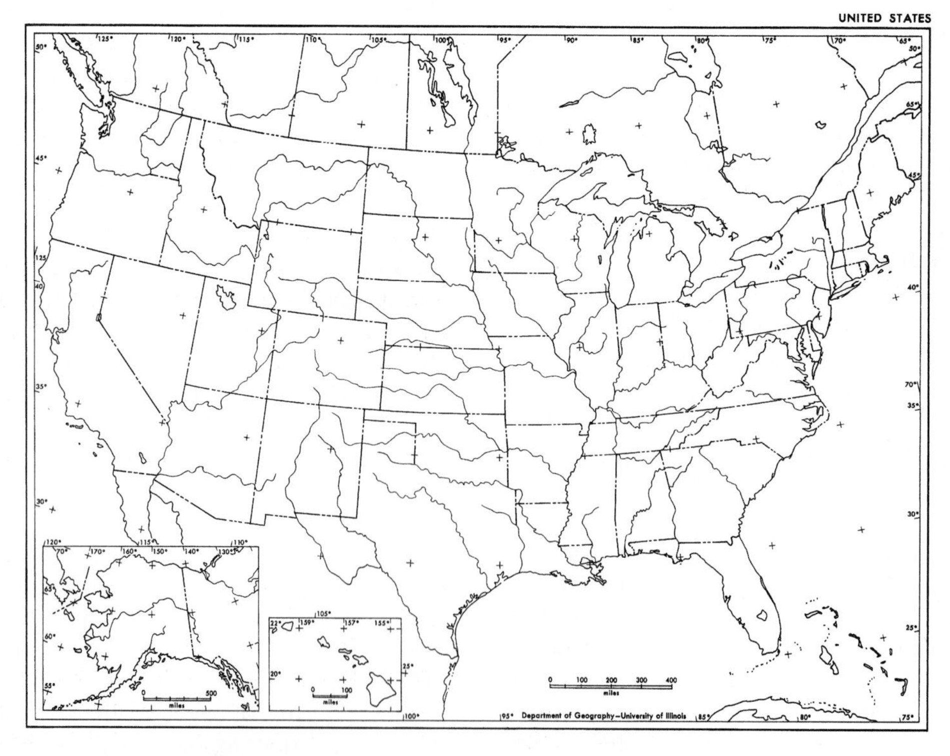

Figure 1.6 Outline map of the United States.
Source: Produced by the Department of Geography, University of Illinois.

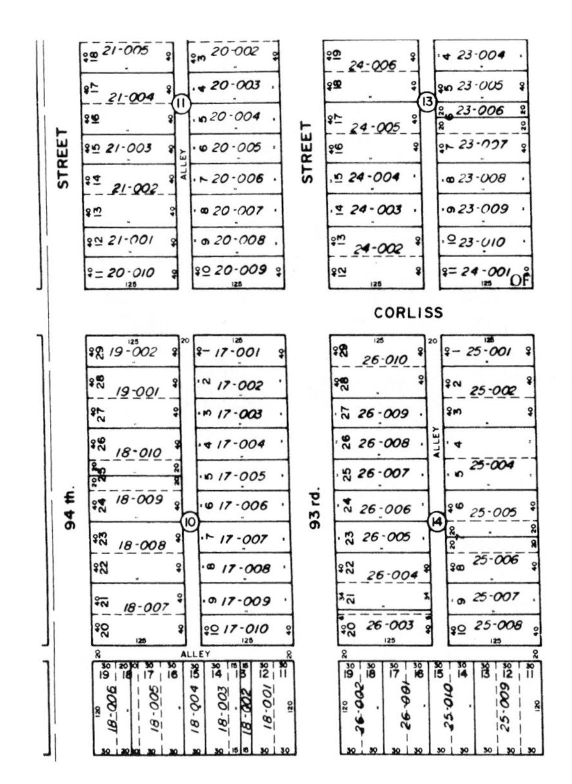

Figure 1.7 Example of a cadastral map. (Village of Sturtevant, Racine County, Wisconsin, 1 inch to 200 feet [1:2400], 1 February 1978.)
Source: Courtesy of Racine County Planning and Development.

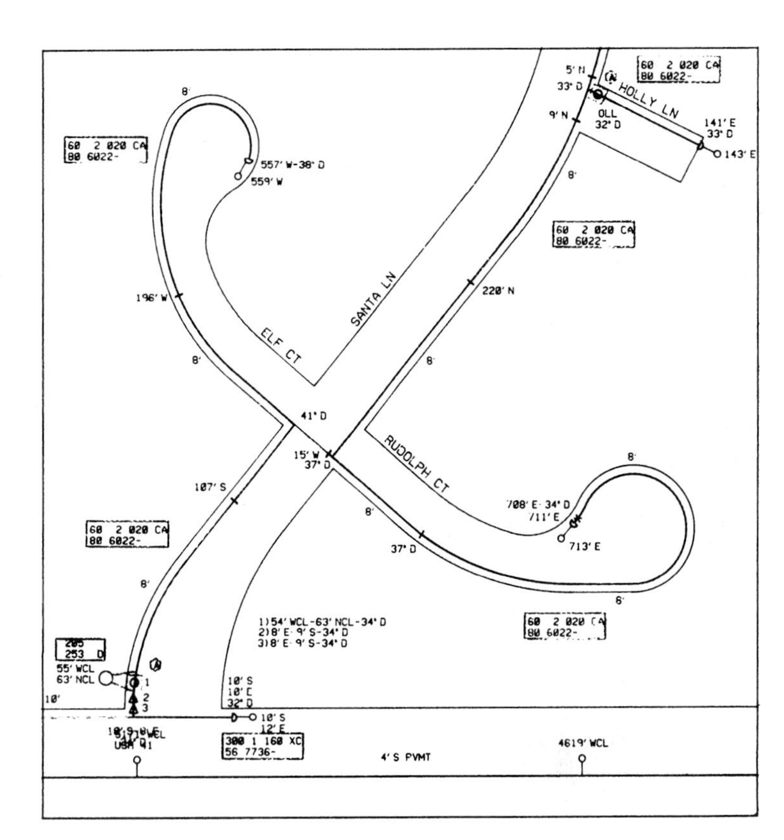

Figure 1.8 Example of a line-route map.
Source: Courtesy of Wisconsin Natural Gas.

you visit your assessor's office, for example, or purchase a piece of real estate, you almost inevitably make use of cadastral maps. Specialized maps of this general type, called **line-route maps,** are used by utility companies to show the routes and rights-of-way of their transmission lines or pipelines (Figure 1.8).

Topographic Maps

Maps that show the shape and elevation of terrain are generally called **topographic maps** (Figure 1.9), to distinguish them from the planimetric type. The term *topographic* is from the Greek *topos* (place) and *graphien* (to describe) and means, therefore, as complete a representation of a visible landscape as scale and symbol limitations allow. Similarly, **bathymetric maps** show water depths and the configuration of underwater topography (Plate 1). Topographic and bathymetric maps have many uses, ranging from helping you plan the route for a hike or sailing outing to detailing the terrain of a piece of property you wish to purchase.

Within the general categories of topographic and bathymetric maps, several types can be distinguished, each produced in a wide range of scales. **Engineering maps** are detailed maps, sometimes called **plans,** used for guiding engineering projects such as bridges or dams and as aids to estimating the construction costs of such projects (Figure 1.10). If you become involved in any significant construction project, you will find it useful to be able to interpret such maps. **Flood-prone area maps** (Figure 1.11), which are used to provide information about areas subject to flooding, are derived from topographic maps. Detailed and accurate terrain information is critical to determining flood risk. It may be of vital importance for you to be able to read such maps when you are evaluating a possible property purchase or other real-estate transaction. Another general type of topographic map—a **landscape map**—provides detailed site information and planting plans for gardens and parks (Figure 1.12). You may want to sketch a map of this type for your own use or you may need to interpret landscaping plans that have been prepared for you by a landscape architect.

Thematic Maps

Thematic maps show information about any one of myriad special topics superimposed on a base map (Figure 1.13). Types of thematic maps include geologic, forestry, soil, land-use, slope, and historical. The

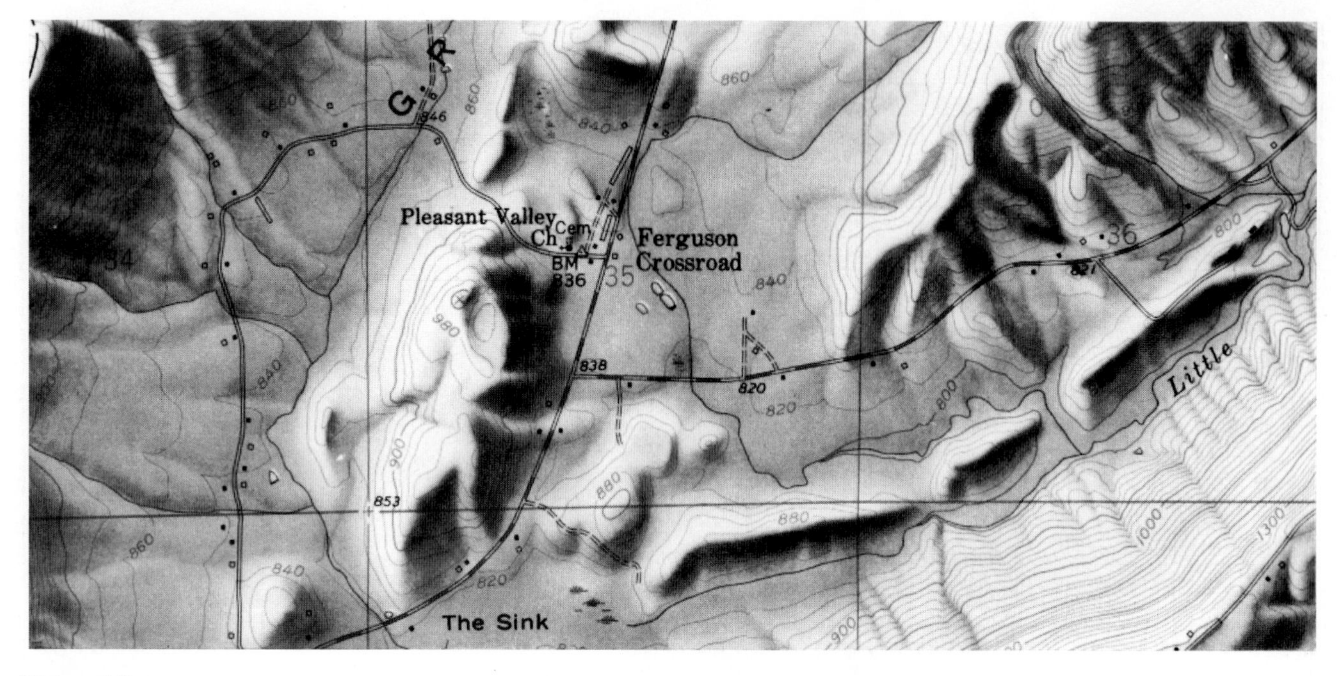

Figure 1.9 Portion of a typical topographic map. (Hyatt Gap quadrangle, Alabama, 1:24,000, 1958.)

Figure 1.10 Example of an engineering map for a landfill.
Source: Courtesy of Racine County Planning and Development.

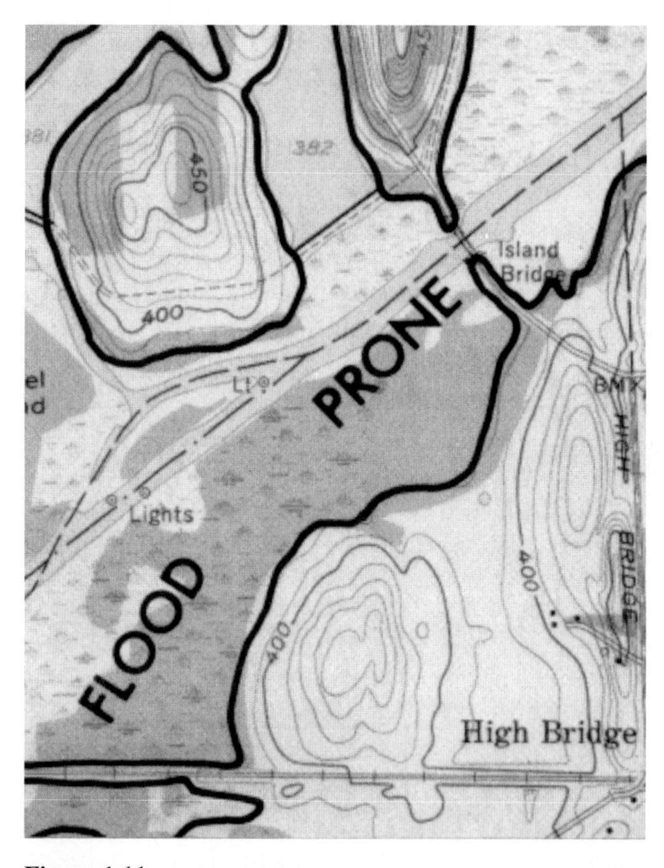

Figure 1.11 Portion of a flood-prone area map. (Montezuma quadrangle, New York, 1:24,000.)

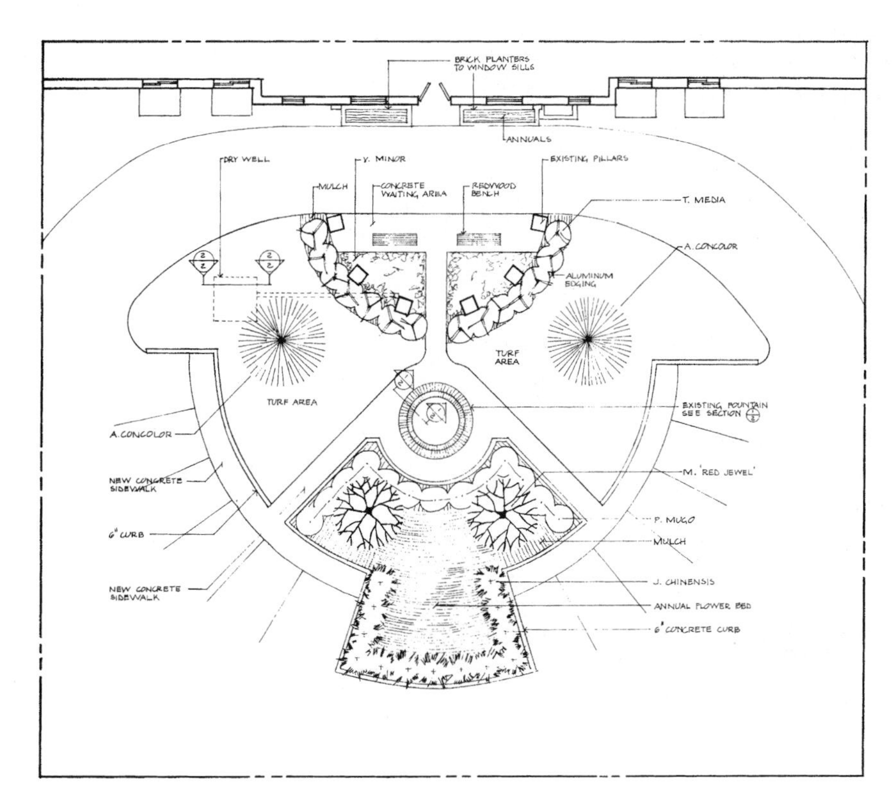

Figure 1.12 Example of a landscape map.
Source: Courtesy of K.E.I. Kujawa Enterprises Inc., Landscape Architecture Division.

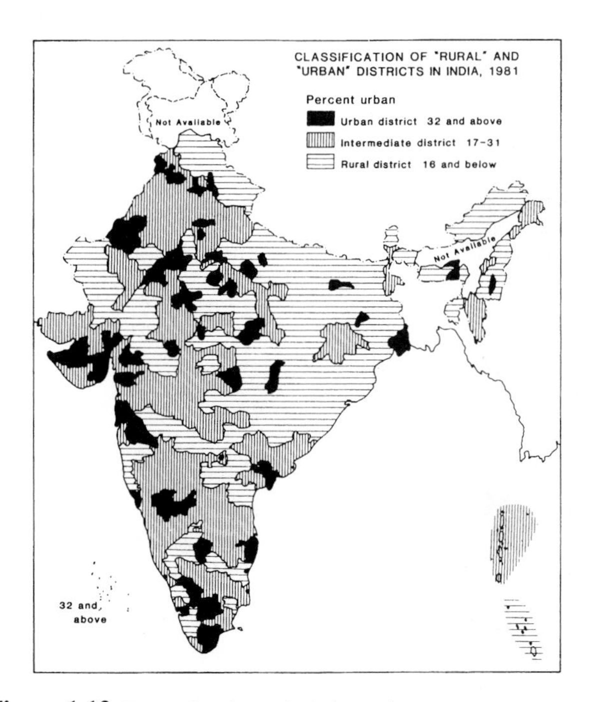

Figure 1.13 Example of a typical thematic map.
Source: From Ashok K. Dutt, Charles B. Monroe, and Ramesh Vakamundi, "Rural-Urban Correlates for Indian Urbanization," *The Geographical Review* 76, no. 2 (April 1986): 173–83. Reproduced by permission of The American Geographical Society.

information on these maps can be presented using a variety of techniques. The study of a selection of thematic maps provides an appreciation for the physical, historical, or social characteristics of a country or region that would be difficult to achieve in any other way. Similarly, thematic maps provide information that allows you to compare the characteristics of two or more different regions.

Thematic maps in which dots are used to represent a specific quantity of a particular variable are called **dot-distribution maps** (Figure 1.14). The goal of such maps is to provide a visual image of the variation in the density of a distribution. The variables presented range from the number of cattle on farms to the distribution of human populations.

Thematic maps that show distributions summarized on the basis of areas delimited by state or county boundaries or other arbitrary boundary lines are called **choropleth maps** (Figure 1.15). On a choropleth map, each area is colored, shaded, dotted, or hatched to appear as a darker or lighter tone. The darkness of the tone is designed to be proportional to the density of the distribution. **Isoline maps,** on the other hand, show numerical values for continuous distributions by means of lines joining points of equal value (Figure 1.16).

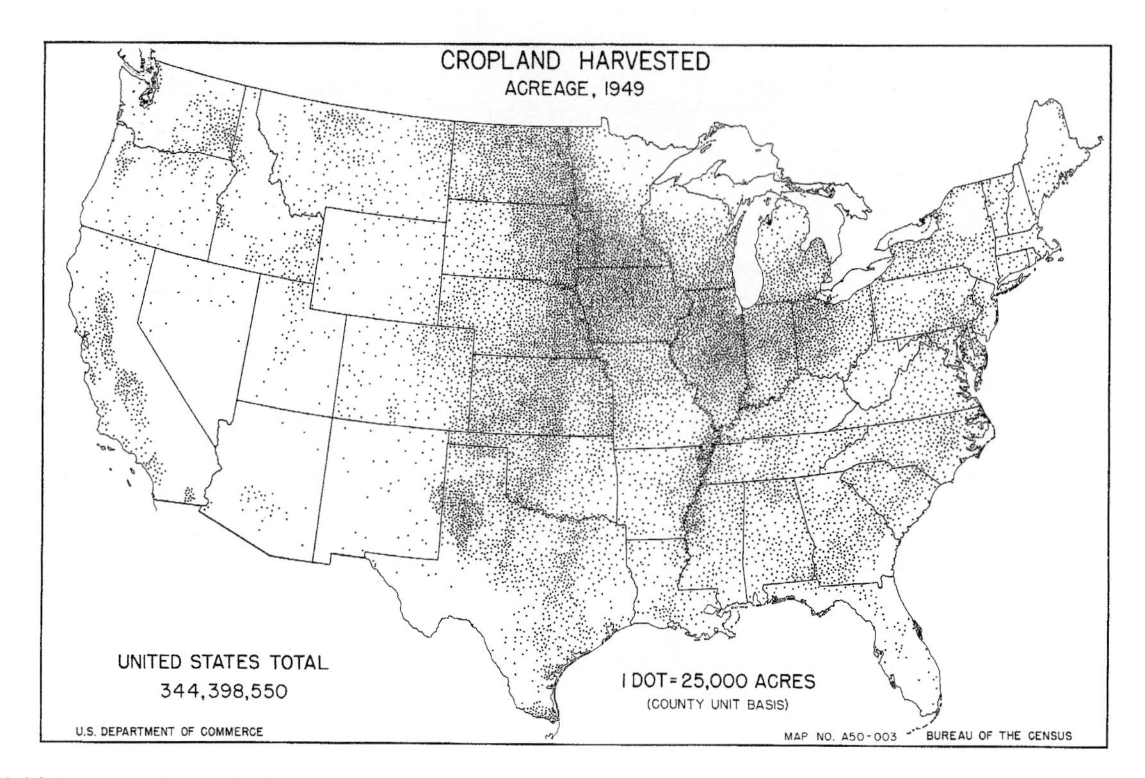

Figure 1.14 Example of a dot-distribution map.

Source: U.S. Department of Commerce, Bureau of the Census, *Portfolio of United States Census Maps* (Washington, D.C.: U.S. Government Printing Office, 1953).

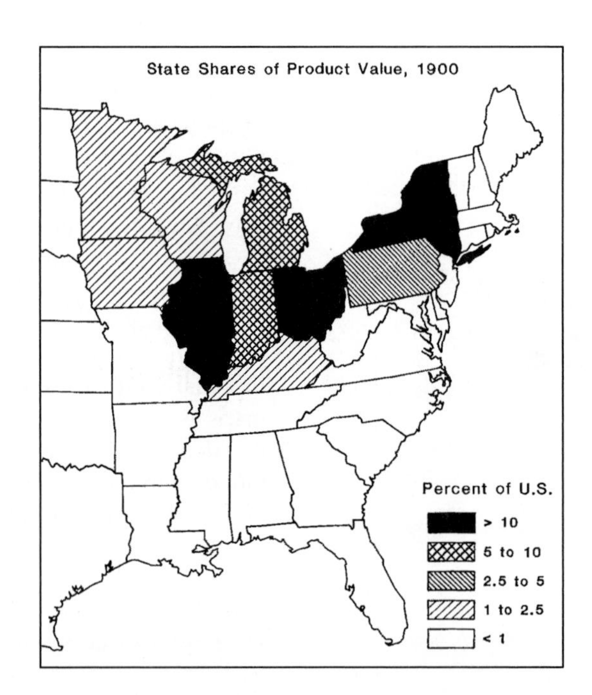

Figure 1.15 Example of a choropleth map, showing state shares of product value, 1900.

Source: From Mary Beth Pudup, "From Farm to Factory: Structuring and Location of the U.S. Machinery Industry," *Economic Geography* 63, no. 3 (July 1987): 203–22. Reprinted by permission.

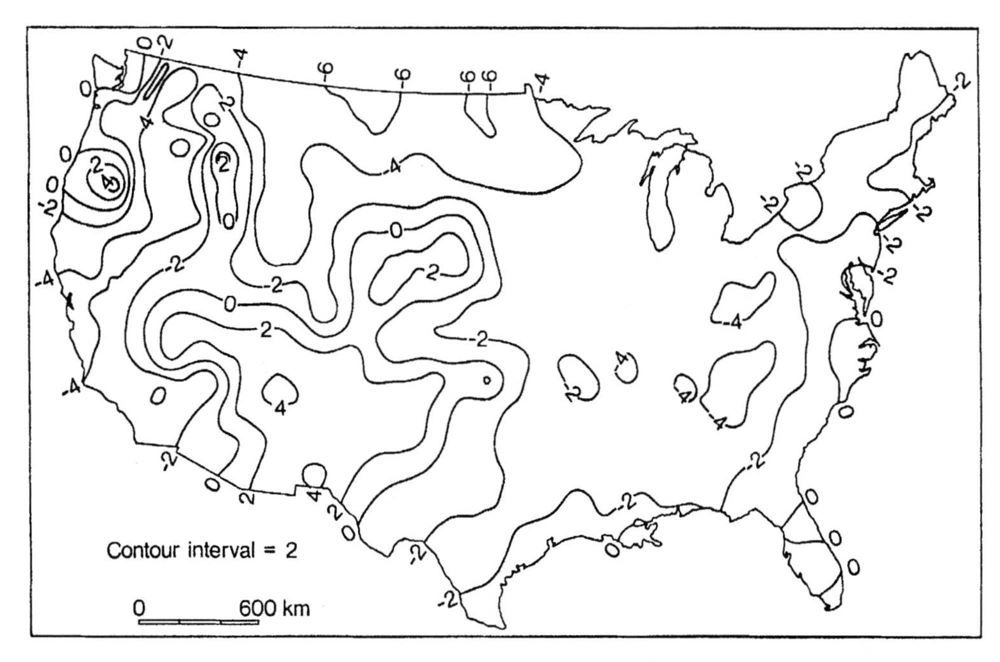

Figure 1.16 Example of an isoline map. Lines join points of equal value on the Palmer Drought Severity Index (PDSI).
Source: From Peter T. Soulé and Vernon Meentenmeyer, "The Drought of 1988: Historical Rank and Recurrence Interval," *Southeastern Geographer* 29, no. 1 (May 1989): 17–25. Reprinted by permission.

Variations in line width or color are used on thematic **flow maps** to show the direction and amount of movement, as well as the types of goods or services moved (Figure 1.17). The placement of the flow lines sometimes indicates the actual route followed by the flow and sometimes simply shows that linkages exist between sources and destinations.

A wide variety of other types of thematic maps present quantitative information through the use of symbols such as proportional circles (Figure 1.18), pie charts, bar graphs, and other graphic methods.

Cartograms

Many representations of distributions are not traditional maps but are, instead, unique representations called **cartograms.** Cartograms are created by substituting a different standard of measurement (time or cost, for example) for the distance measurements customarily used (Figure 1.19). When this is done, sizes, shapes, and distances as we normally think of them are modified or distorted. Although the images thus formed show the world in a different and unfamiliar way, they often provide useful insights precisely because of their different appearance. You should know the strengths and weaknesses of cartograms.

Remotely Sensed Images

Remote sensing consists of gathering information by means of a sensor—human vision is one example—that is not in contact with the objects in the scene being observed. The major types of remotely sensed information in the world of maps are aerial photographs, images from radar and other airborne sensors, and satellite images.

Aerial photographs are the most common remote-sensing product. They are categorized by their angle of view and by the type of photographic emulsion used.

The angle of view of aerial photographs may be either vertical or oblique. **Vertical aerial photographs** are those that are taken with the lens pointed straight down at the ground (Figure 1.20a). **Oblique aerial photographs** are taken with the lens pointed at an angle away from the vertical (Figure 1.20b). It is important to distinguish between these two types because the uses to which they are typically put are quite different. For most mapping purposes, vertical aerial photographs are the most useful. You will find them helpful if you are evaluating real estate, for example, because they clearly show the relationships between various properties, as well as the arrangement of many of the properties' physical characteristics. For some purposes, however, oblique photographs are more suitable. If you are interested in dramatizing the location of a resort area, or showing an industrial site to advantage, an oblique aerial photograph is particularly useful. The techniques involved in the use of aerial photographs are explored in Chapter 18.

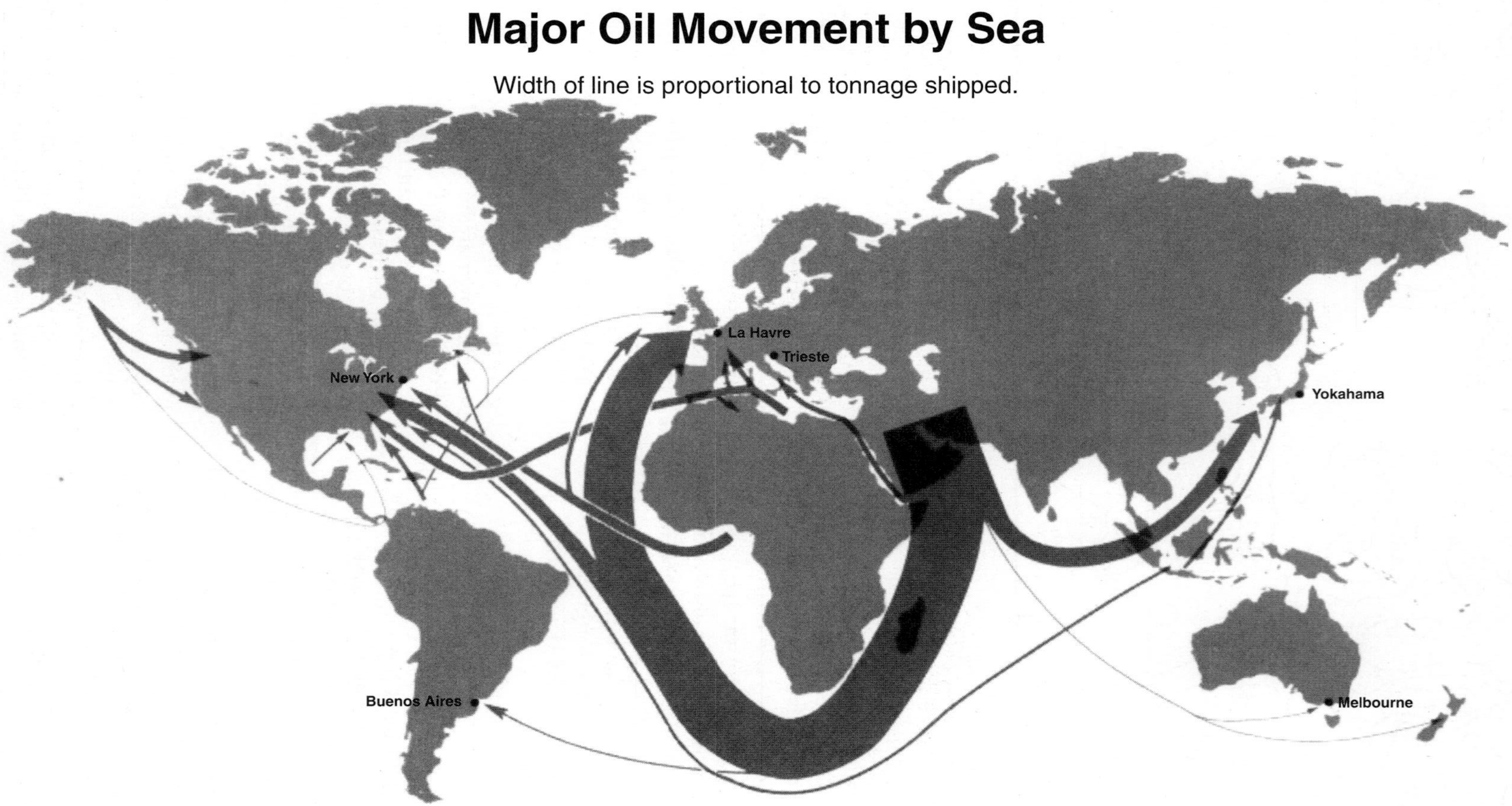

Figure 1.17 Example of a flow map. (1:1,600,000, 1980.)
Source: Portion of Central Intelligence Agency map "The Persian Gulf."

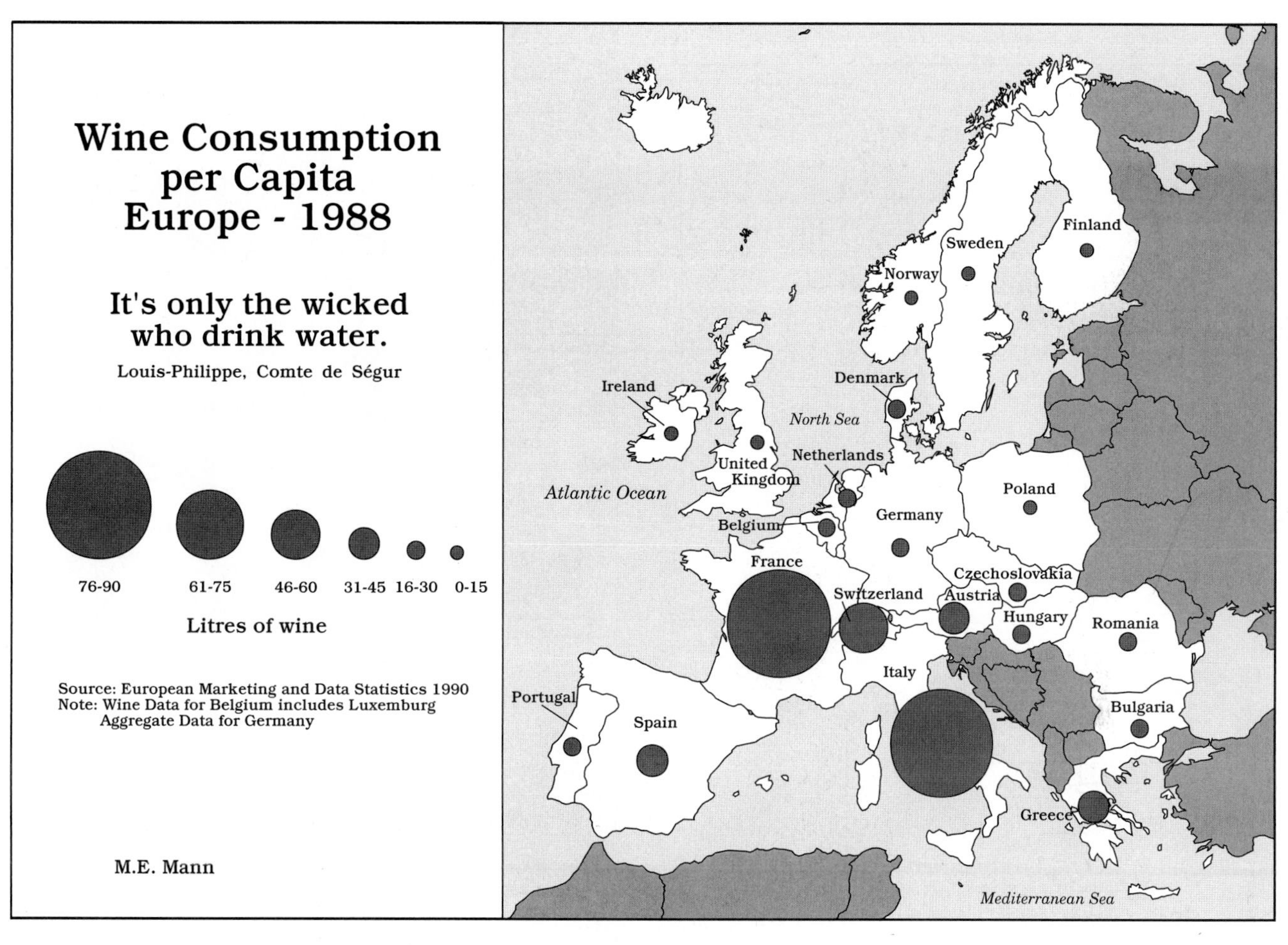

Figure 1.18 Example of a proportional circle map.
Source: Courtesy of Mary E. Mann.

In addition to differing in angle of view, aerial photographs differ from one another in terms of the emulsion on which the image is captured. If a normal, black-and-white (panchromatic) emulsion is used, it is sensitive to a range of wavelengths within the electromagnetic spectrum that is quite similar to that seen by the human eye. The result, however, is a translation of the colors in the original scene into a range of shades of gray in the photograph. If, on the other hand, an emulsion that is sensitive to the ultraviolet wavelengths is substituted for the normal emulsion, the image obtained differs markedly in appearance and can be used for special applications. The use of color emulsions also permits looking at an image in special ways. Knowing the differences introduced into aerial photographs by the various emulsion types is helpful when you are using aerial photographs for a particular application.

Another type of remotely sensed product is the **radar image** (Figure 1.21). To obtain a radar image, a special transmission is beamed from a power source, and the reflected energy is captured and converted into an image of the scene. The geometry and appearance of radar imagery are quite different from those obtained by aerial photography, and there are a number of special applications for which radar imagery is particularly well suited.

Today, images can be captured by instruments taken aloft in spacecraft, with or without a human crew. These images can be returned to earth, whether or not the spacecraft itself ever returns. The technology involved in gathering these **satellite images** from space has been improving rapidly since the early days of space flight (Figure 1.22). A wide variety of images is available today, and the images are used for an even wider variety of applications, some of which may be of interest to you. An appreciation for the characteristics of remotely sensed images will help you to decide if obtaining such coverage is likely to be useful to you.

Now that we have taken a broad overview of the many forms of maps, we can discuss in more detail how best to use them. Chapter 2 provides a basis for this discussion by reviewing some of the earth's important physical characteristics and how some of the information about those characteristics is gathered.

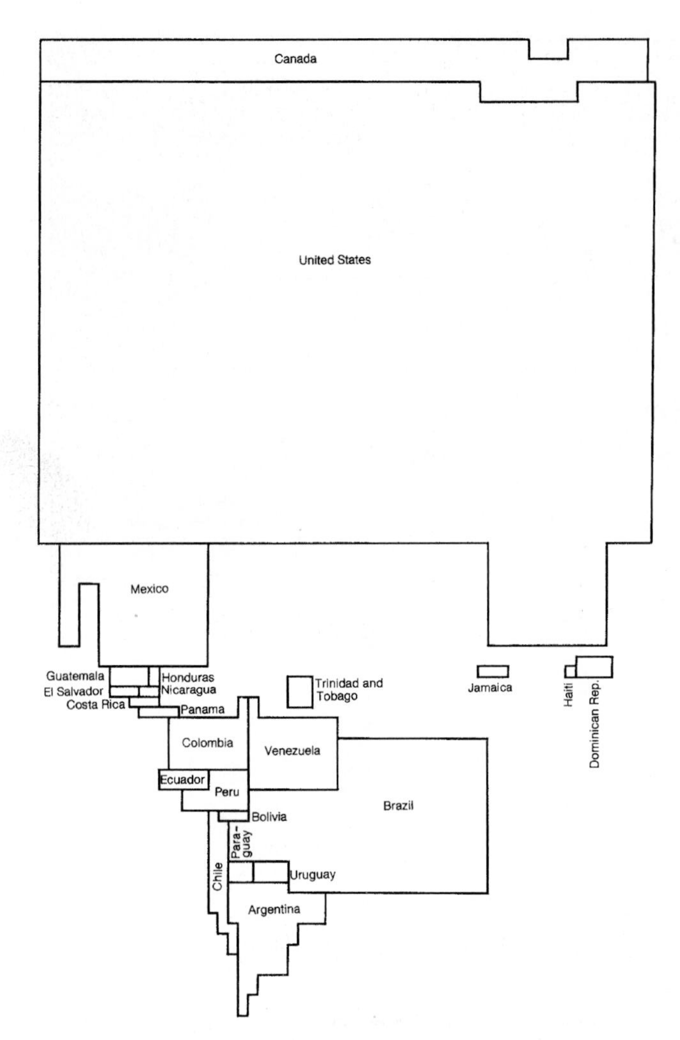

Figure 1.19 Cartogram showing countries of the Western Hemisphere drawn in proportion to their share of world total gross national product.

Source: Adapted and reprinted by permission from International Bank for Reconstruction and Development, The World Bank, *World Development Report 1984* (New York: Oxford University Press, 1984).

(a)

(b)

Figure 1.20 (*a*) Vertical and (*b*) oblique aerial photographs of the same area in downtown Sheboygan, Wisconsin.

Source: Courtesy of Aero-metric Engineering, Inc.

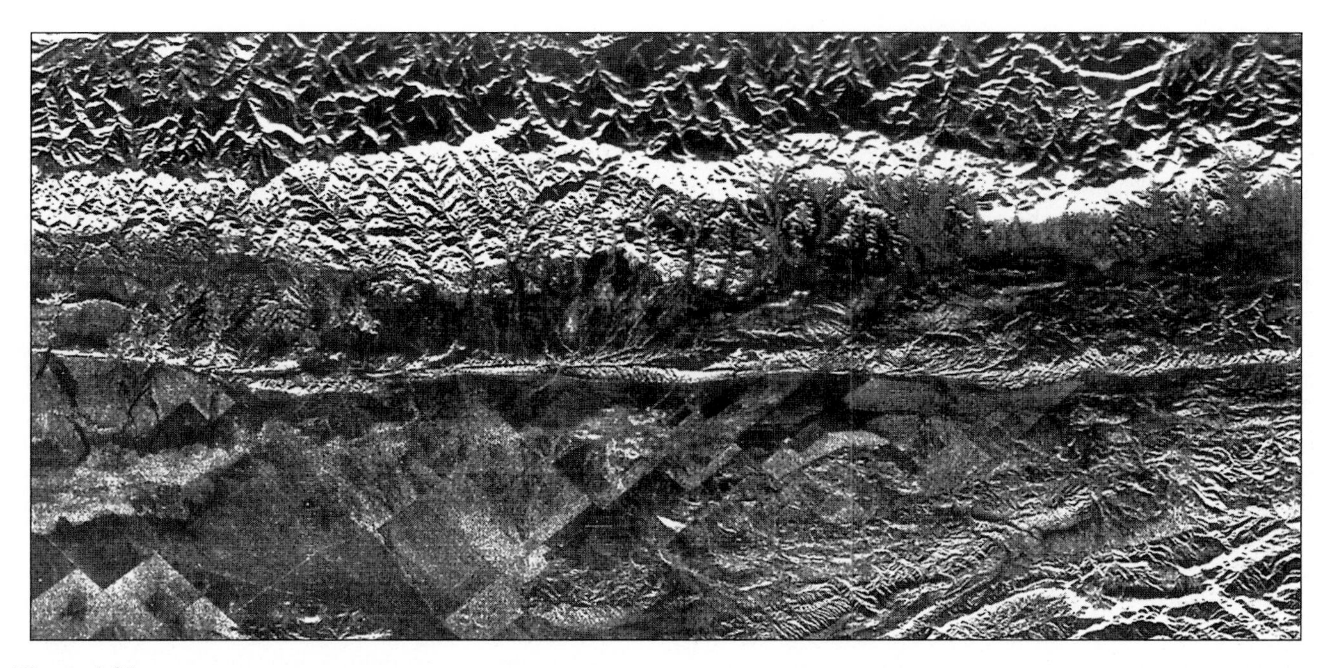

Figure 1.21 Radar image of San Andreas Fault. (Side-looking airborne radar, SLAR, image.)
Source: Courtesy of Westinghouse Electric Corporation.

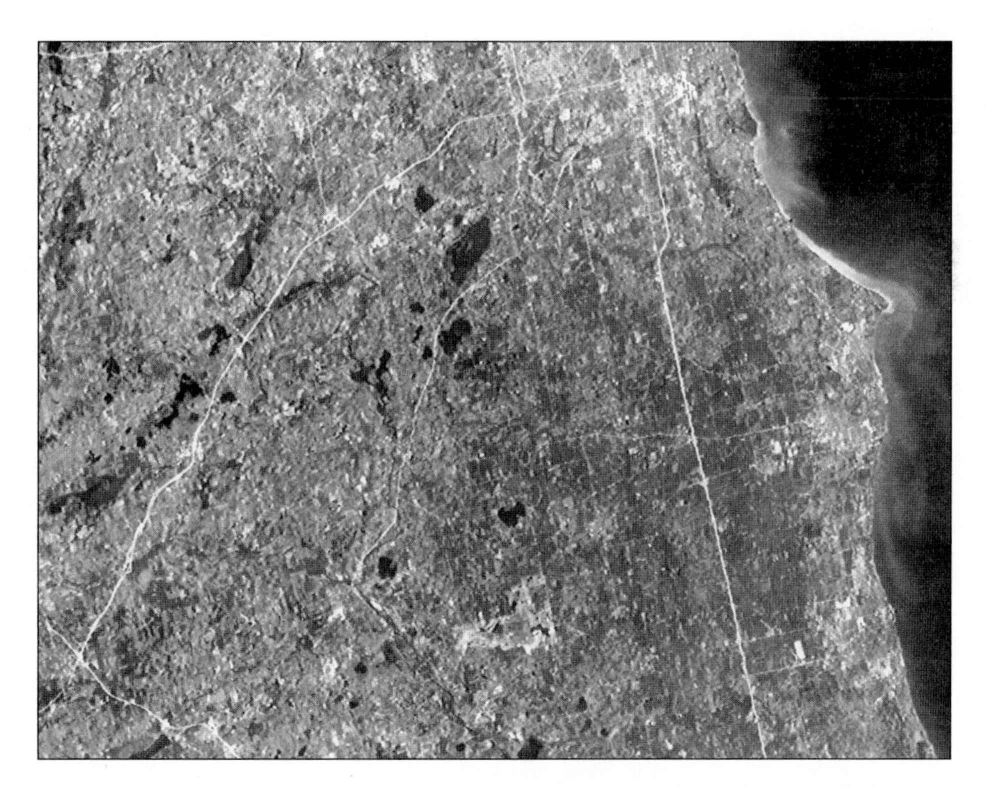

Figure 1.22 Satellite image of Racine, Wisconsin, and vicinity. (Landsat, color composite original, 2 April 1976.)
Source: U.S. Geological Survey EROS Data Center.

SUMMARY

Maps have been produced since ancient times because of the useful functions they perform. They record and communicate information about variations from place to place in the physical and social aspects of the environment. They record distance and direction and are useful tools for exploring, organizing, and analyzing information about the patterns formed by distributions on the earth's surface. They also provide clues for understanding the origins of those patterns. In addition, they provide historical documentation and are useful for regional planning and property-assessment purposes.

Real maps are actual images of the distributions and features that occur on or near the earth's surface. Virtual maps, on the other hand, include images that can be directly viewed but that are not permanent, mental images, and spatial information that is not in standard map form. Mental maps, for example, are images we hold of the exterior physical world. They provide a framework by which we move through and evaluate our environment.

In recent years, computers have allowed speedier, easier, more flexible, and more accurate processing and analysis of vast amounts of remotely sensed data. They have also assisted in the production of maps and in the operation of geographic information systems and have provided the flexibility needed to produce more innovative and effective maps.

Much of this book deals with real maps and related concepts and measures. Real maps take many forms. Planimetric maps, for example, do not try to show relief features in measurable form. Topographic maps, on the other hand, show the shape and elevation of the terrain, and bathymetric maps show water depths and the configuration of the underwater topography. Other maps are used for guiding engineering projects, for planning flood-control measures, and for developing landscapes.

Thematic maps show information about special topics. They use a variety of symbols, including dots to represent the location and quantity of some particular variable, tones or patterns to show the arrangements of variables over areas (choropleth maps), lines to identify locations with the same data values (isoline maps), and lines to show flows and connections between locales.

Cartograms are deliberately distorted representations created by substituting a different standard of measurement (time or cost, for example) for standard distance or area measurements.

Remotely sensed images are gathered by means of a sensor that is not in contact with the objects in the scene being observed. Remotely sensed images include aerial photographs, images from radar and other airborne sensors, and satellite images. These images provide up-to-date information about features on the earth that otherwise might not be available to us.

SUGGESTED READINGS

Downs, Roger M., and David Stea. *Maps in Minds: Reflections on Cognitive Mapping*. New York: Harper and Row, 1977.

______, eds. *Image and Environment: Cognitive Mapping and Spatial Behavior*. Chicago: Aldine, 1973.

Gould, Peter, and Rodney White. *Mental Maps*. Baltimore, Md.: Penguin Books, 1974.

Harley, J. B., and D. Woodward, eds. *The History of Cartography. Volume 1: Cartography in Prehistoric, Ancient, and Medieval Europe and the Mediterranean*. Chicago: University of Chicago Press, 1987.

Lynch, Kevin. *The Image of the City*. Cambridge, Mass.: M.I.T. Press, 1960.

Moellering, Harold. "Real Maps, Virtual Maps, and Interactive Cartography." In *Spatial Statistics and Models*, edited by Gary L. Gaile and Cort J. Willmott. Dordrecht, Holland: D. Reidel Publishing, 1984, 109–32.

______. "Strategies of Real-Time Cartography." *Cartographic Journal* 17, no. 1 (1980): 12–15.

Parfit, Michael. "Putting Our Mental Demons on Maps." *Smithsonian Magazine* (May 1984): 121–31.

Robinson, Arthur H., Joel L. Morrison, Phillip C. Muehrcke, A. Jon Kimerling, and Stephen C. Guptill. *Elements of Cartography*. 6th ed. New York: John Wiley & Sons, Inc., 1995, 20–36.

Thompson, Morris M. *Maps for America*. 3d ed. Washington, D.C.: U.S. Geological Survey, 1988.

Thrower, Norman J. W. *Maps and Man: An Examination of Cartography in Relation to Culture and Civilization*. Englewood Cliffs, N.J.: Prentice-Hall, 1972.

Turnbull, D. *Maps Are Territories; Science Is an Atlas: A Portfolio of Exhibits*. Geelong, Victoria, Australia: Deakin University Press, 1989. Chicago: University of Chicago Press, 1993.

CHAPTER 2

Basic Mapping Processes

This chapter examines methods for measuring the shape and size of the earth, determining locations on the earth's surface, and gathering basic information about objects located on the surface. A general background in these methods will improve your understanding of maps and the information shown on them.

Shape and Size of the Earth

Knowledge of the shape and size of the earth is crucial to understanding the relationships between maps and the world they represent. For this reason, we will discuss the basic attributes of the earth before turning specifically to the study of maps.

Basic Spherical Shape

The earth is spherical, not flat, and its spherical shape is difficult to represent on a flat map sheet. In this section, we will consider the evidence for a spherical earth. Map projections, which are used to represent the earth on maps, introduce many problems into map use. These problems, which must be understood by all map users, are discussed in Chapter 3.

In recent years, astronauts observing the earth from a vantage point in space have seen convincing direct evidence that the earth is a **sphere** (Figure 2.1). Most of us, however, are confined to locations on or near the earth's surface. No matter where we are standing—whether on the shore of the ocean, at the base of a mountain range, on the brink of a canyon, or at the edge of a desert—there seems to be little immediate evidence of the earth's shape. But if we look methodically, we can find clues that lead to the conclusion that the earth is, indeed, spherical rather than flat.

Perhaps the simplest way to gather evidence that the earth has a spherical shape is to watch an object as its distance from us increases or decreases. Such an object, whether it is a ship that is moving out to sea or a building that we are approaching on land, appears or disappears over the horizon in a certain manner, and that is our clue.

Consider, for example, what we see as a ship moves away from us. First, the ship appears to grow smaller as its distance from us increases. At the same time, it appears to sink over the ocean's surface, so that the bottom of its hull disappears first and the top of its mast disappears last (Figure 2.2). These events occur because the ship is moving over the curved surface of the earth. If the surface were not curved, the ship would appear to grow smaller, but its hull would always be in view. The apparent sinking movement just described is uniformly observed, regardless of one's location or the direction of one's movement. The only circumstance that would account for this universal occurrence is a spherically shaped surface of the earth.

Figure 2.1 Astronaut's view of earth rising over the moon's horizon (Apollo photo).
Source: U.S. Geological Survey EROS Data Center.

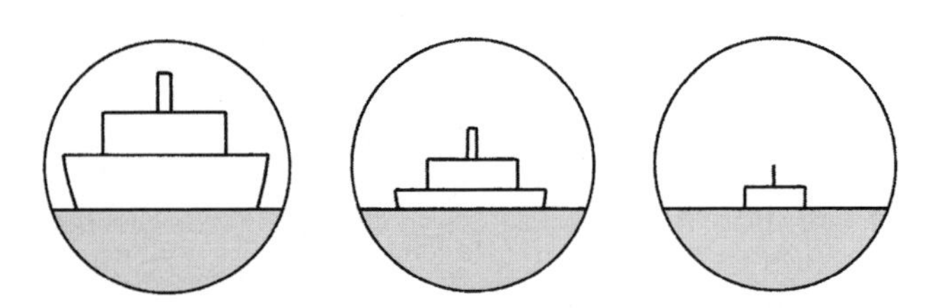

Figure 2.2 Successive views of a ship disappearing over the horizon.

Locational System

Before maps that accurately portray the locations of features on the earth's surface can be prepared, there must be a means of defining those locations. The method that has been adopted for this purpose is the system of latitude and longitude. The chief characteristics of this system are briefly reviewed here because of its fundamental importance in the production and use of maps.

Because the earth is essentially a sphere, there would seem to be no unique location that would provide a natural starting point for a locational system. Fortunately, however, the earth rotates on its axis, and the locations of that axis and of the **poles of rotation** can be determined. The poles provide the starting point for the rest of the system.

Meridians

Visualize a plane passing through the center of the earth. The line defined by the intersection of the plane with the earth's surface is called a **great circle** (Figure 2.3a). An infinite number of great circles can be drawn, but those that pass through the poles are particularly useful. Each half of a great circle that joins the poles is defined as a **line of longitude** or **meridian.** These lines are used to specify locations in the east-west direction (Figure 2.3b). Because all of the meridians are identical, one must be selected as a starting point, or **prime meridian.** Most current maps use the meridian of Greenwich, England, as the prime meridian. The prime meridian is given a value of 0, and the remaining meridians are numbered in degrees, east or west of the prime meridian (Figure 2.4a). The highest numbered meridian is the 180th meridian, which is on the opposite side of the globe from the prime meridian.

Parallels

A second set of locational lines is placed at right angles to the meridians. First, a great circle, called the **equator,** is located midway between the poles (Figure 2.3c). Then a series of **small circles** is defined (a small circle is created when a plane passing through the earth divides the earth into two unequal portions). Selected small circles, which are called **lines of latitude,** or

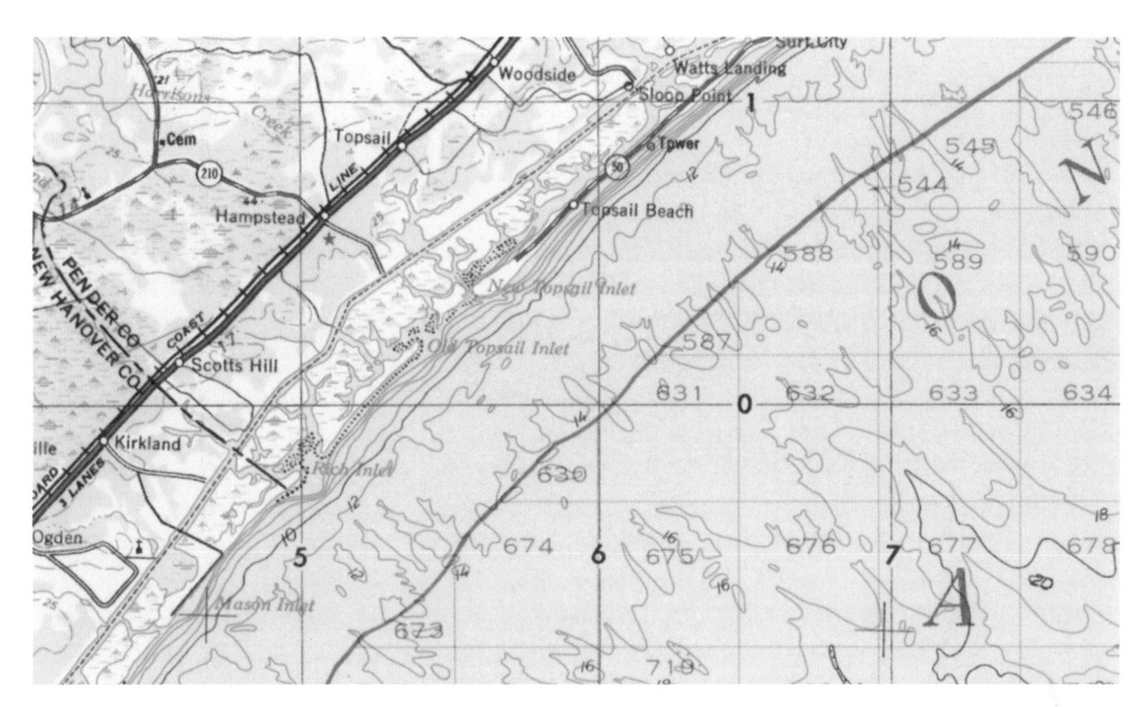

Plate 1 Example of topographic-bathymetric map. Information shown in water areas includes: Universal Transverse Mercator (UTM) grid lines (see Chapter 6); land-management areas (numbered red rectangles) based on Bureau of Land Management survey; and depths based on hydrographic surveys indicated by underwater contours, with depth zones shown by blue tone values.

Source: Portion of U.S. Geological Survey NI 18–4, 1:250,000, 1953 (shoreline revised and bathymetry added, 1972).

Plate 2 Relief shown by layer (hypsometric) tints.

Source: *National Atlas.*

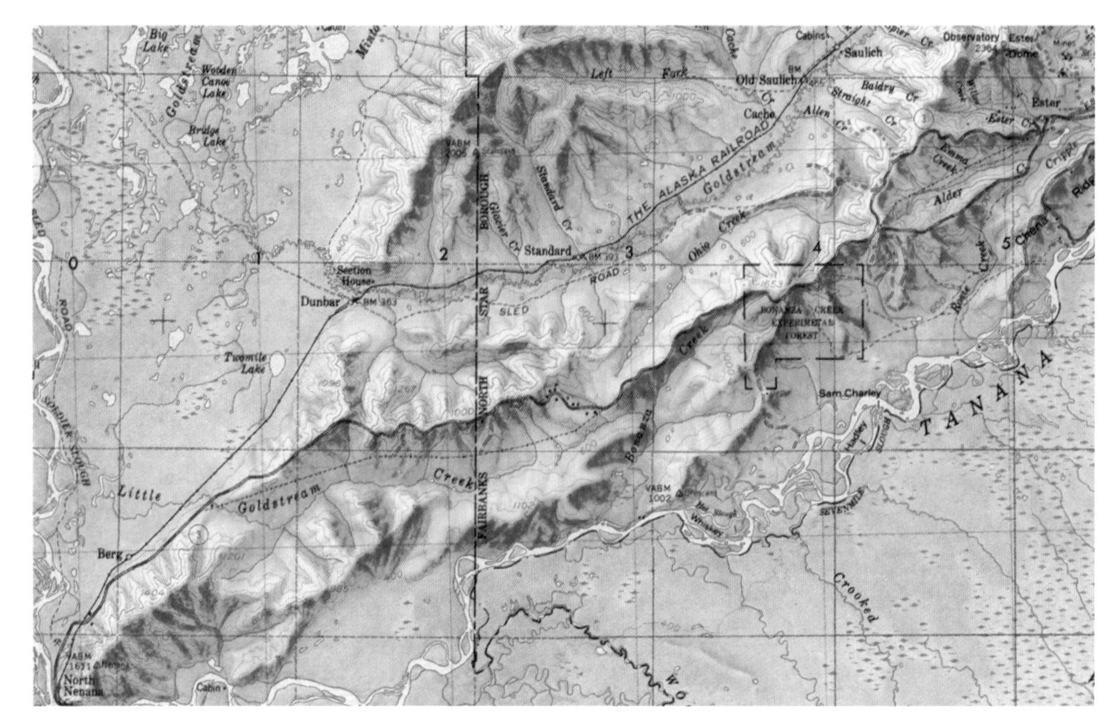

Plate 3 Portion of hill-shaded version of a standard U.S. Geological Survey topographic map (reduced). (Fairbanks, Alaska, 1:250,000, 1956 [revised 1976].)

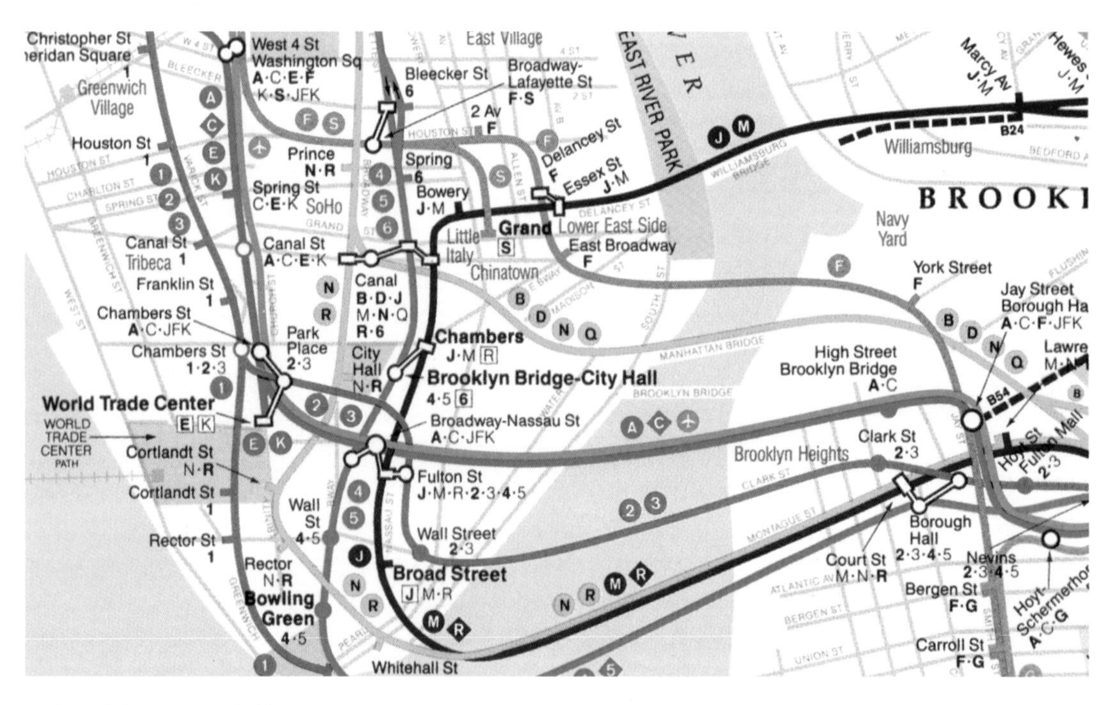

Plate 4 A portion of the New York City subway map.

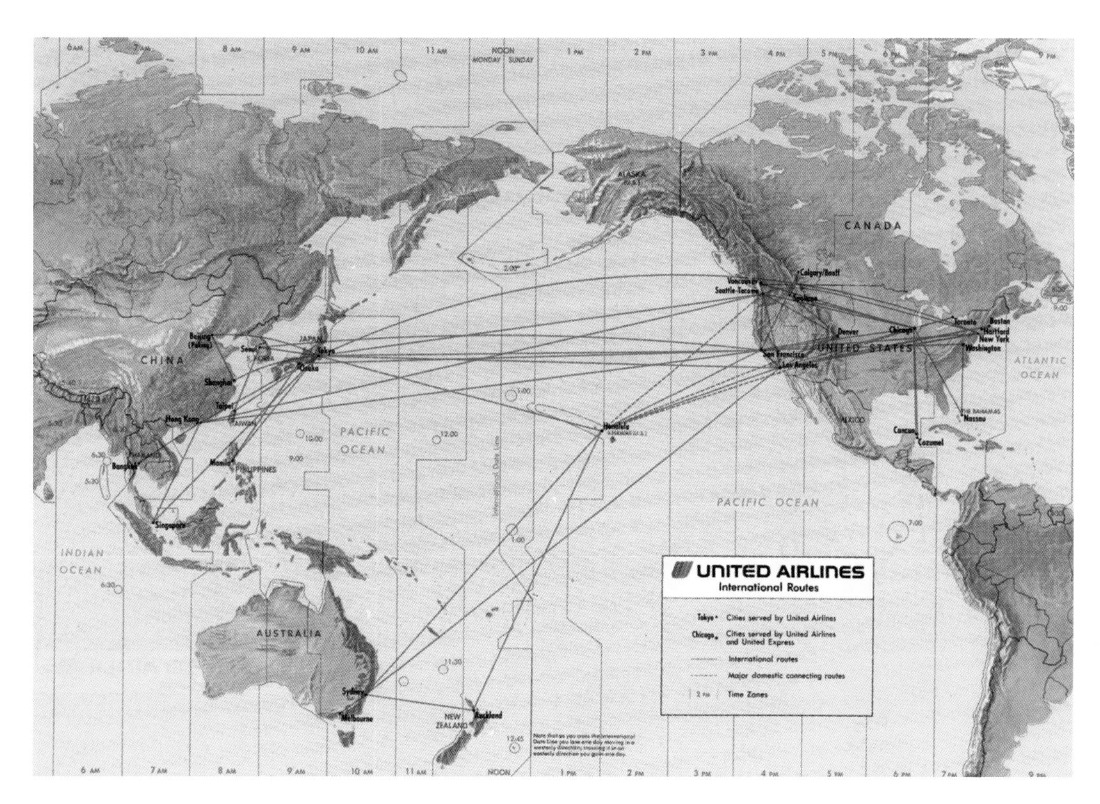

Plate 5 Airline map.
Source: Courtesy of United Air Lines, Inc.

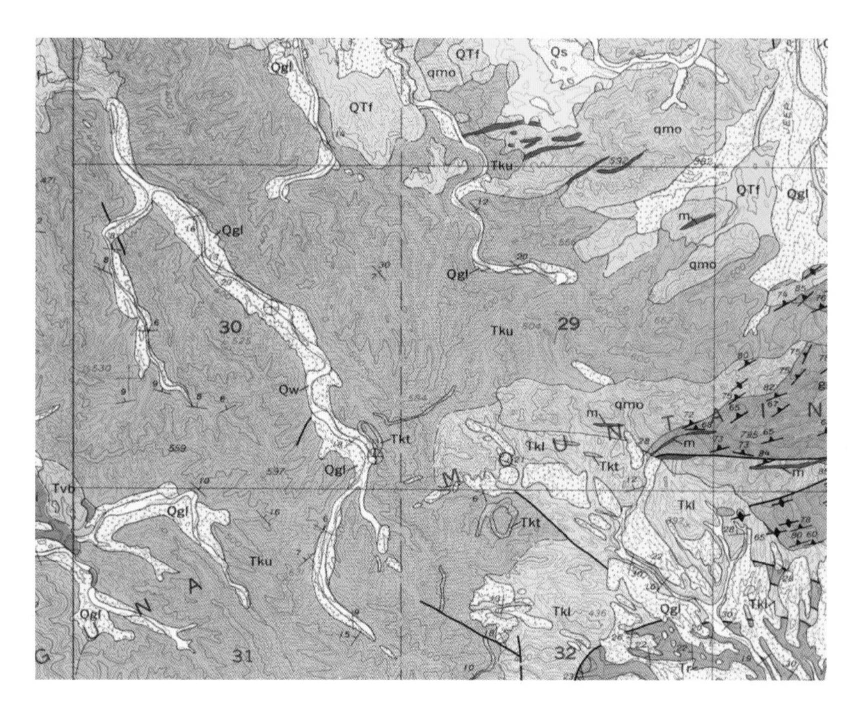

Plate 6 Portion of a U.S. Geological Survey geologic quadrangle map. (F. H. Olmsted, Geologic Map of the Laguna Dam 7.5-Minute Quadrangle, Arizona and California, 1:24,000, 1972.)

Plate 7 Oblique aerial photograph of downtown Chicago.
Source: Courtesy of Aero-metric Engineering, Inc.

Plate 8 Portion of color-infrared aerial photograph, near Footville, Wisconsin, 1986. (Approximately 1:70,000-scale.)
Source: USGS National High Altitude Program (NHAP).

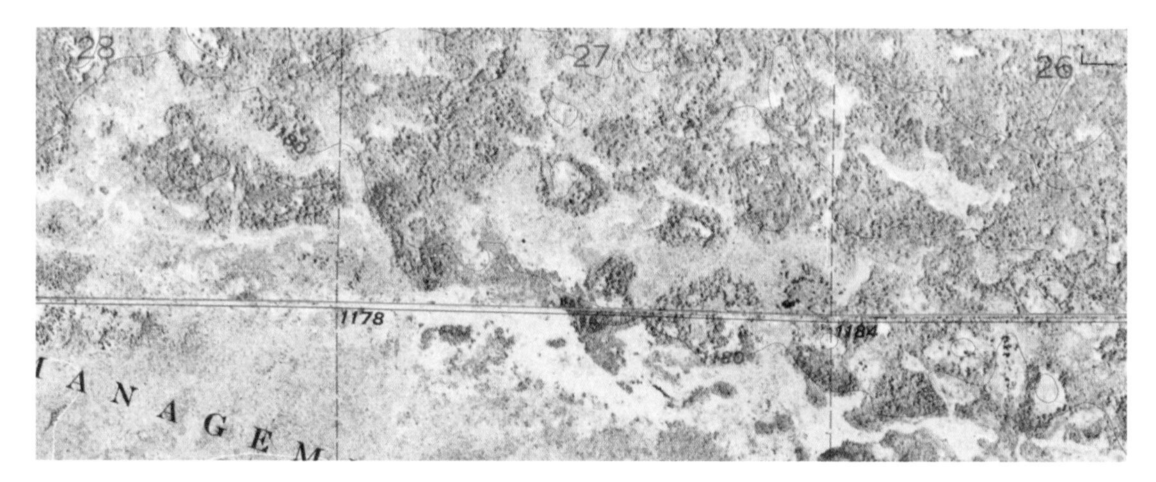

Plate 9 Portion of U.S. Geological Survey orthophotomap. (Randeen Ridge quadrangle, Minnesota, 1:24,000.)

Plate 10 Portion of false-color remote-sensing image, Phoenix, Arizona. (Landsat Thematic Mapper, 30 meter ground resolution, 1:100,000-scale.)

Source: Department of the Interior, U.S. Geological Survey. "Landsat and SPOT Image Display: Phoenix, Arizona Area" Miscellaneous Investigations Series I-1941, 1987.

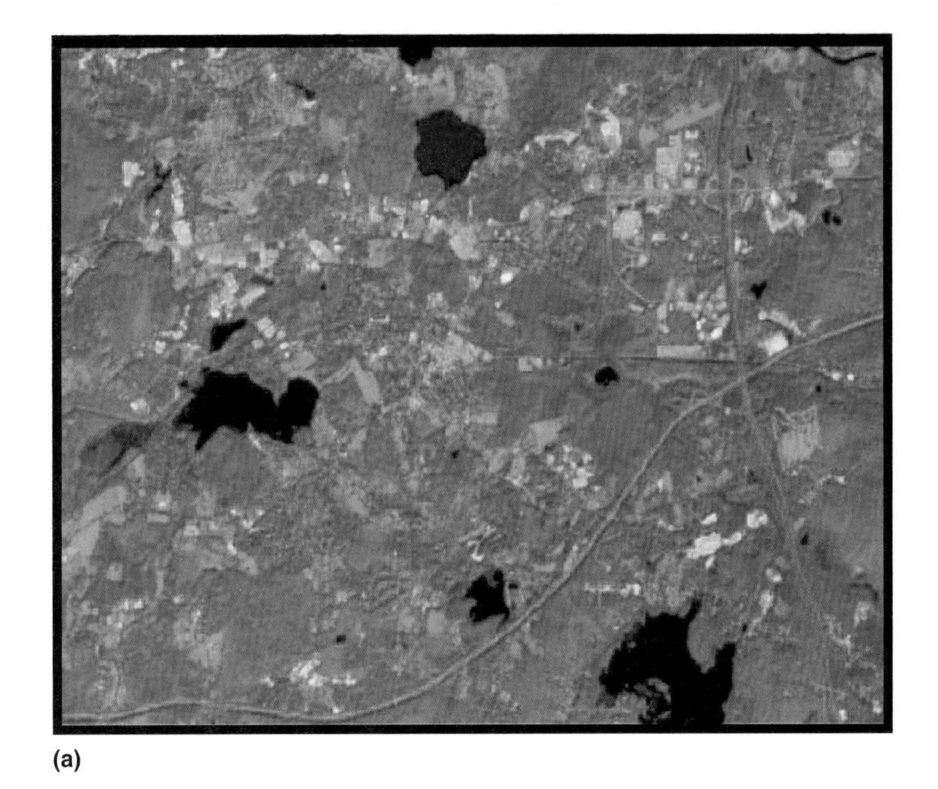

(a)

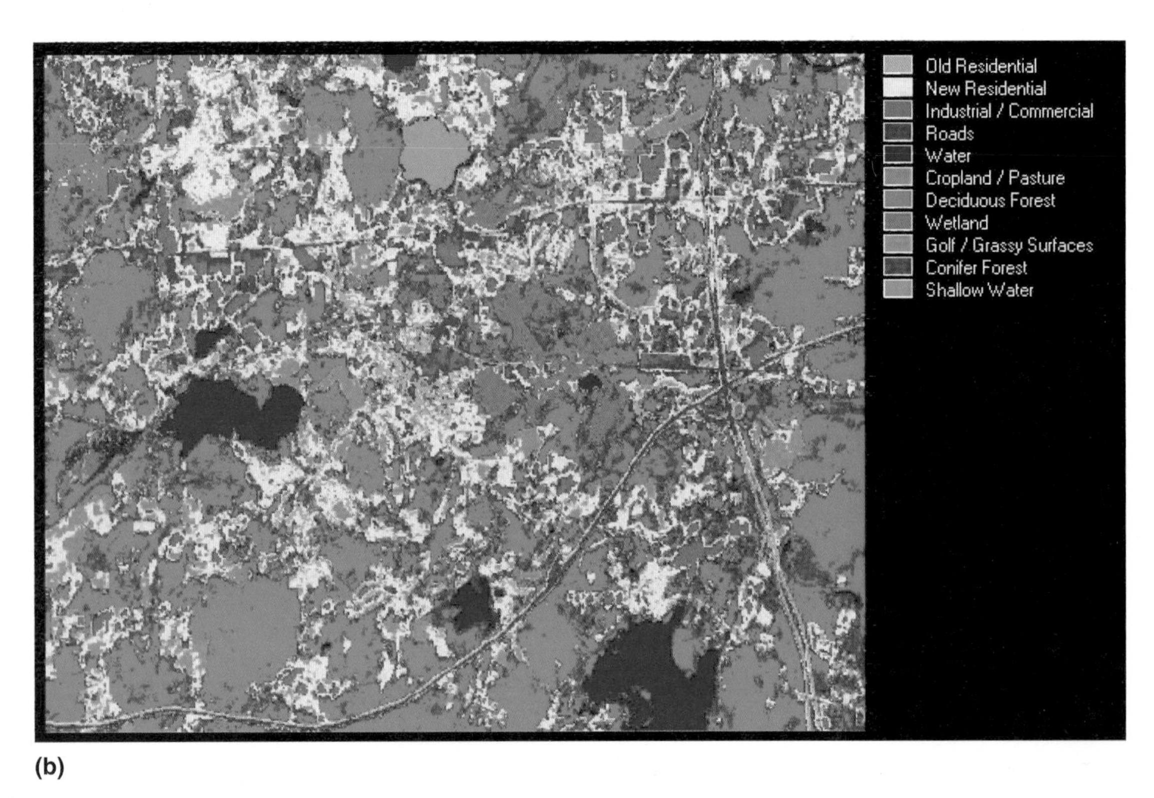

(b)

Plate 11 (*a*) False-color, multispectral SPOT image of Westborough, Massachusetts (bands 1, 2, and 3). The area covered is approximately 10 km by 8 km. Shades of pink and red are areas containing various types of vegetation; shades of bluish-gray are generally in built-up areas; and shades of gray and black are wetlands and water surfaces. (*b*) Maximum-likelihood classification of land cover in the same image area, carried out using IDRISI.

Source: Image Courtesy of Clark Labs, Clark University.

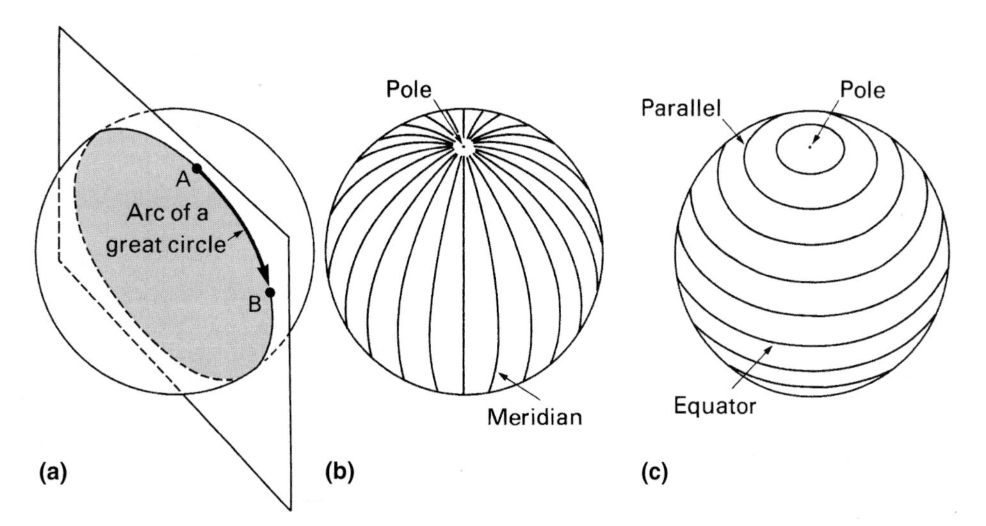

Figure 2.3 Characteristics of the graticule. (*a*) The arc of a great circle is the shortest distance between two points on the earth's surface. (*b*) Meridians (lines of longitude) are arcs of great circles that meet at the poles. (*c*) The equator is a great circle, located midway between the poles. Parallels (lines of latitude) are concentric small circles that are parallel to the equator.

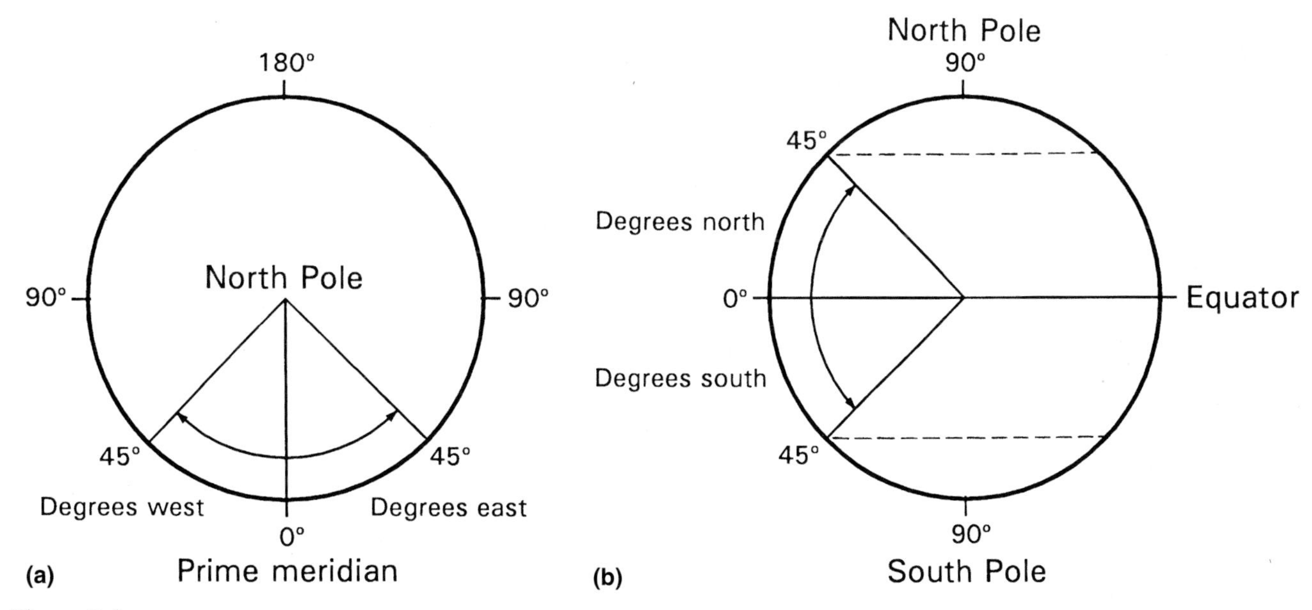

Figure 2.4 Numbering system for the graticule. (*a*) Meridians. (*b*) Parallels.

parallels, are placed parallel to the equator. They are numbered in degrees north and south from the equator, which is given a value of 0, with the maximum value of 90 assigned to each pole (Figure 2.4b). Parallels are used to define locations in the north-south direction.

Graticule

The pattern of meridians and parallels on the earth is referred to as the **graticule.** The following characteristics of the graticule are based on the simplifying assumption that the earth is a sphere:

1. Parallels are equally spaced between the equator and the poles.
2. Parallels are always parallel to one another, so any two parallels are always the same distance apart all the way around the globe.
3. Meridians are spaced farthest apart on the equator and converge to a single point at the poles.
4. Parallels and meridians cross one another at right (90°) angles.

Refinements in Shape

A more detailed examination reveals that the earth's shape is not *exactly* spherical. (See the sources listed in the Suggested Readings at the end of the chapter for more information.) For our purposes, it is sufficient to note the general differences between the true shape of the earth and a sphere.

Ellipsoid

The earth revolves easterly on its axis. This rotation generates centrifugal force, which causes the earth to bulge slightly in the middle and to flatten slightly at the poles, resulting in a shape called an **ellipsoid** (Figure 2.5). The ellipsoid is the shape generated by an ellipse that rotates on its minor axis. It has a slightly greater radius at right angles to the axis of rotation and a slightly smaller radius along the axis (Figure 2.6a). This is called **polar flattening.** The difference is not great. Polar flattening amounts to about 1 part in 298, which means that the earth is very close to the true spherical shape that results from the rotation of a circle (Figure 2.6b). For this reason, many general maps can be based on the assumption that the earth is a sphere. Extremely accurate maps, on the other hand, must be based on the earth's ellipsoidal shape because of its effect on the spacing of the lines of latitude and longitude. The dimensions of the ellipsoid recently adopted for use in maps produced by U.S. government agencies are shown in Table 2.1.

TABLE 2.1 Size of the Ellipsoid

Equatorial radius (in meters): 6,378,137	Flattening (approximate): 1/298

Source: International Union of Geodesy and Geophysics. Adopted for international use and used for North American Datum, 1983 (NAD83).

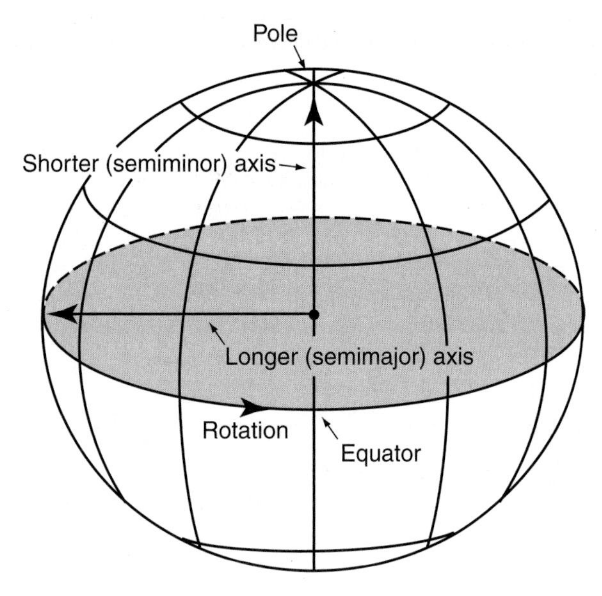

Figure 2.5 An ellipsoid is formed by rotating an ellipse on its shorter axis.

Geoid

Accurate measurements of the earth's size and shape also reveal that it is not a "regular" ellipsoid but that it has a somewhat irregular shape, called the **geoid.** This is not a reference to the mountains and ocean bottoms that wrinkle the earth's surface. The vertical distance from the deepest part of the ocean to the highest mountain peak is about 19,880 meters (65,223 feet). The diameter of the earth is about 12,756,370 meters (41,851,481 feet), so the relief features are quite insignificant in comparison. What is more important, from the standpoint of detailed mapping and geodetic surveying, is that the earth's equigravitational surface undulates because of variations in the density of various portions of the earth's crust. These variations in mass, in turn, introduce variations in the effect of gravity and lead to the irregularities in shape. Again, for

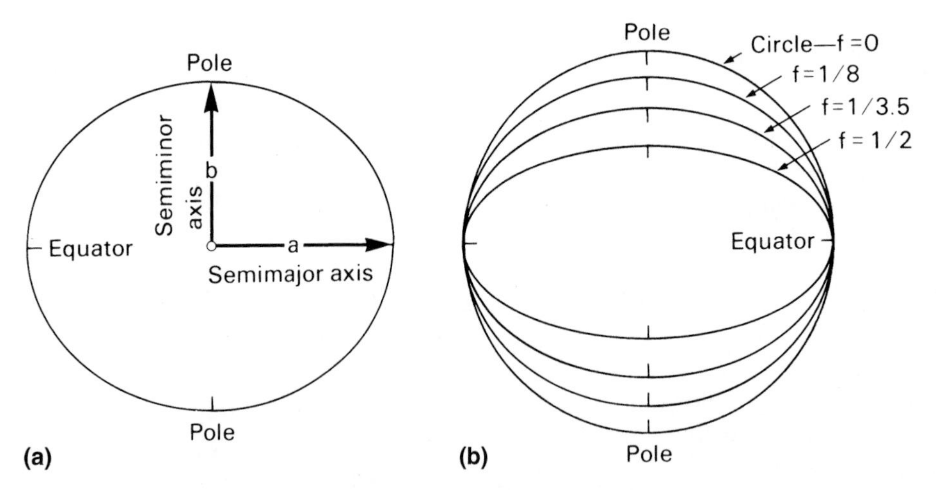

Figure 2.6 Characteristics of an ellipse. (*a*) The shape of an ellipse is determined by the relative lengths of its semiminor and semimajor axes. (*b*) The flattening ratio of an ellipse representing the earth (approximately 1/298) is very close to zero, so the ellipse is almost a circle.

many purposes, the "lumpy" nature of the earth's surface can be ignored, but the effects of these irregularities are taken into account in the geodetic surveys that provide the framework for detailed maps.

Size

How large is the earth? Obviously, knowing the earth's size is as fundamental as knowing its shape. Because maps are a representation of a portion of the earth's surface, the distance from one point to another on the earth must be determined so that it can be represented on maps or globes.

Although visualizing the earth's shape from indirect evidence is difficult, it seems feasible to directly measure distances between points on the earth's surface. For relatively short spans, such as those required for local surveys, distances are measured directly, using surveying techniques. With longer distances, direct measurement is a much more difficult task. It is simply not feasible to take a measuring tape and walk around the world, taking measurements along the way and totaling them at the end. The intervention of oceans, mountains, and other obstacles coupled with a vast array of technical difficulties make such a measurement process difficult to conceive and impossible to accomplish. Small errors that would inevitably be made along the way would accumulate and contribute to an inaccurate size determination. In addition, the earth's spherical shape increases the difficulty of measuring distances over its surface. How, then, can the size of the earth be measured?

The basic approach to measuring the earth is not particularly complex, nor is sophisticated modern equipment required. In fact, about 250 B.C., the Greek mathematician **Eratosthenes** measured the earth, using simple observations and applying his knowledge of basic principles of geometry. The exact methods used by Eratosthenes, as well as the accuracy of his measurements, have been debated many times, but the version presented here covers the major facts.

As a starting point for his measurement of the earth, Eratosthenes observed and measured shadows. He knew that, at the time of the **summer solstice,** the noon sun is directly overhead on the Tropic of Cancer. He then determined that the town of Syene (near present-day Aswan, Egypt) was located on the Tropic of Cancer because vertical shadows were cast there at the solstice.

He also observed that at noon on the occasion of a solstice, the sun cast an angled shadow in the city of Alexandria (which was located north of Syene on approximately the same meridian). He measured the angle of this shadow and found it to be 1/50 of the circumference of a circle. From geometry he knew that the angle between a pair of lines drawn vertically to the center of the earth from Syene and from Alexandria would be the same as the angle of the shadow that he had measured (Figure 2.7). Because the angle between the verticals at Syene and Alexandria amounted to 1/50 of a circle, it was apparent that the distance between the two towns was equal to 1/50 of the earth's circumference.

Eratosthenes next determined that the surface distance between Syene and Alexandria was 5000 stadia, although accounts differ as to whether he had the distance

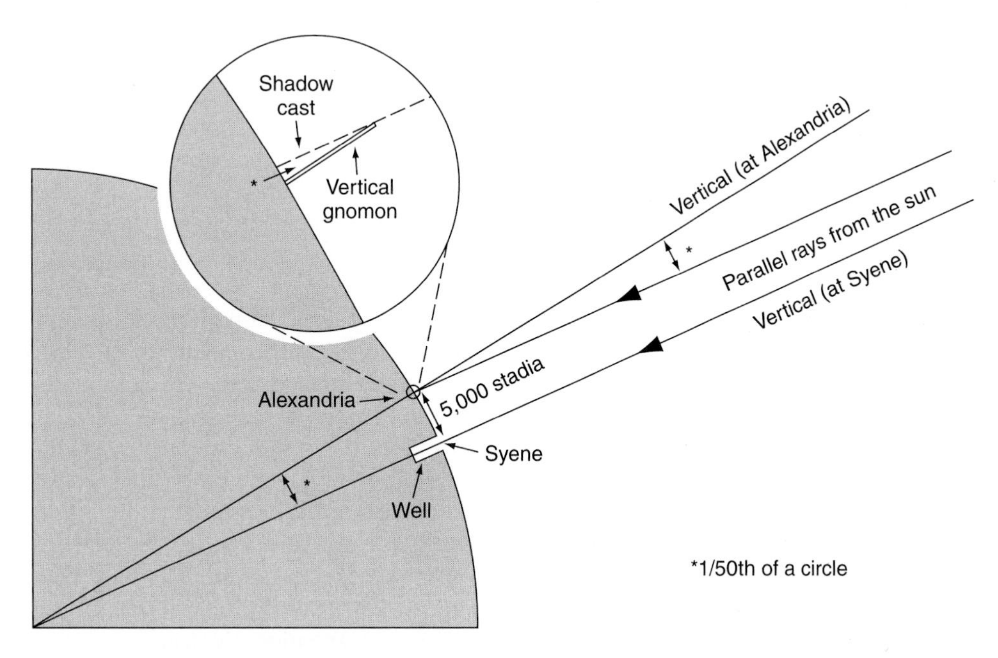

Figure 2.7 Eratosthenes' method of measuring the earth.

carefully measured or whether he estimated it on the basis of travelers' reports. By multiplying 5000 stadia by 50, he estimated the earth's circumference to be 250,000 stadia. Given modern-day estimates of the length of a stadia, the measurement appears to have been rather close to the presently accepted equatorial circumference of the earth, which is 40,075 kilometers (24,901 miles).

Details aside, what is important about the story of Eratosthenes' feat of measurement is his use of a logical, geometric method. Methods in use today employ more sophisticated equipment but the same basic logic.

MAPPING PROCESS

Surveying involves taking measurements to determine the position of points on the earth's surface. A basic description of surveying methods shows how the information used in the production of maps is gathered. The two types of surveying are geodetic control and plane.

Geodetic Control Surveys

Detailed surveys that take into account the curvature of the earth and determine locations in terms of latitude and longitude are called **geodetic control surveys.** The principal reason for considering such surveys is that they provide the overall control, or framework, for maps. Because of the complexity of the subject, our discussion is limited to only a general consideration of geodetic control surveys.

The first step in a geodetic control survey is to establish the latitude and longitude of two starting points on the earth's surface. Such locations are determined by taking astronomic observations.

Latitude Determination

To a present-day observer standing at the **North Pole,** the star **Polaris** is almost exactly overhead.[1] This means that Polaris is an excellent point of reference for determining latitude.

Throughout the Northern Hemisphere, measuring the angle between Polaris and the observer's horizon defines the observer's **astronomic latitude** (Figure 2.8). To understand why this is so, suppose that the observer is standing at the North Pole. Polaris is, then, directly overhead. This means that the angle between the horizon and the star is 90 degrees, which is the latitude of the North Pole. In contrast, suppose that the observer is standing at the equator. At this location, Polaris falls directly on the horizon, at an angle of 0 degrees, which is the latitude of the equator. The same relationships hold, regardless of the observer's position, so that any Northern Hemisphere latitude can be determined by observing Polaris.

Other stars, including the sun, can be observed to determine latitude, and observations also can be made at locations in the Southern Hemisphere. The techniques involved in such observations are similar to those already mentioned, although adjustments are required to take into account differences in the location of the object observed, relative to the earth, at different times and dates.

Longitude Determination

The process of determining longitude is not quite as straightforward as the method for determining latitude because, as already mentioned, longitude must be defined in relation to some arbitrary starting point on the earth's surface. The position of the sun provides the basic information for determining longitude. The moment the sun passes directly over a meridian is defined as **apparent solar noon** on that meridian. When apparent solar noon is observed at a given location, the difference between the observed time and the time of apparent solar noon at the Greenwich meridian provides a direct measure of the difference in longitude of the two locations.

In the past, determining longitude was very difficult. From early maps of the world, it appears that the latitudes of places were rather accurately known at quite an early stage. Longitude, however, was very inaccurately determined. As a result, during the early days of global exploration, navigators constantly faced the problem of not knowing where they were. Indeed, the difficulty in making longitude determinations contributed to Columbus's belief that by sailing west from Europe he would, after a relatively short journey, reach Asia. He was overly optimistic because his estimate of the distance involved, based on longitude, fell far short of the actual distance. By the late 17th and early 18th centuries, commerce on the high seas was highly important, particularly to colonial nations. As the volume of shipping grew, losses of ships that went astray, along with their crews and cargoes, became intolerable. In an effort to mitigate these problems, the British Parliament passed the 1714 Longitude Act, which provided a prize of £20,000—a very large sum for the time—for the development of a practical means of determining longitude.

Inventors who sought the prize offered by the Longitude Act proposed a variety of approaches. For example, one technique was based on observations of the movements of the moons of Jupiter. However, making observations with the accuracy required to implement this method was impractical from the deck of

[1]Polaris is actually about 1 1/2 degrees away from the extension of the earth's axis. Tables that compensate for this difference are available.

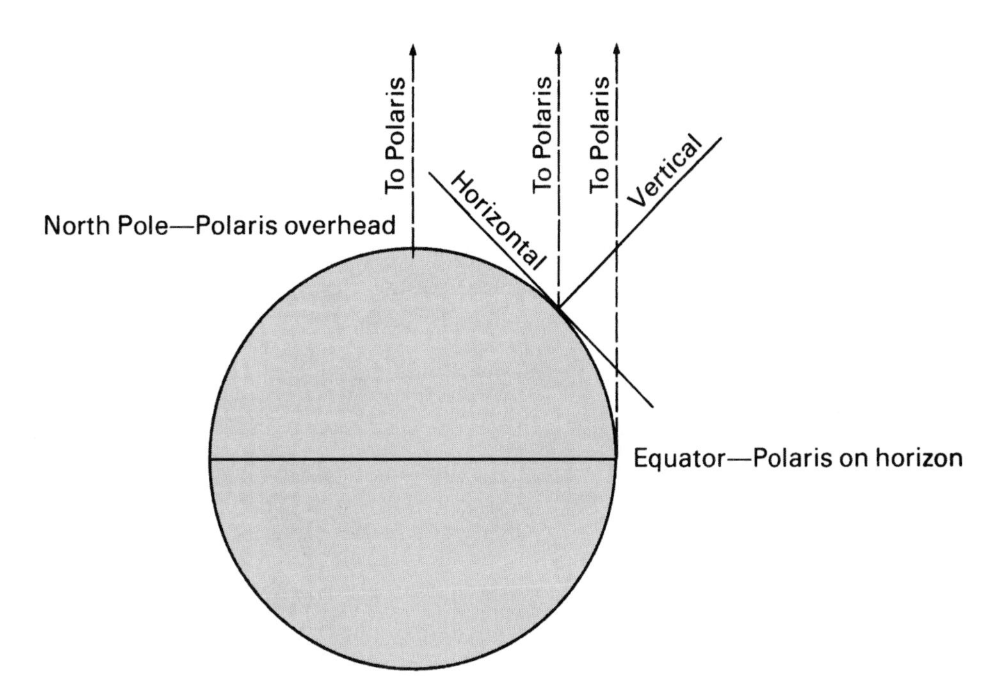

Figure 2.8 Measuring the angle between Polaris and the observer's horizon defines the astronomic latitude (Northern Hemisphere).

a ship, even in calm weather. In addition, the computations involved were long, tedious, and prone to error. Another suggested approach was to anchor a large number of ships at prescribed locations. Each ship would fire a cannon at a prearranged time. A navigator who heard the cannon shot would compare the time at which it was heard with the time at which it should have been fired. The difference between the two times, multiplied by the speed of the passage of sound through the air, would presumably determine the distance of his ship from the navigation vessel. This idea was ruled out because of many impracticalities, such as the vast cost of constantly maintaining a fleet of ships and their crews and the need for precise timekeeping. Yet another suggested method was to observe magnetic declination at a ship's location. (Magnetic declination is the difference between the direction to the north magnetic pole and the direction to the geographic North Pole [see Chapter 7].) Because declination varies somewhat predictably, it was thought that its magnitude would provide locational information. However, problems such as the difficulty of obtaining accurate measurements, and of accurately predicting magnetic variation—which changes over time—led to the abandonment of this method. Many other approaches were proposed, some rational, some not. The one that would eventually claim the prize, however, was the use of a chronometer to improve the measurement of time.

A **chronometer** is simply an extremely accurate clock. It had long been possible to design and construct stationary clocks that met high standards of accuracy. When these clocks were put on moving ships, however, they were subjected to the often-extreme movements caused by wind and waves. These movements created friction or balance problems that introduced errors into the clock's timekeeping. Also, a wide range of temperatures might be encountered as the ship traveled from place to place on a long voyage. These temperature variations also affected the accuracy of ordinary clocks. An English clockmaker, John Harrison, eventually designed a clock mechanism that successfully maintained accurate time despite any environmental problems it might encounter at sea. Harrison's chronometer was finally accepted in 1773, after a long series of delays and machinations introduced by the board that was charged with judging the proposed methods.

The relationship between time and longitude is clear. The sun appears to pass from east to west over a given meridian once every 24 hours. During those 24 hours, the sun passes over the full 360 degrees of the earth's circumference. Thus, every hour, it passes over 15 degrees of longitude (1/24 of 360 degrees). If, for example, it is 3:00 P.M. at the prime meridian when it is noon at the observation point, three hours have passed since the sun passed over the prime meridian. The sun traversed 45 degrees of longitude during that time period, so the observation point is 45 degrees west of the prime meridian. Similarly, if it is 9:00 A.M. at the prime meridian when it is noon at the observation point, the sun will not pass over the prime meridian until three hours later. In this case, the observation point is 45 degrees east of the prime meridian.

Box 1.1 The Analemma

When examining a globe, you may notice a figure-eight-shaped diagram that is usually located, simply as a matter of convenience, in the Pacific Ocean. This diagram is called an **analemma** (Figure 2.9). An analemma gives information about the latitudes at which the sun is directly overhead throughout the year as well as about the *equation of time.* Although this information is provided more exactly in a set of tables, called an **ephemeris,** the analemma gives an interesting visual picture of the relationships.

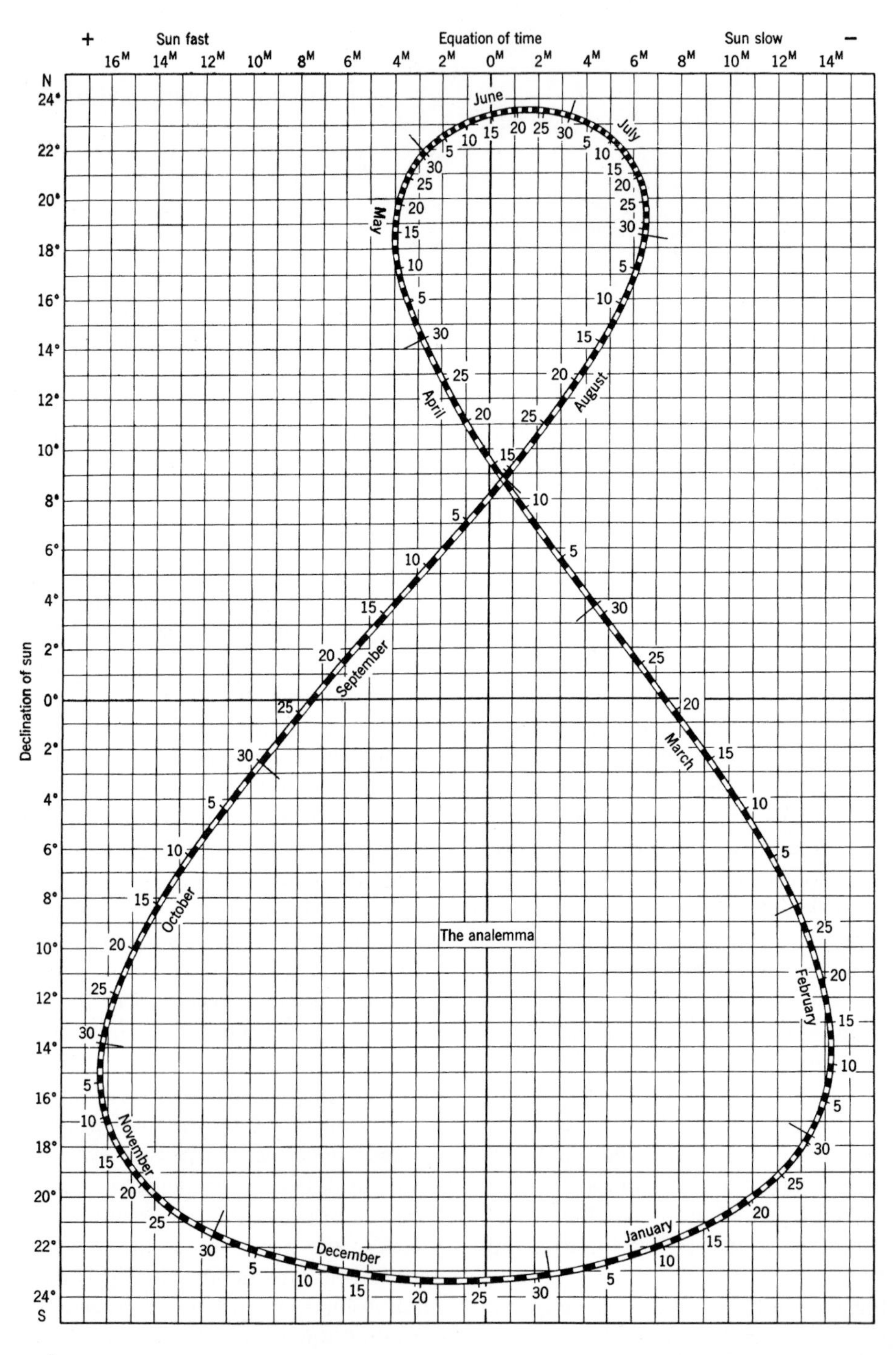

Figure 2.9 An analemma.

The imaginary plane that passes through the sun and the earth at all positions of the earth's orbit around the sun is called the **plane of the ecliptic.** The earth's axis tilts 23 1/2 degrees, relative to a line perpendicular to the plane of the ecliptic (Figure 2.10). This tilt causes the sun to be directly over different latitudes, ranging from the Tropic of Cancer to the Tropic of Capricorn, at different times of the year. The analemma, therefore, stretches between the two tropics on the globe and indicates the parallel at which the sun is directly overhead on any given day. To use the analemma for this purpose, you must locate the date of interest on the diagram. The parallel that intersects it is the required parallel. For example, as shown in Figure 2.9, the sun is almost directly over latitude 16° north on May 5th. This information, which is called the **declination of the sun,** is used to adjust the calculation when an observation of the sun is being used to determine the latitude of a point on the earth's surface.

When the longitude of a point on the earth's surface is determined by observing the passage of the sun, the calculations must take into account the difference between apparent solar time and mean solar time. The reason for this concern is as follows. Noon in **apparent solar time** is the precise moment at which the sun passes over the meridian on which an observer is located. An **apparent solar day** is the period between one noon and the next at the same location. If the earth's orbit around the sun were perfectly circular, every day would be of the same duration. But because of the elliptical shape of the earth's orbit, the speed of the earth's movement varies. As a result, the length of an apparent solar day differs from one part of the year to another. For convenience, however, timekeeping on the earth is based on the average length of the day throughout the year. Thus, every day is taken to be 24 hours long. This is the concept of **mean solar time**. Assume an observer is standing on the central meridian of a time zone. There is usually a difference between the time of passage of the sun over the observer's meridian (apparent solar noon) and the moment of noon on the observer's watch (mean solar time). The difference, which is obtained by subtracting mean solar time from apparent solar time, is called the **equation of time.**

The analemma is used to determine the equation of time on any given date. This is done by locating the desired date on the diagram and extending a vertical line from that date to the horizontal time scale. The intersection of the vertical line with the time scale indicates the number of minutes by which mean solar time is ahead of or behind apparent solar time. On July 10, for example, the line in Figure 2.9 indicates that the sun is *slow* by five minutes, which means that apparent solar noon occurs five minutes after mean solar noon.

The equation of time is used to refine the time difference between an observation location and the prime meridian. This time difference can be important. An eight-minute time differential, for example, makes a difference of two degrees in longitude. Time differentials range from 0 on four occasions throughout the year, to as much as minus 14 minutes in February and more than plus 16 minutes in November.

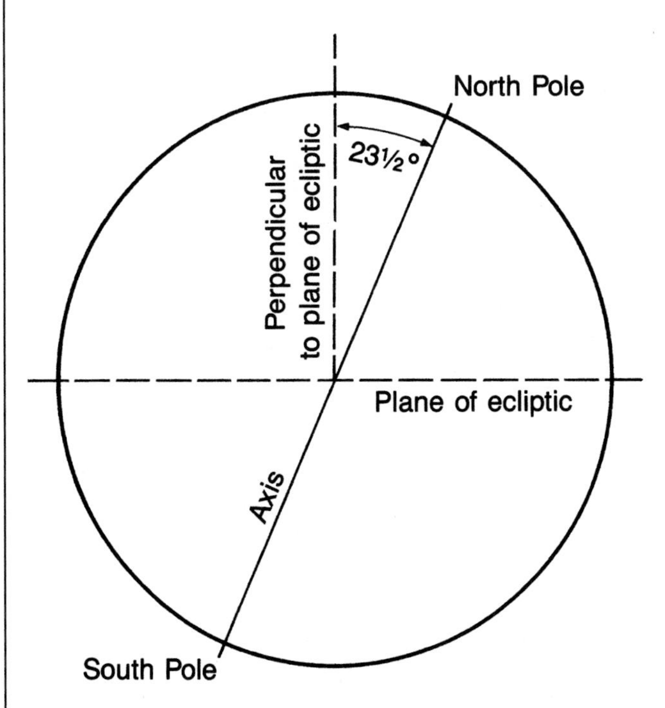

Figure 2.10 Earth's axis is inclined 23 1/2 degrees from a vertical line through the plane of the ecliptic.

DATES AND TIMES

It is noon (solar time) where you are when the sun passes over your meridian. At the same moment, half a world (180 degrees) away from your meridian, it is midnight. At the midnight meridian, one day ends and the next begins—it is no longer Monday, for example, but Tuesday. These facts sound straightforward enough and would probably provide a sufficient set of timekeeping rules if, as in earlier periods of history, transportation and communication between different spots on the earth's surface were slow or nonexistent. The simple rules, however, lead to complications when we can travel or communicate quickly over long distances.

For example, according to the basic rules, if it is midnight Monday at your location, it is 12:15 A.M. Tuesday a short distance to the east and 11:45 P.M. Monday the same distance to the west. The coordination of travel schedules for railroads or airlines or broadcast schedules for radio or television stations would be hopelessly complex if we strictly followed the basic rules. The simple rules are even

—Continued on page 28

Continued from page 27

more confusing over long distances. If it is noon in Chicago, what time and what day is it in London? In Tokyo? In Rio de Janeiro? In the interests of simplifying this potential confusion, two special concepts—the International Date Line and the time zone—were introduced into timekeeping.

INTERNATIONAL DATE LINE

The meridian of Greenwich is generally accepted as the prime meridian (0° longitude). The 180th meridian is halfway around the world from Greenwich. Because the earth turns through 180 degrees in 12 hours, the 180th meridian is also half a day away from Greenwich. Thus, when it is noon at the Greenwich meridian, it is midnight at the 180th meridian. Wherever on earth midnight comes, the old day ends and a new day begins. The 180th meridian has been given a special significance in relation to this event. Because it is half a day away from the prime meridian, it has been designated the **International Date Line.** What this means is illustrated by a complete, 24-hour passage of times and dates (Figure 2.11).

Assume that it is precisely noon at Greenwich (0° longitude) and midnight at the International Date Line (180° longitude) (Figure 2.11a). At this exact moment it is the same day (Monday, for example) everywhere in the world. Then, in the next moment, the day at the date line changes from Monday to Tuesday. As time passes, Tuesday comes to additional areas of the earth. After six hours have passed (Figure 2.11b), Tuesday has spread a fourth of the way around the world, west from the date line, and Monday has become correspondingly smaller. When the midnight line arrives at Greenwich (Figure 2.11c), Tuesday

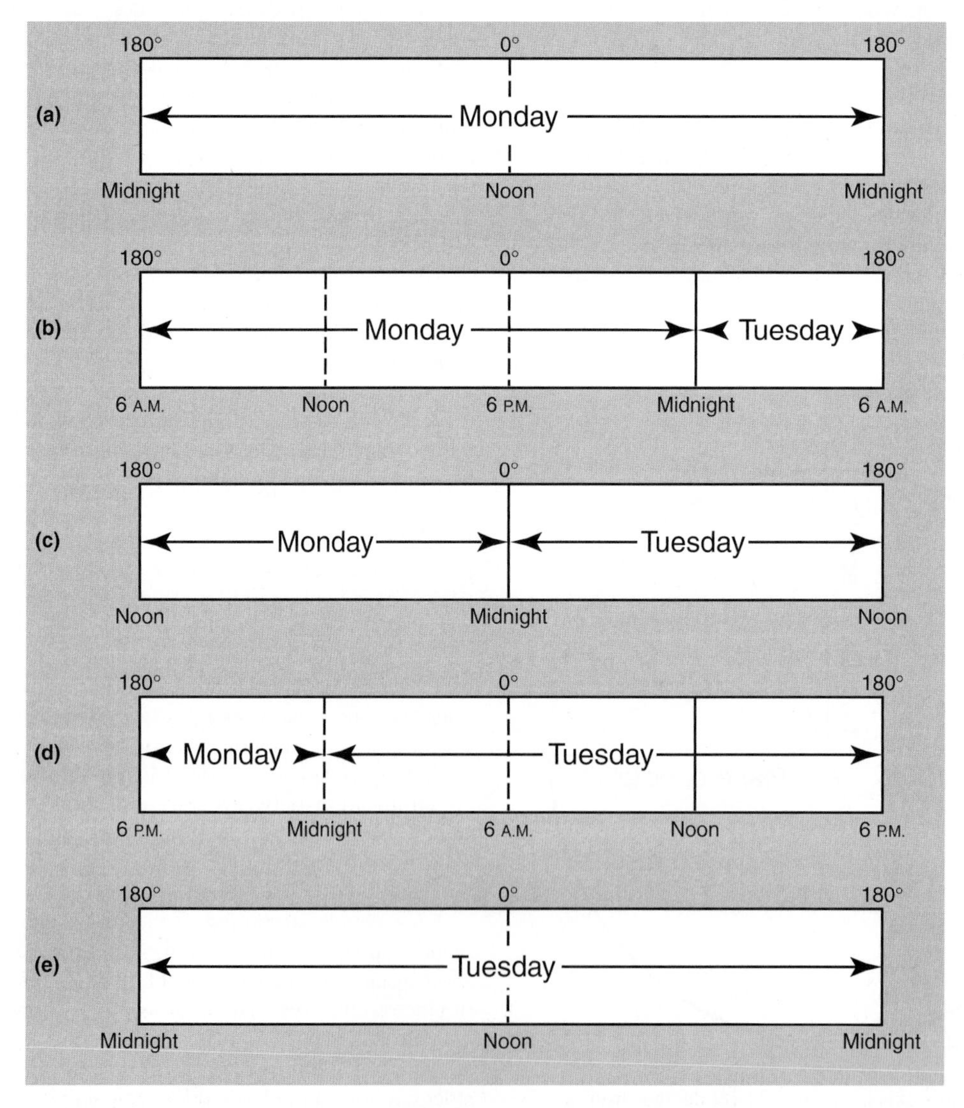

Figure 2.11 The progression of days around the world. See text for details of (*a*) through (*e*).

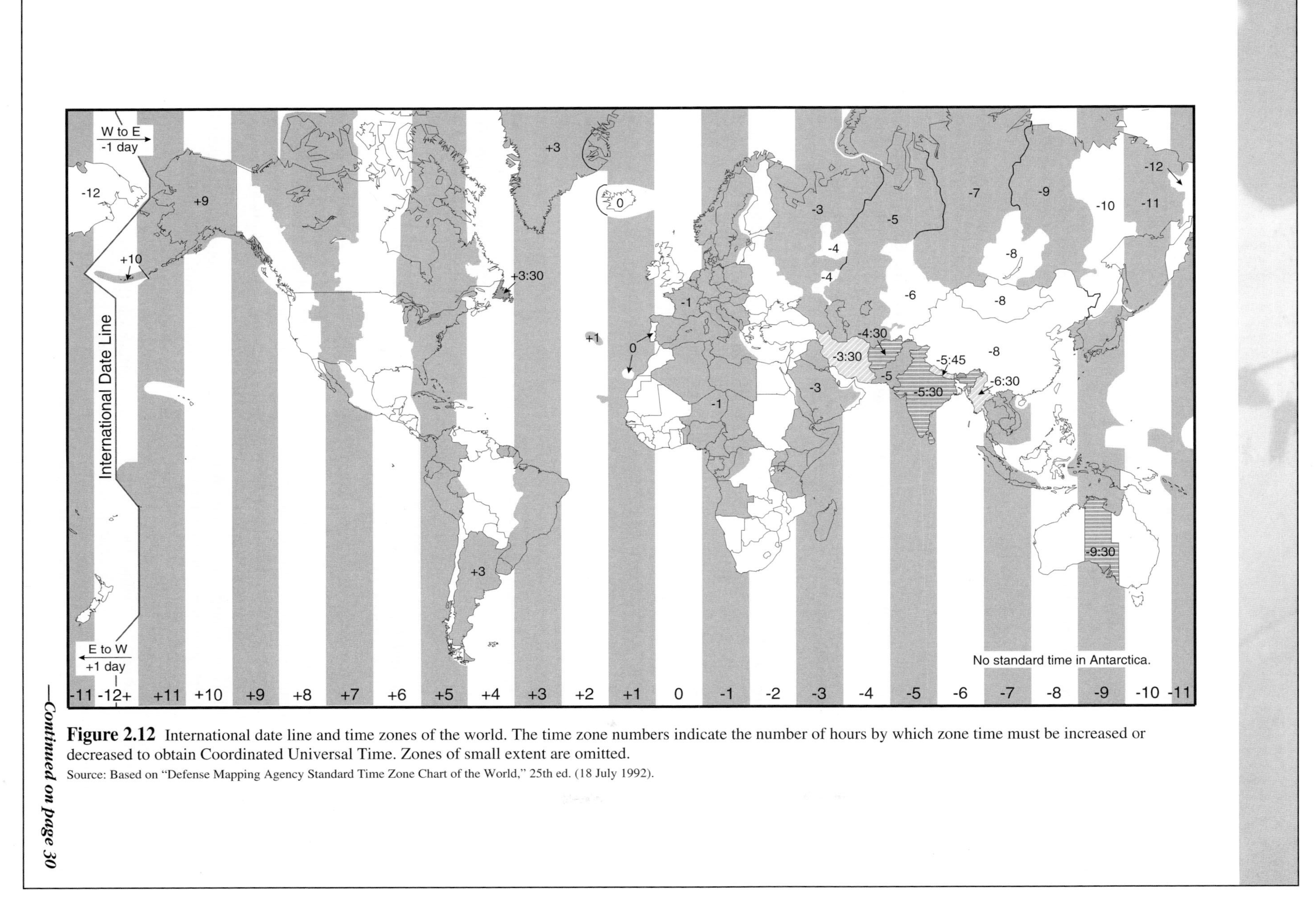

Figure 2.12 International date line and time zones of the world. The time zone numbers indicate the number of hours by which zone time must be increased or decreased to obtain Coordinated Universal Time. Zones of small extent are omitted.

Source: Based on "Defense Mapping Agency Standard Time Zone Chart of the World," 25th ed. (18 July 1992).

—Continued on page 30

Continued from page 29

is 12 hours old and has spread halfway around the world. The process continues (Figure 2.11d) until Monday disappears (Figure 2.11e) and Wednesday begins a new cycle of the same, neverending progression. Except at the stroke of midnight on the date line, there are always two days on the earth at the same time.

The International Date Line introduces a bit of potential confusion into trans-Pacific travel. If you cross the date line, going from east to west (toward Asia from North America, for example), you advance from one day to the next (from Monday into Tuesday) (Figure 2.12). If you cross in the other direction, you go from Tuesday back into Monday. There is a division between one day and the next at the midnight line as well. The midnight line, however, moves around the earth, as we have seen, whereas the International Date Line stays in place.[2]

An additional refinement that should be noted is that the date line does not follow the 180th meridian precisely (Figure 2.12). The deviations from the meridian avoid the arbitrary division of a nation into two different days of the week. Eastern Siberia has the same day as the rest of Siberia, for example, and the western Aleutian Islands have the same day as the rest of that island chain.

[2]Treating the polar view of the earth as a clock face provides a particularly useful way of keeping track of the progression of time and days. See David Greenhood, *Mapping* (Chicago: University of Chicago Press, Phoenix Science Series, 1964), 11–12.

TIME ZONES

The day is divided into 24 hours, and during a 24-hour period, the earth rotates 360 degrees on its axis, at a rate of 15 degrees of longitude per hour. For convenience, the earth is similarly divided into 24 **time zones,** each spanning 15 degrees of longitude (Figure 2.12). The central meridian of each of these zones is designated as a time meridian. In the conterminous United States, for example, the time meridians are 75° west, 90° west, 105° west, and 120° west. Under the time-zone concept, all localities within each zone keep the same time as the zone's time meridian. **Coordinated Universal Time (UTC)** is the basis of the times observed in each time zone. (UTC was formerly known as Greenwich Mean Time [GMT].) For example, the time for the first time meridian east of Greenwich is one hour later than the time at Greenwich. The time for the first time meridian west of Greenwich, however, is one hour earlier than the time at Greenwich. UTC radio time signals are continuously transmitted from carefully regulated observatories and are available worldwide.

The boundaries of the standard time zones have many jogs built in. These jogs tend to follow political boundaries (national and local) or are arranged to fall in sparsely populated areas. These accommodations reduce the necessity for persons engaged in local commerce to constantly adjust their time.

The time differences used in longitude calculations are obtained as follows: First, the navigator determines noon at the observation point by using a sextant, or similar instrument, to observe when the sun is at its highest point in the sky. At the exact moment that the sun reaches its zenith, the time at Greenwich is read from the chronometer. The difference between the two times is then converted into degrees of longitude, as previously described.

In modern times, precise information about the time at the Greenwich prime meridian is available worldwide from radio time signals. These signals provide the time information needed for determining longitude with an accuracy that exceeds that of the most precise chronometer. Thus, to a large extent, the once supremely important invention of the chronometer has been supplanted by modern technology. Nevertheless, the chronometer reigned supreme in the field of navigational timekeeping for over 250 years, and it is still used by navigators who do not have access to the equipment needed to receive radio time signals.

Deflection of the Vertical

The characteristics of the earth present another complication in determining location. The accuracy of the observations of the sun and stars depends in part on a precise determination of the vertical. This is because the instruments used to make the observations must be properly aligned relative to the earth. This alignment is accomplished by observing the pull of gravity using a level, a plumb bob, or a similar device. A plumb bob, for example, points toward the earth's center of gravity. Because of the variation in the earth's density, however, the direction to the center of gravity does not always coincide with the direction to the earth's geometric center. The difference at a given location between a vertical line drawn to the center of the earth and a vertical line drawn to the earth's center of gravity is called the **deflection of the vertical** (Figure 2.13). During a geodetic survey, the effect of the deflection of the vertical must be taken into account.

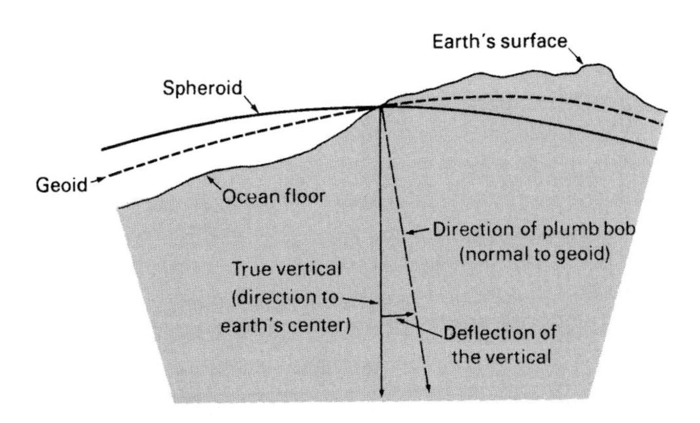

Figure 2.13 Deflection of the vertical.

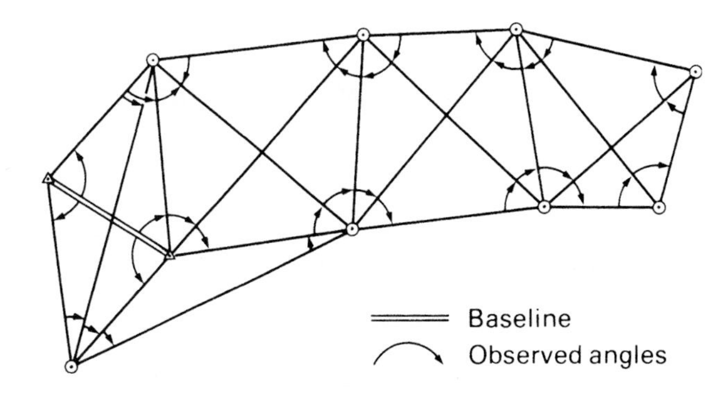

Figure 2.14 Typical triangulation network.

Control Network

After the two starting locations have been determined from astronomical observations, the process of establishing a geodetic control network begins.[3] First, a baseline is carefully and accurately measured between the two known locations to serve as the basis for the remainder of the survey. A system of triangulation is then established by measuring the angles between the baseline and additional observation points (Figure 2.14). From the length and direction of the baseline, and the measured angles from its precisely located endpoints, the surveyor calculates the locations of the newly observed points. Additional angular measurements to new points establish a network of triangles and a series of known locations at which **monuments** are placed. Ultimately, entire countries or continents are covered with geodetic control networks.

Recent technological developments in surveying have increased the accuracy of distance measurements. Many surveys are now based on the measurement of distances instead of the measurement of angles. The **trilateration networks** established by such surveys serve the same purpose as the historically more common **triangulation networks** discussed here.

Horizontal Datum

For many years, maps of North America were based on a vast control network, called the North American Datum of 1927 (NAD27), which originated at Meade's Ranch, Kansas. Positions in the network were recalculated for use in connection with the North American Datum of 1983 (NAD83), which now serves as the standard reference for future maps of the continent. The new **horizontal datum** was necessitated by today's engineering and scientific accuracy requirements and was made possible by satellite observation methods. The datum change resulted in the adjustment of most latitude and longitude locations throughout North America. The adjustments ranged from near zero to as much as 250 meters. Small-scale maps are not affected by such small changes, but large-scale maps require adjustment. For example, relocated corner points are now included on U.S. Geological Survey (USGS) quadrangles. Future maps will be based entirely on the new datum.

[3]In actual practice, the astronomic coordinates of a number of points are determined, and the relationships between them are mathematically adjusted to provide a framework that takes into account the variations from place to place of the deflection of the vertical. The methods by which this is accomplished go beyond the basic concepts presented here.

Plane Surveys

Even though the earth has a basically spherical shape, limited local surveys can be conducted without accounting for shape because minimal inaccuracies occur when surveying over short distances. When a city's sewer system or a county's highways are surveyed, for example, the measurements and computations are made without considering the earth's spherical shape. The term **plane surveying** is used to identify such surveys.

Horizontal Locations

Plane surveying uses a combination of distance and angular measurements to determine the relationships between feature locations. Plane surveys of local areas are tied into the geodetic network by using the geodetic monuments as precisely defined starting and ending points. Given a known, identifiable ground point, such as a geodetic monument, the distances and directions to other points can be measured and recorded (Figure 2.15). From the information collected by the survey, maps are constructed to accurately show the location of selected features, which can range from the property lines of real-estate subdivisions to the alignments of highways or power lines.

Elevation Determination

Because the surface of the earth is not flat, maps often must also incorporate information about the variations

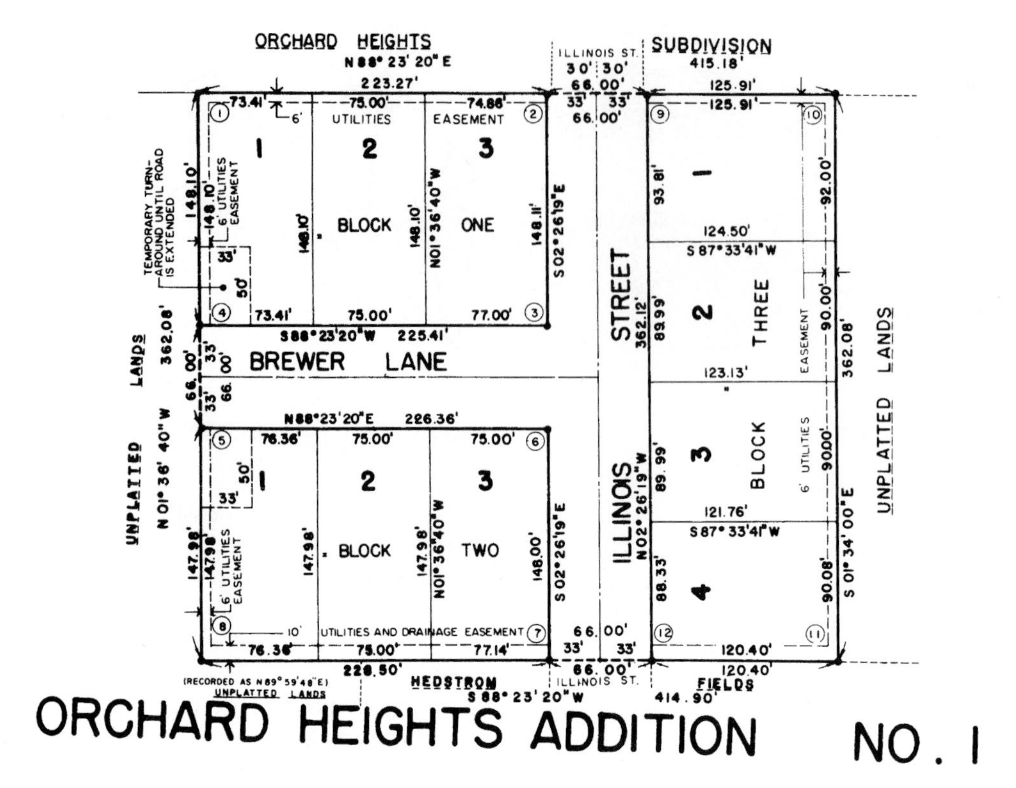

Figure 2.15 Plane survey, subdivision plat. (Town of Mount Pleasant, Racine County, Wisconsin, 28 July 1976. Reduced from original.)
Source: Courtesy of Racine County Planning and Development.

in its elevation. Measuring the difference in elevation between two points is not difficult, especially if the points are relatively close to one another. Using a level and a measuring rod, surveyors start at a point whose elevation is known and compare the elevations of other points to it. A level is basically a telescope that can be aimed in a perfectly horizontal direction to take readings on a measuring rod scaled in the desired units.[4]

One way in which differential **leveling** is used is as follows: First, the level is set up halfway between the starting and ending points. The starting point is preferably a previously surveyed monument of known elevation. (More will be said shortly about the methods used to establish this starting level.) The ending point is the point whose elevation relative to the starting point is to be determined. The measuring rod is placed vertically on top of the starting monument, and the surveyor sights on the rod through the level. The crosshairs in the level intersect the measurements on the rod. The measurement at the intersection of the crosshairs indicates the elevation of the instrument above the starting monument (Figure 2.16). The rod is then moved into position at the ending point. The level is turned, and a sighting is again taken on the vertical rod. This measurement is the elevation of the instrument above the ending point. The difference between the two readings represents the elevation difference between the starting point and the ending point.

Vertical Datum

If the survey is purely local, the elevations of any number of points may be determined, all relative to the elevation of the starting point. However, the elevations in one survey will not necessarily have any relationship to the elevations in the other surveys. This problem can be overcome only if a common elevation, known as a **datum,** is used as the starting point for all of the surveys.

The most commonly used datums are related in a systematic way to the level of the ocean because, in general, sea level meets the requirements for a universal starting level. The determination of sea level requires a considerable amount of effort because the oceans do not maintain a constant elevation relative to the land masses that surround them. This is due to tidal movement caused by variations in the gravitational attraction of the sun and the moon as their positions change relative to the earth. This is not an easily understood set of relationships, because the influence of each of these bodies varies over time in its own pattern. The distance of the sun and the moon from the earth changes, and the attitude of the earth in relation to the direction of their gravitational pull also differs. The combination of these two separate patterns of change results in an extremely complex pattern of variation.

[4]A horizontal sighting is perpendicular to the direction of gravity at the observation point.

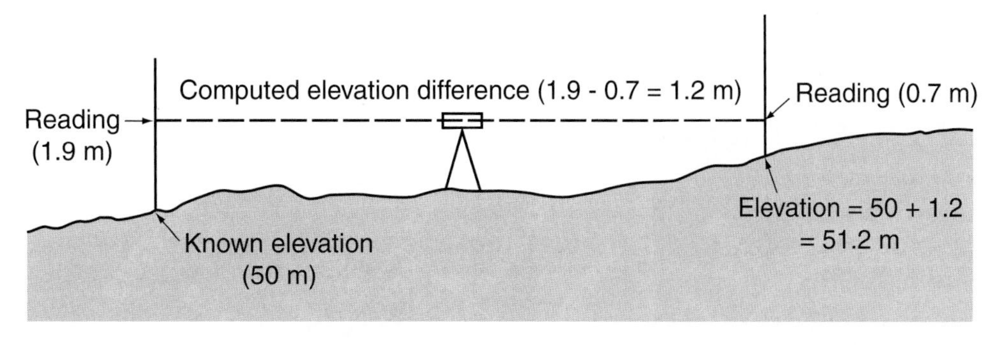

Figure 2.16 Determining elevation difference by differential leveling.

Because of the variation in tidal levels, an averaging concept, called the **mean datum,** is used to establish the desired common starting level. The averaging is done by taking tidal observations at hourly intervals. When a sufficient number of observations has been taken to account for all of the tidal variations and fluctuations, the average is calculated. The total time period involved in this process, if maximum accuracy is required, is 18.6 years! The complete cycle of relationships between the sun, moon, and earth, called the **metonic cycle,** evolves over that period of time.

Other problems must be overcome in the process of determining mean sea level. For example, winds disturb the surface of the water, resulting in waves that can reach major proportions. Obviously, a "surface" existing in a body of water churned by waves is difficult to measure. These difficulties are overcome by creating a **stilling basin,** cut off from the direct influence of wind and waves. A tunnel below the depth of direct wave influence allows water pressure to establish a more easily measured level within the basin.

When a datum is established for large areas, variations *between* stations due to differences in gravity from place to place must also be taken into account. An averaging process based on a leveling survey between tidal measuring stations accomplishes this. The results of this averaging is used to establish the final mean datum. The standard North American Datum, which is used in the United States and the rest of North America, was adjusted in 1929 and again in 1988 (NAD88). Other levels, such as mean low tide or mean high tide, may be calculated for special purposes such as navigational charts or boundary surveys.

TOPOGRAPHIC MAPPING

When the required control-survey work has been done, topographic mapping is built upon the framework that the survey provides. The topographic mapping is usually accomplished by combining surveying with techniques based on aerial photography. Topographic maps are discussed here because they are important in two major ways. First, the maps themselves have many practical uses. Second, they provide information that is the basis of the other types of maps that we will be considering.

Aerial photography allows the rapid acquisition of accurate, up-to-date information about the size, location, and character of objects on the earth's surface. It also reveals the configuration of the surface itself. For these reasons, aerial photography is essential to modern topographic mapping. Since World War II, it has been used extensively in the gathering of information needed to produce new maps and to update existing maps. Also, in many instances aerial photographs substitute for maps. In the remainder of this chapter, we will discuss how aerial photographs are used in the process of creating maps. In Chapter 18, we will consider the use of aerial photographs as map products.

Aerial Photography

Vertical aerial photographs for mapping purposes are taken with an **aerial survey camera,** which typically takes photographs that are 23 centimeters (9 inches) square. This large format permits the capture of a considerable amount of detail over a large image area. As the name suggests, vertical photographs are taken with the camera in as perfectly vertical a position as possible, so that the geometry of the resulting photo can be easily handled. If a photo is not perfectly vertical, corrections must be introduced to make it usable, which adds to the complexity of the process.

Stereoscopic Images

When we look at the world in the course of our normal activities, we see it in three dimensions. The major reason we have this ability is that our eyes are slightly separated from one another. Thus, one eye sees a scene from a somewhat different angle than does the other. This situation is called **parallax.** When parallax is present, the brain is able to interpret the slightly differing views so that we "see" a three-dimensional image.

(a)

(b)

Figure 2.17 Successive, overlapping aerial photographs used for stereoscopy. (Reduced from 9 1/2-by-9 1/2-inch originals.)
Source: U.S. Geological Survey photos.

Unfortunately, individual aerial photographs do not have a three-dimensional quality but simply show a flat image of features on the earth's surface. For this reason, aerial photographs used for topographic mapping applications are taken in a series, with each photo overlapping the preceding photo (Figure 2.17). Overlapping photos provide two slightly differing views of each portion of the earth's surface. Photos taken in this manner are called **stereo pairs.** In use, stereo pairs are placed in a simple instrument called a **stereoscope** that allows one eye to see one view and the other eye to see the other (Figure 2.18). Parallax is thus obtained, and the result is an apparently three-dimensional view of the earth's surface—a **stereoscopic image.**

When stereo pairs are used for mapping purposes, they are typically viewed in a more complex instrument, called a stereoplotter. Many types of stereoplotters are available. Some of them are simple, and some are highly automated, so operator intervention is minimal. Whatever type is used, their common characteristic is that they allow the operator to see one of the stereo-pair images with one eye and the other image with the other eye. This creates an apparent three-dimensional view of the terrain, called a **stereomodel** (Figure 2.19). The discussion that follows is designed to describe the general process involved and not the details related to any particular machine.

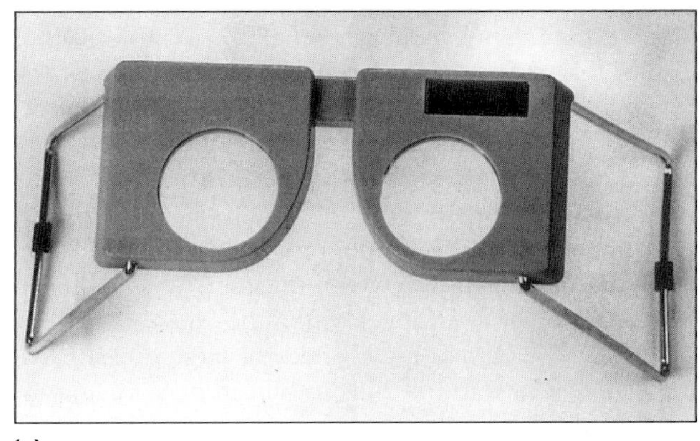

(a)

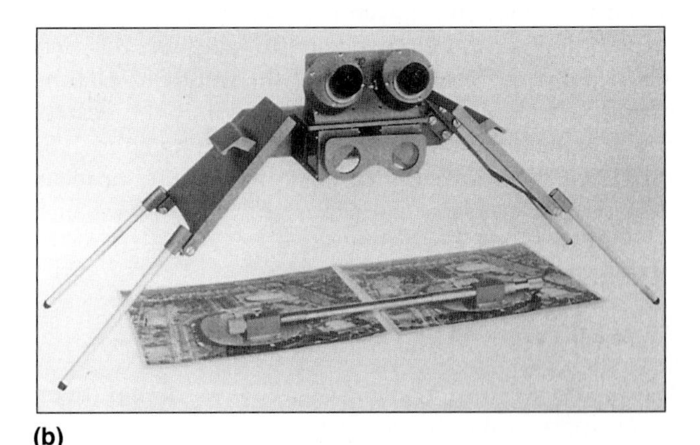

(b)

Figure 2.18 Two types of stereoscopes. (*a*) Pocket type. (*b*) Folding-mirror type with photos and parallax bar in place.
Source: Courtesy of the Sokkia Corporation.

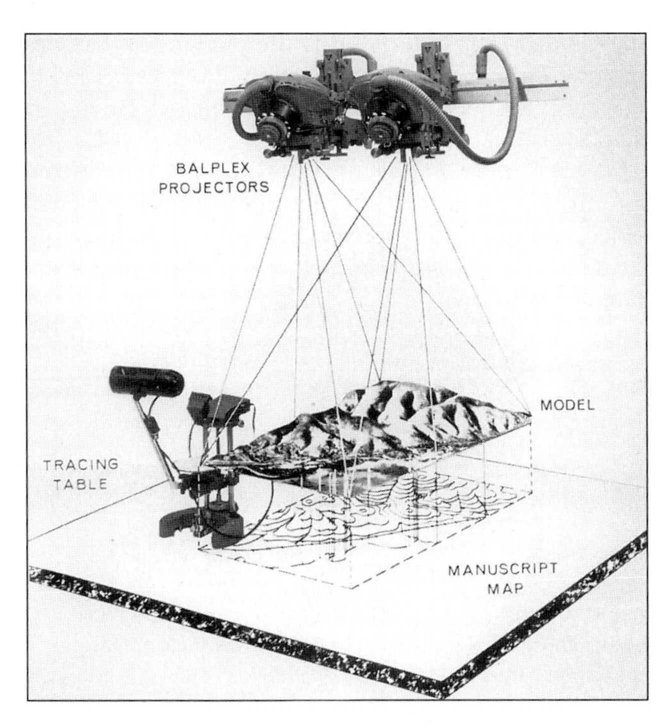

Figure 2.19 Relationship between projectors, stereomodel, tracing table, and control diagram (manuscript map) in a multiplex-type plotter.
Source: Courtesy RFM Associates Inc.

Ground Control

In conjunction with aerial photography, **ground control points** are established at locations that are easily identifiable, both on the ground and in the aerial photos. A survey establishes the precise locations of the control points, and a map, called a **control diagram,** is prepared on the basis of the survey. The operator of the stereoplotter places the control diagram on the base of the instrument. The relationship between the stereomodel and the control diagram is then established by matching the control points on the model to the corresponding points on the diagram.

Plotting

The steroplotter uses a point of light, called a **floating mark,** that appears to be suspended over the stereomodel. Vertical data, such as the location of a contour line (a line joining points of the same elevation), are obtained from the stereomodel by locking the floating mark at the desired elevation. It is then moved so that it remains in contact with the apparent surface of the stereomodel and, therefore, follows a contour line. As the mark is moved, a planimetric line-image of the contour is traced on a manuscript map. Additional contours are established by raising or lowering the floating mark to the desired elevations and tracing the line of contact at each elevation.

Planimetric data, such as the alignment of a road, are obtained by moving the floating mark until it is just touching the apparent surface of one end of the feature in the stereomodel. The mark is then moved so that it follows the center line of the feature. Because the surface of the feature varies in elevation, the mark is moved up and down to keep it in contact with the surface. The planimetric location of the horizontal path followed by the floating mark is recorded on the manuscript map.

This process is repeated, as necessary, until all of the features to be mapped are recorded. Then the manuscript map is checked for accuracy and the necessary drafting, photography, and printing are done to produce a completed map. A digital file of *x-y* coordinate values is often recorded when a map is drawn. This type of file may be used later in conjunction with a plotter to create a hard-copy map (see Chapter 19).

SUMMARY

Knowing the shape and size of the earth is crucial to understanding the relationships between maps and the earth. The earth's basically spherical shape is particularly important because it is difficult to represent on a flat map sheet.

Locations on the earth are specified on the basis of the latitude and longitude graticule. Great circles that pass through the poles are the lines of longitude (meridians) used to designate east-west locations. Other graticule lines are placed at right angles to the meridians. One of these is the equator, a great circle located midway between the poles. The others are the lines of latitude (parallels), which are small circles parallel to the equator.

Because the earth rotates on its axis, it bulges at the equator and is flattened at the poles, resulting in a shape called an *ellipsoid.* The earth is not a regular ellipsoid but is a somewhat irregular shape called a *geoid.*

Surveying determines the position of points on the earth's surface. Geodetic control surveys take into account the curvature and irregular shape of the earth and are used to ascertain latitude and longitude locations. Latitude is determined by measuring the angle between Polaris, for example, and the observer's horizon. Longitude determination, on the other hand, requires knowledge of the difference in the time of occurrence of apparent solar noon at the location to be determined and at the Greenwich meridian.

The establishment of accurate locations is complicated by variations in the earth's density. The difference at a given location between a vertical line to the center of the earth and a vertical line to the earth's center of gravity is called the *deflection of the vertical*. This variation affects the readings of surveying instruments because they respond to the pull of gravity.

A geodetic control survey starts with an accurately measured baseline. A system of triangles is established by measuring the angles between the baseline and additional observation points, at which monuments are established. Plane surveys of local areas are tied into the geodetic network by using the monuments as starting and ending points. Information about variations in elevation is added by leveling techniques, relative to a vertical datum.

Aerial photography provides accurate, up-to-date information about the size, location, and character of objects on the earth's surface as well as about the configuration of the surface itself. Particularly useful are three-dimensional images obtained by using stereoscopic pairs of vertical photos, which are tied to the surveyed control points.

SUGGESTED READINGS

Andrewes, William J. H., ed. *The Quest for Longitude*. The Proceedings of the Longitude Symposium, Harvard University, Cambridge, Massachusetts, November 4–6, 1993. Cambridge, Mass.: Harvard University, 1996.

Campbell, John. *Introductory Cartography*. 2d ed. Dubuque, Iowa: Wm. C. Brown Publishers, 1991, 61–106.

Dickinson, G. C. *Maps and Air Photographs*. 2d ed. New York: John Wiley & Sons, Inc., A Halsted Press Book, 1979, chapter 4.

Robinson, Arthur H., Joel L. Morrison, Phillip C. Muehrcke, A. Jon Kimerling, and Stephen C. Guptill. *Elements of Cartography*. 6th ed. New York: John Wiley & Sons, Inc., 1995, 42–58.

Smith, J. R. *Basic Geodesy*. Rancho Cordova, Calif.: Landmark Enterprises, 1988.

Snyder, John P. *Flattening the Earth: Two Thousand Years of Map Projections*. Chicago: The University of Chicago Press, 1993.

______, and Philip M. Voxland. *An Album of Map Projections*. U.S. Geological Survey, Professional Paper 1453. Washington, D.C.: USGPO, 1989.

Sobel, Dava. *Longitude: The True Story of a Lone Genius Who Solved the Greatest Scientific Problem of His Time*. New York: Walker Publishing Co., 1995.

CHAPTER

3

MAP PROJECTIONS

As discussed in Chapter 2, the earth's shape is basically spherical. Therefore, a **globe,** which shows the correct relationships between features on the earth's surface, is virtually an ideal model of the earth. Why, then, are maps needed at all? Why not use globes for the functions maps typically serve?

Several considerations make maps a necessity. First, globes are expensive to make. The processes involved are time-consuming and require meticulous handwork, such as fitting and gluing many tapered strips of paper, called **gores,** to the surface of the globe (Figure 3.1). Therefore, good-quality globes are difficult to produce in quantity and their prices remain high. Second, globes are cumbersome and awkward to use. Even a classroom-size globe, typically less than a meter in diameter and showing little detail, consumes a large amount of space and is hard to store and handle. Third, to show any significant part of the world in detail, an extremely large globe, perhaps many meters in diameter (or at least a section of such a large globe), is required. Such globes can be built, but the costs and awkwardness are multiplied many times over. Because of all these factors, very few large globes exist.

Plotting locations and routes on the curved surface of a globe is also much more awkward than doing the same tasks on a map. In addition, a viewer can see only part of the globe's surface at once (at most, slightly less than half). For many purposes, therefore, the use of globes is impractical or undesirable. Finally, and perhaps most important, a person looking at a globe sees a **perspective view** (from a single point), rather than the **orthographic view** commonly used for maps (directly overhead at all points).

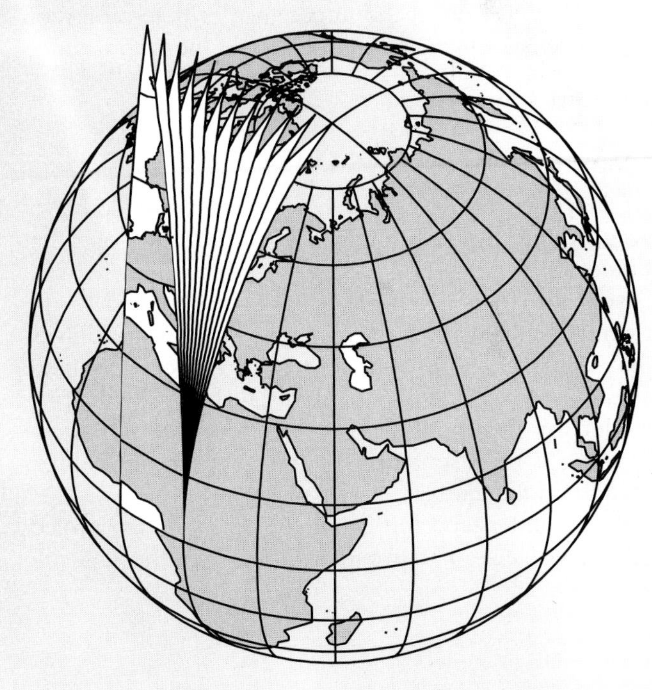

Figure 3.1 Application of a gore to a globe.

Flat maps, in contrast to globes, are inexpensively produced in large quantities. They are easily handled and stored. In addition, it is a simple matter to measure distances, plot paths, and otherwise make practical use of maps, provided the proper projection is used. For these reasons, flat maps are truly a necessity. On the other hand, distortions are inherent when the features that appear on the earth's surface are shown on a flat map sheet.

One can begin to appreciate map distortion problems by simply trying to apply a sheet of paper so that its entire surface is touching the surface of a globe. It can't be done without severely wrinkling the paper. This illustrates the map projection problem. When constructing a map, the cartographer would like to flatten the curved surface of the globe onto a flat sheet of paper without stretching or tearing. Both tasks are equally impossible. Therefore, in practice, the three-dimensional globe is translated to a two-dimensional map representation by map projection techniques. The problem with this approach is that the projection process inevitably involves compromises in the characteristics of the map.

Knowledge of the effects of the map-projection process is an important aspect of understanding maps. This chapter first discusses the general principles that lie behind the various types of map projections. Commonly encountered map projections and their characteristics are then examined. The chapter concludes with a discussion of the importance of map-projection characteristics from the perspective of the map user.

Map-Projection Concepts

A **map projection** is simply a systematic rendering of a **graticule** of lines of latitude and longitude on a flat sheet of paper. (The term *grid,* which is sometimes used instead of *graticule,* should be reserved to describe rectilinear locational systems such as the Universal Transverse Mercator and State Plane Coordinate systems described in Chapter 4.) The skeleton provided by the graticule is converted into a map by adding a representation of the geographic features found on the earth.

To understand the geometric basis of map projections, visualize a transparent globe with the graticule and the coastlines and other features drawn on it. When a light source is placed inside this globe, the various lines drawn on it cast shadows on any surface placed nearby. The outline of these shadows forms the map projection.

Many so-called map projections are actually systematically arranged graticules designed to have particular characteristics. They could not be obtained with the transparent globe-and-shadow approach. The globe-and-shadow idea is useful, nevertheless, as a means of understanding the basic concepts involved in map projections.

Projection Surfaces

Three physical surfaces are commonly used for the construction of map projections. These are the plane and two developable surfaces—the cylinder and the cone.

Plane

The simplest projection surface is a **plane** (flat) surface. Projections onto such a surface are referred to as **azimuthal projections.** The most significant aspect of a plane surface is that an outline projected onto it does not have to undergo further distortion or manipulation because it is already a flat map.

Developable Surfaces

Some projection surfaces are not flat at the time the projection is created but can be flattened later by making an appropriate cut in the surface and unrolling (developing) it. When this unrolling is accomplished without stretching or tearing and, therefore, without distortion of the surface or of the patterns drawn on it, the surface is called a **developable surface.**

The **cone** and the **cylinder** are developable surfaces commonly used for mapping purposes (Figure 3.2). A cylinder is developed by cutting along its length and unrolling it. A cone is developed by making a cut from its base to its apex before unrolling it.

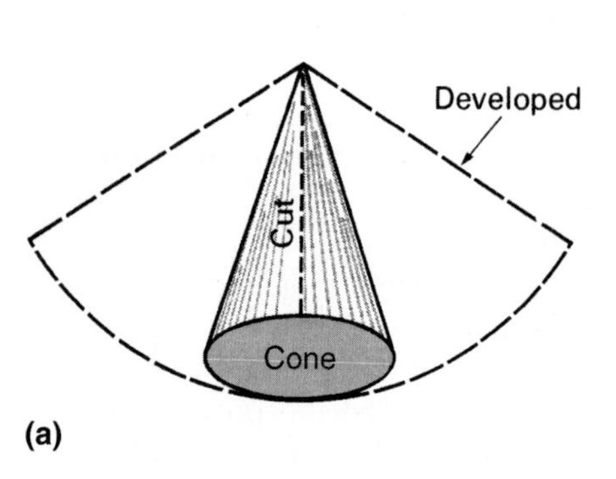

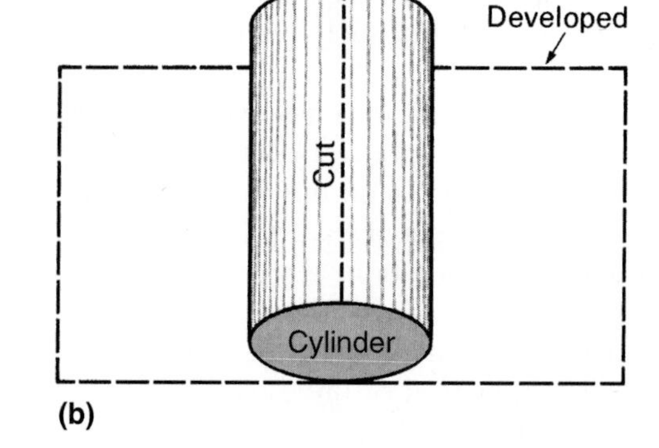

Figure 3.2 Developable surfaces. (*a*) Cone. (*b*) Cylinder.

Light-Source Location

Thus far, this discussion has assumed that the map-projection light source is located at the center of the globe. The light source, however, can be located at any desired point. Three common locations for the projection light source are (1) the **gnomonic position,** at the center of the globe; (2) the **stereographic position,** at the antipode (the point exactly opposite the point of tangency of the projection surface); and (3) the **orthographic position,** at infinity (Figure 3.3). Changing the location of the light source, even when the position and type of projection surface are held constant, alters the characteristics of the resulting projection. The specific effects of these changes are discussed when the appropriate projections are described.

The idea of using different locations for the light source strains the notion of shadows being cast. Sometimes, for example, the light would have to pass through two portions of the globe before reaching the projection surface. In other cases, part of the projection surface may lie between the globe and the light source, so it could not actually have a shadow cast on it. These examples illustrate some shortcomings of the transparent globe as the source for different map projections. The basic idea is clear enough, however, so there should be no confusion when situations arise in which allowances must be made for these difficulties.

Orientation and Tangency

The **orientation of a projection surface** may be changed as desired, although certain orientations are defined as **normal,** or regular (Figure 3.4). A cylinder, for example, is normally oriented so that it is tangent along the equator. A cone is normally oriented so that it is tangent along a parallel, with its apex over the pole, in alignment with the axis of rotation. The normal orientation for a plane is tangent at the pole (polar azimuthal).

If the orientation of the projection surface is changed, the location of the point or line of **tangency** between the surface and the globe also changes. When the projection surface is turned 90 degrees from normal, the result is called a **transverse projection.** In a transverse cylindrical, for example, the cylinder is tangent along a meridian. In the case of a transverse projection on a plane, the plane is tangent at the equator. A transverse conic, which is not frequently seen, has the apex of the cone on the plane of the equator. An **oblique projection** results if the projection surface lies at an angle somewhere between the normal and transverse positions.

The line of tangency between a projection surface and the surface of the globe is called the **standard line** of the projection (if it is along a parallel, as is often the case, it is called the *standard parallel*). Along the standard line, the map has no distortion because there is a one-to-one relationship between the projection and the globe. Away from the standard line, however, distortions occur. The characteristics of these distortions vary according to the specific projection. When the projection surface intersects the globe, the result is called a **secant projection.** If the projection surface of a secant projection is a cone or cylinder, the resulting projection has two standard lines. When the projection surface is a tangent plane, there is distortion everywhere except at the point of tangency. However, if the projection plane intersects the globe, forming a secant projection, the small circle formed along the line of intersection is the standard line of the projection. Some special types of projections may have more than two standard lines.

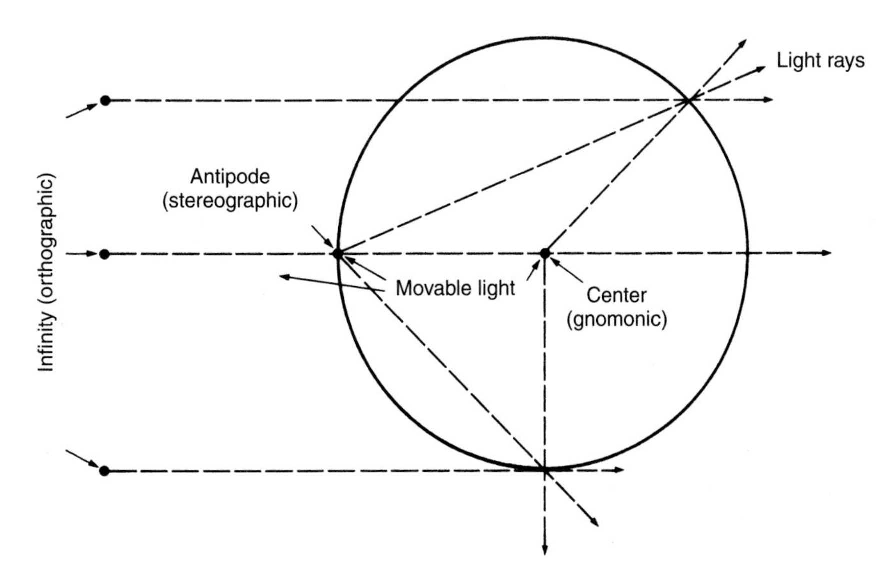

Figure 3.3 Three common locations for the projection light source.

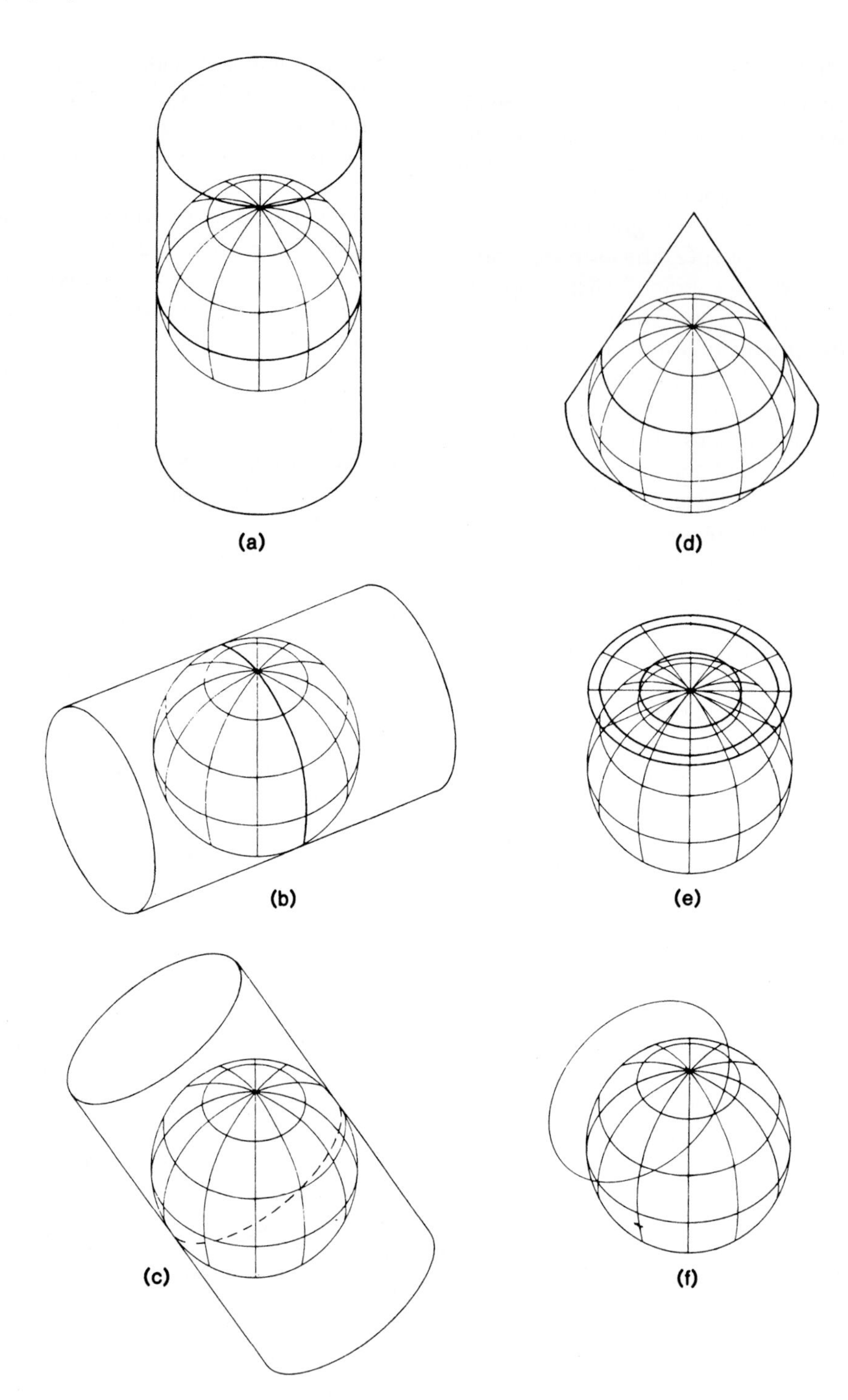

Figure 3.4 Projections on three different surfaces with differing orientations. (*a*) Regular cylindrical. (*b*) Transverse cylindrical. (*c*) Oblique cylindrical. (*d*) Regular conic. (*e*) Polar azimuthal (plane). (*f*) Oblique azimuthal (plane).

Source: John P. Snyder, "Map Projections Used by the U.S. Geological Survey," 2d ed., *Geological Survey Bulletin* 1532 (Washington, D.C.: U.S. Government Printing Office, 1982).

The different orientations of the projection surface each produce a distinctive graticule and pattern of distortion. Although each graticule has a different appearance, those that are produced with the projection light source at the same location and with the same projection surface often retain the same general characteristics. The normal (polar), transverse (equatorial), and oblique cases of the stereographic projection, for example, all maintain correct shapes in small areas. A straight line drawn from the center of any of them represents a great circle. On the other hand, a transverse Mercator (like a normal Mercator) maintains correct shapes, but straight lines drawn on it do not represent great-circle routes (unlike on a normal Mercator). (The Mercator projection is discussed later in the chapter.)

GLOBE CHARACTERISTICS

Because of its similarities to the earth, a globe is a useful reference tool. When features from the earth's surface are shown on a globe, their shape, the area they occupy, and the distances and directions between them are correctly shown. An ideal map projection would retain these characteristics and translate them to the map. Unfortunately, making the transition from a three-dimensional globe to a flat map inevitably involves the loss of some globe properties. If a particular projection retains equal areas, for example, it cannot retain correct shape.

A globe is the only correct representation of the relationships between features on the earth's surface. Therefore, although it suffers from the problems mentioned at the beginning of this chapter, a globe is a useful reference tool. Comparisons between the globe and the map help in resolving questions about the characteristics of a given map projection. Properties of the globe are conveniently classified into two groups: major and minor.

Major Properties

Major properties are those than can exist at all points on certain projections. There are two properties of this type: conformality and equivalence.

Conformality

The retention of correct angles on a map is called **conformality.** The importance of conformality is that map features can be recognized by their distinctive shapes. Conformality also allows the accurate recording of direction. The requirements for a conformal map are that the lines of latitude and longitude cross one another at right angles and that the scale is the same in all directions at any given point. Both conditions exist on a globe.

The term *conformality* is somewhat misleading, because the condition can only be achieved for small areas on a map. The shapes of large mapped areas, such as continents, are different from the shapes of those areas as shown on the globe. This is true even when small areas retain their shape and the map is classified as conformal.

Equivalence

Equivalence is the condition in which a unit area drawn anywhere on the map always represents the same number of square units on the globe's surface. A map with this characteristic is called an **equal-area (equivalent)** projection. Retaining areal relationships is especially important on maps used to represent the areal extent of various phenomena on the earth's surface. Assume, for example, you are interested in the amount of arable land in various countries. It would be very misleading to visually compare the area of arable land in one country with the area in another country if the map showed different-sized areas for the same actual amount of land.

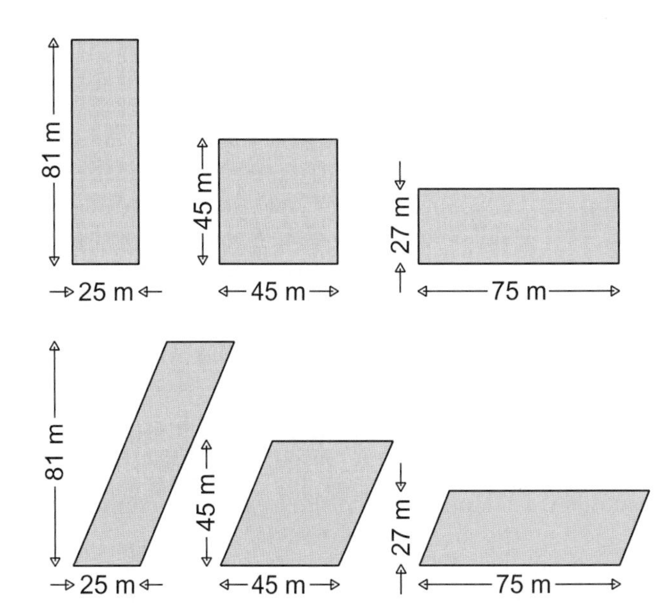

Figure 3.5 Figures with different dimensions and shapes, but with equal areas.

To retain equivalence, scale changes that occur in one direction must be offset by suitable changes of scale in the opposite direction. For example, the three rectangles shown in Figure 3.5 are equal in area, even though their dimensions differ. Compared to the square, the increased height of the tall rectangle offsets its decreased width, whereas the decreased height of the short rectangle offsets its increased width.

The right-angle crossing of parallels and meridians is lost in many projections. The shape changes that result do not affect equivalence; in fact, retaining equivalence inevitably distorts shapes. The fact that different crossing-angles of parallels and meridians do not affect area is shown by comparing the areas of the skewed quadrilaterals in Figure 3.5. Although the shapes of the figures differ, their areas are the same because their height and base dimensions are the same.

Minor Properties

Minor properties are those than can exist in relation to only one, or sometimes two, points or lines on certain projections. There are two properties of this type: distance and direction.

Distance

Correct distance relationships require that the length of a straight line between two points on a map represents the correct **great-circle distance** between the same points on the earth. (The path of a great circle is the shortest distance between two points on the earth's surface.) Maps that have this characteristic can be designed but, even on such **equidistant maps**, correct distances can only be measured along great circles from one or, at most, two points. Distances between other points are incorrect, often by a substantial amount. One should not make the mistake of thinking that measurements between all points, even on an equidistant map, will yield correct distances. Because of the scale distortions involved in many projections, measuring distances on small-scale maps that are not specifically equidistant is not advisable.

Direction

If correct **direction** is retained, a straight line drawn between two points on the map shows the great-circle route and azimuth between the points. An **azimuth** is defined by the angle formed at the starting point of a straight line in relation to a baseline, often a meridian. The angle is usually, but not always, measured in a clockwise direction, starting from north. When north is defined by true north (the direction of a meridian), the azimuth is called a *true azimuth*. Measuring an azimuth with reference to magnetic north or grid north is also possible.

Representing directions accurately from the center of the map to all other points is possible on certain projections. Remember, however, that the azimuths of great circles that do *not* pass through the center of such projections are not shown correctly. Creating a map of the world on which all directions are correct is not possible.

Compromise Projections

Many commonly used projections do not preserve any of the globe properties just described. This might seem at odds with the goal of preserving one or another of the properties. Remember, however, that preserving one globe property (equivalence, for example) results in the distortion of another (conformality). Furthermore, many situations do not demand the use of a map that preserves some particular property. When this is the case, a cartographer may select a **compromise projection** that does not result in extreme distortion of any of the globe properties.

PROJECTION EFFECTS

Each map projection has its own set of characteristics. Ideally, a map user should understand these characteristics in order to realize the advantages and limitations of the map for the purpose for which it is selected. Unfortunately, there are far too many projections to allow a thorough coverage in this book. Instead, seven examples are given. The first four examples each show a projection that retains a specific globe characteristic. The remaining examples show three compromise projections, none of which retains any globe characteristics.

The first example projection is the **Mercator,** which retains conformality (Figure 3.6a). The Mercator is not obtained by geometrical construction but, instead, is mathematically designed so that changes in its north-south scale exactly offset changes in its east-west scale. This means that the scale on the Mercator is the same in all directions at any given point on the map. (Note that this does *not* mean that its scale is the same everywhere). Actually, the scale of a map drawn on a standard Mercator projection is the same as that of an equivalent globe only at the equator (i.e., along its standard line). At 60° North or South, the east-west distance between two meridians on the earth is half the distance between the same two meridians at the equator. On the Mercator projection, however, the spacing between the meridians is the same at 60° as it is at the equator (or at any other latitude). This means that the east-west scale along the 60th parallel of a Mercator map is twice as great as the east-west scale at the equator. These scale differences become increasingly exaggerated away from the equator and toward the poles. In addition to maintaining equality of scale at each point, the meridians and parallels of the Mercator cross one another at right angles. Because these characteristics are also true on the globe, shapes on the Mercator are correct, and the Mercator projection is classified as a conformal projection. It must be reemphasized, however, that conformality does not mean that shapes are correct over large areas, such as continents. What it does mean is that the shapes of smaller, individual features are correct, or nearly so. This is the most that can be expected of a conformal projection, because only on the globe can the scale be the same everywhere.

Other qualities of the Mercator projection, which make it particularly useful for navigational purposes, are discussed in Chapter 7.

Other examples of conformal projections include the Lambert conformal conic (which is often used for maps of the United States) and the stereographic and transverse Mercator.

The second example is the **sinusoidal,** which is an equivalent projection (Figure 3.6b). In the case of the sinusoidal, the scale is true along the central meridian and each parallel, which gives the projection its equal-area characteristic. The right-angle crossing of parallels and meridians is lost in many equivalent projections, including the sinusoidal. The meridians of the sinusoidal projection are sine curves (hence the name sinusoidal). Because of this, the angles at which the meridians cross the parallels become increasingly acute toward the periphery of the projection. The result is an increasing distortion of shapes in those areas. These shape changes do not affect the ability to retain

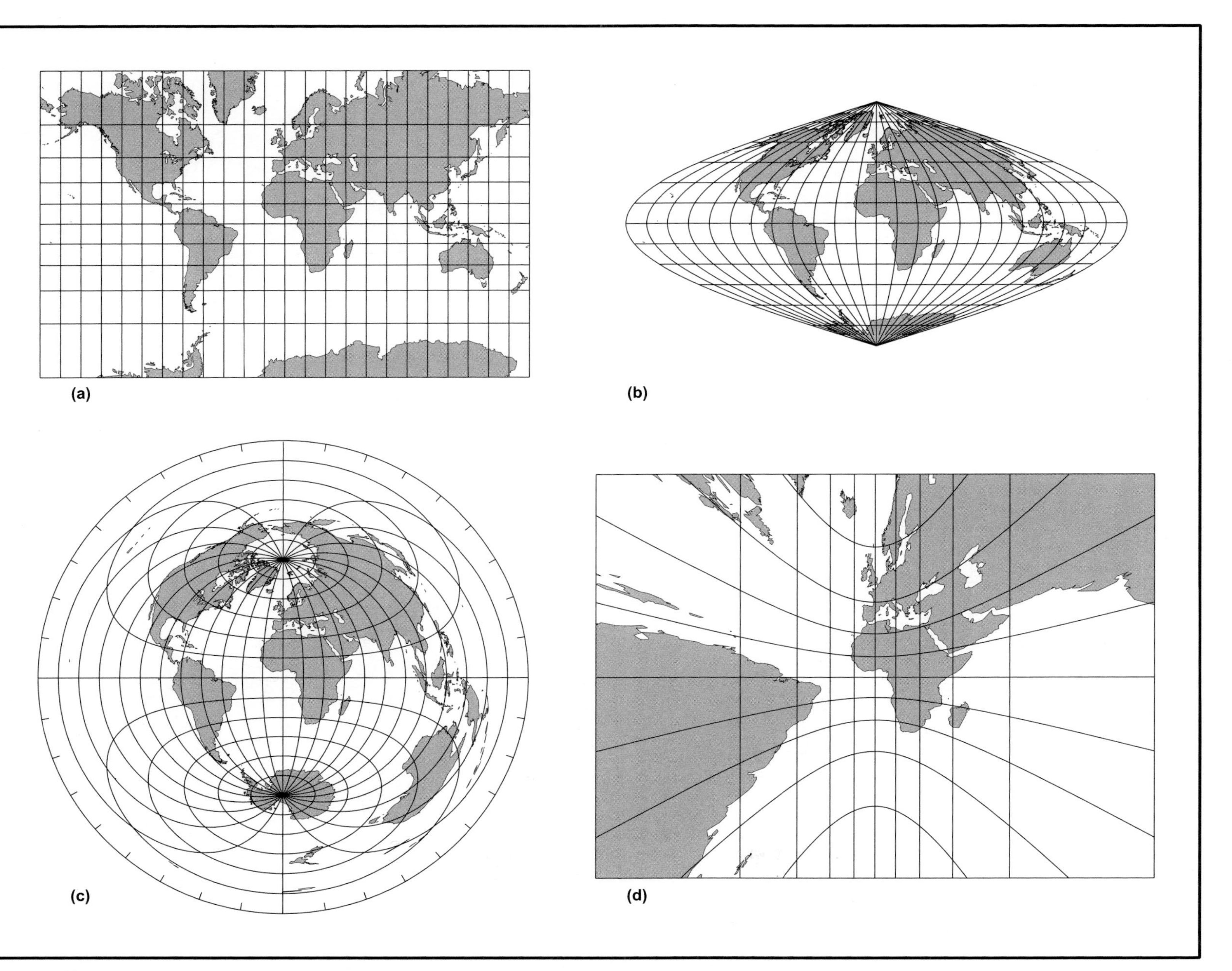

Figure 3.6 Example projections. (*a*) Mercator. (*b*) Sinusoidal. (*c*) Azimuthal equidistant. (*d*) Gnomonic.

equivalence, and, in fact, retaining equivalence inevitably distorts shapes.

Other commonly encountered equivalent projections include the Albers equal-area (which is often used for maps of the United States), the homolographic, the Goode interrupted homolosine, and the Bonne.

Because of their shape distortions, equal-area projections are often presented in interrupted form. **Interruption** involves separating the projection along several dividing lines. This allows the continents (or the oceans, if they are of more interest for a particular application) to be shown with correct areas but with less distorted shapes. The Goode interrupted homolosine is an example of this approach (Figure 3.7). This projection is also unique in that it joins the sinusoidal and the homolosine projections at about 40° North and South to further improve the shapes.

The third example projection is the **azimuthal equidistant** (Figure 3.6c). This projection, as its name states, preserves the globe quality of equidistance. All points on azimuthal equidistant projections are plotted at their true distance from the center of the projection and are in their true global direction, or azimuth, from the center (hence the name azimuthal). Area and shape relationships are not retained anywhere on this projection and are increasingly distorted away from the center.

On the azimuthal equidistant projection, the antipodal point to the center of the projection (the point exactly opposite, on a line drawn through the center of the earth) becomes a circle with a diameter equal to the earth's circumference. This illustrates very forcefully that distance relationships, other than those measured from the center of the projection, are increasingly distorted away from the center point. Measurements between any and all points, even on an equidistant map, do *not* necessarily yield correct distances. Because of this, azimuthal equidistant projections are often centered on some point of particular interest, such as a capital city. This allows distances to be measured from that point to any other point.

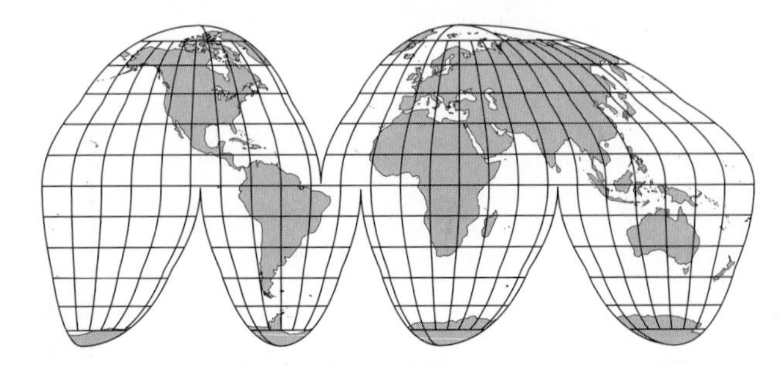

Figure 3.7 Goode interrupted homolosine projection.

The fourth example is the **gnomonic,** on which all straight lines represent portions of great circles (Figure 3.6d). This is so because the light source is at the center of the globe. Imagine that multitudes of planes are passed through the center of the earth. Each plane cuts the globe along the path of a great circle, where it intersects with the globe's surface (Figure 3.8). The planes continue outward until they intersect with the plane on which the map is being created. Because the intersection of two planes always occurs along a straight line, the result is that all great-circle routes appear as straight lines on the map. Applications of this characteristic of the gnomonic are discussed in the section of Chapter 7 that deals with navigation.

Because the light source of the gnomonic is at the center of the earth, the spacing of the graticule increases rapidly away from the center of the projection, resulting in an increasingly exaggerated scale. This means that it is not practical to use the gnomonic projection to cover more than a portion of a hemisphere on a single map.

Compromise Projection Examples

Three compromise projections are introduced here to illustrate the genre: the Miller cylindrical, the Robinson, and the Winkel Tripel.

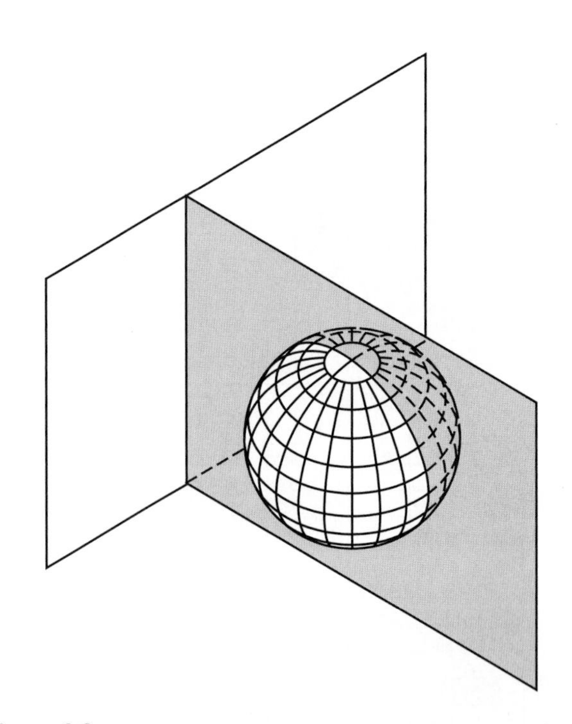

Figure 3.8 The intersection of two planes always occurs along a straight line. In the gnomonic projection, all projection planes pass through the center of the earth. The resulting great-circle plots are therefore straight lines on the map.

Miller Cylindrical

The **Miller Cylindrical** shows the world in a rectangular format (Figure 3.9a). It is modified from the Mercator, so that it distorts the size of areas less, especially in the higher latitudes. It is not conformal, nor does it have the navigational usefulness of the Mercator. Despite the fact that it does not retain any globe qualities, the Miller cylindrical has often been used in atlases, classroom maps, and other applications. One reason for its popularity may be the simple fact that it has a rectangular outline and, therefore, it is easy to fit on an atlas page.

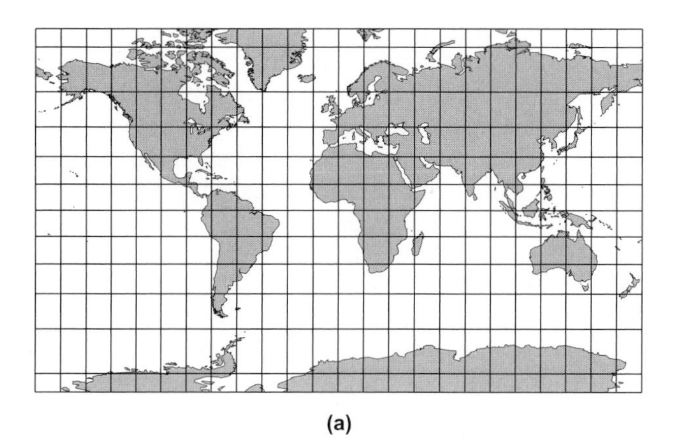

(a)

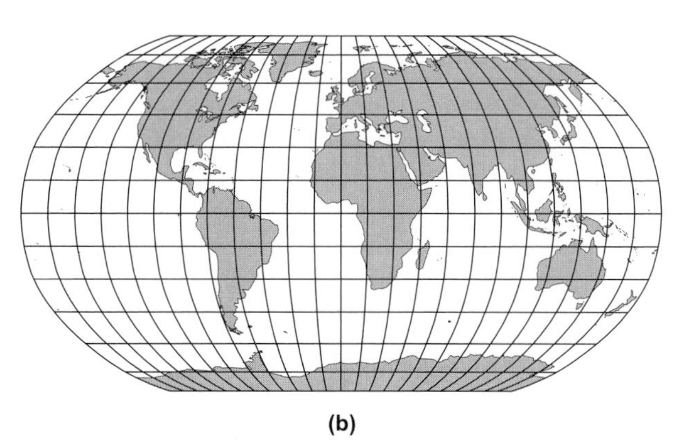

(b)

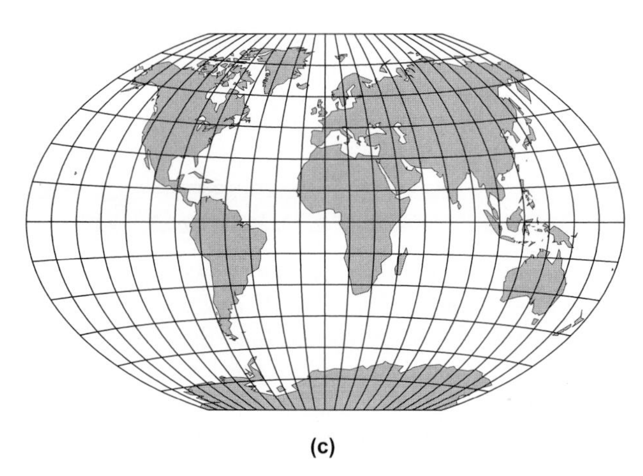

(c)

Figure 3.9 Example compromise projections. (*a*) Miller cylindrical. (*b*) Robinson. (*c*) Winkel Tripel.

Robinson

The **Robinson** was designed by the well-known academic cartographer Arthur H. Robinson (Figure 3.9b). It has the virtue of minimizing visually disturbing distortions that make many world projections unattractive for general uses.

Winkel Tripel

The **Winkel Tripel** is used by the National Geographic Society for their world maps (Figure 3.9c). Its selection for this purpose was based on the balance it provides between size and shape.

PROJECTION SELECTION

As discussed earlier in the chapter, the characteristics of a map are determined by the projection on which it is plotted. The retention of shape, equivalence, direction, or distance may vary in importance, depending on the planned purpose of the map. The projection on which the map is drawn is selected accordingly. Retaining a particular characteristic inevitably results in the distortion of other characteristics, so the choice always involves some give-and-take. When selecting a map for a particular purpose, therefore, the map user must keep the characteristics of the various types of projections in mind.

The projection choices for large-scale maps, such as topographic sheets and navigation charts, have been made by the producing agencies, usually major government agencies, and the map user can be assured that the choices are rational. The Mercator and gnomonic projections are used for navigation charts, for example. The gnomonic charts are used for selecting great-circle routes and the Mercator charts for determining compass headings. Measuring distances on either the gnomonic or the Mercator is difficult because of the constant change in scale. Therefore, an azimuthal equidistant projection is usually better for distance-measurement purposes. Unfortunately, finding azimuthal equidistant projections centered on specific locations is often difficult, so the map user may have to contend with the variable scales of the gnomonic or Mercator.

The Lambert conformal conic projection is frequently used for midlatitude air charts because straight lines drawn on it very closely approximate great-circle routes. Such charts can be used for route selection and distance determination. Similarly, in recent years, conformal projections such as the transverse Mercator have been increasingly used for topographical maps. With the methods described in Chapter 6, the map user can measure short distances, determine directions between nearby points, and calculate areas on such topographic sheets.

In the case of medium- to small-scale maps (in contrast to the situation with large-scale maps), a wide range of producing agencies, companies, and individuals is involved. In these cases, projection selection may be less rational. Thus, the effect of the projection on any such map should be considered before use.

The considerations involved in the selection of map projections include, for example, the size of the area of interest. If a relatively small area is involved, such as one of the counties in the United States, there is no real difficulty. The amount of distortion on virtually any projection would be difficult to detect on map sheets even two or three feet in size, especially if the projection is centered on the area to be mapped. When an area the size of a country or a continent is involved, however, the amount of distortion can become extreme, so the selection is much more important.

The latitude of the area of interest is another important consideration in projection selection. In areas near the equator, for example, cylindrical projections tend to be favored because the cylinder, in the normal orientation, is tangent at the equator, and there is minimal distortion. The lines of tangency of conic projections, on the other hand, are along the parallels, usually in the middle latitudes. This makes conic projections a likely choice for maps of that part of the globe. Pole-centered azimuthal projections are suitable for areas near the poles because the amount of distortion at the poles is minimized and increases toward the lower latitudes. Different versions of this class of projection have special characteristics that may also be selected, including correct distances from the center or equivalence.

The shape of the area of interest also influences the choice of projections because of the different areas of minimal distortion of each projection. When the area is wide in one direction and narrow in the other, one should choose a map that has its area of least distortion aligned with the long axis of the area of interest. For example, the Bonne, sinusoidal, or transverse Mercator projections are usually preferred when the area that is mapped is long in the north-south direction but has little east-west extent, such as Chile. This is because the areas of least distortion of these projections lie along their central meridians. Conic projections, on the other hand, are particularly useful for maps of areas with considerable east-west extent and relatively less north-south extent because their area of minimum distortion is along their standard parallels. The Lambert conformal and Albers equal-area projections, for example, are very suitable for mapping the United States.

Maps of areal distributions are best done on equal-area projections. This is because the area occupied by a particular phenomenon in one part of the world may need to be compared with the areas it occupies in other parts of the world. Or the areal extent of one phenomenon may need to be compared with that of another phenomenon, no matter where in the world it occurs. The equal-area projections, by definition, allow such comparisons, but the map user must be careful to select maps of the same scale.

Conformal projections, on the other hand, are best when compass directions between locations are important. The Mercator, for example, is useful when it is necessary to record direction accurately, as in plotting winds and ocean currents.

Four azimuthal projections have particular applications. The azimuthal equidistant, for example, is useful when it is centered on a point from which the airline distance and direction to any other place in the world are to be determined. The oblique aspects of the **orthographic** projection, on the other hand, are used to represent views of the earth from space. The difference between the projection and the actual perspective of an observer in space is so slight that it is virtually unnoticeable. The stereographic projection is often used for plotting purposes when radiating patterns are involved. Any circle plotted from the globe appears as a circle on a stereographic projection. Finally, the gnomonic projection is useful for radio or seismic work because the waves involved in such studies travel in approximately great-circle directions. They are, therefore, plotted as straight lines on the gnomonic projection.

SUMMARY

Although a globe shows the correct relationship between features on the earth's surface, flat maps are more practical and useful for many applications. Representing the spherical earth as a flat image requires a systematic rendering of the latitude and longitude graticule, called a *map projection*. Although some projections are obtained by the use of the transparent globe-and-shadow approach, most are systematically arranged graticules designed to have particular characteristics. An ideal map projection would retain the characteristics of the globe and translate them to the map. But the transition from a three-dimensional globe to a flat map inevitably involves the loss of some globe characteristics.

Three physical surfaces are commonly used for the construction of projections. These are the plane and two developable surfaces—the cylinder and the cone. Three common locations for the projection light source are (1) at the center of the globe, (2) at the antipode (the point exactly opposite the point of tangency of the projection surface), and (3) at infinity. The orientation of a projection surface may be changed as desired, although certain orientations are defined as normal. A cylinder is normally oriented when it is tangent along the equator, a cone when it is tangent along a parallel (with its apex over the pole), and a plane when it is tangent at the pole. When the projection surface is turned 90 degrees from normal, the result is a transverse projection. An oblique projection results if the projection surface lies at an angle somewhere between the normal and the transverse positions. The importance of the different aspects of the projection surface is that each one produces a distinctive graticule and a different pattern of distortion.

The line of tangency between a projection surface and the surface of the globe is called the standard line of the projection. Along the standard line, the map has no distortion. Away from the standard line, however, distortions occur. When the projection surface intersects the globe, the result is a secant projection, which usually has two standard lines (a secant plane results in only one standard line).

Retaining correct shapes is called *conformality,* whereas retaining equal-area characteristics is called *equivalence.* Correct distance is obtained when the length of a straight line between two points on a map represents the correct great-circle distance between the same points on the earth. Direction is retained when a straight line drawn between two points on the map shows the correct azimuth of the line.

Projections are selected on the basis of the retention of globe characteristics or of some compromise in relation to them. Retaining a particular characteristic inevitably results in the distortion of other, potentially important, characteristics, so the choice involves some give-and-take.

SUGGESTED READINGS

Bugayevskiy, Lev M., and John P. Snyder. *Map Projections: A Reference Manual.* London: Taylor & Francis Ltd., 1995.

Campbell, John. *Introductory Cartography.* 2d ed. Dubuque, Iowa: Wm. C. Brown Publishers, 1991, chapter 2.

Canters, Frank, and Hugo Decleir. *The World in Perspective: A Directory of World Map Projections.* Chichester, England: John Wiley & Sons, Inc., 1989.

Committee on Map Projections of the American Cartographic Association. *Choosing a World Map: Attributes, Distortions, Classes, Aspects.* Special Publication No. 2 of the American Cartographic Association. Falls Church, Va.: American Congress on Surveying and Mapping, 1988.

______. *Matching the Map Projection to the Need.* Special Publication No. 3 of the American Cartographic Association. Falls Church, Va.: American Congress on Surveying and Mapping, 1991.

______. *Which Map Is Best? Projections for World Maps.* Special Publication No. 1 of the American Cartographic Association. Falls Church, Va.: American Congress on Surveying and Mapping, 1986.

Fisher, Irving, and O. M. Miller. *World Maps and Globes.* New York: Essential Books, 1944.

Grime, A. R. *The Earth Grid.* Scarborough, Ontario: Bellhaven House Limited, 1970.

Kellaway, George P. *Map Projections.* London: Methuen, 1949 (reprinted 1970).

McDonnell, Porter W., Jr. *Introduction to Map Projections.* New York: Marcel Dekker, 1979.

Pearson, Frederick II. *Map Projections: Theory and Applications.* Boca Raton, Florida: CRC Press, Inc., 1990.

Richardus, Peter, and Ron K. Adler. *Map Projections for Geodesists, Cartographers and Geographers.* Amsterdam: North-Holland Publishing Company, 1972.

Robinson, Arthur H., Joel L. Morrison, Phillip C. Muehrcke, A. Jon Kimerling, and Stephen C. Guptill. *Elements of Cartography.* 6th ed. New York: John Wiley & Sons, Inc., 1995, chapter 5.

Snyder, John P. *Flattening the Earth: Two Thousand Years of Map Projection.* Chicago: University of Chicago Press, 1993.

______. *Map Projections Used by the U.S. Geological Survey.* 2d ed. Geological Survey Bulletin 1532. Washington, D.C.: U.S. Government Printing Office, 1983.

______, and Philip M. Voxland. *An Album of Map Projections.* Geological Survey Professional Paper 14–53. Washington, D.C.: U.S. Government Printing Office, 1989.

Steers, J. A. *An Introduction to the Study of Map Projections.* 14th ed. London: University of London Press, 1965.

CHAPTER

4 Locational and Land-Partitioning Systems

Humankind has always needed to describe the locations of features on the earth's surface. In the earliest times, primitive and relatively immobile humans could use word descriptions to convey the locations of hunting or fishing sites, attractions or dangers, or other features that were of importance to them. Their descriptions could be simple word pictures of the location, such as "the hill with snow on top" or "the hole of gushing waters." As mobility and the complexity of life increased, however, it became increasingly necessary to identify more and more features and to differentiate between many features of the same type. This caused a great increase in the number of place names and feature names.

Today we must be able to describe the location of thousands of places on earth in accurate and unequivocal terms. Commonly, we must locate some object or locale of interest on a map, given its locational description. In addition, we frequently need to provide such locational information to others so that they can find their way. Regardless of which task we face, there are a number of methods of specifying locational information. The commonly used locational methods described in this chapter range from the latitude and longitude coordinate system to the State Plane Coordinate (SPC) system. They include the World Geographic Reference System (GEOREF) and the Universal Transverse Mercator (UTM) and Universal Polar Stereographic (UPS) systems.

As land-ownership concepts evolved, land-ownership description systems were developed to accurately describe the limits of property tracts. Situations in which we need to convey land-ownership information to someone else or to plot a land-ownership description onto a map are common. Both of these abilities are vital to anyone involved in real-estate transactions, even when the transaction is as relatively simple as the purchase of a home, and are useful in many other situations as well. This chapter, therefore, also describes two major types of land-partitioning systems. The first is the unsystematic subdivision of land using metes-and-bounds descriptions. The second is the systematic subdivision system utilized by the U.S. Public Land Survey (USPLS) and the similar Canada Land Survey System (CLSS).

Locational Systems

Latitude and Longitude Positions

The latitude and longitude graticule on a map may be used for two opposite purposes. First, the location of a specific map feature may be measured and recorded for future use. Second, the known latitude and longitude of a feature may be used to find and map its location. In this section, we discuss how to accomplish the first of these tasks. Although the second task is not specifically discussed, it is simply the reverse of the first.

If a location falls on the lines of latitude and longitude that are printed on the map, it is a simple matter to read off the coordinates. The difficulty comes in the more usual situation in which the location lies somewhere *between* the graticule lines. When this occurs, a process of **interpolation** is used to estimate the intermediate coordinate value.

Interpolation

A simple case illustrates the interpolation process. Assume that a feature is halfway between 10° and 20° North, for example. Clearly, its latitude is 15° North. When features do not fall in such convenient locations, however, interpolation becomes more difficult.

Obviously, interpolation requires a graticule as a starting point. Some maps do not contain a full graticule but use short lines, called **grid ticks,** to indicate the intersection of the graticule with the map outline (called the **neatline**). When using such a map to determine locations, you must first establish the graticule by joining the ticks with a lightly drawn pencil line. If the graticule of the map projection that you are using is curved, joining the ticks with straight lines introduces some error into the estimation process. To partially alleviate this problem, some maps contain additional reference ticks that indicate the intersection points of the graticule within the map. USGS 1:24,000-scale quadrangle maps, for example, use this method. If the map you intend to use lacks both a graticule *and* grid ticks, you must obtain a more suitable map before proceeding.

One complicating factor in the measurement of latitude and longitude is that the graticule spacing varies from place to place. This is particularly apparent in the case of the lines of longitude, which converge toward the poles (Table 4.1). It is also true, however, that the parallels are slightly farther apart toward the poles because of the slight flattening of the earth. The effect of this flattening is sufficient to be apparent with large map scales (Table 4.2). Other variations in graticule spacing may occur, depending on the projection used.

The variation in spacing between the graticule lines often means that a simple scale that works accurately for subdivisions of a degree of longitude at 40° latitude, for example, will not work at 50°. Nor can it be used at other latitudes or for subdividing latitude readings. This means that different scales must often be used for each latitude and longitude subdivision. Three methods flexible enough to deal with this complication are described here.

A simple ruler or engineer's scale, subdivided in either inches or metric units, can be used for interpolation.

TABLE 4.1 Length of Degrees of Longitude

Latitude	Kilometers
0	111.319
5	110.899
10	109.639
15	107.551
20	104.647
25	100.950
30	96.486
35	91.288
40	85.394
45	78.847
50	71.696
55	63.994
60	55.800
65	47.176
70	38.187
75	28.902
80	19.393
85	9.735
90	0.000

Note: This table shows the length of a degree of arc along the latitude named.
Source: Computed from data supplied by National Geodetic Survey, based on the Geodetic Reference System 1980 ellipsoid.

TABLE 4.2 Length of Degrees of Latitude

Latitude	Kilometers
0	110.574
5	110.583
10	110.608
15	110.649
20	110.704
25	110.773
30	110.852
35	110.941
40	111.035
45	111.132
50	111.229
55	111.324
60	111.412
65	111.493
70	111.562
75	111.618
80	111.660
85	111.685
90	111.694

Note: The table shows the length of a degree of arc centered on the latitude named.
Source: Computed from data supplied by National Geodetic Survey, based on the Geodetic Reference System 1980 ellipsoid.

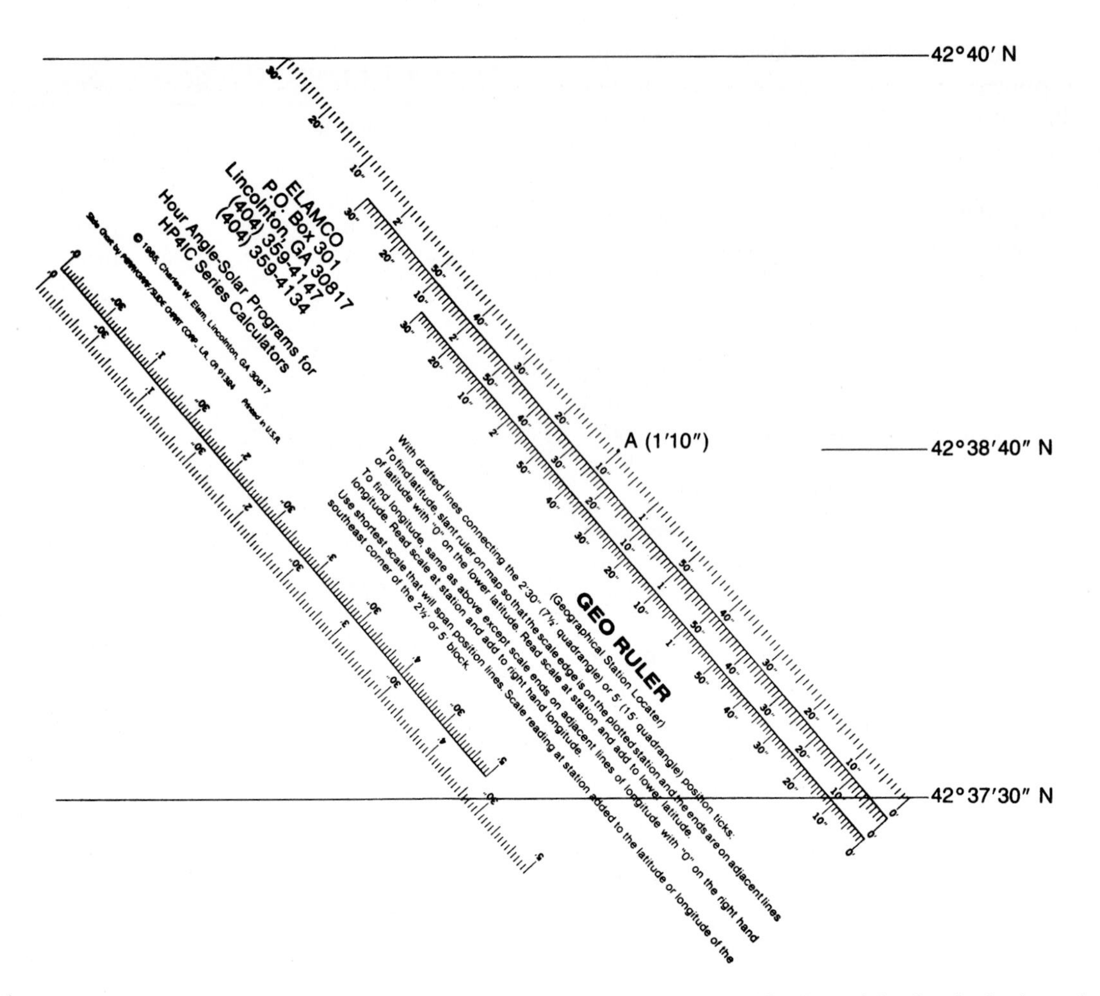

Figure 4.1 Use of a ruler for graphically interpolating latitude. The same approach is used for determining longitude. Any ruler with an appropriate number of subdivisions can be used, but this Geo Ruler is conveniently divided into minutes and seconds.
Source: Courtesy of Elamco.

(An even more useful device for this purpose is the special ruler shown in Figure 4.1.) Assume, for example, that the latitude of point *A* in Figure 4.1 is to be determined. As is the case with USGS 7.5-minute quadrangles, the lines of latitude are 2 minutes and 30 seconds apart (42°37′30″ and 42°40′ North). First, select a scale that comfortably spans the distance between the two lines of latitude. You will probably have to place the ruler on a slant to accomplish this. Next, align the ruler so that it passes through point *A*, with the 0 mark on the lower graticule line and the 2′30″ mark on the other. Each unit on the ruler represents 1 second of latitude. In this case, point *A* is at the 1′10″ mark, which means that it is that amount north of 42°37′30″ North, or 42°38′40″ North.

Another method of expediting the interpolation process, in the case of USGS 7.5-minute and 15-minute topographic quadrangles, is to use another type of commercially available transparent template. This template allows for the subdivision of degrees of latitude or longitude, regardless of the latitude range involved. It is used as shown in Figure 4.2.

Other scale selections are convenient for different graticule spacings. As already mentioned, on USGS 1:24,000-scale quadrangle maps the spacing between grid ticks is 2 minutes and 30 seconds, which totals 150 seconds of arc. In this case, if a scale with 150 units is used, each unit on the scale is equivalent to 1 second on the map.

Another method of interpolation involves using **spacing dividers** (Figure 4.3). Suppose, for example, that the latitude of the dot in Figure 4.3 is to be determined and that the graticule lines for 10° North and 20° North are on the map. The spacing dividers have eleven points, so when the endpoints are placed on the two parallels, the space between them is divided into ten equal parts. In this example, then, the dot is four spaces north of 10° North, at 14° North.

Spacing dividers have the advantage that they can be adjusted to any length (up to their maximum reach). In addition, division into anywhere from two to ten units is easily done by selecting a different number of points. Using five points, for example, gives a division into four units.

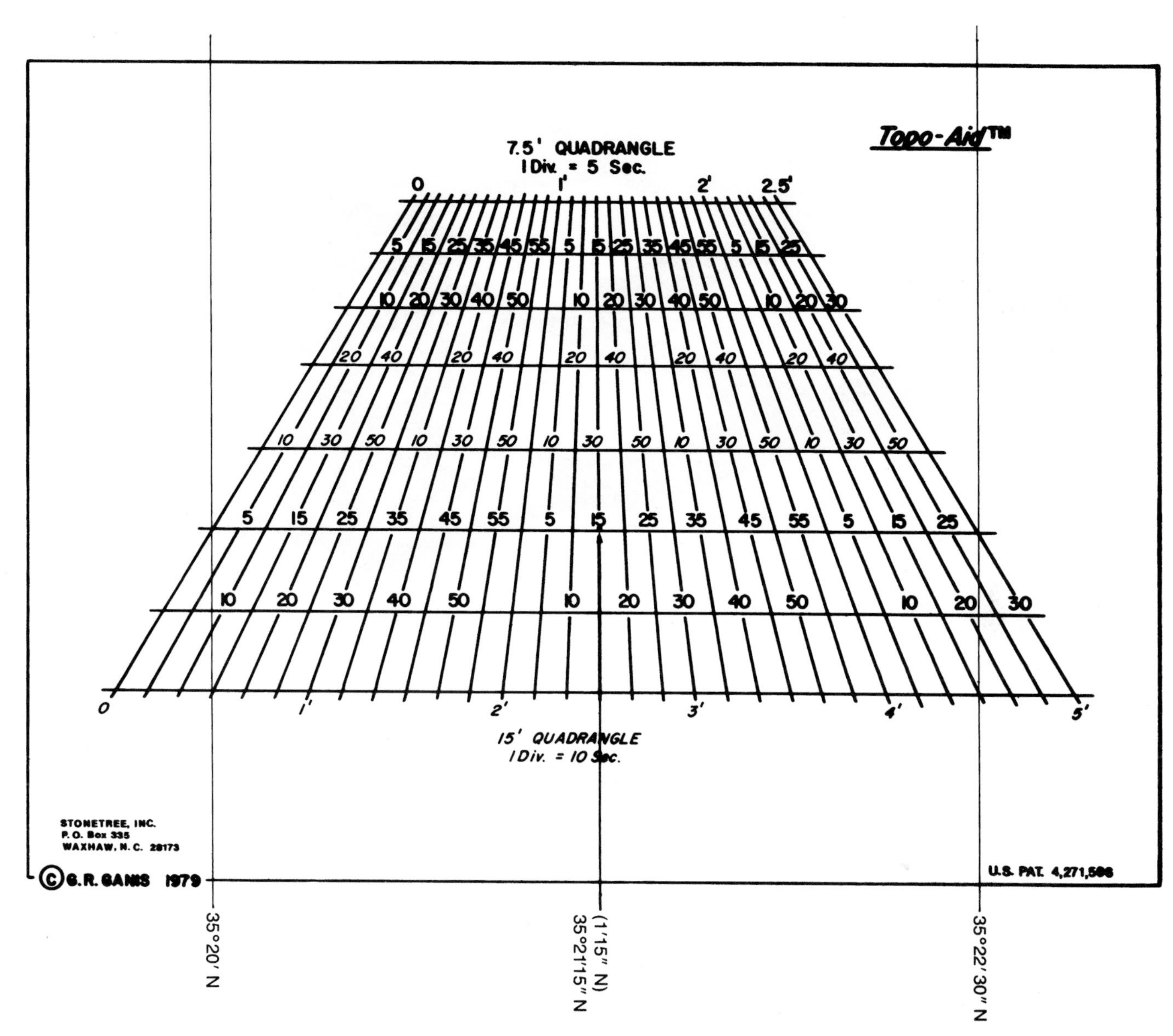

Figure 4.2 Use of a Topo-Aid™ transparent subdivision template.
Source: Courtesy of Stonetree Inc.

Subdivisions of Degrees

The system of degrees, minutes, and seconds used for latitude and longitude designations is awkward because of the subdivision of each degree into 60 minutes and of each minute into 60 seconds. One way to avoid this complexity is to use decimal fractions of a degree instead. When this method is used, 10°30′, for example, is converted to 10.5°.

When decimals are not used, minutes and seconds can be located by subdividing in a sequential manner. Locating 57°10′, for example, would require first locating 57° and 58° and then interpolating between them to locate 10′. Since 10′ is one-sixth of the 60 minutes that make up 1 degree, it would be sufficient to divide the space into six parts. When this is done, the first space north of 57° is 57°10′ north. If accuracy to the level of seconds is required, a further subdivision of the 1-minute spacing is done in a similar manner.

World Geographic Reference System

Specifying a location on the earth's surface in terms of degrees, minutes, and seconds of north and south latitude and of east and west longitude is somewhat cumbersome. The **World Geographic Reference System (GEOREF)** allows more convenient and rapid reporting and plotting of locations.

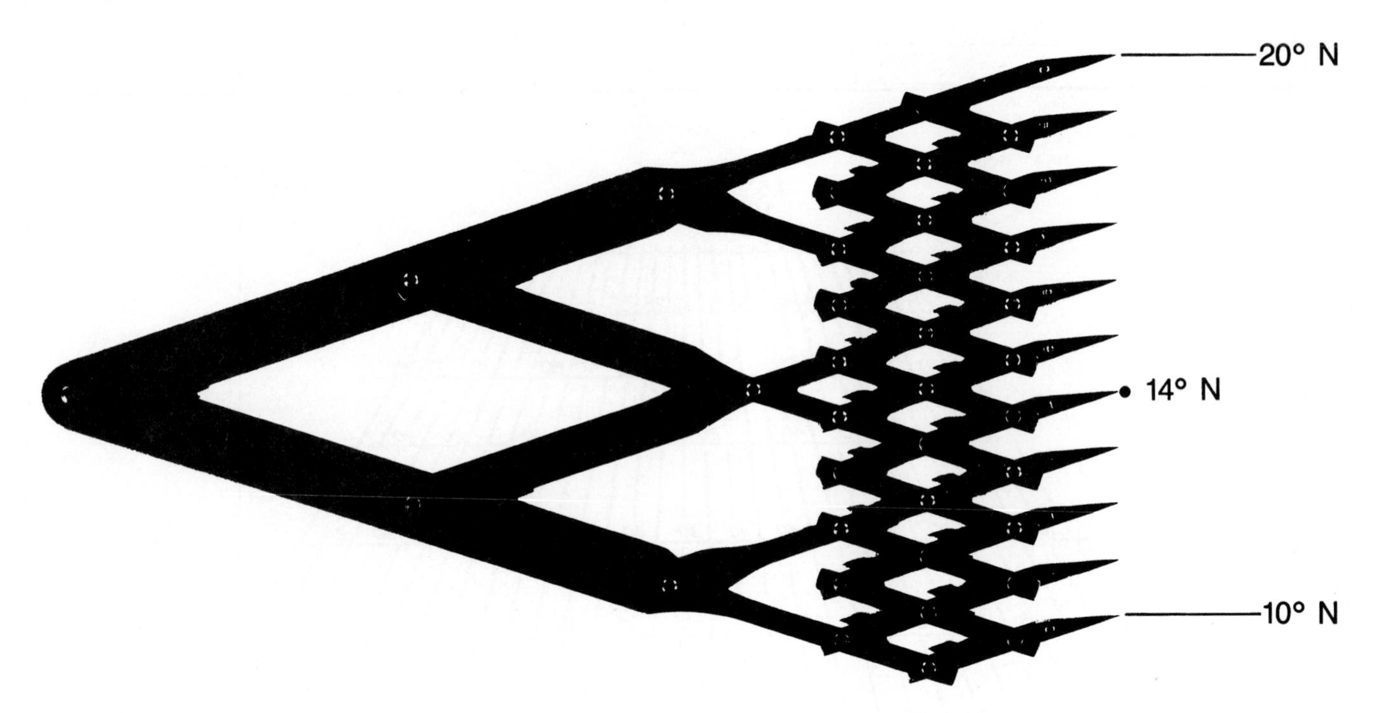

Figure 4.3 Use of spacing dividers for latitude/longitude interpolation.
Source: Courtesy of Charroz.

In GEOREF, which is used primarily for military air operations, latitude and longitude designations are replaced by a simple set of letters and numbers. The world is first divided into 15-degree-wide north-south and east-west bands, using the lines of longitude and latitude (Figure 4.4a). The 24 longitudinal zones are lettered in an eastward direction, beginning at the 180th meridian. The first zone to the east of 180° is identified by the letter *A*. The identifying letters continue in alphabetical order with the letter *Z* applied to the zone immediately to the west of the 180th meridian (*I* and *O* are omitted). The 12 latitudinal bands begin with *A*, for the first band north of the South Pole, and end with *M*, for the last band south of the North Pole (*I* is omitted).

Establishing the 15-degree zones and bands results in the formation of 288 quadrangles, each uniquely identified by a pair of letters. The first letter of the pair identifies the vertical zone and the second letter identifies the horizontal band. This follows the general rule of grid systems: "Read right and then up." The British Isles, for example, fall largely within square *MK* using this system.

When more precise location descriptions are required, additional subdivisions are used. The first increase in precision is provided by subdividing each of the 15-degree quadrangles into 225 1-degree quadrangles (Figure 4.4b). Each of these quadrangles is identified by adding a third and fourth letter to the original pair of letters. The north-south zones are identified by the third letter, using *A* through *Q* (omitting *I* and *O*), beginning at the western edge of the 15-degree quadrangle. The east-west bands are identified by the fourth letter, using the same letter sequence and beginning at the southern edge of the 15-degree quadrangle.

The use of the four identifying letters described so far allows the full identification of any 1-degree quadrangle in the world.

More precise identification of locations within a 1-degree quadrangle is accomplished by adding a numerical component to the letter codes. In the case of a four-digit number, the first two digits specify the number of minutes the location is to the east of the indicated degree line. The last two digits specify the number of minutes the location is to the north of the indicated degree line. In effect, this divides each 1-degree quadrangle into 3600 1-minute quadrangles. Further precision is provided by using decimal designations. Thus, a GEOREF coordinate with four letters and six numbers identifies any 0.1-minute quadrangle in the world. The first three digits indicate easting and the last three digits indicate northing (the decimal point is not actually written) (Figure 4.4c). Similarly, any 0.01-minute location is identified by four letters and eight numbers.

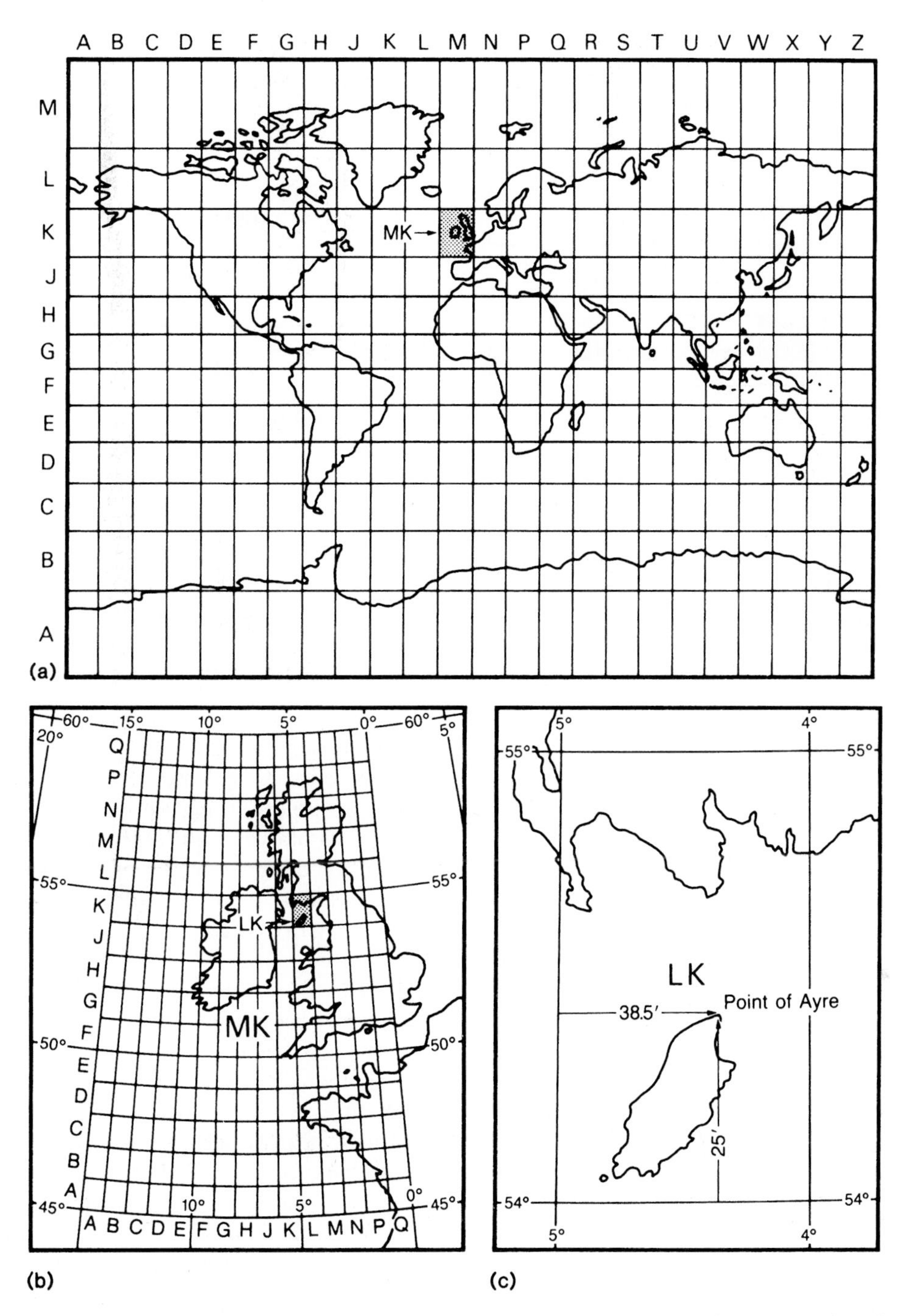

Figure 4.4 Determination of a reference using the GEOREF system. (*a*) Worldwide 15-degree-wide zones and bands. (*b*) Isle of Man: 1° square MK LK. (*c*) Point of Ayre: MK LK 385250.

Grid Systems

Reference Grids

Maps often include special reference grids. Some grid systems are very easily understood because they are designed simply as a supplement to a map index. Examples of this type are commonly encountered in atlases and on street maps. The index lists cities, towns, rivers, streets, and other features and designates the grid square in which each feature appears (in the case of an atlas, the page number is given as well). To find a feature, you first locate the designated square, using the identifying letters and numbers printed around the outside border of the map (Figure 4.5). You then search within the square for the feature. Reference grids of this type often use letters to designate the horizontal bands

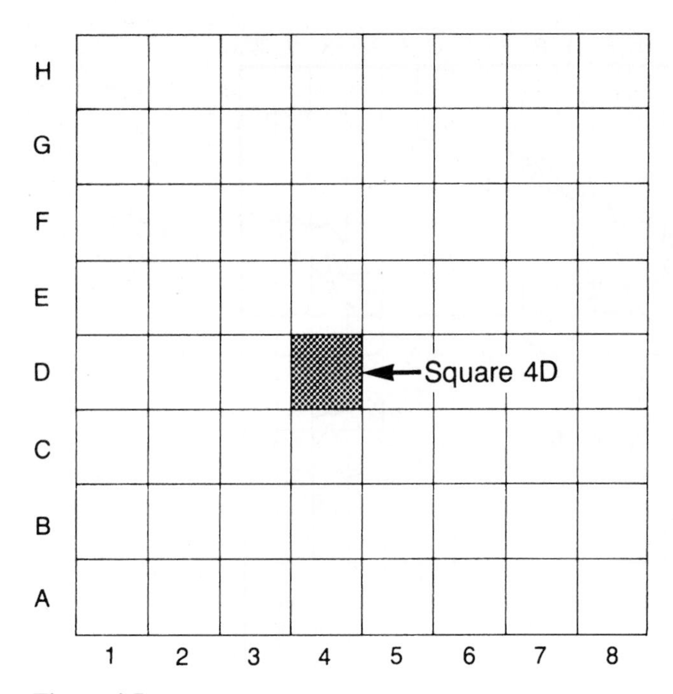

Figure 4.5 Simple reference grid.

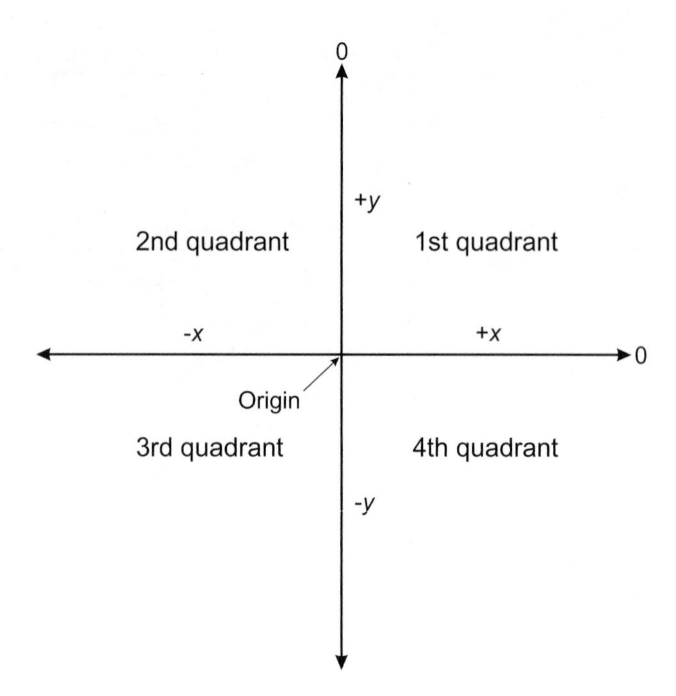

Figure 4.6 Cartesian coordinate system.

and numbers to designate the vertical ones, with both series beginning at the lower-left corner of the map. The numbering and lettering patterns are not standardized, however, so each system must be examined individually. The major shortcoming of such grids is that they are specific to each map and, therefore, do not tie into any overall locational system.

A more important type of grid system allows the user to specify any desired point in relation to other locations within a relatively large region. The general characteristics of such systems are described next, and two specific examples are examined.

Plane Rectangular Grids

One or more **plane rectangular grid systems** may be superimposed on a map after it has been constructed on a latitude and longitude graticule. The purpose of such grids is to provide an easily used locational system that does not require the introduction of corrections for the curvature of the earth.

Although details vary, plane rectangular grids tend to have a number of elements in common. In the typical case, an origin is established first. Two sets of parallel lines are then established, beginning at the origin, with one set perpendicular to the other, thus forming the rectangular grid. The grid lines are not directly associated with the latitude and longitude graticule, although they may be aligned with it at the grid origin; they are simply arbitrary lines, superimposed over the map. In addition, a systematic labeling system is established to allow users to specify locations within the grid system.

In rectangular coordinate systems, measurements of locations are specified in terms of distances away from the origin, along the grid lines. Typically, measurements above (north) or to the right (east) of the origin are designated as positive numbers, and those below (south) or to the left (west) are negative. Because negative designations are awkward to work with, most plane coordinate grid systems are established in the equivalent of the first quadrant of a **Cartesian coordinate system,** so that all of the coordinate measurements are positive (Figure 4.6). In other words, the origin is usually established south and west of the area to be mapped, and distances are measured eastward and northward from that point.[1] There are exceptions to this rule, but they are relatively easy to deal with once the more usual arrangements are understood. One simply has to be careful when measuring distances to increase or decrease the numbers in the correct directions.

Reference lines for plane coordinate grids are often drawn on maps. They are usually at a regular spacing, such as 1000 yards, 1000 meters, or 10,000 feet. The spacing is selected so that the grid does not interfere with other information on the map. If the scale is relatively large, the grid lines are spaced at a closer interval than if the scale is relatively small. Other than convenience, there is nothing to prevent the map placement of grid lines at *any* arbitrary interval.

[1]Such an origin is often called a **false origin** because it lies outside the area being mapped.

A numbering system is needed so that the location of a grid reference can be quickly determined. A variety of systems is used involving letters, Arabic numbers, Roman numerals, or some combination of letters and numbers.

The general rule for plane coordinate grid systems is that locations are defined by giving the east-west coordinate first and the north-south coordinate second. Usually, the coordinates are simply distance values, measured from the system origin. They are frequently written as a single number block, with the division into east-west and north-south components merely implied. Suppose, for example, that a point is located 642,000 meters east and 195,000 meters north of the origin. One way to represent the coordinates for this point is by the number 642000-195000. Alternatively, the hyphen is omitted, so that the number is shown as 642000195000. In addition, trailing zeros are often omitted on a grid reference, but always as an equal number of digits from each half. When this is appropriate, the example coordinate is conveniently shortened to 642195. An abbreviated coordinate of this type indicates that the feature is located in the grid square that lies to the east and north of the point that is 642,000 meters east and 195,000 meters north of the origin.

Leading digits in coordinate strings are also sometimes omitted when work is being done in a limited region. Continuing with the previous example, a location that is 1,642,000 meters east and 2,195,000 meters north might normally be listed as 16422195. When the leading digits are omitted, the reference is shortened to 642195. However, this same set of numbers is the full reference for the point located at 642,000 meters east and 195,000 meters north. It could also be the shortened reference for other points located at 1,000,000-meter intervals. Obviously, this shorthand approach depends on an agreement by all parties involved that prevents a shortened reference from being confused with other possible locations. In some systems, the preceding digits and trailing zeros that may be omitted are printed as smaller-sized numbers on the face of the map.

More accurate designations of points to be located are often given by using decimal places to indicate subdivisions between grid lines. Thus, a point located at 642,500 meters east and 195,300 meters north could be designated as 642.5-195.3. Other systems call for running decimal numbers together, again without specifically dividing them into east-west and north-south components. In this case, the decimals are implied rather than written (64251953). Obviously, the user of any grid system must know that system's numbering conventions to interpret a particular coordinate.

Universal Transverse Mercator System

The **Universal Transverse Mercator (UTM) system** is based on the transverse Mercator projection, with two standard lines. It covers the earth's surface between 80° South and 84° North. The UTM system is used in conjunction with the **Universal Polar Stereographic (UPS) system,** which covers the polar caps.

The UTM system involves establishing 60 north-south zones, each of which is 6 degrees of longitude wide. In addition, each zone overlaps 1/2 degree into the adjoining zones. This allows easy reference to points that are near a zone boundary, regardless of which zone is used for a particular project.

Two methods of location identification are used in the UTM and UPS systems. The first is intended for civilian use and the second for military use. The basic framework, however, is the same for both schemes. Because the civilian system is most commonly encountered, it is described first, and the military system is compared to it.

Civilian System

First, a false origin is established 500,000 meters west of the central meridian of each UTM zone (Figure 4.7a). In the Northern Hemisphere, this origin is on the equator; in the Southern Hemisphere, it is 10,000,000 meters south of the equator. A square grid, with the lines extended north and east from the origin, provides a basic locational framework. With this framework, any point on the earth's surface, within each zone, has a unique coordinate.

A locational grid system is imposed on the two polar (UPS) projections as well (Figure 4.7b). Grid north is arbitrarily established so that the prime meridian is at the top-center of the south zone, and the 180th meridian is at the top-center of the north zone. The origins of these grids are located at the lower-left corner of each grid, 2,000,000 meters to the left and 2,000,000 meters down from the pole.

The first step in locating a feature, using the UTM system, is to identify the zone and hemisphere in which it lies. Assume, for example, that the location of the Kenosha, Wisconsin, airport is to be specified. Kenosha is located in zone 16 in the Northern Hemisphere, so this is the first portion of the locational description. The specific location in relation to the origin is then determined. This location is designated in terms of *eastings* (distances measured from west to east) and *northings* (distances measured from south to north).

UTM coordinates are shown on the edges of many topographic maps. A portion of the USGS 1:24,000 Pleasant Prairie quadrangle is shown in Figure 4.8. A considerable amount of extraneous marginal information is

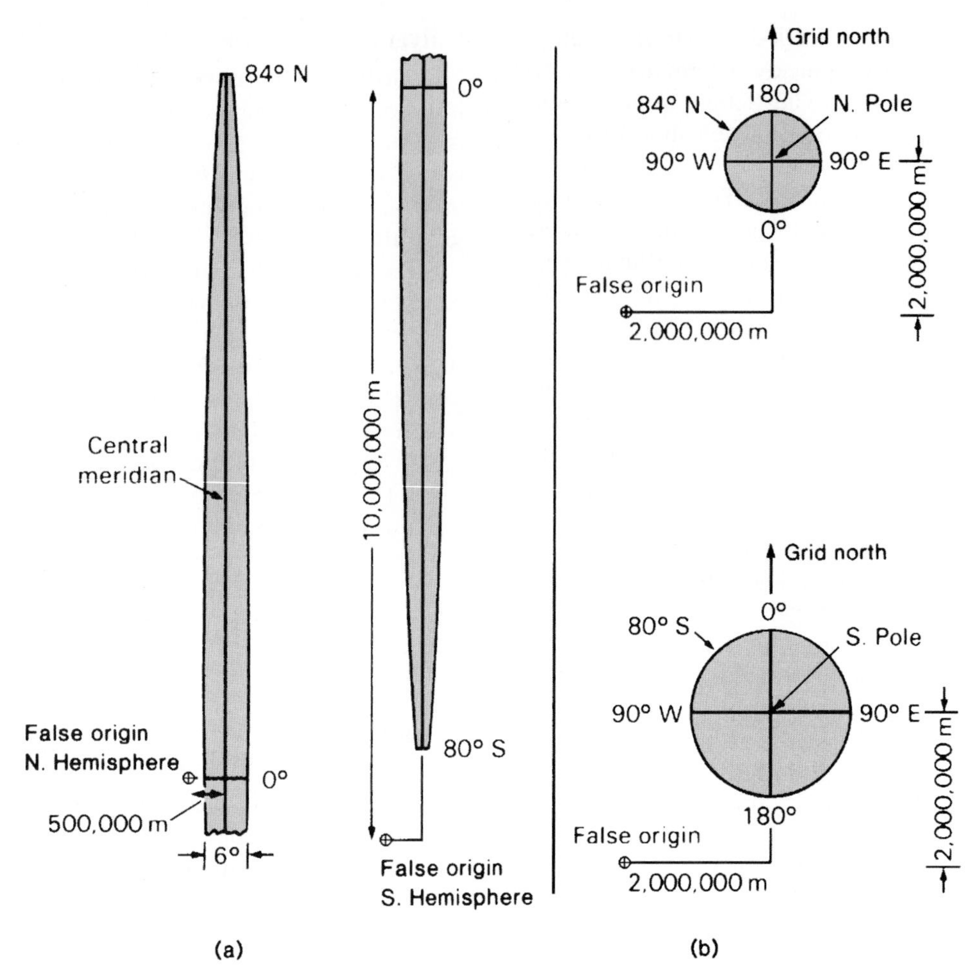

Figure 4.7 (*a*) Universal Transverse Mercator (UTM) zone. (*b*) Universal Polar Stereographic (UPS) zones.

deleted from this figure so that the UTM system is clearly shown. Note that the UTM grid designations are abbreviated in most locations by omitting the last three zeroes and showing designations of 100,000 or more in smaller type. In Figure 4.8, for purposes of clarity, the full identification numbers for a portion of the grid are added. The grid itself is also shown, although it is indicated only by short blue ticks in the margins of the original map.

The northernmost tip of the runway of the Kenosha Airport is identified in Figure 4.8. It is located in the UTM grid square that lies between 423,000 meters east and 424,000 meters east and between 4,716,000 meters north and 4,717,000 meters north, as is indicated by the grid lines. In this case, the exact location of the point of interest, within the nearest 10 meters, is 870 meters east and 510 meters north of the south-west corner of the grid square. The full UTM coordinates of the location, then, are 423,870 meters east and 4,716,510 meters north (Northern Hemisphere, zone 16).

Determination of UTM coordinates is facilitated by the use of a Metric Coordinate Reader, which is available from the U.S. Geological Survey (Figure 4.9). The device can be used with any map containing a UTM grid. Most USGS maps, for example, have UTM grid ticks around the map margin. A full grid is constructed by drawing lines that connect these ticks. After the full grid has been drawn, the UTM coordinates of any given point on the map are obtained by using the coordinate reader in the following manner.

In this example, the feature of interest is the building symbol shown in Figure 4.10. First, note the UTM grid zone in which the feature lies, as given in a marginal note on the map. In this example, it is zone 16. Then draw in the coordinate grid in the area of interest and locate the southwest corner of the grid square that surrounds the building symbol. The designations of the grid lines that form the southern and western edges of this grid square provide the general coordinates

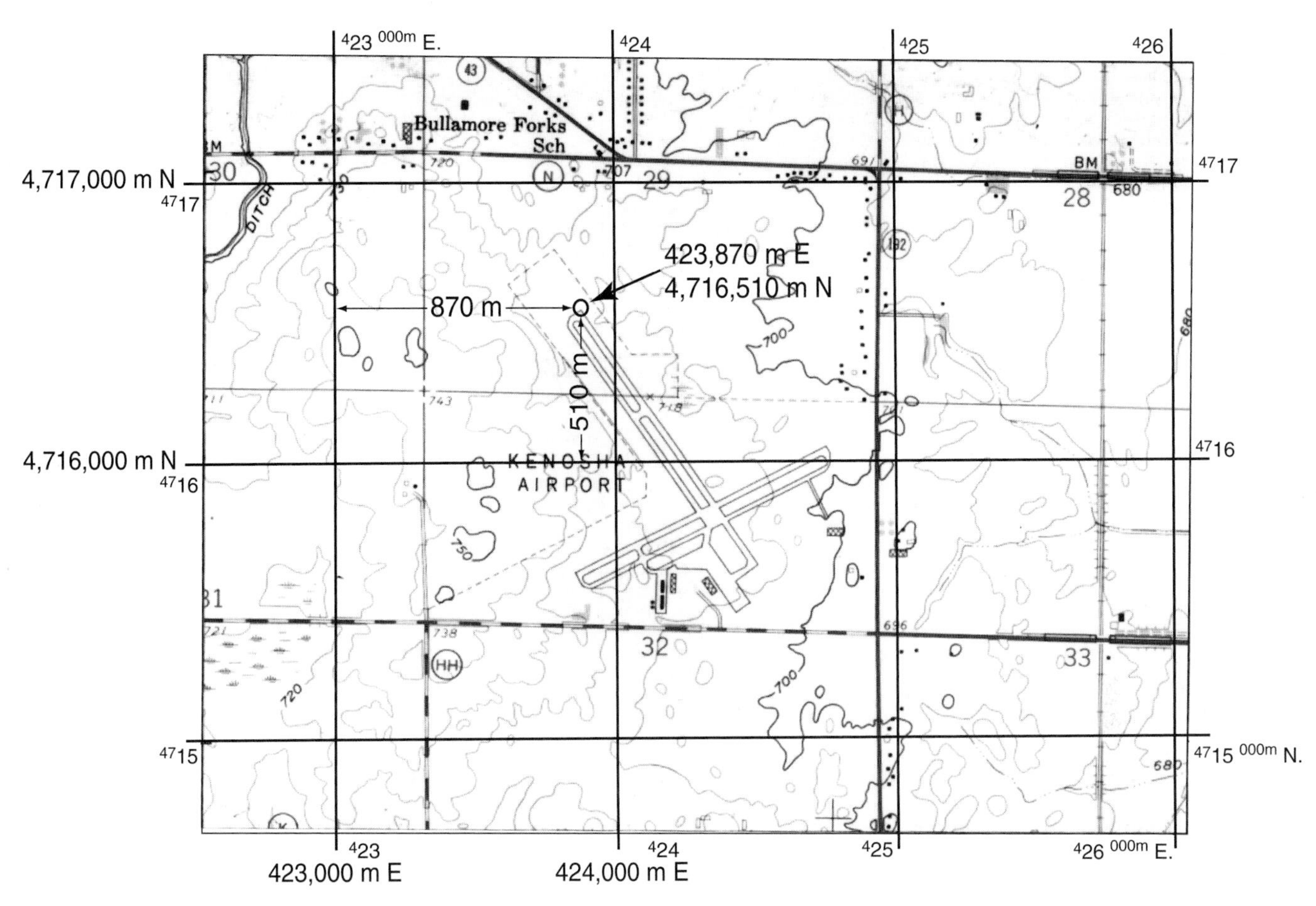

Figure 4.8 Use of a UTM coordinate grid. (Portion of U.S. Geological Survey Pleasant Prairie quadrangle, Wisconsin, 1:24,000, 1971.)

(429,000 meters east and 4,720,000 meters north). Next, select the appropriate scale on the coordinate reader and hold the reader so that the scale designation (1:24,000 in this case) is in normal reading orientation. Place the bottom edge of the scale so that it is aligned on the southern edge of the selected grid square. Slide the scale right or left, until the vertical scale line passes through the center of the school symbol. Then read the fine grid designations from the scales. In this case, each small tick indicates a distance of 20 meters, so distances can be accurately estimated to the nearest 10 meters. The east-west distance between the grid line and the building symbol is approximately 450 meters, and the north-south distance is 150 meters. Add these values to the general coordinates to obtain the final UTM designation of the location of the building symbol: 429,450 meters east, 4,720,150 meters north (Northern Hemisphere, zone 16).

Because the civilian location reference uses only numbers, it is easily handled by computer systems, which is an advantage in many applications.

Military Grid Reference System

The military versions of the UTM and UPS systems are identical to the civilian systems already described, except for the method of identifying grid zones. The Military Grid Reference System is used for this purpose. Because this system avoids the long strings of numerals that occur in the civilian systems, it is believed to be preferable for military use.

In the military system, the area of the world lying between 84° North and 80° South is subdivided into grid zones that are 6 degrees east-west by 8 degrees north-south (with some exceptions north of 56° North). Each grid zone is identified by a column number and a row letter (Figure 4.11). The column numbers, 1 through 60, are the UTM zone numbers. The zone letters, *C* through *X* (with *I* and *O* omitted), start at 80° South and continue to 84° North. The grid zone for a given location is identified by its column and row designation, such as 3N. In addition, grid zones *A* and *B* are assigned to the south zone of the UPS system and *Y* and *Z* to the north zone.

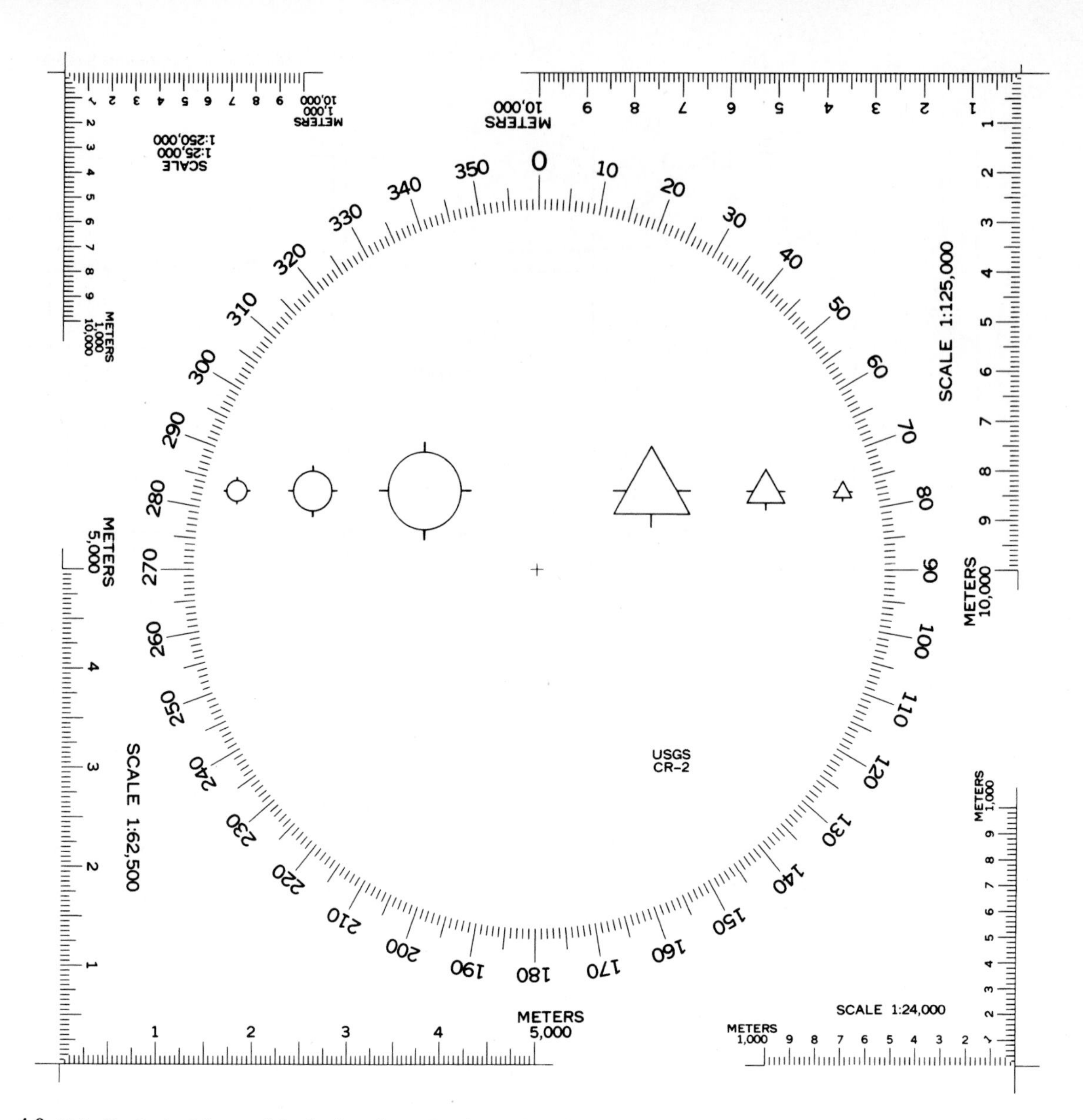

Figure 4.9 U.S. Geological Survey Metric Coordinate Reader (reduced).

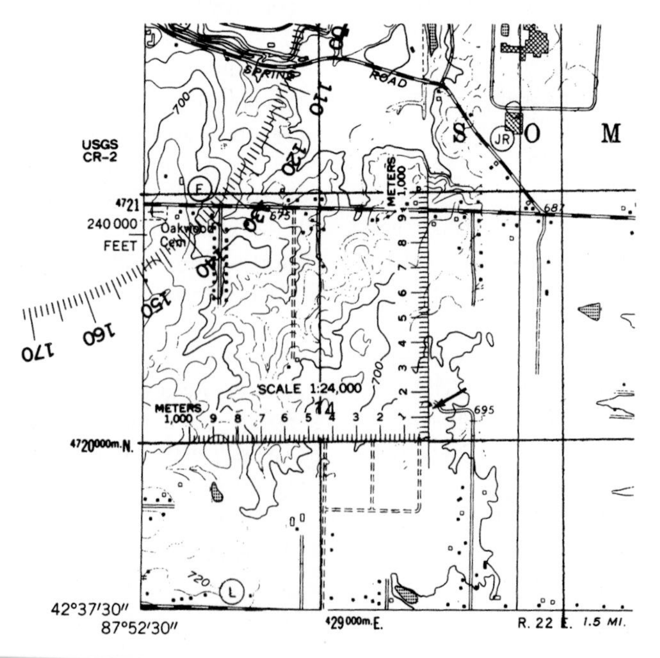

Figure 4.10 Use of a Metric Coordinate Reader. (Superimposed on portion of U.S. Geological Survey Racine South quadrangle, Wisconsin, 1:24,000, 1958. UTM grid lines added. Illustration reduced.)

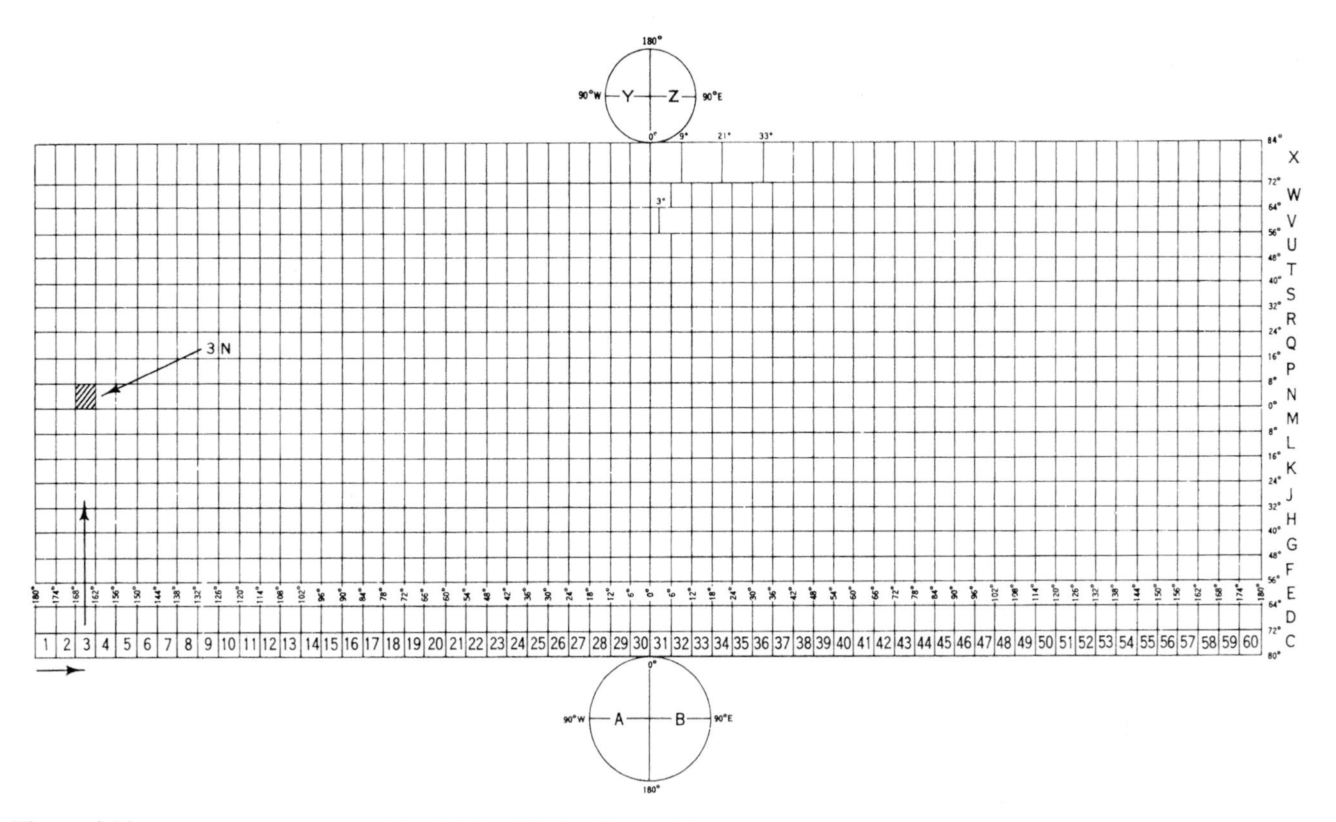

Figure 4.11 Grid-zone designations of the UTM and UPS military grid system.
Source: Department of the Army, *Grids and References,* TM 5-241-1 (Washington, D.C.: Department of the Army, March 1983).

The grid zones are subdivided by a pattern of 100,000-meter squares, each of which has a two-letter identifier (Figure 4.12). The identification pattern is set up so that repeated combinations of the identifying letters are as far apart as possible. The first identifying letter indicates the column, and the second letter designates the row. The pattern is not completely regular. Details of the variations, which are numerous, are found in the Department of the Army Training Manual TM5-241-1. (See the Suggested Readings at the end of the chapter.)

The column identification letters for the 100,000-meter squares apply both north and south of the equator. The first group begins with *A,* east of the 180th meridian, and continues easterly for 18 degrees (*I* and *O* are omitted), after which it is repeated every 18 degrees. As a result, in the 6-degree UTM zones 1, 2, and 3, for example, the letter *A* is located in the westernmost column of zone 1, *Z* is located in the easternmost column of zone 3, and the first column in zone 4 begins again with the letter *A*. The tapering of the UTM zones toward the poles affects the columns on the western or eastern sides of the zones. As the tapering of the zones progresses toward the poles, the outer columns gradually become narrower or are eliminated.

The pattern for the row letters is somewhat more complex, and, again, there are variations that are not described here. The 20 letters *A* through *V* (omitting *I* and *O*) are used to identify the rows, with a different pattern applied to even- and odd-numbered UTM zones. In odd-numbered zones, the first row north of the false origin (the equator in the Northern Hemisphere and 10,000,000 meters south of the equator in the Southern Hemisphere) is designated as *A*. In even-numbered zones, the same pattern is used, except that the sequence is advanced five letters, so that squares with the same identification are farther apart than they would be if the offset were not introduced. Thus, the first row in an even-numbered zone is *F* instead of *A*. Thereafter, the sequence of letters is the same as that already specified. In each case, the sequence of letters progresses northward and is repeated every 2,000,000 meters.

The UPS zones also have two-letter grid-square designations. The pattern for the north zone is shown in Figure 4.13; the south zone has a similar pattern with different letter designations.

Locations within each 100,000-meter square are given in meters as coordinate distances from the southwest corner. The location is defined by the easting, which is given first, and by the northing, which is

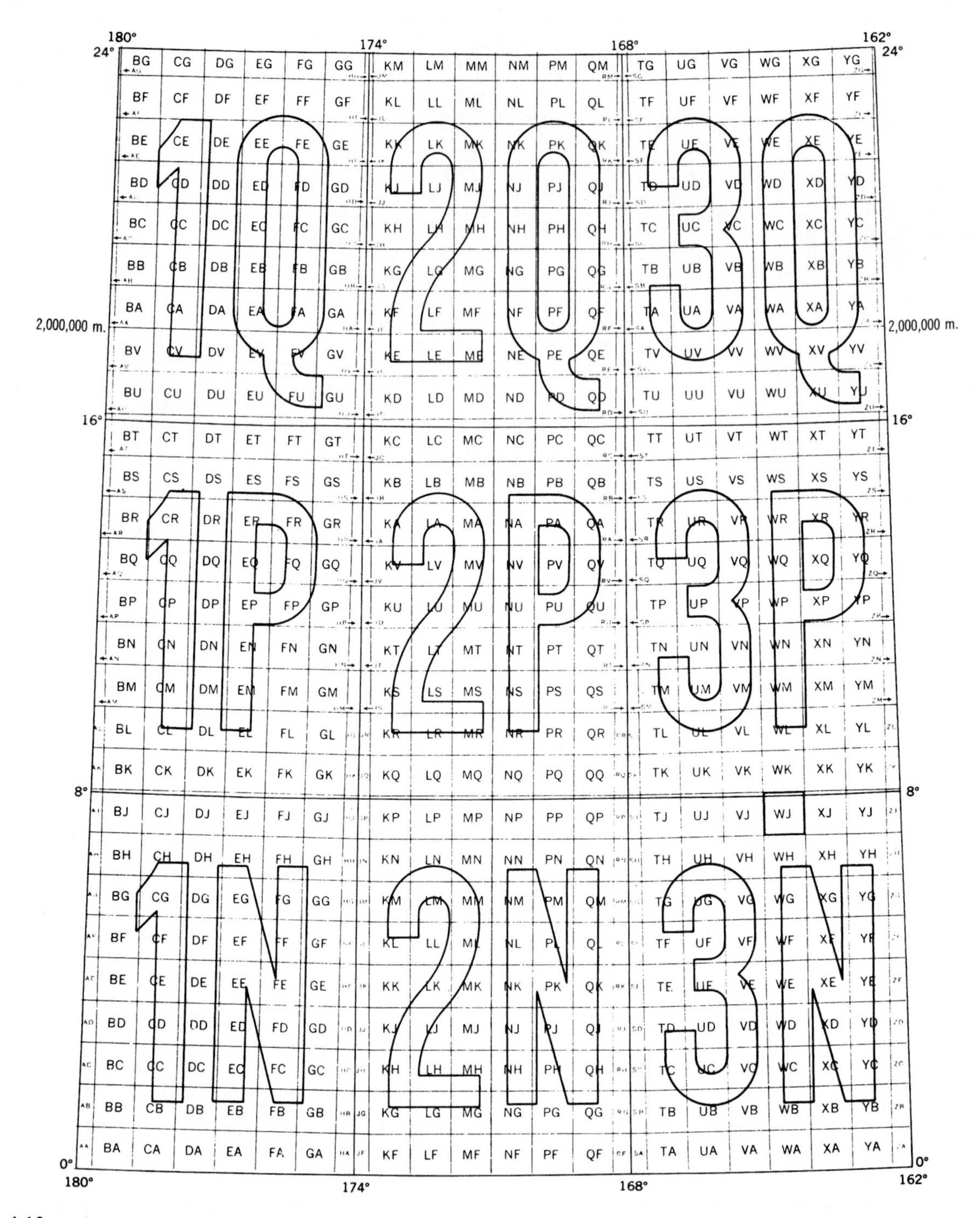

Figure 4.12 Basic plan of the 100,000-meter square identifications of the UTM military grid system zones.

Source: Department of the Army, *Grids and References,* TM 5-241-1 (Washington, D.C.: Department of the Army, March 1983).

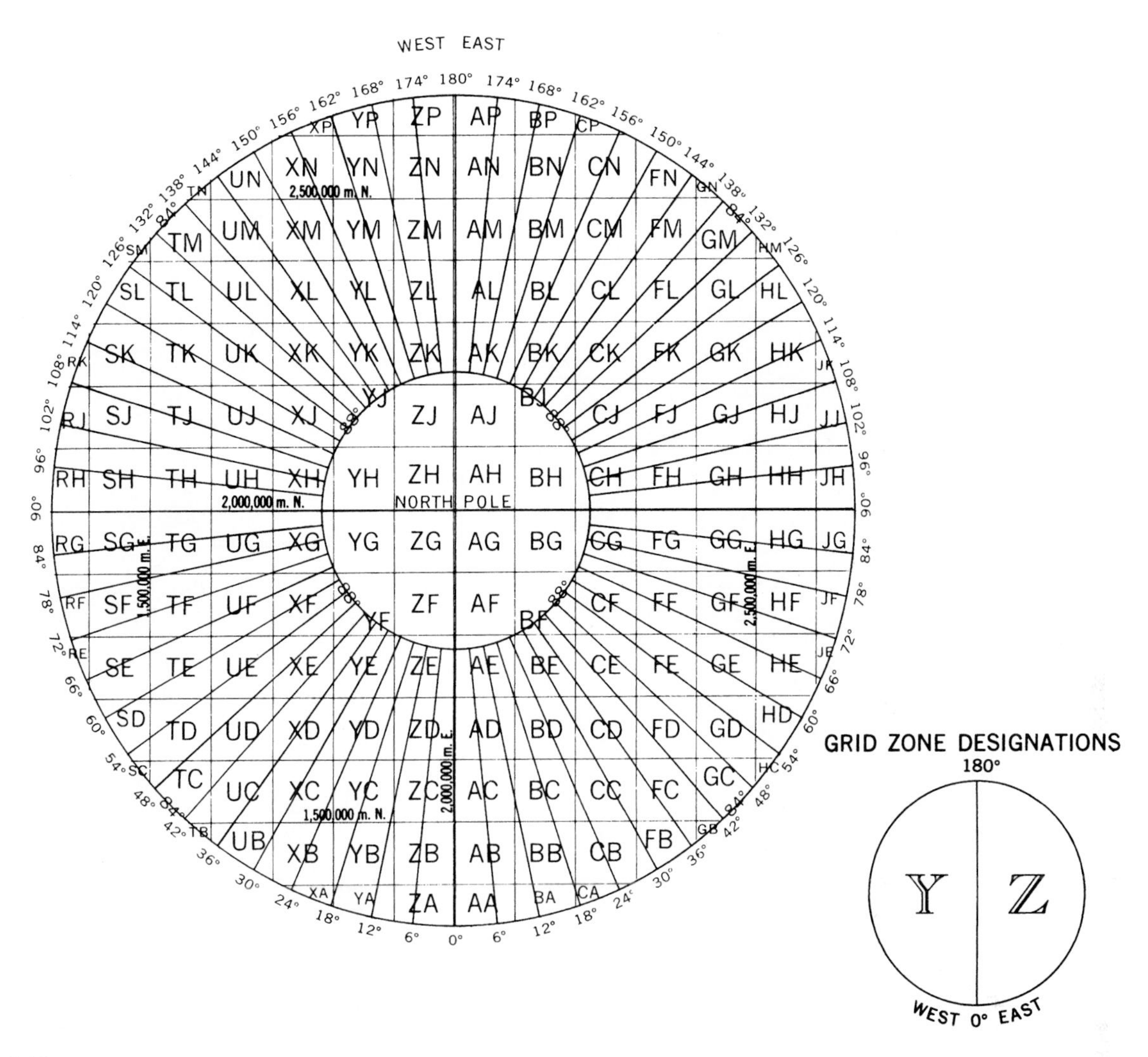

Figure 4.13 Two-letter identifications of the 100,000-meter squares and partial squares of the UPS military system grid (north zone).
Source: Department of the Army, *Grids and References,* TM 5-241-1 (Washington, D.C.: Department of the Army, March 1983).

given second. The digits are written in sequence without punctuation. The number of digits used in the coordinates depends on the accuracy that is required or that is possible, given the scale of the map. A ten-digit number is the maximum, and it indicates a location within a 1-meter square. This level of accuracy is seldom reached. At the other extreme, if no grid coordinates are given, the location simply lies somewhere within the designated 100,000-meter square.

A full grid reference consists of the grid-zone designation, the 100,000-meter square identification, and the grid coordinates (written to the desired degree of accuracy). The reference (within 100 meters) for Kent Point, in Figure 4.14, for example, is 18SUT916091. If a less accurate location is required, the coordinate portion of the reference is shortened accordingly. Within 1000 meters, the reference is 18SUT9109, and within 10,000 meters, it is 18SUT90. If the scale of the map permits greater accuracy, coordinates are adjusted accordingly. Within 10 meters, for example, they might be 18SUT91620914, and within 1 meter, 18SUT9162309143.

If there is no danger of confusion, as would be the case if operations were being carried out within a limited area, the full grid reference is not required. If the area is restricted to one grid zone, for example, the grid square and its coordinates are sufficient (UT916091) or, if it is restricted to one 100,000-meter square, the coordinates alone can be used (916091).

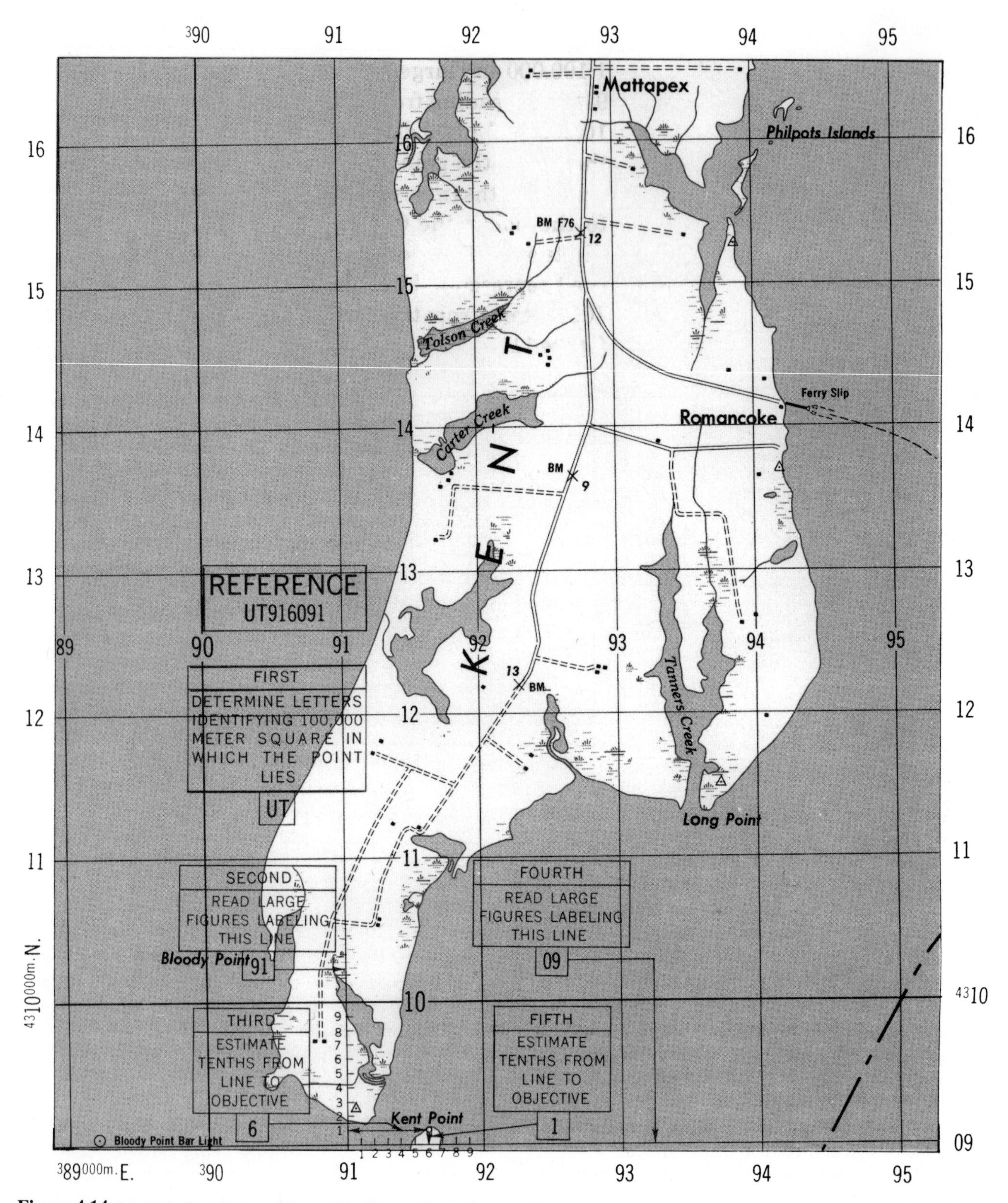

Figure 4.14 Method of reading a military grid reference from a large-scale map.

Source: Department of the Army, *Grids and References,* TM 5-241-1 (Washington, D.C.: Department of the Army, March 1983).

State Plane Coordinate System

The **State Plane Coordinate (SPC) system** is a rectangular coordinate system that is individually applied to each of the United States. The system is designed to simplify local surveys and at the same time tie them into the national geodetic network. As the name indicates, the system uses plane coordinates. This is done so that the surveyor does not have to take the curvature of the earth into account, which greatly simplifies the survey computations. This simplification is acceptable for the small areas usually involved in local surveys because the amount of error introduced is quite small.

A separate SPC is established for each state. The larger states are subdivided into zones, with the number of zones determined by the distortion limits that have been established for the SPC (Figure 4.15). A map of each zone is then drawn on a projection selected to reduce distortion. With some exceptions, the Lambert conformal projection is used for east-west trending states and the transverse Mercator is used for north-south trending states.[2]

[2]A revised SPC system, based on the transverse Mercator projection and with distances expressed in meters, is in use in some states. The defining constants for the 1983 State Plane Coordinate system are presented in the National Oceanic and Atmospheric Administration Manual NOS NGS 5, *State Plane Coordinate System of 1983* (Rockville, Md.: National Geodetic Center, 1989) and in Appendix B of Borden D. Dent, *Cartography: Thematic Map Design,* 4th ed. (Madison, Wis.: Brown & Benchmark, 1996).

For purposes of designating locations, a false origin is established for each zone. This false origin is located outside the zone and is placed so that all of the measurements based on it are positive numbers. The typical false-origin location is west of the central meridian of the zone. The location is 2,000,000 feet west in the case of the Lambert projection and 500,000 feet west for the transverse Mercator projection. It is south of the southern edge of the zone in both cases. Location coordinates are measured from the false origin in feet. They are measured to the right (called *false eastings*) and upward (called *false northings*). A full SPC location designation consists of the easting, the northing, the name of the state, and the designation of the zone, such as north, south, east, west, or central. The designation of the identified point in Figure 4.16, for example, is 2,557,910 feet east, 225,360 feet north, Wisconsin, South Zone.

LAND-PARTITIONING SYSTEMS

Two basic types of surveys used to determine land-ownership boundaries and property descriptions in the United States and Canada are described in this section.[3]

[3]The survey terms used here follow the glossary in *Definitions of Surveying and Associated Terms,* revised, American Congress on Surveying and Mapping and the American Society of Civil Engineers, 1978.

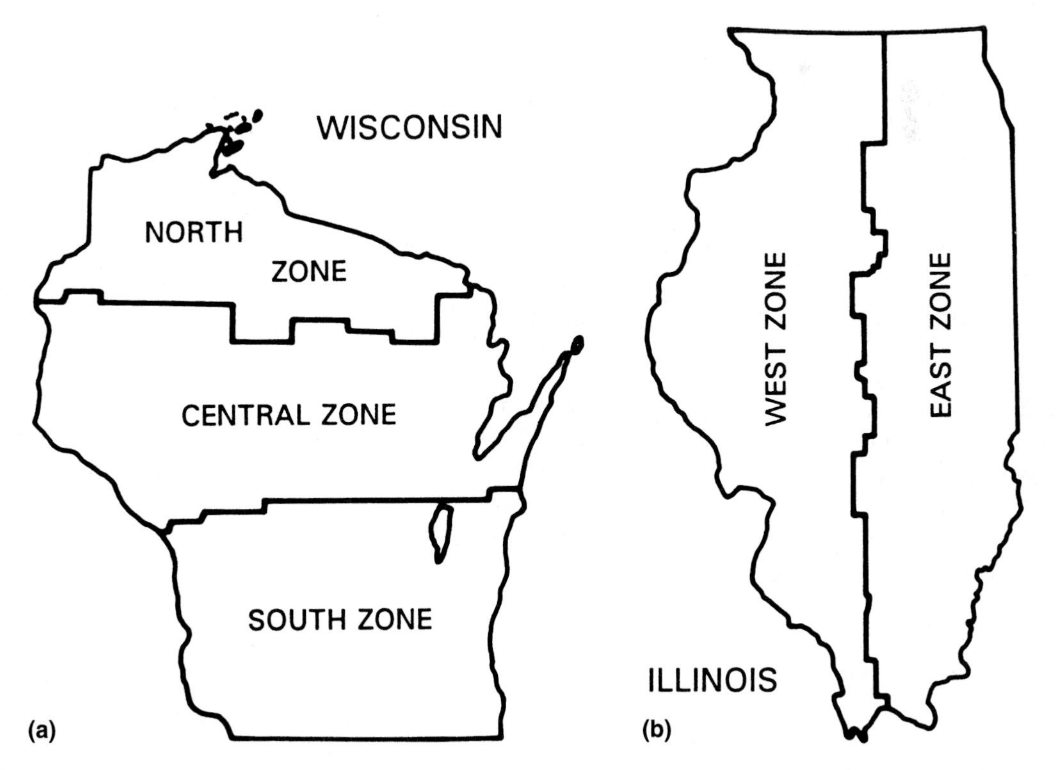

Figure 4.15 Examples of State Plane Coordinate System (SPC) zones. (*a*) In a state mapped on the Lambert projection. (*b*) In a state mapped on the Transverse Mercator projection. Zone boundaries fall on county boundaries.

Source: Joseph F. Dracup, *Fundamentals of State Plane Coordinate Systems.* (Rockville, Maryland: U.S. Department of Commerce, National Ocean Survey, 1974.) Reprinted 1977.

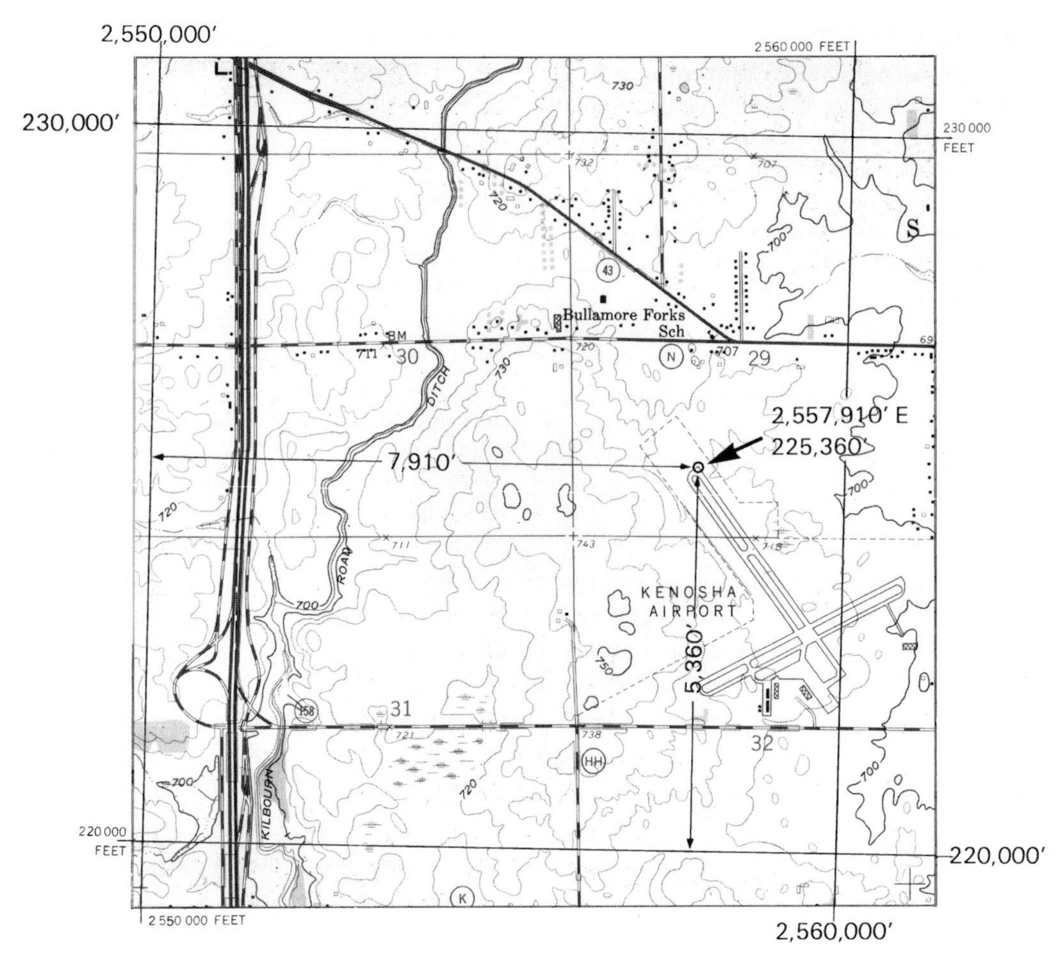

Figure 4.16 Use of a state plane coordinate grid. (Portion of U.S. Geological Survey Pleasant Prairie quadrangle, Wisconsin, 1971. Original scale 1:24,000. Illustration reduced.)

Surveys that serve these purposes are generally referred to as **cadastral surveys.** In the United States, however, the use of the term *cadastral survey* is preferably reserved to describe land-ownership surveys of the public lands. The term *land survey* is used to describe such surveys outside the public lands.

Unsystemic Subdivision

In most of the originally colonized portions of the eastern United States, land-ownership patterns developed in an unplanned manner called **unsystematic subdivision.** Property boundaries were established on the basis of the claims of the settlers who had occupied, cleared, and farmed specific plots of land. Typically, the boundaries of these properties were defined first, and surveys were later carried out to record the ownership claims.

Metes and Bounds

The method of land survey used to describe these unplanned land ownerships is called **metes-and-bounds surveying.**[4] A metes-and-bounds survey records the direction and length of each portion of the boundary line of the property in sequence, as though one were walking around the perimeter of the parcel. Distinctive markers, such as trees of particular species, are often used for such surveys (Figure 4.17).

Bearings

Directions in metes-and-bounds surveys are indicated by the use of bearings. A **bearing** is the angular difference between a north or south reference and the direction of the surveyed line. The reference line, and hence the bearing, may be expressed as a true, magnetic, or grid direction. The listing of a bearing consists of three parts: the designation of the reference direction (either north or south), the angle between the reference direction and the surveyed direction (not to exceed 90 degrees), and the direction of the angle (east

[4]Metes-and-bounds surveys are also used in certain situations elsewhere in the United States, including areas with systematic subdivisions.

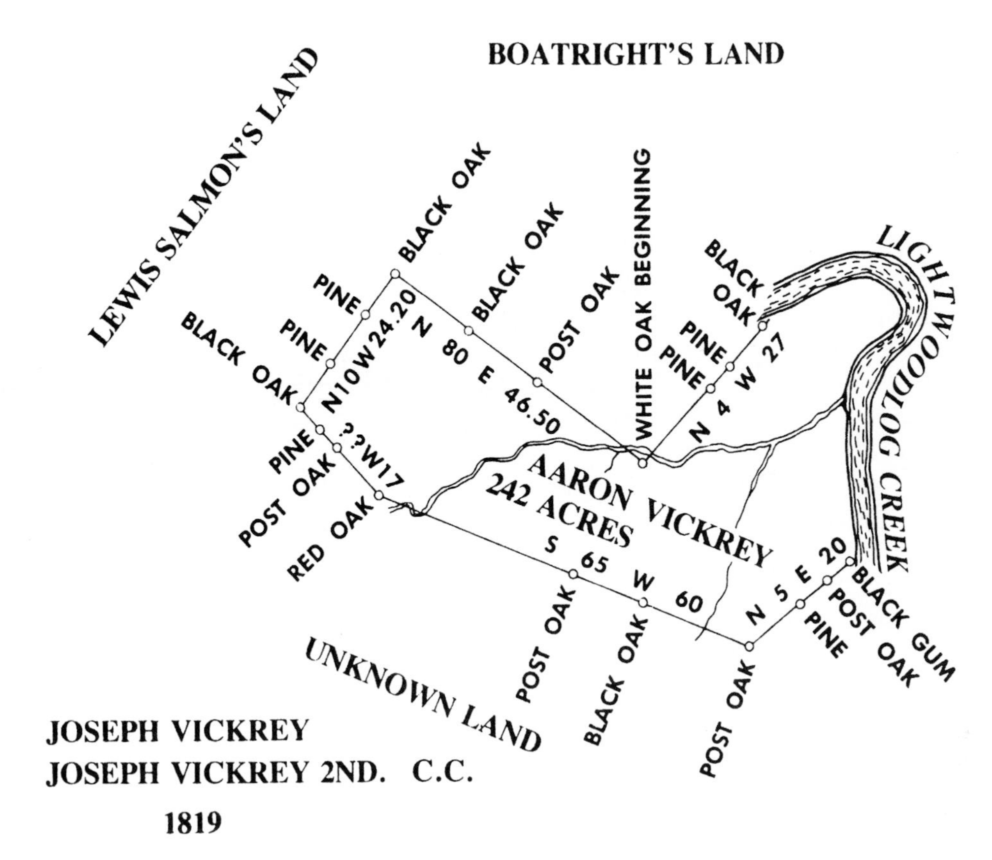

Figure 4.17 Metes-and-bounds survey.
Source: On file at the Georgia Department of Archives and History.

or west). If, for example, the line being described lies at a 45-degree angle to the east of a north reference line, its bearing is north 45° east (see Figure 7.6).

Systematic Subdivision (U.S. Public Land Survey)

In New England, in contrast to other colonial areas, there were instances of organized land-ownership plans. Towns were often surveyed prior to settlement. They were usually laid out along riverbanks or some other transportation arteries, were of a fairly uniform size, and were divided into lots that were mapped prior to settlement. The lots were of variable shape and size but were frequently bounded by straight lines. New England town plans were the predecessors of the public land-survey system that was later extended throughout much of the United States.

U.S. Public Lands

The Treaty of Paris in 1783, which ended the Revolutionary War, gave control of the territory east of the Mississippi River and south of the Great Lakes (except Spanish Florida) to the United States. In addition, during the period between 1780 and 1802, the original 13 colonies ceded their western land claims to the newly formed United States. The impending sale and settlement of these public lands created a need for a systematic method of determining and recording land ownership. As a result, in 1785, Congress passed a land ordinance that established a procedure for disposing of lands in the Western Territory. This act provided the initial framework for the **rectangular survey system** that was begun when the settlement of the western portion of the nation was in its early stages. This system, which has come to be called the **United States Public Land Survey (USPLS),** was started in Ohio and ultimately was extended throughout most of the western states and parts of the south.[5]

Many variations in the details of the USPLS were introduced as the system evolved. For this reason, different specifics may be encountered when actual surveys are examined. When variations occur, it is necessary to consult other sources, such as those listed in the Suggested Readings at the end of the chapter, to determine their origins. The basic outline of the system as it is described here has nevertheless remained quite constant.

[5]The rectangular survey system is not used in Connecticut, Delaware, Kentucky, Maine, Maryland, Massachusetts, New Hampshire, New Jersey, New York, North Carolina, Pennsylvania, Rhode Island, South Carolina, Tennessee, Texas, Vermont, Virginia, or West Virginia.

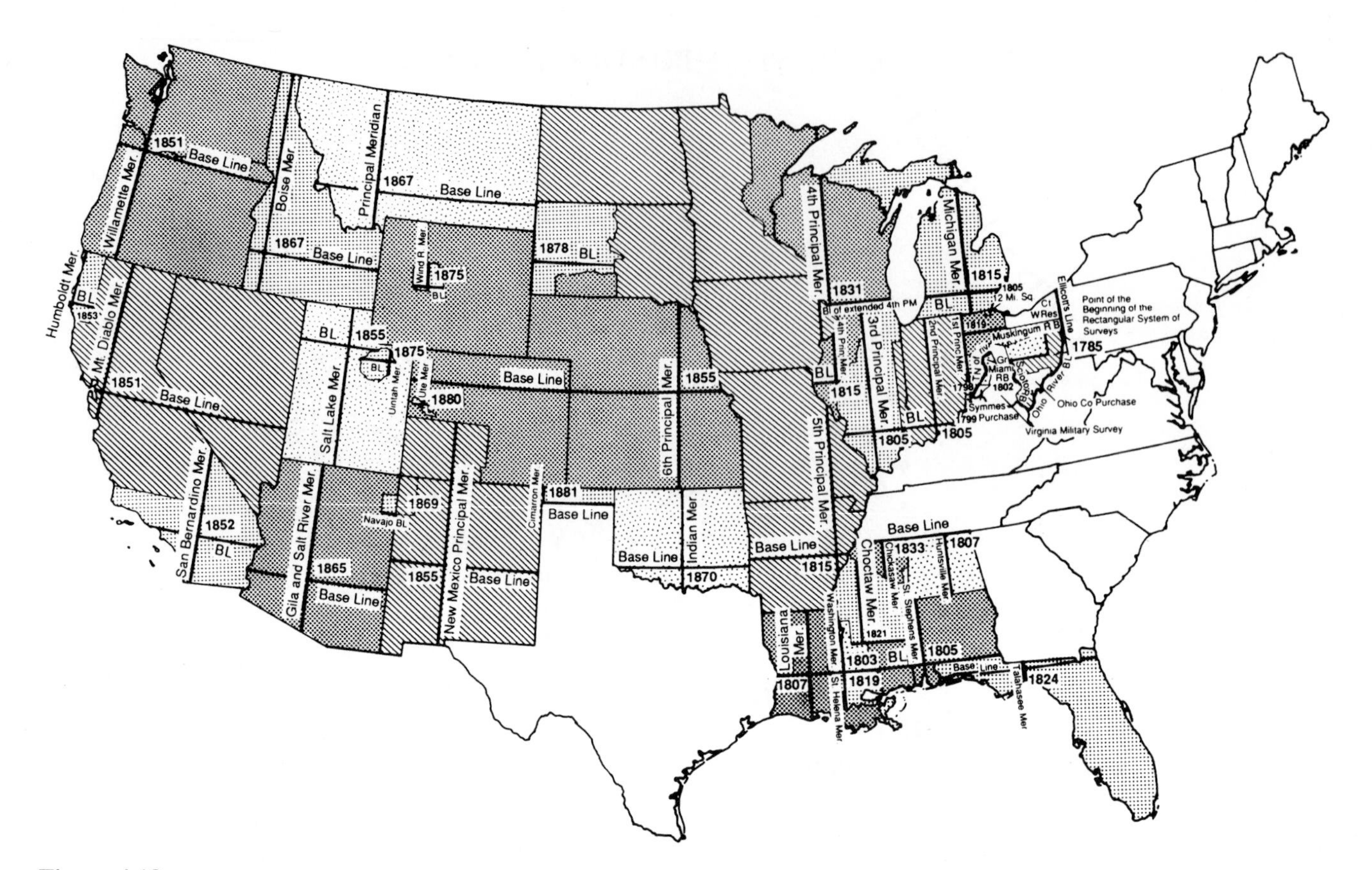

Figure 4.18 USPLS principal meridians and baselines and the territory governed by each.
Source: U.S. Department of the Interior, Bureau of Land Management, *Surveying Our Public Lands* (Washington, D.C.: U.S. Government Printing Office, 1980).

Township, Range, and Section

In the USPLS, an initial point is first established for the particular region to be surveyed. Astronomical observations are used to determine the location of this initial point. A north-south **principal meridian** is then established, and an east-west **baseline** is set at right angles to it, with both lines passing through the initial point. The actual arrangement of the thirty-four principal meridians and primary baselines and the territory governed by each are shown in Figure 4.18.

A grid of lines, spaced at 6-mile intervals, is established next. The east-west lines in this grid form rows, called **townships.** They are numbered in sequence, both north and south of the baseline (Figure 4.19). The north-south lines in the grid form columns, called **ranges.** They are also numbered in sequence, both east and west of the principal meridian. The intersection of the township rows and the range columns creates a pattern of squares, also called **townships** (an unfortunate repetition of the word). Townships are subdivided into thirty-six **sections,** each approximately 1 square mile (640 acres); the exact size of actually surveyed townships often differs from this standard, however. The sections are numbered in alternating rows, beginning at the northeast corner of the township. The reasons for the variation in township size are discussed and the subdivision of sections is explained next.

Plats

Detailed maps that show the boundaries of USPLS land ownerships within a township are called **township plats.** The subdivision of the sections, as recorded on the plats, varies considerably. The basic system, however, uses the so-called **aliquot parts.** Aliquot parts result from the division of a section into quarters (160 acres) and, subsequently, of the quarters into quarter-quarters (40 acres). In addition, half-sections are sometimes used (either northern and southern or eastern and western halves).

Special units, called **lots,** are used when the normal system of fractional sections cannot be applied. Lots frequently occur along the northern and western edges of townships, for example, because measurement errors are adjusted along these edges. They are also used when a navigable stream or a lake that is 25 acres or more in extent is encountered. This is because a settler purchasing public land would generally not want to pay for a full section when a considerable area was under water and not usable for farming.

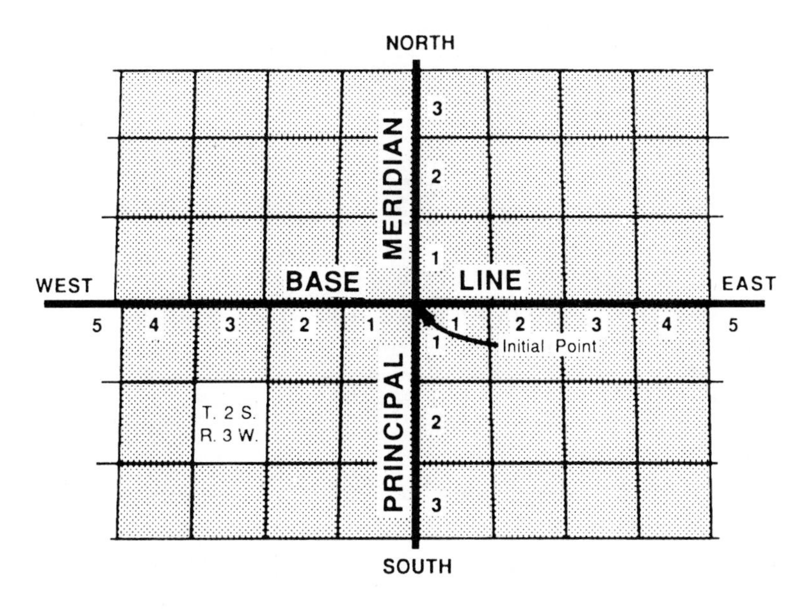

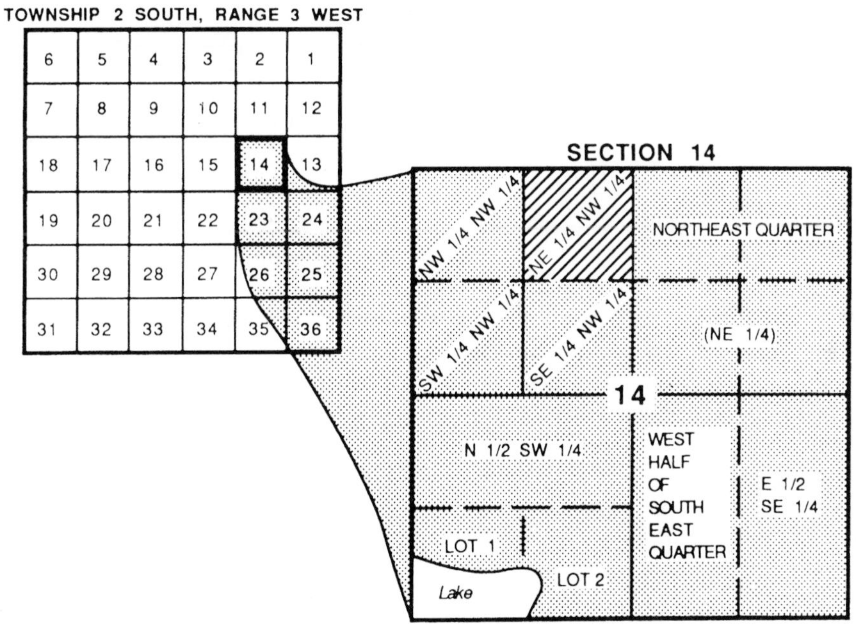

Figure 4.19 Townships, ranges, sections, and further divisions of the USPLS.

Source: Modified from the U.S. Department of the Interior, Bureau of Land Management, *Surveying Our Public Lands* (Washington, D.C.: U.S. Government Printing Office, 1980).

When lots are established, a line called a **meander line** is surveyed along the water body's mean high-water elevation. Then, as many standard quarter-quarter sections as possible are designated, and the remaining land is subdivided into lots. The lots are kept as close to 40 acres as possible, although subdivisions as small as 2.5 acres are sometimes encountered. The lots are identified by sequential numbers within each section. Although meander lines are used to determine acreages, high-water lines are used to establish property boundaries.

There are also instances in which preexisting claims were incorporated into the USPLS, such as in portions of California, Oregon, and Florida. Because these claims were usually irregular in shape and alignment compared to the units of the USPLS, they were incorporated into the system as lots. Other special units that were established were the so-called French Tracts in Orleans Territory (now Louisiana). These tracts were arranged to provide frontage on navigable waterways, important to the French settlers in the region (Figure 4.20). More information about this type of

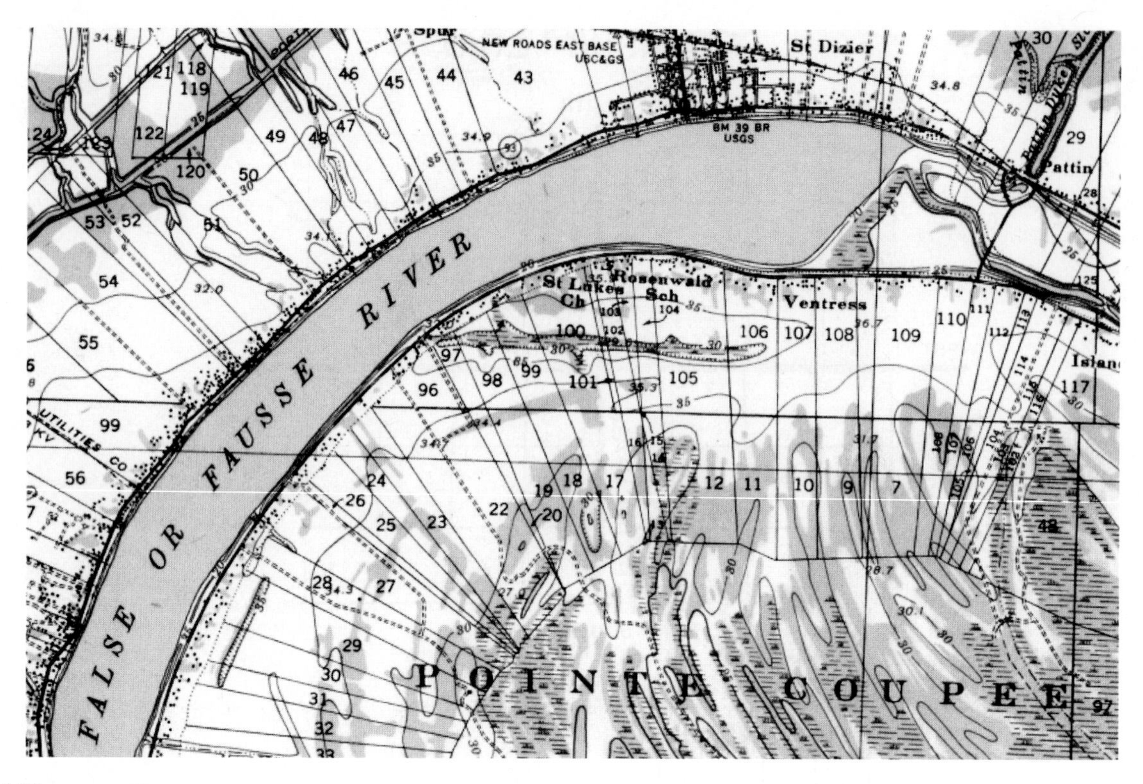

Figure 4.20 French Tracts in Louisiana. (Portion of New Roads quadrangle, Louisiana, 1:62,500, 1939.)

tract is provided in the discussion of Canadian land-survey systems, later in this chapter.

Property Descriptions

Rectangular survey system designations are used to describe a tract of land for legal purposes. Consider, for example, the 40-acre tract marked with the diagonal line pattern in Figure 4.19. This tract covers the northeast quarter of the northwest quarter of Section 14 and is located in the township that is at the intersection of Township 2 South, Range 3 West, within the area of survey that is based on a particular principal meridian. This description would normally be written in abbreviated form as NE 1/4, NW 1/4, Sec. 14, T 2 S, R 3 W, __P.M.

Assume, then, that you wish to locate the same tract, on the basis of its description. This task involves working from the broadest category down to the narrowest. You first determine the area covered by surveys based on the particular principal meridian. Then you need a map showing the intersection of the second township south of the origin and the third range west. Within that township, you first find Section 14, then the northwest quarter of the section, and, finally, the northeast quarter of the north-west quarter. This system is valuable because such a brief description allows a specific small piece of land, within the extensive area of the United States that is covered by the USPLS, to be unambiguously defined and located.

Irregularities

The curvature of the earth affects the size of townships in the USPLS because lines of longitude converge toward the poles. As a result, although range lines are spaced 6 miles apart along a baseline, they come closer together as they are extended north along their meridian. For this reason, corrections must be made to prevent spacing between the range lines from deviating too much from the 6-mile standard. This is accomplished by establishing correction lines, called *standard parallels* and *guide meridians*. Generally, a **standard parallel** is established every 24 miles north or south of a baseline. Along this standard parallel, north-south **guide meridians** are surveyed, at a starting spacing of 24 miles. The southern, western, and eastern boundaries of each of the blocks formed by the guide meridians and the standard parallels are nominally 24 miles long. Surveying discrepancies almost always cause the actual measurements to deviate somewhat from the desired standard. In addition, the guide meridians converge, so that the northern boundaries of the blocks are less than 24 miles long.

The imposition of a rectangular survey system onto the spherical surface of the earth, then, faces fundamental problems that result in unavoidable deviations from a strictly right-angled system of property lines. In addition, for a variety of reasons including inaccurate instruments, difficult terrain, poor work, and, unfortunately, occasional dishonesty, irregularities are often found in the pattern of USPLS lines. In the early years, for example, some surveyors lacked the skills or the instruments with which to measure horizontal distances between points, instead of distances over the irregular surface of the ground (see Chapter 6). Such discrepancies introduced inaccuracies into the system, especially in areas with steep terrain.

Similarly, the early use of the magnetic compass, which is an unreliable tool for determining directions, often resulted in survey errors. Standard USPLS practice, therefore, excludes the use of the magnetic compass for establishing the direction of survey lines. Instead, readings from a special instrument called a *solar transit* or observations of Polaris are used to establish true north. Other directional readings are related to the reference direction thus established.

Land Survey in Canada

Different land-ownership systems have been used at various stages in Canadian history. Each system has resulted in a distinctive landscape pattern in the region to which it was applied. It is not feasible to describe these systems in detail here, but some of their major characteristics will be noted.

Patterns in Areas of French Settlement

In the seventeenth century, the region that is now Quebec and the Maritime provinces, except Newfoundland, was under French domination. The French government granted large land holdings, called **seigneuries,** often to noblemen but sometimes to yeomen or farmers. The **seigneurs,** as the land holders were called, had the obligation to promote settlement and development. To this end, they subdivided the seigneuries into lots, called **rotures,** that were allocated to individual farmers.

The pattern of land subdivision in the seigneuries was largely controlled by the rivers, the principal highways and communication links of the time. In the valley of the St. Lawrence River, for example, the general pattern consisted of tiers of long, narrow lots. These lots were arranged in a northwesterly southeasterly alignment, at right angles to the trend of the river (Figure 4.21). The first tier of lots fronted on the river. Although there was considerable variation, the typical river frontage of a lot was three arpents[6] (approximately 575 feet). The lots usually had a depth of thirty arpents (5755 feet) or more. The favorably located riverfront lots were usually settled first, and the tiers of lots away from the river were settled later.

[6]One arpent equals 191.838 feet.

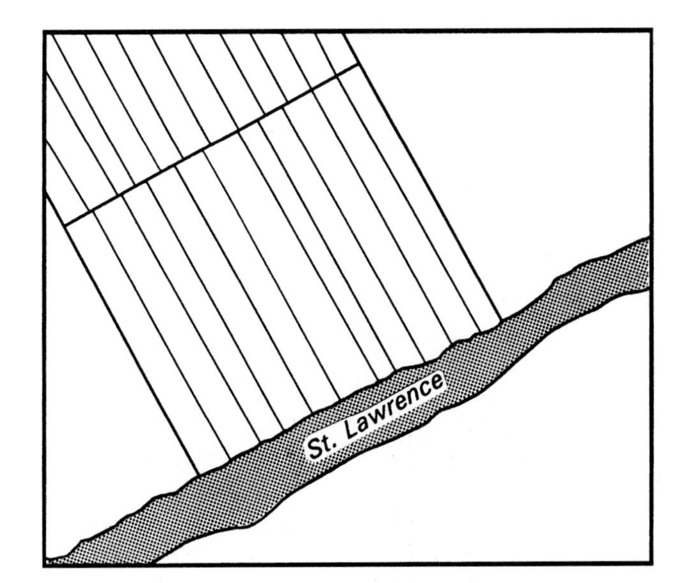

Figure 4.21 Example of lots (rotures) within a seigneury, in an area of French settlement in Canada.

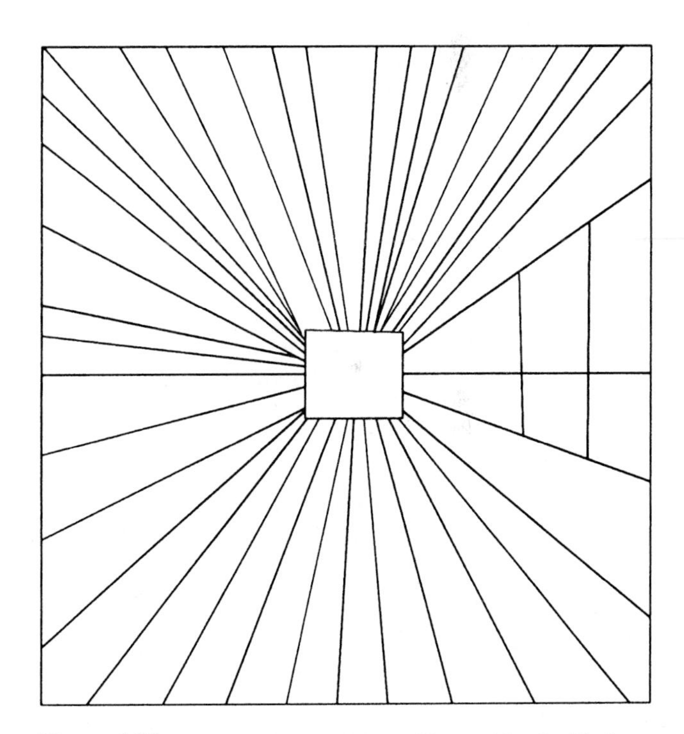

Figure 4.22 Example of lots within a village with a fortified central square, in an area of French settlement in Canada.

The pattern of lots differed when a village was laid out. Villages often consisted of a square, perhaps 40 arpents (approximately 1 1/2 miles) on a side. Forty farms, each containing 40 square arpents (34 acres), might be laid out in a radial pattern around the fortified central square (Figure 4.22).

Patterns of long lots were also established by French settlers in the United States. They were found, for example, in areas around Detroit, Michigan, and Green Bay, Wisconsin, and in Louisiana.

Patterns in Areas of English Settlement

In later years, in areas settled by the English, lots of 100 to 300 acres were often surveyed in large, rectilinear blocks. As with the seigneuries, the patterns of these blocks frequently began with a first tier of lots approximately paralleling the irregular bank of a river. The lots in the other tiers were then arranged in rows parallel to the rear line of the first tier. A variety of systems was used at different times, so township and lot sizes are found in a variety of forms.[7] In one example, the township contained 36 square miles (6 by 6 miles) (Figure 4.23). It consisted of 25 lots along the river bank, each 19 chains wide (1254 feet). The pattern extended inland for seven tiers, with each tier 68 chains 25 links (4356 feet) deep, plus a road allowance of two chains 25 links. The total township contained 175 lots of 120 acres each, plus provision for roads.

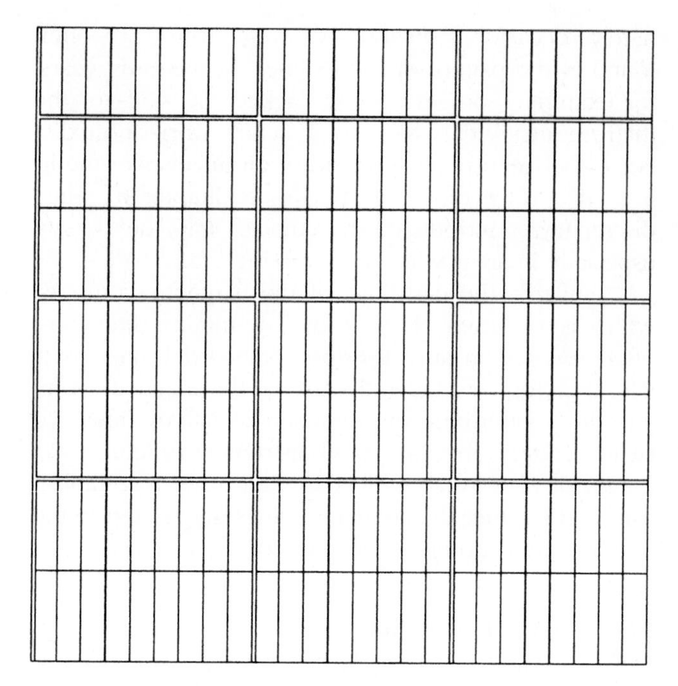

Figure 4.23 Example of a township and lots in an area of English settlement in Canada.

Westward Expansion

Still later, in the 1860s and 1870s, Canada was faced with a rapid expansion to the west. The situation was similar to that in the United States about 80 years earlier. During the early part of this period, townships were surveyed on the north shore of Lake Huron, following U.S. practices.

In 1870, to provide an orderly framework for expansion into the prairies and the mountains of the west, the Dominion Lands Survey System was established. This system, now called the **Canada Land Survey System (CLSS),** uses principal meridians and baselines, townships, ranges, sections, and subdivisions of sections. It is basically the same as the USPLS pattern. The measurements of these features are the same as in the USPLS, although serious consideration was given to making the townships larger. In 1869, for example, a recommendation was made that townships contain 64 sections, each containing 800 acres, plus 40 acres for roads. The system extends from the eastern boundary of Manitoba to the Pacific Coast. Within its area of coverage, however, already occupied and surveyed lands are usually omitted from the new system.

In Canada, the survey system that was finally adopted provides a more organized overall framework than is the case in the United States. The first element of the CLSS is a pattern of regularly spaced principal meridians. This contrasts with the heterogeneous scattering of principal meridians used in the USPLS. The initial principal meridian, called the *First* or *Prime Meridian,* is located approximately 15 miles west of Winnipeg, Manitoba, along 97°27′28.41″ West. A round-numbered meridian, such as 98° West, would have been simpler. However, the historical sequence of events led to the adoption of an already surveyed meridian. Other principal meridians were established with a regular, 4-degree spacing (Figure 4.24). In addition, there is an irregular meridian called the *Coast Meridian*. This meridian is located near Vancouver, British Columbia, along 122°45′39.6″ West.

Another departure from the USPLS style is that only one baseline is used. This baseline is 49° North, the parallel that defines much of the Canada-United States boundary. For this reason, there are no ranges with "south" designations. Additional standard parallels are established at regular intervals to correct for the convergence of the meridians (Figure 4.25). At these parallels, the north-south dividing lines jog to compensate for the convergence of the meridians.

The numbering of ranges also deviates from the USPLS pattern. Starting with the Principal Meridian, ranges are numbered in sequence to the west, until the next initial meridian is reached. This pattern is repeated over the span of each initial meridian. Therefore, to the west of the First Meridian, there are no ranges with "east" designations. The opposite holds true to the east of the First Meridian, where ranges are numbered in sequence to the east, until the eastern boundary of Manitoba is reached.

The final differences from the USPLS involve the numbering sequence and the subdivision of the sections. In the CLSS, the numbering of the 640-acre sec-

[7]Don W. Thomson states that seven land surveying systems were used in present-day Ontario (*Men and Meridians,* vol. 1, 237).

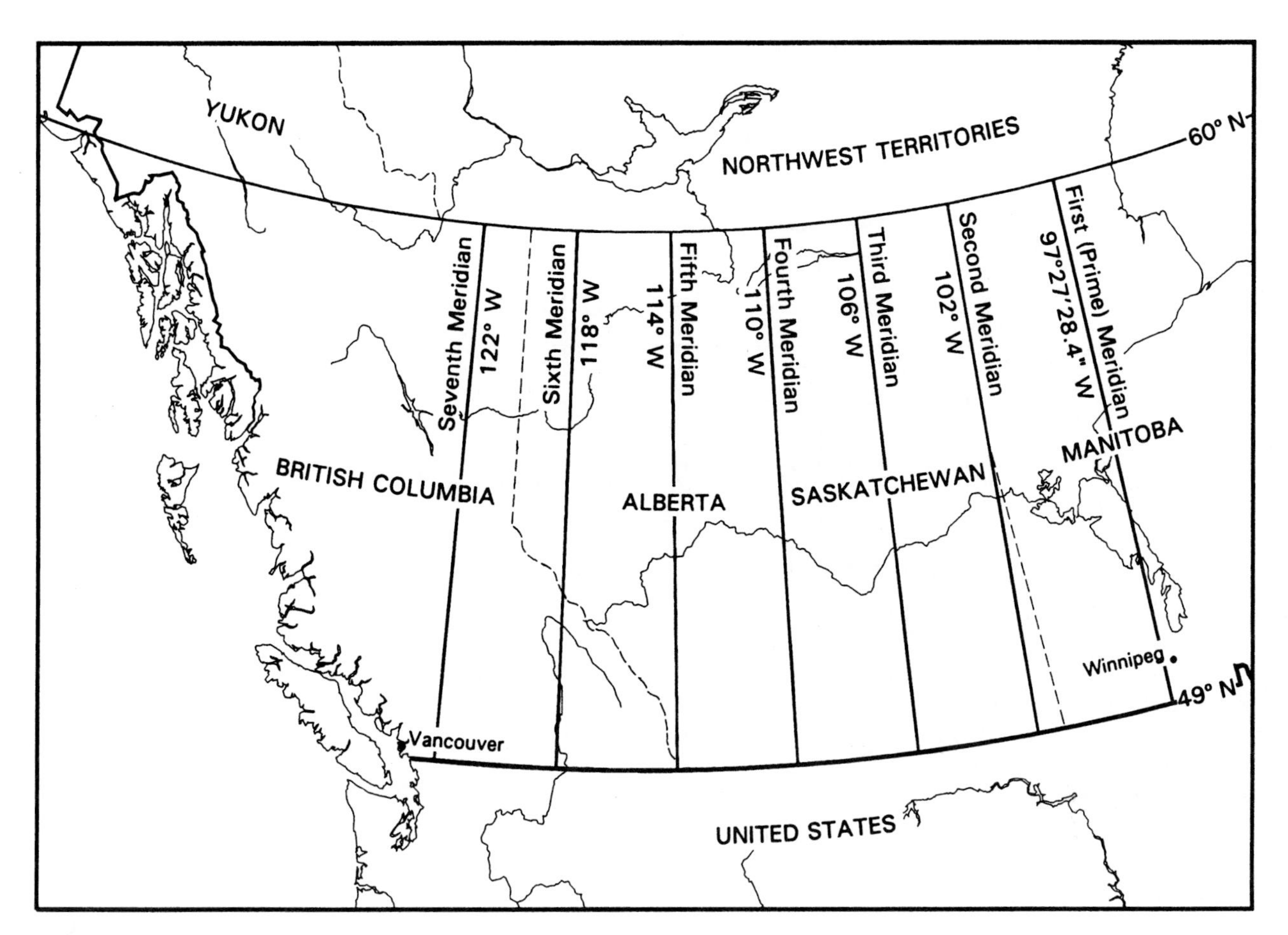

Figure 4.24 Principal meridians of the Canada Land Survey System (CLSS).

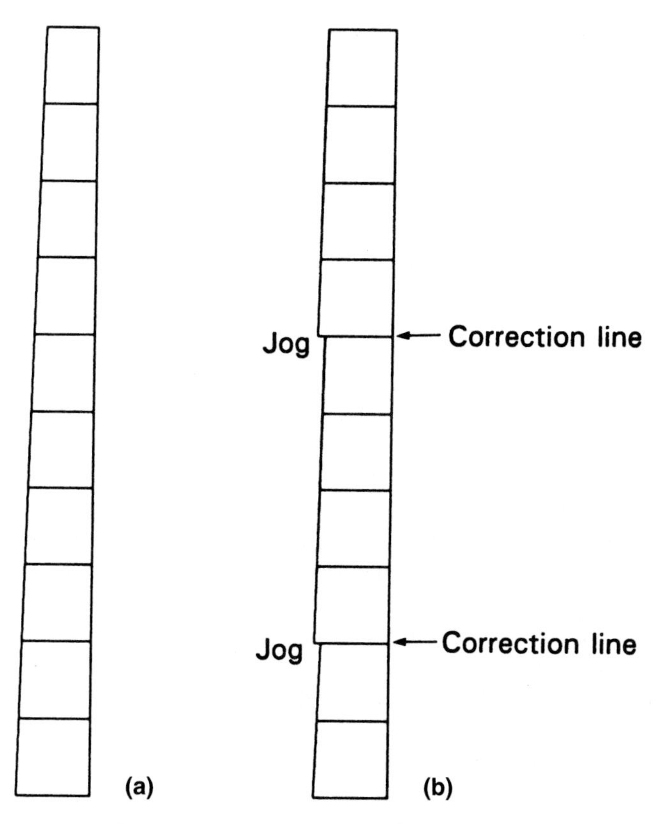

Figure 4.25 (*a*) Convergence of meridians without correction. (*b*) Correction lines, as used in the CLSS, to compensate for convergence.

tions zigzags from the southeast corner of each township (Figure 4.26), instead of from the northeast corner as in the USPLS. Sections are sometimes divided into 160-acre quarter-sections, as in the USPLS, but may be divided into legal subdivisions instead. When they are used, there are sixteen 40-acre **legal subdivisions** in a section, numbered as shown in Figure 4.26. The legal subdivisions are the equivalent of quarter-quarter sections in the USPLS. Finally, legal subdivisions are infrequently subdivided into 10-acre quarters.

There are many variations on the general plan just outlined because of the special circumstances or history of particular locales. It is not possible to describe these variations here, but they are explained in the publication *Understanding Western Canada's Dominion Land Survey System,* which is listed in the Suggested Readings.

Cultural Influences

Workers in many fields, including geography, history, archaeology, forestry, park and recreation planning, landscape architecture, land use planning, and conservation are concerned with aspects of both the human and the natural environment—often referred to as *cultural landscapes.* The concept of cultural landscape is important in the understanding and management of

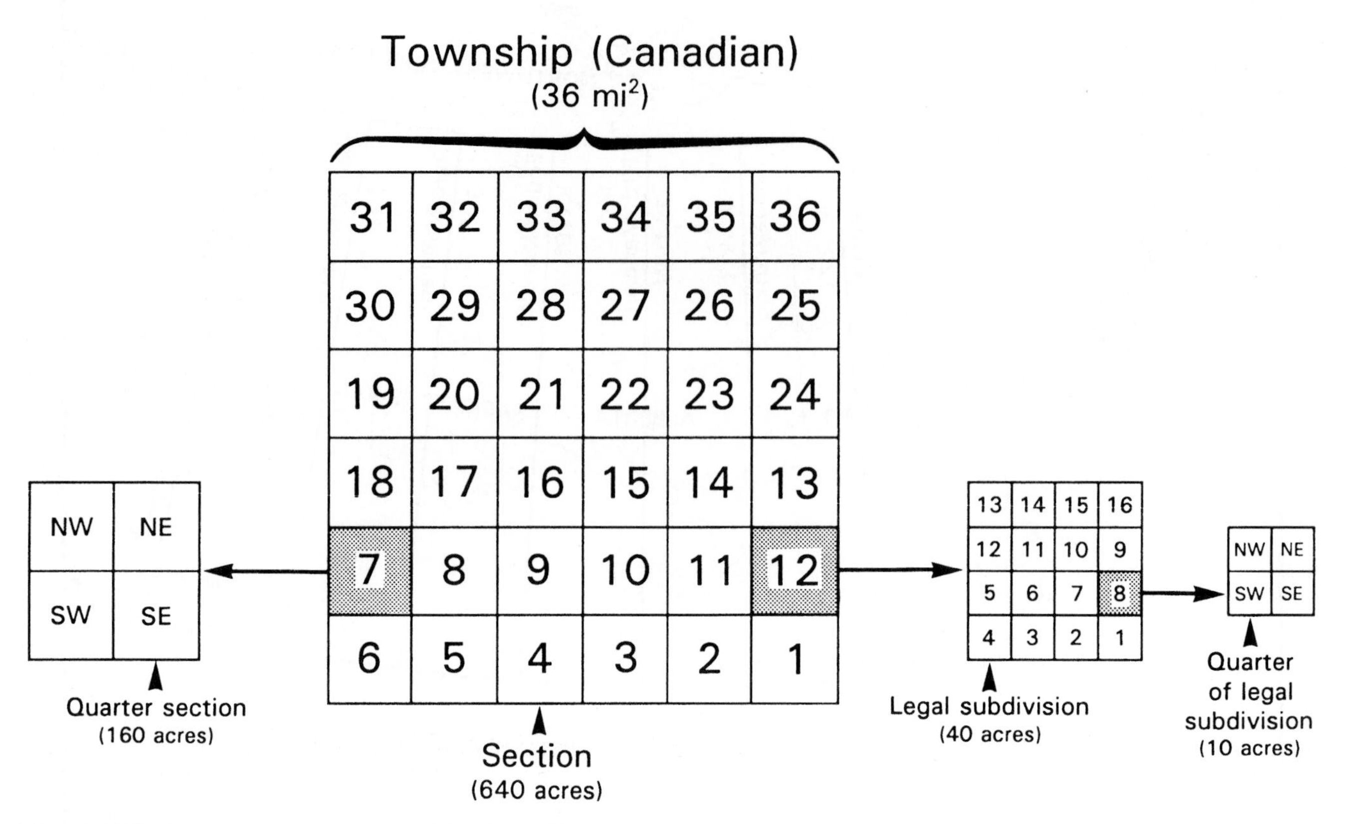

Figure 4.26 Standard subdivision of a township in the CLSS.

human interactions with the land. Knowledge of the elements of the cultural landscape provides insights that help reveal the way humans directly and indirectly shape the land. These insights relate to aesthetics, conservation, pollution, land use, zoning, and myriad other concerns. Therefore, they influence decisions about the preservation or transforming of modern landscapes, whichever is desired. These decisions may range from efforts to preserve historic districts or protect ordinary neighborhoods, to the development of farms or the conservation of forests. As we have seen, maps can record present-day or historical information about both human-built and natural features. These include, but are not limited to, features of the cultural landscape such as the topography, the extent and type of vegetation, the size and location of buildings, the extent of built-up areas, and the location of transportation facilities. Therefore, maps provide essential tools for anyone concerned with the study of cultural landscapes.

The existence of the rectangular survey system has had a significant influence on the cultural landscape of the western United States and Canada. Most of the world was settled without a preexisting land-ownership system. For that reason, the size and arrangement of property ownerships, the alignment of roads and railroads, the boundaries of political jurisdictions, and the alignment of similar cultural features in such areas has a relatively random appearance. The appearance of such areas is more attuned to the relatively free-form alignments of the physical landscape than to the artificial rectilinear arrangements of a human-made system. This difference in appearance is readily recognized by anyone who has traveled to different portions of the country, especially by air. The rectilinear patterns of the west and the "natural" patterns of the east are in distinct contrast with one another (Figure 4.27).

The effects of land subdivision on the landscape are long lasting and pervasive. We can, for example, see traces of the Roman method of land division, called *centuriation,* in the landscape of rural central Italy. There is, even today, a "persistent orthogonal layout of roads, fences, and irrigation canals and even in the orientation of houses and barns" that has persisted for some two thousand years."[8]

Although not easily documented, direct influences on human activities of these differences in cultural landscapes certainly exist. The influences are as subtle as in the use of a phrase such as "plowing the south forty," which indicates the pervasiveness of the system. Such a phrase would have no meaning in much of the world, but it is easily understood by farm dwellers in areas settled under the influence of the USPLS or CLSS.[9]

[8]Samuel Y. Edgerton Jr., "From Mental Matrix to *Mappamundi* to Christian Empire," In *Art and Cartography: Six Historical Essays,* ed. David Woodward (Chicago and London: University of Chicago Press, 1987), 21.

[9]For more extensive examples of cultural influences, see the books by Norman J. W. Thrower and John F. Rooney, Jr., Wilbur Zelinsky, and Dean R. Louder, listed in the Suggested Readings at the end of the chapter.

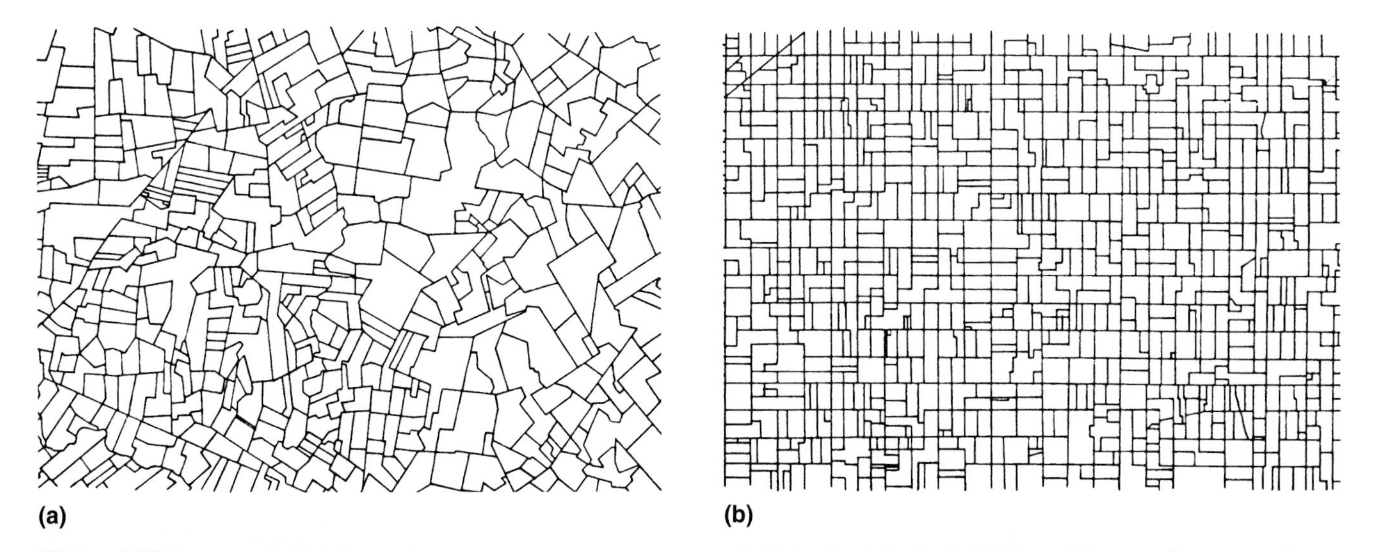

Figure 4.27 Contrasting land-survey patterns of physically similar areas in Ohio. The irregular boundaries of the area of unsystematic survey (*a*) contrast sharply with the rectilinear pattern that developed under the influence of the systematic USPLS (*b*).

Source: From Norman J. W. Thrower, *Original Survey and Land Subdivision* (Chicago: Published for the Association of American Geographers by Rand McNally and Company, 1966). Reprinted by permission of the Association of American Geographers.

SUMMARY

The locational systems commonly used to describe locations on the earth's surface include latitude and longitude, the World Geographic Reference System (GEOREF), the Universal Transverse Mercator (UTM) and Universal Polar Stereographic (UPS) systems, and the State Plane Coordinate (SPC) system. Other systems, including metes-and-bounds surveys and the U.S. Public Land Survey (USPLS), are used to record land-ownership information.

Positions on the latitude and longitude graticule are rather easily determined by interpolation. The main requirement is to keep in mind the division of the circle into 360 degrees, each degree containing 60 minutes that are further divided into 60 seconds.

The World Geographic Reference System (GEOREF) uses 15-degree-wide north-south and east-west bands. Each quadrangle formed by the intersection of these bands is uniquely identified by a pair of letters. The quadrangles are each subdivided into 1-degree quadrangles, identified by a second pair of letters. Thus, four letters allow the full identification of any 1-degree quadrangle in the world. Closer identification of locations is provided by subdividing each 1-degree quadrangle into 1-minute quadrangles. These are identified by a four-digit number and can be further subdivided into decimal parts, if necessary.

Simple reference grids that do not tie into any overall locational system are commonly encountered in atlases and on street maps. More important grid systems, such as the UTM system and its companion UPS system or the SPC system, designate locations within a large region.

The UTM system extends between 80° south and 84° north, and the UPS system covers the polar caps. The UTM system involves 60 north-south zones, each 6 degrees wide. Two false origins are established in each zone: one for the Northern Hemisphere and the other for the Southern Hemisphere. A square locational grid, measured in meters, is extended north and east from each origin, so that any point within each zone has a unique coordinate. A similar grid is used in the UPS system. The full UTM or UPS coordinates of a location contain its easting and northing, the hemisphere, and the zone number.

The Military Grid Reference System is a modification of the civilian UTM and UPS systems. The area covered by the UTM system is subdivided into grid zones that are generally 6 degrees east-west by 8 degrees north-south. Each grid zone is identified by a column number and a row letter and is subdivided into 100,000-meter squares, each of which has a two-letter identifier. Locations within each 100,000-meter square are given as coordinate distances, in meters, from the square's southwest corner. A full grid reference consists of the grid zone designation, the 100,000-meter-square identification, and the grid coordinates, written to the desired level of accuracy.

The SPC system is a rectangular coordinate system that is individually applied to each state in the United States, with the larger states divided into zones. It is designed so that it is unnecessary to take the earth's curvature into account. The amount of error thus introduced is usually acceptable for the small areas involved in local surveys. In this system, a false origin is established for each zone, and location coordinates are measured east and north from it, in feet. A full SPC location designation consists of the easting, northing, name of the state, and designation of the zone (such as north, south, east, west, or central).

Land partitions are based on either unsystematic or systematic subdivision. The method used to describe unplanned land ownerships is metes-and-bounds surveying. In this system, the direction (bearing) and length of each portion of the boundary line of the property is recorded in sequence, as though one were walking around its perimeter.

The predominant type of systematic subdivision in the United States is the USPLS. In this system, an initial point is established, by astronomical observations, for each of 34 regions. A north-south principal meridian is then established, and an east-west baseline is set at right angles to it, both passing through the initial point. Six-mile-wide east-west rows (townships) are established next, as are similar north-south columns (ranges). The intersection of the townships and ranges creates a pattern of squares, also called *townships*. Townships are subdivided into 36 1-square-mile (640-acre) sections. The sections are numbered in alternating rows, beginning at the northeast corner of the township, and may be subdivided into halves, quarters, quarter-quarters, and lots. Detailed maps of these USPLS land ownerships are called *township plats*.

SUGGESTED READINGS

Department of the Army. *Grids and Grid References*. TM5–41–1. Washington, D.C.: Headquarters, Department of the Army (latest ed.).

______. *Universal Transverse Mercator Grid*. TM5-241-8. Washington, D.C.: Headquarters, Department of the Army, April 1973.

Department of Energy, Mines and Resources. *The Universal Transverse Mercator Grid: As Applied to National Topographic System Maps of Canada*. Ottawa, Ont.: Department of Energy, Mines and Resources, 1969.

McKercher, Robert B., and Bertram Wolfe. *Understanding Western Canada's Dominion Land Survey System*. Saskatoon, Saskatchewan: University of Saskatchewan, Division of Extension and Community Relations, 1986.

Rooney, John F., Jr., Wilbur Zelinsky, and Dean R. Louder, eds. *This Remarkable Continent: An Atlas of the United States and Canadian Society and Cultures*. College Station, Tex.: Texas A & M University Press, for the Society for the North American Cultural Survey, 1982.

Rubinstein, James M. *The Cultural Landscape*, 6th ed. Englewood Cliffs, N.J.: Prentice-Hall, 1999.

Sebert, L. M. "The Land Surveys of Ontario, 1750–1980." *Cartographica* 17, 3 (1980): 65–106.

Stewart, Lowell O. *Public Land Surveys*. New York: Arno Press, 1979.

Thomson, Don W. *Men and Meridians: The History of Surveying and Mapping in Canada:* Vol. I, Prior to 1867, Ottawa: Queen's Printer and Controller of Stationery, 1966; Vol. II, 1867 to 1917, Ottawa: Information Canada, 1967; Vol. III, 1917–1947, Ottawa: Information Canada, 1969.

Thrower, Norman J. W. *Original Land Survey and Land Subdivision*. Chicago: Rand McNally, for The Association of American Geographers, Monograph 4, 1966. Compares patterns of cultural features, such as land-ownership units; administrative districts, such as counties and school districts; and transportation facilities in two areas of Ohio—one settled under the influence of the USPLS and the other not.

White, C. Albert. *A History of the Rectangular Survey System*. Washington, D.C.: U.S. Department of the Interior, Bureau of Land Management, 1983.

CHAPTER

5

Scale and Generalization Concepts

After gathering information about the earth and its surface features, the cartographer must decide how the information is to be presented on a map. Two particularly important questions to ask and answer are (1) What is an appropriate scale for this map? and (2) How much information should be included? As a map user, you need to be aware of the effects that these decisions have on the final map.

Scale

Physical distances on the earth's surface are relatively easy to visualize because of our experience with them. We may have to commute 7 miles (11.3 kilometers) to work, the nearest major-league baseball park may be 200 miles (321.9 kilometers) away, and we may be used to watching (American) football teams play on a field that is 100 yards (91.44 meters) long. Recognizing that a half-inch (1.27-centimeter) measurement on a map may, depending upon the particular map scale, represent any one of these ground distances or, indeed, any other distance is not as easy. This variable relationship between the size of actual features on the earth and the size of the map symbols that represent those features is discussed in this section.

Methods of Expressing Scale

The **scale** of a map is the ratio between map distances and earth distances. This scale relationship is expressed in one of three different, but equivalent, ways: (1) as a word statement, (2) as an arithmetic ratio (representative fraction), or (3) as a graphic symbol.

Word Statement

A **word statement** of map scale gives a quick idea of size relationships, but it is in a form that is awkward to use in many applications. A word statement might say, for example, that one map is at a scale of "1 inch to 1 mile," and another is at a scale of "1 inch to 2 miles." The second map then will show considerably less detail than the first, because 2 miles instead of 1 are represented by every inch.

Representative Fraction

A **representative-fraction (RF) scale** is the ratio between the map distance and the ground distance between equivalent points. It is expressed either as a true ratio, such as 1:100,000, or as a fraction, such as 1/100,000. Both means of notation have the same meaning; in this example, one unit of distance on the map represents 100,000 of the same units on the earth's surface.

A representative-fraction scale is "unit free." That is, the ratio between the map-distance values and the earth-distance values remains the same, regardless of whether the measurements are expressed in feet, miles, meters, kilometers, or any other unit of distance. Another way of saying this is that both portions of the ratio (or the numerator and denominator of the fraction) are always expressed in the same units of measurement.

Box 1.1 Units of Measurement and Map Scales

Units of measurement must be defined before features on the earth's surface can be mapped. In the United States, we are used to working with units of the U.S. Customary System, such as inches, feet, yards, miles, and acres. These units are familiar, but they are extremely awkward for calculation purposes. Most of the rest of the world works with metric units such as centimeters, meters, kilometers, and hectares. Although the metric system is unfamiliar to many of us, it has great advantages in ease of computation.

The metric system has been approved for use in the United States since 1866. Despite this long period of approval, the process of converting to metric units continues to move slowly in this country. Nevertheless, many maps are now produced using metric units, and this trend is expected to continue. From a practical point of view, therefore, you should be familiar with metric units so that you will be able to use both foreign and domestic metric maps properly. As a learning aid, most of the measurements in this book are given in both metric units and their U.S. Customary System equivalents.

LINEAR MEASUREMENTS

U.S. CUSTOMARY UNITS

U.S. Customary Units of measurement are based on the units of the older English system (Table 5.1). This system developed over a long period of history. Most of the units were established arbitrarily and have been subject to change over the centuries. The **foot,** for example, is a measure descended from numerous ancient measurements. Those measurements, which were based on different ideas about the length of a human foot, varied from about 25 to 34 centimeters (approximately 9.8 to 13.4 inches). The division of each foot into 12 equal inches was begun by the Romans.

In the United States today, there are two different definitions of a foot. From 1893 until 1959, the foot was defined as 1200/3937 meter (1:0.30480061). In 1959, the **international foot** was defined as 1:0.3048 meter. At the same time, it was decided that within the United States any data expressed in feet and derived from geodetic surveys would continue to use the foot as it was defined in 1893. This foot is called the **U.S. survey foot**. In day-to-day usage, the difference between the two measures is not very important, but in accurate measurements of long distances, the difference between them can be significant. For example, a measurement of 800,000 meters is 2,624,666.67 U.S. survey feet, but 2,624,671.92 international feet. Given such potential confusion, it is little wonder that surveyors are encouraged to use meters instead of feet.[1]

[1]James Stern, "Do You Know Which Foot to Use?" *ACSM Bulletin* (August 1987): 36.

TABLE 5.1 U.S. Customary Units—Linear Measure

Land Units
12 inches = 1 foot
3 feet = 1 yard
16.5 feet* = 1 rod
66 feet* = 1 chain
40 rods = 10 chains = 1 furlong = 660 feet*
80 chains = 8 furlongs = 1 mile* = 5280 feet*
Nautical Units
1852 meters = 6076.11549 feet (approx.) = 1 international nautical mile

*Survey units, not international units.

TABLE 5.2 Metric Units—Linear Measure

1 millimeter	= 1/1000 meter
1 centimeter	= 1/100 meter
1 decimeter	= 1/10 meter
1 meter	
1 decameter	= 10 meters
1 hectometer	= 100 meters
1 kilometer	= 1000 meters

Note: This table reflects part of the International System of Units (SI), a modernized metric system established by international agreement in 1960.

The concept of a **mile** is descended from a Roman measurement called, in Latin, *mille passus* (1000 paces). A Roman pace, as it was originally defined in England, was approximately 5 feet long, so a 1000-pace mile covered 5000 feet. In 1593, however, a law was passed in England that brought a shorter statute foot into use. Because the physical length of a mile remained unchanged, the number of feet in the newly defined **statute mile** was increased to the familiar but awkward 5280 feet (1609.3 meters).

METRIC UNITS

The **metric system** has the major advantage of being a decimal system. This is a great convenience when conversions between units are required because they are related to one another by simple multiples of ten (Table 5.2).

In addition to offering computational convenience, metric measurements are, theoretically, particularly well-suited to mapping applications. This is because the **meter** was originally defined by the French Academy of Sciences in 1791 as one ten-millionth of the great-circle distance from the earth's equator to its pole. A major problem with using the earth as the standard for determining the length of a meter, however, is that determining the precise measurement of the earth is extremely difficult. The length of the original meter, which was finalized by an international conference in Paris in 1875, for example, was established on the basis of a detailed survey that was as accurate as the instruments of the period allowed. As newer, more accurate methods of measurement were introduced, however, the presumed size of the earth had to be adjusted, so that the precise relationship of the meter to that size was altered.

For this reason, the modern definition of the meter is based on a specific physical measurement that does not change over time. Since 1983, the meter has been defined as the distance that light travels in 1/299,792,458 of a second. This measurement, although it is undoubtedly impossible to visualize, can be precisely determined whenever necessary, anywhere in the world, provided that the necessary equipment is available.

The modern definition of the meter, then, is no longer based on an exact relationship between the length of the unit and the size of the earth. But the initial earth-referenced length is still influential because the standard for the modern meter was based on that earlier measurement. In practice, however, a measure established on the basis of a particular relationship to the size of the earth is really no more satisfactory than any other arbitrary measure. What is most important is the measure's practicality in use.

AREAL MEASUREMENTS

The areal units of the U.S. Customary System are a mixture of types (Table 5.3). Some basic units are simply derived from linear measurements—9 square feet = 1 square yard, or 144 square inches = 1 square foot, for example. Other areal units have their origins in land surveying and are the result of specially defined units that may be puzzling, until one realizes the relationships between them.

One common unit of areal measurement in land surveying, for example, is an **acre.** The size of an acre is related to the structure of the U.S. Public Land Survey (USPLS) system, as follows. Each township in the USPLS is divided into 36 sections (see Chapter 4). Each section, in turn, measures 1 mile on a side and, therefore, encompasses 1 square mile. At the time of the establishment of the USPLS, the standard method for measuring a mile was to stretch a 66-foot-long chain, end-to-end, 80 times (66 ft. × 80 = 5280 ft. = 1 mile). This meant that each section contained 6400 square chains (80 × 80 = 6400). However, by definition, a section also consisted of 640 acres. This meant that each acre contained 10 square chains (6400/640 = 10). It also meant that an acre equaled 43,560 sq. ft. (10 sq. chains = 10 [66 ft. × 66 ft.] = 43,560 sq. ft.). Therefore, the very unusual-appearing relationship, 43,560 square feet in an acre, is simply the outcome of applying a series of definitions and standard surveying measures.

TABLE 5.3 U.S. Customary Units—Areal Measure

144 sq. inches	= 1 sq. foot
9 sq. feet	= 1 sq. yard = 1296 sq. inches
1 sq. rod	= 272.25 sq. feet*
160 sq. rods	= 10 sq. chains = 1 acre = 43,560 sq. feet*
640 acres	= 6400 sq. chains = 1 sq. mile*
1 mile by 1 mile*	= 1 section of land
6 miles by 6 miles*	= 1 township = 36 sections = 36 sq. miles*

*Survey foot or mile, not international foot or mile.

TABLE 5.4 Metric Units—Areal Measure

1 are	= 100 sq. meters (10 meters on a side)
1 decare	= 10 ares = 1000 sq. meters
1 hectare	= 100 ares = 10,000 sq. meters

Note: This table reflects part of the International System of Units (SI), a modernized metric system established by international agreement in 1960.

Given the complex set of relationships between the areal measurements of the U.S. Customary System, it is easy to see why metric units of areal measure are an attractive alternative (Table 5.4).

NAUTICAL MEASUREMENTS

Another element of complexity in linear measurements is that the **nautical mile** used for measuring distances at sea is different from the statute mile used on land. The original reason for this was navigational convenience. Navigators determine the position of their vessels in terms of latitude and longitude. It therefore made sense to measure distances between locations in units that are related to the graticule.

The concept of an earth-related, nautical measurement has been accepted for a long time. For example, the former **U.S. nautical mile,** sometimes called the **geographical mile**, measured 6080.2 feet (1853.24 meters). This definition was based on the length of an arc of 1 minute, measured along the equatorial great circle of the earth. Confusingly, the British used a unit called the **admiralty nautical mile,** which was equal to 6080 feet (1853.18 meters). The two nautical miles differed in size by this small amount because they were based on slightly different measurements of the earth.

continued on page 78

Continued from page 77

The **international nautical mile** was established in 1959 to eliminate the confusion caused by different nations using different units of measurement. This unit, which is 1852 meters (6076.12 feet), is based on an idealized concept. First, a perfect sphere with a surface area equal to the surface area of the earth was defined. The international nautical mile was then determined by calculating the length of an arc of 1 minute along a great circle of this perfect sphere. Thus, the relationship between a nautical mile and a degree of latitude is, for practical purposes, 1 nautical mile per minute, or 60 per degree. Some error is involved in this relationship, however, because the earth is not a perfect sphere, and the lengths of actual degrees of latitude vary slightly, depending upon where they are located between the equator and the pole. Also, of course, degrees of longitude vary in length, from a maximum at the equator to zero at the pole, so the desired relationship between nautical miles and degrees of longitude only exists at the equator.

The nautical term *knot* is often misused. A **knot** is defined as 1 nautical mile per hour, so it is a measure of speed, not of distance. That is why it makes no sense to say that a ship is traveling at 15 knots per hour; the correct statement is that it is traveling at 15 knots.

Nautical depth measurements have also had, traditionally, their own unique measure—the fathom. A **fathom,** which equals 6 feet (approximately 1.83 meters), was defined by the amount of sounding line that could be encompassed between the outstretched hands of an average seafarer. Many nautical charts record depths in fathoms, but recently there has been a transition to using feet or meters. When using a nautical chart, you must be certain which of the various units is in use or grievous errors can result.

TABLE 5.5 Conversion Factors (U.S. Customary Units to Metric Units and Metric Units to U.S. Customary Units)

Given	Multiply by	To Yield
Linear Measure		
Nautical miles	1852.0	Meters
Statute miles	1609.3	Meters
Yards	0.9144	Meters
Feet	0.3048	Meters
Inches	0.0254	Meters
Meters	0.00054	Nautical miles
Meters	0.00062	Statute miles
Meters	1.094	Yards
Meters	3.281	Feet
Meters	39.37	Inches
Areal Measure		
Given	**Multiply by**	**To Yield**
Sq. miles	2.59	Sq. kilometers
Acres	0.405	Hectares
Sq. yards	0.836	Sq. meters
Sq. feet	0.0929	Sq. meters
Sq. inches	6.452	Sq. centimeters
Sq. kilometers	0.386	Sq. miles
Hectares	2.471	Acres
Sq. meters	1.196	Sq. yards
Sq. meters	10.764	Sq. feet
Sq. centimeters	0.155	Sq. inches

Note: For conversions to and from other linear metric units, simply shift the decimal in the multiplication factor the appropriate distance and direction. For example, given centimeters, multiply by 0.3937 to derive inches or, given inches, multiply by 2.54 to derive centimeters.

UNIT CONVERSIONS

As you work with maps, you are likely to encounter many situations that require you to convert from metric units to U.S. Customary Units, or the reverse. The conversion factors in Table 5.5 will assist in this task.

Units of measurement have a strong influence on the scales at which maps are drawn. A very common scale for topographic maps, especially in Great Britain and in other parts of the world that have come under British influence, is 1:63,360 (or a multiple or fraction thereof). These scales were adopted because there are 63,360 inches in a mile. At the scale of 1:63,360, then, 1 inch represents 1 mile, 1/4 inch represents 1/4 mile, 1/10 inch represents 1/10 mile, and so on, so it is easy to measure distances with a regular ruler that is subdivided into conventional fractions of an inch.

In the United States, on the other hand, the base scale for topographic mapping was for a long time 1:62,500. This base provided a convenient progression of scales, each of which was an equal multiple of the others (1:62,500, 1:125,000, 1:250,000, 1:500,000, and 1:1,000,000). On maps drawn at these scales, however, inches do not exactly match multiples or fractions of a mile on the scale, so a ruler cannot be easily used for direct scaling.

In recent years, the scale 1:24,000 has become dominant for topographic maps in the United States. This scale is convenient because it is equivalent to 1 inch to 2000 feet, so a ruler is easily used for direct scaling.

In parts of the world that use the metric system, common map scales are 1:25,000, 1:50,000, 1:100,000, and 1:1,000,000. Direct scaling, using a metric ruler, is convenient at these scales. At a scale of 1:100,000, for example, 1 centimeter on the scale represents 1 kilometer on the earth's surface.

A distance of 1 centimeter on a map with a scale of 1:100,000 represents 100,000 centimeters on the surface of the earth; or 1 inch on the same map represents 100,000 inches on the ground. To say that a scale of 1:100,000 means that 1 inch on the map represents 100,000 feet on the earth, or some other mixed measurement, is obviously incorrect. However, a correct word statement of the scale of a map with an RF of 1:100,000 could be "1 centimeter represents 100,000 centimeters," or "1 centimeter represents 1000 meters," or "1 centimeter represents 1 kilometer." These are all equivalent statements.

Sometimes, especially on engineering plans, a different type of numerical scale statement is used. This engineering style takes the form 1″ = 800′. This statement can be read as "1 inch equals 800 feet," which is literally impossible. What is actually meant is that 1 inch on the drawing *represents* 800 feet on the ground. It is clear, then, that the representative fraction 1:9600 is equivalent to 1″ = 800′. Care must be taken to interpret these different forms of scale expression correctly.

Graphic Scale

A line drawn on a map and subdivided into units appropriate to the scale of the map is called a **graphic scale** (or **bar scale**). A graphic scale allows a distance measured on the map to be translated directly into the correct earth distance by comparing it to the scale (depending, as already discussed, on the map projection). Alternatively, the graphic scale allows the transfer of a desired earth distance directly onto the map.

Graphic scales on a single map often include a variety of scale units (such as kilometers, miles, feet, yards, meters, and so on). This arrangement allows measurements to be taken directly in the desired units without the necessity of converting from one unit of measure to another.

Graphic scales are particularly useful because they change size in proportion to changes in map size. If a map is photographically enlarged to twice its original size, for example, its scale has been altered proportionately. For this reason, neither the original word statement nor the RF scale relationship is valid for the enlargement. If a graphic scale is included with the rescaled map, on the other hand, its length is changed in the same proportion as the map, and it continues to be correct. This characteristic is beneficial if a photostatic or photographic reduction or enlargement of a map needs to be made, assuming that the map is not copyrighted or that proper permission has been obtained (copyright guidelines are presented in Appendix C).

There are two basic styles of graphic scales. The first style consists simply of a series of major units. In addition, the first major unit of such a scale is sometimes subdivided to allow greater accuracy in measuring relatively short increments of distance (Figure 5.1a). Because this arrangement is rather awkward when measuring fractional distances, a second type of scale, which has a subdivided extension added in front of the first major unit, is more easily used for that purpose (see Figure 5.1b).

For the sake of accuracy and ease of use, specially drawn scales (instead of simple rulers) are almost always used for map measurements. One exception is a map drawn at a scale of 1:63,360, on which 1 inch is equivalent to 1 mile. Another exception is a map drawn at a scale of 1:100,000, on which 1 centimeter represents 1 kilometer. In these cases, an inch or centimeter ruler can be used to measure distances directly on the respective maps. (Multiples or fractions of these scales are also conveniently measured with the appropriate ruler.)

Variable graphic scales are sometimes drawn on maps on which the scale ratio varies considerably. Variable graphic scales are simply a collection of individual scales with a separate, subdivided scale line for each specified portion of the map and with some method of designating where each scale line should be used (Figure 5.2). For example, in a typical situation, each scale line is associated with a specific parallel. In addition, it is normal to use the specified scale line in a zone on either side of the appropriate parallel (usually halfway to the next parallel). Because scale changes on maps are continuous and do not occur in steps or increments, some error is introduced by not having continuously changing graphic scales. This error, however, is likely to be within a reasonable range for most purposes.

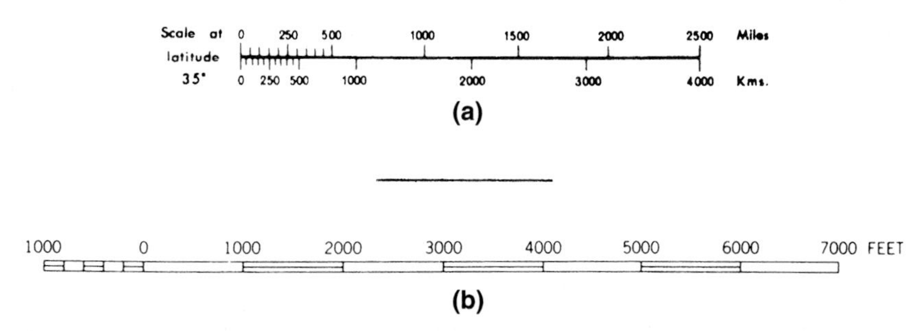

Figure 5.1 Types of graphic scales. (*a*) First major unit subdivided. (*b*) Subdivided extension added in front of first major unit.

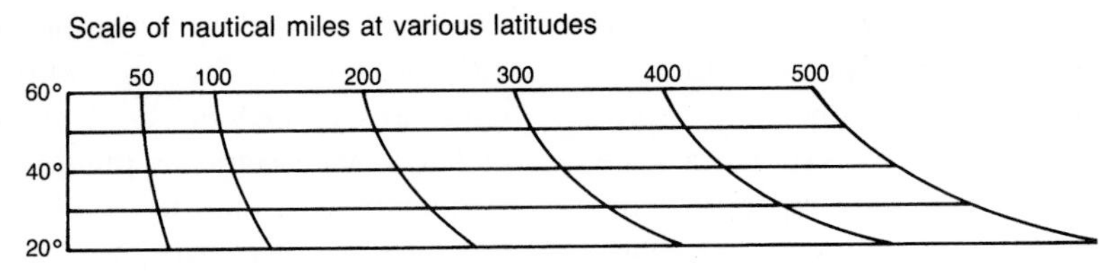

Figure 5.2 Variable graphic scale, designed for use with a polar stereographic projection.
Source: Redrawn from U.S. Weather Service, Surface Weather Map.

Large and Small Scale

It is not unusual to hear in a discussion of maps that one map is **large scale** and another is **small scale.** Unfortunately, the use of these terms is often a source of confusion, in part because there are no universally accepted definitions of a large-scale map and a small-scale map. The following definitions are generally acceptable, although many lists with different divisions may be found.[2] Large scale is 1:50,000 or larger (such as 1:10,000); intermediate scale is 1:50,000 to 1:250,000; and small scale is 1:250,000 to 1:7,500,000.

City plans may be at very large scales, such as 1:2500, and maps in some atlases, books, and reports may be at scales of 1:100,000,000 or smaller and be classified as very small scale.

Another frequent source of confusion regarding large and small scales involves a misunderstanding of what is implied by the terms. This type of concern can be dealt with by memorizing the following rule: "The smaller the number in the denominator, the larger the scale." This means that a map with a representative-fraction scale of 1/1000, for example, is a large-scale map, as compared with a smaller-scale 1/1,000,000 map. A small-scale map of a given size, then, shows a relatively large portion of the earth's surface, with relatively little detail, which is useful for general orientation purposes. A large-scale map of the same size, on the other hand, shows a relatively small portion of the earth's surface in relatively greater detail, which is useful for determining more specific information, such as street widths or exact distances between map features.

These large- and small-scale definitions are only suggestions. Actual scales, rather than these categories, should be used to establish the scale in a particular case. The terms are useful, however, for comparing two or more maps ("This map is at a larger scale than that map") or for deciding the relative scale appropriate for a given task ("We need a very large-scale map for this job").

[2]Some U.S. Geological Survey publications use the following subdivisions: large scale, 1:25,000 or larger; intermediate scale, 1:50,000 to 1:100,000; and small scale, 1:250,000 to 1:7,500,000. This list's usefulness is limited because it leaves gaps between categories and also does not define very small or very large scales.

In special applications, the concepts of vertical and areal scale are often encountered. For purposes of greater clarity, these concepts are discussed elsewhere—areal scale in conjunction with areal measurement (Chapter 6) and vertical scale in conjunction with profiling (Chapter 9).

Scale Determination

Occasionally, a map will not include a scale. This problem may appear on individual aerial photographs or with maps that have been copied on an office copy machine. The absence of a scale greatly limits a map's usefulness, because you usually want to know the size of features shown or the distances between them. When faced with such a situation, you can often create a reasonably accurate scale of your own, using whatever clues the map provides. The specific clues that you can use vary, depending on the type of map and the amount of detail it shows. The basic technique used involves proportionality.

Use of Known Features

The use of proportionality involves comparing the size of a reference feature on the map to the size of the same feature on the earth (or on a map of known scale). For example, this type of comparison is easily visualized in the case of an aerial photograph that shows a familiar object with a well-known standard size, such as a baseball diamond, tennis court, or football field. Objects of this type can be used for scaling purposes if you are confident from other clues, such as their relationship to parking, the type of stadium, and so forth, that they are likely to conform to standard dimensions. A sandlot baseball diamond will obviously not provide dimensions as reliable as those taken from a major-league ballpark.

Assume, then, that a standard baseball diamond appears on a particular photograph. The first step in determining the scale of the photograph is to measure the distance between bases on the photograph using a convenient scale. You then determine the standard distance between bases (90 feet) and establish the ratio between the distance on the photograph and this standard distance. If, in this case, the distance between bases on the photograph is 0.3 inch, the computation is as follows:

$$\frac{0.3''}{90'} = \frac{0.3''}{90 \times 12''} = \frac{0.3}{1080} = \frac{0.3/0.3}{1080/0.3}$$

$$= \text{RF } 1/3600$$

Notice that the photo (or map) distance is reduced to a value of 1.

Unfortunately, there is seldom anything as convenient as a baseball diamond or football field on the segment of photograph or map that you are trying to scale. You must usually be prepared to be ingenious or to work a little harder to determine scale. The map or photo may, for example, contain section lines (see Chapter 4). The standard distance between section lines is 1 mile, so if such lines are on average spaced 2.65 inches apart on a map, the scale of the map is as follows:

$$\frac{2.65''}{1 \text{ mile}} = \frac{2.65''}{1 \times 63,360''} = \frac{2.65/2.65}{63,360/2.65}$$

$$= \text{RF } 1/23,909$$

This is, of course, very close to the standard 1/24,000 U.S. Geological Survey quadrangle scale. Indeed, it is likely in this example that the map is actually at the standard scale. The slight difference may arise because the spacing between section lines is frequently not exactly the prescribed 1-mile distance. Because of such considerations, it is good practice to take several measurements of similar map features and to use the average measurement for scale-calculation purposes. In addition, in this case care must be taken to be certain that quarter-section lines, or other regularly spaced divisions, are not mistaken for section lines.

Use of Lines of Latitude and Longitude

Another useful scaling reference often shown on maps is the latitude and longitude graticule. The map distance between a pair of latitude or longitude lines that are spaced a known number of degrees apart can be compared to the earth distance between the same lines to determine the map's scale ratio. On an atlas map, for example, if the map distance between the lines of latitude for 40° and 41° North is 24.5 millimeters, the map scale is as follows:

$$\frac{2.45 \text{ mm}}{111.133 \text{ km}} = \frac{245 \text{ mm}}{111.133 \times 1,000,000 \text{ mm}}$$

$$= \frac{24.5/24.5}{111,133,000/24.5}$$

$$= \text{RF } 1/4,536,040$$

$$(\text{or, for practical purposes, } 1/4,500,000)$$

You must be careful to use the correct figures for the spacing between graticule lines. In this case, 111.133 kilometers is the average length of a degree of latitude. If the lengths of degrees of longitude are used, however, the convergence of the meridians toward the poles must be taken into account. This is done by multiplying 111.133 kilometers by the cosine of the latitude (the cosines of angles are listed in standard mathematical tables). For example, at 30° North the result would be as follows:

$$\cos 30° \times 111.133 \text{ km} = 0.8660 \times 111.133 \text{ km}$$

$$= 96.24 \text{ km (at 30°)}$$

If greater accuracy is required, the exact lengths of degrees of latitude and longitude should be used (see Tables 4.1 and 4.2).

Use of Map Comparison

Maps can also be compared to calculate an unknown scale. A map of known scale that covers at least a portion of the area covered on the map of unknown scale is required. In this case, select a pair of features that are likely to be accurately located on both maps, such as railroad crossings, city symbols, drainage features, section lines, or lines of latitude or longitude. The choice depends partly on the scale of the maps and partly on the types of features common to both maps.

To use the map-comparison method, measure the distance between the same pair of features on each of the maps, using any convenient units. The unknown map scale is then obtained by multiplying the known RF scale by the ratio of a measurement on the map of unknown scale to the equivalent measurement on the map of known scale. For example, assume that two features are 50 millimeters apart on a map of scale 1:125,000, and the same features are 100 millimeters apart on a map of unknown scale:

$$\frac{100}{50} \times \frac{1}{125,000} = \frac{2}{1} \times \frac{1}{125,000} = \frac{2}{125,000}$$

$$= \frac{2/2}{125,000/2}$$

$$= \text{RF } 1/62,500$$

Scale Conversion

It is often helpful to convert a map scale from one form to another. You may wish to make measurements from a map that lacks a graphic scale but whose representative-fraction (RF) scale you know. Or the map may have a graphic scale that you wish to express as a word statement for ease in conversation. Other **scale conversions** may also be helpful for a variety of reasons. The procedures for making such changes are quite straightforward, because each form is simply an alternative way of expressing the same relationships.

Scale conversions are facilitated if you keep certain standard units in mind. For example, if you are dealing in metric units, it is helpful to remember that there are

1000 meters in 1 kilometer, 100 centimeters in 1 meter, 10 millimeters in 1 centimeter, and so on. If you are dealing in English units, on the other hand, you should remember that there are 5280 feet (and 63,360 inches) in 1 mile. (Maps are often produced at a scale of 1:62,500 because the multiples of 62,500, such as 125,000, 250,000, and so on, are mathematically convenient.)

Representative-Fraction (RF) and Word-Statement Conversions

If the map scale is expressed as a representative fraction, it may be more meaningful to you or to someone with whom you are speaking to put the scale into words. A map with an RF of 1:100,000, for example, has a word-statement scale of "1 centimeter per kilometer" (because there are 100,000 centimeters [1000 meters] in 1 kilometer). Similarly, if a map has an RF of 1:63,360, its scale in words is "1 inch to 1 mile."

The conversion of a word-statement scale to an RF is the mirror image of the conversion from an RF to a word statement. Thus, the RF equivalent of "2 inches per mile" is 1:31,680 (2/63,360 = 1/31,680). Several common RF/word-statement relationships are shown in Table 5.6. Note that maps at the scale of 1:62,500, or its multiples, are identified as "1 inch to approximately 1 mile," and so on. Note also that alternative expressions are in common use, such as the equivalent statements "1/4 mile to an inch" and "4 inches per mile." Determining which style of expression to use is simply a matter of personal choice.

Equivalencies for values not included in Table 5.6 can be easily determined. To determine how many inches to the mile, for example, divide the denominator of the RF of the map into 63,360. Thus, on a map with an RF of 1:40,000, there are approximately 1.58 inches per mile (63,360/40,000). Conversely, to determine how many miles to the inch, divide 63,360 into the denominator of the RF. Thus, on a map with an RF of 1:400,000, there are approximately 6.3 miles to the inch (400,000/63,360). If you are working in metric units, substituting 100,000 for 63,360 in the previous formulas yields centimeters per kilometer and kilometers per centimeter, respectively. You may wish to check these relationships by recalculating some of the entries in Table 5.6.

TABLE 5.6 Scale Relationships

Representative Fraction (RF)	Word Statement
1:1200	100 feet to an inch
1:4800	400 feet to an inch
1:7920	⅛ mile to an inch (or 8 inches per mile)
1:9600	800 feet to an inch
1:15,840	¼ mile to an inch (or 4 inches per mile)
1:20,000	⅕ kilometer per centimeter (or 5 centimeters per kilometer)
1:24,000	2000 feet to an inch
1:25,000	¼ kilometer per centimeter (or 4 centimeters per kilometer)
1:31,680	½ mile to an inch (or 2 inches per mile)
1:50,000	½ kilometer per centimeter
1:62,500	1 mile to an inch (approx.)
1:63,360	1 mile to an inch
1:100,000	1 kilometer per centimeter
1:125,000	2 miles to an inch (approx.)
1:126,720	2 miles to an inch
1:250,000	4 miles to an inch (approx.)
1:253,440	4 miles to an inch
1:500,000	5 kilometers per centimeter
	8 miles to an inch (approx.)
1:506,880	8 miles to an inch
1:1,000,000	10 kilometers per centimeter
	16 miles to an inch (approx.)
1:1,013,760	16 miles to an inch

Representative-Fraction (RF) and Graphic Conversions

If a map includes an RF scale statement, you can create a graphic scale by deciding the number of map units you wish to represent, drawing a line of the proper length, and applying the appropriate label. Assume, for example, that the RF scale of a map is 1:100,000, and you wish to draw a 10-kilometer scale. Because there are 100,000 centimeters in 1 kilometer (1000 meters, each containing 100 centimeters), a 10-centimeter line would represent 1,000,000 centimeters, or 10 kilometers. Similarly, if the RF is 1:63,360, a 1-inch line would represent 1 mile.

To convert a graphic scale to the RF form, select a convenient scale distance, such as 1 mile or 1 kilometer, and measure its length as precisely as possible. The measurement should be done in appropriate units, such as inches for miles and centimeters for kilometers. Then divide the measured distance into the distance indicated on the graphic scale, both in the same units, to obtain the denominator of the RF. If 1 mile (63,360 inches) on the graphic scale measures 2.64 inches, for example, the RF is 1:24,000 (63,360/2.64). Similarly, if 1 kilometer (100,000 centimeters) on the scale is 2 centimeters long, the RF is 1:50,000 (100,000/2).

One of the most convenient aspects of maps drawn to metric scales becomes apparent when you are making scale-conversion measurements such as those just described. On a map with a scale of 1:100,000, each centimeter represents 1 kilometer; at a scale of 1:1,000,000, each centimeter represents 10 kilometers; and so on. This means that a standard centimeter scale is easily used for scaling distances on metric-scale maps.

Word-Statement and Graphic Conversions

Conversions between word-statement scales and graphic scales are accomplished by combining the procedures just described. To go from a word statement to a graphic scale, for example, simply convert the word statement to an RF and then convert the RF to graphic form. Converting from a graphic scale to a word statement is easily done by first determining the equivalent RF and then converting the RF to a word statement.

GENERALIZATION

Every map has, to some extent, been simplified or **generalized.** Even large-scale topographic maps cannot possibly show every detail of the real world they represent. Because of this, cartographers adjust the content of each map to make it a useful and recognizable representation of the real world.

One constraint on the amount of generalization that is introduced is the scale of the map. The smaller the scale of the map, the greater the amount of generalization required, because the amount of space available to show any given feature decreases as scale decreases.

Generalization is also strongly influenced by the purpose for which the map was designed. Two maps of a given area may vary significantly in content, even if they are of similar scales, depending upon their purpose. For example, a highway map will emphasize the road system, possibly including the number of lanes, the surface type, and the name or route number of each road. Other features, such as streams and topographic features, may be shown in very general form or not at all. On the other hand, a topographic map will emphasize surface characteristics, perhaps using contour lines and terrain shading, and is likely to show the drainage system in some detail. Such a map, however, may not include any significant information about the road system. Both maps could be considered valid maps; their difference in content is simply a reflection of their differing purposes.

If generalization is properly done, the distinguishing characteristics of the mapped features will still be effectively represented. An awareness of the techniques that cartographers use to accomplish generalization will help you to interpret and understand the level of detail presented on any given map.

Selection

One means of generalization that cartographers use is the **selection** and retention of the more important features in an area and the elimination of the less important ones. If a series of lakes is shown, for example, some of the smaller lakes in the group are eliminated as scale is reduced, so that the larger ones can be retained (Figure 5.3a). This means that you must be aware of the possibility that some of the smaller features in a given area may simply not be shown on the map.

Simplification

A second technique of generalization is the **simplification** of the shapes of the features retained on the map. To continue the previous example, this approach means that the shorelines of the lakes are made less complex at smaller scales. Ideally, enough detail is retained so that the major characteristics of the features are identifiable. A coastline with many bays and headlands, for example, should not be reduced to a smooth curve. Instead, sufficient complexity should be retained so that the nature of the actual coastline can be realized, even though the individual features are not all retained (Figure 5.3b).

Combination

Another step that cartographers take in the generalization process is the **combination** of two or more similar features into a single symbol. If there are many small wooded areas in a region, for example, two or more of them may be combined to show as one. If the combination of features is carried too far, however, there is a danger that the map user will not be aware that the region is characterized by scattered areas of woodland, separated by unwooded areas (Figure 5.3c).

Locational Shift and Size Exaggeration

The generalization of a particular map may also require the cartographer to **shift** the location of some features. For example, a road, a railroad, and a river may all be crowded into a narrow valley, but the purpose of the map makes it necessary to show all of them. Because of the need for legibility, however, the weight of the linework on the map must be sufficient to be distinguishable, and there must be sufficient space between the lines to keep them from running together. Faced with the incompatible requirements of retaining features as well as legibility, the cartographer may **exaggerate** the size of the features to improve visibility and may shift their location slightly to gain the necessary space. This combination of techniques retains legibility, even though the features lie in slightly shifted positions and occupy more space than they would if they were represented to precise scale (Figure 5.3d). Keep in mind the possibility—indeed the likelihood—that the size of some of the features on a map may be exaggerated and that not every feature is likely to be shown at its precise location.

In most instances, the cartographer uses one or more methods of generalization on a given map. If sound judgment has been exercised, the result is a

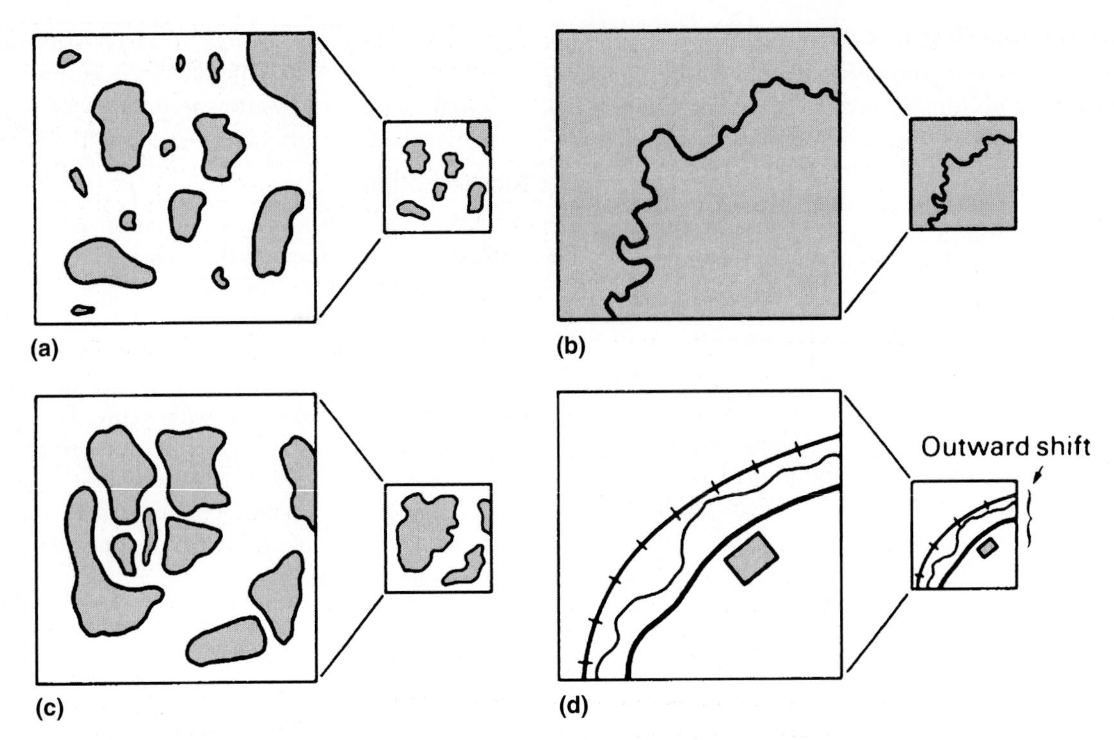

Figure 5.3 Generalization techniques. (*a*) Selection. (*b*) Simplification. (*c*) Combination. (*d*) Locational shift and size exaggeration.

level of generalization that allows appropriate detail for the particular map scale to be shown. Figure 5.4 shows portions of maps centered on the same area but drawn at five different scales. The variations in generalization apparent in these examples are typical of those generally encountered.

You must be aware of the existence of generalization and try to select the most appropriate map for the purpose at hand. Because judging the validity of a particular map's level of simplification is difficult, the best safeguard is to consult a variety of maps. Maps drawn at larger scales than the map you are currently using are likely to be the most useful. A comparison between the details shown on several maps, even maps of the same scale, will improve your knowledge of the mapped area and allow you to better judge the accuracy of any given map.

SUMMARY

The representation of the earth's features on a map is affected by scale and generalization.

The scale of a map is the ratio between map distances and earth distances. It is expressed in one of three equivalent ways: as a word statement, representative fraction, or graphic symbol. Although maps are sometimes referred to as *large scale* or *small scale,* the terms are generally more useful for comparative purposes.

Because scale involves proportionality, an unknown scale may be determined by comparisons. For example, the size of a reference feature on the map can be compared to the size of the same feature on the earth. Comparisons between maps also can be used to calculate an unknown scale. All that is required is a map of known scale that shows at least some of the same features as the map of unknown scale.

Because the three methods of stating scale are equivalent, conversions from one method of expression to another are feasible. In addition, such conversions may be recommended for certain tasks.

Every map is, to some extent, simplified or generalized, because it is impossible to show every detail of the real world it represents. The content of each map must be adjusted to make it a useful and recognizable representation of the real world, within the limits of the space available and the scale of the map. The smaller the scale of the map, the greater the amount of generalization required. One method of accomplishing generalization is to retain the more important features in an area and eliminate the less important ones. A second technique is the simplification of the shapes of the features retained on the map. Another method of generalization is the combination of two or more similar features into a single symbol. Finally, the size of some features may be exaggerated, and some features may be shifted in location.

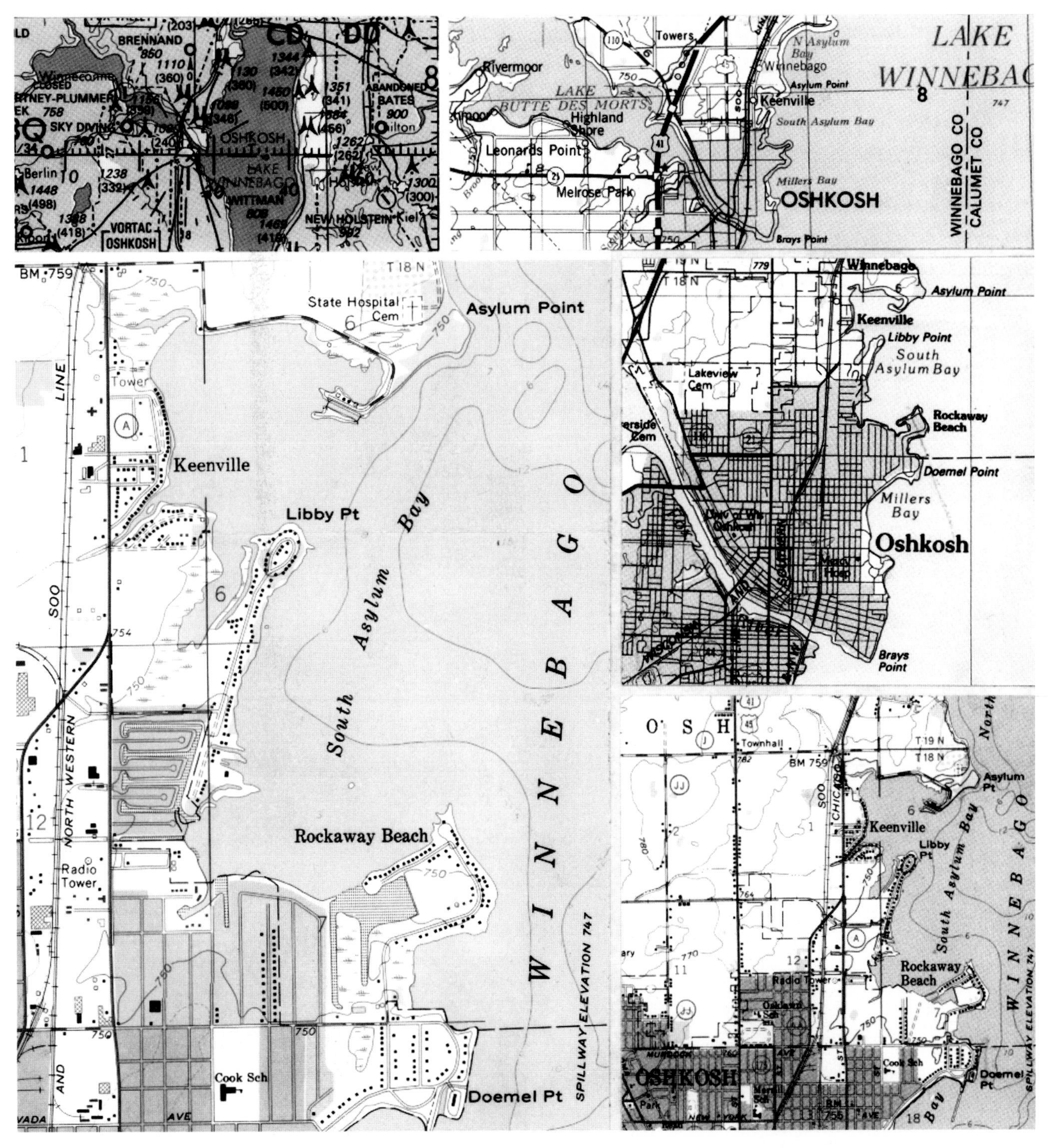

Figure 5.4 Effects of generalization at five different scales. The map in the lower-left corner is at the largest scale, 1:24,000. Moving in a counter-clockwise direction, the scales of the other maps are 1:62,500; 1:100,000; 1:250,000; and 1:1,000,000.

SUGGESTED READINGS

Cuff, David J., and Mark T. Mattson. *Thematic Maps: Their Design and Production*. New York: Methuen, University Paperbacks, 1982, 91–94.

Dickinson, G. C. *Maps and Air Photographs*. 2d ed. New York: John Wiley & Sons, Inc., A Halsted Press Book, 1969, 143–49.

Greenhood, David. *Mapping*. Chicago: University of Chicago Press, Phoenix Science Series, 1964, 42–50.

Hodgkiss, A. G. *Maps for Books and Theses*. New York: Pica Press, 1970, 37–43.

Muehrcke, Phillip C. *Map Use: Reading, Analysis, and Interpretation*. 3d ed. Madison, Wis.: JP Publications, 1992, 519–28.

Robinson, Arthur H., Joel L. Morrison, Phillip C. Muehrcke, A. Jon Kimerling, and Stephen C. Guptill. *Elements of Cartography*. 6th ed. New York: John Wiley & Sons, Inc., 1995, 92–95.

CHAPTER

6

MEASUREMENT FROM MAPS

MAP ACCURACY

The accuracy with which a map is constructed is critical when measurements are taken from it. The difference in elevation between two points on a topographic map, for example, cannot be precisely determined by merely noting the locations of the points relative to the contour lines. The level of accuracy with which the contours are drawn limits the accuracy that can legitimately be expected. (This is explained in more detail in the box on "National Map Accuracy Standards.")

A variety of information about the earth and about the features distributed on its surface is obtained by direct measurement from maps. Two measurements of primary importance are (1) the distances between locations and (2) the areas of regions. This chapter presents some of the techniques by which map measurements of these values are made. The determination of other factors, such as direction, shape, and slope, is discussed in later chapters. This chapter also examines the importance of map accuracy in relation to measurements taken from maps.

The discussion of the taking of measurements in this chapter assumes that the appropriate map projection and map scale have been chosen. If the projection is unsuitable, any measurements can be inaccurate, sometimes by a very large margin. Therefore, the issues discussed in the map selection section of Chapter 3 should be reviewed before deciding which map to use for taking measurements. In addition, if the map is at an inappropriate scale, accurate measurements cannot be expected. This is true simply because of the limitations imposed by map generalization. It would not be realistic, for example, to use a small-scale map of the world to try to measure distances between towns or cities, even if they were visible on the map. A related concern is the fact that map symbols are usually disproportionately larger than the actual scale size of the object on the ground. Because of this, measurements taken between symbols may not correctly show the distance between the actual features. In most cases, the center of the symbol is taken as the correct ground location. Using this assumption will reduce, but not necessarily eliminate, the problem of disproportionate symbols.

Three more issues should be considered when measurements are taken on maps. First, the map must be sufficiently up-to-date to reflect any changes in the mapped region. Natural events, such as fault-line movement, floods, or landslides, often cause significant changes in the landscape. Similarly, human activities, such as urban expansion, building dams or canals, or constructing new buildings, often introduce changes that should be reflected on maps. Depending on the types of measurements being made, any of these changes could affect the accuracy of measurements taken from an out-of-date map. Second, many maps are printed on paper, or other material, that expands and contracts with changes in atmospheric moisture. When accurate measurements are needed, changes due to shrinkage or expansion can introduce significant error.

Box 6.1 National Map Accuracy Standards

The accuracy of topographic maps has two components: horizontal and vertical. The **U.S. National Map Accuracy Standards**[1] provide the framework by which maps produced by federal government mapping agencies are judged. These standards were revised on 17 June 1974 and are still in use.

Maps tested for accuracy are selected by the producing agency. Those that meet or exceed the established standards are published with the statement, "This map complies with National Map Accuracy Standards," included in the legend.

HORIZONTAL ACCURACY

The **horizontal accuracy** of a map produced by a U.S. government mapping agency is measured by testing the placement of "well-defined points" on the map. This involves first the selection of a number of points on the map for testing. The points selected for this purpose are limited to those that are "visible or recoverable on the ground." Examples of such points are bench marks and property boundary monuments, road and railroad intersections, the corners of large buildings or structures, and the center points of small buildings.

The correct location of each of the selected points is then determined, using a survey of greater accuracy than that of the map being checked. These "correct" locations are then checked against the actually plotted locations of the same points on the map being tested, and the amount of error at each location is measured. Assuming that the map being tested is at a scale larger than 1:20,000, no more than 10 percent of the points tested can be in error by more than 1/30 inch. At a scale of 1:10,000, then, an error of 1/30 inch would represent 27.78 feet (8.47 meters) on the ground. By comparison, at a scale of 1:24,000, an error of 1/50 inch would represent 40 feet (12.19 meters) on the ground.

VERTICAL ACCURACY

The standards for **vertical accuracy** apply to contour maps at all scales. The points to be tested are selected and their actual elevations are determined, again from more accurate surveys. The elevations of the same points are then determined on the map being tested, and the differences between the two elevations are calculated. The rule applied is "Not more than 10 percent of the elevations tested shall be in error more than one-half the contour interval."

An additional complication involved in the testing of vertical accuracy is that the tester may, in effect, deliberately shift the map location of the point being tested by as much as the permissible horizontal error for a map of that scale. Such a shift may serve to offset some or all of the vertical error that would otherwise occur.

COMMENTARY

Although the National Map Accuracy Standards are the current standards for printed maps, they have certain deficiencies. For example, they do not specify how many horizontal or vertical points should be tested nor where they should be located on the map. In addition, the standards do not specify what constitutes a survey of higher accuracy. Because of the general recognition of the inadequacy of the national standards, most mapping agencies have their own procedures that go beyond them.[2] In addition, the national standards do not deal directly with accuracy standards for digital mapping. This aspect is discussed in Chapter 19.

[1]Morris M. Thompson, *Maps for America,* 3d ed. (Washington, D.C.: U.S. Department of the Interior, U.S. Geological Survey, 1987), 104.

[2]E. Terrence Slonecker and Nancy Tosta, "National Map Accuracy Standards: Out of Sync, Out of Time," *Geo Info Systems* 2, no. 1 (January 1992): 20, 24–26.

Third, some maps may be inaccurate because of poor compilation or production techniques. The best protection from this type of error is to use maps produced by reputable companies or agencies whenever possible.

DISTANCE MEASUREMENT

Direct Methods of Measurement

This discussion of direct methods of map distance measurement assumes that one of two conditions is met so that the measurements taken are valid. One condition is that the distances are short, so that the earth's curvature is not an important factor in the measurement. This condition is usually met when a large-scale map is used to measure distances of a few miles. On the other hand, when distances of several hundreds of miles or more are measured on small-scale maps, the other condition must be met. This requires a map on an equidistant projection, with the measurements taken along radials from the center point of the map (see Chapter 3). To take an extreme example, it would not be valid to directly measure the distance between New York and London on a Mercator projection using the methods discussed here. A large-scale Mercator projection might be used, however, to measure the distance between the downtown and the suburbs of a city, a matter of only a few miles.

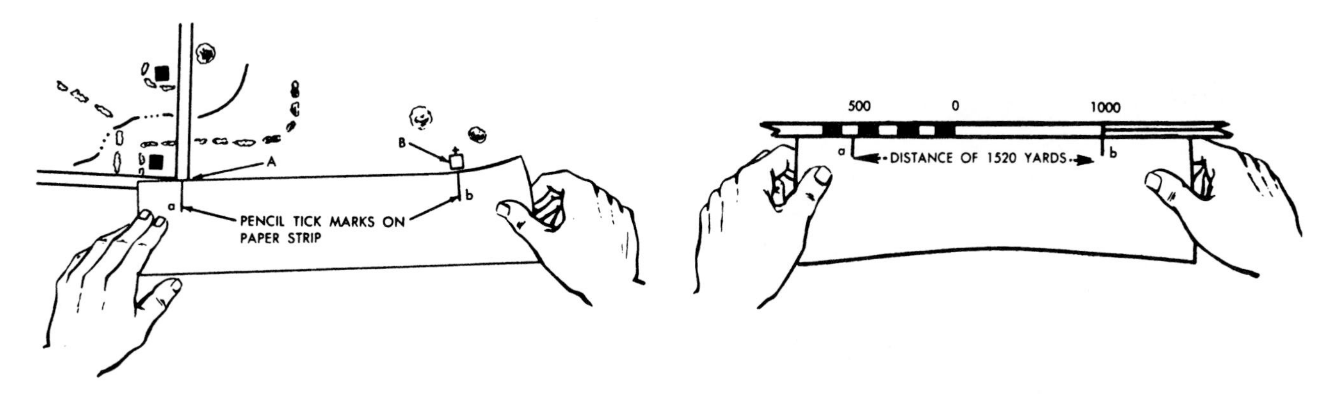

Figure 6.1 Measuring straight-line distance using a bar scale.
Source: Department of the Army, *Map Reading,* FM 21–26 (Washington, D.C.: Department of the Army, January 1969).

Scaling

The simplest method of determining the straight-line distance between two points on a map is to measure the map distance directly with an engineer's scale, a pair of dividers, or marks on the edge of a strip of paper. The earth distance is then determined by checking the measurement against the appropriate map bar scale (Figure 6.1). If the measured distance exceeds the length of the bar scale, several partial measurements are combined.

The paper-strip method has an advantage in that distances other than straight-line distances are easily accommodated. The distance along an irregular route is measured by individually marking sections of the route, end to end, on the paper strip (Figure 6.2). These sections are kept as short as is necessary to ensure that the length of each one closely approximates the curved-line distance between the marked points. Once the total length of the route is marked on the strip, the distance is scaled by the same technique as is used for straight-line distances.

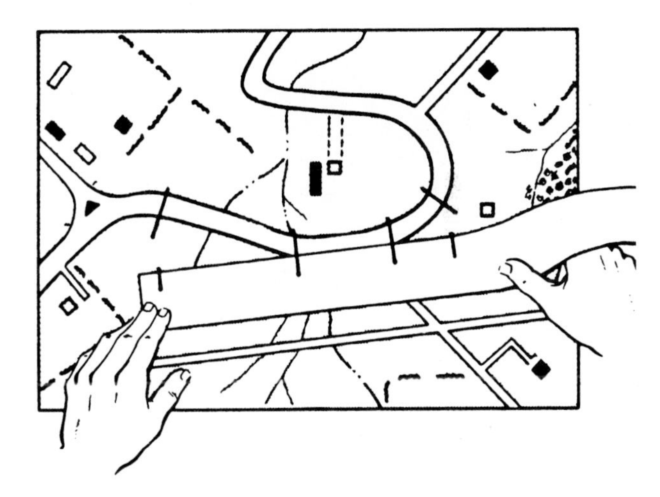

Figure 6.2 Measuring the length of an irregular route.
Source: Department of the Army, *Map Reading,* FM 21–26 (Washington, D.C.: Department of the Army, January 1969).

Map Measurer

Another convenient method of measuring distances on a map is to use a special device called a **map measurer** or **opisometer** (Figure 6.3). The scale on the dial of the map measurer is set to zero, and the tracing wheel is placed at the desired starting point on the map. The wheel is then guided along the route to be measured, and the pointer on the dial indicates the number of centimeters or inches the wheel travels. The swivel handle of the device makes it possible to trace an undulating route, although it is a tricky maneuver to measure a path that contains numerous curves, especially if they are tightly spaced.

The map distance is converted to earth distance by multiplying the map scale denominator by the distance shown on the wheel and converting to the desired units. If the map distance is 3 centimeters (approximately 1.18 inches), for example, and the map scale is 1:100,000, the earth distance is as follows:

3 cm × 100,000 = 300,000 cm = 3000 m = 3 km
(or 1.18 in. × 100,000 = 118,000 in. = approx. 1.86 mi.)

Alternatively, the wheel of the map measurer is placed at the 0 point on the map's graphic scale and is rolled along that scale (in the direction opposite to the direction it was moved on the map), until the pointer once again rests on 0. The graphic scale reading at the stopping point gives the ground distance represented by the length of the path on the map. If the distance measured is longer than the scale, the process must be repeated as often as necessary to return the scale on the map measurer to zero, and the accumulated values indicate the total distance.

Computer Assisted

Computer-assisted methods for direct distance measurement typically involve the use of a digitizing table

Figure 6.3 Map measurer.
Source: John Kehoe photo, by permission of Keuffel and Esser Company.

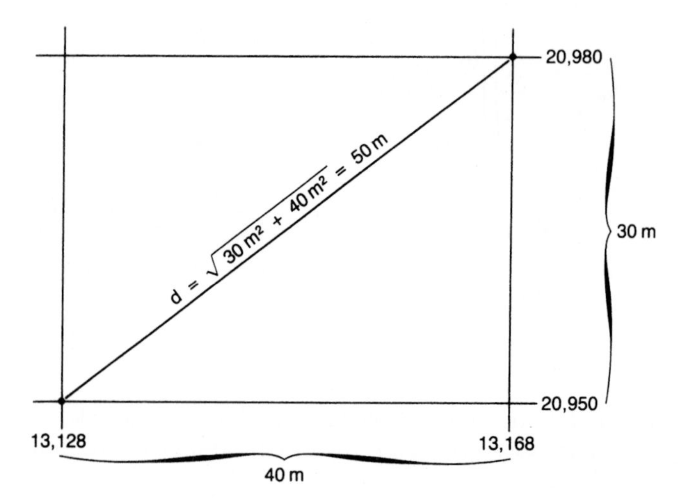

Figure 6.4 Measuring distance using the Pythagorean theorem.

and appropriate software. In the simplest mode, the map is fastened to the digitizing table, and the map scale is entered. The starting and ending points are then selected, and the software computes the direct, straight-line distance between them. In a more complex mode, significant points are selected along the desired route, and the distance between the endpoints is calculated, taking the twists and turns of the route into account.

Indirect Methods of Measurement

When it is inconvenient or impractical to measure distances directly from a map, indirect methods of measurement are useful.

Short Distances

When two points are located within a relatively short distance of one another, within the confines of a medium-sized state, for example, the distance between them may be indirectly measured without correcting for the curvature of the earth's surface. If the UTM or State Plane Coordinate locations of the points are known or can be determined, the distance between them can be easily calculated, provided both points lie within a single grid zone. The technique used in these circumstances is an application of the Pythagorean theorem. The east-west distance between the points is obtained by finding the difference between the east-west coordinates (Figure 6.4). The same process is carried out for the two north-south coordinates. The required distance is then calculated by taking the square root of the sum of the squares of the east-west and north-south distances.

Distances that span two or more grid zones can be calculated by taking advantage of the overlap provided between adjoining zones. A "linking point" along the alignment of the path between the two points of interest and within the area of overlap needs to be established. The distance from each point to the linking point is then determined, using the appropriate grid values for each. The two partial distances are then added to obtain the total distance.

Long Distances

Distances of several hundred miles or more are measured in generally the same way as shorter distances. With these longer distances, however, spherical coordinates are used instead of plane coordinates because the shape of the earth must be taken into account.

Spherical triangles have different characteristics than do the more familiar plane triangles, so the straightforward Pythagorean method cannot be used to determine distances on a sphere. Also, the spherical coordinate system involves degrees, minutes, and seconds of latitude and longitude, or decimal degrees, rather than familiar distance units. Both of these factors complicate the computations. The scope of this book does not include computations involving spherical coordinates, so they are not discussed further here.

Computer Assisted

A computer may be used for indirect distance calculations if suitable software is available. A minimal system of this type simply allows the entry of the desired beginning and ending coordinates. In some systems, the coordinates are entered directly by typing them on the computer keyboard. In other systems, the beginning and ending points are identified by moving a cursor to the desired locations on the screen, in sequence, and pressing a designated key on the keyboard or clicking a button on the mouse. The necessary calculations to determine the direct-route distance between the coordinates are then performed, under program control. The program can be designed to take into account as much detail about the spherical geometry of the earth as is desired. This aspect is most important when long distances are calculated. In a more advanced system, a data file of available routes is required. In this case, a specification of the particular route to be followed is included, along with the starting and ending coordinates. This type of calculation is often a feature within a geographic information system.

A simple way to find the accurate great-circle distance between two locations, whether near or far apart, is to use a distance calculation service on the Internet. All that is necessary is to specify the two locations of interest. This can be done by entering a name for most U.S. cities, and for many cities worldwide. Alternatively, the latitude and longitude of either or both locations may be entered. If the name is misspelled, or if there is more than one city with the same name, choices are provided until any ambiguity is clarified. After the locations are entered, the program calculates and reports the great-circle distance between them in miles and kilometers. It will also produce a map, showing the locations of the points. Both results can be printed on an attached printer. (The web site for this chapter provides links).

Error Sources

Three factors influence the accuracy of horizontal distance measurements: (1) measurement error, (2) the level of generalization of the line being measured, and (3) the vertical displacements along the route.

Measurement

Following curved lines accurately when using a map measurer or a paper strip is difficult, but these difficulties are minimized by careful measurement. Error is also reduced when each measurement is repeated at least three times and the average of the three readings is used as the most likely distance value.

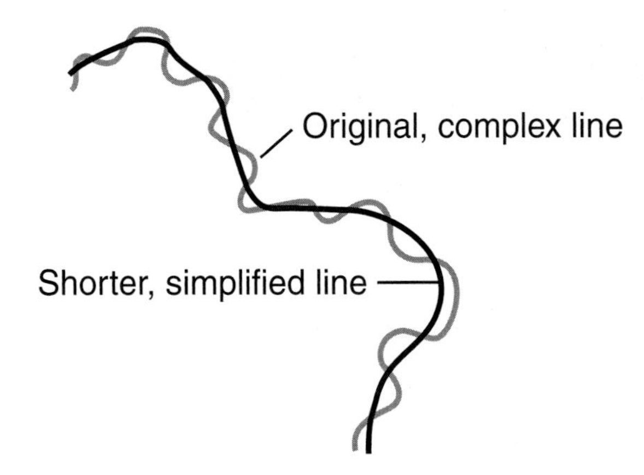

Figure 6.5 Simplification of a line results in shortening its length.

Smoothing

Because of map generalization, routes (other than straight lines) inevitably are represented in somewhat simplified form on maps. The smaller the map scale, the greater the amount of simplification (**smoothing**) that occurs. In addition, simplification of a line results in shortening the distance between its starting and ending points (Figure 6.5). For these reasons, route distance measurements taken from maps tend to underestimate the true distances between features on the earth's surface.

In the case of linear features, the more complex the shape, the greater the effect of map generalization. A straight-line feature, for example, is accurately represented by a simple straight line on a map. A linear feature made up of intricate, closely spaced curves, on the other hand, is not accurately represented by the inevitably smoothed form shown on the map.

In one experiment, the influence of generalization differences was illustrated by measuring two stretches of coastline, one smooth and the other irregular, on maps of three different scales. As shown in Table 6.1, the measured length varied considerably for both stretches of coastline, but the more irregular coastline showed the greatest variation. Also, in general, the measured lengths increased as scale increased, reflecting decreased generalization. This experiment suggests that the only feasible way to avoid serious underestimation of the length of irregular features is to use a source map that is at the largest scale appropriate for the task at hand.

Slope

When distance measurements are taken from a map, **slope errors** may occur because features are shown orthographically on the map. This means that their distances apart are not affected by elevation differences and are the same as they would be if the earth were

TABLE 6.1 Effects of Scale on Measured Distances (Distances in Kilometers)

	Map Scale		
	1:250,000	1:63,360	1:25,000
Smooth Coast	12.5	13.2	13.375
Crenulated Coast	15.0	16.2	21.750

Source: Adapted from J. P. Cole and C. A. M. King, *Quantitative Geography* (London: Wiley, 1968), 60.

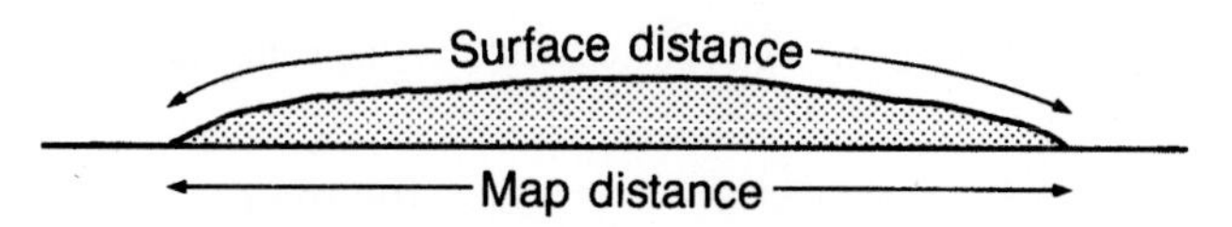

Figure 6.6 When slopes exist, the surface distance is longer than the map distance between two points.

flat. In the real world, the routes between locations almost always involve ups and downs. The actual surface distance between two objects, therefore, is longer than the map distance, except in the exceptional case where the ground surface is perfectly flat (Figure 6.6).

In many circumstances, the difference in distance traveled due to elevation differences is trivial. In traveling from Chicago to Denver, for example, there is an elevation difference of approximately 1432 meters (4698.4 feet). Taking this difference into account increases the 1686 kilometers (1045 miles) travel distance by only 0.6 meter (1.97 feet) (less than 0.1 percent). In this example, the difference obviously can be ignored.

If you are planning a hike in mountainous territory, on the other hand, your origin and destination may be only a few kilometers apart in horizontal distance. The vertical distance that you have to traverse, however, may increase the total hiking distance significantly. In the Grand Canyon, for example, the horizontal distance from Yavapai Point, on the south rim, to Phantom Ranch, at the floor of the canyon, is approximately 4660 meters (15,289.5 feet). The vertical elevation difference between these two points is 1391 meters (4563.9 feet), and the "actual" distance between them is 4863 meters (15,955.5 feet), assuming that you could travel in a straight line from one point to the other. (In actuality, of course, the trail travels a very tortuous path, so that the actual total travel distance is much greater.) The elevation difference, then, adds 4.3 percent (203 meters or 666 feet) to the horizontal distance. This difference is for many purposes not of major importance, although your feet may tell you otherwise after you hike the trail.

Assuming that there is a relatively uniform slope between two points, the calculation of the distance correction for elevation involves a simple application of the Pythagorean theorem. If, for example, there is a horizontal distance of 1000 meters (3281 feet) between two points and a vertical difference of 100 meters (328.1 feet), the total surface distance between the two points is as follows:

$$\sqrt{(1000\ \text{m}^2 + 100\ \text{m}^2)}$$

$$= \sqrt{(1,000,000\ \text{m}^2 + 10,000\ \text{m}^2)}$$

$$= \sqrt{1,010,000\ \text{m}^2} = 1004.98\ \text{m}\ (3297.3\ \text{ft.})$$

Another factor to be considered in distance measurement is that undulations in the earth's surface add additional errors. These differences might be significant in very rough terrain. If great accuracy is required for the measurement of distance over such terrain, the route must be broken into a number of sections, each with a relatively uniform up- or down-slope. The surface distance for each section must then be calculated, and the total distance must be obtained by totaling the figures for the individual sections.

AREA DETERMINATION

Areal Scale

The **areal scale** of a map is sometimes used to describe the relationship between the *area* of a feature plotted on a map and the *area* of the same feature on the earth's surface. The ratio between the area of a region on the map and the area of the same region on the earth is the square of the map's linear scale. On a map with a linear scale of 1:24,000, for example, each *linear* unit on the map represents 24,000 *linear* units on the earth. At the same time, however, each *square* unit on the map represents 576,000,000 ($24,000^2$) *square* units on the earth. This areal scale is best expressed as $1:24,000^2$.

A number of other methods are available for the measurement of map areas. These range from quick, relatively inaccurate estimation techniques to more accurate methods that utilize special instruments and often require a longer period of time or special data to accomplish.

Direct Methods of Measurement

Scaling and Computation

The area occupied by some relatively small features can be determined with a linear graphic scale. The length and width of rectilinear features, such as a building or a city block, are easily measured. Determining the area that the feature occupies is then a matter of simple arithmetic. This method is reasonably accurate for small areas if the feature has a relatively regular geometric shape. It is not satisfactory, however, if the feature has a very irregular outline.

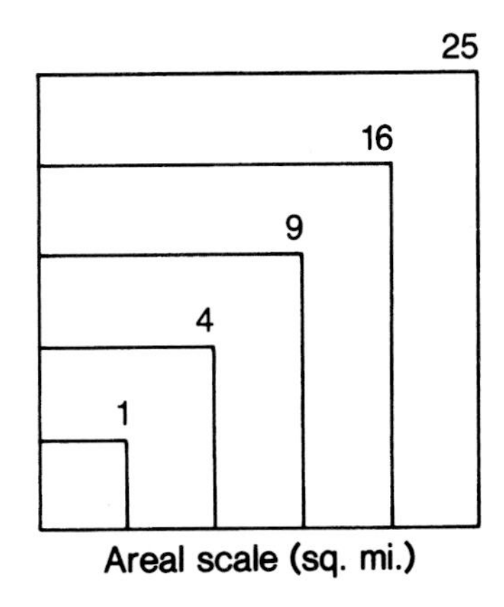

Figure 6.7 Graphic areal scale.

The size of features can be estimated directly with a graphic areal scale (Figure 6.7). Ideally, such scales are provided on transparent material so that they can be laid directly over the map. This form is particularly easy to use; it is simply a matter of totaling the unit areas occupied by the feature of interest. The area of a county in square miles, the area of wheat land in hectares, or the size of a lake in square kilometers, for example, can be determined by this method, if a scale with the desired units is available. This method is inaccurate if the feature has an irregular shape.

Polar Planimeter

A **polar planimeter** is an instrument that measures areas on a map in terms of square inches or square centimeters (Figure 6.8). These measurements are then arithmetically converted to earth measurements.

The pivot point of the planimeter is set up so that the area to be measured can be traced. (The pivot point may be either inside or outside the area to be measured, although the location affects the conversion factor that must be used.) The planimeter dial is set to zero, and the tracing point is placed at a predetermined starting point. The tracing point is then carefully guided around the perimeter of the area, returning to the starting point. When the tracing is completed, the reading on the dial is recorded. At least three readings should be taken because of the difficulty of tracing the boundary exactly. The average value of the three map area readings is then converted into the area on the ground in the desired units. This is done by using either of the following formulas:

1. To determine ground area in square kilometers:

$$\text{ground area (km}^2) = \frac{\text{map area (cm}^2) \times \text{RF}^2}{100{,}000^2}$$

2. To determine ground area in square miles:

$$\text{ground area (mi.}^2) = \frac{\text{map area (in.}^2) \times \text{RF}^2}{63{,}360^2}$$

Figure 6.8 Polar planimeter.
Source: John Kehoe photo, by permission of Keuffel and Esser Company.

where the RF value, in each case, is the denominator of the map scale.

Some planimeters are available with an adjustable tracer arm that allows direct scaling in ground units. Also, some elaborate instruments such as digitizing boards may be connected to a computer that is programmed to automatically convert instrument readings directly into the desired areal units.

Indirect Methods of Measurement

Grid Squares

The **grid-squares method** of measurement involves overlaying the area of interest with a right-angled grid of lines, spaced an appropriate distance apart (Figure 6.9). It is convenient if the spacing between the lines is scaled so that the desired areal units are obtained without further calculation. If the lines are spaced 1 scale kilometer apart, for example, each grid square represents 1 scale square kilometer.

The cells that fall within the area to be measured are then counted. If a full cell falls within the area, it is counted as one. If only a portion of a cell falls within the area, one of two methods of counting is adopted.

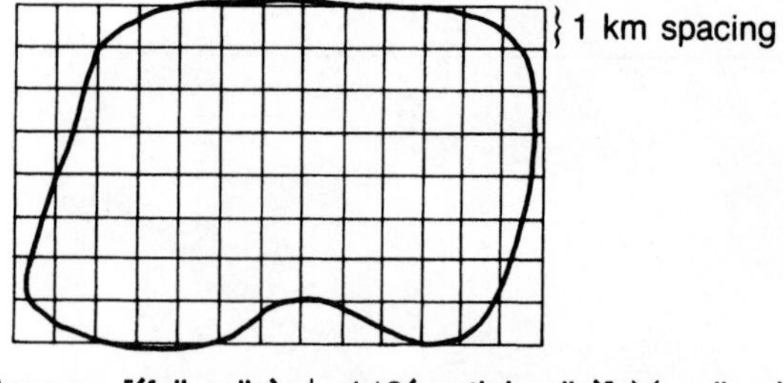

Area = [(full cells) + 1/2(partial cells)] × cell value
= [71 + 1/2(26)] × 1 km² = 84 km²

Figure 6.9 Grid-squares method of areal measurement.

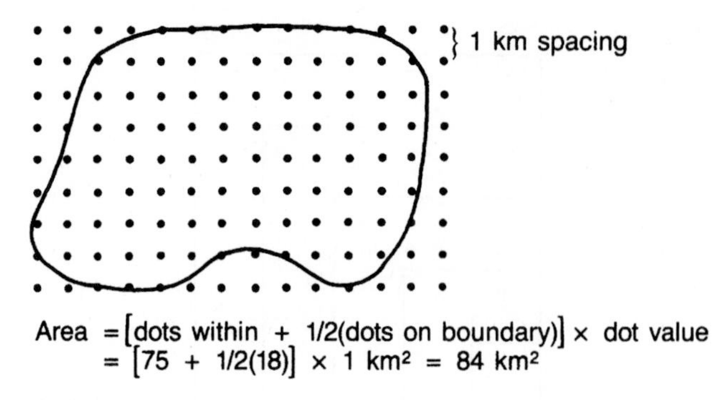

Figure 6.10 Dot-planimeter method of areal measurement.

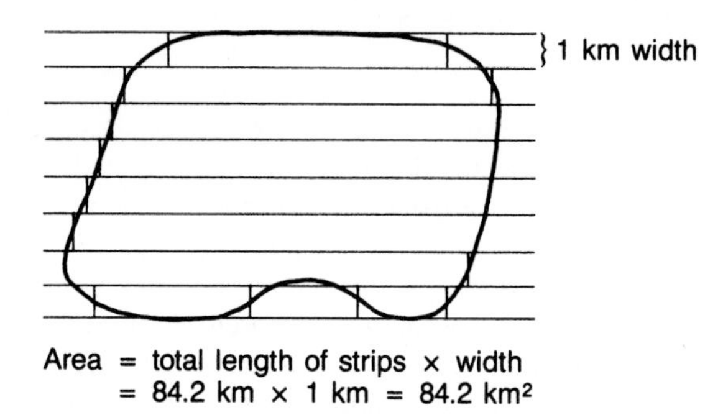

Figure 6.11 Strip method of areal measurement.

In the first approach, the total number of partial cells is divided by two and added to the total number of full cells to obtain the total area. The assumption that lies behind this approach is that, by chance, some cells fall in such a way that a large portion is within the area and others fall so that only a small portion is within the area, so that the two categories will balance one another, and the average value of one-half cell will be relatively accurate. If the grid of lines is fine enough, this method is likely to be satisfactory. In any case, the grid-square method should not be used when extreme accuracy is required.

If the grid pattern is somewhat coarse (or if the measurer does not trust random processes), a second approach is taken. In this method, an estimate is made of the fraction of each partial cell that falls within the area being measured. The total of these fractional values is added to the number of full cells to determine the total area.

Dot Planimeter

With the **dot-planimeter method** of areal measurement, which is a variant of the grid-square method, a regularly spaced pattern of dots is placed over the map, and each dot represents the center of an imaginary grid square. A total dot count is obtained by adding the number of dots that fall fully within the area to be measured and one-half of the number of dots that fall on the boundary of the area to be measured (Figure 6.10).

Next, a dot unit value is determined, based on the spacing between the dots and the scale of the map. Assume, for example, that the dots are spaced 1 centimeter (0.394 inches) apart and the map scale is 1:10,000. The imaginary square that surrounds each dot is 1 centimeter (0.394 inches) long on each side, which, at the scale of 1:10,000, represents 10,000 centimeters or 100 meters (3940 inches or 109.4 yards). The area of each square, therefore, represents 10,000 square meters (11,968 square yards). The total dot count is then multiplied by the dot unit value, and the result is the desired areal measurement. In contrast to the grid-square and strip methods, which are exhaustive counting methods, the dot-planimeter method is a sampling procedure.

Strip

In the **strip method,** a series of parallel lines is drawn over the area to be measured. As with the grid-square method, it is helpful if the spacing between the lines is scaled so that the results of the measurement are obtained directly in the desired units.

The ends of each of the strips are established by drawing vertical lines at the edge of the area being measured. Because the edge of the area is uneven and not likely to cross a strip at right angles, each line is located so that it balances the portion of the area that is outside the line with the portion that is inside (Figure 6.11).

The length of each strip is then measured using the same units that were used to establish the strip's width. Adding the individual lengths and multiplying the total length by the width of a single strip yields the total area. Assume, for example, that the strip is 1 centimeter (0.394 inches) wide and the map scale is 1:10,000. If the length of the strip is 91 centimeters (35.850 inches), it represents, at the map scale, a width of 100 meters (3940 inches or 109.4 yards) and a length of 9100 meters (3585 inches or 9955.4 yards). Its area, therefore, represents 910,000 square meters (1,089,121 square yards).

Polygon

The **polygon method** of areal measurement is based on direct measurements. It is most suitable for use when the boundary of the area to be measured is made up of a series of relatively straight lines (or can be approximated by a series of such lines). The method involves subdividing the study area into a number of regular geometric figures (rectangles, triangles, and trapezoids) (Figure 6.12). The dimensions of these figures are determined, using the desired map units, and their areas are obtained by standard procedures (height × width for rectangles 1/2 height × base for triangles, and height × 1/2 [$side_1$ + $side_2$] for trapezoids). The total of the individual areas gives the overall measurement.

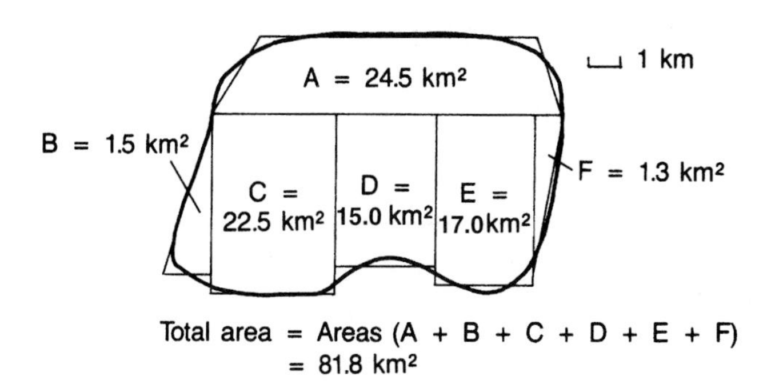

Figure 6.12 Polygon method of areal measurement.

Computer-Assisted Methods

A computer can be used to calculate areal measurements if suitable software is available. This greatly reduces the tedious calculations necessary using regular computational methods and usually speeds up the operation as well. As is the case with many computer operations, the preparation of a suitable data file is the major task required for the calculation of areas.

Small Areas

For relatively small areas, the curvature of the earth can be ignored. The data file in this case consists of a list of x,y coordinates (UTM or State Plane Coordinates, for example) and is done with a standard digitizing tablet (see Chapter 19). Usually, critical points on the outline of the area to be measured are selected, and a digital file of their coordinates is created.

The physical process of digitizing is very similar to tracing the outline of an area with a standard planimeter. The advantage over using a planimeter is that a file of coordinates is created, and this is potentially useful for other operations such as creating a map of the area. Indeed, if a computer mapping coordinate file already exists, it can be used directly, and the creation of a new file is unnecessary.

The coordinate file, whether it is newly created or already exists, is simply read into the appropriate software package, which computes the area and reports the result. The task is accomplished in somewhat different ways by different software packages, but the general approach is as follows.

Assume that we wish to compute the area of the region defined by the vertices *A, B, C,* and *D* in Figure 6.13. This region can be related to the vertical axis in the figure to form four trapezoids (*HABGH, GBCEG, HADFH,* and *FDCEF*). The data file for this region would contain x and y coordinates for each of the vertices in these trapezoids.

In general, the area of a trapezoid is determined by multiplying its height by the average length of its two parallel sides, or height × ½($side_1$ + $side_2$). The coordinates in the data file are used in the following manner to compute the area of each trapezoid: Take trapezoid HABGH as an example. In this case, its height is found by subtracting the y coordinate of vertex B from the y coordinate of vertex A. The average length of its parallel sides is found by adding the x coordinate of vertex A (which is the length of side HA) to the x coordinate of vertex B (the length of side BG) and dividing the result by 2.

After the areas of the four trapezoids have been determined, they are used to calculate the area of the region of interest on the basis of the relationship of each trapezoid to that region. Some of the trapezoids encompass the region (and more), whereas others fall entirely outside it. The procedure, therefore, is to add the areas of the trapezoids that encompass the desired region and to subtract from that total the areas of the trapezoids that fall outside it. This process is shown in Figure 6.13. Here, the region of interest (*ABCDA*) is shown, as is the reference framework (Figure 6.13a). The region *ABCDA* is included in two trapezoids (*HABGH* and *GBCEG*) (Figure 6.13b), but excess areas are also enclosed. This is resolved by excluding the area encompassed by the two outside trapezoids (*HADFH* and *FDCEF*) (Figure 6.13c), which leaves the area of interest.

This may seem to be a rather roundabout method of calculation, especially when dozens, or perhaps hundreds, of trapezoids are involved. With computer calculations, however, the results are obtained very speedily.

Large Areas

A different procedure is needed for relatively large areas, because the curvature of the earth cannot be ignored. In such cases, a computer-assisted method that uses a data file consisting of geodetic coordinates is employed. The article "Area Computation from Geodetic Coordinates on the Spheroid," by A. J. Kimerling, which is listed in the Suggested Readings at the end of this chapter, gives the details of this method.

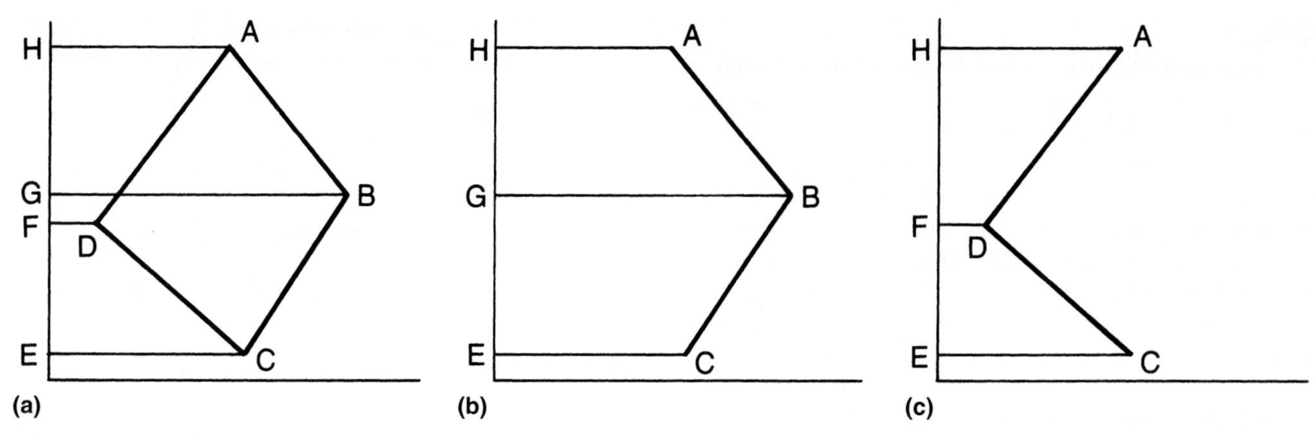

Figure 6.13 Computer-assisted method of areal measurement. (*a*) Area of interest (*ABCDA*) and reference framework. (*b*) Trapezoids encompassing the area of interest (*HABGH* and *GBCEG*) but including the excess area. (*c*) Trapezoids excluding the area of interest (*HADFH* and *FDCEF*).

SUMMARY

Distances and areas are often directly measured from maps, although the accuracy with which the map is constructed must be kept in mind when this is done.

Ideally, distance measurements are taken over short distances on large-scale maps so that the earth's curvature is not an important factor in the measurement. Over long distances on small-scale maps, distance measurements should be taken along radial alignments, on a map drawn on an equidistant projection. If these requirements are met, direct scaling using the map's graphic scale and dividers, scales, paper strips, or a map measurer is feasible, as are computer-assisted methods.

Indirect distance measurement methods are sometimes preferable. These include the application of the Pythagorean theorem to the grid coordinates of the endpoints of the distance to be measured. Over longer distances, computations using spherical coordinates are used.

Three factors influence the accuracy of distance measurements: measurement error, map generalization, and vertical displacements along the route. Measurement error stems from simple difficulties encountered when using the measuring devices and is minimized when measuring is done with care. Map generalization inevitably simplifies linear features, which normally results in shortening the distance between starting and ending points. Slope errors occur whenever a route involves ups and downs, and the actual surface distance between two locations is longer than the map distance, except where the ground surface is perfectly flat. The underestimation of distances that results from generalization and slope can be corrected if it is judged to be important enough.

Area determination can involve quick-estimation techniques or more time-consuming, but often more accurate, methods. Direct methods include simply scaling the dimensions of features and computing their area accordingly. Alternatively, the sizes of features can be estimated directly by using a graphic areal scale. Also, a polar planimeter can be used to measure areas on a map in terms of square inches or square centimeters and the measurements converted to earth measurements. Indirect methods of measurement include the use of grid squares, dot planimeters, and the strip method, all of which are relatively inaccurate. The polygon method can be more accurate, especially if the outline of the study area is relatively regular in shape. Computer-assisted methods can also be used to calculate areal measurements. These reduce the calculations and usually speed up the operation.

SUGGESTED READINGS

Kimerling, A. J. "Area Computation from Geodetic Coordinates on the Spheroid." *Surveying and Mapping* 44, no. 4 (1984): 343–51.

Lopshits, A. M. *Computation of Areas of Oriented Figures*. Boston: D.C. Heath, 1963.

Maling, D. H. *Measurements from Maps: Principles and Methods of Cartometry*. New York: Pergamon Press, 1989.

Slonecker, E. Terrence, and Nancy Tosta. "National Map Accuracy Standards: Out of Sync, Out of Time." *Geo Info Systems* 2, no. 1 (January 1992): 20, 24–26.

Thompson, Morris M. *Maps for America*. 3d ed. Washington, D.C.: U.S. Geological Survey, 1987, 104.

U.S. Bureau of the Budget. *United States Map Accuracy Standards*. Washington, D.C.: U.S. Government Printing Office, 1947.

CHAPTER

7 Route Selection and Navigation

A primary use of maps is as an aid to route selection and navigation; maps provide indispensable information about locations, obstacles, directions, and distances. We may need to decide which roads to travel to reach an adjoining town, to lay out a compass path for an orienteering outing, to plot the course for a ship to follow to reach harbor, to pick the radio beacon that will guide an airplane to the next airport, or to perform any one of a multitude of similar tasks. Regardless of the simplicity or complexity of our route-selection problem, a map of some sort is likely to be an important part of the solution, as is a means of establishing direction.

Although this chapter treats land, water, and air applications separately, they overlap a great deal. The basic considerations involved in planning a course by dead reckoning, for example, apply equally to all three, even though the discussion of dead reckoning has been placed in the nautical section. Therefore, keep an open mind when considering which techniques may be applicable in a particular situation.

Basic Considerations

Direction Determination

Direction simply refers to the path along which something is pointing or moving. It leads from one location to another. We may say, for example, that we plan to travel toward the tree on the hill, or in the direction of the waterfall. Such general statements are of little use in navigation and mapping, however, so the concept of direction must be developed into something more useful. A navigator traveling over the ocean, for example, needs to know in which direction to steer the ship. Initially the cardinal directions (north, south, east, and west) were used as guides. In the Northern Hemisphere one simply faced north, which was the direction of the North Star. Then, south was behind, east was to the right, and west was to the left. These cardinal directions were only a rough guide, however, and the North Star was not visible during the daytime or in the Southern Hemisphere. Fortunately, at a time lost in antiquity, the magnetic compass was discovered. This simple device used a free-floating magnetized indicator, which aligned itself with the earth's magnetic field and, therefore, pointed generally north. This provided a fixed reference to which other directions could be related, at any time of the day or night and in either hemisphere. (The complications introduced by the difference between true or geographic north and magnetic north are discussed later in this chapter.)

An early refinement of the compass was to divide the directions into finer intervals than the four cardinal directions. This was accomplished by the use of a compass rose. Subdivisions were established halfway between the cardinal directions to establish eight standard directions. These subdivisions were subdivided twice more into 32 directions (Figure 7.1a). Each of these directions was given a designation, and seamen committed them to memory. Calling off the designations in sequence, in the clockwise direction from north, was called "boxing the compass" (Table 7.1).

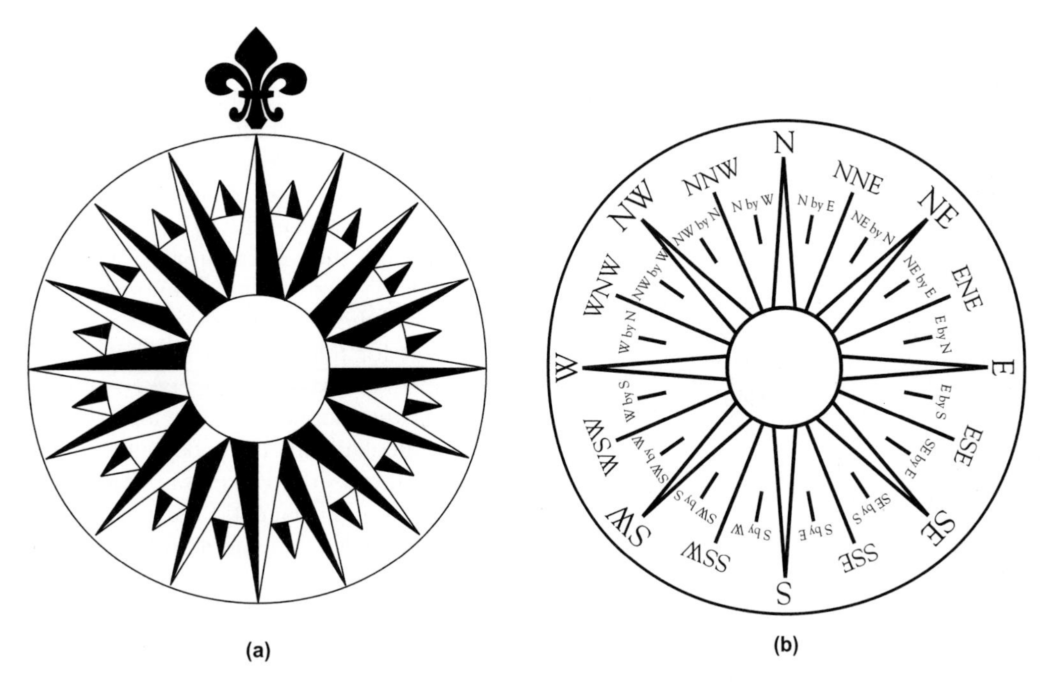

Figure 7.1 Compass roses. (*a*) 32-point rose. (*b*) With directional names.

For a long time, compass roses were simply diagrams, with no labels. Later, it was common to place the directional names on the compass (Figure 7.1b). Today, compasses are usually divided into degrees, and the names of the points of the compass are seldom used. The use of degrees in designating azimuths and bearings is discussed later in this chapter.

The determination of direction requires a frame of reference. A baseline is established first, then direction is expressed in terms of angular measurements in relation to the baseline. The baseline usually used for direction finding is a north-south line that passes through the viewer's position and establishes "north." Because more than one north reference can be established, however, the map user must decide which one to use.

True North

The most commonly used north for direction-finding purposes is **true north** (sometimes called **geographical north**). True north is the northerly direction along a line of longitude.

As mentioned earlier, true north is easily determined, on a clear night in the Northern Hemisphere, by sighting on Polaris. Polaris is a star located almost directly over the geographical north pole.[1] With a little more difficulty, observations of other stars, including the sun, may be used for this purpose.

A **gyrocompass** provides another means for determining true north. It has an advantage in that it can be used at any time of the day or night and at any point on the earth's surface or above or below the earth's surface. Also, because a gyrocompass aligns itself with the spin of the earth's axis, its north arrow points steadily true north and is not affected by the vagaries of the earth's magnetic field, as is the more common magnetic compass.

Magnetic North

The north arrow of a magnetic compass aligns with the earth's magnetic lines of force. This means that it points roughly toward the north magnetic pole, a direction called **magnetic north.** The north magnetic pole currently is located in northern Canada, approximately 1227 kilometers (762 miles) from the geographic pole (the pole of rotation). The two poles are not in direct alignment with one another in relation to most locations on the earth's surface. As a result, the magnetic compass does not usually point directly toward the geographic north pole and, therefore, does not indicate true north. The difference between true north and magnetic north at any given location is called **magnetic declination.**[2]

A special map, called an **isogonic chart,** shows magnetic declination (Figure 7.2). The lines drawn on this chart to show locations with equal declination are called **isogonic lines.** In the western United States, as

[1]Because Polaris is not directly on a line extended from the earth's axis of rotation, a discrepancy of approximately 1½ degrees must be taken into account when true north is determined very accurately. The required correction is obtained from the appropriate tables in a reference publication for the current year. Standard references, for this and other astronomical observations, include *The Astronomical Almanac* (Washington, D.C.: U.S. Naval Observatory); the *Solar Ephemeris* (Morristown, N.J.: Keuffel and Esser); *The Nautical Almanac* (joint publication of the U.S. Naval Observatory and H. M. Nautical Almanac Office, Royal Greenwich Observatory); and the *Air Almanac* (by the publishers of *The Nautical Almanac*) for the current year.

[2]Navigators often use the term *variation* for the same concept.

TABLE 7.1 Points of the Compass

Point	Name	Azimuth
0	North	0°0′
1	North by East	11°15′
2	North-northeast	22°30′
3	Northeast by North	33°45′
4	Northeast	45°0′
5	Northeast by East	56°15′
6	East-northeast	67°30′
7	East by North	78°45′
8	East	90°0′
9	East by South	101°15′
10	East-southeast	112°30′
11	Southeast by East	123°45′
12	Southeast	135°0′
13	Southeast by South	146°15′
14	South-southeast	157°30′
15	South by East	168°45′
16	South	180°0′
17	South by West	191°15′
18	South-southwest	202°30′
19	Southwest by South	213°45′
20	Southwest	235°0′
21	Southwest by West	236°15′
22	West-southwest	247°30′
23	West by South	258°45′
24	West	270°0′
25	West by North	281°15′
26	West-northwest	292°30′
27	Northwest by West	303°45′
28	Northwest	315°0′
29	Northwest by North	326°15′
30	North-northwest	337°30′
31	North by West	348°45′

the chart shows, declination is generally eastward, which means that the compass points east of true north. The reverse is true in the eastern United States. The isogonic line that joins points of zero declination is called the **agonic line.**

The location of the magnetic pole changes with time. The changes are somewhat predictable over a period of several years. As a result, the change in declination expected in the near future can be predicted. An isogonic chart, therefore, is drawn to show the situation at a particular time. It also usually includes an indication of the expected direction and amount of annual change. This information allows the adjustment of the declination values on the chart for current use, even though the chart may be several years old.

Deviations are another complication involved in the determination of compass direction. **Compass deviation** is the difference between the direction to the north magnetic pole and the direction that the compass needle actually points. One source of deviation is the presence of ore bodies that distort the earth's magnetic field. Some such deviations are relatively well-known and predictable and affect areas of considerable extent. These deviations are often included in the information from which isogonic charts are drawn. Other deviations, such as those due to power lines, are not so predictable and their effects not so widespread. This idiosyncratic type of deviation is not shown on an isogonic chart, and the user of a magnetic compass must be on the alert for the deviation's effects when operating in the field. Compass readings taken near metallic objects, such as automobiles or fence lines, or near power lines are particularly likely to be inaccurate. Indeed, even keys in the user's pocket can lead to significant compass deviation. Magnetic compasses installed in ships are carefully located and adjusted in an attempt to minimize deviation problems, although these efforts are never entirely successful.

Grid North

Any map may have one or more specialized grids, such as the Universal Transverse Mercator (UTM) or State Plane Coordinate (SPC) grids, superimposed on it (see Chapter 4). The direction of the north-trending grid lines of any such grid is referred to as **grid north.**

Meridians, which are aligned with true north, converge as one moves away from the equator and toward either pole, whereas the lines of a rectangular grid do not. Therefore, although the north-south lines of rectangular grids may coincide with true north at the grid origin (or at some other special point), they usually do not align at other locations. Thus, there is almost always a difference between the direction of the nominally north-south lines of a particular grid and *either* true *or* magnetic north. In addition, the grid north of two or more grids seldom matches, so there may be more than one grid north on a given map.

Declination Diagram

A **declination diagram** is placed on a map to show the relationships between the three "norths" that have been described (Figure 7.3). True north is usually marked by a line with a star at the end, suggesting Polaris; magnetic north is often shown by a line with a partial arrowhead, representing a compass needle, or with the letters *MN;* and the grid-north line is typically indicated by the abbreviation *GN.* The angular difference between the three north directions is indicated on the face of the diagram. The date of the diagram is also usually indicated, sometimes with the expected annual amount and direction of change in declination.

Declination diagrams vary from map to map. Some diagrams show magnetic north with respect to true north, for example, whereas others show magnetic north with respect to UTM grid north. Still other diagrams

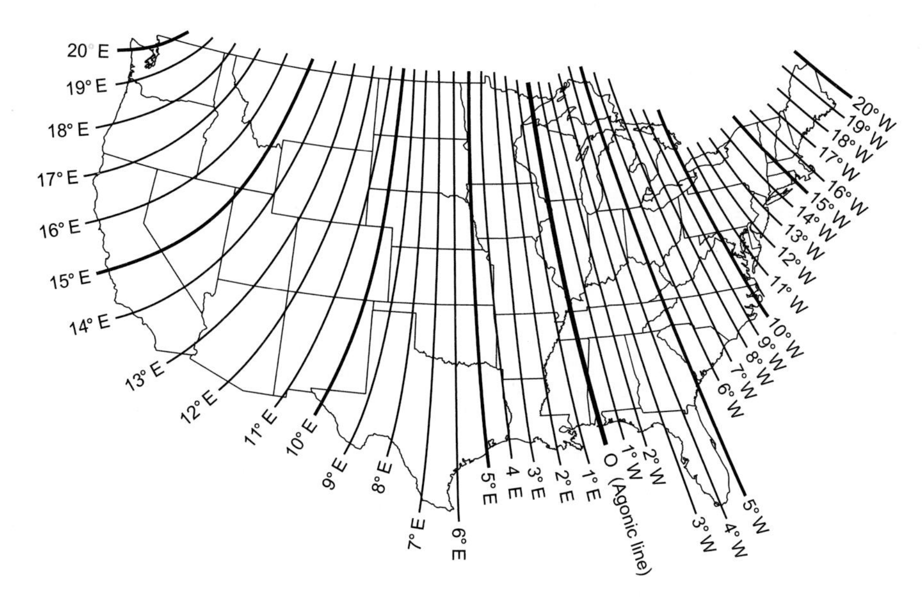

Figure 7.2 Chart showing isogonic lines, which join points of equal compass declination.
Source: Defense Mapping Agency Chart 42, *Magnetic Variation: Epoch 1995.0,* (11th edition, July 27, 1996).

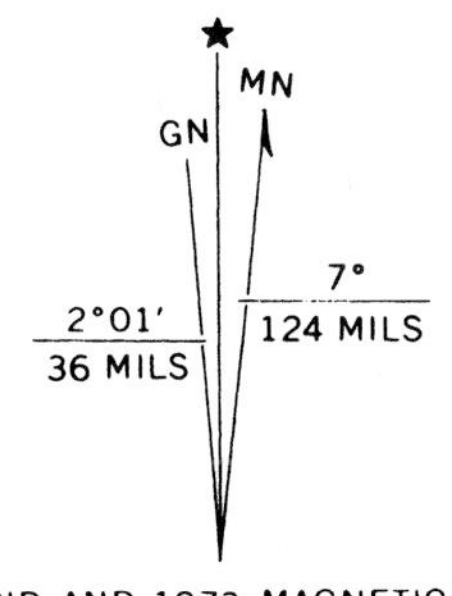

Figure 7.3 Declination diagram. The star indicates true north, *GN* stands for grid north, and *MN* and the arrow designate magnetic north. Angles on the diagram show only relative directions; actual angles are plotted when making adjustments. (Randeen Ridge quadrangle, Minnesota, 1:24,000, 1973.)

show magnetic north with respect to both true and UTM grid north. It is obviously important to know the frame of reference before using any declination diagram.

Declination diagrams on maps are sometimes drawn diagrammatically, rather than exactly. This is especially true in the case of small angles of declination, which are difficult to reproduce accurately. For this reason, exact conversions from one reference direction to another are often done mathematically, rather than graphically. Nevertheless, a declination diagram is useful for keeping track of the calculations involved in converting a directional reference from true to magnetic, from grid to true, and so on. If an accurate graphical conversion is needed for direct-measurement purposes, a larger version of a declination diagram can be drawn using exact measurements. Such a diagram is often drawn directly on the face of the map so that angular measurements can be made more easily in relation to the desired base direction.

Direction Designation

Directions are usually designated as either *azimuths* or *bearings*.

Azimuth

An **azimuth** is a directional designation that is usually measured in a clockwise direction from north (although south is the reference direction in some cases). Depending upon the reference direction, azimuths are stated as orthodromes (based on true north), magnetic (based on magnetic north), or grid (based on grid north). Azimuth measurements are usually stated in degrees, minutes, and seconds or in degrees, minutes, and decimal fractions of minutes. They range from 000° (north), through 090° (east), 180° (south), 270° (west), and 359° (approximately north, again).[3]

Over a short distance, an azimuth is simply the angle between two sightings, one taken on the reference direction and the other on the target, or destination. It can be directly plotted, with relative impunity, when the distance between the origin and destination is relatively short. When directions over long distances are involved, however, the curvature of the earth must be taken into account because, depending upon how one defines *direction,* an azimuth from a starting point may or may not remain constant. The concepts *orthodrome* and *constant azimuth* are involved here and require some clarification.

Orthodrome (sometimes called the **true azimuth**) refers to the direction of a great circle (see Chapter 2).

[3]Occasionally, angular measurements are expressed in *grads* (see Appendix B). Also, in military applications such as aiming of artillery weapons, *mils* are used to measure angles. In this system, a circle is divided into 6400 mils, so that each mil represents approximately 1/18 of a degree.

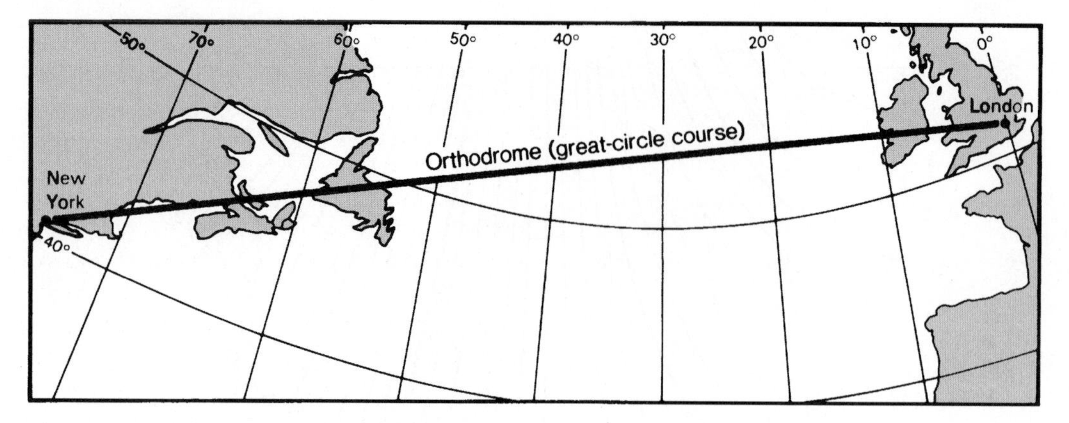

Figure 7.4 The orthodrome (true azimuth or great-circle route) crosses successive meridians at constantly changing angles.

This is because, when one starts out in a given direction and follows a great circle, one eventually returns to the point of beginning. In most cases, the angle at which an orthodrome crosses successive meridians constantly changes (Figure 7.4). One exception to this rule occurs when the orthodrome is along the equator. In this case, it crosses all of the meridians at a right angle. Another exception occurs when the orthodrome is along a meridian. In this case, of course, no other meridians are crossed. In contrast to an orthodrome, a **constant azimuth,** or **rhumb line,** is defined as a directional line that crosses each succeeding meridian at a constant angle. The difference between an orthodrome and a rhumb line is particularly important in long-distance navigation.

A simple **conversion diagram** may be used to assist in the change of azimuths from one reference direction to another. A conversion diagram is a sketch that shows two reference lines (Figure 7.5). One reference line represents the original base direction (true north, for example), and the other represents the new base direction (magnetic north, for example). The diagram also shows the azimuth being converted to the new base in its correct relationship to the reference directions. The angles between each of the lines are indicated, but they need not be drawn to the exact measurements because the sketch is used simply as an organization aid, not for actual plotting.

The examples that follow illustrate two possible conversion combinations. Assume, for example, that a magnetic azimuth of 045° is to be converted to an orthodrome and that the declination is 004° East (Figure 7.5a). The relationships between magnetic azimuths and orthodromes in this situation require *adding* the declination to the magnetic azimuth to obtain the orthodrome of 049°. Obviously, the reverse would be true when converting from orthodrome to magnetic. Figure 7.5b, shows that a west declination requires that the amount of the declination angle be *subtracted* from the magnetic azimuth to obtain the orthodrome.

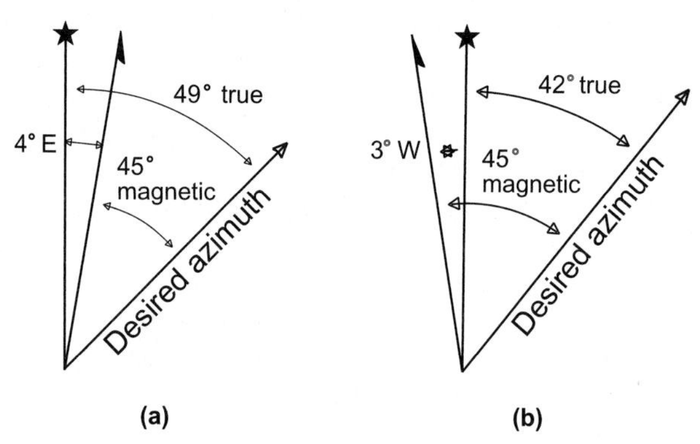

Figure 7.5 Use of a conversion diagram, showing the conversion of magnetic azimuth to true azimuth (orthodrome). (*a*) East declination. (*b*) West declination.

Conversions from grid to orthodrome or from grid to magnetic (or the reverse of either) require the use of the angle between grid north and true north or between grid north and magnetic north, as appropriate. Addition or subtraction is then carried out, as required.

Back-Azimuth

A **back-azimuth** is the exact reverse of an azimuth. (The regular azimuths just described are sometimes called **forward azimuths** for the sake of clarity.) The value of a back-azimuth is obtained by adding 180 degrees to an azimuth that has a value of less than 180 degrees or subtracting 180 degrees from an azimuth that has a value of 180 degrees or more.

Bearing

A **bearing**[4] is a type of direction designation often used in metes-and-bounds surveys. Bearings are mea-

[4]In marine navigation, the term *bearing* is used in the same sense as the term *azimuth* is used in land navigation. This potentially confusing difference in terminology results from historical usage; the context in which the terms are used usually clarifies their specific meaning.

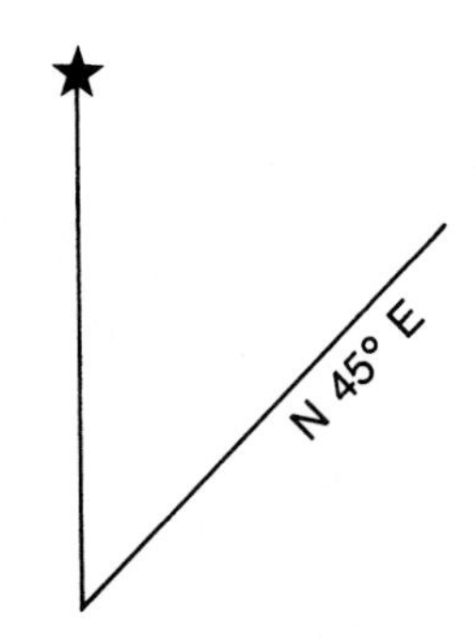

Figure 7.6 Bearing, North 45° East.

sured from either a north or south baseline, whichever is nearer to the direction being designated. Bearings may be relative to either magnetic, true, or grid north, so the baseline should be designated on the map.

Because of the manner in which they are measured, bearings have a maximum value of 90 degrees. A bearing is designated with the following form: North 45° East, for example (Figure 7.6). The base direction is stated first (in this case, north). The numerical designation indicates the number of degrees that the bearing differs from the reference direction (here, 45°). If the accuracy of the measurement is sufficient, the bearing direction includes minutes, seconds, and fractions of seconds. The last portion of the designation indicates whether the bearing is to the east or west of the reference direction (east, in this case).

Back-Bearing

Just as a back-azimuth is the opposite of an azimuth, a **back-bearing** is the exact reverse of a bearing. The value of a back-bearing is obtained by changing both of the directional references of the bearing. For example, assume that the bearing to a position from an observation point is North 45° East. The back-bearing from the position to the observation point, then, is South 45° West.

LAND OPERATIONS

Movement over land is generally the simplest form of navigation. Features that can be used for orientation are usually abundant, and the ground provides a stable platform for observations. These circumstances are in marked contrast to conditions at sea or in the air. Heavy tree cover or adverse weather conditions, on the other hand, can hide reference points and make the land navigator's task considerably more difficult.

Map Orientation

An important aspect of organizing route-finding activities on land is to orient the map being used. A simple method of **map orientation** is the process of **inspection.** Inspection involves selecting easily recognizable features on the landscape and locating the same features on the map. The map is then turned so that the landscape features are in the proper relationship to the mapped features (Figure 7.7).

A more exact method of map orientation involves the use of a compass. In this approach, a compass is placed on the map near the declination diagram. Because the compass arrow points to magnetic north, the map and compass are rotated as a unit until the direction of the north arrow on the declination diagram matches the direction of the compass needle (see Appendix D). If the rate and direction of change in declination are known, any necessary correction of the north arrow on the declination diagram is made prior to starting the orientation process.

Position Determination

If a person does not know his or her exact location, a properly oriented map provides the means for determining it, using a method called **resection.** Resection can be done either with or without a compass. In the compass method of resection, two or three known positions on the ground are located and marked on the map. The magnetic azimuth to one of the known positions is determined by taking a compass sighting. The back-azimuth of this sighting is then drawn on the map from the known position toward the unknown position, using the compass to establish the direction (see Appendix D). Preferably, this process is repeated for two other known locations. The crossing of the back-azimuths marks the ground position (Figure 7.8).

The use of three known references for resection is preferred because, ideally, the lines from all three will intersect at the same location. If they do not, a triangle, called the **triangle of error,** is formed. A large triangle of error indicates that the measurements need to be rechecked. A small triangle of error, however, indicates that the center of the triangle is the best estimate of the correct location.

If a compass is not available, resection can be accomplished using the straightedge method. In this approach, the map is oriented by inspection. Two or three known positions are then located and marked on the map. Next, a straightedge is laid on the map, with its center at the map location of one of the known positions. With this map location as a pivot point, the straightedge is turned until its edge is aligned with the same known position on the ground. (This alignment is determined by sighting along the edge of the straightedge.) A line is then drawn along the straightedge through the map location of the known position and toward the unknown position. This operation is repeated with one or two additional points, and the intersection of the lines is the unknown position. Again,

Figure 7.7 Map orientation by inspection.
Source: Department of the Army, *Map Reading,* FM 21–26 (Washington, D.C.: Department of the Army, January 1969).

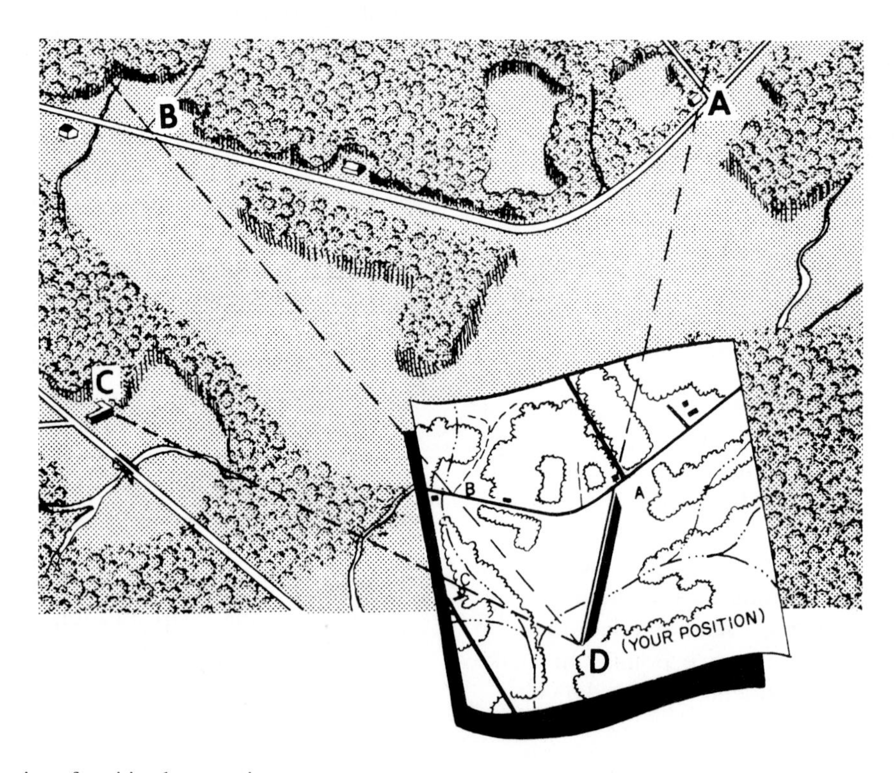

Figure 7.8 Determination of position by resection.
Source: Department of the Army, *Map Reading,* FM 21–26 (Washington, D.C.: Department of the Army, January 1969).

the accuracy of the method is indicated by the size of the triangle of error; the more accurate the orientation, the better the result that will be obtained.

Route Selection

Once the map is oriented and your location determined, you can plan the route to your desired destination. Plotting your proposed path on the map allows you to select a travel route that avoids steep gradients, hazardous areas, and other difficulties, while steering you to as many points of interest as possible. You can also make note, in advance, of reference points for monitoring your progress. Your selected route may be followed by using clues provided by reference to the map, by use of a compass, or by a combination of both methods.

When no compass is available, you must use landscape features, such as hilltops or water towers that also appear on the map, as reference points. When such a reference point lies in line with your desired route, for example, you can simply keep it dead ahead and be certain that you are on the proper path. You can check your progress by periodically reorienting your map. This will verify that other visible reference points lie in the correct relationship to your estimated position. When you have reached the end of a particular line of travel, you simply select a new target that will guide you in the appropriate direction for the next portion of your journey.

When you are using a compass, the magnetic azimuth of each portion of the route is obtained by measuring its angle from magnetic north. Aligning the compass with the desired route on a properly oriented map allows a direct reading of compass direction.[5] The compass is then sighted in the desired direction, and a target that lies on that azimuth is selected. In this case, the target can be much less prominent and need not be shown on the map. A distinctively shaped tree or a large rock, for example, often makes an excellent target. You then move in the desired direction by keeping the selected target directly ahead. When you reach one target, you select another that is on a heading appropriate to the next portion of your route and continue as before.

Water Operations

The special maps used for navigational purposes are usually called **charts.** The two principal types of charts are nautical charts for navigation at sea and aeronautical charts for navigation in the air. Each type incorporates special characteristics suited to its particular applications.

[5]Compass use is discussed in appendix B.

Nautical Charts

Many different types of **nautical charts** are published to serve the varying needs of chart users (Figure 7.9). The National Ocean Service (NOS), for example, publishes four major types of conventional nautical charts of U.S. waters: (1) harbor charts, (2) coast charts, (3) general charts, and (4) sailing charts.[6] These vary in scale from harbor charts at scales of 1:50,000 and larger to sailing charts at 1:600,000 and smaller. The larger scales show the detail necessary for operation in relatively confined harbor and coastal situations, whereas the smaller scales are suitable for operations in less-confined, offshore areas. Some of the more important general characteristics of nautical charts are described here.

Publication Date

A chart's publication date is vitally important because both navigational hazards and aids change over time. Using a chart on which such changes have not been recorded is extremely dangerous. For this reason, charts are revised at regular intervals. To ensure that the chart you intend to use is the most recent version, you should check the pamphlet *Dates of Latest Editions,* which is issued quarterly by NOS.

Some countries publish *Notices to Mariners,* for their own waters, to provide the information necessary for updating charts. The user obtains, from the appropriate agency, any notices that affect the area covered by the chart and that have been published since its effective date. The changes are then recorded on the face of the chart.

A ***Notice to Mariners (NTM)*** for U.S. waters is published weekly by the Marine Navigation Department of the National Imagery and Mapping Agency (NIMA).[7] These notices specify changes to aids to navigation and other information of importance for waters likely to be used by large oceangoing vesses. A *Local Notice to Mariners (LNM)* is published, as required, by U.S. Coast Guard District Offices, through the Coast Guard Navigation Center. These notices, which are more extensive than the *NTMs,* specify changes of importance to all vessel operators. The Marine Navigation Services of the Canadian Coast Guard publishes *Canadian Notices to Mariners (NOTMAR).*

Recently, many *Notices to Mariners* have been made available on the World Wide Web (the web site for this chapter provides links). Also, some notices are available by means of a radio/computer interface so that ships at sea have immediate access to the most up-to-date navigational information.

[6]Small-craft and intracoastal waterway charts and a variety of other charts and materials are also published by NOS. In addition, nautical charts and related products for inland waterways are published by the U.S. Army Corps of Engineers, and charts are published on a worldwide basis by the Defense Mapping Agency (DMA).

[7]Notices for the Great Lakes are published weekly by the U.S. Coast Guard.

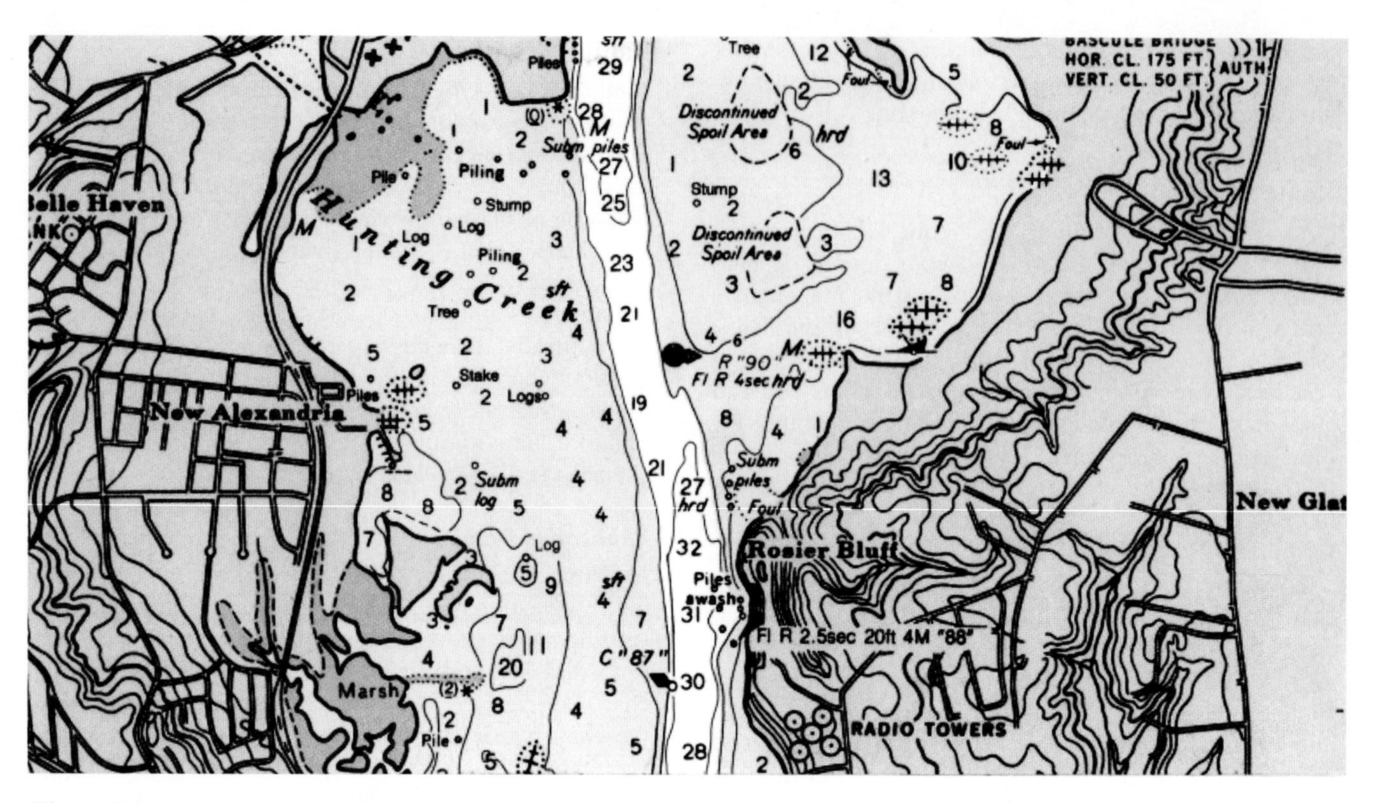

Figure 7.9 Portion of a typical nautical chart. (Not for navigational purposes.)
Source: "Example of a National Ocean Service Nautical Chart: Potomac River," NOS Educational Publication no. 14.

Projections

A useful characteristic of the Mercator projection is that a straight line drawn between any two points represents a line of constant heading (known as a *loxodrome* or *rhumb line*). Because of this characteristic, many nautical charts are drawn on the Mercator projection. Great-circle planning charts, however, are drawn on the gnomonic projection, and other projections are used for particular purposes.

Units of Measurement

Navigators must often measure distances on nautical charts. For this reason, some charts contain graphic scales. Because of strong historical precedent, these scales are often in nautical miles. Other scales are sometimes used, however, including statute miles, meters, or feet. These scales are used in the same manner as any other graphic scale (see Chapter 5).

When no graphic scale is provided, a question naturally arises as to how distance measurements are made. The answer to this question is provided by the information in the box entitled "Units of Measurement and Map Scales" in Chapter 5. As stated there, the international nautical mile is based on the measurement of an imaginary perfect sphere with a surface area equal to the surface of the earth. Specifically, it is the length of 1 minute of arc along a great circle of the sphere.

When you examine a chart, you will find that its neatline is subdivided into minutes and fractions of minutes. This subdivision provides a built-in scale on which each minute of latitude represents 1 nautical mile and fractions of minutes represent equivalent fractions of a nautical mile. To measure a distance on this scale, the chart distance between two points is recorded, usually by spreading a pair of dividers between them. The dividers are then carried to the neatline and the nautical mile distance is read from the marks representing the lines of latitude. If distances are needed in units other than nautical miles, conversions are easy to carry out (see Chapter 5). For example, 1 nautical mile = 1852 meters, 6076.12 feet, or approximately 1.151 statute miles.

Most nautical charts are drawn on the Mercator projection. This is convenient for navigational purposes, because a straight line on the Mercator represents a constant compass heading. However, it also means that the scale of the chart constantly changes as the latitude changes. Because of this change in scale, two limitations must be observed when taking and scaling measurements. First, both operations must be done at the same latitude. Second, degrees of latitude must be used, not degrees of longitude.

Depth Indications

A vital piece of safety information for anyone operating a vessel is the depth of the water under the vessel's keel. The depth of water bodies such as lakes and oceans varies as remarkably as the elevations on the earth's surface, but the undulations are hidden from view by the water. For this reason, the depth information on navigational charts provides vital information about the conformation of the bottom.

On most charts, depth information is indicated in two ways. First, individual depths, or **soundings,** are specified by printing a number that is the equivalent of a spot height. There is usually no dot or other symbol to indicate the location of the sounding. The number that specifies the depth is simply centered at the correct location. Second, points of equal depth are indicated by lines called *depth curves*. Sometimes, hypsometric coloring between the contours indicates areas that fall within a particular range of depths.

Depths on charts are traditionally shown in **fathoms** (6 feet), although feet and meters are also used. The units used on a particular chart are clearly indicated on the face of the chart. One must be cautious, nevertheless, because of the potential danger of confusing depth values. If a ship draws 2 fathoms, areas that have a depth of less than 2 fathoms (12 feet), plus a margin for error, must be avoided. If the depths are marked in meters, however, the critical depth value is 4 meters (4 meters is slightly more than 2 fathoms). Heights of objects or natural features are usually recorded in feet. Both ocean and Great Lakes charts, however, are gradually being changed to a metric standard, so that depths and heights will both be recorded in meters.

The datum used for recording depths is an important variable that must also be taken into consideration when using depth indications on navigational charts. Information regarding the datum used on a particular chart is printed with the chart title. In general, depths are measured from the level of some form of average low water, in tidal waters. This means that the depths shown provide safe operating margins under most conditions. At extreme low tide, however, depths are less than those indicated on the chart. The wise mariner, therefore, takes tidal conditions into account when determining safe operating conditions.

Special Symbols

The symbols used on nautical charts are selected to meet the special needs of chart users. These symbols deal in particular with such matters as water depth, the characteristics of the ocean floor or lake bottom, and the locations and types of navigational aids such as channel markers. For obvious reasons, such important navigational information is presented in exhaustive detail on nautical charts. Information about land features is quite limited and is usually confined to the immediate coastal areas and to those features such as radio towers, bridges, railroad tracks, and cities and towns that provide locational clues to the navigator. *Chart No. 1* is a reference booklet showing the symbols used on the nautical charts of the National Ocean Service.[8] Selected examples of these symbols are shown in Figure 7.10. One important additional method of symbolization on charts is the use of upright lettering for features that are dry at higher water and italics for submerged or floating features.

Position Determination

The determination of your vessel's position is a critical part of navigation because knowing where you are is fundamental to selecting the route you must follow to reach a new destination. Also, once your position has been accurately determined, you can check for any potentially hazardous conditions in the vicinity.

Early methods of position determination depended on visual observations assisted by various types of relatively simple instruments. Modern methods, on the other hand, often depend on sophisticated radio and electronic gear. This section describes some of the more fundamental approaches to locational determination and provides a basic understanding of how they relate to navigation.

Astronomical Observations

The traditional method by which navigators determine their position is by observing the sun and stars (see Chapter 2). This method, although relatively time consuming, is accurate and is particularly useful for navigation out of sight of land, provided atmospheric conditions allow taking the necessary observations. Depending on circumstances, a variety of other methods can supplement or take the place of direct observations.

Lines of Position

Navigators operating within sight of land maintain close watch on the location of their vessel by using **lines of position.** The use of lines of position is simply an application of the method of resection described earlier in the chapter. Lines of position may be **bearings** on known objects, such as church spires or water towers (Figure 7.11a). A *bearing* is the nautical term for an azimuth: the angle, in degrees, between a reference direction and the direction to an observed feature. Nautical bearings, therefore, range from 000 degrees to 360 degrees, in contrast to surveyor's bearings, which were described earlier in the chapter.

Lines of position may also be **ranges,** which are alignments of two known objects (Figure 7.11b), or they may be *distances* from known objects (Figure 7.11c). When lines of position are plotted on a navigation chart, their point of crossing establishes the vessel's location,

[8] *Chart No. 1, United States of America, Nautical Chart Symbols and Abbreviations* (Washington, D.C.: Department of Commerce, National Oceanic and Atmospheric Administration). The booklet also contains a glossary of principal foreign charting terms and information regarding aids to navigation in U.S. waters.

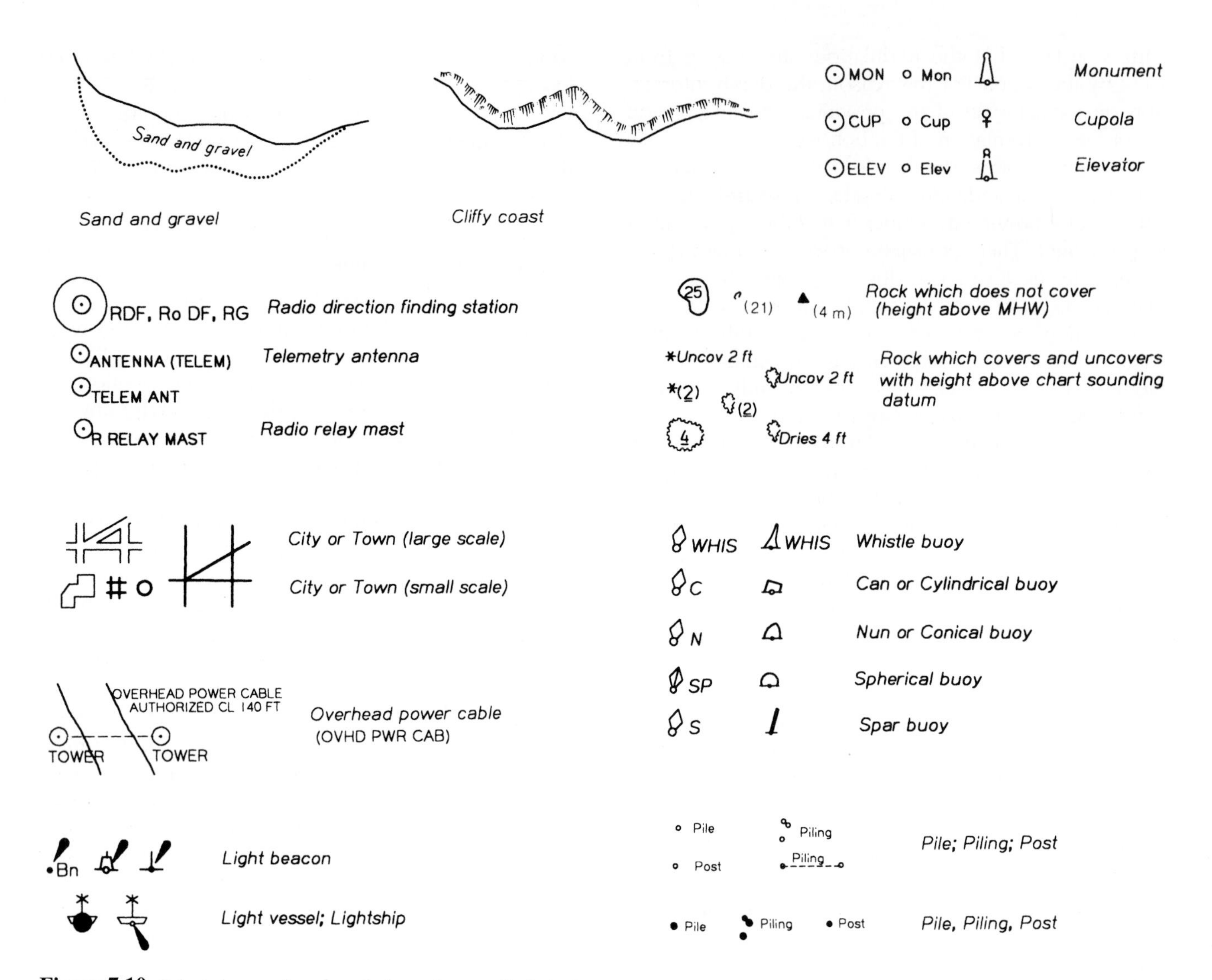

Figure 7.10 Selected examples of symbols used on nautical charts.

Source: *Chart No. 1, United States of America, Nautical Chart Symbols and Abbreviations,* 8th ed. (Washington, D.C.: Department of Commerce, November 1984).

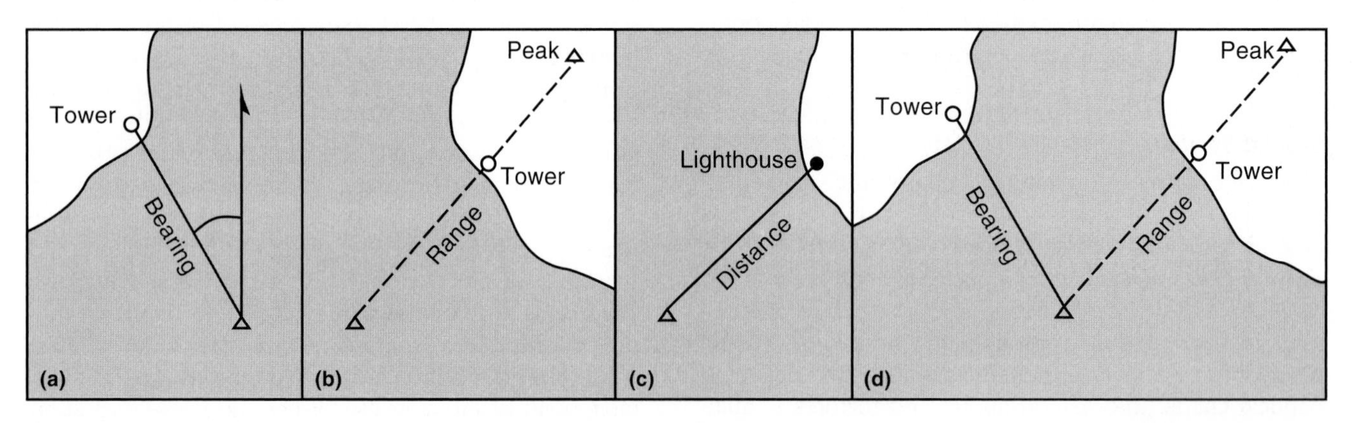

Figure 7.11 Lines of position. (*a*) Bearing. (*b*) Range. (*c*) Distance. (*d*) Fix.

which is called a **fix** (Figure 7.11d). Obtaining a fix is the most accurate and trustworthy method of determining a vessel's location.

Radar

Radar is an active remote-sensing technique involving the transmission, reflection, and detection of radio waves that are only a few centimeters in length. It can be used regardless of weather and lighting conditions. The same general principles apply whether the transmitter is located on a ship, on an airplane, or at a land base, but this section is written from the perspective of a ship-based installation.

A radar transmitter sends out pulsed signals, which are focused into a narrow beam by a special antenna. The antenna rotates as the signal is transmitted, so that a complete circle is swept every few seconds. When a pulse strikes the land surface, a tree, a building, another ship, a navigational aid, or any other object, it is reflected back to the radar antenna and is detected by a receiving unit. The returned signals are then used to create a display on a computer monitor. Objects that reflect signals show up on the screen as points of light, with the transmitter located at the center of the screen (Figure 7.12). Each point of light is located in the correct direction and at the correct scale distance from the ship. The appropriate location is determined by the direction from which the radar beam was reflected and by the length of time it took for it to reach the target and return.

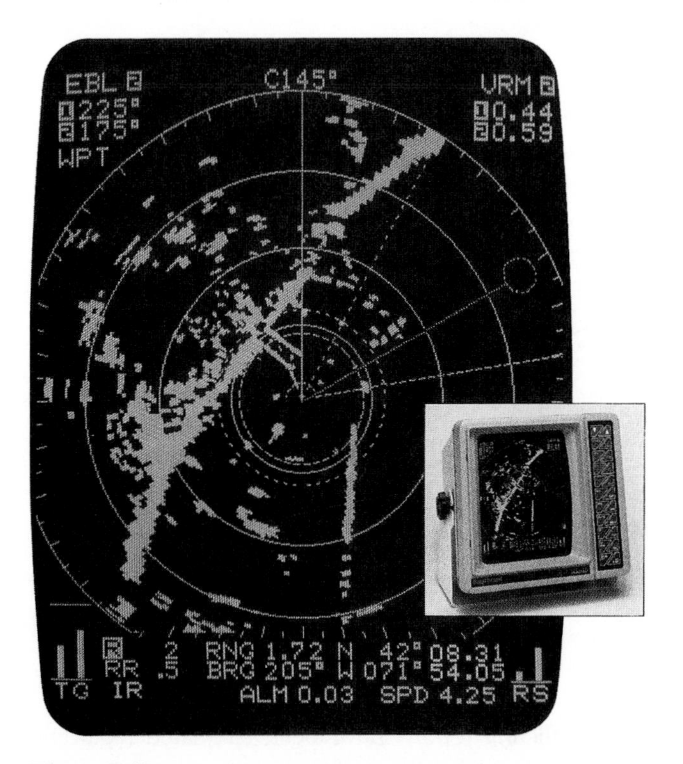

Figure 7.12 Photograph of an image on a radar screen. (Inset shows radar unit.)
Source: Courtesy of Raytheon Marine Company.

The radar image is used to provide navigational information. Bearings are taken on identifiable objects that appear on the radar screen, for example. These bearings are plotted on the navigator's chart, in the same manner as other lines of position, to establish the vessel's location. In addition, the location, speed, and direction of movement of other ships and the location of navigational hazards can be determined by careful observation, so that collisions and groundings can be avoided.

A radar observer must have a considerable amount of training, experience, and skill. The image on the radar screen consists simply of points of light, whose meaning must be interpreted. Different objects, at different orientations, for example, reflect signals with varying strength. As a result, their images on the screen vary in brightness. In addition, because the resolution of the radar image depends on the width of the radar beam, an object may sometimes look larger than it actually is. Also, the length of the pulse transmitted affects the ability to differentiate objects. Features located close to shore, for example, may merge with the shoreline. Signals may also be reflected from the surface of the water, under certain wave conditions, producing a confusing "clutter." In addition, atmospheric conditions may adversely affect the reception of the radar signals. All of these factors make the interpretation of a radar image as much an art as a science, and although radar provides a vital navigational aid, other navigational methods must be used in conjunction with it.

Global Positioning System

Recently, the **Global Positioning System (GPS)** has made accurate, virtually automatic position determination possible. The system is worldwide and operates on a 24-hour basis, in all types of weather. Although a major use of the system is for navigation at sea, it is equally useful for land and aerial navigation and for surveying.

The basis of GPS is a constellation of 24 satellites, operated by the U.S. Department of Defense. These satellites orbit about 10,900 nautical miles above the earth (Figure 7.13). Their orbits are designed to ensure that at least four satellites are visible from almost any location on earth, at any time of the day or night. As the GPS satellites move in precisely controlled orbits, they each transmit complex radio codes, an important component of which is an extremely accurate time signal. The time signals are monitored and controlled from earth and are accurate to better than 1 nanosecond (one billionth of a second). The time signals radiate in all directions from the satellites, at the speed of light.

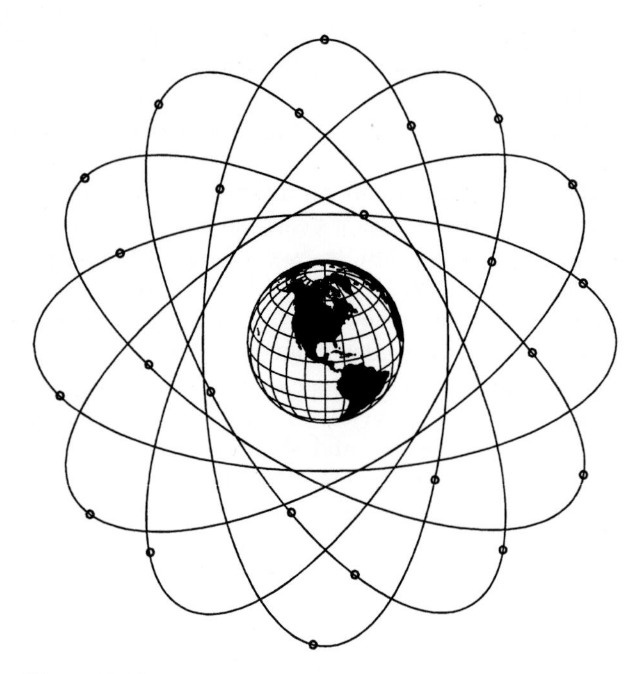

Figure 7.13 Global Positioning System (GPS) satellite orbits.

To use GPS, an appropriate receiver is required. Receivers range from very small, waterproof, battery-powered units to large, survey-grade instruments. When in operation, the receiver searches for signals from the satellites. When it receives a signal of sufficient strength, it identifies the transmitting satellite, based on a code that is part of its signal. The receiver has its own internal clock, and this enables it to determine the signal's travel time from the satellite to the receiver.[9] Based on the speed of light, this time is converted into the equivalent distance traveled. Also, at each moment, the receiver has information regarding the precise location of each satellite.

The information obtained about the satellites is used as follows. First, for each satellite, the position of the receiver must be on the surface of a sphere centered on the location of that satellite. The diameter of this sphere is determined by the analysis of the time signal (Figure 7.14a). When the information from two satellites is considered, the location must be somewhere on the circle formed where the two distance spheres intersect (Figure 7.14b). Taking a third distance into account, the location is narrowed to one of two possibilities, where all three spheres intersect (Figure 7.14c). Finally, the use of a fourth satellite allows for the correction of certain timing problems, and the location is calculated in terms of latitude, longitude, and elevation.

[9]The actual computation is more complex, but this description should be satisfactory for introductory purposes.

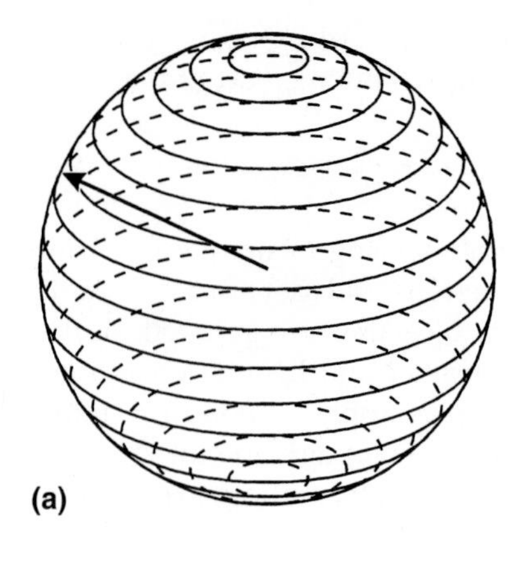

(a)

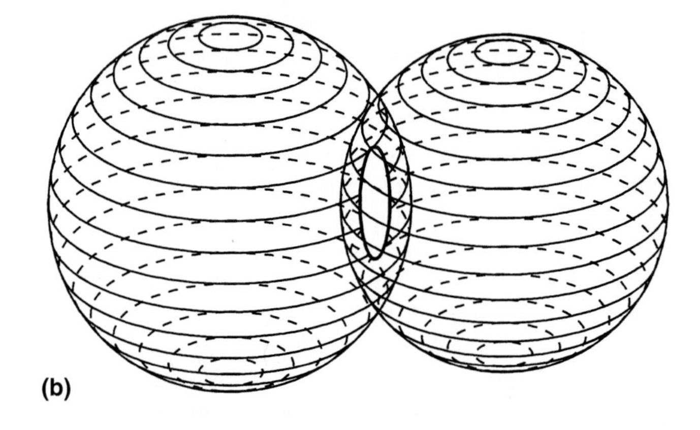

(b)

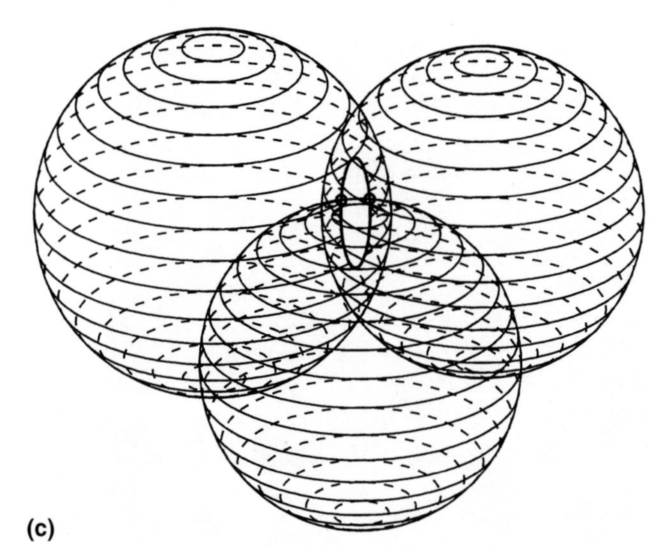

(c)

Figure 7.14 GPS location determination. (*a*) Distance from a single satellite is on the surface of a sphere. (*b*) Distances from two satellites are on a circle. (*c*) Distances from three satellites are at one of two points on a circle.

Conflicting reports have been circulated regarding the accuracy of GPS position determination. One source may say that positions are determined within fractions of a centimeter and another may suggest that accuracy is within a range of ±100 meters. Actually, either statement can be correct, because the actual accuracy depends on several variables, including the quality of the receiver, the length of the observation period, atmospheric conditions, and so on. Expected accuracy with a hand-held, battery-powered receiver, without additional correction information, is in the ±100-meter range. In general, the accuracy of vertical measurements is less than the horizontal accuracy.

The main reason for the 100-meter accuracy expectation is that the Department of Defense deliberately degrades the accuracy of the GPS time signals. This degradation, which is called **Selective Availability (S/A),** is designed to prevent the system from being used against our national interests by a hostile force. Methods are available to offset the effects of S/A, such as using information from a fixed reference location. This approach is called *differential GPS (DGPS)*. DGPS improves position determination to ±10 meters, or better, but requires the use of a special, additional receiver.

As mentioned, a standard piece of information provided by a GPS receiver is a readout of the calculated latitude and longitude of a position. This is usually reported in degrees, degrees and minutes, or degrees, minutes, and seconds. (Universal Transverse Mercator [UTM] and other grid coordinates may also be available.)

Because the earth is not a perfect sphere, it is usually represented for large-scale mapping applications as an ellipsoid (see Chapter 2). Because there are many different ellipsoids, designed to work for specific areas of the world, the GPS receiver must be set for the reference ellipsoid appropriate for the use to which the information will be put. For example, for use with current U.S. navigational charts, the most common reference ellipsoid is the North American Datum of 1983 (NAD83), so the GPS receiver should be set to refer to that ellipsoid (which is identical to the World Geodetic System of 1984 [WGS84]).

Latitude and longitude positions obtained from GPS can be plotted on a chart or map in the standard way, and navigation can be done as described elsewhere in this chapter. However, other GPS applications make the system far more useful.

One of these additional uses of GPS is way-finding. That is, the location of a destination can be indicated, either by entering its latitude and longitude manually or by recalling locational information recorded during an earlier visit. To use the way-finding feature, after the destination is selected, the system calculates the direct course heading from the current location to the destination. As the boat moves along the course, the unit monitors and provides continuous information about its progress. First, it tells the boat's speed and how long it will take to reach its destination at that speed. Second, it tells the direction of travel and shows the appropriate correction to come back to the direct course.

The way-finding feature can also be used by entering a series of way-points, such as the destinations for several legs of a trip. The system then monitors progress along the course and provides directions to the way-points, in sequence.

Another task that can be done by a GPS unit is logging the actual route of a trip. To retrace this track, the appropriate command is entered and the system provides, in sequence, information about the course legs to steer.

Some GPS navigation units include digital chart information. This allows direct checking for hazards, aids to navigation, and so on without the need of a regular chart.

Radio Direction Finding

Radio direction finding (RDF) is a method of position determination that involves the use of directional information obtained from radio signals. It is used in place of visual observation of lighthouses, points of land, and other landmarks.

RDF requires a series of separately identifiable radio transmitters placed at specified locations. These transmitters operate continuously, and their signals radiate in all directions from their transmission towers. A radio receiver with a rotating-loop antenna is needed aboard the ship whose location is to be determined. The strength of the signal picked up by the receiver varies as the antenna is rotated. By relating the orientation of the antenna to the strength of the signal, the operator determines the directions to at least two transmitting stations. The back-bearings from the known locations of those stations are then plotted to determine the position of the ship, in the same manner as was shown in Figure 7.8.

An alternative method of RDF is for the ship to transmit a signal. When this signal is picked up by two or more receiving stations, the directions from those stations to the ship are determined by the use of rotating-loop antennas. When the bearings are plotted, the ship is located at the crossing points of those lines. Some receiving stations are also equipped to measure distances. One disadvantage of this system is that the locational information obtained must be relayed to the ship, which introduces a time delay, as well as a potential for miscommunication.

RDF can be utilized in weather, lighting, and distance conditions that prevent visual sightings. Unfortunately, various types of radio-wave interference sometimes make it difficult to obtain accurate readings, so the method is not totally dependable.

Loran

Another method used to determine the location of a ship or an airplane is the radio navigation system called **loran** (an acronym for *lo*ng-*ra*nge *n*avigation). The loran system is based on radio-transmission travel times.

Loran employs pairs of radio-transmitting stations, called *master (M)* and *slave (S) stations,* located some distance apart. Each pair of stations transmits readily identifiable, synchronized radio pulses. These pulses are received by a special radio receiver on the ship or airplane whose location is to be determined. This receiver measures the time interval between the two pulses, which is used to determine the location of the receiver.

When the *M* and *S* stations simultaneously transmit signals, a radio located at a point equidistant from them, such as along line *A* in Figure 7.15, receives both transmissions simultaneously. On the other hand, a receiver at point *B* is closer to *M* and, therefore, receives the signal from *M* earlier than it receives the signal from *S*. The same timing difference occurs at many other points which, when plotted, fall on the hyperbolic curve *CD*. Sets of these curves, based on expected time differences between the reception of *M* and *S* signals, are preplotted on specially constructed loran charts (Figure 7.16). Each individual curve on the chart represents locations at which the signals from the designated pair of stations are received at a specified time interval.

To use the loran chart, the operator first tunes to a given pair of stations. The time reading thus obtained indicates that the receiver is located somewhere along the curve that represents that interval. The measurement and plotting process is then repeated with a second pair of stations and the locational curve that corresponds to the observed time interval from those

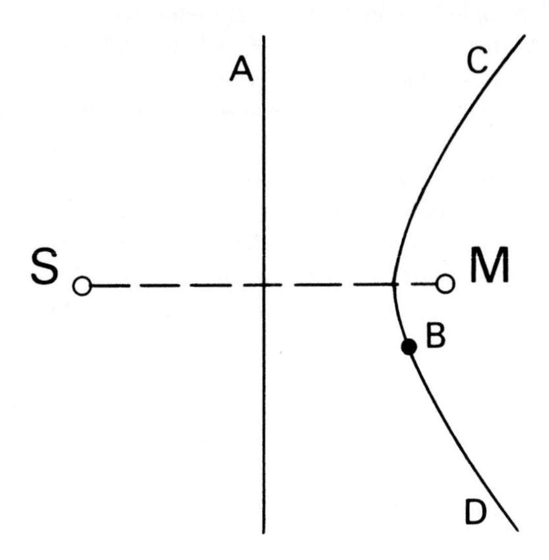

Figure 7.15 Relationship between loran slave (*S*) and master (*M*) stations and lines of position.

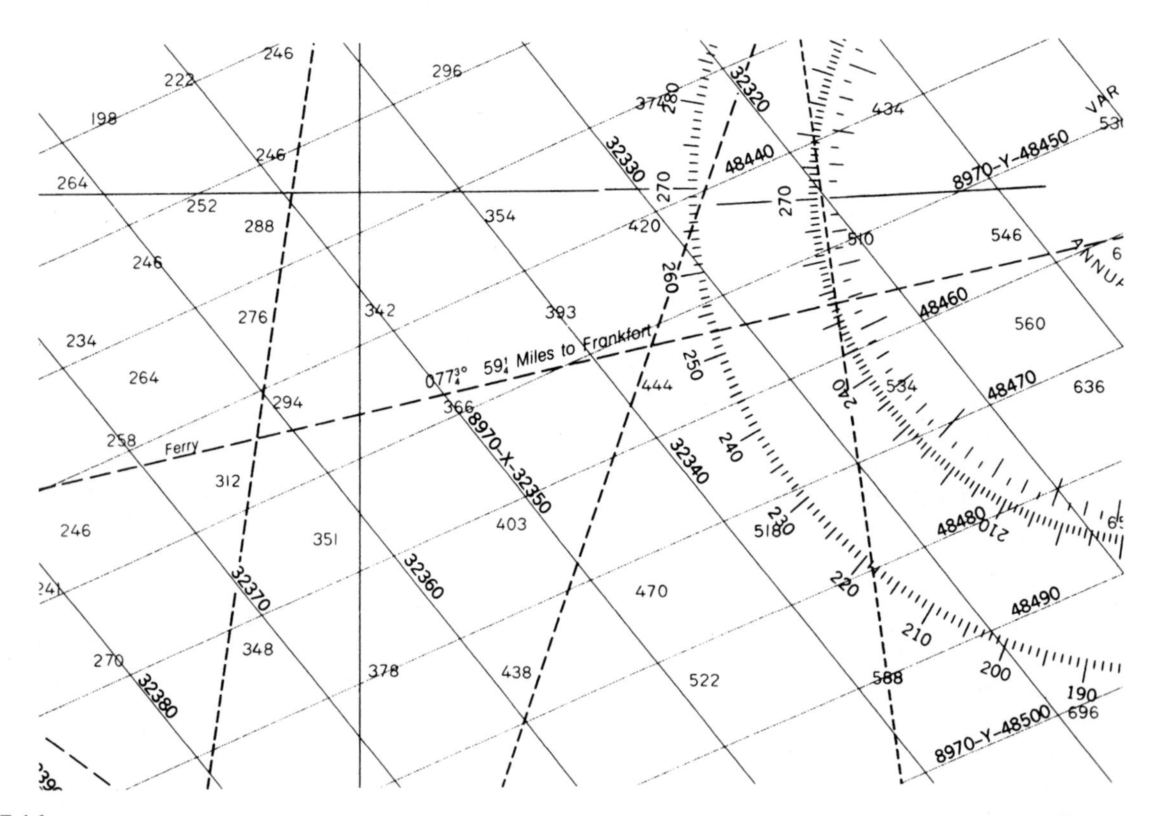

Figure 7.16 Portion of a loran chart. Numbers printed beside the loran curves are time delays from designated stations. (Not for navigational purposes.)

Source: National Ocean Service Chart 14903 (1984).

stations is identified. The intersection of the two curves indicates the location of the receiver.

The accuracy of the loran technique partly depends on the existing atmospheric conditions and the manner in which they affect radio transmissions. Generally, for example, transmission conditions are better at night than in the daytime. Also, the location of the transmitter pairs relative to the observer affects the angle at which the location curves intersect one another. If the angle is rather flat, the amount of inaccuracy due to an error in measuring the time interval between signals is greater than if the curves intersect at an angle that is closer to a right angle (Figure 7.17). Finally, the effective range and level of accuracy of the system depend on the particular frequency used and the vintage of the equipment. In general, loran accuracies are sufficient for navigational purposes on the high seas but not for determining locations within confined waterways.

Loran stations provide coverage for much of the north Pacific Ocean, the north Atlantic Ocean, and the Mediterranean Sea. The Decca and Omega systems are worldwide locational systems. The transmission frequencies and some other details of these systems are different, but they are based on the same basic principles as loran.

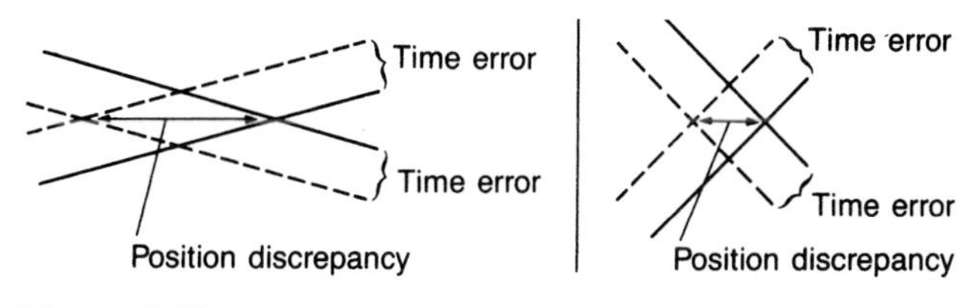

Figure 7.17 Angle differences affect the accuracy of the loran plot.

Dead Reckoning

Once position has been determined, the navigator's task is to plan and track the vessel's movements. The most straightforward approach to these tasks is to use a process known as **dead reckoning.**[10] This technique is used either for advance route planning or for keeping track of progress toward a destination. Dead reckoning depends on knowing the starting point and the direction, speed, and time of travel away from that point. Although the same general technique is applicable to travel by ship, airplane, land vehicle, or foot, the description here is based on travel by ship.

To estimate current position with the dead-reckoning method, the route that has been traveled from a known starting point is divided into separate components, called **legs.** Information about the direction of travel, as well as the length of time and the speed for each leg of the journey, is recorded. Speed of travel divided by elapsed time equals distance traveled. Thus, one simply locates the starting point and plots, in sequence, the direction and distance of each leg of the journey away from that point. The theoretical location of the ship at each stage of the voyage is shown by the resulting plot (Figure 7.18).

[10]The derivation of the term *dead reckoning* is uncertain and has resulted in some confusion between estimated location, which takes current and wind into account, and dead reckoning, which does not. See *American Practical Navigator: An Epitome of Navigation* (Washington, D.C.: Defense Mapping Agency Hydrographic/Topographic Center, Pub. no. 9, 1984), 59. [Originally by Nathaniel Bowditch, LL.D., vol. 1.]

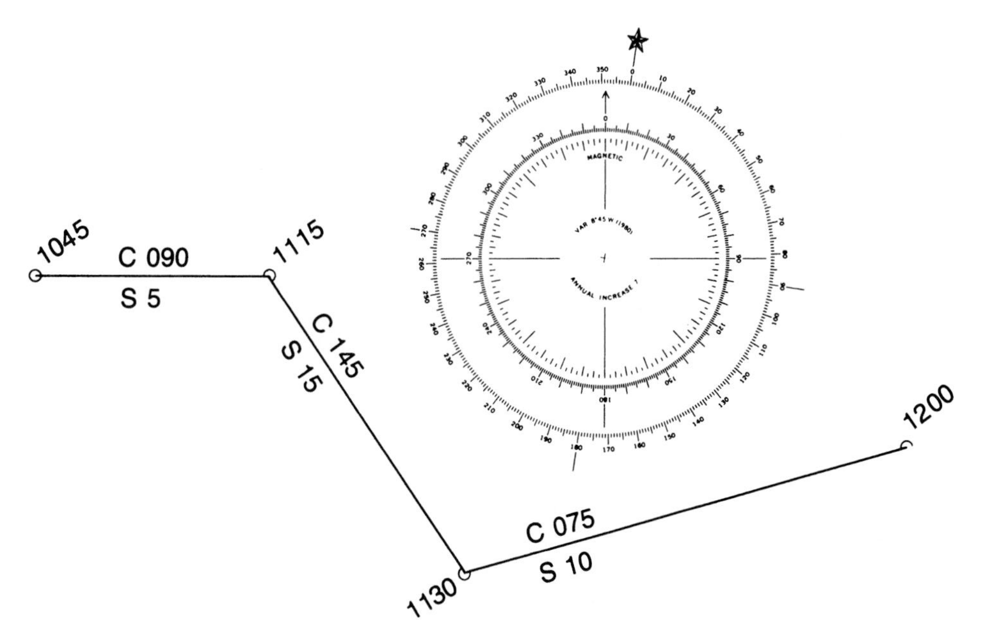

Figure 7.18 Example of a dead-reckoning plot. Angled notations are times of observations; *S* indicates speed; *C* indicates course heading (magnetic).

Many inaccuracies can creep into the dead-reckoning process. For example, directions may be inaccurately recorded because of compass deviations. In addition, speeds and directions may not be accurate because of the effect of winds, currents, and tides on the motion of the vessel. If either wind or water movement works opposite to the direction of travel, for example, progress is slowed, and the distance traveled per unit of time decreases. On the other hand, influences in the direction of travel increase speed and, therefore, the distance traveled per unit of time. Similarly, a cross-wind or movement of water causes a drift away from the planned course, as well as a change in speed. The exact influence of these factors on the motion of the vessel is difficult to determine.

For these and other reasons, positions obtained by dead reckoning are ordinarily used only if other sources of position information are not available. When more accurate information is obtained, such as an astronomical fix or a fix based on observed lines of position, a correction is made, and a new series of dead-reckoning position estimates is started, based on the updated location. If an accurate locational fix cannot be obtained, the navigator may estimate the effects of winds and tides and use them to adjust the vessel's dead-reckoning location. A location based on this type of adjustment is called an **estimated location.**

Dead reckoning is also useful as a method of advance route planning. In this case, the desired route of travel is laid out on the chart. Headings for each leg are then determined, and this data, combined with expected speeds, allows the calculation of estimated elapsed time en route, arrival times, and similar information.

Aviation Operations

Earlier sections of this chapter that deal with land and water navigation are equally relevant to aerial navigation because the principles and techniques involved are virtually identical. The chief difference between aerial navigation and the other types is that locations must often be determined and courses selected very rapidly. Also, specialized charts and navigational aids have been developed for aerial navigators.

Aeronautical Charts

Aeronautical charts are special maps designed to meet the needs of aviators. They are published at a variety of scales to suit different users. Charts published in the United States range from planning charts at scales of 1:2,000,000 or smaller to airport obstruction charts at scales as large as 1:12,000.

Aids to aeronautical navigation are subject to frequent change. For this reason, aeronautical charts, like nautical charts, must be as current as possible to avoid confusion or disaster, and they are frequently revised. In addition, notices called ***Notice to Airmen (NOTAM)*** are regularly published by the Federal Aviation Administration (FAA) and distributed by the U.S. Government Printing Office. Between chart revisions, *NOTAMs* provide up-to-date information about changed navigational aids.

Because of the variety of specialized charts available, the descriptions given here are limited to two commonly encountered series: the World Aeronautical Charts and the visual flight rule version of the sectional charts. These series provide good examples of the general characteristics of aeronautical charts.

World Aeronautical Charts

The **World Aeronautical Chart (WAC) series** is based on the Lambert conformal conic projection. WAC charts are published at a scale of 1:1,000,000, which is convenient for navigation by moderate-speed aircraft. Coverage is limited to land areas. Sheet edges conform to lines of latitude and longitude, but the number of degrees covered by each sheet varies somewhat, as the index map shows (Figure 7.19).

WACs show cities and towns as well as principal roads, railroads, and distinctive landmarks (Figure 7.20). The terrain representation includes spot elevations, contours, and elevation tints, as well as the stream-drainage pattern. Information regarding aeronautical features is overprinted on the base map in dark blue or magenta ink. This includes the location of radio navigation information, airports, runways, restricted areas, obstructions, and so on. The symbols used on WACs are virtually identical to those described in the section that follows on sectional charts.

Sectional Charts

Sectional Charts are based on the Lambert conformal conic projection and are published at a scale of 1:500,000. Each chart covers an area that averages 4 degrees of latitude and 6 to 8 degrees of longitude (Figure 7.21). These charts are designed for the visual navigation of slow- to medium-speed aircraft (Figure 7.22).

The major classes of data shown on sectional charts include topographic data, radio aids to navigation and communication, airport traffic and airspace information, and airport information. Topographic information includes features that are useful checkpoints for visual navigation, such as cities and towns, roads and railroads, and power transmission lines, as well as coastlines and drainage patterns (Figure 7.23a). This type of information is printed in black or dark blue. Built-up city areas are shown in bright yellow, and elevation zones are printed in relatively subdued hypsometric tints. A detailed latitude and longitude graticule is also included.

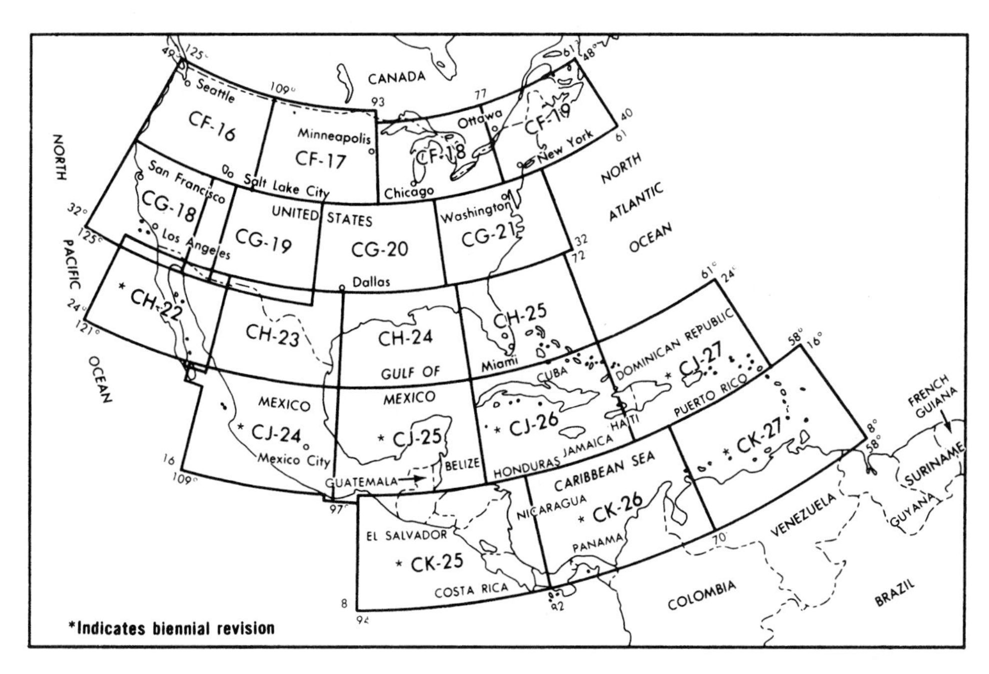

Figure 7.19 World Aeronautical Chart (WAC) series index for the conterminous United States, Mexico, and the Caribbean.
Source: National Ocean Service, *Aeronautical Charts and Related Publications.*

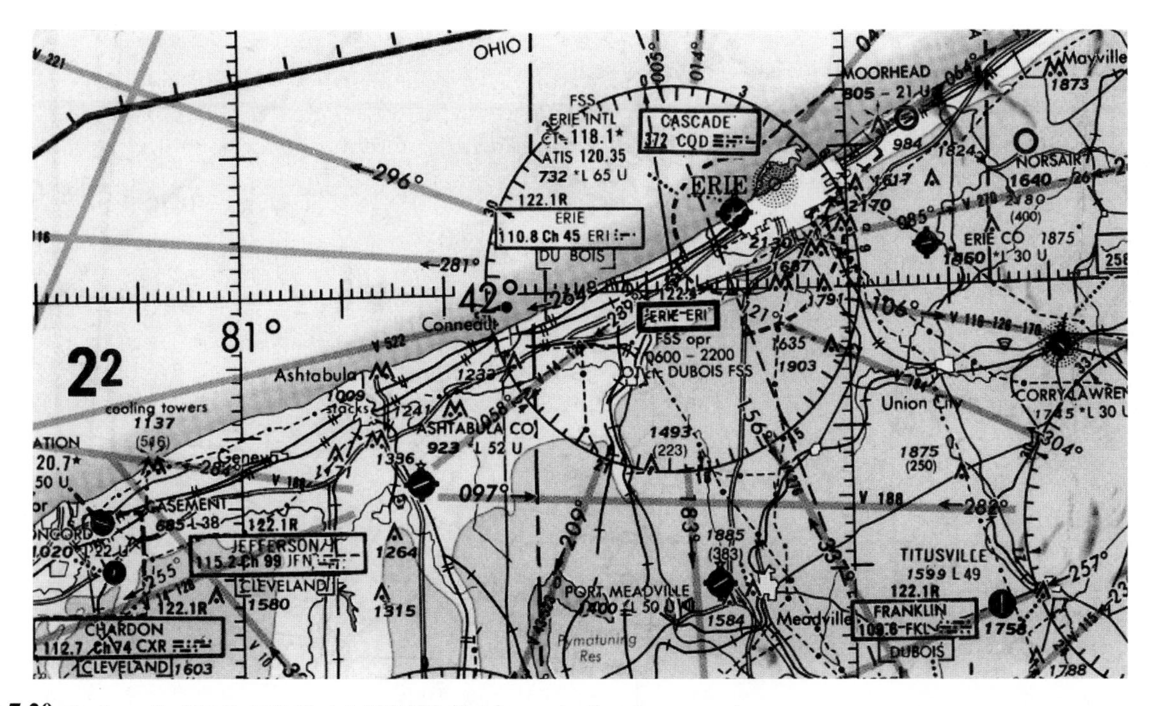

Figure 7.20 Portion of a WAC. (CF-18, 1:1,000,000. Not for navigational purposes.)

Just as marine navigators must pay particular attention to the depth of the water under the keels of their ships, aerial navigators must be aware of the altitude of their aircraft above the earth. The elevation of the terrain must be considered before safe operating altitudes can be determined. Flying 10,000 feet above sea level, for example, is not recommended when terrain features reach the vicinity of 10,000 feet or above. For this reason, the elevation of the highest terrain elevation within each latitude and longitude quadrangle on a sectional chart is printed in boldface type (Figure 7.23b). The large numbers indicate elevation above sea level in thousands of feet, and the small numbers represent hundreds of feet. Additional elevation information is

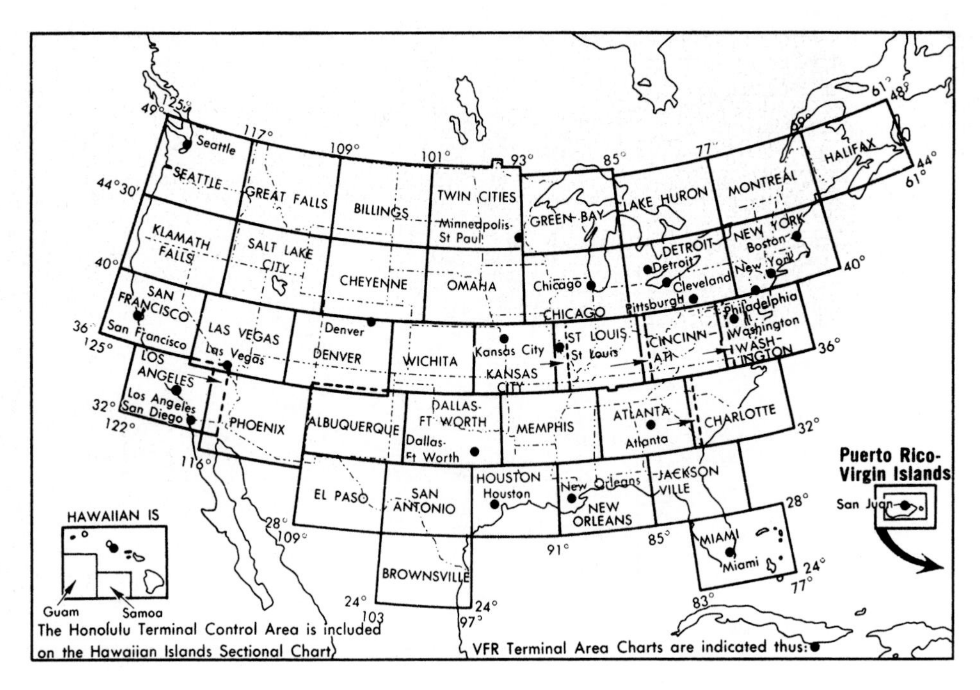

Figure 7.21 Sectional chart series index for the conterminous United States, Hawaiian Islands, Puerto Rico, and the Virgin Islands.
Source: National Ocean Service, *Aeronautical Charts and Related Publications.*

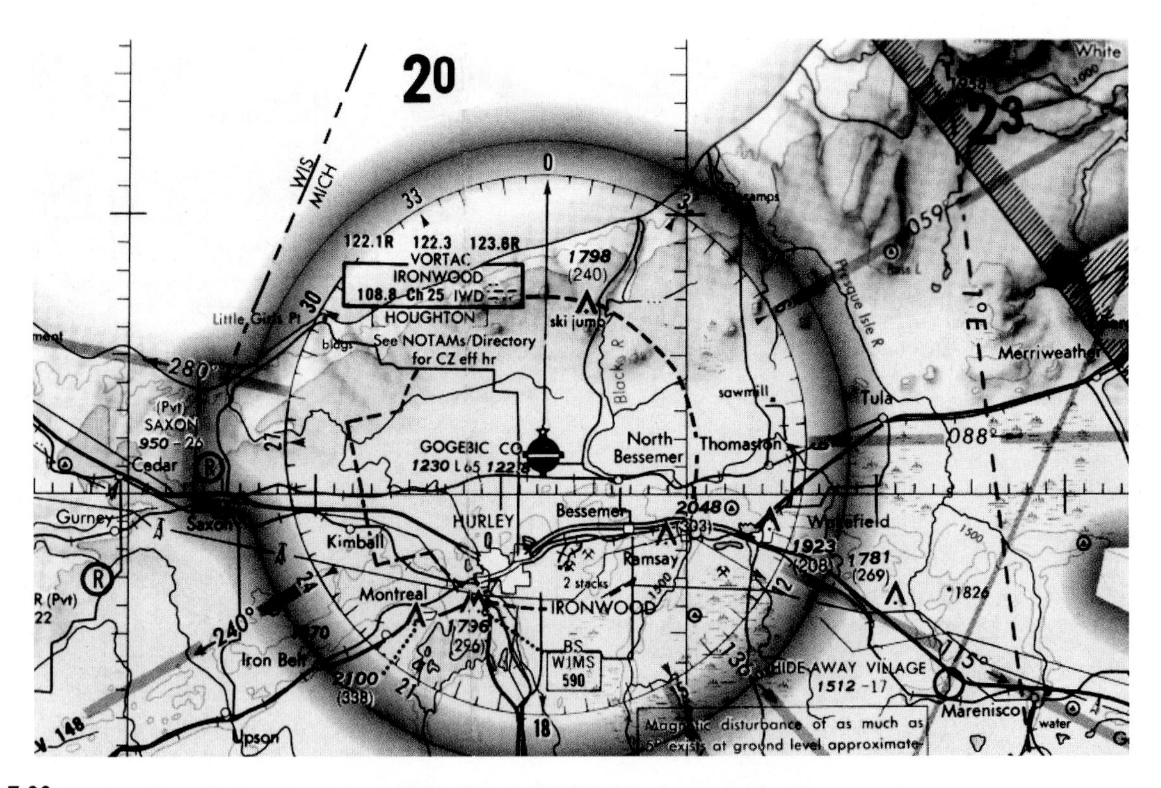

Figure 7.22 Portion of a typical sectional chart. (Green Bay, 1:500,000. Not for navigational purposes.)

provided in the form of contour lines and hypsometric tints.

Aircraft typically use **aneroid altimeters.** As altitude increases, barometric pressure decreases, and these altimeters convert the current pressure into a direct altitude reading. The altimeter must be calibrated so that it reflects current barometric conditions. This means that, before departure, for example, it is set so that it shows the elevation of the airport *above sea level.* Adjustments are made during a flight if conditions warrant.

The remaining symbols on sectional charts are printed in dark blue or magenta. This color scheme as-

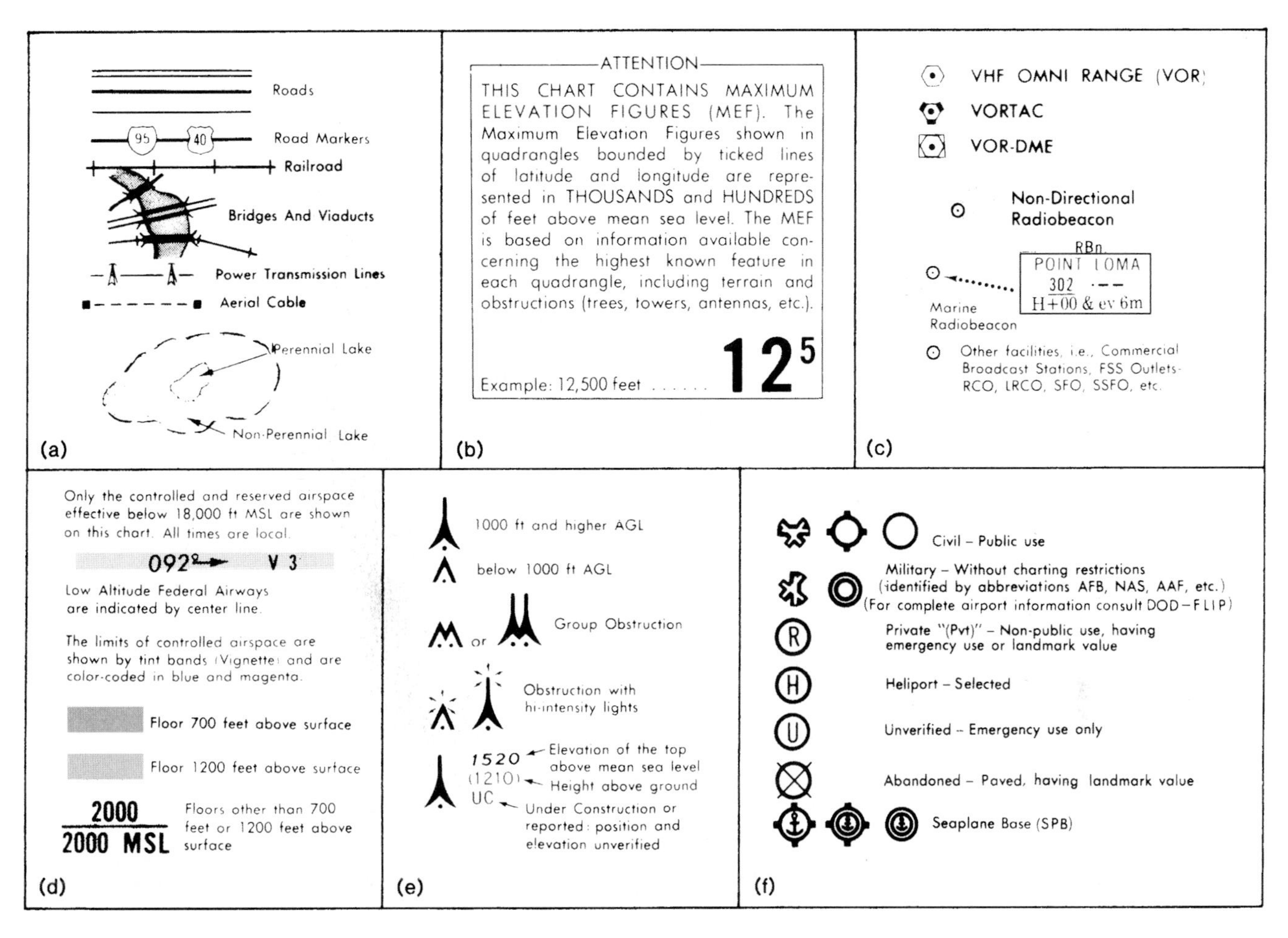

Figure 7.23 Examples of the symbols used on WACs and sectional charts. (*a*) Topography. (*b*) Elevation markings. (*c*) Radio aids to navigation and communication. (*d*) Airport traffic service and airspace information. (*e*) Major obstructions. (*f*) Information about airports.

sures maximum visibility, even under the subdued red lighting used in aircraft at night.

A variety of different types of radio aids to navigation and communication are provided on sectional charts (Figure 7.23c). Among these are various types of radio beacons that are of direct use in navigation. By homing in on such beacons, it is possible to follow designated routes between locations. Other radio transmissions provide traffic-control information and weather broadcasts.

Airport traffic service and airspace information includes designated airways with directional indications. Also shown on sectional charts are operational floors, which are designated minimum elevations above the surface, and a variety of control areas and restricted areas (Figure 7.23d).

In addition, sectional charts indicate the elevations of major obstructions such as tall buildings, smokestacks, and radio towers (Figure 7.23e). The elevations of such features are shown relative to sea level, and their height above the ground is also indicated. The latter information is particularly useful when a **radio altimeter** is available. This type of altimeter provides direct information about the **absolute altitude** of the aircraft—that is, its altitude above the earth's surface instead of above sea level.

Finally, sectional charts provide information about airports. This includes runway surfaces and lengths, services available, elevation, military or civilian status, hours of operation, lighting facilities, and other vital details (Figure 7.23f).

Flight Planning

Two distinct types of rules govern aircraft flights. The simplest type governs flights under favorable weather conditions, in which visual references to landmarks can be used to reliably control the operation of the plane. Such flights operate under **visual flight rules.** When some or all of a flight occurs under conditions that prevent reliable visual contact, instruments must be used to determine location, direction, and altitude. Under such conditions, **instrument flight rules** are observed. If you are interested in actual aircraft navigation, you must be thoroughly trained in the use of navigational aids and equipment, have a detailed knowledge of flight rules, and learn a great deal of technical information. The discussion here is designed simply to indicate the importance of map-use skills for aerial navigation purposes.

Because there is often very little time to work out navigational matters while airborne and because of the need for control of the use of limited airspace, air navigators usually work out a **flight plan** prior to departure. In addition, in many cases, the flight plan must be filed with the appropriate authorities.

Flight planning is based on dead-reckoning procedures. It includes determining the courses that will be flown, usually using both true and compass directions. It also notes landmarks that will be important checks on the correct route, as well as distance, estimated flying time, and altitude for each leg.

When a visual flight is under way, roads, railroads, power lines, shorelines, and other easily distinguished features are used to determine the course to be followed. In particular, the visual alignment of two objects, called a range, is often used to provide route alignment.

Different types of charts are used at various stages of flight planning and during the flight itself. In the planning phase, small-scale charts are used, because the major concern is the selection of the best general route to follow. In the absence of other factors, this is usually the great-circle route, especially for long-distance flights. When in flight, larger-scale charts are needed so that detailed information about landmarks, hazards, and aids to navigation can be easily seen. These charts should not be at too large a scale, however, as the speed of the plane carries it across the mapped area rapidly. On an approach to a terminal, a larger-scale chart may be needed to provide sufficient detail. Indeed, under some conditions, a large-scale, detailed landing chart may be used during the final stages of the landing procedure.

Other aeronautical chart series, in addition to the World Aeronautical Chart series, include Jet Navigation (JN) (1:2,000,000 scale) and Global Navigation (GN) (1:5,000,000 scale), which are designed to provide similar functions to those served by WAC. All of these are readily available (either for sale or in map libraries), provide virtually worldwide coverage, are carefully compiled, and are frequently updated and revised. Because of these characteristics, this set of map resources is also valuable to the general map user who is not involved directly with air navigation.

SUMMARY

Maps are used extensively for route selection and navigation. The basic considerations involved apply equally to land, water, and aerial navigation, although there are special aspects to each.

Direction refers to the path along which something is pointing or moving. The cardinal directions (north, south, east, and west) provide a rough guide to direction. However, the magnetic compass, which aligns itself with the earth's magnetic field, provides a fixed reference to which other directions are related. A compass rose is divided into 32 directions, with a directional name.

Direction is expressed in terms of angular measurements from a north-south baseline. The most common reference is to true (geographical) north, the direction to the geographical north pole. In contrast, magnetic north is the direction to the north magnetic pole. Geographical and magnetic north differ at most locations by the amount of magnetic declination. Isogonic charts show declination and, usually, the expected direction and amount of annual change in declination. Points of equal magnetic declination are shown by isogonic lines, and the line of zero declination is the agonic line.

Compass deviation is the difference between magnetic north and the direction that the compass needle actually points. If this is well-known and predictable, it is included on isogonic charts. However, compass readings taken near metallic objects are likely to suffer from unpredictable deviation.

The direction of the north-trending lines of a grid, such as Universal Transverse Mercator (UTM) or State Plane Coordinate (SPC) grids, is grid north. Grid north almost always differs from true or magnetic north. In addition, the grid north of two or more grids is seldom the same.

A declination diagram shows the relationships between the three "norths." A form of declination diagram called a *conversion diagram* is used to assist in the change of direction from one reference to another.

Directions are designated by either azimuths or bearings based on true, magnetic, or grid north. An azimuth is usually measured clockwise in degrees, minutes, and seconds from north. An orthodrome (true azimuth) is the direction of a great circle and, in most cases, crosses successive meridians at a constantly changing angle. A constant azimuth (rhumb line), in contrast, crosses each succeeding meridian at the same angle. A back-azimuth is the exact reverse, by 180 degrees, of an azimuth. A bearing is measured from either a north or a south baseline, whichever is nearer to the direction being designated. Bearings have a maximum value of 90 degrees.

Orientation involves aligning a map with the terrain by inspection or by compass. Position determination begins with a properly oriented map. Resection is then used to locate an unknown position by marking three known ground positions on the map. The magnetic azimuths to these positions are determined, and back-azimuths are drawn from the known positions toward the unknown position. Alternatively, visual sightings to the known positions are drawn on the map. The crossing of the back-azimuths, or sightings, marks the unknown position. If back-azimuths do not meet exactly, a triangle of error is formed. The center of this triangle is an estimate of the correct location.

Route selection involves the identification of reference points that can be used to monitor location. The route is followed by using clues provided by the map, by using a compass, or by a combination of both.

Nautical charts are designed to serve the needs of waterborne navigators. Because hazards and aids to navigation change over time, the latest edition of a chart, posted with changes listed in the *Notice to Mariners,* should always be used.

Water depths are traditionally shown in fathoms, although feet and meters are also used. The units used on a particular chart, as well as the chart's datum, must be determined to prevent confusion.

The determination of a vessel's position is critical to navigation. Navigators determine position by observation of the sun or stars or, if they are operating within sight of land, they may obtain a fix by resection, using lines of position. Radar, GPS, radio direction finding, and loran are also used to determine location.

Dead reckoning is used to plan and track the movements of a vessel. The method depends on knowing the starting point and the direction, speed, and time of travel along the various legs of the voyage. The effect of winds, currents, and tides on the motion of the vessel may be used to adjust a dead-reckoning position to obtain an estimated location.

Aeronautical charts are specially designed to meet the needs of aviators. Because aids to aeronautical navigation are subject to frequent change, the most recent chart and information from the latest *Notice to Airmen* must be used.

Generally, aeronautical charts show cities and other distinctive landmarks and terrain features. Aeronautical information, including the location of radio navigation aids, airports, runways, restricted areas, and obstructions, is overprinted in dark blue or magenta. Elevation information is emphasized, and the highest elevation within each latitude and longitude quadrangle is printed in boldface type.

Two distinct sets of rules govern aircraft flights. Visual flight rules govern flights under favorable weather conditions. When some or all of a flight occurs under conditions that prevent reliable visual contact, instruments are used to determine location, direction, and altitude, and instrument flight rules are observed.

SUGGESTED READINGS

American Practical Navigator: An Epitome of Navigation. Washington, D.C.: Defense Mapping Agency Hydrographic/Topographic Center, Pub. No. 9, 1984. [Originally by Nathaniel Bowditch, LL.D., vol. 1.]

Andresen, Steve. *The Orienteering Book.* Mountain View, Calif.: Anderson World, 1977.

Department of the Army. *Map Reading.* Department of the Army Field Manual FM 21–26. Washington, D.C.: Department of the Army, Headquarters (current edition).

Gardner, A. C. *A Short Course in Navigation.* New York: Funk and Wagnalls, 1968.

Haug, Moir D., Francis H. Moffitt, and James M. Anderson. "A Simplified Explanation of Doppler Positioning." *Surveying and Mapping* 40, no. 1 (March 1980): 29–45.

Hurn, Jeff. *Differential GPS Explained.* Sunnyvale, Calif.: Trimble Navigation, 1993.

———. *GPS: A Guide to the Next Utility.* Sunnyvale, Calif.: Trimble Navigation, 1989.

Kjellstrom, Bjorn. *Be Expert with Map and Compass.* New York: Charles Scribner's Sons, 1976.

Maloney, Elbert S. *Dutton's Navigation and Piloting.* 14th ed. Annapolis, Md.: Naval Institute Press, 1985.

Monmonier, Mark S. *Technological Transition in Cartography.* Madison, Wis.: University of Wisconsin Press, 1985, 15–45.

Muehrcke, Phillip C. *Map Use: Readings, Analysis, and Interpretation.* 3d ed. Madison, Wis.: JP Publications, 1992, 273–88, 291, 584–90.

Points and Positions. Torrance, Calif.: Marine and Survey Systems Division of Magnavox Advanced Products and Systems Company (latest issue).

Snufeldt, H. H., and G. D. Dunlap. *Navigation and Piloting.* Annapolis, Md.: Naval Institute Press, 1970.

Stewart, John Q., and Newton L. Pierce. *Marine and Air Navigation.* Boston: Ginn and Company, 1944.

CHAPTER

8 Terrain Representation

Because the earth's surface is seldom absolutely flat, mapmakers have always been challenged to convey information about its undulations to map users. And because maps are drawn or printed on flat sheets of paper, they are not particularly well-suited to representing a three-dimensional surface. Three-dimensional physical models and a number of other techniques provide partial answers to the problem. Each technique has difficulties of its own, however, and does not provide the total answer. The purpose of this chapter is to describe the advantages and disadvantages of some of the terrain-representation techniques that have been used with varying success over the years.

Symbols

Spot Heights

The simplest way to represent relief is to use numbers called **spot heights.** A spot height on a map indicates the elevation of the earth's surface at that point relative to some datum (see Chapter 2). Sometimes, the number appears without a symbol, but often, a point symbol of some type indicates the location to which the elevation value refers (Figure 8.1a).

Spot heights are typically located at the crest of hills, road intersections, railroad grade crossings, or some similar, easily identifiable points. A **bench mark,** placed by geodetic surveyors, is an ideal location for a spot height because it can be precisely located in the field and then used as a reference point for additional survey work (Figure 8.1b). On a map, a special symbol is usually used to distinguish a bench mark from other types of spot heights.

Land Surfaces

The advantage of spot heights as a means of representing land surfaces is that they provide an exact value at an identifiable location. Their main disadvantage is that they do not provide useful information about the shape of the terrain or about the elevations *between* the spot heights. Nor do they provide any visual three-dimensional effect. For these reasons, spot heights are often combined with one or more other methods of terrain representation.

Water Surfaces

Water surfaces provide readily identifiable reference levels. The elevation of the water surface above a datum is, therefore, often indicated on maps. Usually, the elevation value is simply printed on the water surface, without any additional symbols. On USGS quadrangle maps, for example, blue italic numerals indicate water-surface elevations. Different symbols may be encountered on maps produced by other agencies.

Contours

A **contour** is an imaginary line that joins points of equal elevation above or below some datum (Figure 8.2). The contours drawn on topographic maps are extremely important because they provide the basis for

Horizontal control:	
Third order or better, permanent mark	Neace △
With third order or better elevation	BM △ 148
Checked spot elevation	△64
Coincident with section corner	△ Cactus
Unmonumented	Not Shown
Vertical control:	
Third order or better, with tablet	BM × 53
Third order or better, recoverable mark	× 394
Bench mark at found section corner	BM + 61
Spot elevation	× 17

(a)

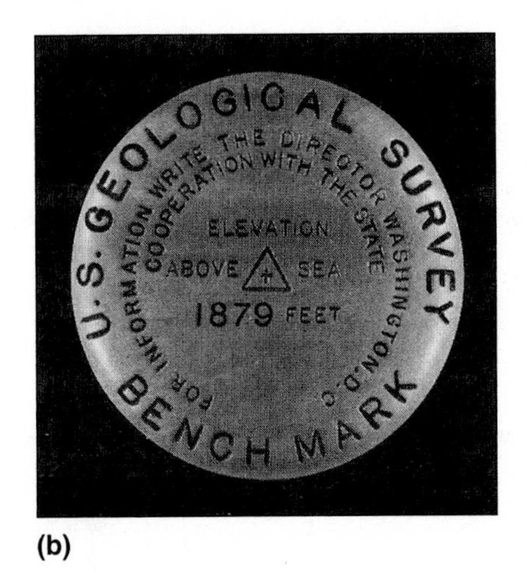

(b)

Figure 8.1 (*a*) Examples of spot height symbols used on U.S. Geological Survey topographic maps. (*b*) Bench mark monument.
Source: (*b*) Morris M. Thompson, *Maps for America,* 3d ed. (Washington, D.C.: U.S. Department of the Interior, Geological Survey, 1987).

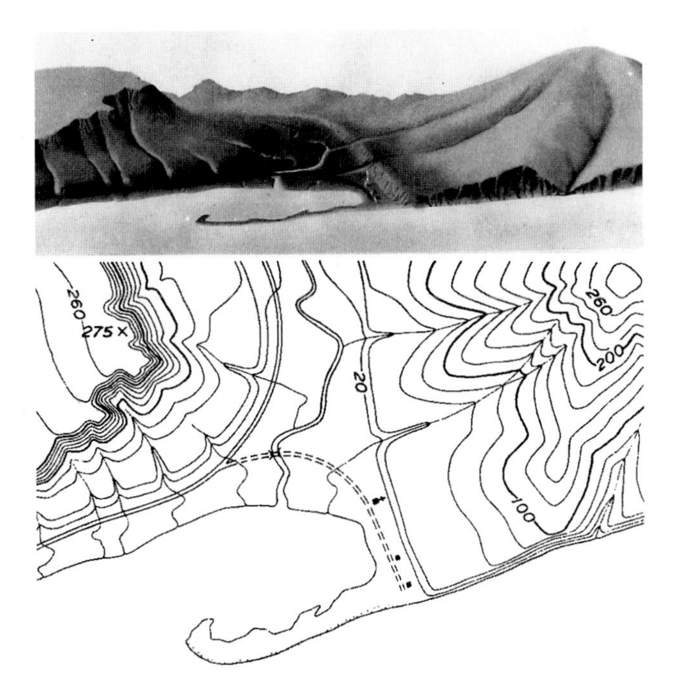

Figure 8.2 Relationship between surface and contours.
Source: U.S. Geological Survey.

the development of a great deal of helpful information about the terrain.[1]

Contour Interval

An infinite number of contours could theoretically be drawn on any given map. Contours could be drawn at sea level, at 1 meter above sea level, at 2 meters above sea level, and so on, for every meter of difference in elevation. In the extreme, a contour could conceivably be drawn to represent every centimeter of elevation, or a similar, very fine interval. Closely spaced intervals are often used for detailed engineering or landscape maps. On the other hand, contours may be drawn at only 10 meters, 50 meters, or even more widely spaced intervals on smaller-scale maps.

The basic **contour interval** is the vertical distance between contours. It is selected by the mapmaker to clearly display for the user the predominant terrain features in the mapped area. If the terrain is relatively flat, a close interval is used so that the small vertical differences can be seen (Figure 8.3a). If the terrain is more mountainous, on the other hand, a wider interval is used so that map details are not overwhelmed by detailed contour information (Figure 8.3b).

The selection of the basic contour interval also takes into account the scale of the particular map. A closer contour interval is generally selected for large-scale, detailed maps so that the density of contour lines is sufficient to provide a good representation of the shape of the terrain. A wider interval is used for small-scale, generalized maps, to avoid overcrowding.

Two basic contour intervals can be used on one map. This is usually avoided, but it may be necessary when an area contains abrupt contrasts in relief (see Figure 9.2). In such a case, a closer interval is used in the areas that have relatively flat terrain and a wider interval is used in the steep areas.

Contour Types

In an area of relatively complex terrain, mapmakers commonly accentuate every fourth or fifth contour by drawing it in a wider line weight. The paths of these **index contours** are then easier to identify and follow. Index contours usually have numbers inserted at intervals along their length to indicate the elevations that

[1]The contour definitions used here are based on Morris M. Thompson, *Maps for America,* 3d ed. (Washington, D.C.: U.S. Department of the Interior, U.S. Geological Survey, 1987), 26–28.

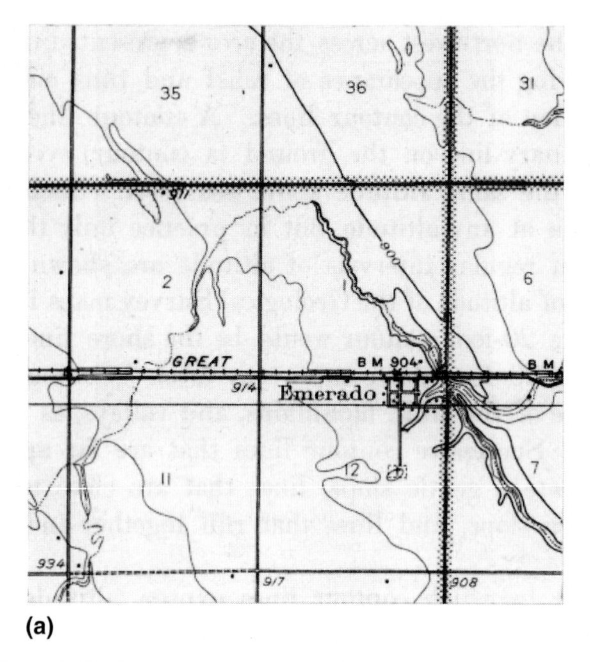

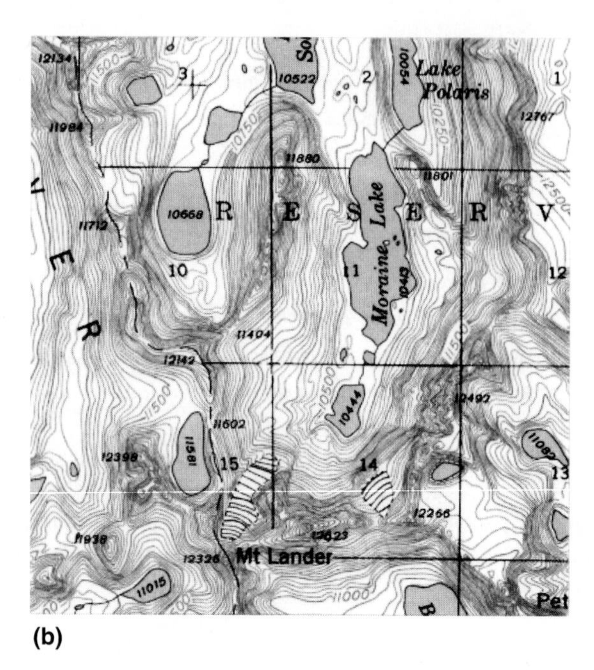

Figure 8.3 Contour intervals vary, depending on the characteristics of the terrain. (*a*) If the terrain is relatively flat, a close interval is used so that the small vertical differences can be seen. (Ten-foot contour interval, Emerado quadrangle, North Dakota, 1:62,500, 1936. Reprinted 1943.) (*b*) If the terrain is steep, a wider interval is used so that map details are not overwhelmed by the contours. (Fifty-foot contour interval, Mount Bonneville quadrangle, Wyoming, 1:62,500, 1938.)

they represent (Figure 8.4a). Index contours are usually selected so that convenient numbering sequences result, such as 5, 10, 15 or 50, 100, 150, and so on.

Regular contours, spaced at the normal interval and drawn with a finer line weight, lie between the index contours. These **intermediate contours** may or may not be numbered, depending in part on the space available (Figure 8.4b).

Supplementary contours are additional contours usually drawn at intervals that are some regular fraction (often half) of the basic contour interval (Figure 8.4c). Supplementary contours are appropriate in areas of flat terrain, where contours drawn at the basic interval would be spaced relatively far apart. They are usually drawn as dashed or dotted lines to distinguish them from contours drawn at the basic interval.

When using a contour map, examine the map legend for information regarding the contour interval. At the same time, check for the possibility of two intervals or of supplementary contours. This precaution reduces the possibility of misinterpreting the contours.

When an area lies at a lower elevation than all of the surrounding terrain, it forms a depression. A depression that fills with water forms a lake. The land surface of a depression not normally filled with water, however, is mapped with contours on its surface. These **depression contours** are distinguished from regular contours by short ticks, at right angles to the contour line (Figure 8.4d). The ticks point downslope, toward the bottom of the depression.

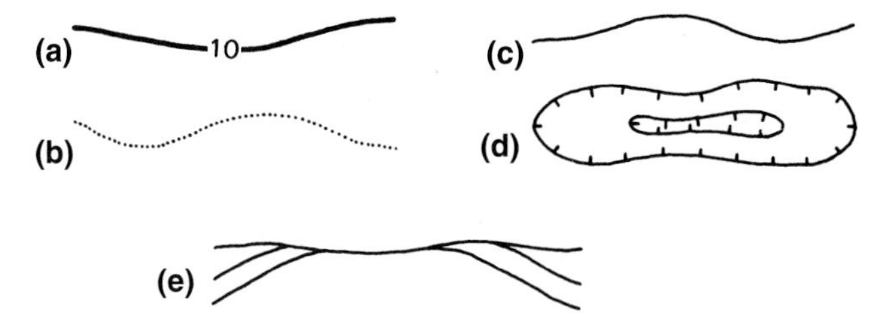

Figure 8.4 Types of contours. (*a*) Index. (*b*) Intermediate. (*c*) Supplementary. (*d*) Depression. (*e*) Carrying.

It is not always possible to draw accurate contour lines because of the inaccessibility of an area, the presence of a particularly heavy forest cover that obscures the view of the terrain in aerial photographs, or some similar problem. In these situations, contour lines may be drawn to represent the likely surface of the terrain. Such **approximate contours** are usually dashed or dotted to indicate their uncertain nature.

If a series of contours falls extremely close together, as may happen in a particularly steep area, they may not all be shown as individual lines. A single contour drawn to represent the several contours that would be drawn if space allowed is called a **carrying contour** (Figure 8.4e). This type of representation does not occur frequently because, if the basic contour interval causes extensive crowding on a given map, a wider interval is usually substituted.

When a dam is constructed, a portion of the land surface behind it becomes inundated. The contours that represent the land surface that was exposed before the flooding may be retained, however. Such contours are called **underwater contours.** Underwater contours are *not* the same as depth curves, which show the depth of the water (and are discussed in the next paragraph). Instead, underwater contours indicate elevations above the same datum as the land-surface contours on the map.

When contours are drawn to represent the bottom configuration of a water body, they are called **depth curves,** or **isobaths.** Depth curves are measured *down* from the specific water surface and, therefore, are not directly related to the overall map datum. However, depth curves in ocean areas are frequently the exception because they are normally related to the mean low-tide level.

Layer Tints

Layer tints, sometimes called *hypsometric tints,* are often used to fill in between contour lines or isobaths to give a stronger visual indication of general elevation or water depth. Suppose, for example, that there are three contour lines on a particular map, at 100, 200, and 300 meters, in addition to the coastline. Layer tints added to this map would probably consist of four different tones or colors: one between sea level and 100 meters, another between 100 and 200 meters, a third between 200 and 300 meters, and a fourth above 300 meters.

Layer tints are also commonly used in conjunction with **form lines.** Form lines are generalized contours that are frequently used on small-scale, less detailed maps. When layer tints are added to form lines, the area of the map that lies within each general elevation level is overprinted with a specific tone or color.

Sometimes, layer tints are shades of gray or of a single color, such as brown. When tones of a single color are used, the light tones are usually assigned to the lower elevation ranges and the darker tones are used at the higher elevations. When the map is printed in multiple colors, however, a color sequence may indicate the elevation zones (Plate 2). One commonly used sequence begins with green in the lower elevations and proceeds through buffs and yellows to oranges, reds, purples, and, finally, white, as higher elevations are reached. The idea behind this gradation is that the warmer colors (oranges, reds) "advance" toward the eye, while the cooler colors (greens, buffs) "retreat." Presumably, then, the color sequence tends to produce a visual three-dimensional effect. Unfortunately, little evidence supports this view, and the color-sequence effect is probably of little or no help. In fact, because the colors have different *values* (that is, some are dark and some are light), the eye may even be confused by the mixture of effects due to color differences and those due to value differences.

Another problem with layer tints is that users may fall into the trap of thinking that the colors represent vegetation or climatic types. For example, map users who are unfamiliar with the layer-tint color sequence may assume that the green colors represent well-watered, fertile areas. That the Sahara Desert is usually green on this type of map illustrates the fallacy of this view.

Despite its shortcomings, this type of color sequencing is used extensively. Some mapmakers use a very effective color sequencing that *does* represent the vegetation types of different areas, which, again, points out the importance of consulting each map's legend to know the meaning of the symbols and colors being used.

Because the information provided by layer tints is usually very generalized, the tints are often used in conjunction with other techniques. Most commonly, this means that contours, spot heights, or both may be encountered on layer-tinted maps.

Illuminated Contours

Illuminated contours bring a more effective visual impression of relief to what is, otherwise, a standard contour map.[2] In this method, the overall map is covered with an intermediate value of gray or of some single color, such as blue or brown. An imaginary source of light is placed at the upper-left corner of the map, and the contours on the side of the landform facing the light source are shown as white lines. The contours on the side away from the light source, on the other hand, are shown as dark lines, as though they are in shadow. The result is a strong impression of the region's surface relief (Figure 8.5).

An advantage of the illuminated-contour method, in addition to the visual effect of relief, is that the contours can be read in the usual way. It is still possible to measure and estimate elevation values using conventional techniques. Unfortunately, the method gives the unrealistic impression that the terrain surface is made up of layers instead of continuous slopes.

[2]Kitirô Tanaka, "The Relief Contour Method of Representing Topography on Maps," *The Geographical Review* 40, no. 3 (July 1950): 444–56.

Figure 8.5 Illuminated contours.

Source: From Kitirô Tanaka. "The Relief Contour Method of Representing Topography on Maps," *The Geographical Review* 40, no. 3 (July 1950): 444–56. Reprinted by permission of The American Geographical Society.

HILL SHADING

One visually effective means of indicating the shape of the terrain is to use **hill shading.**[3] Hill shading is the application of gray values to the map so that the modeling of the surface is apparent.

The idea of hill shading can be visualized by imagining that a light has been placed so that it shines on a three-dimensional terrain model (see Figure 8.6). The light is angled so that it strikes obliquely, from the upper-left corner of the model, and shadows are cast toward the lower right. If a vertical photograph is taken of this illuminated model, the result is similar to a hill-shaded map. In reality, construction of this type of illuminated model requires considerable time and expense, and the model is difficult to properly light. Because hill shading is less expensive and easier to control, it is usually preferred to the illuminated model. Hill shading is traditionally rendered by means of an airbrush (Plate 3). More recently, computer-assisted techniques have been used (Figure 8.7).

One difference between an illuminated model and a hill-shaded map is that the shadows cast on the

[3]This method is also known by other names, including *plastic shading, relief shading,* and *terrain shading.*

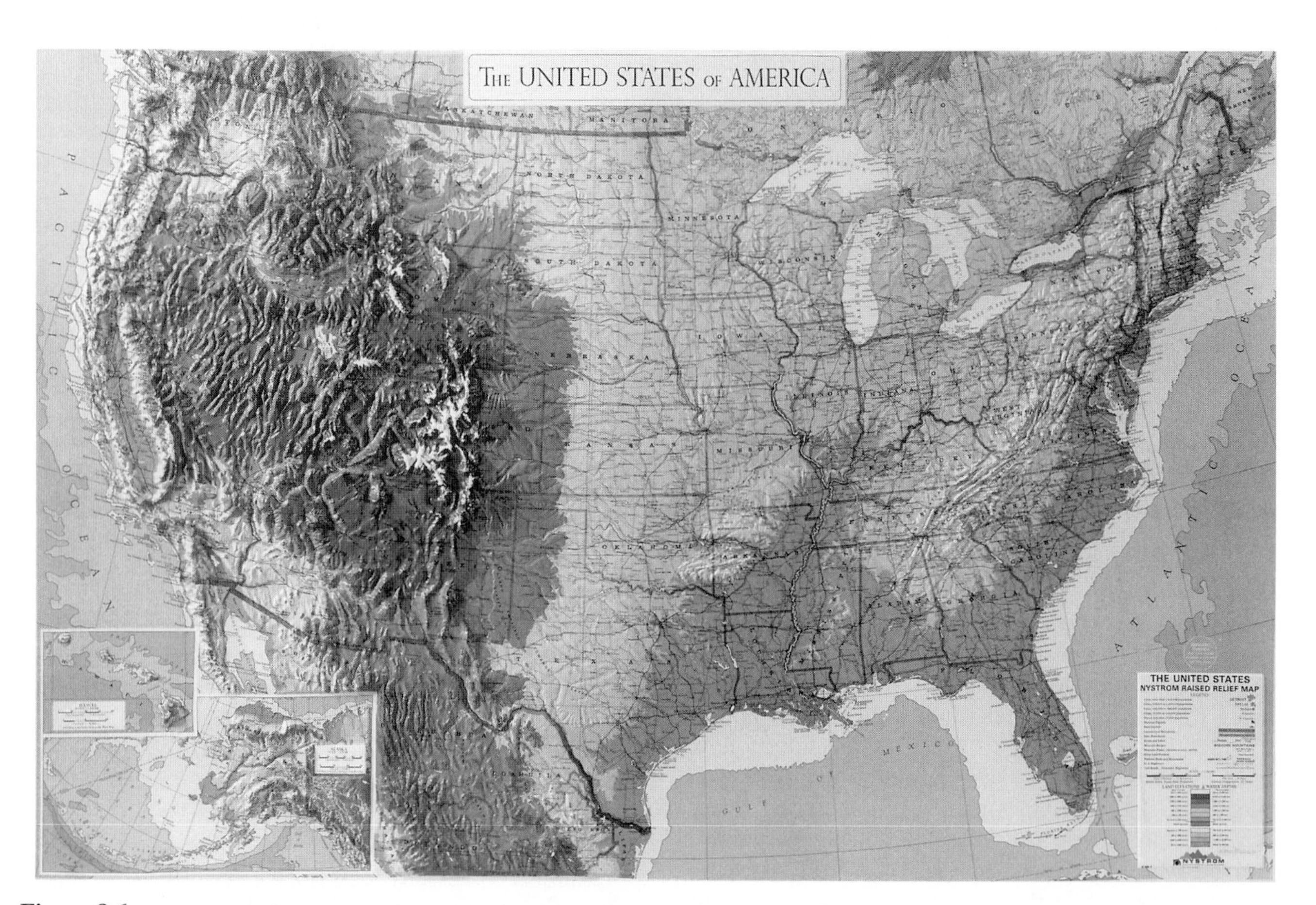

Figure 8.6 Terrain model in the form of a vacuum-formed relief map.

Source: Courtesy of Nystrom, Division of Herff Jones Inc.

model always fall directly in the direction of the light. This means that features that trend in the same direction as the light falls often do not cast sufficient shadow to show their actual relief. With hill shading, on the other hand, the gray values are determined on the basis of angle of illumination and the amount of light striking the surface, assuming that nothing else is shadowing it. Also, in some cases, the light cast on a model can fall across the floor of a valley and cover a portion of the slope on the opposite side. This is especially true in the case of rather steep-walled, deep valleys. This results in some visual confusion, because the map user tends to lose track of the valley floor. When hill shading is used, the shadows are stopped at the bottom of the valley, thus eliminating the confusion. It may seem paradoxical, but the "artificial" method (hill shading) can result in a more effective map than the "natural" method (the illuminated model).

One shortcoming of hill shading is that the successful creation of a strong three-dimensional effect may produce very dark tonal values in some areas. Such dark tones reduce the visibility of other symbols printed over them, symbols that may be equally important to the map user.

Hill shading also has the disadvantage of not providing the accurate determination of elevations. The technique of the artist who does the shading and the characteristics of the reproduction process used greatly affect the map's final appearance. If the shading is relatively dark, for example, the map user may interpret the terrain slope as being steeper than it actually is, whereas relatively light shading is likely to have the opposite effect. The interpretation of the shading, in other words, is largely subjective. For this reason, hill shading is often used in combination with spot heights, contours, or both so that accurate elevations can be determined.

PHYSICAL MODELS

A **physical model** is the most straightforward means of showing the three-dimensional aspect of the earth's surface (Figure 8.6). In a physical model, the third dimension does not have to be simulated, as it does on

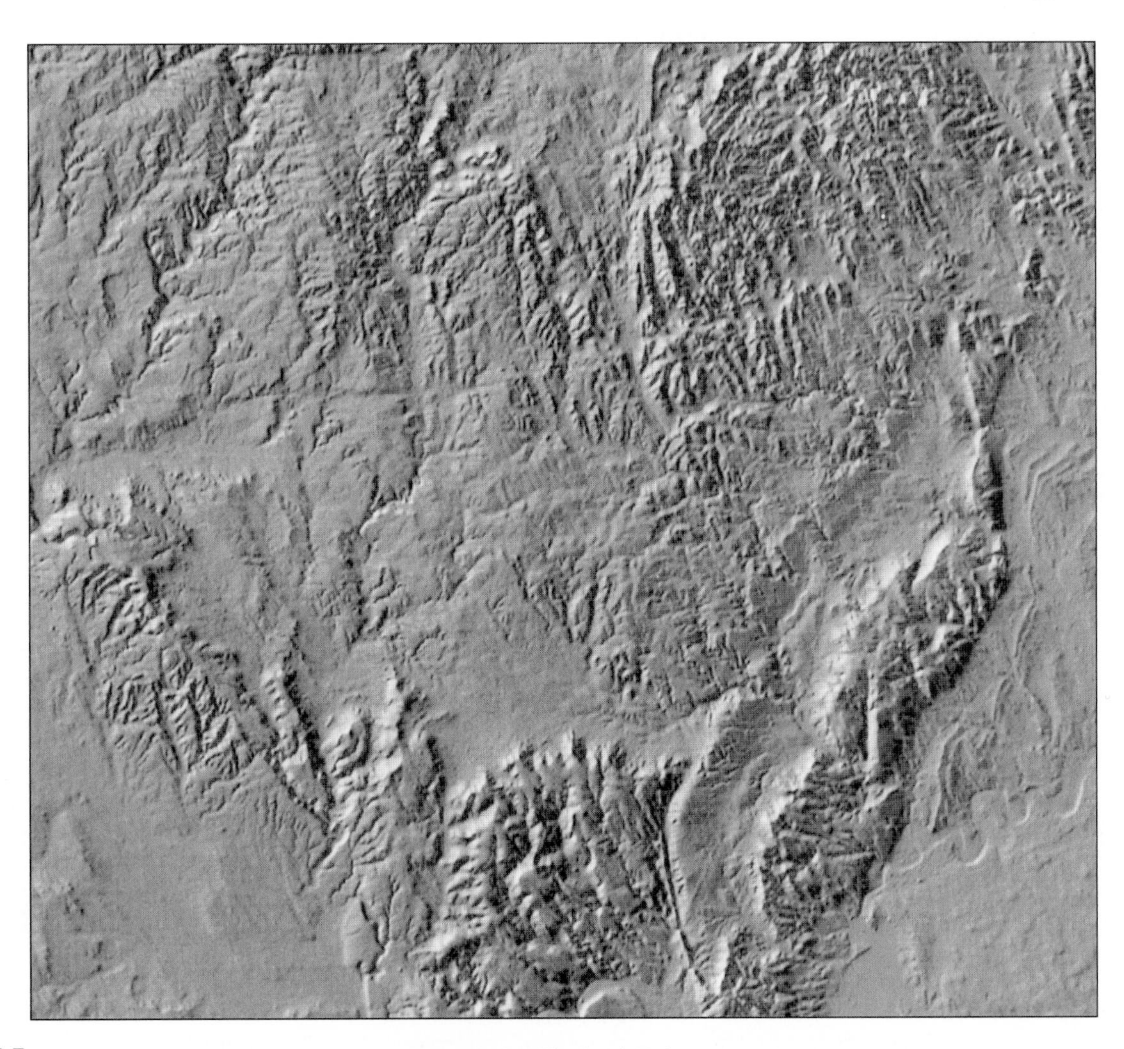

Figure 8.7 Computer-assisted hill shading of part of the Black Hills, South Dakota.
Source: Courtesy of EROS Data Center.

maps, because it actually exists, and the hills and valleys of the terrain are directly seen and understood.

Despite their visual appeal, terrain models are not used extensively. One reason is that they are expensive to build. Another reason involves handling and storage. Many models are built of layer upon layer of heavy materials, although lighter materials are often used today. Even when lighter-weight materials are used in their construction, however, the thickness and inflexibility of models make them extremely difficult to handle and store. Some models are made of vacuum-formed plastic sheets on which the appropriate map has been printed. These plastic-relief models are lighter and easier to handle but must be stored individually because stacking damages them.

Another problem with physical models is that, while they are created on the basis of quantitative measures, they are somewhat difficult to interpret. The main reason for this is that, even in areas of what would be considered severe relief, the elevations of the earth's surface are relatively slight in comparison to the immense horizontal distances usually involved. **Vertical exaggeration** is usually introduced into models to overcome this difficulty.[4] Vertical exaggeration involves using a different scale to measure vertical distances than is used, on the same model, to measure horizontal distances. The result is that the vertical relief of the area *appears* to be more realistic, even though it is not to the same scale.

Finally, the exact surface elevation at any specific point on a terrain model is difficult to determine, although the relative elevations of different areas may be quite clear. Spot heights and contours may be added to the model, but this is relatively rare.

[4]See also the section on profiles in chapter 9.

PHYSIOGRAPHIC DIAGRAMS

Physiographic diagrams are pictorial representations of the earth's surface. In general, they consist of an oblique view of a portion of the surface from a position high above the ground (Figures 8.8a and 8.8b).

There are strong elements of artistic license in the design of physiographic diagrams. In particular, the base map on which the diagram is drawn is a conventional, vertical view, but the symbols that represent the terrain features are drawn from an oblique point of view. This means that there is a problem with planimetric displacement. If the bottom of a mountain is placed at its correct location, for example, its top is

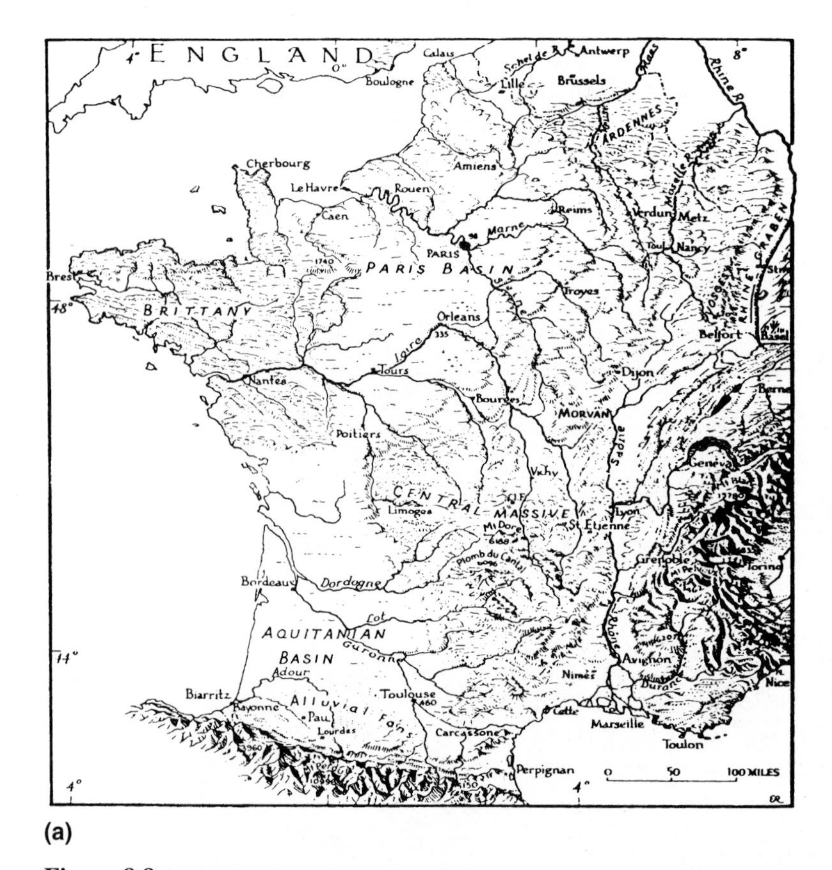

(a)

(b)

Figure 8.8 Physiographic diagrams. (*a*) France. (*b*) The coastal plain of England and France.

Source: (*a*) From Erwin J. Raisz, "The Physiographic Method of Representing Scenery of Maps," *The Geographical Review* 21, no. 2 (April 1931). Reprinted by courtesy of The American Geographical Society. (*b*) From Armin K. Lobeck, *Geomorphology, An Introduction to the Study of Landscape* (New York: McGraw-Hill, 1939).

necessarily misplaced. In addition, the oblique viewpoint may cause features that lie behind the higher elevations to be hidden from view. Despite these problems, the general effect of a well-drawn physiographic diagram is quite realistic.

The second element of artistic license in physiographic diagrams is that the features are drawn in a generalized, somewhat idealized, manner. In many cases, the landscape features are classified into groups, and standardized symbols are used to represent each type. Thus, each and every feature of the real-world landscape is not realistically represented in a physiographic diagram, although major features are sometimes explicitly shown. In general, a physiographic diagram gives an overall idea of the extent and types of terrain found in a region.

BLOCK DIAGRAMS

A **block diagram** is a pictorial representation of a portion of the earth's surface, drawn as though it were a picture taken at an oblique angle from the window of a high-flying aircraft (Figures 8.9 and 8.10). The oblique angle of view contributes to the three-dimensional impression that is the hallmark of such diagrams. The terrain is drawn as though it were a piece of the surface, sliced from the earth in a block, and the sides of the block are usually shown in the drawing. In addition to the surface features, block diagrams often include geological profiles drawn on the sides of the block. These profiles provide information regarding the underlying geological structure of the area.

Although block diagrams are visually attractive and provide an effective three-dimensional appearance, they have disadvantages similar to those of physiographic diagrams. First, planimetric displacement occurs. The top of a hill, for example, appears to be shifted away from its map location toward the back of the diagram. Second, the oblique viewing angle causes features with high elevations to block the view of lower features located behind them.

In addition, block diagrams are not designed for the measurement of distances or elevations. This is because of the isometric or perspective construction methods used, as well as the vertical exaggeration usually introduced.

For ease of drawing, many block diagrams use an isometric construction method (Figure 8.11a). In an isometric drawing, the corner of the drawing faces the front, and the two visible edges are aligned at 30-degree angles with the horizontal. The result is that the block appears to tilt up at the back. Because of this visual shortcoming of the isometric method, many block diagrams are drawn with a more realistic perspective view (Figure 8.11b). In a perspective view, the edges taper toward the back. If one-point perspective is used, as in the example shown in Figure 8.11b, one edge is parallel to the viewer, and the other tapers toward the vanishing point. If two-point perspective is used, both sides taper, each toward its own vanishing point. The use of perspective removes the appearance of tilt that occurs in an isometric view, but more-distant points are relatively smaller than those closer to the viewer.

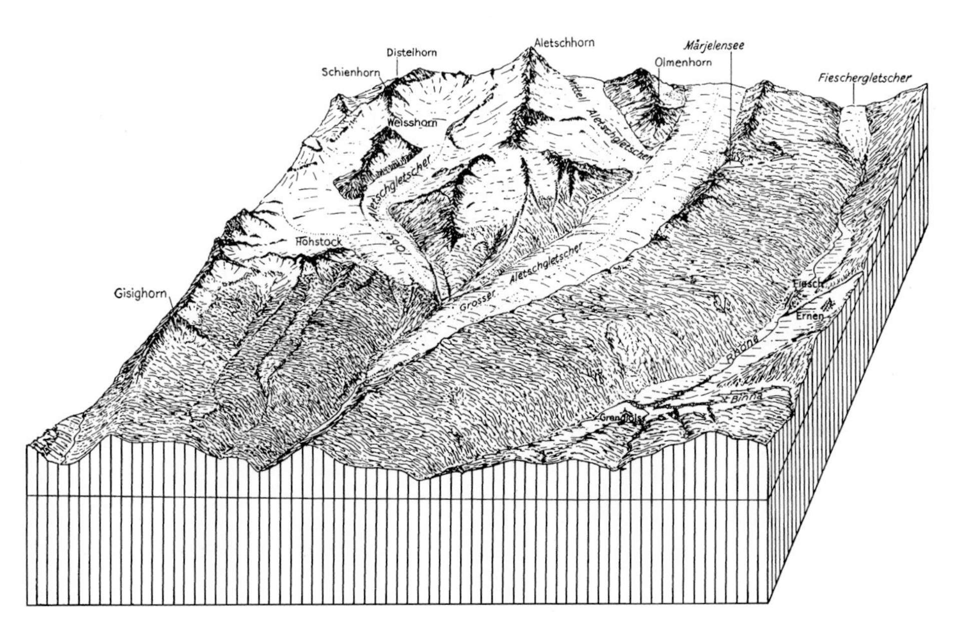

Figure 8.9 Block diagram of a portion of the Swiss Alps.
Source: From Armin K. Lobeck, *Geomorphology, An Introduction to the Study of Landscapes* (New York: McGraw-Hill, 1939).

To effectively portray terrain, virtually every block diagram incorporates vertical exaggeration. The amount of vertical exaggeration introduced depends on the local relief and the scale of the diagram.

The primary purpose of block diagrams, then, is to illustrate the relationships between features and *not* to provide measurements, which are better obtained from traditional maps.

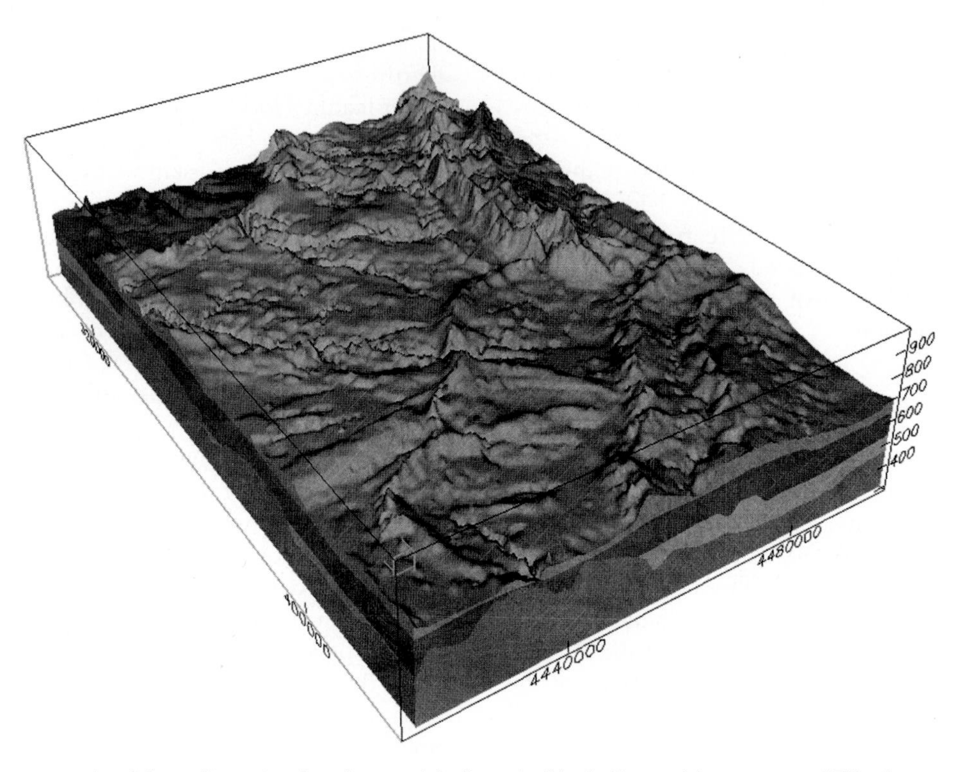

Figure 8.10 Computer-assisted three-dimensional surface model of terrain (block diagram) in east-central Illinois.

Source: U.S. Geological Survey, Reston, Virginia, by David R. Soller, Susan D. Price (U.S. Geological Survey), and Richard C. Berg (Illinois State Geological Survey). Courtesy David R. Soller.

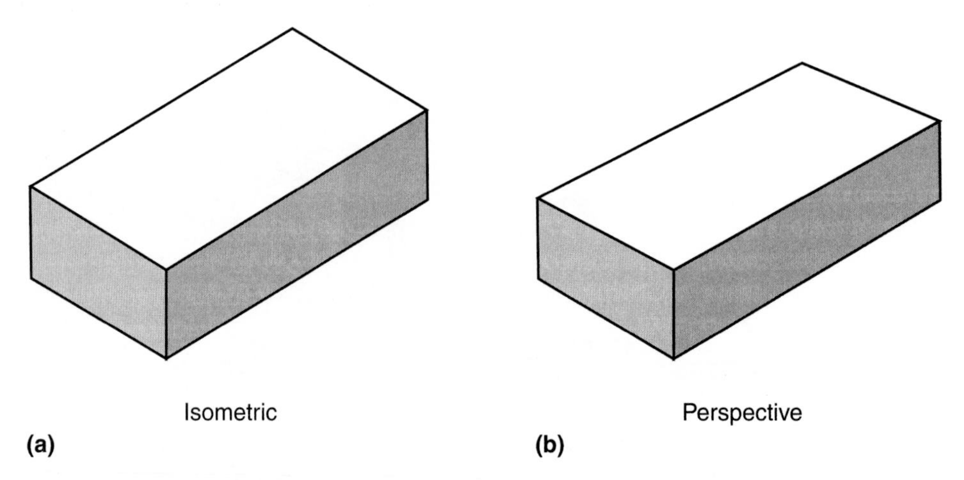

Figure 8.11 Comparison between isometric and perspective blocks. (*a*) Isometric view. (*b*) Perspective view.

SUMMARY

Representing the earth's three-dimensional surface on flat sheets of paper is a challenge to mapmakers. Various techniques are used in an attempt to solve this problem.

The simplest method of relief representation is the use of spot heights that indicate the elevation of points on the earth's surface. This has the advantage that exact values are provided at identifiable locations. It does not, however, provide useful information about the shape of the terrain between the spot heights, nor does it provide any visual, three-dimensional effect.

Another method of relief representation is to use contours, which are imaginary lines that join points of equal elevation. The paths of the contours provide elevation information and also a limited visual effect of relief. To improve the visual effect, contours are sometimes enhanced. One way of doing this is to add layer tints over zones of the same general elevation. Another method is to use illuminated contours. This involves making the contours on the northwest side of the landform white, as though they are directly illuminated. The contours on the side away from the light source, on the other hand, are shown as dark lines, as though they are in shadow.

The most visually effective method of relief representation is hill shading. Hill shading is simply the application of gray values to the map to simulate shadows that make the modeling of the surface apparent. This method is often combined with contours and spot heights so that measurements are available, in addition to the visual effect.

Physical models are the most straightforward means of showing the earth's surface. In a model, the third dimension does not have to be simulated, as it does on maps, because it actually exists. Models are expensive and awkward to handle and store, however. Also, even in areas of severe relief, the elevations of the earth's surface are relatively slight in comparison to the horizontal distances. This means that vertical exaggeration is usually required to produce a visually satisfactory result.

Two pictorial methods of representing the earth's surface are also used. Both methods present an oblique view of the terrain from a position high above the ground. Physiographic diagrams involve drawing stylized representations of the terrain features on a conventional map base. Block diagrams, on the other hand, present a realistic picture of a portion of the earth's surface. Both methods involve planimetric displacement, which means that a feature is not always located in its correct position. Also, their oblique viewpoint causes features that lie behind the higher elevations to be hidden from view.

SUGGESTED READINGS

Keates, J. S. "Techniques of Relief and Representation." *Surveying and Mapping* 21, no. 4 (December 1961): 459–63.

Lewis, Peirce. "Introducing a Cartographic Masterpiece: A Review of the U.S. Geological Survey's Digital Terrain Map of the United States, by Gail Thelin and Richard Pike." *Annals of the Association of American Geographers* 82, no. 2 (June 1992): 289–304. Also discusses the Erwin Raisz landform map of the United States and the Raven Maps and Images versions of the USGS map.

Lobeck, Armin Kohl. *Block Diagrams.* 2d ed. Amherst, Mass.: Emerson-Trussell, 1958.

Raisz, Erwin J. "The Physiographic Method of Representing Scenery on Maps." *Geographical Review* 21 (1931): 297–304.

Thompson, M. M. *Maps for America.* Washington, D.C.: U.S. Department of the Interior, Geological Survey, 1979.

Yoeli, Pinhas. "Relief Shading." *Surveying and Mapping* 19, no. 2 (June 1959): 229–32.

CHAPTER 9

Contour Interpretation

In many situations, knowing the terrain characteristics of a particular area in advance is valuable. Engineering projects and military operations benefit from such prior knowledge, as do such day-to-day activities as hiking and camping. A number of map-analysis techniques are used to determine terrain characteristics. Although maps are seldom the only source of information for a terrain-analysis project, much can be learned from their systematic examination.

Topographic maps are particularly useful for terrain-analysis projects. Contour lines are the fundamental source of terrain information on such maps, and contour analysis and interpretation are discussed in this chapter. The application of contour interpretation to the classification of terrain types is covered in Chapter 10.

The general approaches discussed in this chapter lend themselves to many practical applications. Computer calculations often are more rapid, flexible, and accurate than many of the more traditional approaches described here. Nevertheless, an understanding of the general characteristics of the applications is useful, whether or not computers are used, because the computer approaches are typically based on, or are the equivalent of, the more traditional ones.

Contours

Characteristics

One characteristic of contours is that they all close by returning to the point from which they started. This is always the case, although the closure does not necessarily occur on a given map sheet. Contours that close on a given sheet represent either hills or depressions of relatively limited size. Depressions are differentiated from hills by the use of depression contours, with their distinctive tick marks (Chapter 8).

The horizontal spacing of contour lines reveals the nature of the slope that the lines represent. The vertical interval between contour elevations is constant (except in the infrequent situation where two intervals are used on the same map). Variations in horizontal spacing, therefore, reflect variations in slope. If the contour lines are evenly spaced on the map, the slope is uniform. Widely spaced lines indicate a relatively flat slope (Figure 9.1a), whereas closely spaced lines indicate a steep slope (Figure 9.1b). If the contours are closely spaced at the top and widely spaced at the bottom, the slope has a concave shape (Figure 9.1c). If the contours are closely spaced at the bottom and widely spaced at the top, on the other hand, the slope has a convex shape (Figure 9.1d). Variations on these combinations can be similarly interpreted.

As already mentioned, two basic contour intervals are sometimes used on the same map. The differences in horizontal spacing and, consequently, in appearance that result from the different intervals influence the map user's interpretation of the terrain. In Figure 9.2, for example, apparently steep terrain in the area with closely spaced contours appears to abruptly change to relatively flat terrain in the area with widely spaced contours. In actuality, terrain characteristics have not

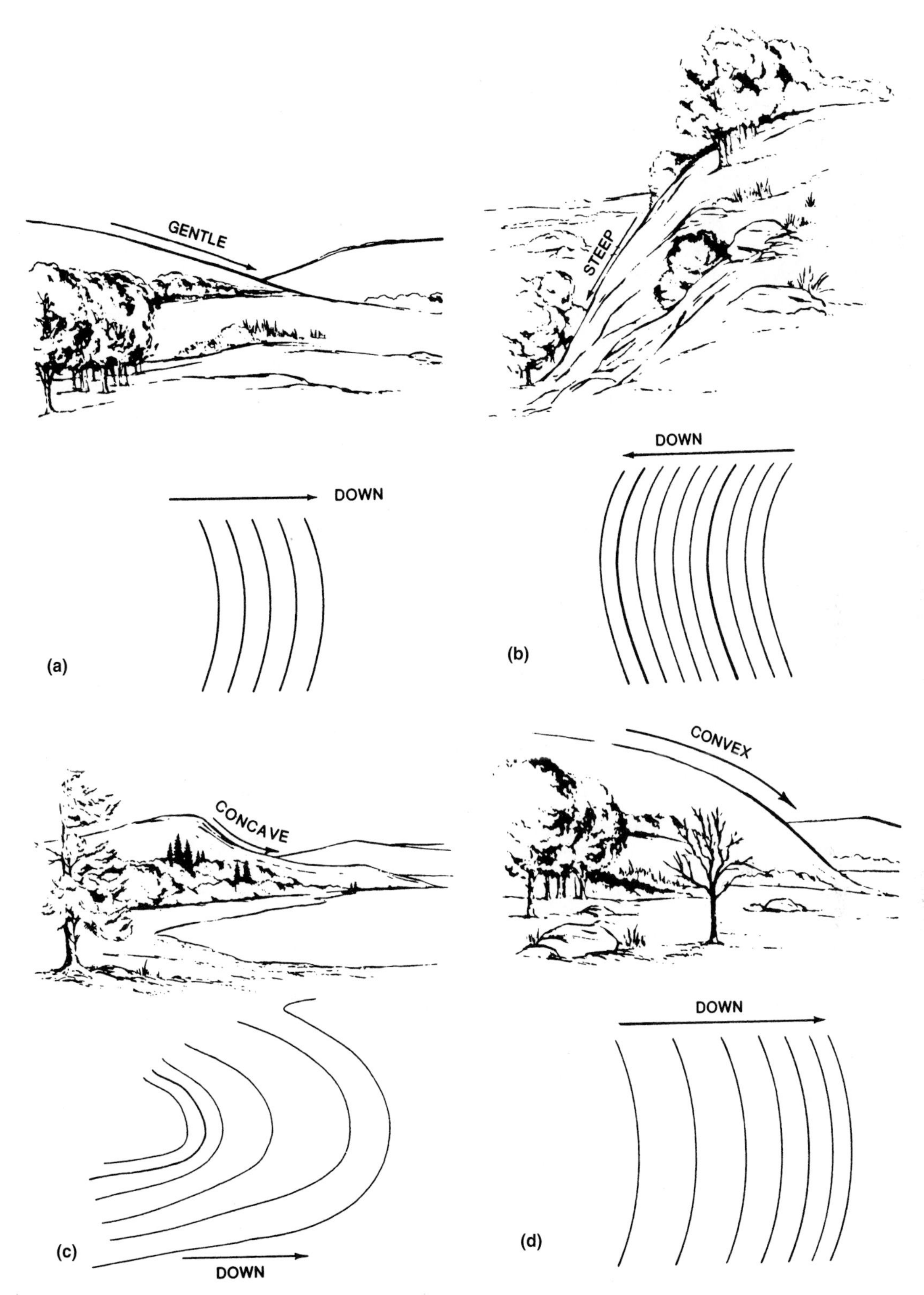

Figure 9.1 Characteristics of slopes are shown by contour spacing. (*a*) Uniform, gentle slope. (*b*) Uniform, steep slope. (*c*) Concave slope. (*d*) Convex slope.

Source: Department of the Army, *Map Reading,* FM 21–26 (Washington, D.C.: Department of the Army, January 1969).

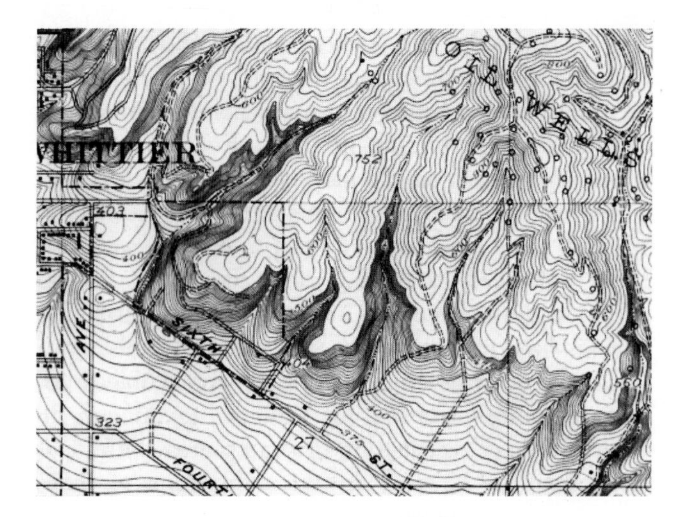

Figure 9.2 Two basic contour intervals used on the same map (interval is 5 feet up to 500 feet, 25 feet above 500 feet). The change in contour interval causes an exaggerated, apparent sharp change in slope. (Whittier quadrangle, California, 1:24,000, 1925. Reprinted 1932.)

changed at all; only the basic contour interval has changed. In such circumstances, a map user who is not aware of the use of the two different basic contour intervals incorrectly interprets the terrain characteristics. Obviously, the marginal information or legend on any map should be checked to determine the contour interval (as well as other important information).

The shape of contours reflects other aspects of the terrain. For example, water-cut stream valleys frequently have what is characterized as a V-shaped cross section, whereas glacial valleys are rounded into a U-shaped cross section (see examples in Chapter 10). The contours that cross valleys point in the upstream direction (point *b* in Figure 9.3). The closed ends of contours that represent ridges or drainage divides, on the other hand, point in the downslope direction (point *c* in Figure 9.3). The relative sharpness or rounded form of the contours as they cross the ridge line reflects the form of the ridge itself (point *a* in Figure 9.3).

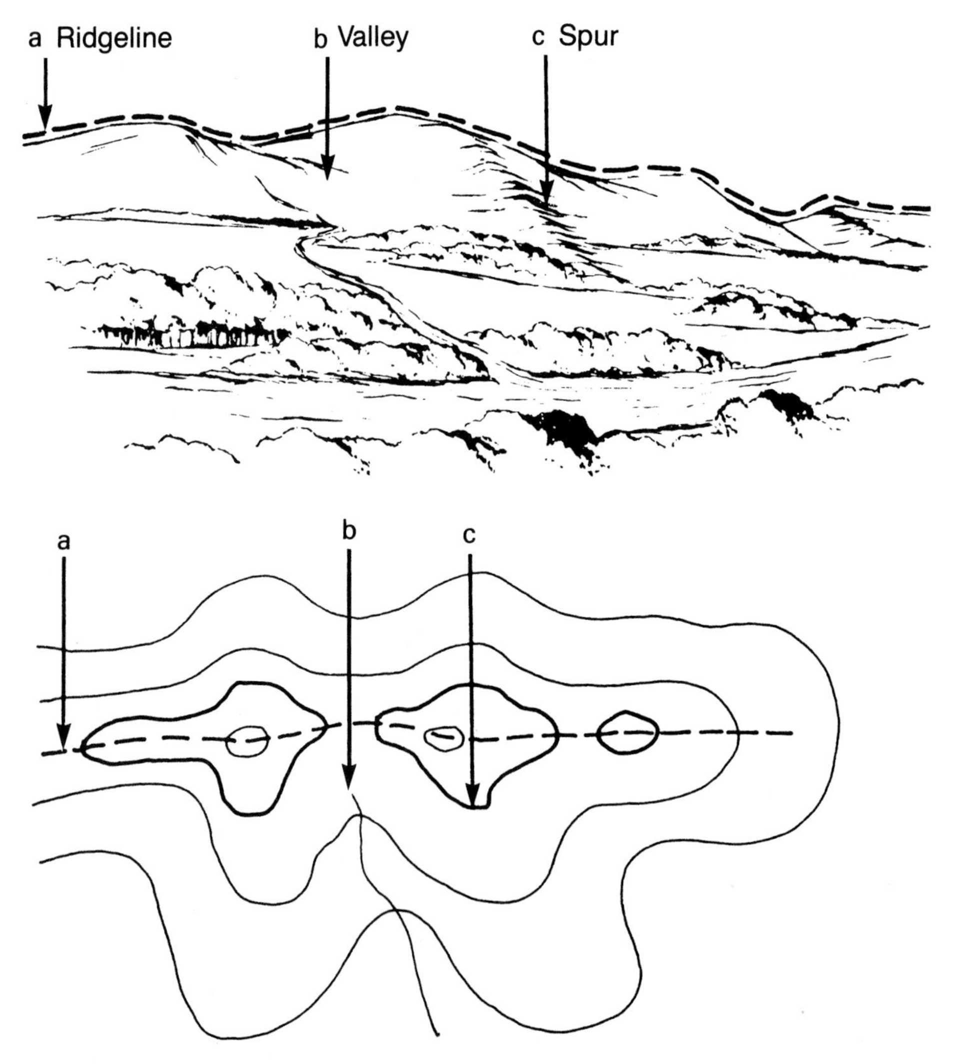

Figure 9.3 Contour patterns indicate locations of valleys and ridges. Contours that cross valleys point upstream, whereas those that cross ridges or spurs point downslope.

Source: Modified from Department of the Army, *Map Reading,* FM 21–26 (Washington, D.C.: Department of the Army, January 1969).

Thus, the shapes of ridges, valleys, hills, and depressions, as well as the characteristics of slopes, are indicated by the arrangement of contour lines on a map. The sections that follow illustrate some of the uses of contours. The use of contours for the identification of distinctive landforms is discussed in Chapter 10.

Applications

Elevation Estimation

On contour maps, the elevations of bench marks and other selected locations, such as road crossings, are often given exactly. Elevations of other locations must be estimated.

If the point whose elevation is to be estimated falls on a contour line, the elevation of that contour is the best estimate of the elevation of the point. If the point falls between contours, however, a technique called **linear interpolation** may be used to arrive at the estimated elevation.

Linear interpolation is based on the assumption that the slope between contours is constant. This assumption, which may not actually be correct, introduces potential error into the process. In the absence of additional information, however, linear interpolation is a logical procedure that involves using rules of proportionality to estimate. If, for example, a point is located halfway between two contour lines, its elevation is estimated as being halfway between the elevation values of those lines.

Figure 9.4 illustrates linear interpolation for estimating elevation at any location. First, draw a line through the point whose elevation is to be obtained, in the direction of maximum slope (that is, at a right angle to the contour lines). Next, determine the point's distance from the lower contour as a proportion of the total distance between the two contours. Then apply that proportion to the contour interval. Finally, add the resulting fraction of the contour interval to the lower contour elevation to obtain the point's estimated elevation.

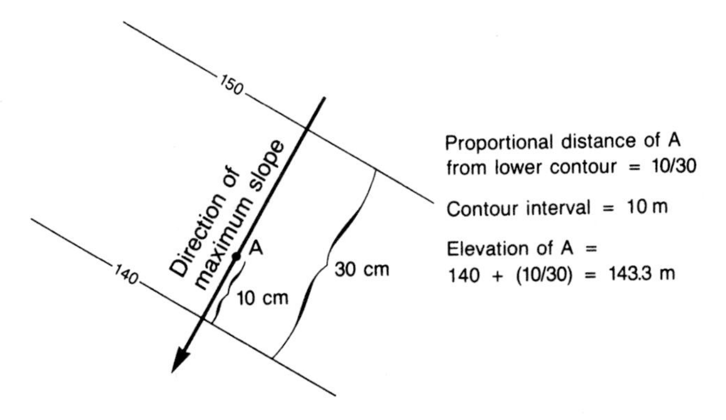

Figure 9.4 Linear interpolation for elevation estimation.

Elevation estimates based on contour lines, whether or not the values are interpolated, usually are not completely accurate. This is because there is a certain amount of permissible error in the placement of contours, even on maps that meet national accuracy standards, and contours are often smoothed, or generalized, especially on small-scale maps. Any elevation estimate based on contours, then, usually is not completely accurate. Plus or minus one-half the contour interval is usually an accepted estimate of the possible error.

Reservoir Capacity Estimation

Often-encountered applications of contour information are the calculation of the capacity of a proposed reservoir, the calculation of the volume of a pile of stockpiled coal, and other similar problems. Such calculations can be made with the use of detailed contour maps of the features involved. Photogrammetric techniques are frequently used for these purposes and often provide more up-to-date and accurate information. In the absence of the necessary photos and equipment, however, existing maps are a useful source of information.

When estimating reservoir capacity, the area enclosed by the contour line that represents the eventual surface level of the reservoir is measured, as are the areas encompassed by each lower contour. These areas are then used to determine the approximate reservoir capacity, given the following formula:

$$V = i\left(\frac{A_1}{2} + A_2 + A_3 \ldots + A_n\right)$$

where i is the contour interval and each A (A_1, A_2, etc.) is the area contained within the given contour outline and where A_1 is the *surface* contour area.

The area of the surface contour is divided in half because the formula is based on the assumption that the contour interval is centered on the elevation of each contour line, and therefore, half of the volume is above the contour and half is below. In the case of the topmost contour, then, only half of the volume is below the water surface. Another assumption of the formula is that the effect of the sloping sides of the feature being measured will average out. This means that the greater volume above the contour, where the reservoir is usually wider, will be balanced by the smaller volume below the contour. The actual variation due to slope can be calculated if necessary but is not likely to be important if the contour interval is sufficiently close.

Flood-Zone Maps

Around the banks of lakes, rivers, and other water bodies, areas that lie at relatively low elevations are likely to be subject to flooding. Despite the threat of flooding, people have made and continue to make extensive use of the lands surrounding water bodies. Agricultural activity, settlements, industry, and transportation facilities are often situated in such areas. Information regarding flood danger, then, is extremely important.

The specific areas subject to flooding vary from time to time, depending on the water volume present and the other conditions that prevail. The factors that affect susceptibility to flooding include natural conditions, such as temperature, soil types, and vegetation, and human-made factors, such as paving and diking.

Of particular importance to flooding, of course, is the shape of the terrain surface. A topographic map, therefore, provides a starting point for the production of a **flood-zone map.** The volume of water expected to overflow the banks is calculated based on the flood level selected for planning purposes. The shape of the river valley determines the level of flooding from the predicted water volume. If the valley is broad and flat, the level of flooding from a given volume of water will be less than if the valley is narrow and steep sided. These considerations allow the selection of the probable maximum level of the expected flood, along the length of the stream. Other factors that affect the rate of water flow out of the valley must be taken into account in refining the basic map.

The Federal Emergency Management Agency (FEMA) produces a specific type of large-scale flood-zone map in support of the national flood insurance program. These maps show, as closely as possible, the areas subject to flooding under what are called **100-year-flood** conditions, as well as other, related information, such as floodways (Figure 9.5).

A 100-year-flood is defined as a major level of flooding that has one chance in one hundred of being equaled or exceeded in any one-year period. The term is simply a statement of probabilities and does *not* mean that such conditions occur at 100-year intervals, or, indeed, that a 100-year flood will actually occur within a given 100-year period. Although the 100-year flood is commonly used as a standard, it is not necessarily the *maximum* flood that could occur. More serious conditions could arise, although they would be extremely rare. If greater safety is desired, the probable extent of a 500-year flood, for example, has to be taken into consideration.

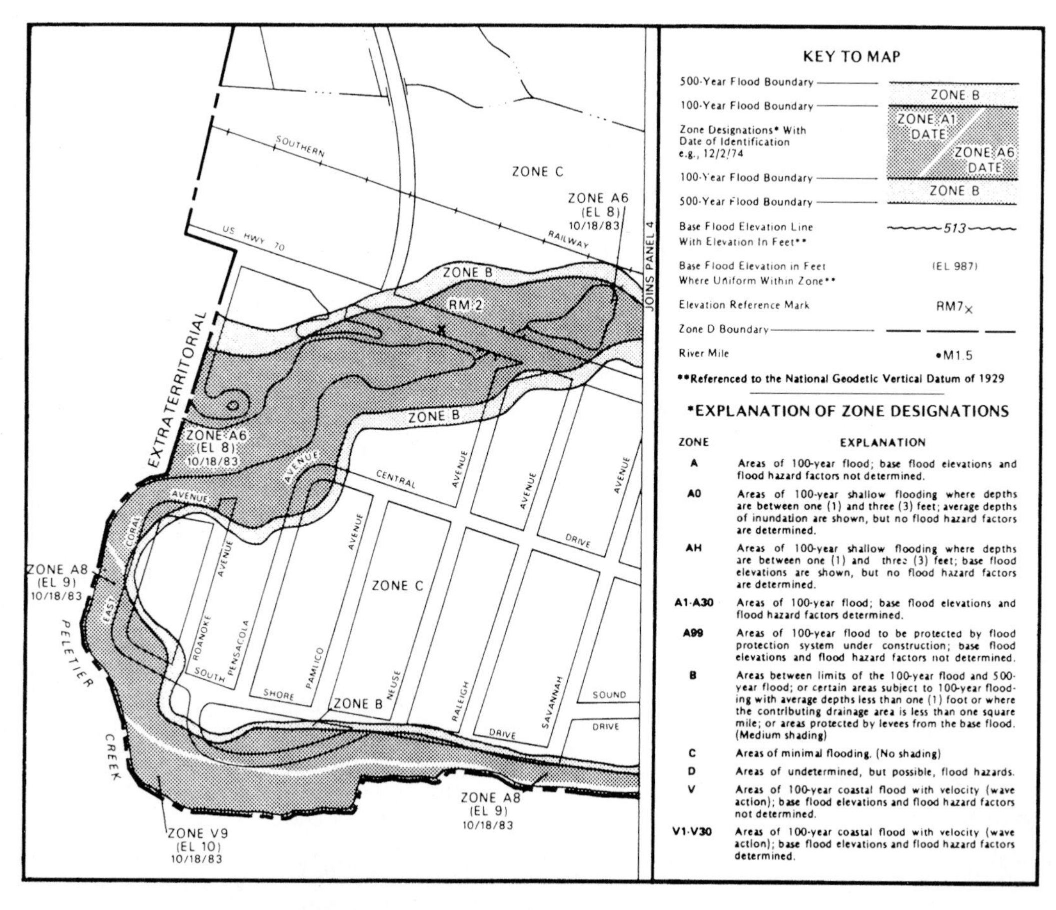

Figure 9.5 Portion of a flood insurance rate map produced by the Federal Emergency Management Agency (FEMA) in support of the national flood insurance program. (Town of Morehead City, North Carolina. Reduced scale.)

Source: Courtesy of Federal Emergency Management Agency.

PROFILING

A **profile** is simply a cross-sectional view through a particular piece of terrain. Profiles provide a relatively quick and accurate means of determining such useful information as the relative steepness of slope of the terrain at given locations, the form of hills, the relationships of hilltops to depressions, the intervisibility of points, and the determination of hidden areas. Profiles are also used in the planning of construction projects involving earth moving, such as road, railroad, and pipeline construction, because they provide information regarding the amount of material that must be moved in grading the right-of-way.

Methodology

Cut-Line

Drawing a profile begins with the selection of starting and ending points that fit the needs of the task at hand. Once these points are selected, a **cut-line** joining them is drawn on the map (or on a map overlay if the map surface must be protected) (Figure 9.6).

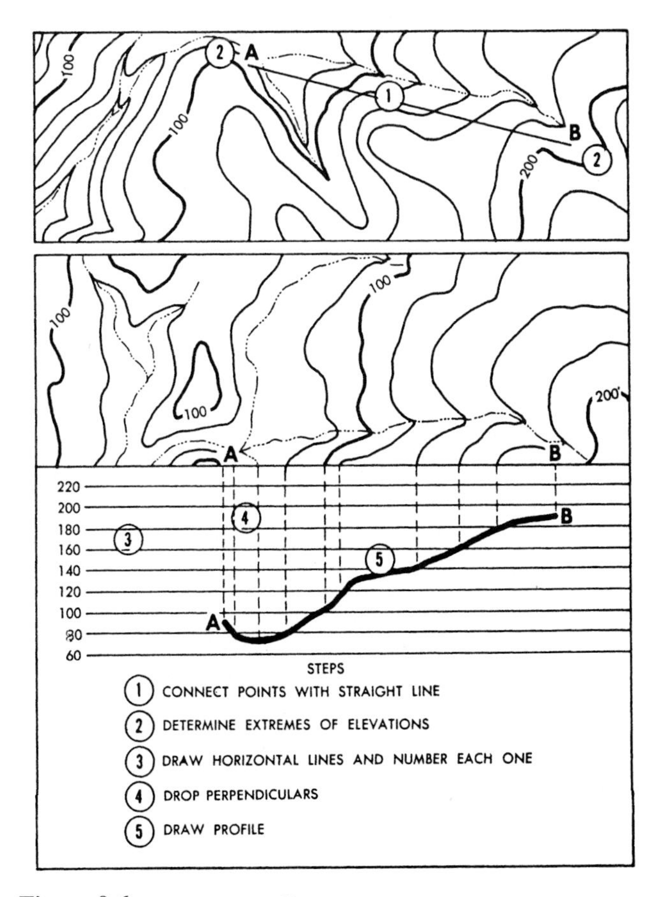

Figure 9.6 Drawing a profile.

Source: Department of the Army, *Map Reading*, FM 21–26 (Washington, D.C.: Department of the Army, January 1969).

Although the cut-line is often a simple, straight line between two points, it is perfectly feasible to draw a line that follows any desired route. It might be useful, for example, to draw a profile along a curving highway route or a twisting hiking trail. The cut-line in such a case is simply drawn on the centerline of the route. The subsequent step in the drawing process, which consists of transferring elevations to the profile drawing, is slightly more difficult. The balance of the operation is the same as for a profile drawn along a straight-line route.

Vertical Scale

The next step in creating the profile involves the selection of the vertical scale. Assume, as an example, that the map from which the profile is being taken has a scale of 1:10,000, which is a relatively large scale. This means that each horizontal centimeter (0.3937 inch) on the map represents 10,000 centimeters or 100 meters (approximately 328 feet) on the ground. If the same scale is used for the vertical dimensions on the profile, a rise of 100 meters (328 feet) is similarly represented by a vertical dimension of 1 centimeter (0.3937 inch). This may be satisfactory if the hills in the area rise several hundreds, or even thousands, of meters or feet above the valleys. However, if this local vertical relief is in the range of 20 meters (65.6 feet), for example, the profile's vertical dimensions will not exceed 2 millimeters (0.08 inch), and the surface characteristics will be very difficult to measure and interpret.

This type of problem is likely to be even more pronounced at the commonly encountered scales of 1:24,000 or 1:62,500. At 1:24,000, a rise of 1500 meters (4921.5 feet) is only 6 centimeters (2.36 inches) high, and at 1:62,500, the same map-profile distance represents a rise of 3906 meters (12,816 feet).

Implicitly, the vertical scale of a map is the same as its horizontal scale, because there is no *direct* representation of vertical distances on a map. Instead, some indirect method is used to indicate the elevations at various locations (as is discussed in Chapter 8). With profiles, however, vertical distances are directly visible (as they are on terrain diagrams and physical models). In such cases, the vertical scale may *not* be the same as the horizontal. The cartographer may decide, instead, to exaggerate the vertical scale.

The use of a vertical scale that is greater than the horizontal scale of a cross section or model is called **vertical exaggeration.** Cartographers often use vertical exaggeration to give a visually more satisfactory appearance to the features shown or to obtain an elevation difference that can be easily measured. Generally, less exaggeration is needed in areas of greater relative relief, whereas more is needed in areas that are relatively flat.

Vertical exaggeration is both a help and a problem to map users. While it increases the visual effectiveness of the presentation, it also has the potential to give *too* strong an impression and, therefore, to be misleading. Noting the amount of relief exaggeration that has been incorporated into a presentation is helpful in interpreting the profile's appearance.

Formally, vertical exaggeration is defined by comparing the vertical earth distance represented by a unit of map distance with the horizontal distance represented by the same measurement. Assume, for example, that a profile has a horizontal scale of 1:50,000. This means that 1 horizontal centimeter (0.3937 inch) on the model represents 50,000 horizontal centimeters (19,685 inches) on the earth. A centimeter (0.3937 inch) measured vertically on the profile, however, may represent 10,000 centimeters (3937 inches). In this case, the vertical exaggeration is five times (50,000/10,000 or 19,685/3937).

Vertical exaggeration causes the slope of a hillside to look steeper than it actually is. The same thing occurs with regard to the apparent dip of a surface in the case of a geologic cross section. Slope information must be interpreted carefully when the slope is drawn with vertical exaggeration.

Drawing the Profile

When the desired vertical exaggeration has been chosen, the next step is to prepare the profile itself (Figure 9.6). The profile is usually drawn on a separate piece of paper of the required length. A series of equally spaced, parallel, horizontal lines is drawn on the paper, or prepared cross-section paper may be used. The spacing between these lines is based on the contour interval and the desired vertical scale so that the correctly scaled vertical interval is represented.

The elevation lines are labeled with the appropriate elevation values so that there is one line for each contour line on the map. The bottom elevation line is usually designated at least one contour interval below the lowest point on the profile and the upper line at least one contour interval above the highest point. This procedure provides clearance for the tops of the hills and the bottoms of the valleys.

The prepared profile overlay is then taped over the map, with its upper edge aligned with the cut-line on the map. It is usually convenient at this stage to tape the map/overlay unit onto a drawing table and to align the horizontal lines using a T square.

The cut-line on the map intersects the contour lines along its path. These intersections are used in drawing the profile. The contour intersection closest to one end of the profile line is located, and its elevation is determined. A light vertical line is then drawn from the contour intersection to the correct horizontal elevation line. The intersection of the vertical line with the elevation line is marked with a dot. This process is continued until each contour crossing on the map is transferred to the equivalent point on the profile.

The final step in creating the profile is to draw a line joining the points plotted on the profile. The path of this line represents the ground surface. The profile should not be drawn as a series of straight lines joining the designated elevation points. Instead, the ground surface is assumed to be rolling or fluctuating. The lines between the points, therefore, are drawn as relatively smooth curves that take into account the general direction of the slope on both sides of each intersection. The result is that sharp breaks in the surface are generally avoided. When hilltops or valley bottoms are encountered, the profile line should not abruptly flatten out to follow the horizontal line. Instead, the profile should be allowed to rise above or drop below the contour line, as appropriate to the terrain's general slope.

Applications

Cut and Fill

Engineers often need to establish road or railroad routes and gradients that minimize the movement of earth. It is generally most economical to balance the amount of material removed from the high areas (cut) with the amount of material required to fill low areas (fill).

A profile drawn along the proposed route assists in making cut-and-fill determinations. A proposed gradient is drawn on the profile (Figure 9.7), and the areas above and below it are measured, using one of the methods described in Chapter 6. These measurements are used as a rough estimate of the relative amounts of **cut and fill** needed along that route.[1] The gradient is then adjusted, as necessary, to come as close to balancing cut and fill as feasible. Alternative routes, with various gradient schemes along each, can be tested to determine the best choice from the standpoint of balancing cut and fill. Other factors, such as gradient limits and the directness of the route, are also considered in the final route selection.

Intervisibility

A profile is useful for determining whether two points are **intervisible.** The two points of interest are located, and a profile is drawn between them. A straight line is then drawn on the profile between the locations of the two points. If the line does not encounter any obstructions, the points are intervisible (Figure 9.8).

[1]Accurate determinations depend on the characteristics of the terrain for some distance on each side of the proposed route, so multiple profiles may be needed.

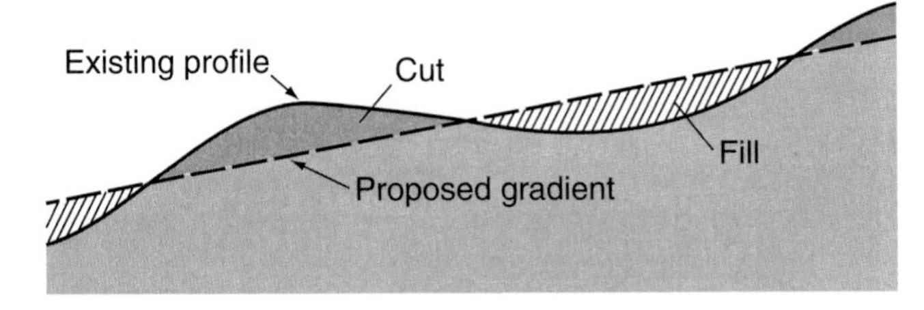

Figure 9.7 Proposed gradient drawn on a profile.

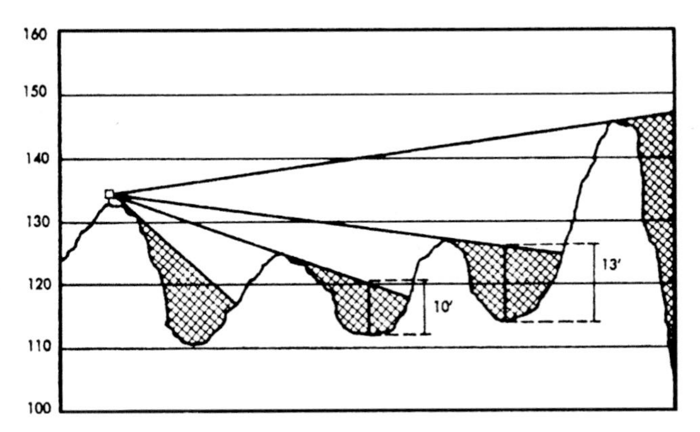

Figure 9.8 Intervisibility profile. Locations not visible from the viewpoint are shown by cross-hatching.

Source: Department of the Army, *Map Reading*, FM 21–26 (Washington, D.C.: Department of the Army, January 1969).

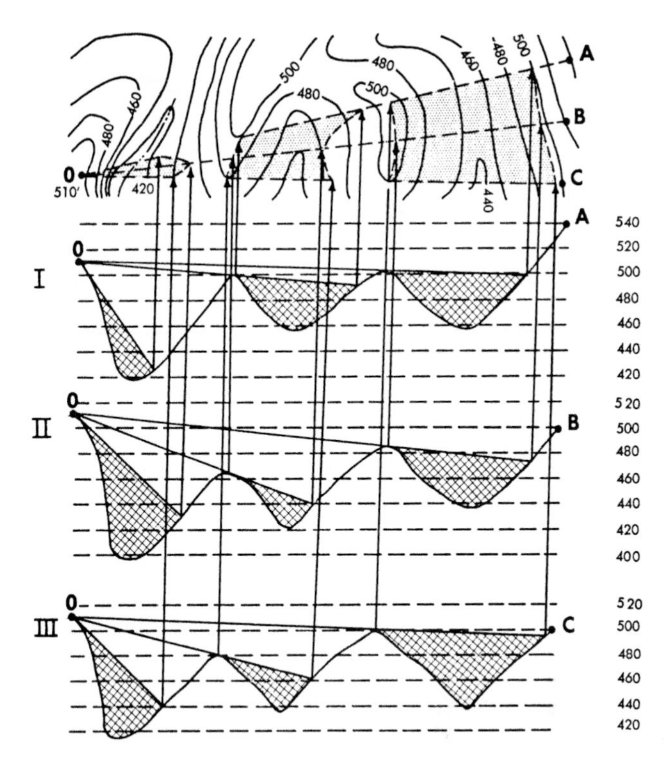

Figure 9.9 Plotting masked areas for an intervisibility map.

Source: Department of the Army, *Map Reading*, FM 21–26 (Washington, D.C.: Department of the Army, January 1969).

An extension of the same technique is used to map areas hidden from the view of a given observation point. Such areas are said to be in **defilade.** A series of profiles is drawn that radiates from the proposed viewpoint, with the accuracy of the final product governed by how close together the profiles are placed. The limits of visibility for each profile are then plotted on the map and are linked together to outline areas that are masked from view (Figure 9.9).[2]

In addition to its obvious importance in military operations, intervisibility information is useful for the evaluation of proposed sites for forest-fire lookouts, the planning of logging and other activities so that they interfere as little as possible with recreational and scenic values, and similar civilian applications.

SURFACE INCLINATION

The surface of the terrain typically rises and falls, and variations in its inclination are important in determining the likelihood of landslides, determining the feasibility of building a logging road or installing a septic-tank sewage system, or many similar concerns. A slope map graphically depicts the steepness of the land and is useful for planning such projects. It shows, by colors or patterns, the various areas that fall within specific ranges of surface slope.

Two terms are commonly used to describe surface variations: slope and gradient. These terms are often used synonymously. However, in mapping applications it is sometimes desirable to distinguish between them.

Slope

Slope is a measure of the vertical difference in the elevation of a surface at two different points, in relation to the horizontal distance between the same points. Slope, therefore, does not exist *at* a point; it only exists *between* two points.

It may seem paradoxical, but the calculation of slope does not take into account the actual shape of the surface between two points. The surface may be uniform, concave, or convex (see Figure 9.1), or it may be irregular and complex—the slope will still be the same. What determines the slope is the relationship between the end points of a line, not what happens between them.

All expressions of slope are derived from the relationship between the vertical distance between two points, known as the **rise,** and the horizontal distance

[2]A computer-assisted method of preparing intervisibility maps is described in Pinhas Yoeli, "The Making of Intervisibility Maps with Computer and Plotter," *Cartographica* 22, no. 3 (1985): 88–103.

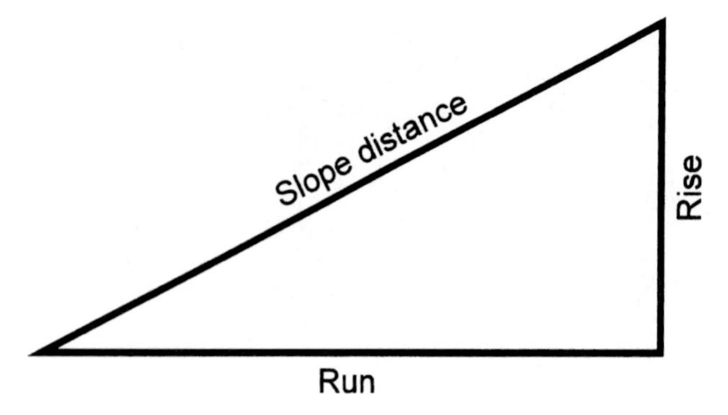

Figure 9.10 Relationship between rise, run, and slope distance.

between them, known as the **run** (Figure 9.10). (Notice that the run is the map distance between the points, not the slope distance as measured on the earth's surface.) The measure of this relationship is expressed in one of four ways: (1) as a ratio, (2) as a simple fraction, (3) as a percent, or (4) as an angle, in degrees. Slope can be expressed in any of these forms, and the values are easily converted from one form to another. The form chosen is simply a matter of convenience in relation to the task at hand.

Slope Ratio

A **slope ratio** is established by locating a starting point and an ending point on the map. The elevation of each point and the horizontal distance between them are determined by the measurement techniques described earlier. Both rise and run must be measured in the same units, whether they are feet, meters, inches, yards, or any other units. The measurements are written as a ratio between the rise and the run, with the rise reduced to 1. If the vertical distance is 3 meters (9.843 feet), for example, and the horizontal distance is 30 meters (98.43 feet), the relationship between them is 3:30 (9.843/98.43). This reduces to 1:10, which means that there are ten units of run per unit of rise. In words, this is usually stated as "a slope of one in ten." A slope of 1:10 is equally valid when interpreted to mean 1 meter to 10 meters, 1 foot to 10 feet, and so on.

Slope Fraction

A **slope fraction** is derived in the same manner as a slope ratio, but is written as a fraction, rather than as a ratio. Thus, a slope ratio of 1:10 would simply be written as the fraction 1/10.

It should be noted that the units used in a slope fraction are not always the same. For example, it is not unusual to see the slope of a stream expressed as 1 meter/kilometer or 1 foot/2 miles. In words, these would be stated as "one meter of rise per kilometer of run," or "one foot of rise per two miles of run." Note that there is a potential confusion when this done. For example, 1 meter/kilometer yields the fraction 1/1,000, whereas 1 foot per 2 miles yields the fraction 1/10,560.

Percent

Slope is expressed as a **percent** by simply determining the rise per 100 units of run (divide the rise by the run and multiply the resulting decimal fraction by 100). In the previous examples, the rise was 3 units, and the run was 30 units. This is equivalent to a 10% slope ($3/30 \times 100 = 0.1 \times 100 = 10\%$).

Expressing slopes in percentage terms can be unintentionally misleading. If someone were asked whether 100% is the steepest possible slope, for example, the likely answer would be yes. However, 100% simply means that the rise of the slope is equal to its run. For example, 40 units of rise and 40 units of run equals 100% ($40/40 \times 100 = 100\%$). This slope has a gradient of 1/1 or, as described in the next section, a slope angle of 45 degrees—certainly not the steepest possible slope.

In hilly or mountainous terrain, slopes are often steeper than 100%. A slope with a rise of 200 units and a run of 100 units, for example, is a 200% slope ($200/100 \times 100 = 200\%$), and steeper slopes have even larger percentage values. For example, a cliff might have a rise of 800 feet and a run of 10 feet. Its percentage slope, therefore, is 8000% ($800/10 \times 100 = 8000\%$). Finally, a vertical slope has a large rise and a run of zero, which yields an infinite percentage measure (e.g., 500 feet/0 feet $\times$ 100 = ∞%). Because such extremely large values result, the use of percentage measurements is seldom practical for steep slopes.

Degree

The fourth way to express slope is as an angle, in **degrees** from the horizontal. The angle of slope may be measured directly from a profile drawn with no vertical exaggeration. More commonly, however, the angle is calculated using the same rise and run information used to determine fractional or percentage slopes.

Dividing the rise by the run results in a decimal value defined as the tangent (tan) of the slope angle. The slope angle can be determined by locating this value in a table of trigonometric functions or by using the $\tan^{-1}$ key on a calculator. Continuing the previous examples, the slope 3/30 has a tangent of 0.1, and the angle with this tangent value is approximately 6 degrees.

Gradient

The most common use of the term **gradient** is simply as a synonym for slope.[3] At times, however, the term is used in a more specialized way, to define the maxi-

[3]See, for example, Robert L. Bates and Julia A. Jackson, eds. *Glossary of Geology*, 2 ed. Falls Church, Va.: American Geological Institute, 1980.

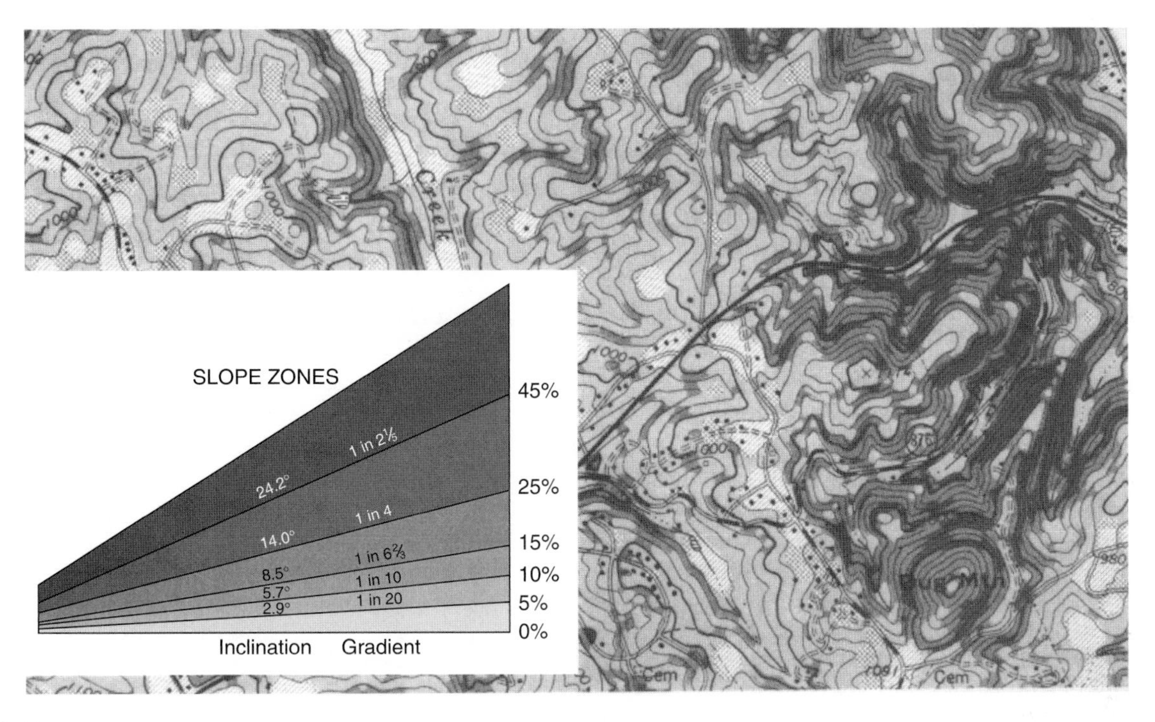

Figure 9.11 Slope-zone map. (Hershey, Pennsylvania, 1:24,000.)
Source: Melvin Y. Ellis, ed., *Coastal Mapping Handbook* (Washington, D.C.: U.S. Government Printing Office, 1978), 165.

mum slope of a surface. In this usage, gradient is thought of as the rate of change of slope *at* a given point. This is because the inclination of the surface can vary in different directions. On a contour map, the steepest gradient occurs where the contours are closest together. Its direction is at right angles to the contours at that location. A gradient is expressed by any of the measures used for slope measurement (slope ratio, slope fraction, percentage, or degree), with the addition of an azimuth or other directional indicator to indicate the direction of the maximum slope.

A further extension of the specialized gradient idea is the **gradient path.** In this case, a series of maximum slopes is combined to define the steepest path down a hill.

Applications

Slope-Zone Maps

The slope of the land surface is often an important factor in making decisions about land-use restrictions or the general desirability of an area for cultivation. Slope is also a major factor in the relative ease or difficulty of moving equipment or personnel over the terrain. A specialized map, called a **slope-zone map,** presents slope information for such purposes.

A slope-zone map shows, by means of coloring or shading, areas of relatively homogeneous slopes (Figure 9.11). The process of establishing slope zones is time consuming because it requires repeated slope determinations over the entire area of interest, and the greater the accuracy required, the more closely spaced the determinations must be. One way to create a slope-zone map is to draw a series of profiles throughout the area of interest. The profiles are then examined to determine the relative steepness of the slopes at different locations. This information is transferred back to the map, and areas with similar slopes are identified by appropriate symbols. Because of the expense of such detailed work, the U.S. Geological Survey usually prepares slope-zone maps only at the request of a government agency. In addition, some private firms produce such maps on a contract basis. Computer-assisted techniques can expedite the production of slope-zone maps.[4] As such techniques are more frequently applied, slope-zone maps may become easier to obtain.

You can prepare a slope-zone map, if only moderate reliability is required, with a simple visual method. First, you must prepare a set of templates (Figure 9.12). Mark each template with a set of lines spaced to represent a specific slope at the scale and contour interval of the map to be analyzed. The number of templates required is determined by the number of zones and the fineness of classification needed and is usually limited to a few easily distinguishable categories. To produce

[4]See, for example, D. A. Sharpnack and G. Akin, "An Algorithm for Computing Slope and Aspect from Elevations," *Photogrammetric Engineering* 35, no. 3 (1969): 247–48.

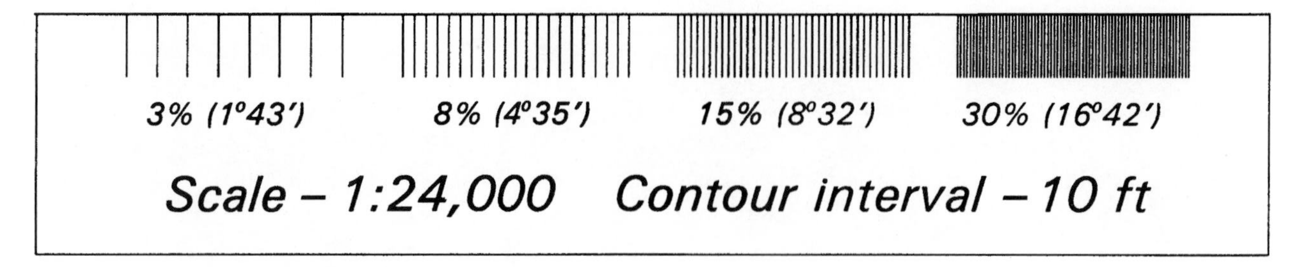

Figure 9.12 Template for determining slope zones on a 1:24,000-scale map with a 10 ft contour.

the slope-zone map itself, visually match the templates to the contour lines on the base map and draw an outline around areas of comparable slope. Add identifying colors or patterns to complete the project.

Maximum Slope Limitations

A route between two locations usually must meet certain grade restrictions. Road or railroad alignments, for example, cannot be too steep because of the limited capabilities of the vehicles that use them. Similarly, vehicles that can move directly across the terrain, such as those often used in military operations, search-and-rescue missions, logging operations, or similar activities, have maximum grade restrictions. Even hikers prefer to limit the steepness of their routes to avoid too much stress. The determination of the slopes along the path between a given starting point and destination is one method of ensuring that the capability of the particular equipment or personnel involved in a planned movement is not exceeded. If the objective is to be reached, the route must not include grades that are steeper than the operational limit.

Slopes may be determined by examining a profile taken along the planned route, as was already discussed. If no profile is available and it is not considered worthwhile to draw one, a contour map can be used to check slope limits. Both vertical distances (based on the contour interval of the map) and horizontal distances (between contours) can be determined directly from the map. These two pieces of information are sufficient for checking that the **maximum slope limit** is not exceeded.

First, the horizontal distance between contours that will satisfy the maximum slope limit is calculated. Suppose, for example, that the maximum permissible slope is 4%, and the contour interval of the map is 10 meters (32.81 feet). The 4% slope limit (4 meters [13.124 feet] of rise per 100 meters [328.1 feet] of horizontal distance) means that 25 meters (82.025 feet) of horizontal travel is required for each meter (3.281 feet) of rise. Given the contour interval of the map, a minimum horizontal distance of 250 meters (820.25 feet) must be traveled for each contour interval that is overcome.

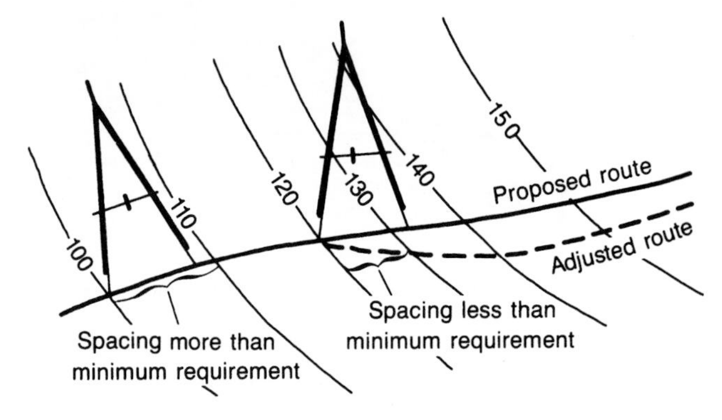

Figure 9.13 Checking maximum slope limitations on a planned route of travel by measuring the minimum horizontal distance requirement.

Once the minimum horizontal distance requirement is determined, the process of checking to see that the maximum slope is not exceeded is simple and mechanical. The example explained here assumes a continuous uphill gradient, but it is not difficult, once the basic procedure is understood, to accommodate uphill or downhill variations.

The desired route is first plotted on the map. Then a pair of dividers is set to the correct scale equivalent of the minimum horizontal distance requirement. The first leg of the route is checked by placing one divider tip at the starting point (assuming, for simplicity's sake, that it is on a contour line). The dividers are then aligned so that the other tip lies on the uphill path (Figure 9.13). If the dividers do not reach beyond the next contour line, the slope does not exceed the limit. The dividers are then moved to the next contour crossing, and the process is continued until the final destination is reached. If the distance between any pair of contour lines is less than the minimum horizontal distance for which the dividers are set, the maximum slope limit is exceeded. The route at that point is adjusted accordingly, and checking of the realigned portion is continued.

Trafficability

The ability of an area to support vehicles of a certain type is called **trafficability.** Trafficability is of particular importance in military operations because the selection of routes able to support specific types of vehicles can assist in meeting a military objective. Trafficability information is also important in peacetime engineering and forestry projects.

A trafficability map shows zones of relative ease or difficulty of travel for various modes of transportation. The determination of trafficability is based on multiple criteria. Slope is usually a major factor, but other common criteria include types of vegetation cover, surface materials, and weather conditions. Geographic information systems, which are discussed in Chapter 21, are particularly useful for the analysis necessary to generate trafficability maps.

SUMMARY

Contour lines are particularly useful for terrain analysis. The horizontal spacing of contours reveals the nature of the slope. If the contours are evenly spaced, for example, the slope is uniform. Widely spaced contours indicate a relatively flat slope, and closely spaced contours a steep one. If contours are closely spaced at the top of a slope and widely spaced at the bottom, the slope has a concave shape. If the contours are closely spaced at the bottom and widely spaced at the top, the slope is convex.

Contours reflect other aspects of the terrain. Water-cut stream valleys, for example, usually have a V-shaped cross section, whereas glacial valleys are rounded into a U-shaped cross section. Contours that cross valleys point in the upstream direction, but those that cross ridges or drainage divides point downslope. The relative sharpness or rounded form of the contours as they cross the ridge line reflects the form of the ridge itself. Thus, the shapes of ridges, valleys, hills, and depressions, as well as the characteristics of slopes, are indicated by the arrangement of contour lines.

Elevations of locations are estimated by means of linear interpolation between contours. This process is based on the assumption that the slope between contours is constant.

Practical applications of contour analysis include volume estimation and the development of flood-zone maps. In addition, profiles, which are cross-sectional views through the terrain, provide information about slope steepness, intervisibility, and so on and are used in planning construction projects.

Slopes are expressed as ratios, fractions, percentages, or degrees. The measure of a slope is based on the relationship between the rise and the run, expressed as a ratio or a fraction. Slope, in percentage terms, is given by the rise per 100 units of run. Slope may also be denoted as an angle, in degrees from the horizontal.

Gradient is usually synonymous with slope. It is sometimes defined as the maximum slope of a surface, or as the rate of change of slope at a given point. A gradient path is the steepest path down a hill.

Slope-zone maps show areas of relatively homogeneous slope.

A route between two locations must usually meet grade restrictions. Slopes along the proposed route may be determined by examining a profile taken along its path. Alternatively, the horizontal distance between contours that will satisfy the maximum slope limit is calculated and stepped off along the planned route. If the limit is exceeded, the minimum horizontal distance is used to discover an alternative route.

Trafficability is the ability of an area to support vehicles of a certain type. A trafficability map shows zones of relative ease or difficulty of travel for various modes of transportation, with slope a major consideration.

SUGGESTED READINGS

Department of the Army. *Map Reading*. Department of the Army Field Manual, FM 21–26. Washington, D.C.: Headquarters, Department of the Army (current edition).

Muehrcke, Phillip C. *Map Use: Reading, Analysis, and Interpretation*. 3d ed. Madison, Wis.: JP Publications, 1992.

Speak, P., and A. H. C. Carter. *Map Reading and Interpretation*. 2d ed. London: Longman Group, 1970, 14–18.

Tyner, Judith. *The World of Maps and Mapping*. New York: McGraw-Hill, 1973, 12–21.

CHAPTER

10 Topographic Features

Topography is the surface configuration of the earth resulting from complex interactions of natural processes, working throughout the eons of geologic time. One set of processes involves movements of the earth's crust, called tectonic activity. These crustal movements create the basis for the resulting landforms by causing the broad uplift or depression of the earth's crust, as well as its folding and faulting. In addition, external flows and internal intrusions of volcanic material introduce unique details into the landscape. A second set of processes sculpts the features created by the first. These involve a variety of geologic, hydrologic, atmospheric, and biotic forces, including the physical and chemical action of surface water, groundwater, ice, and wind, and even the action of plants and animals. These forces bring about weathering, mass movement, erosion, and deposition, which create the details of the landscape.

As a major environmental element, the earth's topography has many significant and diverse effects on human life and endeavors. Some of these effects are direct. Mountain ranges, for example, are barriers to transportation, but they also harbor minerals used by industry, and they provide scenic attractions and enjoyment. Valleys shelter agricultural activities. They also hold the rivers that create disastrous floods which, at the same time, bring renewal to the soils. Some effects of topography on human activity are indirect, operating through topography's interrelationship with other environmental elements. Deserts develop in the lee of mountain ranges, for example, and rain forests are created on mountains' windward flanks. Each of these dramatically different environments has differential impacts on the activities of the human population. In addition, there are more subtle effects on plant and animal life and, therefore, on human activity. Different varieties of vegetation grow at different elevations and exposures, for example. In these and many other ways, topography influences human life and activities. An examination of landforms also helps us to develop an understanding of and appreciation for the past and for the ongoing processes that have led to where we are and that will continue to mold our environment.

A full understanding of how a specific topographic feature developed requires a knowledge of the nature and history of the underlying geological structures. It also requires an appreciation of a variety of physical processes, as well as of the modifying effects of climate and vegetation. These matters go well beyond the scope of this chapter, which is limited to presenting typical examples of local landforms. The purpose of providing these examples is simply to illustrate that it is possible to recognize many topographic features from their appearance on a map. The book by Victor Miller and Mary Westerback listed in the Suggested Readings at the end of this chapter provides an excellent reference for more extensive study.

DRAINAGE PATTERNS

Drainage patterns, the arrangements of streams on the landscape, are indicators of underlying geology. Although the patterns that streams form on maps do not allow extensive analysis of underlying geology, preliminary drainage-pattern classifications can be generated in preparation for more extensive work. Drainage patterns can be classified in two ways: type of dissection and texture of dissection.[1]

Type of Dissection

The classification of drainage patterns into characteristic types is a subjective process, but several generally adopted major categories can be identified. Knowledge of these categories provides some initial clues about underlying geological structure when a map is being analyzed.

Dendritic

The term *dendritic* means treelike, an apt analogy for this common drainage pattern that is characterized by treelike branching (Figure 10.1). The tributaries (lower-order streams) of a dendritic drainage pattern join the main (higher-order) streams at acute angles. Dendritic patterns usually form in areas of homogeneous (uniform) soil and rock, often consisting of horizontal strata or unjointed rock masses, which exert no predominant directional control on the drainage.

[1]Douglas A. Way, *Terrain Analysis: A Guide to Site Selection Using Aerial Photographic Interpretation,* 2d ed. (Stroudsburg, Pa.: Dowden, Hutchinson and Ross; New York: McGraw-Hill, 1978), 49.

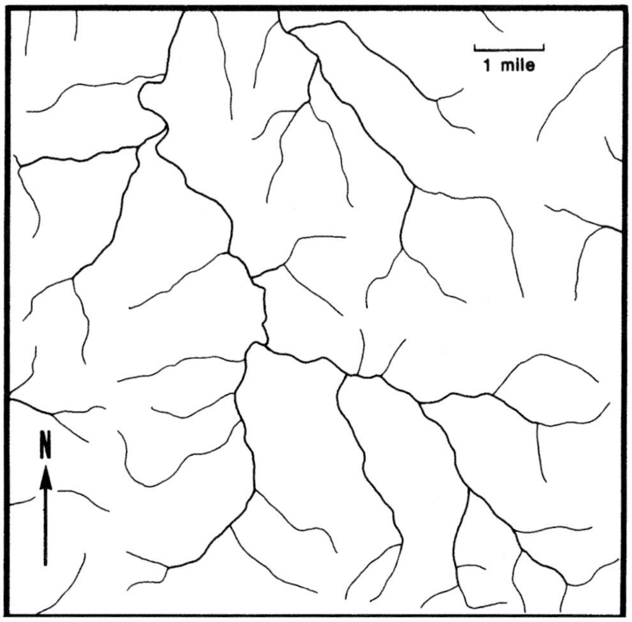

Rectangular

Rectangular drainage patterns (Figure 10.2) are a variation of the dendritic type. In this case, however, the next-highest-order tributaries join highest-order streams at approximately right angles. The rectangular pattern of these major streams originates with bedrock, whose jointing or fracturing controls the pattern's development.

Trellis

Another modification of the dendritic pattern is the trellis drainage pattern, which has parallel tributaries. This pattern often forms in areas of alternating parallel ridges and valleys (Figure 10.3).

Radial

A circular pattern of channels flowing away from, or toward, a central point is called a radial drainage pattern. When the streams flow outward from the feature that controls the pattern, such as a volcano or structural dome, the pattern is referred to as centrifugal (Figure 10.4). In this type, the radial streams often flow

Figure 10.1 Dendritic drainage pattern.
Source: Traced from St. Paul quadrangle, Arkansas.

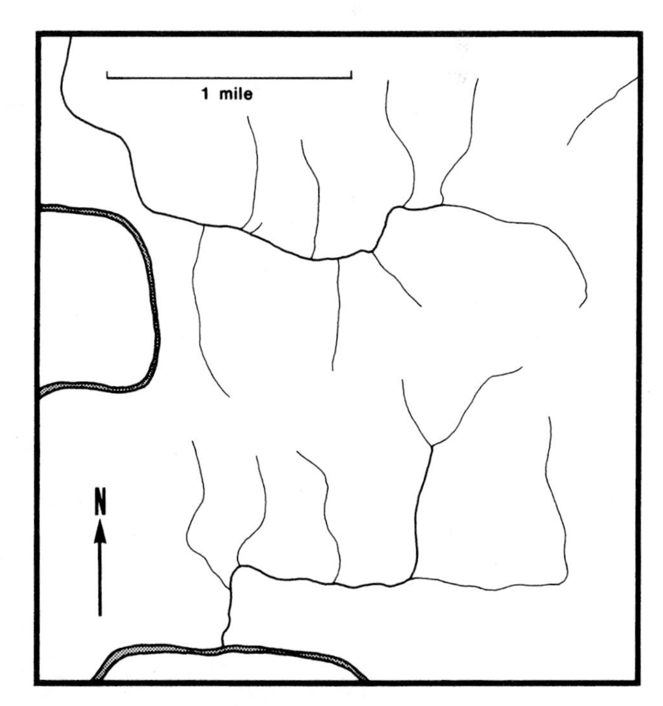

Figure 10.2 Rectangular drainage pattern.
Source: Traced from Hillsboro quadrangle, Kentucky.

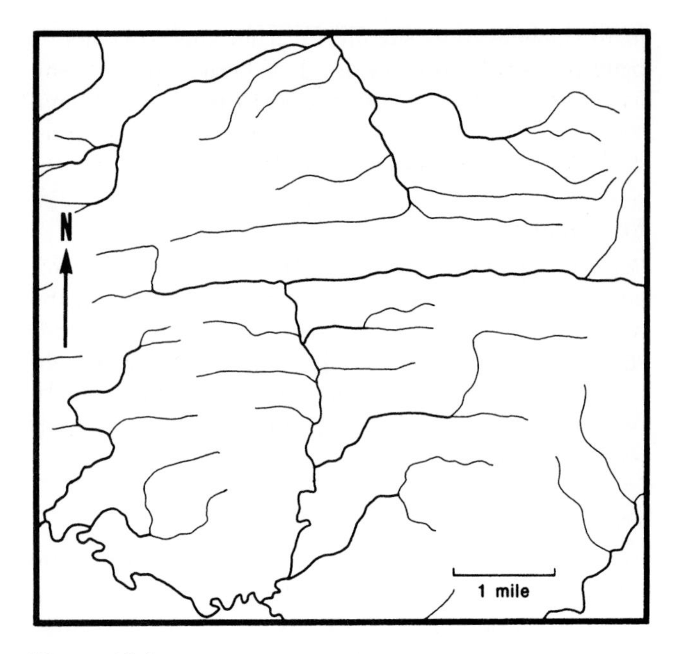

Figure 10.3 Trellis drainage pattern.
Source: Traced from Waldron quadrangle, Arkansas.

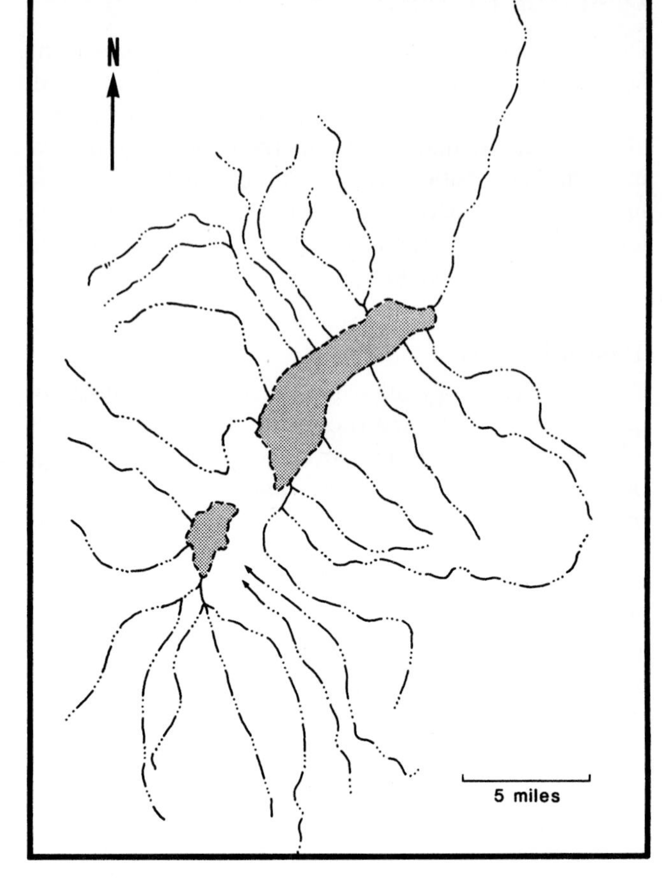

Figure 10.5 Radial drainage pattern (centripetal). Clayton Valley, Nevada (intermittent streams and lakes).
Source: Traced from USGS State of Nevada map.

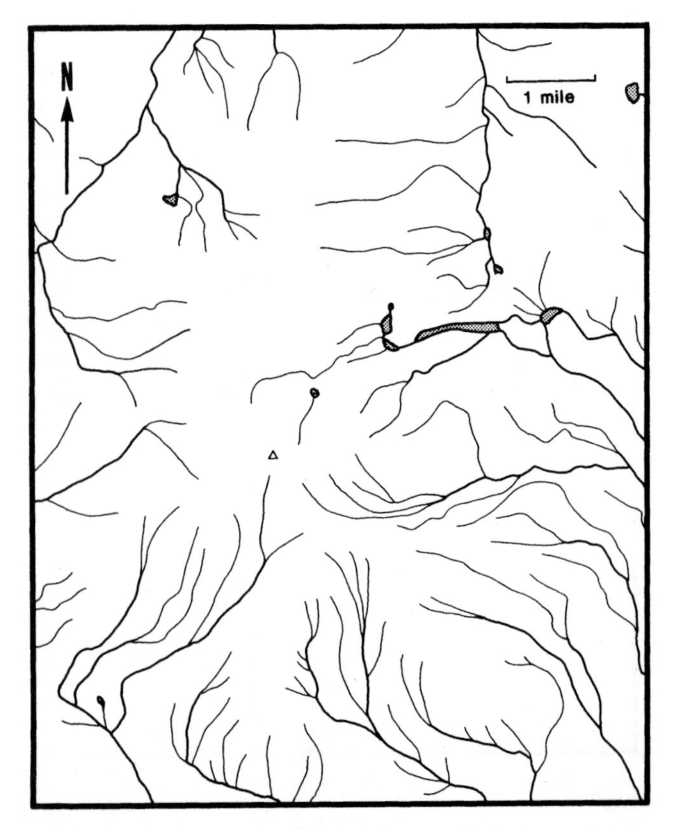

Figure 10.4 Radial drainage pattern (centrifugal).
Source: Traced from Katahdin quadrangle, Maine.

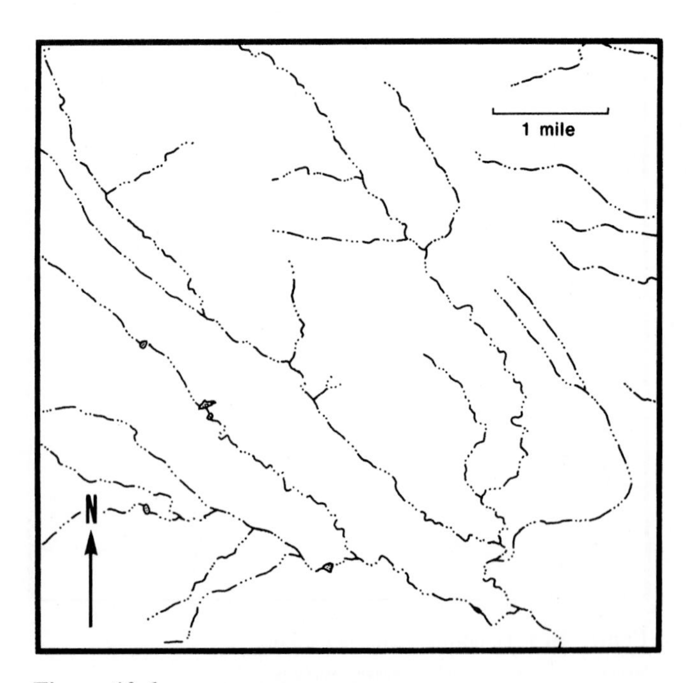

Figure 10.6 Annular drainage pattern.
Source: Traced from Maverick Spring quadrangle, Wyoming.

into a collector stream around the base of the feature. When the streams flow inward into a crater or similar structure, the resulting pattern is called centripetal (Figure 10.5).

Annular

The underlying topography control that leads to a radial pattern is similar to the control that results in an annular drainage pattern (Figure 10.6). The annular characteristic is introduced by the presence of joints or fractures that create elements of parallel drainage.

Miscellaneous

A great variety of other stream patterns form in special circumstances. For simplicity's sake, these are grouped here into a *miscellaneous* category. The following examples illustrate the diversity: A disrupted pattern is characteristic of areas of poor drainage resulting from glaciation (see the discussion of glaciation later in the chapter). Disappearing streams and internal drainage occur in areas where bedrock solution results in the development of underground drainage channels (see Figure 10.25). A fanlike pattern of streams often spreads out over the base of an alluvial fan (Figure 10.7). A braided, meandering pattern is characteristic of a stream that is close to the level of a water body, such as a lake or the ocean (Figure 10.8). The pattern evolves because the water body establishes a limit on the stream's downward cutting. As a result, the stream wanders from side to side and creates a broad, flat valley. An artificial pattern is imposed when there is human intervention in the form of channels and ditches for navigation, irrigation, or drainage (Figure 10.9).

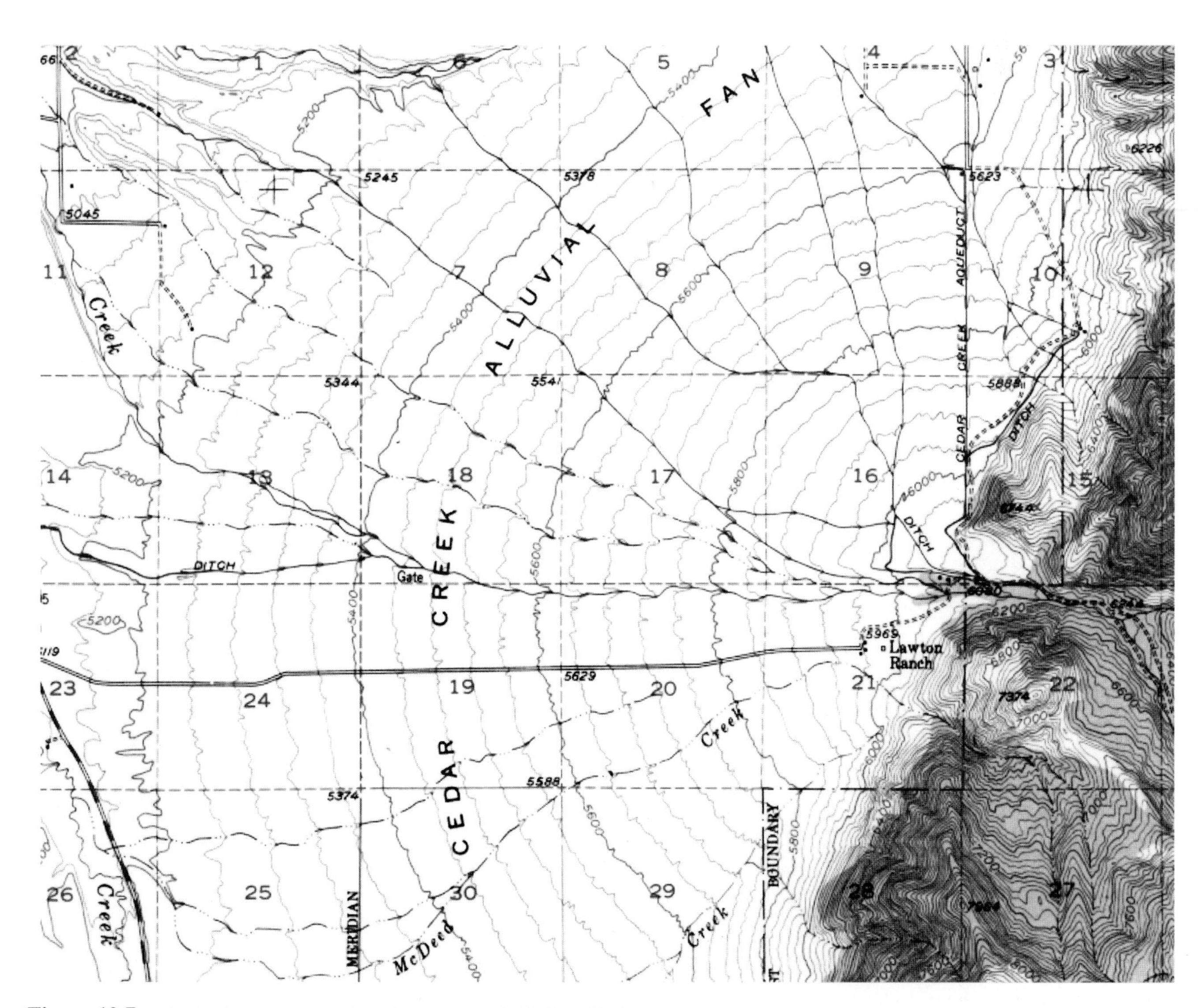

Figure 10.7 Alluvial fan. (Ennis quadrangle, Montana, 1:62,500, 1949.)

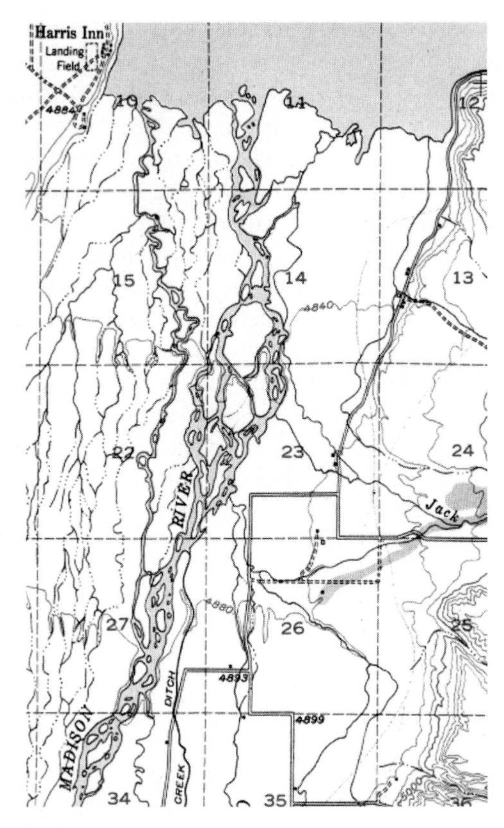

Figure 10.8 Braided, meandering stream. (Ennis quadrangle, Montana, 1:62,500, 1949.)

Texture of Dissection

Texture of dissection describes the relative fineness of the stream pattern.[2] The categories typically established to describe texture—fine, medium, and coarse—are indicative of its rather subjective nature. The concept, however, offers some insight into the possible nature of the underlying geology and soils.

The definition of texture is based on the spacing of first-order streams (see Chapter 13). If the first-order streams are very close together, say on the order of 100 meters or less, the texture is classified as fine. If, on the other hand, the first-order streams are spaced 400 or more meters apart, the texture is classified as coarse. A pattern with medium texture would, of course, fall in the 100- to 400-meter range.

Fine texture is likely where there is relatively weak rock because in such a circumstance a small volume of water is sufficient to establish a drainage channel. An impermeable soil type, which results in a greater volume of runoff from the surface, also can be expected to contribute to a fine texture. A coarse texture, on the other hand, is more likely where there is resistant bedrock because a large volume of runoff is needed to establish stream channels. If the soil is permeable, runoff is reduced, and any tendency toward a coarse

[2]See, for example, Way, *Terrain Analysis,* 49.

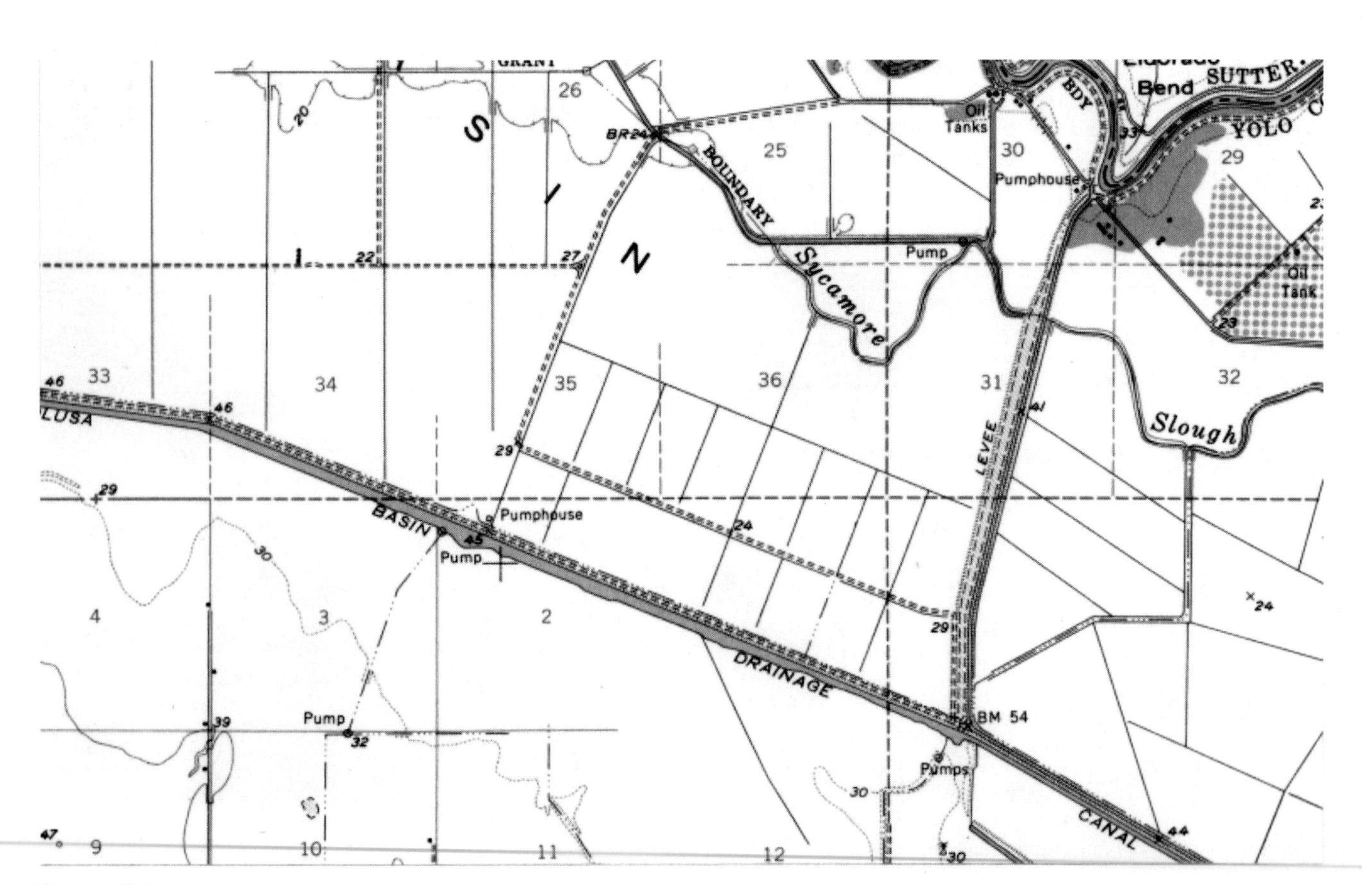

Figure 10.9 Human intervention: irrigation and drainage canals. (Dunnigan quadrangle, California, 1:62,500, 1953.)

texture is reinforced. In addition, the presence of vegetation protects the surface and increases the tendency toward coarseness, because runoff is decreased.

SELECTED EXAMPLES OF LANDFORMS

For descriptive purposes, landforms may be grouped into categories based on their location or origin. In this section, we discuss the following landform groupings: coastal features and shorelines, escarpments, glaciation (alpine and continental), mountains, plains, plateaus, solution features, valleys, volcanic features, water features, wind features, and the inevitable miscellaneous features.

The tremendous variety of landforms found around the world cannot be addressed in any detail here, nor can their underlying geologic origins be adequately explained. A textbook on physical geography or geomorphology should be consulted for additional examples, as well as for discussion of the influence of subsurface structure and other geologic factors on the development of landforms. In addition, the examples shown in the U.S. Geological Survey collection *A Set of 100 Topographic Maps Illustrating Specified Physiographic Features* provide an invaluable source of further information regarding typical landforms.[3] Although the examples in this collection are from the United States, similar features are found throughout the world.

Coastal Features and Shorelines

The features found in coastal areas and along shorelines develop in a zone where land and water meet. Sand, for example, is moved by wave and current action until it reaches a location where those actions cannot carry it any longer. Deposits of sand and other materials on the shoreline are shaped by the complex wave and current action into distinctive features, such as sand spits (Figure 10.10).

[3]Extracts from the collection are presented in Richard DeBruin, *100 Topographic Maps Illustrating Physiographic Features* (Northbrook, Ill.: Hubbard Scientific, 1970).

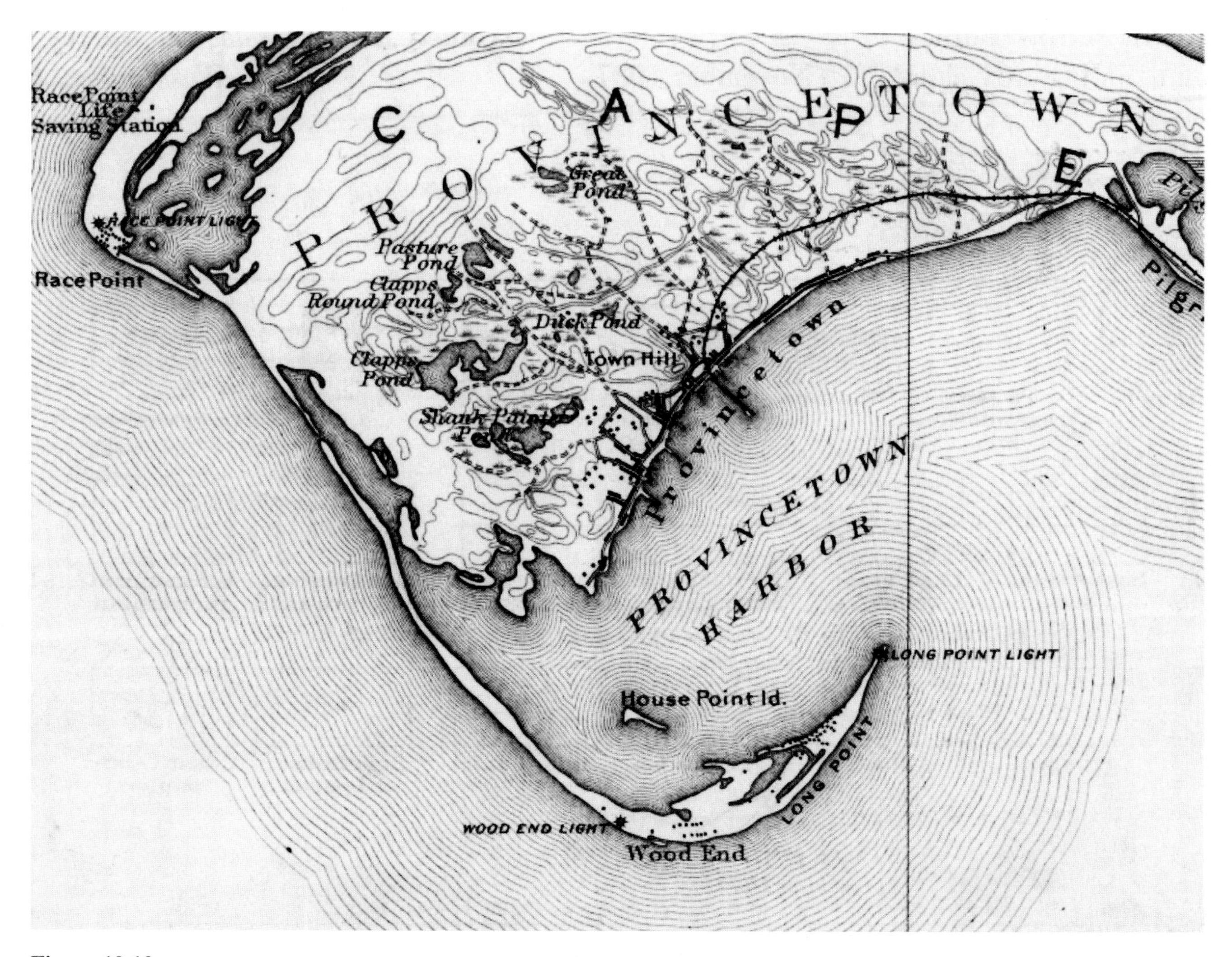

Figure 10.10 Sand spit. (Provincetown quadrangle, Massachusetts, 1:62,500, 1889. Reprinted 1940.)

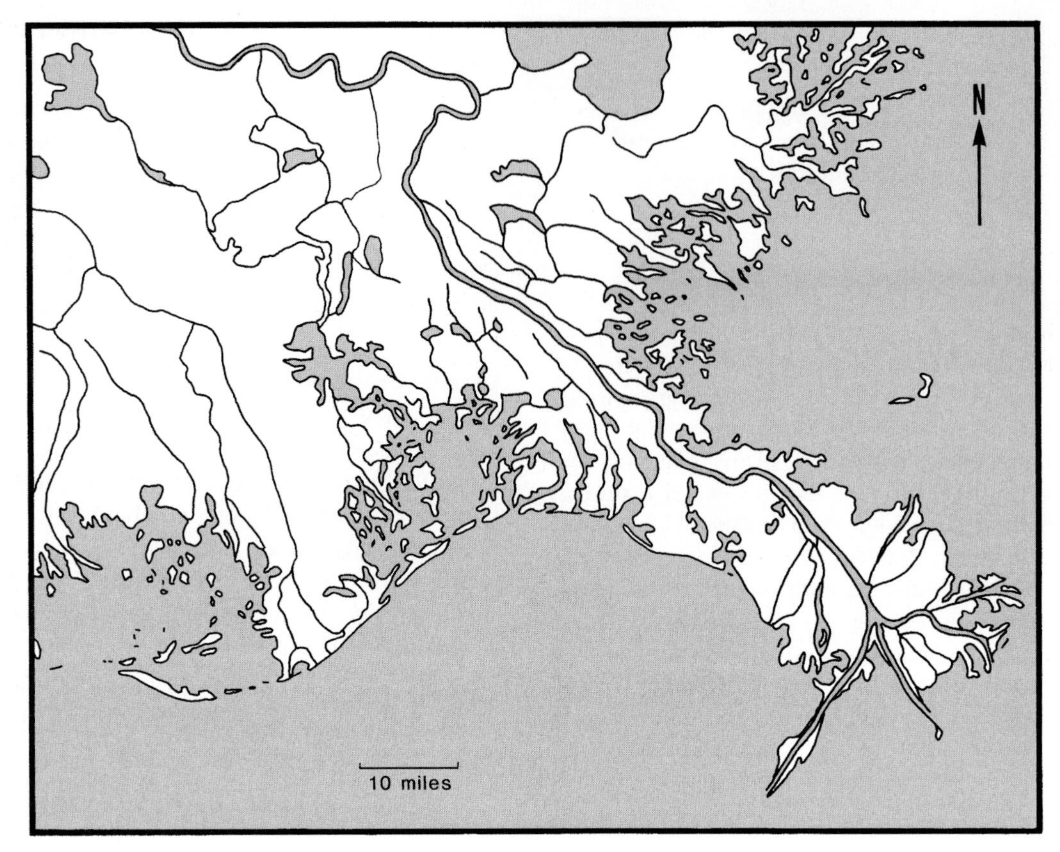

Figure 10.11 Delta of the Mississippi River.

Source: Traced from Army Map Service Strategic Map (undated).

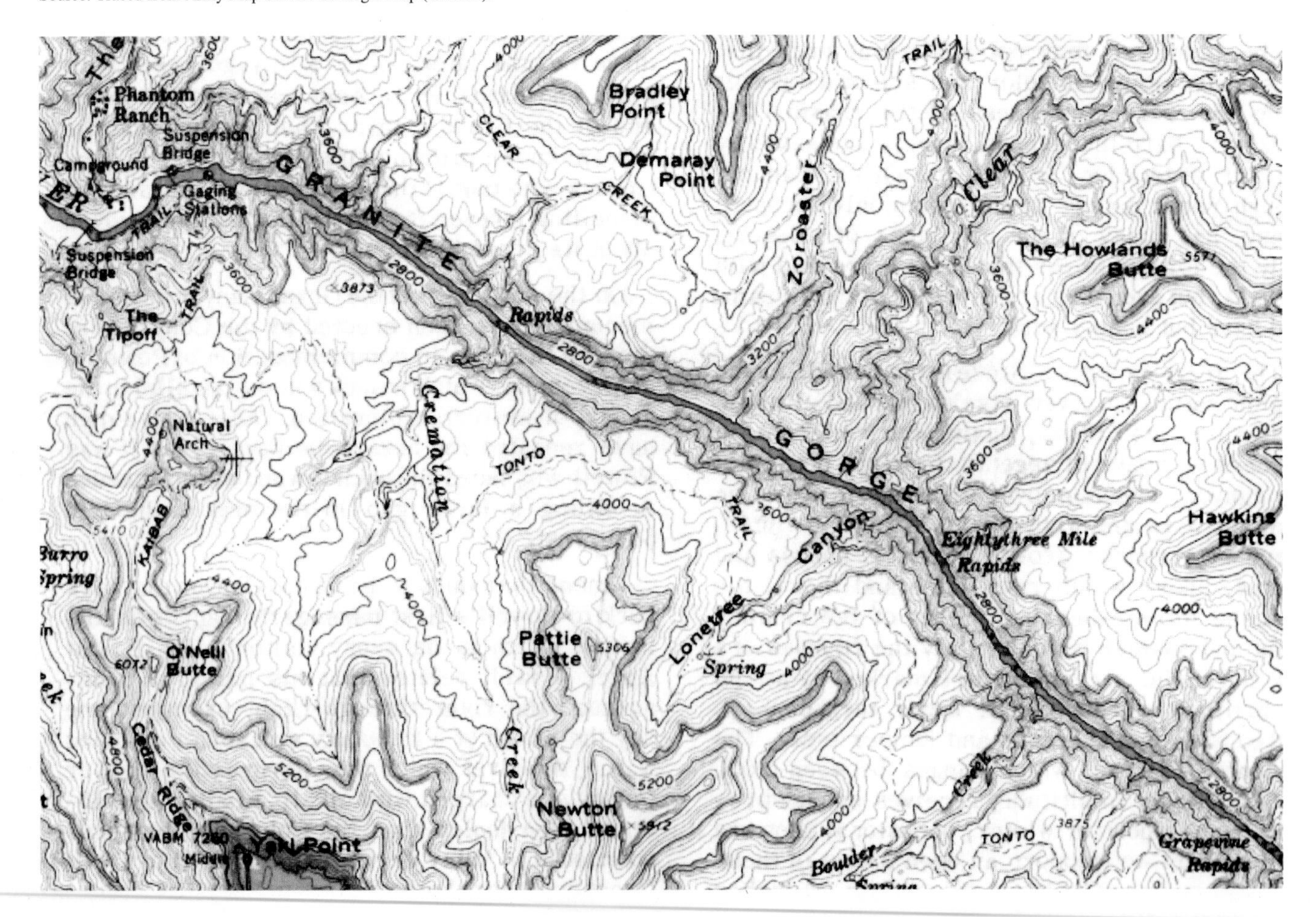

Figure 10.12 Cliffs. (Bright Angel quadrangle, Arizona, 1:62,500, 1962.) Cliffs abound in the Grand Canyon but are especially notable in the Granite Gorge.

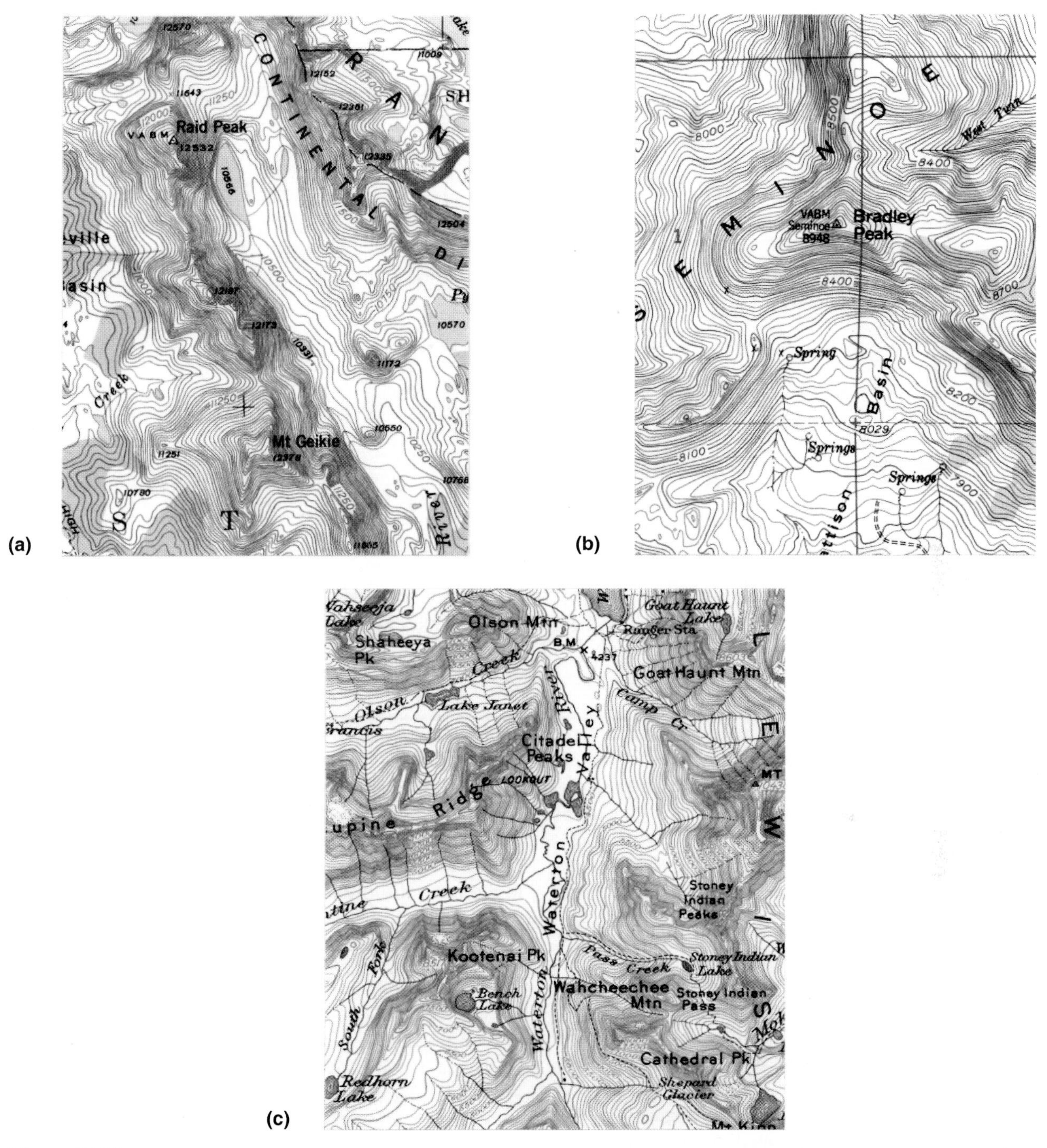

Figure 10.13 Typical alpine glaciation features. (*a*) Arête between Raid Peak and Mount Geikie, with col just north of Mount Geikie. Both Raid Peak and Mount Geikie are horns. Cirque, with tarn, southeast of Raid Peak. (Mount Bonneville quadrangle, Wyoming, 1:62,500, 1938.) (*b*) Matterhorn (Bradley Peak). (Bradley Peak quadrangle, Wyoming, 1:24,000, 1953.) (*c*) U-shaped valley (Waterton Valley). (Chief Mountain quadrangle, Montana, 1:125,000, 1938.)

As a river enters a larger body of water, such as an ocean, a sea, or a lake, the speed of water flow decreases, and materials that had been suspended and carried by the stream are deposited. If conditions are suitable, such deposition forms a delta (Figure 10.11).

Escarpments

Escarpments occur where erosion has worked on the underlying structures and resulted in a marked, extensive difference in elevation, often in the form of cliffs (Figure 10.12).

Glaciation

Alpine

Alpine areas are often subjected to glaciation. The actions of the ice have carved a variety of alpine glaciation features in the mountains (Figure 10.13), including

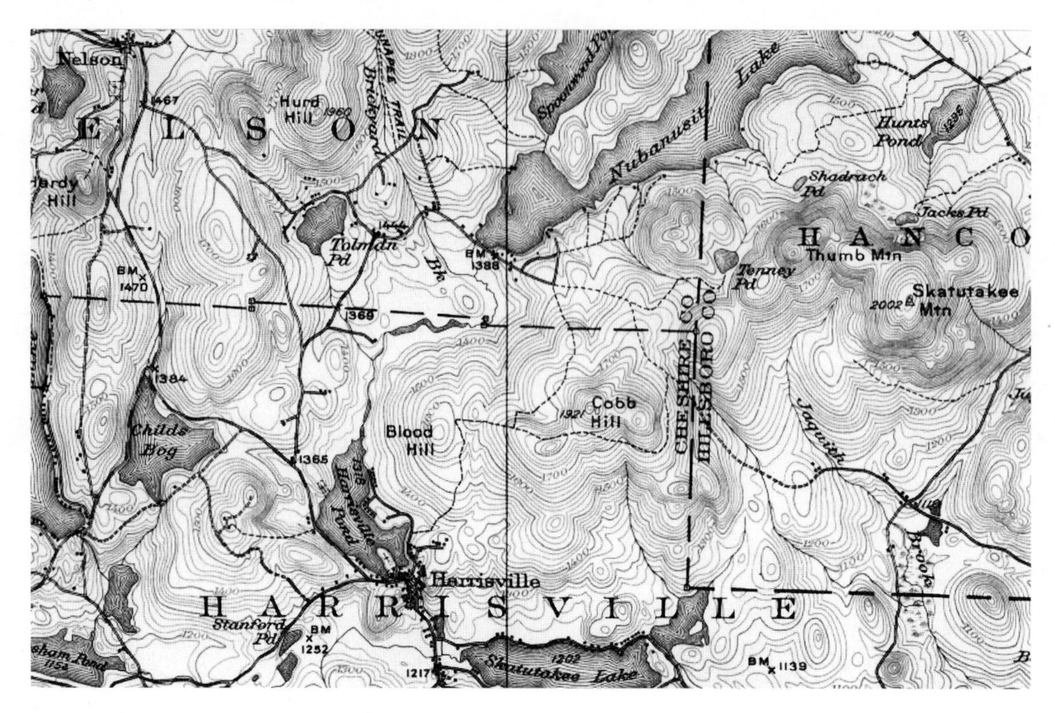

Figure 10.14 Continental glaciation: erosion to bedrock, with interrupted drainage, rounded hills, and depressions, often with lakes or marsh. (Monadnock quadrangle, New Hampshire, 1:62,500, 1936. Reprinted 1950, with corrections.)

basins at the heads of valleys (cirques) that sometimes contain lakes (tarns), sharp-edged ridges (arêtes) with passes or saddles (cols), and mountain peaks (horns or matterhorns). In addition, U-shaped valleys typically result from ice action.

Continental

Many distinctive features result from either the removal or deposition of materials by the action of the extensive ice sheets associated with continental glaciation. One frequently encountered aspect of areas subjected to glaciation is an interrupted drainage pattern, in which streams run in many directions without visible pattern, often through areas of lake and marsh. (Similar features are often found in areas of alpine glaciation.) In areas where material was removed, the bedrock left behind formed into smoothed depressions and hills (Figure 10.14). In areas where deposition was dominant, numerous depressions and small lakes called kettles formed when large fragments of ice lying within the glacial debris (till) melted and left cavities behind. Also, small hillocks called kames were formed by the deposition of debris in hollows within the ice (Figure 10.15). Drumlins, which often occur in groups called drumlin fields, are lozenge-shaped hills with steep slopes facing in the direction from which the ice advanced and more gradual slopes in the opposite direction (Figure 10.16). Terminal moraines were left at the margins of the farthest advanced of the glaciers (Figure 10.17), and lakes frequently occupied the space behind them. The resulting terrain features may include a line of small hills and hummocks at the location of the moraine, with a broad, flat lacustrine plain in the area of the former lake bottom (Figure 10.18).

Mountains

The features related to mountains include a great variety of ridges, hills, faults, and folds. Extensive concave or convex slopes may occur (see Figures 9.1c and 9.1d), or a distinctive formation may result from the erosion of particular types of folded bedrock. A hogback, for example, is a sharp-crested ridge formed by the eroded edge of steeply inclined strata (Figure 10.19). In other locations, streams may have eroded water gaps (Figure 10.20). Where the river has not been able to keep up with the uplift of the bedrock, partial gaps, called wind gaps, are left (Figure 10.20).

Plains

Features associated with plains may occur where lakes or streams have deposited materials to form an alluvial plain (Figure 10.21). Features of plains also may be formed in desert conditions or as a result of glaciation.

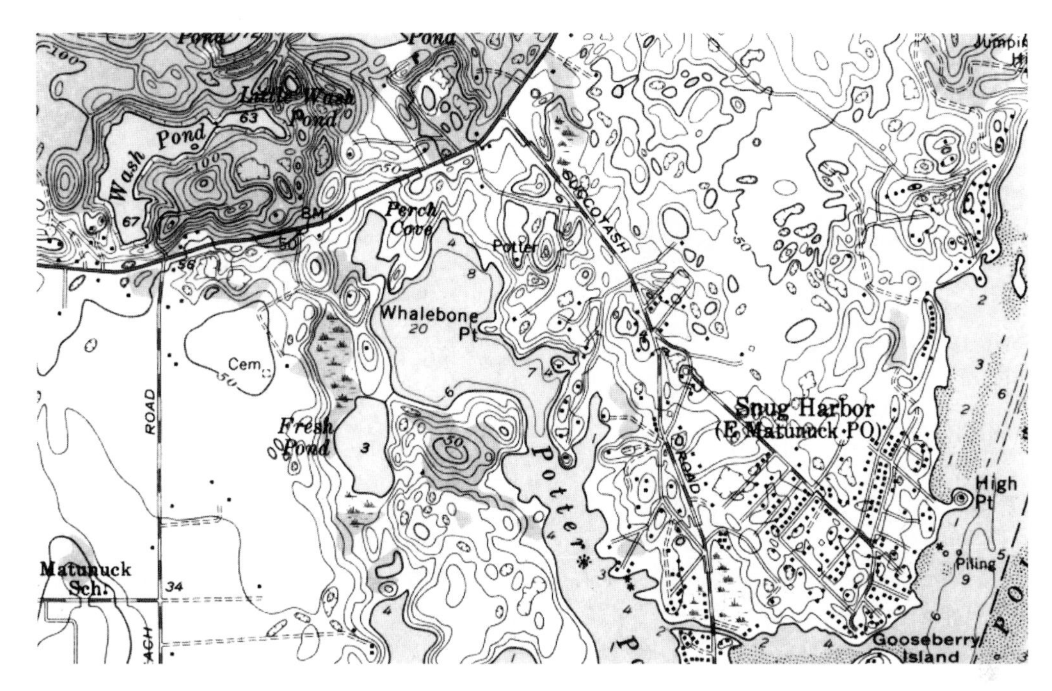

Figure 10.15 Continental glaciation: kames (hills) and kettles (depressions). (Kingston quadrangle, Rhode Island, 1:24,000, 1957.)

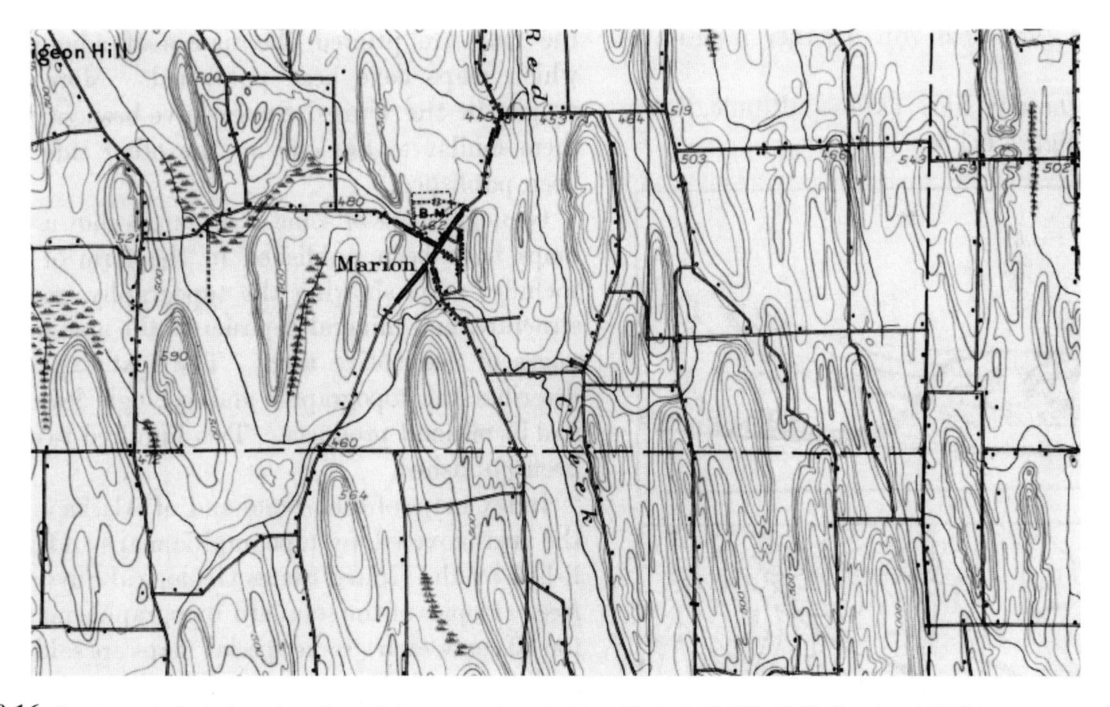

Figure 10.16 Continental glaciation: drumlins. (Palmyra quadrangle, New York, 1:62,500, 1902. Reprinted 1948.)

Plateaus

Plateaus often result when a relatively horizontal resistant rock layer is gradually eroded away, and less resistant materials are removed more rapidly from its edges (Figure 10.22). A variety of related features may become apparent as the edges of the plateau recede. A mesa is similar to a plateau, although less extensive, whereas a butte is an isolated remnant with very small surface area and clifflike sides (Figure 10.23). Where the resistant layer is tilted, a low ridge called a cuesta

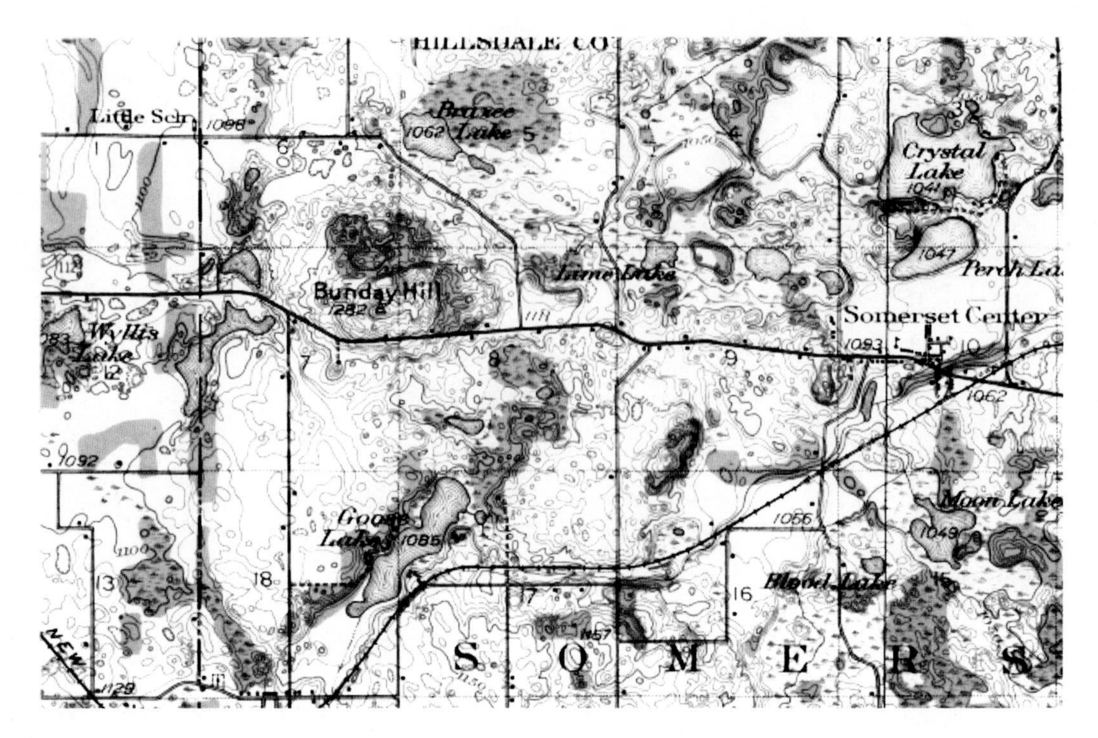

Figure 10.17 Continental glaciation: terminal moraine, with interrupted drainage, kames, and kettles with marshes and lakes. (Jackson quadrangle, Michigan, 1:62,500, 1935. Enlarged.)

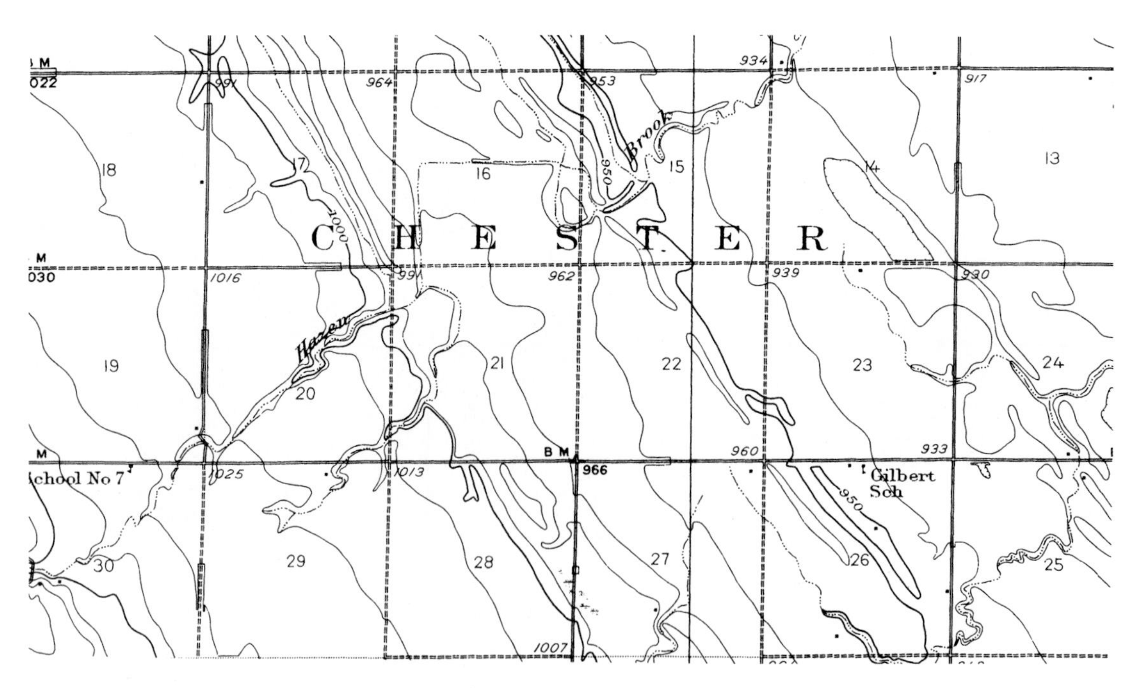

Figure 10.18 Continental glaciation: lacustrine plain, with characteristically gentle slopes and strand lines of glacial Lake Agassiz. (Emerado quadrangle, North Dakota, 1:62,500, 1936. Reprinted 1943.)

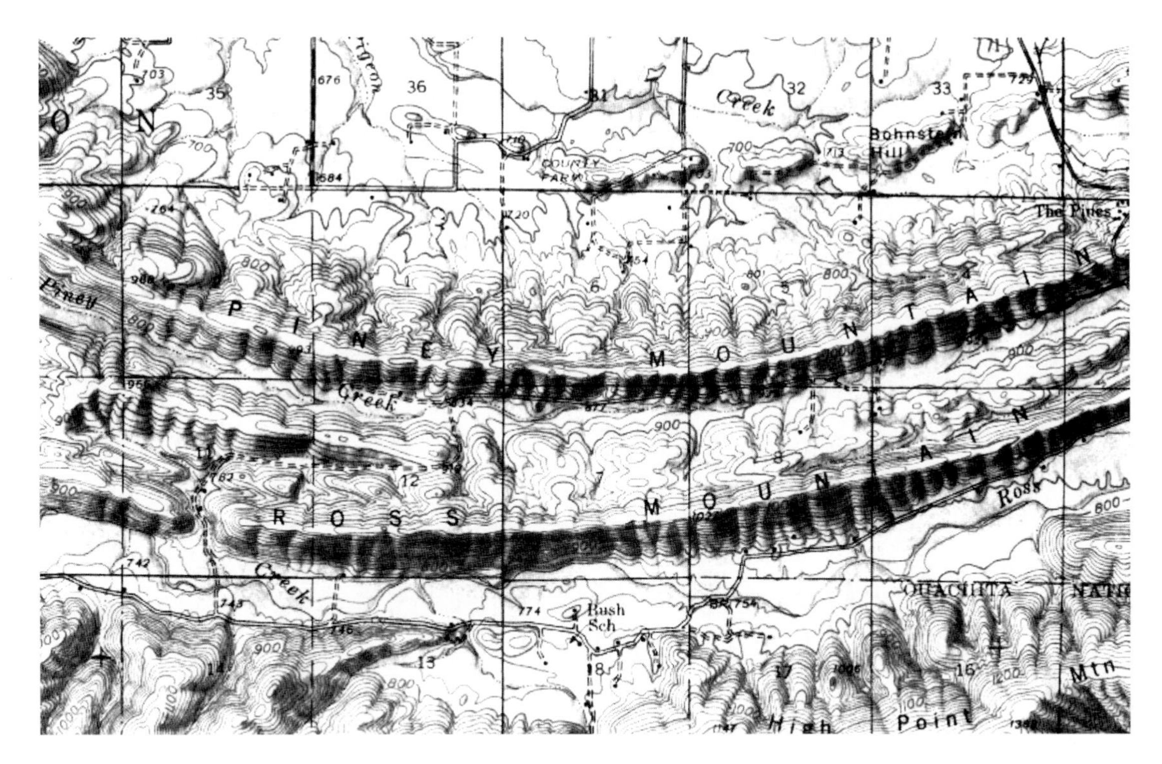

Figure 10.19 Hogbacks (Piney Mountain and Ross Mountain). (Waldron quadrangle, Arkansas, 1:62,500, 1939.)

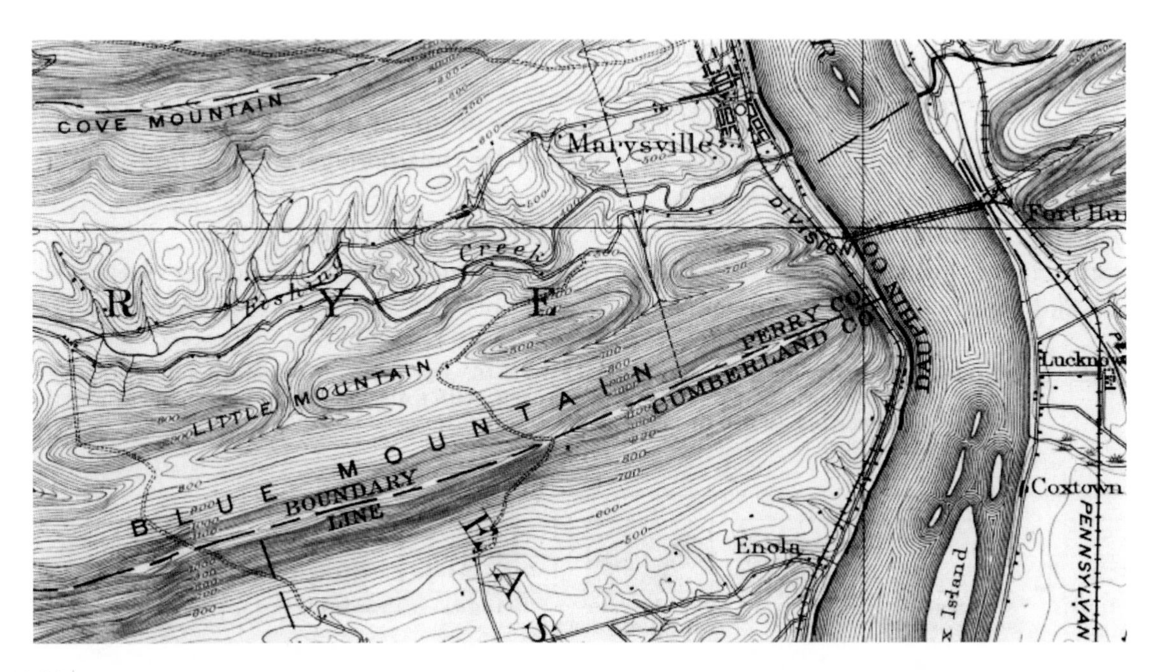

Figure 10.20 Water gap (stream at Little Mountain) and wind gap (road at Little Mountain). (Harrisburg quadrangle, Pennsylvania, 1899. Reprinted 1948.)

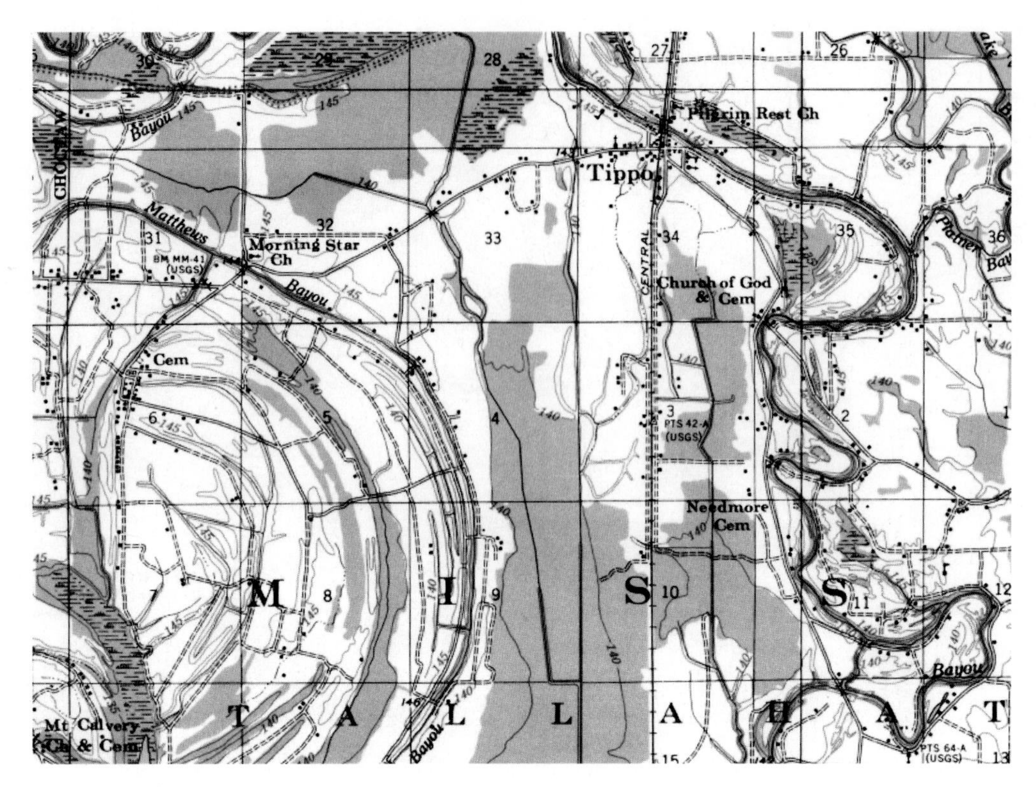

Figure 10.21 Alluvial plain, with low relief, lakes, meandering streams, abandoned meanders, and swampy areas. (Philipp quadrangle, Mississippi, 1:62,500, 1957.)

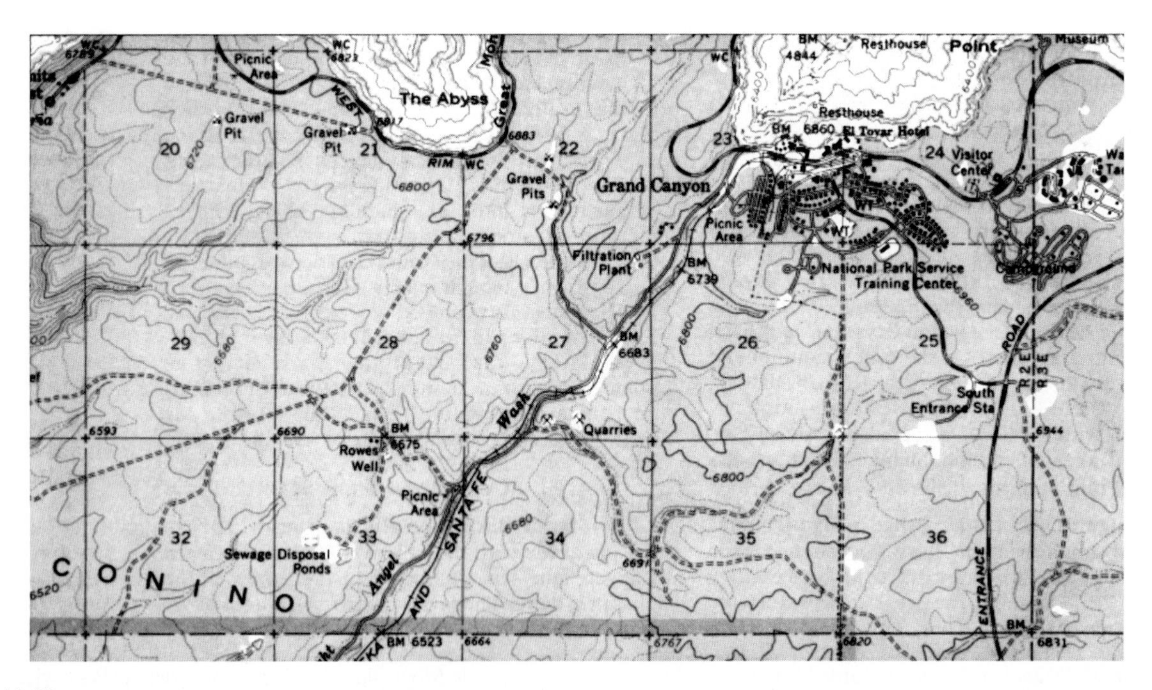

Figure 10.22 Portion of Coconino Plateau, at the edge of the Grand Canyon. (Bright Angle quadrangle, Arizona, 1:62,500, 1962.)

may form instead of a plateau. A cuesta has a steep slope on one face and a more gently sloping surface (the equivalent of the plateau surface) on the other (Figure 10.24).

Solution Features

Certain types of rock, such as limestone, are susceptible to solution by water. Where such underlying bedrock is dissolved by water action, underground drainage may

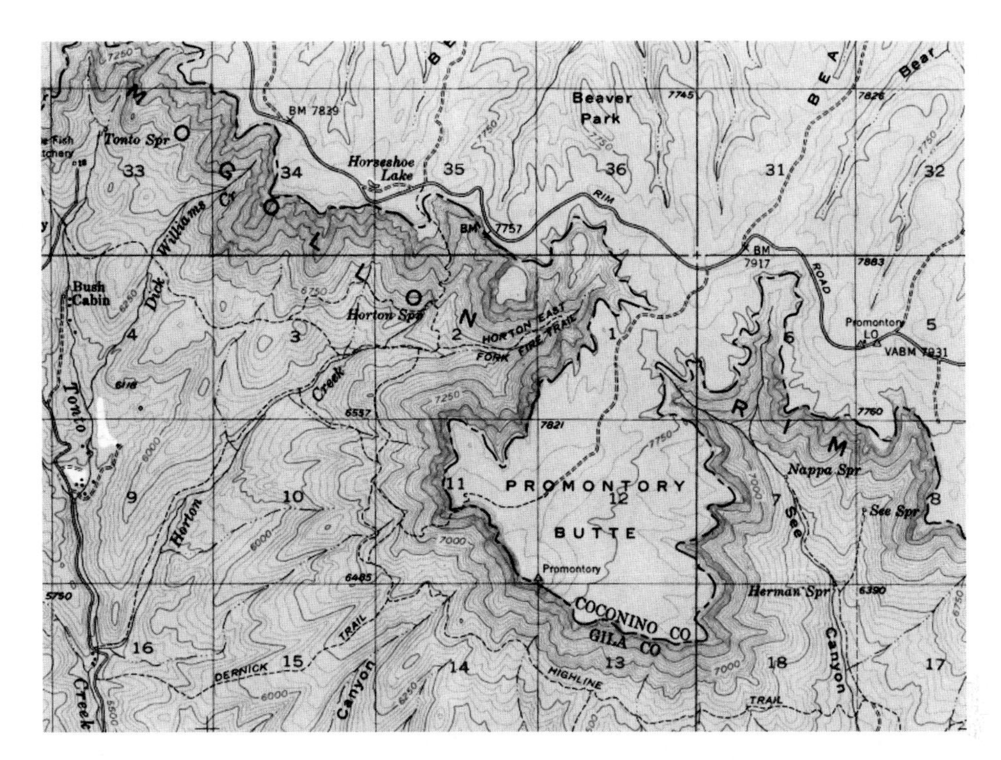

Figure 10.23 Butte (Promontory Butte) and mesa (tableland north of Promontory Butte). (Promontory Butte quadrangle, Arizona, 1:62,500, 1952.)

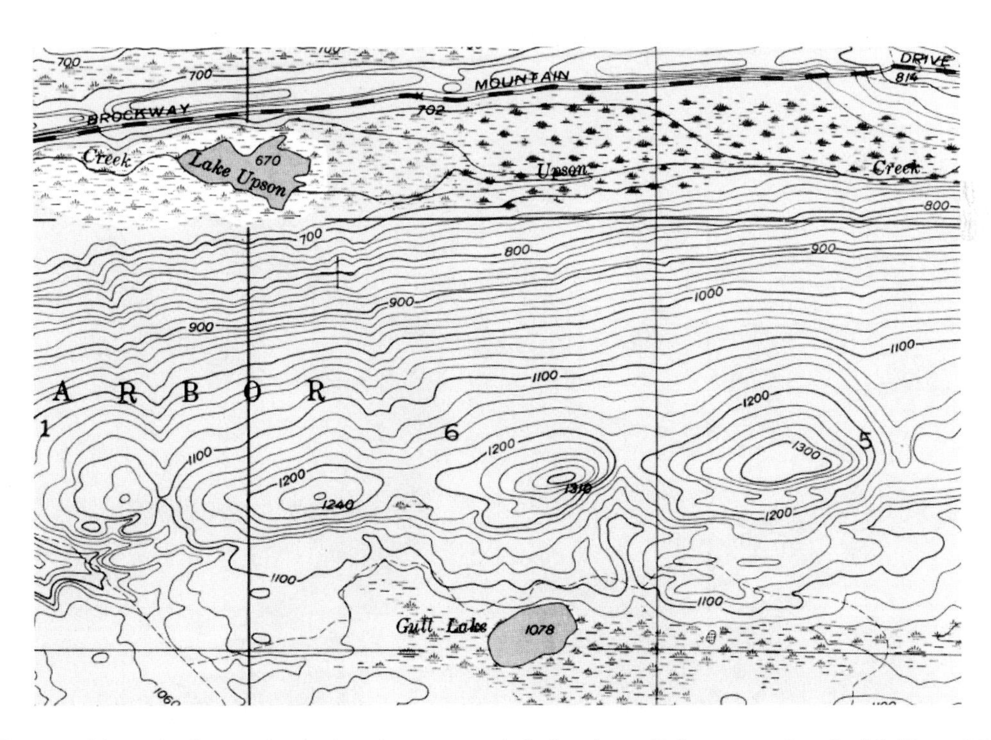

Figure 10.24 Cuesta, with gentle slope to the north and steeper south-facing slope. (Delaware quadrangle, Michigan, 1:24,000, 1948. Photo-revised 1975.)

develop. As a result, the surface drainage pattern in such areas is irregular and disrupted, with streams sometimes sinking from view. In addition, sinks may appear as the roofs of the underground streams and caverns collapse, and some sinks may be deep enough to form lakes (Figure 10.25). This situation emphasizes the need for more knowledge than can be gained from a topographic map alone. The kettles mentioned in the

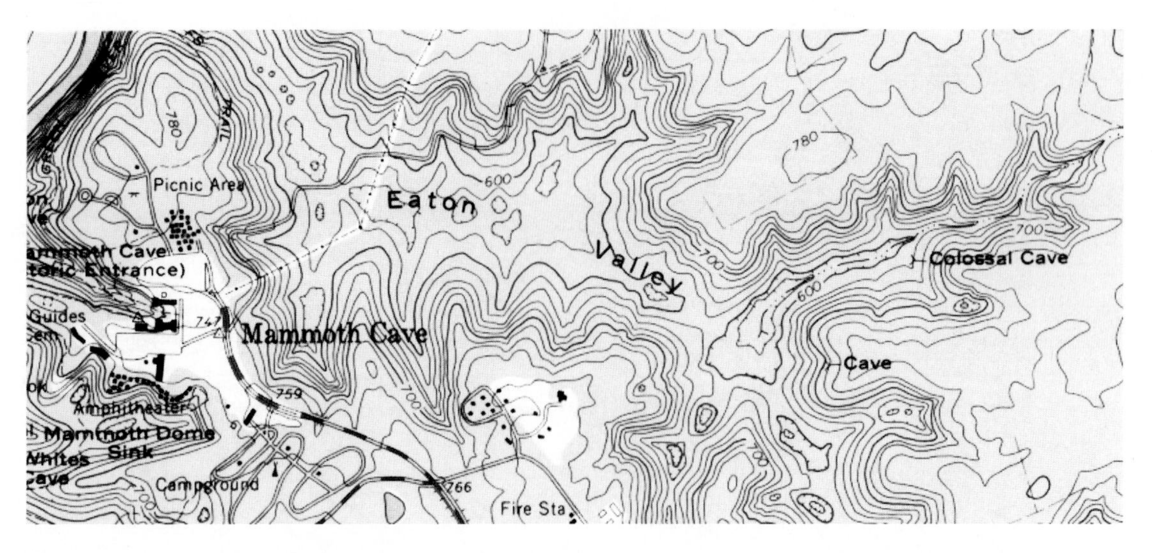

Figure 10.25 Solution features: depressions (sinks) (in Eaton Valley), cave entrance (Colossal Cave), and disappearing (sinking) stream (adjacent to Colossal Cave). (Mammoth Cave quadrangle, Kentucky, 1:24,000, 1965.)

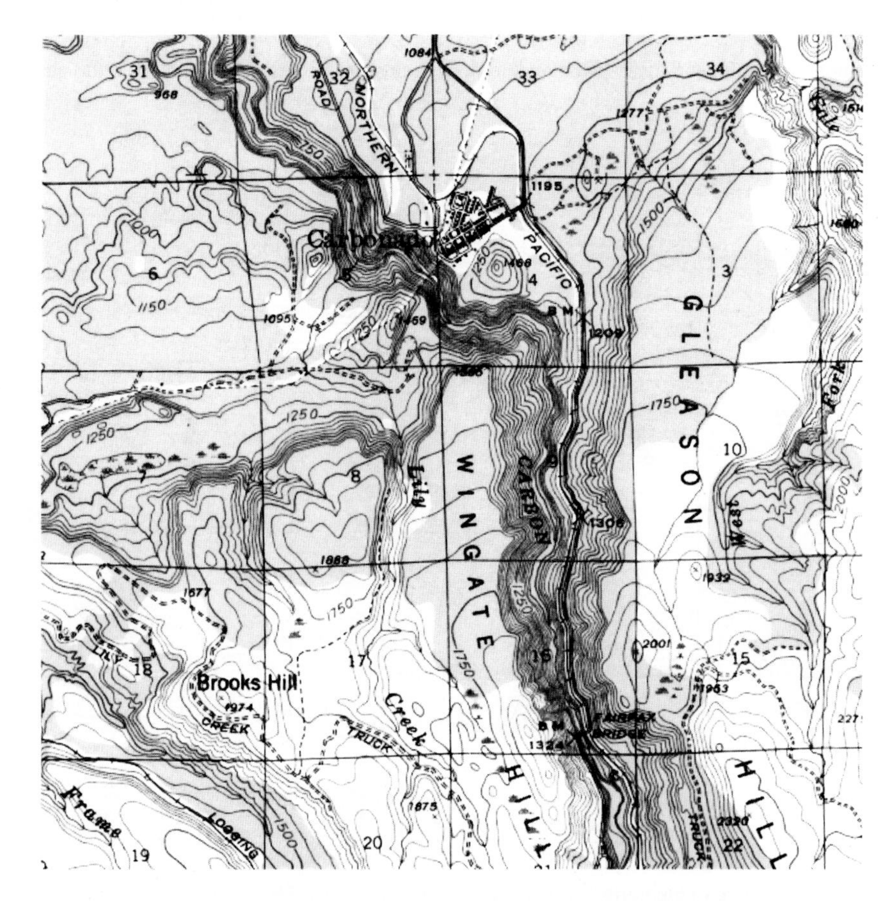

Figure 10.26 V-shaped valley (valley of Carbon River). (Lake Tapps quadrangle, Washington, 1:62,500, 1942.)

discussion of glacial features, for example, may take a form that looks very similar to solution features, even though their origins are entirely different.

Valleys

There is a tremendous variety of valley features resulting from the action of water in either removing or depositing materials. For example, a typical stream valley is V-shaped where downward cutting is active (Figure 10.26). Where streams are near to grade and downward cutting is minimal, however, meanders and natural levees form, and oxbow lakes develop in remnants of the changing river channel (Figure 10.27). In other areas, water action may deposit alluvial fans as water movement slows where it leaves steep, restricted channels and empties into a valley bottom (see Figure 10.7).

Volcanic Features

Features produced by volcanic action that are of sufficient size to be visible on a contour map include cinder cones, sometimes with craters at the top (Figure 10.28). Large volcanic peaks dominate many landscapes (Figure 10.29).

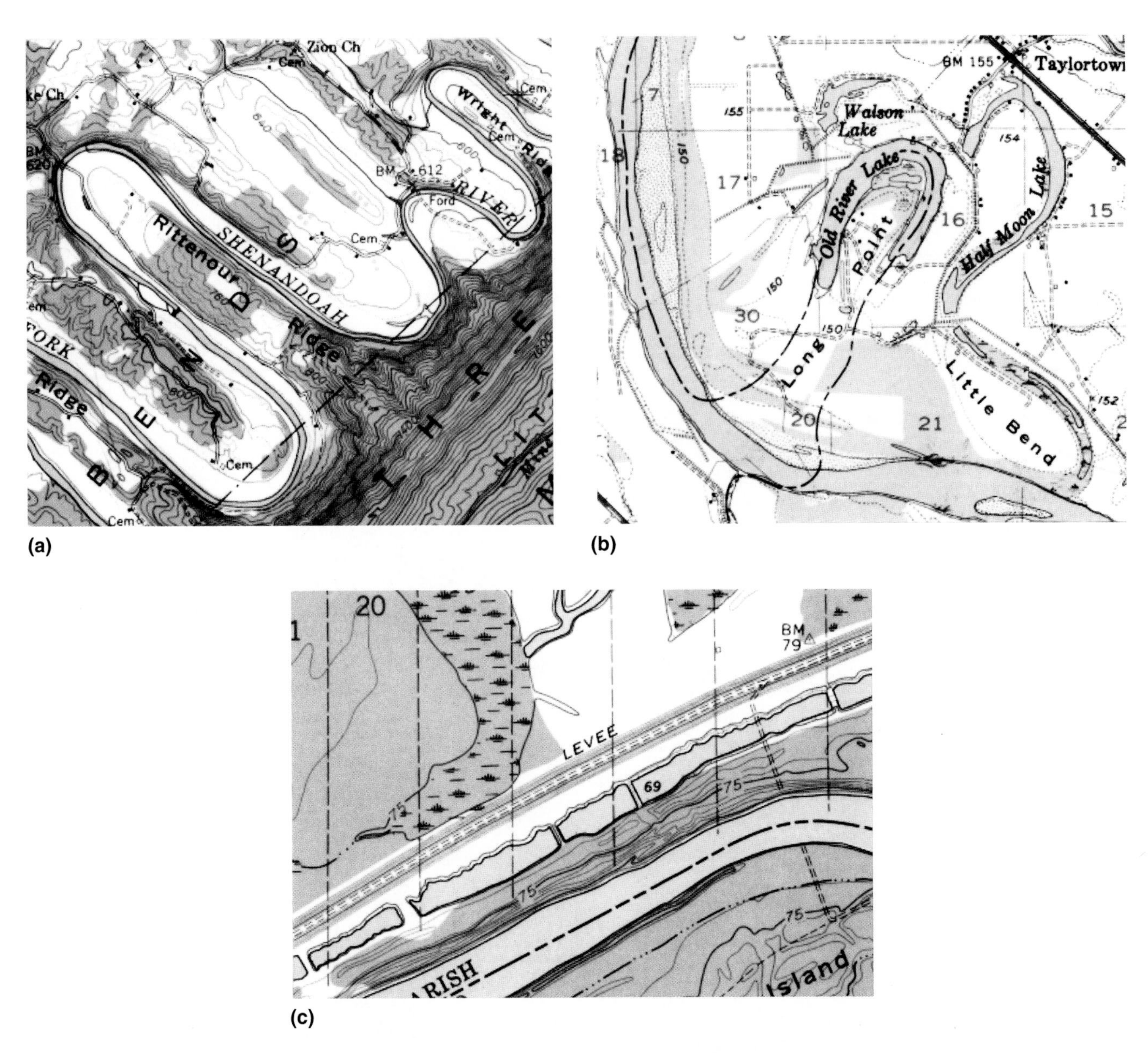

Figure 10.27 Stream features. (*a*) Meanders. (Strasburg quadrangle, Virginia, 1:62,500, 1947.) (*b*) Oxbow lakes. (Caspiana quadrangle, Louisiana, 1:62,500, 1955.) (*c*) Natural levees on the riverbanks, supplemented by man-made levees (labeled "LEVEE"). (Somerset quadrangle, Louisiana-Mississippi, 1:24,000, 1963.)

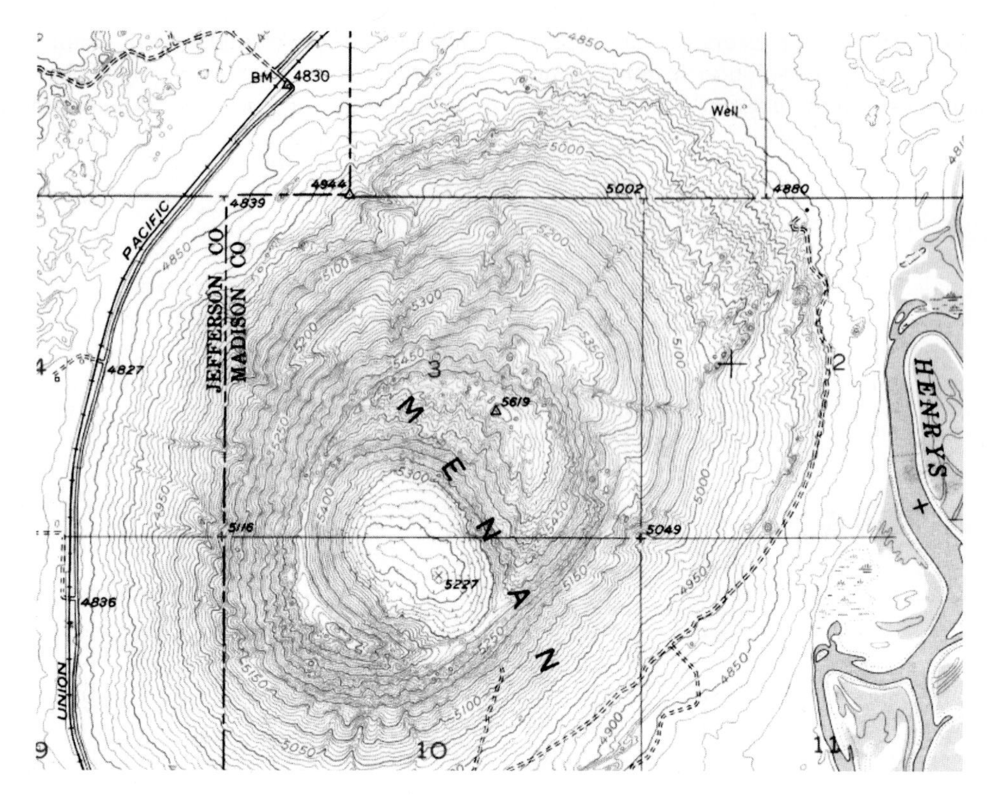

Figure 10.28 Cinder cone, with crater. (Menan Buttes quadrangle, Idaho, 1:24,000, 1951.)

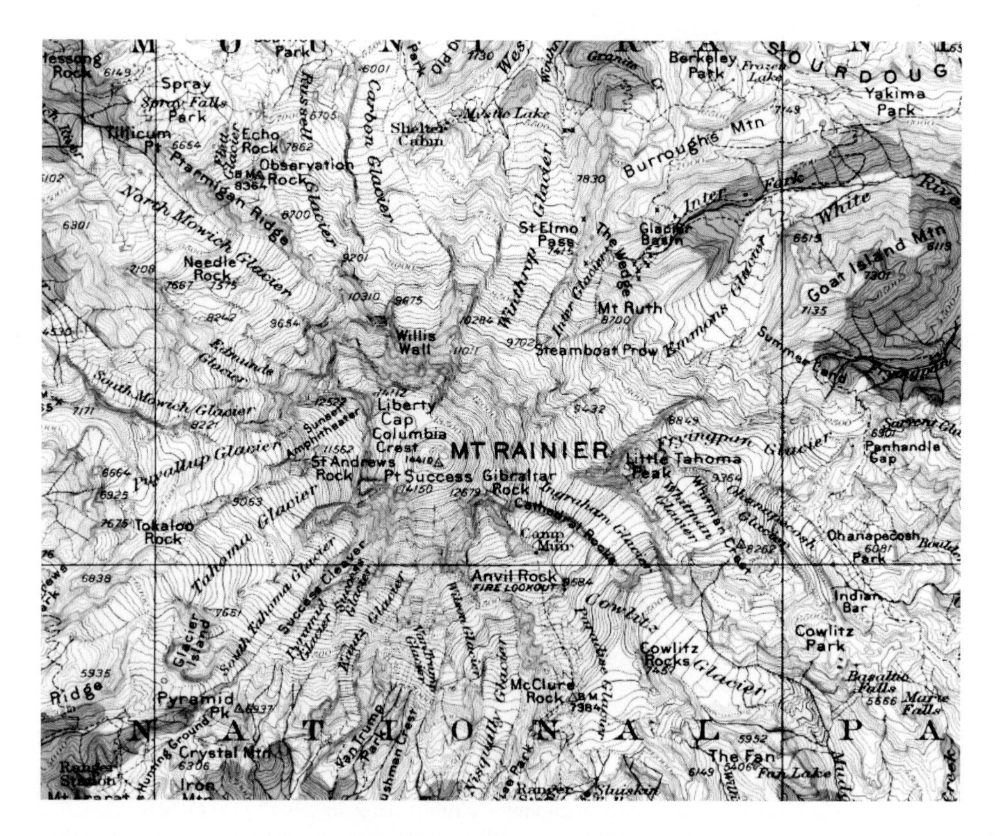

Figure 10.29 Volcanic peak, with radial drainage pattern. (Mount Rainier quadrangle, 1:125,000, 1924.)

Water Features

Another broad category of features comprises those distinctive forms created by the action of water. Many of the forms associated with lakes and streams are strongly interrelated with features already discussed, such as oxbow lakes, disappearing streams, and water gaps. Among the other forms are rapids and springs (Figure 10.30).

Wind Features

The wind is a powerful force for the transportation and deposition of materials. Features formed by the action of wind include dunes (Figure 10.31a) and ancient sand hills (Figure 10.31b).

Miscellaneous

Because of the enormous variety of landforms, some types fall into a category that must be labeled *miscellaneous*. Features in this category range from natural bridges (Figure 10.32) to fall lines (Figure 10.33).

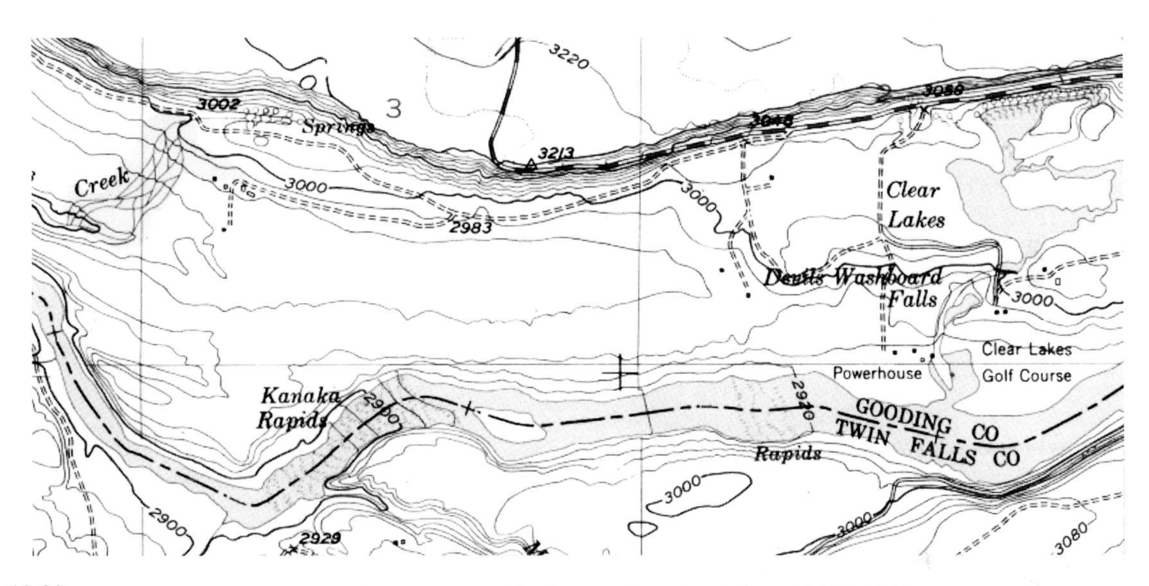

Figure 10.30 Water features: rapids and springs. (Thousand Springs quadrangle, Idaho, 1:24,000, 1949.)

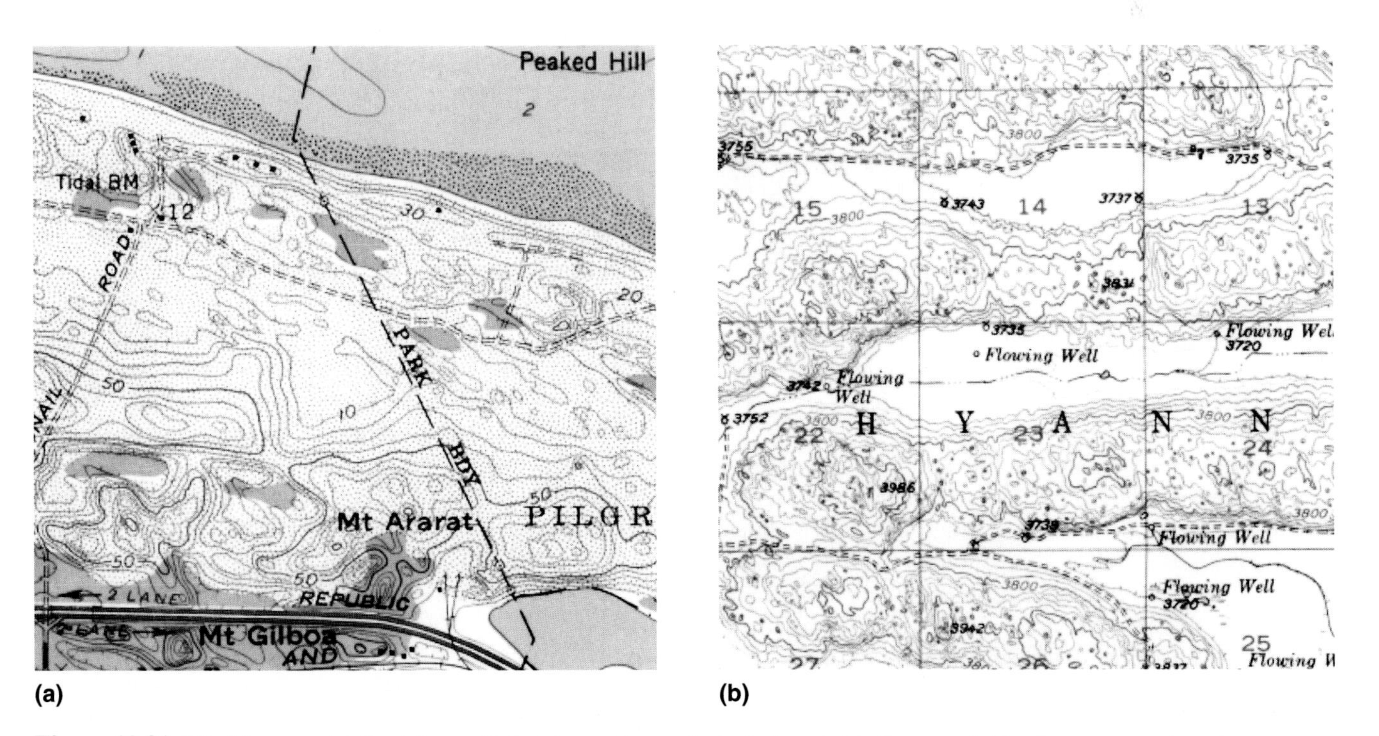

Figure 10.31 Wind features. (*a*) Sand dunes. Undulating surface and stippled symbol indicate sand. (Provincetown quadrangle, Massachusetts, 1:24,000, 1958.) (*b*) Sand hills. (Ashby quadrangle, Nebraska, 1:62,500, 1948.)

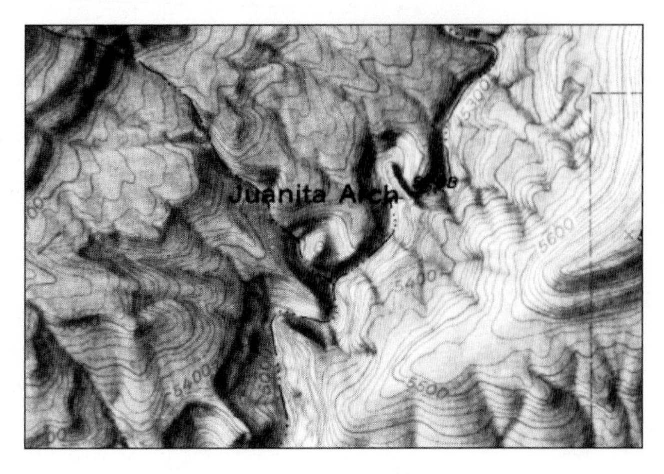

Figure 10.32 Juanita Arch natural bridge. An intermittent stream flows under the arch. (Juanita Arch quadrangle, Colorado, 1:24,000, 1960.)

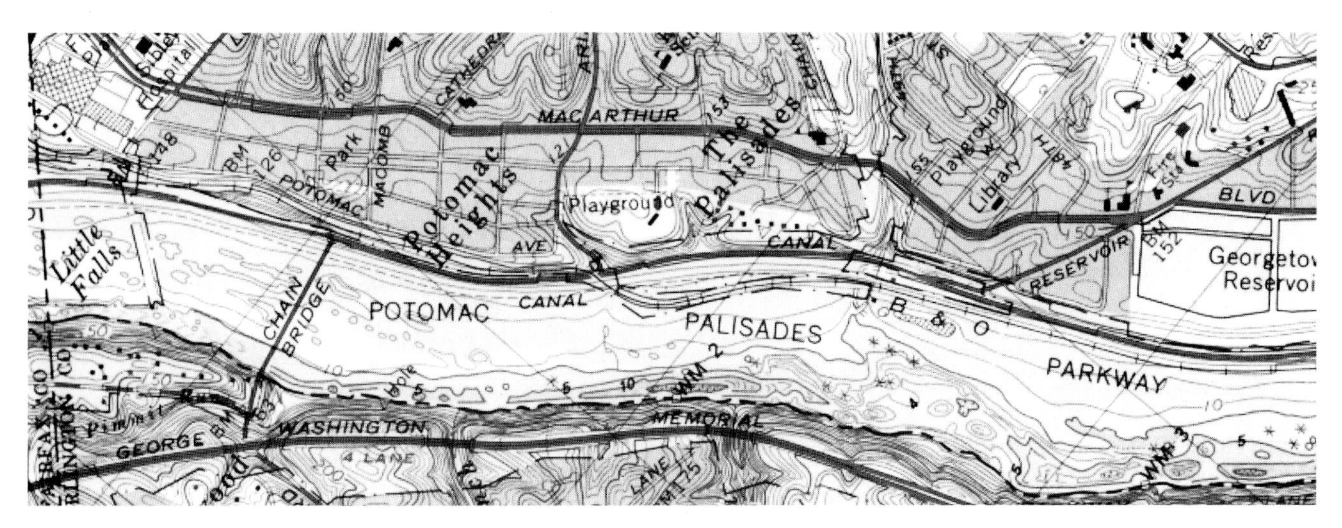

Figure 10.33 Fall line. (Washington West quadrangle, D.C.-Maryland-Virginia, 1:24,000, 1965. Photo-revised 1983.)

Summary

Landforms are the result of the complex interactions between the tectonic movement of the earth's crust and the geologic, hydrologic, atmospheric, and biotic forces that bring about weathering, mass movement, erosion, and deposition. As a major environmental element, topography has direct and indirect effects on human life.

Drainage patterns are classified in terms of the type and texture of dissection. Types of dissection include the treelike dendritic pattern and the related rectangular and trellis types. The circular patterns of channels flowing away from, or toward, a central point are radial patterns; those that flow outward are centrifugal, and those that flow inward are centripetal. Annular patterns are circular patterns modified by the presence of joints or fractures that create elements of parallel drainage.

A great variety of stream patterns form in differing circumstances. Interrupted patterns, for example, are characteristic of areas of poor drainage resulting from glaciation. Disappearing streams and internal drainage occur in areas of bedrock solution. Fanlike patterns of streams often spread out over the bases of alluvial fans. Braided, meandering patterns are characteristic of streams that are close to the level of a water body that establishes a limit on their downward cutting. Also, human intervention is seen in the artificial patterns of irrigation and drainage ditches and navigation canals.

Texture of dissection describes the relative fineness of the stream pattern. The concept offers some insight into the nature of the underlying geology and soils.

For descriptive purposes, landforms are grouped into categories based on their location or origin. The features described in this chapter represent only a sample of the tremendous range and variety of existing landform features. In particular, the challenging study of the influence of geologic structure and other factors on the formation of these features has been largely ignored because of lack of space; a separate book would be needed to do justice to the topic. The goal of this chapter, therefore, is simply to show that the contours on topographic maps provide useful information for the identification and initial study of distinctive landform features.

SUGGESTED READINGS

Brunsden, D., and J. Doornkamp, eds. *The Unquiet Landscape.* Bloomington, Ind.: Indiana University Press, 1974.

DeBruin, Richard. *100 Topographic Maps Illustrating Physiographic Features.* Northbrook, Ill.: Hubbard Scientific, 1970.

Garner, H. F. *The Origin of Landscape.* New York: Oxford University Press, 1974.

Miller, Victor C., and Mary E. Westerback. *Interpretation of Topographic Maps.* Columbus, Ohio: Merrill Publishing Co., 1989.

Pitty, W. F. *The Nature of Geomorphology.* London: Methuen, 1982.

Way, Douglas A. *Terrain Analysis: A Guide to Site Selection Using Aerial Photographic Interpretation.* 2d ed. Stroudsburg, Penn.: Dowden, Hutchinson and Ross; New York: McGraw-Hill, 1978.

CHAPTER

Qualitative and Quantitative Information

The modern postindustrial world thrives on the collection and dissemination of statistical information. Indeed, so much information of this type is gathered that finding the time to interpret and understand much of it is difficult. Graphics help to organize and interpret statistical information. Maps are a form of graphics, not totally dissimilar to graphs and statistical charts. The main feature that distinguishes maps from other graphics is that maps automatically incorporate locational information.

Various symbolization schemes are used to present quantitative information on maps. This chapter deals with the characteristics of some of the more common approaches. Other aspects of the presentation of statistical information are included in the discussions of cartograms (Chapter 14) and graphs (Chapter 15).

Point Symbols

Many variables shown on maps occur at single locations (points) on the earth's surface. The symbols used to represent such data are called **point symbols.** Point symbols are used to show two different characteristics of points: (1) the type or category of the phenomenon that occurs there and (2) the value or importance (population, amount produced, and so on) connected with it. The first type of information is nominal, whereas the latter is quantitative (see the discussion of "Levels of Measurement" in this chapter).

The use of the term *point symbol* must be interpreted rather broadly, because most objects mapped as points obviously are not actually dimensionless. A town, for example, is often represented on a map as a single dot. This does not imply that the town occupies no space. It simply means that, at the scale of the map, the space the town takes up is very small, compared with the area covered by the map. Maps drawn at small scales necessarily provide less room for symbol placement than maps drawn at large scales.

Nominal Information

When a phenomenon occurs at an identifiable point, its location can be shown by a **nonproportional point symbol.** This type of symbol is called *nonproportional* because it is not varied in size. Thus, it simply records a location and does not suggest any difference in relative importance from place to place. Placing a single dot on a map at the location of each person's home in a city, for example, results in what is called a **dot-distribution map.**

Most dot-distribution maps do not use a simple, one-dot-for-one-occurrence approach. A map might become overcrowded if a large number of dots are required, especially if they are kept large enough to be readable. Assume, for example, that a population map of the United States is being drawn. If one dot is used to represent each person, more than 270 million dots are needed, which is obviously impractical. Such a large set of data is inevitably condensed, or grouped, before it is mapped. As a result, a single dot may represent 20 people, factories, cows, production dollar

Box 11.1 Levels of Measurement

The measurements used to present statistical information on maps provide different levels of information. The possible types of interpretation differ, depending on the information level of a particular set of measurements. The four levels of measurement discussed here are nominal, ordinal, interval, and ratio.[1]

NOMINAL

Sometimes the information available simply divides a group into subgroups. This type of **qualitative information** carries no implication that one subgroup is more or less important than another. The level of measurement involved in qualitative information is called **nominal measurement.**

A simple example of nominal measurement is the separation of 30 people in a room into groups on the basis of eye color. If there are some people with blue eyes, some with brown eyes, some with black eyes, and some with green eyes, there are, then, four eye-color groups in the room. There is no logical basis, however, for saying that the green-eyed people should be ranked ahead of the brown-eyed and behind the blue-eyed. Any ordering in such a case is purely arbitrary. What you *can* do with the groups with different eye colorings is count the members of each group so that you know the number of people with eyes of each particular color. From these simple counts, you can also calculate the proportion or percentage of the total group that falls into each category. Nominal information of this type is used to create maps showing, for example, city classifications, types of pipelines, crop types, or soil categories.

ORDINAL

Assume that you line up the people in a room so that the tallest person is at one end of the line and the shortest is at the other. In addition, you make certain that each person is standing between someone who is taller and someone who is shorter. When you do this, you have classified the members of the group by an **ordinal measurement,** using relative height as a logical ordering criterion. Notice, however, that you are only comparing one person with another to see who is taller and who is shorter. You are not obtaining measurements, in centimeters, for example, so that you know each person's exact height.

Many types of ordinal information are used on maps. Boundaries, for example, are usually classified according to the level of the political entity that they define. County boundaries are less important than state boundaries, and national boundaries are more important than either state or county boundaries. Similarly, roads are classified according to their level of jurisdiction: county road, state highway, or federal interstate highway. Any such arrangement is organized around the concept of more or less important but does not include a quantitative measurement.

INTERVAL AND RATIO

The data levels beyond ordinal measurement are interval and ratio. These levels provide the exact dimensions, on some standard scale, of the particular phenomenon of interest. Interval and ratio levels of measurement are very similar, but the difference that exists between them affects the way in which each type of information can be used. We will discuss the ratio type first and then compare the interval type to it.

Taking a **ratio measurement** requires the use of a regularly spaced scale that begins at zero and extends out without any specific limit. For example, distance is measured using a scale such as millimeters, centimeters, or meters (or inches, feet, or yards). A meter stick, with a zero point at one end and with millimeter subdivisions marked, is a kind of scale typically used for such measurements. The zero mark on its scale indicates the total absence of length. If the scale is appropriately subdivided, an object of very small dimension can be measured and its size recorded. Such a scale can also be extended to measure objects of greater and greater size. Theoretically, at least, the scale can be extended to any length necessary, conceivably to include interplanetary distances.

A metric distance scale allows you to say that *B* is, for example, 10 meters east of *A* or that *C* is 20 meters south of *A*. In this case, you can also say that *A* is twice as far from *C* as it is from *B*. This latter capability gives the measure its name, because it allows the ratio between two measurements to be determined.

The characteristic that distinguishes an interval scale from a ratio scale, then, is that an **interval scale** allows you to determine values according to a particular measure but does not allow you to determine the ratio between two values. Such scales are relatively rare, but one type that commonly occurs on maps is temperature information on either the Celsius or Fahrenheit scales. If two temperature readings, such as 20° and 40°, are taken using the Celsius scale, we can say that 40° is warmer than 20° (an ordinal statement). We can also say that there is a 20-degree difference between the two readings, which is the same as can be done using a ratio scale. What we *cannot* say is that 40° is *twice* as warm as 20°. The reason that the latter statement makes no sense is that the starting point of the Celsius scale is an arbitrary zero point. That is, 0° Celsius represents the freezing point of water but *does not* represent an absolute zero, which would occur in the complete *absence* of heat.[2]

[1]Adapted from S. S. Stevens, "On the Theory of Scales of Measurement," *Science* 103 (1946): 677–80, and Jacques Bertin, *Semiology of Graphics* (Madison: University of Wisconsin Press, 1983), 34–39.

[2]The Kelvin scale, on the other hand, starts at absolute zero, which *does* represent the absence of heat. It is, therefore, a ratio scale.

values, or whatever. Or one dot may represent a thousand items or even a million or more (Figure 11.1). The number of items represented by each dot depends both on the scale of the map and on the size of the population involved.

Dot maps do not always show the distribution of a single phenomenon, such as cows. Suppose, for example, that the locations of various types of alcoholism treatment programs are plotted on a map of Oklahoma, with the location of each program indicated by a point symbol. The types can be nominally differentiated by varying the shapes of the point symbols (Figure 11.2). Besides shape variations, variations in color are also often used to distinguish nominal categories. On an

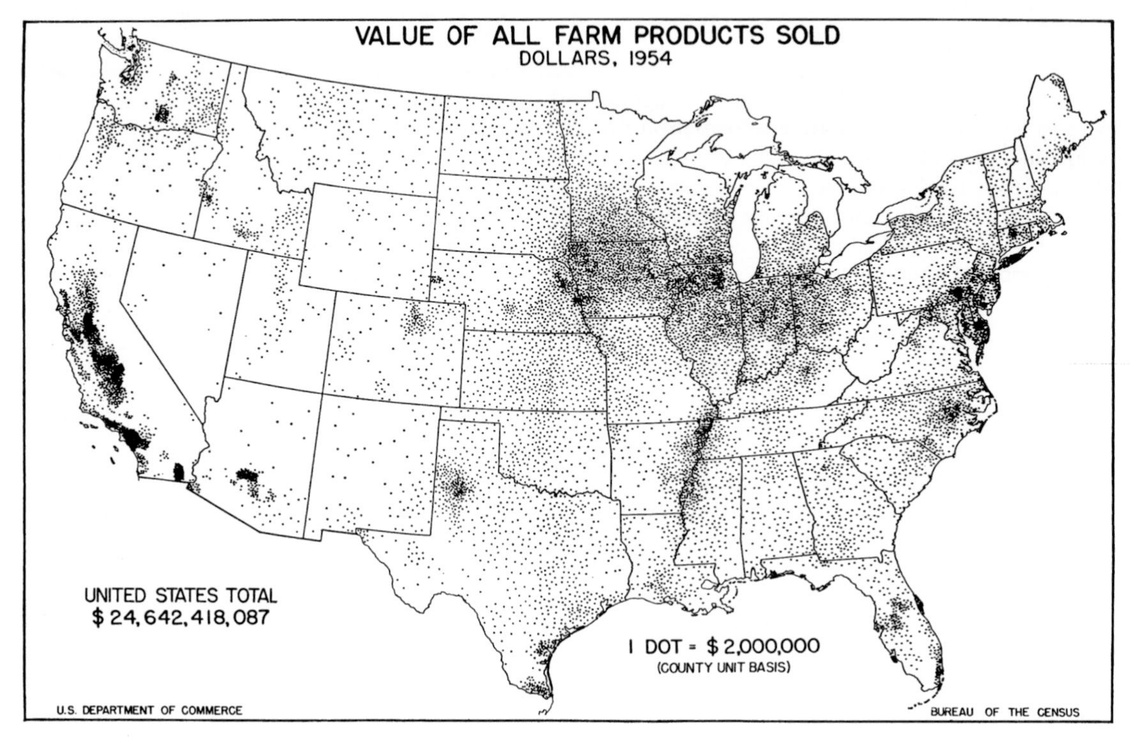

Figure 11.1 Dot-distribution map.
Source: Bureau of the Census.

Figure 11.2 Nominal point symbols used to show types of programs providing alcoholism treatment centers in Oklahoma, 1980.
Source: From Christopher J. Smith, "Locating Alcoholism Treatment Facilities," *Economic Geography* 59, no. 4 (October 1983): 368–85. Reprinted by permission.

agricultural map, for example, blue dots may represent milk cows, and red dots may stand for beef cattle.

Remember that the nominal symbols on a given map are kept to the same size, whether they represent a specific number of occurrences or a specific type of occurrence. This is so they do *not* suggest any quantitative differences. Thus, if the legend shows that a single symbol represents five cows, every symbol of that type represents five cows wherever it occurs on the map. Similarly, a symbol representing a residential treatment center is kept to the same size as one representing an outpatient center, because, for the purpose of the map, both types are equally important. Showing differences in quantity or importance requires at least an ordinal level of information and the use of some type of proportional symbol.

Quantitative Information

Graduated point symbols are often used to indicate the values associated with various locations. The population range of the cities within a region, for instance, can be shown by appropriately varying the size of the dot that represents each city.

When dot-size variations are used, the cartographer provides a key that indicates their meaning. Usually, a large dot indicates a city with a relatively large population, whereas a small dot indicates one with a smaller population (Figure 11.3). However, whether the symbol simply reflects the *extent* of the geographical area encompassed by the city or whether it reflects the city's population may be unclear. The specific populations involved and whether one city's population is only slightly larger than the others' or, perhaps, much larger may also be uncertain. For these reasons, the map legend must be read carefully to determine the specific population or population range indicated by each size symbol.

Proportional circles, which are an extension of the graduated point symbol idea, are discussed in Chapter 15.

Figure 11.3 Map using graduated point symbols to show sizes of known Old Order Amish settlements in the United States, 1976.

Source: From William K. Crowley, "Old Order Amish Settlement: Diffusion and Growth," *Annals of the Association of American Geographers* 68, no. 2 (June 1978): 249–64. Reprinted by permission.

LINEAR SYMBOLS

Because of their nature, many features such as railroads, roads, telephone or power lines, political boundaries, and rivers are best represented on maps by **linear symbols.** Two general types of linear symbols are used. One type differentiates between nominal categories of features, and the other presents quantitative information.

Nominal Information

Nominal linear symbols simply allow the map user to recognize that one line represents a river and another a railroad, for example. Differences in line color (or gray value), shape, or pattern are the most common nominal linear symbols. When color is used, the line representing a river may be printed in blue, whereas a black line of similar width may represent a highway. When shape is used, two lines may be the same color, but a dashed black line may indicate a trail and a solid black line a road. Combining colors and shapes produces a great variety of symbols, so that a variety of features can be uniquely identified.

The map legend specifies what the nominal linear symbol—for example, the solid red line, the dashed green line, or the dotted black line—represents on that particular map. Consulting the legend is necessary because the same symbol may stand for entirely different things on different maps.

Quantitative Information

In addition to indicating what is being represented, linear symbols often represent quantitative information.

Ordinal Symbols

Ordinal linear symbols are often used to differentiate the relative importance of what they represent. For example, different line symbols may represent the various levels of importance of political boundaries. There is obviously a range from less important county boundaries to increasingly important state

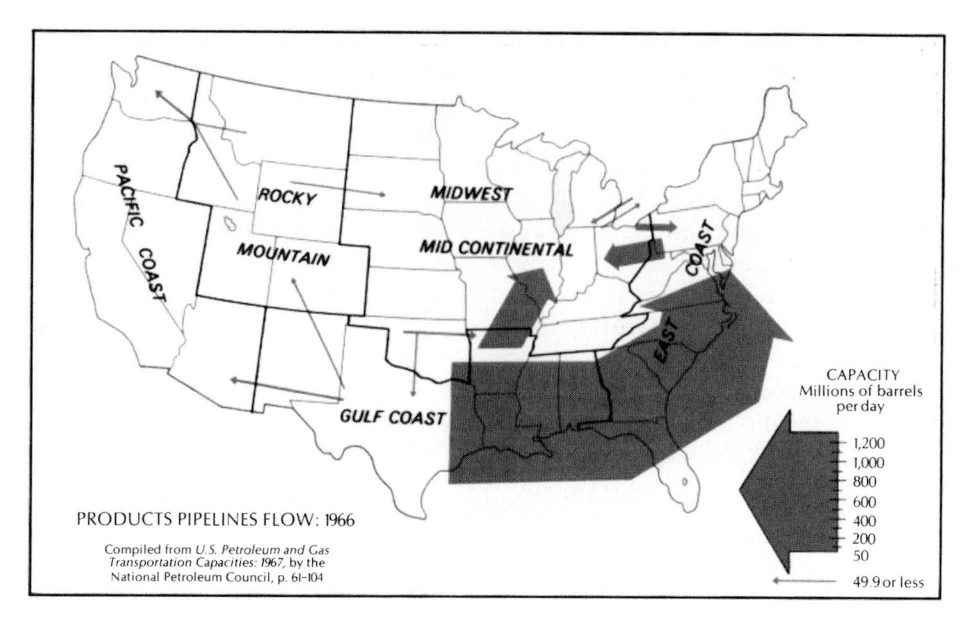

Figure 11.4 Flow map with proportional line widths.
Source: *The National Atlas,* 235.

and national boundaries. A similar hierarchy often applies to road designations, from local through county and state to national levels. When hierarchies such as these are mapped, the style and weight of the lines used to represent each level is typically altered to represent the appropriate level of prominence. Flow maps also often use ordinal symbolization, but they frequently are extended to show ratio levels of information.

Flow-Line Symbols

Quantitative linear symbols are found on **flow maps,** which show connections of various types that exist between locations on the earth's surface. A trade exchange between two cities, for example, may involve the movement of a certain quantity of products worth a given amount of money. Either of these pieces of quantitative information—volume or value—can be the basis of a flow map.

The usual convention regarding the use of flow lines to convey ordinal or quantitative information is to vary the width of the line in relation to the amount or value of the flow, with a wider line indicating a greater quantity. Flow-line symbols may also indicate nominal information, such as the types of goods or services involved in the exchange, in combination with an indication of the value.

Line widths on flow maps are often drawn in direct proportion to the value represented. A scale in such cases might show, for example, that a line 1 millimeter wide represents 10,000 tons of product, and a line 10 millimeters wide represents 100,000 tons of product (Figure 11.4). This approach is feasible when the range of values is not too great.

When the flows range from very small to very large values, the data are often divided into several categories. These categories may simply be small, medium, and large when ordinal information is being presented. The various line widths represent specific data ranges such as 1–10, 11–100, and over 100, when ratio information is involved. In either case, a particular line width is used to represent each range of values, so the symbols are referred to as **range-graded**[3] (Figure 11.5). When symbols are range-graded, the line that represents ten times the value may only be five times as wide (or some other multiple). When range-graded symbols are used, therefore, it is particularly important to check the legend to determine the values represented by each width.

Symbols are sometimes used instead of width variations, to represent different quantities in flow lines. For example, a dotted line may indicate a minimal-value flow, a dashed line a greater flow, and a solid line the maximum flow.

One additional style variation in flow maps has to do with the placement or routing of the flow lines themselves. In some cases, the placement of the lines indicates the actual routes followed. The lines may follow the alignments of the highway or the railroad along which the goods are moved, for example. In

[3]This is the same approach used for classed choropleth maps and range-graded circle symbols.

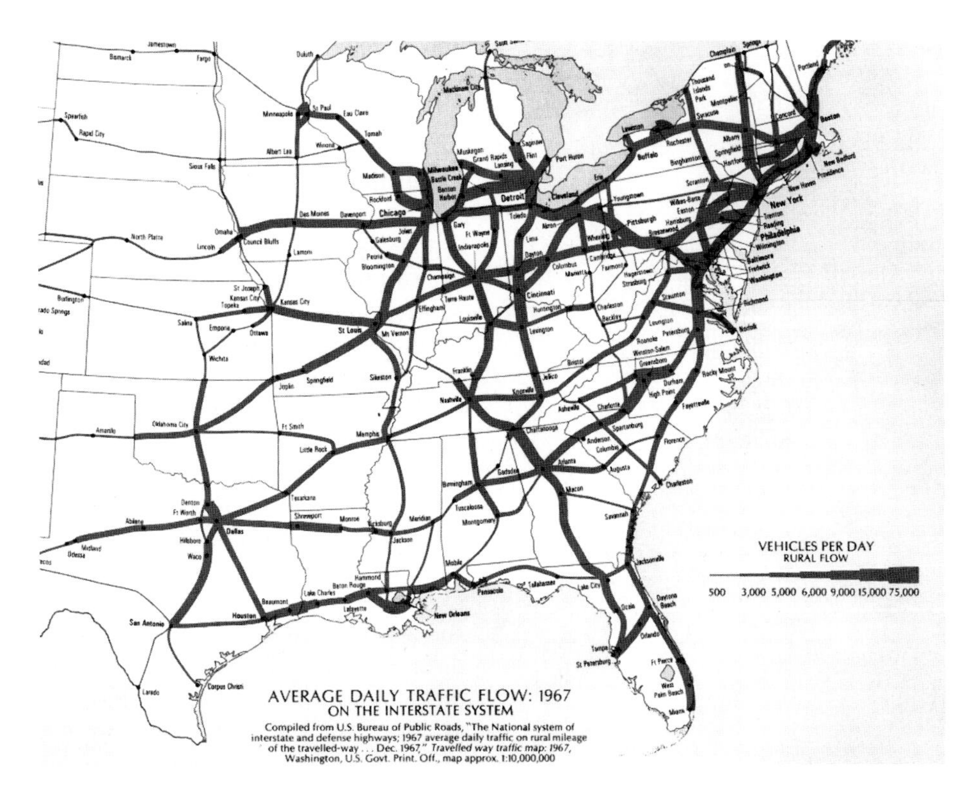

Figure 11.5 Flow map with range-graded line widths.
Source: *The National Atlas,* 227.

many instances, however, the scale of the map relative to the widths of the lines involved does not permit this type of exact alignment. When this occurs, the flow lines may simply indicate the general route followed. The annual monetary value of ocean trade between countries, for example, can be shown with flow lines, but the flow lines probably won't correspond to the actual routes used. In such a case, the lines drawn on the map simply indicate an interchange of some sort between place *A* and place *B* and a certain value associated with that interchange. For these reasons, the meaning of the routes shown on flow maps must be interpreted with caution.

AREAL SYMBOLS

Areal symbols are used to represent phenomena that are spread out over the earth's surface. A vast array of subjects falls into this category, ranging from climatic-zone data to areas where certain crops predominate, or from regions occupied by specific ethnic groups to counties with specific income levels. The common characteristic of these subjects is that they occupy some more or less precisely defined area on the earth's surface. Similarly, the symbols used to represent them are spread out over the two-dimensional surface of the map. As is true with other types of symbols, areal symbols can represent either nominal, ordinal, or quantitative information.

Nominal Information

With **nominal areal symbols,** the appropriate area on the map is covered with a pattern or color that differentiates it from the surrounding areas (Figure 11.6). The map may indicate the different types of crops grown in a county, for example. The area in which wheat is the predominant crop may be shown as a stippled pattern, whereas the area in which corn is predominant may be striped, and the soybean area may have another pattern. The same goal may be accomplished on a color map by using three different colors, one for each of the categories. Any differences in the textures or colors used to designate the different areas, however, do not imply any hierarchy or order of importance. A red area representing potato cropland is no more or less important, for example, than a green area representing turnip cropland.

Some level of generalization is to be expected on maps of nominal distributions. For example, an area identified as a wheat-growing region may contain subareas that support other crop types. Depending upon the scale of the map, however, these subareas may be omitted so that only the predominant crop type is indicated. If the number of subareas of different types is

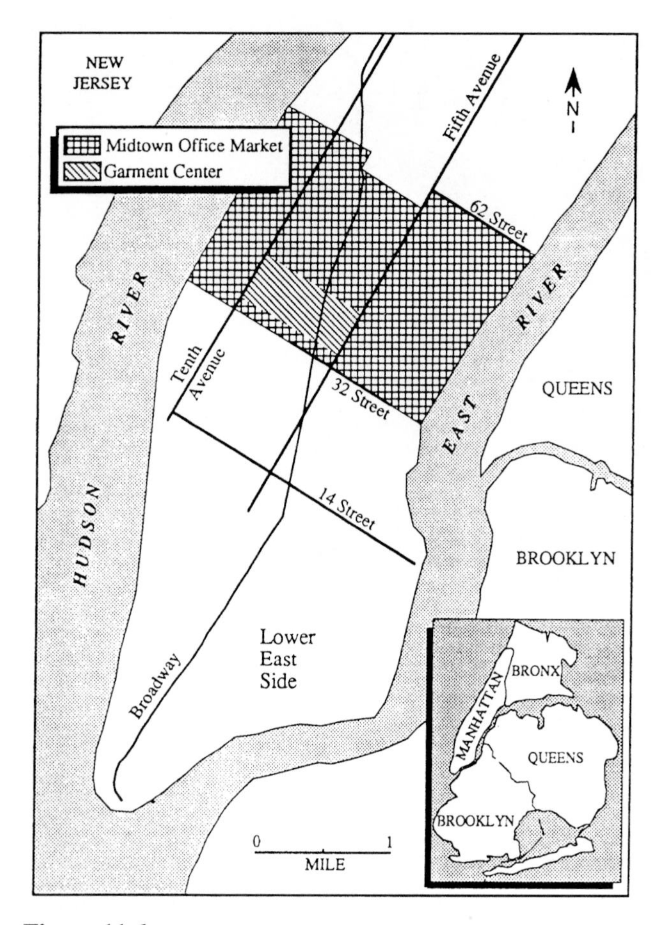

Figure 11.6 Map using nominal areal symbols.
Source: From Andrew Herod, "From Rag Trade to Real Estate in New York's Garment Center: Remaking the Labor Landscape in a Global City," *Urban Geography* 12, no. 4 (1991): 324–28. Reprinted by permission.

significant, the classification scheme should include suitable mixed categories.

Quantitative Information

The presentation and understanding of ordinal, interval, or ratio areal data is more complex. The symbolization for this purpose must indicate either the relative importance of one area compared to another or the actual values involved. Also, the symbols must often simultaneously present nominal information regarding the various types of variables being shown.

When **quantitative areal symbols** are used, subareas on the map are distinguished from one another by different color shades or different gray values that indicate the rank or numerical value of the phenomenon in each subarea.[4] Commonly, darker shades indicate greater values, and lighter shades indicate lesser values. In the case of rank information (ordinal data), the symbols indicate the relative importance of some phenomenon in each area on the map. The information is limited to an indication of a range from low to high, without any precise values attached (Figure 11.7). When specific numerical values (interval or ratio data) are associated with the areas, the information is presented on a choropleth map.

[4]Different hues (colors) usually indicate differences in kind, not differences in value.

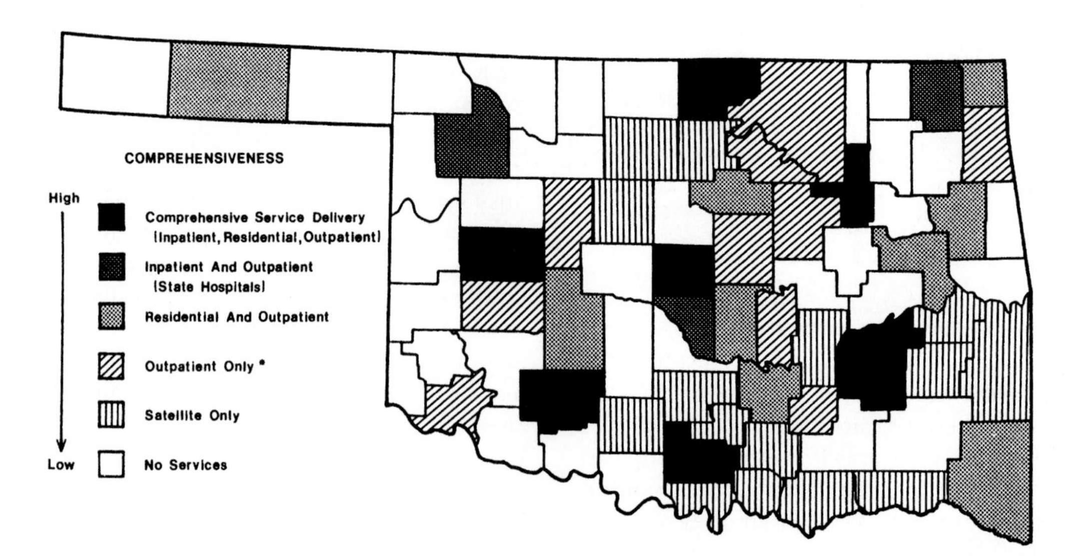

Figure 11.7 Ordinal areal symbols used to show alcoholism treatment comprehensiveness in Oklahoma, 1980.
Source: From Christopher J. Smith, "Locating Alcoholism Treatment Facilities," *Economic Geography* 59, no. 4 (October 1983): 368–85. Reprinted by permission.

Choropleth Maps

A common type of areal quantitative map is the **choropleth map.** On a choropleth map, the subdivisions are preexisting units; census tracts and county, state, or national boundaries are frequently used. If the map shows information about textile and apparel workers by county, for example, each county is printed in a pattern that indicates the particular level of employment associated with it (Figure 11.8).

The numerical values used on choropleth maps are based on **areal averaging.** Areal averaging means that the information presented is *not* a simple total value. Total values require another means of presentation, such as a value-by-area cartogram (see Figures 14.3 and 14.4) or a proportional symbol map (see Figure 15.24). On choropleth maps, the average value for each areal unit is calculated and symbolized. In general, ratio values, such as population density, yield per acre, or average income, are used for choropleth maps, instead of absolute values such as total population, total volume of agricultural production, or total income. An example will clarify the reasons for this practice.

Figure 11.9 provides examples of both proper and improper use of the choropleth technique. In Figure 11.9a, the gray-value applied to each state is based on the state's *absolute* population. (This example uses five classes, determined by the minimum-variance classification method.) California has the largest total population and, therefore, is black. New York and Texas, with almost as large total populations, receive the darkest gray-value, and so on. Wyoming with the smallest total population is white, along with other states with relatively small populations. The overall impression of the U.S. population distribution gained from this map is of heavy population concentrations in the three most populous states, with lesser concentrations in Florida, Illinois, Michigan, Ohio, and Pennsylvania. New England appears to have a very sparse population, along with most of the west.

The map of population density (Figure 11.9b) conveys a very different impression of the U.S. population distribution. Here, the very dense population concentrations in the northeast stand out. New Jersey and

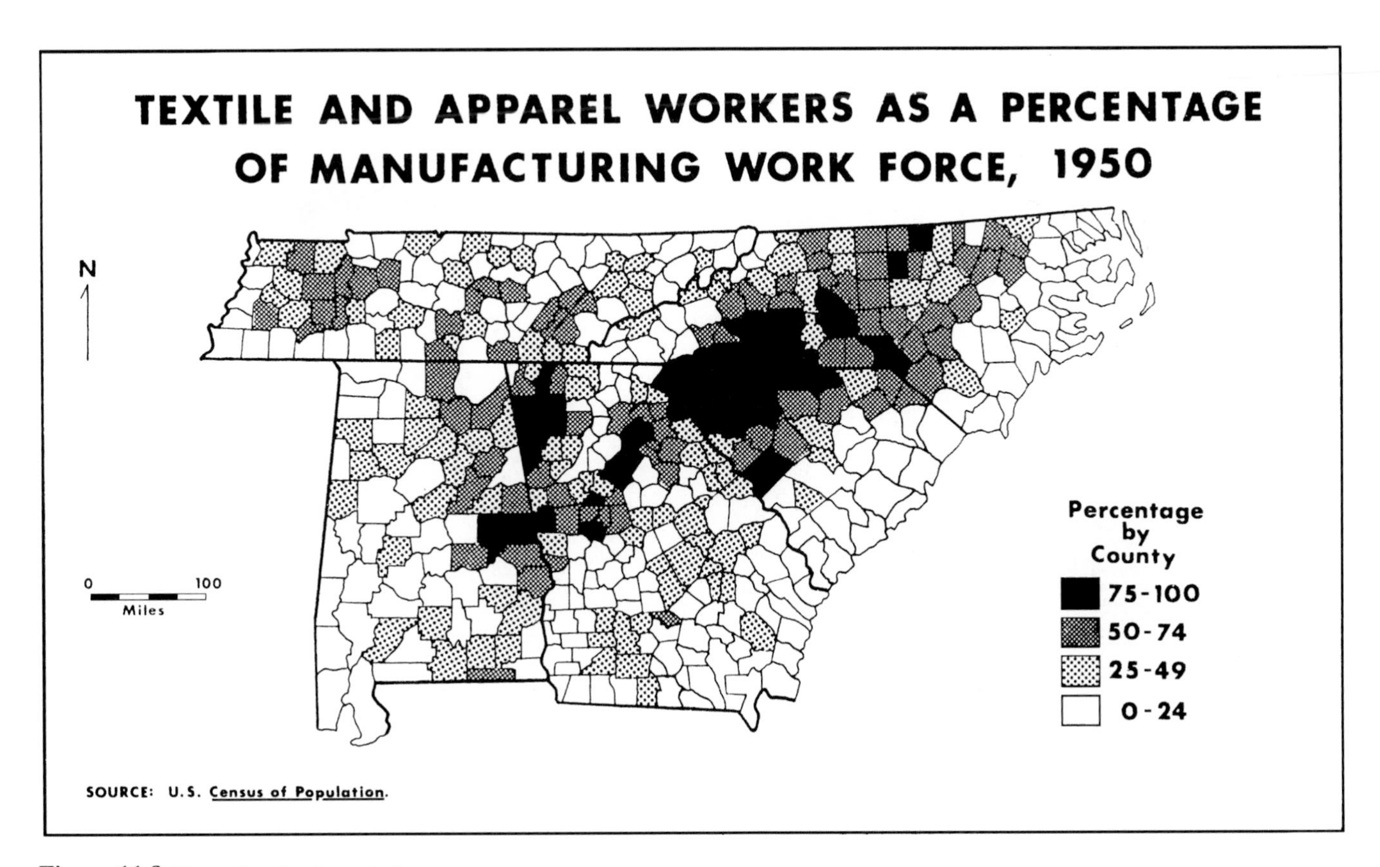

Figure 11.8 Example of a choropleth map.

Source: From Merrill L. Johnson, "Postwar Industrial Development in the Southeast and the Pioneer Role of Labor-Intensive Industry," *Economic Geography* 61, no. 1 (January 1985): 46–65. Reprinted by permission.

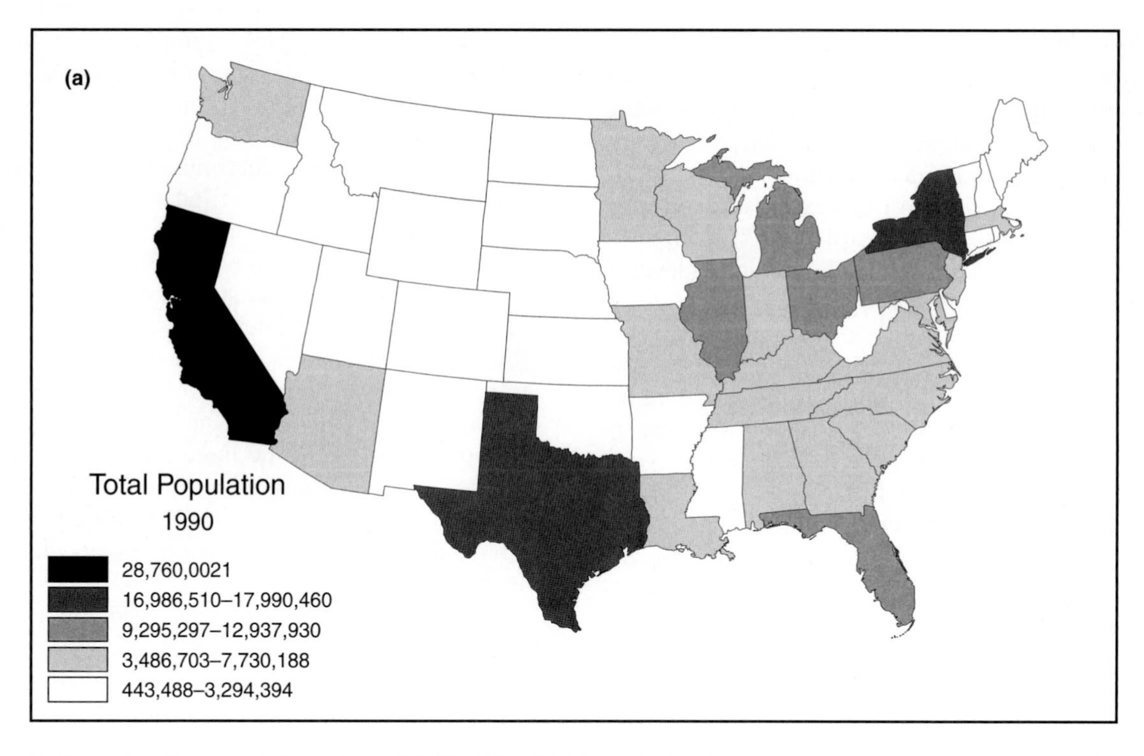

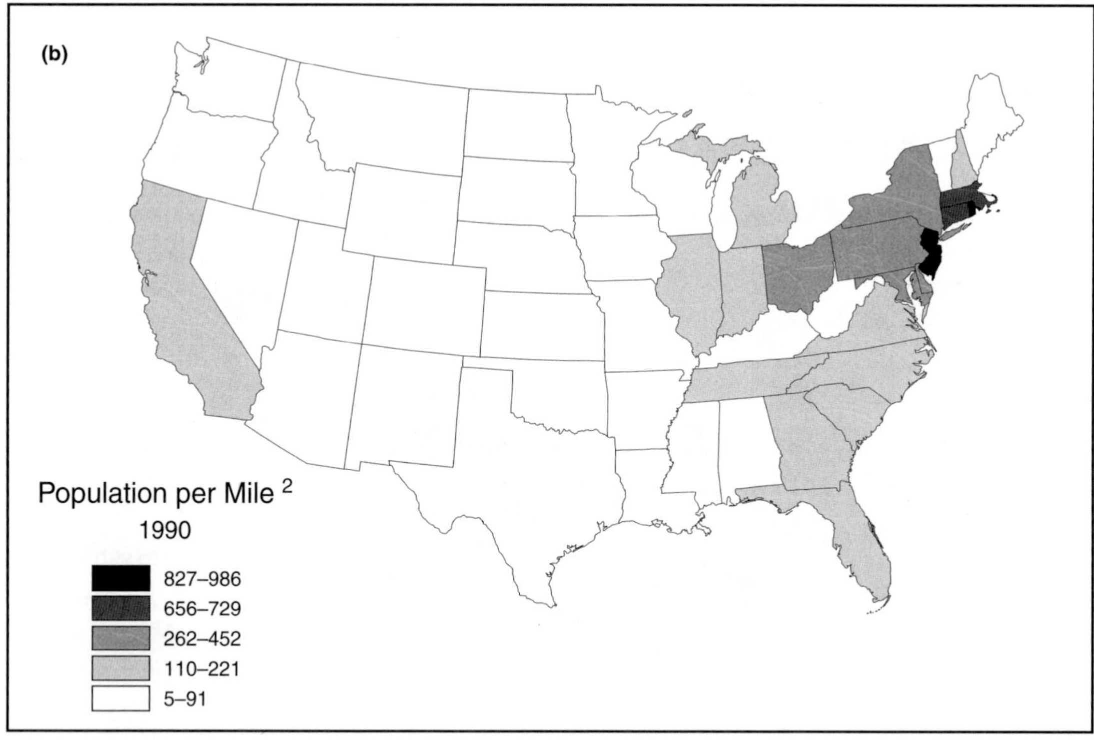

Figure 11.9 Examples of choropleth maps. (*a*) Improperly based on absolute values. (*b*) Properly based on density values.
Source: (*a*) 1990 U.S. Census. (*b*) Derived from 1990 U.S. Census.

Rhode Island are in the highest density class, followed by Connecticut and Massachusetts. The third class includes Ohio, Pennsylvania, Maryland, Delaware, and New York. California, in contrast, falls in the second-lowest density class, along with several midwestern and eastern states. Texas, which stood out as a population concentration in the absolute value map, falls into the least-dense category, along with most of the nation west of the Mississippi River and the remaining midwestern, southern, and eastern states.

This example shows that choropleth maps based on absolute values should be interpreted with caution: Usually, they convey a false impression. The exception is where all the areas mapped are of the same size. Then, the distortions introduced by size differences, such as in the case of Texas in this example, are eliminated.

It should also be emphasized that even properly drawn choropleth maps may still convey a misleading impression, because they are based on averages for the areas mapped. Here, for example, the presentations of New York and California in Figure 11.9b are still not ideal. In both cases, some parts of the state have heavy population concentrations and other areas are sparsely settled. Thus, the map would convey a better impression of the situation if the area around New York City were darker gray and the rest of the state lighter gray. Similarly, in California, the impression would be improved with dark values in the Los Angeles, San Francisco, and San Diego areas and with lighter values in the rest of the state. This type of problem is avoided by dasymetric maps, which are discussed later in this chapter.

Classed Choropleth Maps

One method of portraying data regarding states, counties, and other divisions involves combining the values into only a few groups so that a range of values is included in each group. The desired goal of such a **classification** system is to produce groupings that minimize the differences within each established category and maximize the differences between the categories. When grouped data are used, the result is a conventional choropleth map that makes it easier for the map user to understand the nature of the distribution pattern because fewer symbols are needed.

When classification is done, the cartographer first decides how many groups to use. This is a necessary decision but frequently an arbitrary one that greatly affects the resulting map. Map users must be sensitive to the choices made by the cartographer to effectively evaluate the resulting map.

The cartographer's dilemma in trying to decide how to group the data on a map can be illustrated by considering two extreme possibilities. One extreme would be to use a separate symbol for *every* individual value. This approach would result in a large number of symbols on the map.[5] In fact, so many symbols would be needed that some cartographers feel that map users might be unable to tell the symbols apart, no matter how conscientiously they try. The other extreme would be to show only one group; that is, the data could be completely aggregated and show, for example, *all* of the counties with the same symbol. The resulting map would be useless, of course, because map users would be unable to differentiate the counties at all.

Rather than take either extreme, cartographers frequently try to find some satisfactory middle ground. This involves establishing enough categories to provide meaningful distinctions between groups but not so many that the amount of information provided is overwhelming and confusing. A typical solution is to settle on some *manageable* number of groups or divisions, frequently somewhere between five and eleven, which allow the various gray values, or colors, used as symbols to be distinguished.

Classification is important because the class intervals selected determine the appearance and meaning of the final map, and a poor selection results in a misleading product. Figure 11.10 shows four maps that look very different from one another. What is not apparent is that all of these maps are based on exactly the same set of data. The differences in appearance are the result of the variations in data classes that resulted from using different classification schemes. Although many classification methods have been developed, the four shown in this figure provide typical examples of the problems that can occur. For the sake of comparability, each method is used here to create five data classes.

In this example, the map user should be aware that many counties in Indiana have sparse population densities, a few have intermediate densities, and one has a very high density. The mapping problem, then, consists of selecting a classification system that reflects the characteristics of the density distribution (Figure 11.11). When other types of distributions are encountered, the goal of correct representation is the same, although the details of the problems involved may vary.

The *equal steps* method entails dividing the data range into numerically equal intervals. Population densities of Indiana counties range from 9.50 to 745.80 persons per square kilometer. This range of 736.30 units is divided into five equal parts, making each range 147.26 units wide. When this is done, 86 counties fall into the lowest range (9.50–156.76), four counties fall into the next range (156.77–304.02), one falls into the third range (304.03–451.28), and one falls into the highest range (598.55–745.80). None of the counties fall into the fourth range (451.29–598.54), which is called an *empty set*. The map is an ineffective representation of the data, largely because it makes most of the state look virtually unpopulated. On the other hand, one county looks very densely populated (which is true), and a few counties fall somewhere between the extremes.

The *quintiles*[6] method avoids the problem of empty sets, because it requires equal numbers of counties in each class. The values then range from the lowest density in the group to the highest density in

[5]This type of unclassed choropleth map is discussed later in the chapter.

[6]This general approach is called the *quantile* approach. In this case, because it divides the data into five equal parts, each part is called a *quintile*. If there were ten classes, each would be called a *decile*, if four, a *quartile*, and so on.

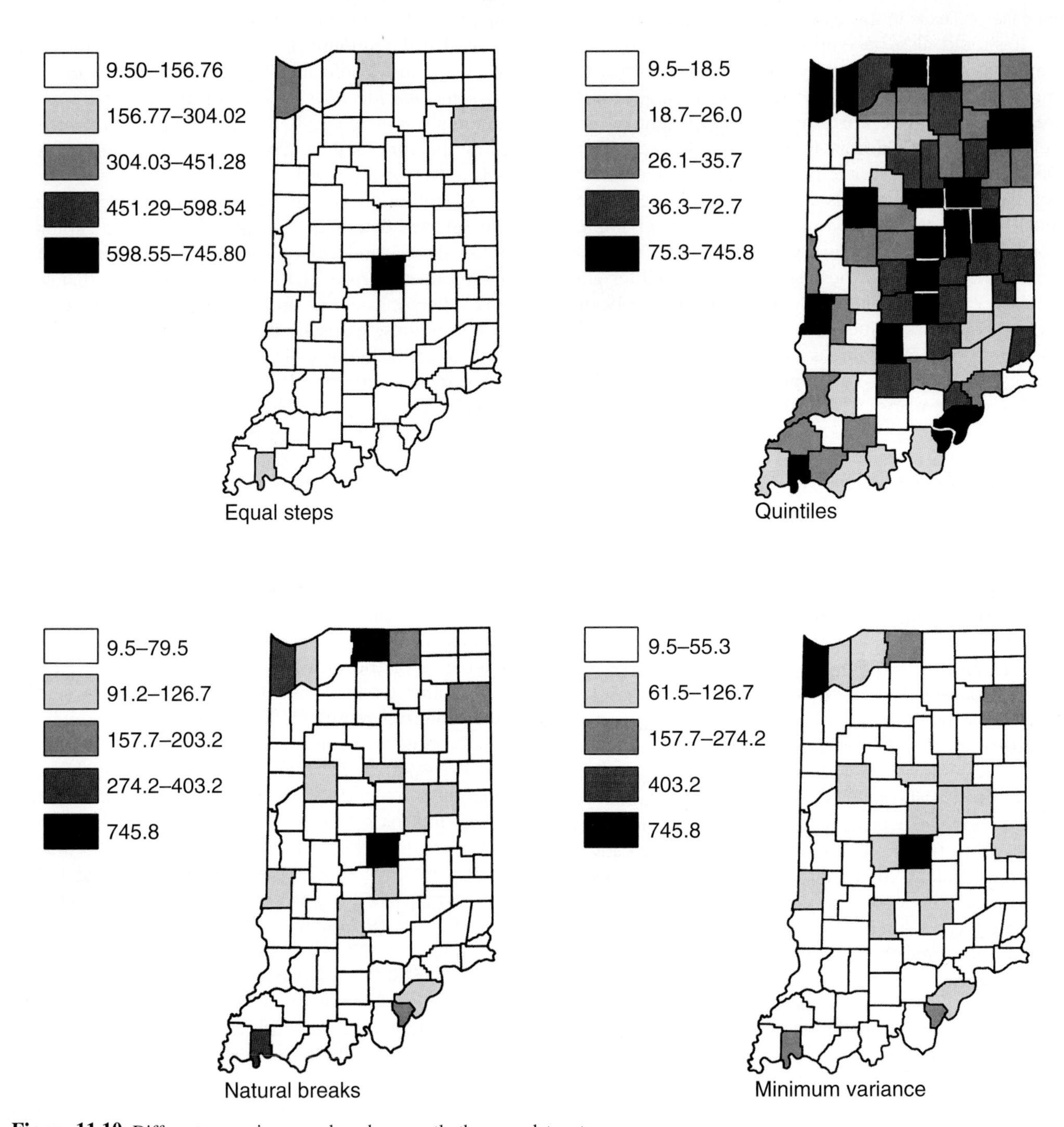

Figure 11.10 Different-appearing maps based on exactly the same data set.

the group. One problem with this approach is that it may result in placing very similar occurrences into separate categories. Here, for example, the narrow range from 9.5 to 72.7 persons per square kilometer is divided into four categories, whereas the very wide range from 75.3 to 745.8 is placed in one category. Here, the resulting map is ineffective because so many counties are included in the densely populated category, although some of them are only about one-tenth as densely populated as the highest-density county. Also, many remaining counties appear to have a significant population density whereas, in actuality, the densities of some of them are relatively sparse.

Using the *natural breaks* approach involves direct examination of the data plotted on the histogram. Natural breaks are represented as points where apparent

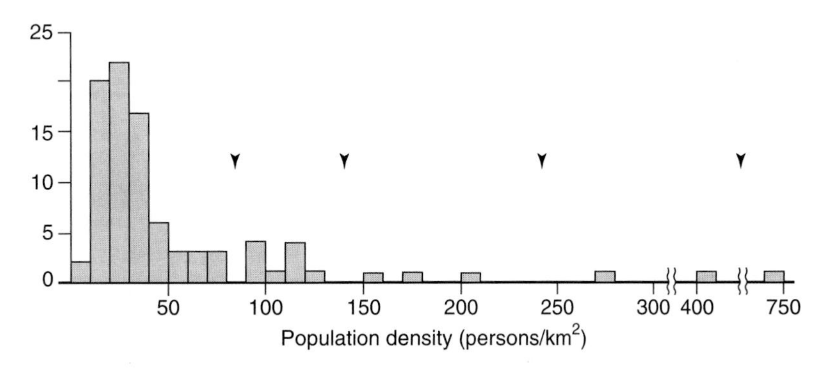

Figure 11.11 Histogram showing the distribution of population densities in Indiana counties.

gaps in the distribution are selected as dividing points for the classes. Here, the results of this approach are breaks in the classes at 79.6–91.1, 126.8–157.6, 203.3–274.1, and 403.3–745.7 (arrows in Figure 11.11). Because this is a subjective process, other breaks might have been chosen, with perhaps equal justification. Thus, 76 counties are placed in the lowest-density class and only one in the highest. The remaining classes subdivide the intermediate-density counties into relatively low, medium, and high groups. Overall, this is a better representation of the data than given by the two previous methods. As mentioned, however, it is subjective, and, therefore, any given results are difficult to defend against other possible outcomes.

The final example shown, the *minimum variance* method, attempts to improve on the natural breaks method by using a more objective, mathematical process to produce the classes. In this approach, the classes are established so that they have maximum internal homogeneity and yet are as different from each other as possible. The result is that the two counties with the highest population densities are each assigned to a separate class. This is because they are so dissimilar to all of the other counties and to each other—the highest density county has 745.8 persons per square kilometer, the next highest value is 403.2, and the highest density in the remaining counties is only 274.2. The counties that remain have low densities but are divided into three classes. This preserves the widespread occurrences of counties with low densities in the state. Also, the quite different intermediate density areas show up distinctly but are differentiated from one another.

The variation in the four maps illustrates the importance of the choice of the classification method to the final appearance of a map. Unfortunately, judging whether the classification method was appropriate, simply by a map's appearance, is difficult.[7] The following section discusses some considerations involved in such judgments.

[7]For additional information about the classification processes described here, see John Campbell, *Introductory Cartography,* 2d ed. (Dubuque, Iowa: Wm. C. Brown Publishers, 1991), 196–203.

Evaluation of Classed Choropleth Maps

When examining a given map, keep in mind the potential pitfalls. You can, for example, examine the map's classification system by looking at the legend. From the legend, you may be able to visualize how the appearance of the distribution is being affected by the classes selected. To do this effectively, you must go beyond the map itself. You may be able to find other maps based on the same data, for example, so that you can compare the results. Or you may be able to find written descriptions of the distribution. You might be able to find the data that were used to construct the map in question and, in the extreme, might even experiment with your own system of classification.

You might occasionally encounter a choropleth map with overlapping categories, such as 0–10, 10–20, and 20–30. The problem with this system is that some values could be placed in either of two categories. In this example, 10 could be placed in either the 0–10 or the 10–20 range. This type of classification is ambiguous and does not reflect a basic understanding of appropriate procedures. The map on which it is used, therefore, is likely to have minimal value.

Unfortunately, there are no other helpful guidelines for you to follow in evaluating choropleth maps, but, as an educated map user, you can at least maintain a healthy skepticism toward what you see on them.

Classification problems are as important for other types of maps as they are for choropleth maps, although the discussion of the subject has been concentrated in this section for convenience. You should also consider the question of appropriate data subdivisions, however, when using maps that incorporate proportional circles, flow lines, or any other symbols grouped by classes.

Unclassed Choropleth Maps

Imagine a map depicting the average income level of persons in the United States, by county. The income-level value of each county is likely to be at least slightly different from the values associated with the other counties. If each of the counties is shaded with a symbol that indicates its average income level, a very large number of different symbols is needed. Some cartographers feel that using such an "infinitely divided" range of symbols is an effective way to present information.[8] They believe that the method has the advantage of giving a strong visual impression of the distribution of the variable mapped. Maps that show specific values for each area are called **unclassed choropleth maps** (Figure 11.12).

Bivariate and Multivariate Maps

The maps discussed so far in this chapter are **univariate maps;** that is, they show variations from place to place of a single factor, such as wheat production. In many situations, however, there is a geographic relationship between two variables, such as wheat production and average rainfall. An appropriate **bivariate map** illustrates such relationships.

The discussion here focuses on a bivariate map that uses continuous shading created by crossed-line patterns (Figure 11.13). The interpretation of this type of symbolization is based on the characteristics of the network created by the intersecting lines.[9] One aspect of the network is that, where the lines are closer together, a darker gray value is produced and, by implication, the density of the particular variable is greater. Another aspect of the network is that differing relationships between the two distributions result in a different pattern appearance. To illustrate, if both values are high, the lines are close together, and the openings in the network are relatively square. On the other hand, if both values are low, the lines are far apart, but the network openings are still square. These relationships are shown on the diagonal from the lower left to the upper right of Fig-

[8]The seminal paper on this topic is W. R. Tobler, "Choropleth Maps without Class Intervals?" *Geographical Analysis* 5, no. 3 (July 1973): 262–65. A summary of arguments for and against the unclassed approach, as well as useful citations of the discussions in the literature, is contained in Michael P. Peterson, "An Evaluation of Unclassed Crossed-Line Choropleth Mapping," *The American Cartographer* 6, no. 1 (1979): 21–37.

[9]Bibliographic information, as well as research into aspects relating to the symbolization, interpretation, and measurement of such maps, is reported in a number of publications of Laurence W. Carstensen Jr., which are listed in the Suggested Readings at the end of the chapter.

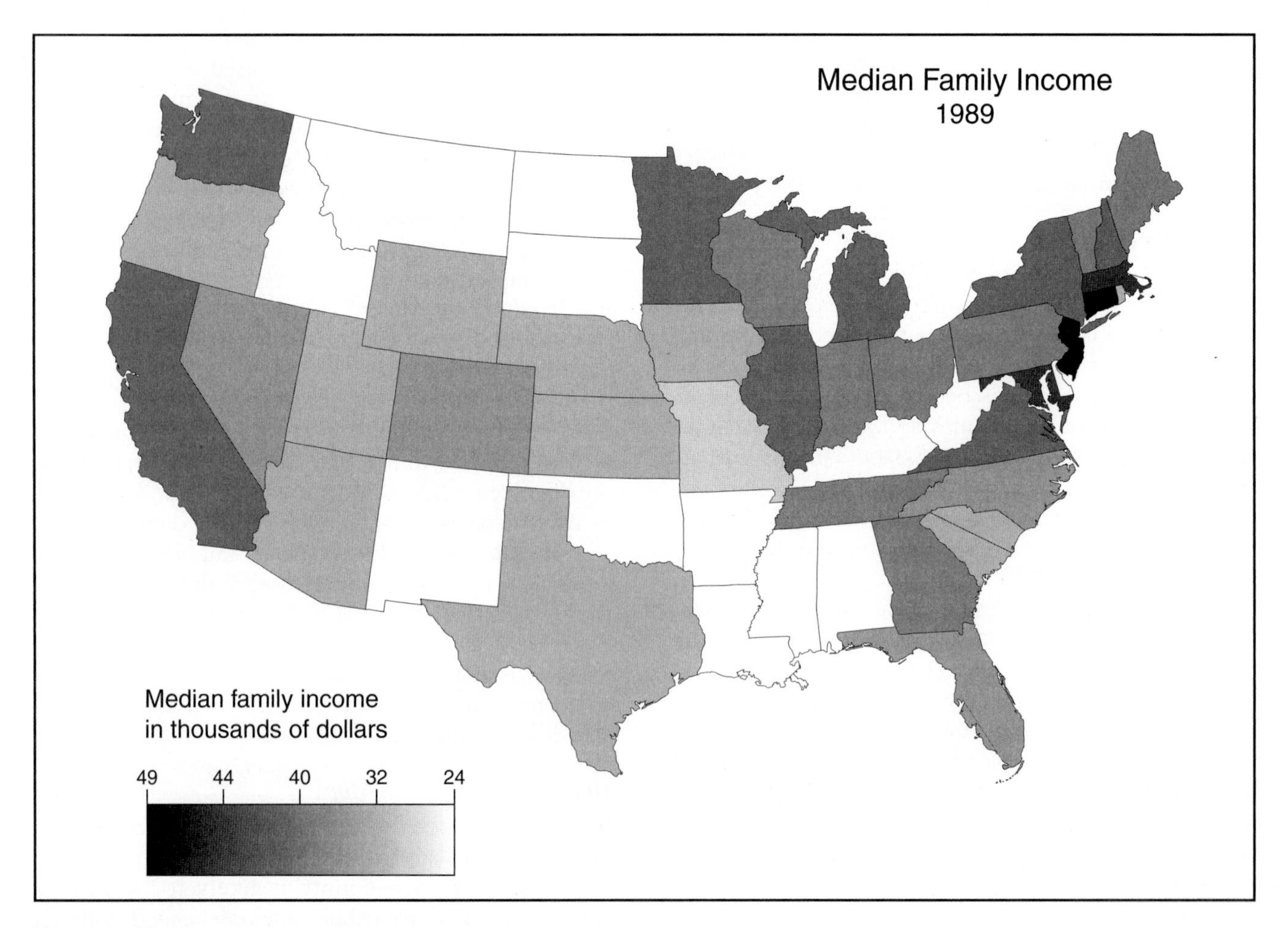

Figure 11.12 Unclassed choropleth map.
Source: 1990 U.S. Census.

ure 11.13. In contrast, if one value is high and the other low, the network openings take on a rectangular appearance, with the wider spacing for the low-value variable and the narrower spacing for the high-value variable. These relationships are shown in the upper left and lower right corners of Figure 11.13.

Although only black-and-white patterns have been discussed here, the same general approach applies to color maps, on which a different color is used to represent each variable (blue for one and red for the other, for example). As usual, lighter tones represent lower data values, and darker tones represent greater data values. Printing one color over the other results in mixed colors that reflect the relationships between the variables. In the case of blue and red, for example, relatively equal values of both variables result in a purple color. Where the values of the variables are low, the purple is light; the purple is dark where the values of both variables are high. Mixtures of high values of one variable and low values of the other produce bluish-red or reddish-blue tints. The specific intensity and tint that result are dependent on the relative importance of each value. **Multivariate maps,** which show the relationships between more than two variables, are sometimes produced by introducing additional colors for the additional variables. The major difficulty with color bivariate and multivariate maps is that the correct interpretation of the meanings of the color combinations is sometimes difficult.

Dasymetric Maps

A major problem with choropleth maps is that the entire area of each subdivision is symbolized the same way because any internal variations in the distribution are not recognized. In fact, as has already been indicated, internal variations are automatically eliminated by the calculation of average values for each unit. This may be satisfactory for some types of information, such as the distribution of rainfall, which may not vary in intensity to a great extent over a relatively small unit area. It may be entirely unsatisfactory, however, for showing the distribution of population or other information that can vary significantly in intensity within a similar unit area.

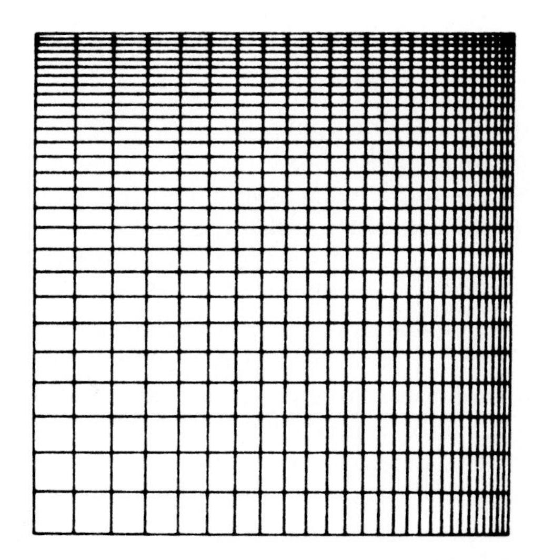

Figure 11.13 Key for bivariate map using continuous shading created by crossed-line patterns. (See text for explanation.)

When they wish to show internal distributions, cartographers sometimes use a technique called **dasymetric mapping** to partially overcome the problem of areal averaging. The dasymetric technique recognizes that what is being mapped varies in intensity within the mapping units. To accommodate this, each mapping unit is divided into a number of smaller areas. Each of these subdivisions is defined so that it is relatively homogeneous in terms of the phenomenon being mapped. Although the value associated with each subdivision is still based on an areal average, the internal variation within each subdivision of a dasymetric map is much less than it would be on a choropleth map. A dasymetric map has a recognizable general appearance. The subdivisions no longer follow county boundaries or similar outlines, so their shapes are likely to be much less uniform or regular than those of a choropleth map (Figure 11.14).

Prism Maps

A **prism map** is based on a bird's-eye view, looking down at the region at an angle, from an elevated viewpoint. Each data unit on the map is raised above the base level an amount proportional to the value attached to it. The result is a series of columns, or prisms, of variable height, each with the outline of the appropriate mapping unit (Figure 11.15).

The heights of the prisms are difficult to judge, especially if their bases are hidden. Furthermore, if the cartographer has not been careful in selecting the viewpoint, some prisms may be hidden from view entirely. Despite these possible difficulties, prism maps are often effective presentations of quantitative information.

SURFACE SYMBOLS

The idea behind a map of a **surface** is perhaps most easily illustrated by a topographic map of the earth's surface. A topographic map shows, by contour lines or some other technique, the variations in the elevation of the earth's surface above a datum level. Maps are drawn of many other, more abstract surfaces, however. In this section, we will discuss the use of surfaces as a means of representing statistical information.

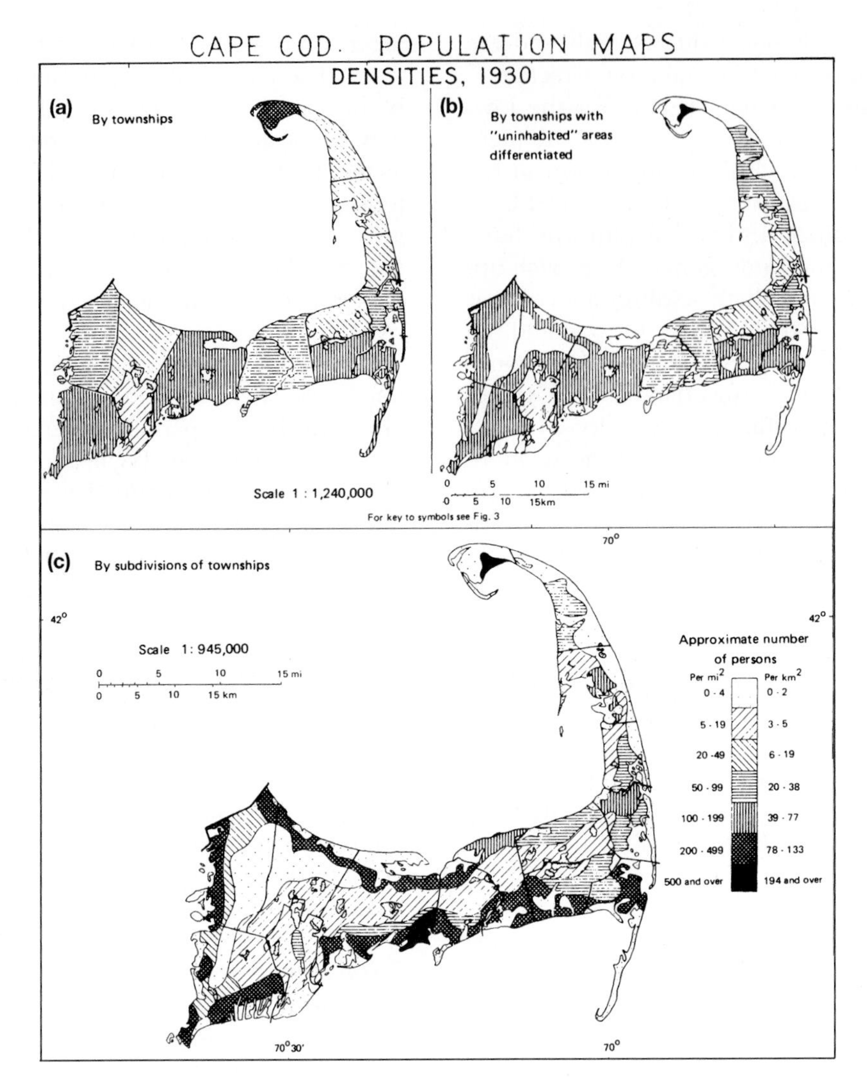

Figure 11.14 Development of a dasymetric map of Cape Cod population, 1930. (*a*) Township population densities. (*b*) Township population densities after elimination of "uninhabited" areas. (*c*) Final dasymetric map after taking into account other factors affecting population distribution.

Source: From John K. Wright, "A Method of Mapping Densities of Population with Cape Cod as an Example," *The Geographical Review* 26, no. 1 (January 1936): 105. Reprinted by permission of The American Geographical Society.

Isolines

It does not require too great a stretch of the imagination to visualize surfaces that are more abstract than the surface of the earth. For example, during a rainstorm, different amounts of precipitation fall at different locations. If the amounts that fall at a number of points are recorded, the values obtained can be used as the basis for mapping. The amount of rain at each location is simply treated as a vertical distance above a datum, and the representation of the values of all of the points taken together creates an abstract "rainfall surface." One way that this surface can be mapped is by using each recorded value as the equivalent of an elevation control point. **Isolines** are then drawn to join points that have received equal amounts of rainfall.

The chief difference between such a rainfall map and a topographic map is that the rainfall does not create an actual *physical* surface to observe and map. It does, however, provide an abstract, conceptual surface. In a similar way, even more abstract measurements, such as population densities, crop yields, and temperature statistics, represent **statistical surfaces.** A contour map, or a map that uses some other appropriate type of symbolization, can be created to represent these abstract concepts.

As is implicit in the preceding discussion, two types of isolines are commonly recognized. These are

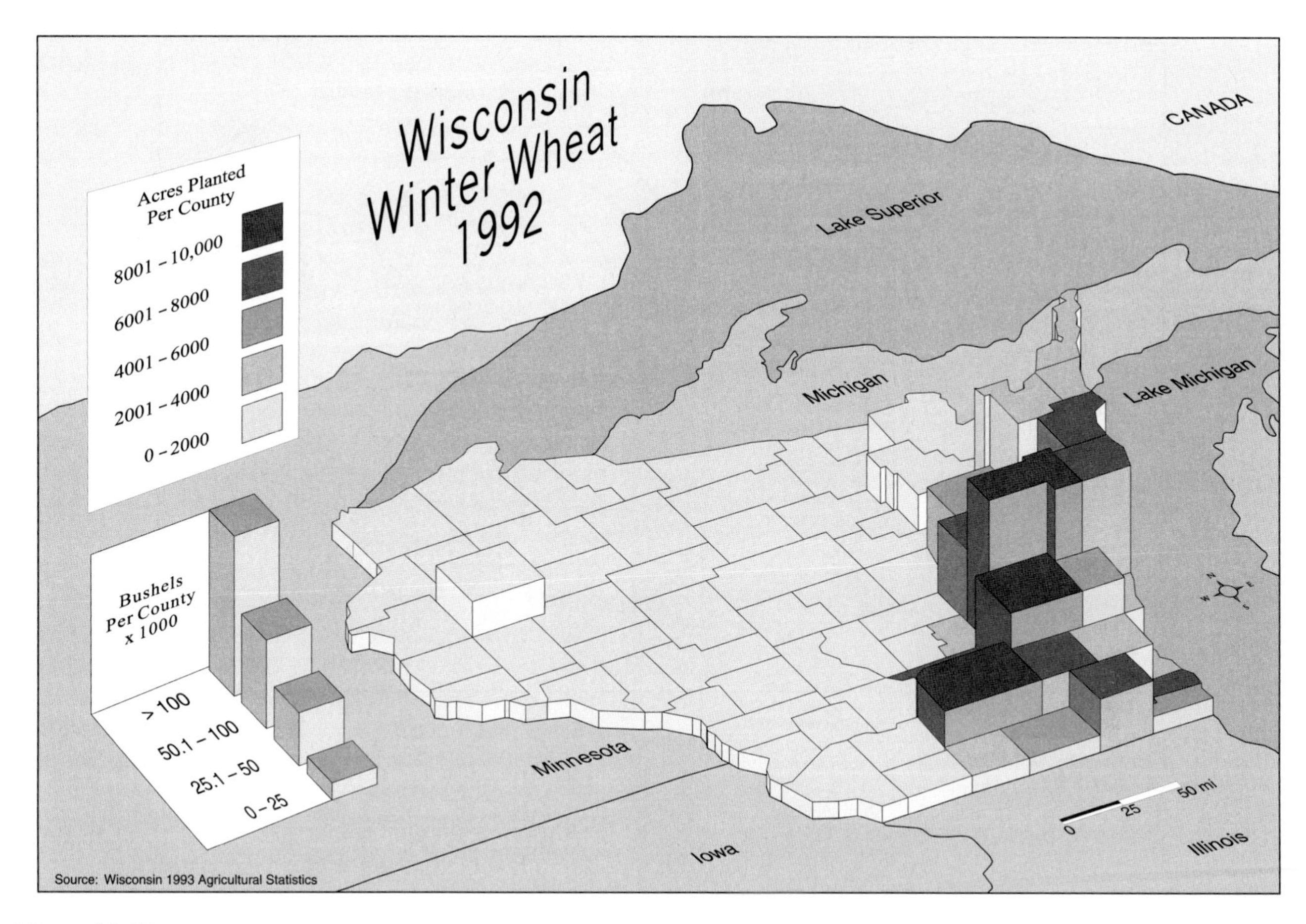

Figure 11.15 Prism map.
Source: Courtesy of Joseph M. Polder.

usually called *isometric lines* and *isopleths,* although other names for the same concepts exist.

Isometric Lines

Isometric lines are isolines based on control points that have actually observed values associated with them. The earth's surface provides an example of this type of situation, because it is higher here or lower there but can be measured everywhere. This means that every point on the surface has *some* value connected to it. Furthermore, the values that exist between two points are fairly predictable, and interpolation techniques can be used, with reasonable confidence, to determine what those values are. Contours, then, are isometric lines, as are isolines showing temperature, rainfall, barometric pressure, or other continuous variables (Figure 11.16).

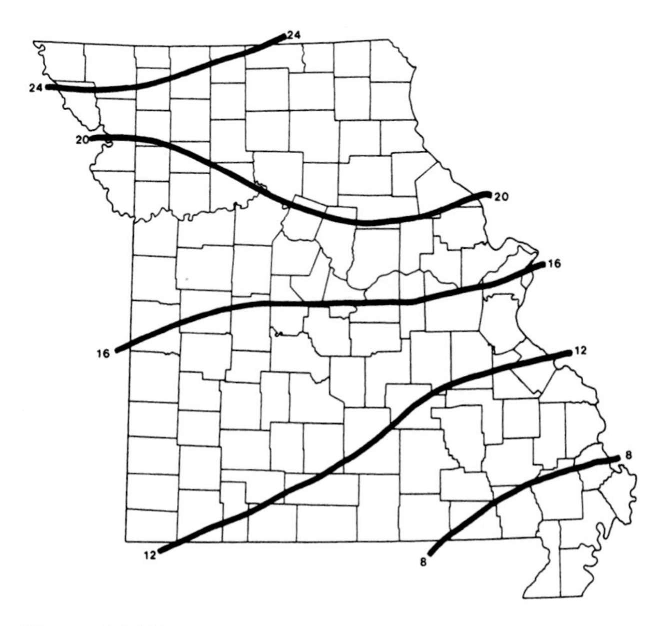

Figure 11.16 Isometric-line map showing average annual snowfall in Missouri.
Source: Walter A. Schroeder, *Missouri Water Atlas* (Jefferson City, Mo.: Missouri Department of Natural Resources, 1982).

Isopleths

Isopleths, on the other hand, are drawn on the basis of areal averages assigned to arbitrary control points and are *not* directly observable. An isopleth map of

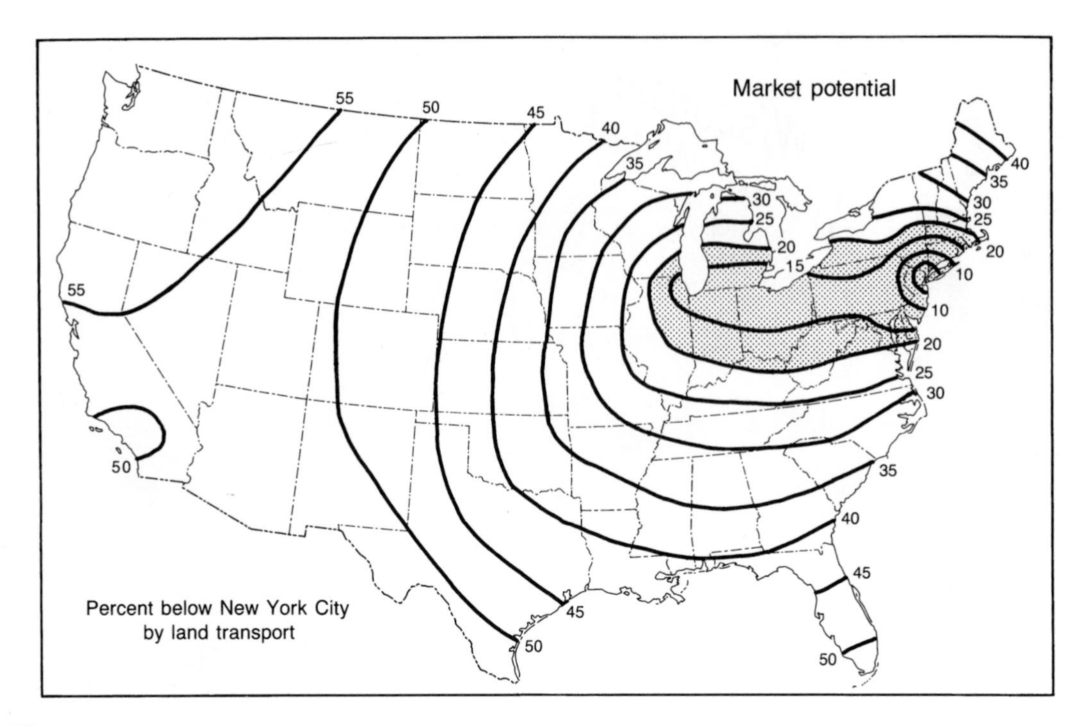

Figure 11.17 Isopleth map of market potential.

Source: Redrawn from Chauncy D. Harris, "The Market as a Factor in the Localization of Industry in the United States," *Annals of the Association of American Geographers* 44, no. 4 (December 1954): 315–48. Reprinted by permission.

population density for a given state, for example, might be based on county-level data. To construct the map, the cartographer has to obtain or compute the average population density for each county. Each of these average values is then assigned to control points, usually located at the centers of each county, and the isopleths are constructed around these control points (Figure 11.17).

The fact that the values used to produce isopleth maps are not directly observable has caused some to conclude that they are not a valid means of representing areal averages.[10] The arguments against isopleths are not universally accepted, however, so you are likely to encounter many maps that fall into the isopleth category. When examining such maps, keep in mind the questions that have been raised and consider how they affect the usefulness of the maps for your purposes.

[10]See, for example, David Unwin, *Introductory Spatial Analysis* (New York: Methuen, University Paperbacks, 1981), 158.

SUMMARY

Various symbolization schemes are used to present qualitative and quantitative information on maps.

Many variables occur at specific locations (points) on the earth's surface. The most straightforward type of point symbolization is the dot-distribution map, in which each dot indicates the location of one item or, in a form of generalization, many items.

When a phenomenon occurs at an identifiable point, its location may be indicated by a nonproportional point symbol. This shows the type of phenomenon that occurs there (qualitative information). Graduated point symbols, on the other hand, indicate the values (quantitative information) associated with various locations. This conveying of quantitative information may be done by varying the size of the symbol in relation to the value it represents.

Linear symbols differentiate linear features of various types (qualitative and ordinal information) or show flows or connections between locations (quantitative information). Flows are symbolized by varying the width of the line in relation to the amount or value of the flow.

Area symbols represent phenomena that are spread out over the earth's surface. In the case of qualitative information, the appropriate area on the map is symbolized so that it is differentiated from the surrounding areas. When quantitative information is mapped, the symbols are designed to indicate the relative importance of the areas. This is done by using gray values or color shades that reflect a hierarchy of importance. One map category of this type is the choropleth map, in which the values presented are average values for preexisting subdivisions, such as counties.

Sometimes, the values shown on choropleth maps are grouped. When this is done, the first decision is how many groups to use in the classification system. A typical solution is to settle on some manageable number of groups or divisions, perhaps between five and eleven. Classification is important because the chosen class intervals determine the appearance and meaning of the final map, and a poor selection results in a misleading product.

When the values on a choropleth map are not grouped, an unclassed choropleth map is produced. Such maps depict an "infinitely divided" range of symbols.

A major problem with choropleth maps is that the entire area of each subdivision is symbolized in the same way. Dasymetric mapping is used to partially overcome this often unrealistic presentation. This technique recognizes that the phenomenon being mapped varies in intensity within the mapping units. Each unit is divided into a number of smaller areas that are relatively homogeneous in terms of the phenomenon being mapped.

Both choropleth and dasymetric maps may be produced in bivariate and multivariate forms that illustrate relationships between two or more distributions at one time.

A prism map is an angled view in which each data unit is raised above the base level an amount proportional to the value attached to it.

Surfaces, both physical and statistical, are often represented by isoline maps. Two types of isolines are recognized. Isometric lines are based on control points that have actually observed values associated with them. Isopleths, on the other hand, are drawn on the basis of areal averages assigned to arbitrary control points and not directly observable. There is some controversy regarding the legitimacy of isopleth maps.

SUGGESTED READINGS

Birch, T. W. *Maps: Topographical and Statistical.* 2d ed. Oxford, England: Clarendon, 1964, part II.

Carstensen, Laurence William, Jr. "A Continuous Shading Scheme for Two-Variable Mapping." *Cartographica* 19, nos. 3 and 4 (Autumn and Winter 1982): 53–70.

———. "Perceptions of Variable Similarity on Bivariate Choroplethic Maps." *The Cartographic Journal* 21, no. 1 (June 1984): 23–29.

———. "Bivariate Choropleth Mapping: The Effects of Axis Scaling." *The American Cartographer* 13, no. 1 (January 1986): 27–42.

———. "Hypothesis Testing Using Univariate and Bivariate Choropleth Maps." *The American Cartographer* 13, no. 3 (July 1986): 231–51.

"Commentary." *Annals of the Association of American Geographers* 70 (1980): 106–8.

Crawford, P. V. "Perception of Grey-Tone Symbols." *Annals of the Association of American Geographers* 61 (1971): 721–35.

Cromley, Robert G. "Classed Versus Unclassed Choropleth Maps: A Question of How Many Classes." *Cartographica* 32, no. 4 (Winter 1995, published October 1996): 15–27.

Dickinson, Gordon Cawood. *Statistical Mapping and the Presentation of Statistics.* London: Edward Arnold. New York: Crane, Russak, 1973.

Dobson, Michael W. "Refining Legend Values for Proportional Circle Maps." *The Canadian Cartographer* 11, no. 1 (June 1974): 45–53.

Flannery, James J. "The Relative Effectiveness of Some Common Graduated Point Symbols in the Presentation of Quantitative Data." *The Canadian Cartographer* 8, no. 2 (December 1971): 96–109.

Jenks, George F. "Contemporary Statistical Maps—Evidence of Spatial and Graphic Ignorance." *The American Cartographer* 3, no. 1 (1976): 11–19.

———. "Optimal Data Classification for Choropleth Maps." *Occasional Paper No. 2.* Lawrence, Kans.: Department of Geography, 1977.

Jenks, George F., and F. C. Caspall. "Error on Choropleth Maps: Definition, Measurement, Reduction." *Annals of the Association of American Geographers* 61 (1971): 217–44.

Mackay, J. Ross. "Dotting the Dot Map." *Surveying and Mapping* 9, no. 1 (January–March 1949): 3–10.

———. "Some Problems and Techniques in Isopleth Mapping." *Economic Geography* 27, no. 1 (January 1951): 1–9.

McCleary, George F., Jr. "How to Design an Effective Graphics Presentation." In *How to Design an Effective Graphics Presentation,* Harvard Library of Computer Graphics, 1981 Mapping Collection, vol. 17. Cambridge, Mass.: Harvard University, Laboratory for Computer Graphics and Spatial Analysis, 1981, 34–48.

Muehrcke, Phillip C. *Thematic Cartography.* Resource Paper No. 19. Washington, D.C.: Association of American Geographers, Commission on College Geography, 1972.

Muller, Jean-Claude. "Perception of Continuously Shaded Maps." *Annals of the Association of American Geographers* 69, no. 2 (1979): 240–49.

Peterson, Michael P. "An Evaluation of Unclassed Crossed-Line Choropleth Mapping." *The American Cartographer* 6, no. 1 (1979): 21–37.

Sibert, John L. "Continuous-Color Choropleth Maps." *Geo-Processing* 1, no. 3 (November 1980): 207–16.

Smith, Richard M. "Comparing Traditional Methods for Selecting Class Intervals on Choropleth Maps." *Professional Geographer* 38, no. 1 (1986): 62–67.

Stevens, S. S. "On the Theory of Scales of Measurement." *Science* 103 (1946): 677–80.

Tobler, Waldo R. "Choropleth Maps Without Class Intervals?" *Geographical Analysis* 5, no. 3 (July 1973): 262–65.

Williams, Robert L. *Statistical Symbols for Maps: Their Design and Relative Values*. New Haven, Conn.: Yale University Map Laboratory, 1956.

Wright, J. K. "A Method of Mapping Densities of Population with Cape Cod as an Example." *Geographical Review* 26, no. 1 (January 1936): 103–10.

CHAPTER

12

CHARACTERISTICS OF MAP FEATURES: SHAPE AND POINT PATTERNS

The features shown on maps often have distinctive characteristics, particularly of shape, pattern, and arrangement. Indeed, maps are frequently used as tools for the discovery of such characteristics. Once the characteristics are identified, various methods can be applied to measure the spatial patterns, to generate hypotheses about their origins, and, ultimately, to test the hypotheses and draw conclusions regarding their validity. This chapter presents some common methods by which shape and pattern are measured.

SHAPE

A feature displayed on a map, whether it is a physical object, a political unit, or the market area of a manufacturer, is often described in terms of its *shape*.

Subjective Description

Some regions or features on maps can be described in terms of regular, geometric shapes, such as squares, circles, ovals, triangles, and octagons. The shortcomings of these descriptors quickly become apparent, however, when irregular shapes need to be described. In these cases, more exotic terminology, such as is used to describe Italy—its "bootlike" shape—may be necessary. Professional map analysts often resort to similar subjective shape descriptions. They have compared various atolls, for example, to a "stretched-out diamond," a "heron," or a "headless goose without legs."[1] Shape descriptions of this type are vaguely defined and undoubtedly mean different things to different observers. Also, this descriptive approach does not provide index numbers that an observer can consistently assign. Such index numbers would objectively distinguish one shape from another and would allow, for example, sorting into groups of similar shapes or analyzing the origins of particular shapes.

Numerical Indices

Investigators in many fields have developed numerical indices for the more scientific description of shape. Boyce and Clark reviewed examples of geographical approaches to the description of urban form, trade areas, political areas, and physical features, and of shape measures in other disciplines, dating from the 1920s.[2] They found that most of the approaches in their review were descriptive, except in physical geography, the biological sciences, and mineralogy, where they found various methods of more objective measurement. The difficulty with all of these methods was their ambiguity. Areas with different shapes could generate identical index values. In addition, these methods did not possess the other desirable characteristics described in the next section.

[1]Quoted in D. R. Stoddart, "The Shape of Atolls," *Marine Geology* 3, no. 5 (1965): 370.

[2]Ronald R. Boyce and W. A. V. Clark, "The Concept of Shape in Geography," *The Geographical Review* 54 (1964): 561–72.

Desired Characteristics of a Shape Index

One requisite of a satisfactory shape index is that all investigators obtain the same value when applying the index to the same task. The methodology should ensure that the outcome is not affected by scale, starting location, orientation, or translation. A second requisite is that the index provide sufficient information to permit the regeneration of the original shape. We will now consider each of these requirements in more detail.

Yielding Identical Value

A satisfactory shape index yields the same value for a city, for example, whether the outline is plotted on a scale of 1:100,000 or on a scale of 1:250,000. The amount of detail in the outline must remain the same at both scales so that we are truly measuring the *same* shape. Although the amount of generalization of two maps at such different scales is likely to vary considerably, this requirement refers to the result that would be obtained if the same image were simply enlarged or reduced without changing the amount of generalization. This is equivalent to saying that the same value would be obtained when comparing similar shapes of different sizes. A large island that has the same outline as a small island, for example, should have the same shape-index value.

When tracing the outline of a shape, you are not likely to find a preordained starting point on its perimeter. On a square, for example, the northeast corner is as logical a starting point as the southwest corner, either of the other two corners, or, for that matter, any other point, such as the middle of one of the sides. A satisfactory shape measure, then, should yield the same value, regardless of what starting point is selected.

When the shapes of two regions on a map are so similar that a tracing of one of them could be placed over the other and would match it perfectly, the shapes may be said to be identical, even if one is "upside down" on the map, as compared to the other—in other words, is in a different orientation. Similarly, if the position of the tracing is changed (translated), the shape of the outline should obviously not be affected. A satisfactory shape index, then, should yield the same result for two identical but differently oriented or translated shapes.

Allowing for Regeneration

A shape index should also contain enough information so a person familiar with its characteristics can *regenerate* the original shape using only the index as a basis. One of the requirements for regeneration is that there be a one-to-one correspondence between shapes and indices. This means that each shape is represented by a unique index value, and each index value describes only one shape.

Specific Measures

Unfortunately, it is not an easy task (in fact, as we shall see, it may very well be impossible) to devise a measure that provides clear and unambiguous definitions of shapes. This section describes some of the existing measures used to define shapes and considers their strengths and weaknesses.

Miller's Measure

One common approach to the measurement of shape is to establish a specific reference figure and to determine how much the figure being measured differs from it. A frequently used reference figure is the circle, because the circle is the most *compact* two-dimensional shape. That is, any other figure of equal area to the circle has a greater perimeter than the circle. A measure of this type is **Miller's measure,** which was developed for the study of the shape of drainage basins.

In Miller's measure, the shape of the drainage basin (C) is the ratio between the area of basin (A_b) and the area of a circle with the same perimeter (A_c)—that is, $C = A_b/A_c$. With this approach, the basin shape indices range from almost 0, in the case of a very elongated basin, to 1, in the case of a perfectly circular basin (Figure 12.1a). While this range in values provides a very convenient index, Miller's measure (and similar methods based on the same general idea) has a major problem. The problem is that the values used in the calculation of the index are based on the areas and perimeters of the two different figures and *not* their shape. The result is that many *different* shapes can yield the same index value (Figure 12.1b).

Bunge's Measure

William Bunge suggested a measure of shape that does not depend upon area. **Bunge's measure** is based on a set of distance measurements taken between systematically placed vertices on the perimeters of the shapes to be studied. In Bunge's example application, the outlines of 97 Mexican communities were measured.

The same number of vertices must be used for all of the shapes involved in a particular study. If the shapes are generally very irregular, a closer spacing is required than if the shapes are uncomplicated. The shapes being measured may be of different sizes, and the influence of size (or scale) is excluded from the measure by giving a value of 1 to the standard distance established between the vertices for each shape. This same distance is then used for all of the measurements of that particular shape.

After the vertices and the standard distances are established (Figure 12.2a), the shortest straight-line distances between different pairs of vertices are measured for each shape. A "lagging," or skipping, technique is

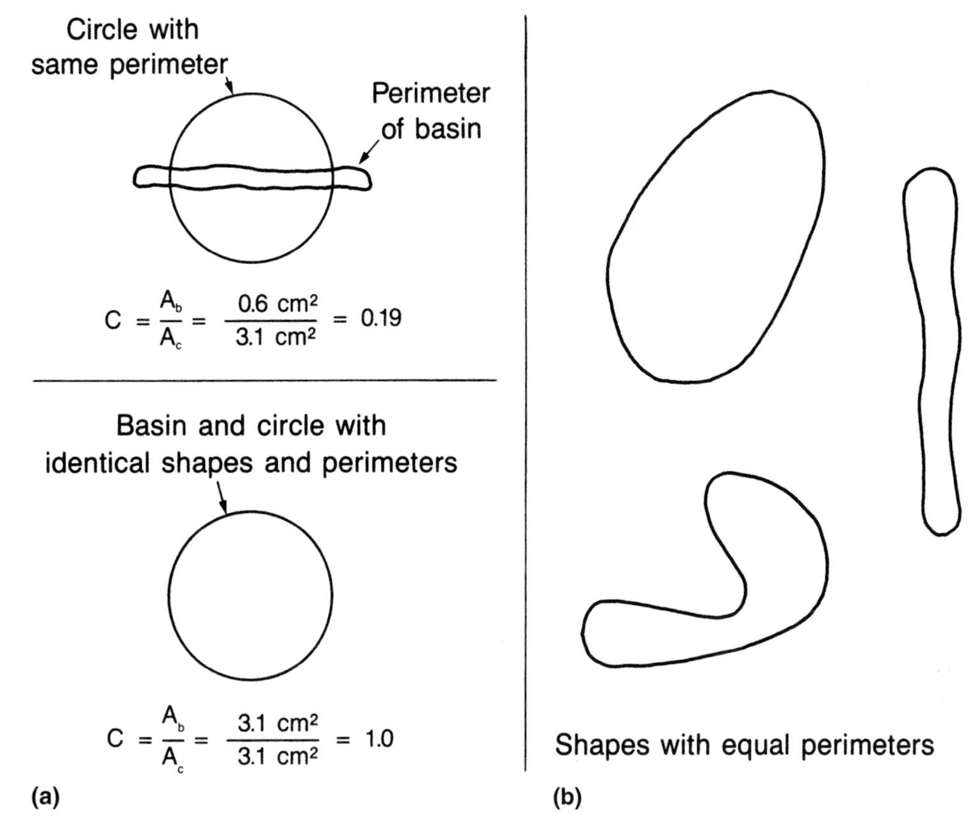

Figure 12.1 Characteristics of shape—Miller's measure. (*a*) Index values range from almost 0, in the case of a very elongated shape, to 1, in the case of a perfectly circular shape. (*b*) Various shapes can have equal perimeters and, therefore, equal index values.

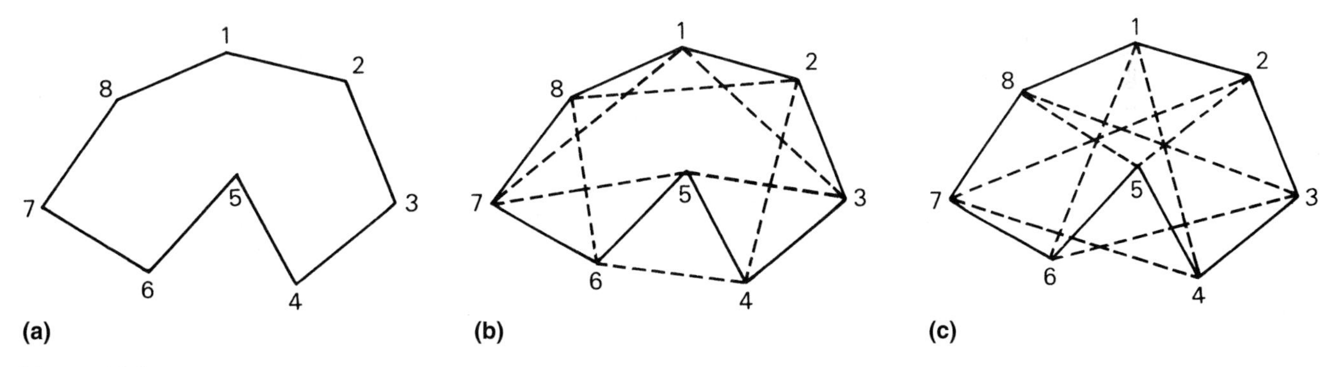

Figure 12.2 Distance measurements for Bunge's measure. (*a*) Numbered vertices of eight-sided polygon. (*b*) Distance between starting vertex and "lag-1" neighbors. (*c*) Distance between vertex and "lag-2" neighbors.

used for this purpose. A starting vertex is selected (any vertex may be used), and measurements are taken in one direction (either clockwise or counterclockwise) around the outline. The first distance measured is between the starting vertex and its "lag-1" neighbor, which means that the starting vertex's immediate neighbor is skipped (Figure 12.2b). The distance is recorded in the standard units previously established. Similar lag-1 measurements are then taken from the vertex next to the original starting vertex. This process is repeated until all of the vertices have served as starting points. The values thus obtained are added to obtain the lag-1 measure.

The distance between each vertex and its lag-2 neighbor is taken next (Figure 12.2c) and added, and the process is continued until all of the unique lagged values (lag-2, lag-3, and so on) have been obtained.[3]

[3]The total number of "laggings" required depends on the number of vertices used, because only unique results are desired. In the case of eight vertices, for example, the maximum number of lags required is three, whereas sixteen vertices require seven lags, and so on.

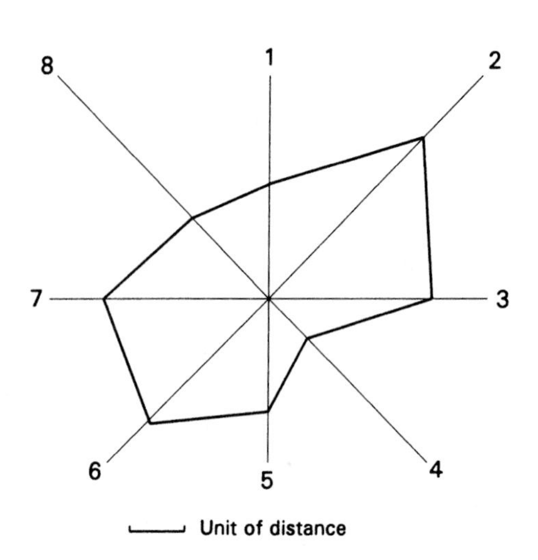

Radial Shape Index

$$\sum_{1}^{n} \left| \left(\frac{r_i}{\sum_{1}^{n}} \times 100 - \frac{100}{n} \right) \right|$$

Radial No. (r)	Length (r_i)	Actual % of Total $\left(\frac{r_i}{\sum_1^n} \times 100\right)$	Theoretical % of Total $\left(\frac{100}{n}\right)$	Absolute Difference in % $\left(\left|\left(\frac{r_i}{\sum_1^n} \times 100 - \frac{100}{n}\right)\right|\right)$
1	2.0	10.0	12.5	2.5
2	4.0	20.0	12.5	7.5
3	3.0	15.0	12.5	2.5
4	1.0	5.0	12.5	7.5
5	2.0	10.0	12.5	2.5
6	3.0	15.0	12.5	2.5
7	3.0	15.0	12.5	2.5
n = 8	2.0	10.0	12.5	2.5
	$\sum_1^n = 20$			Radial Shape Index = 30.0

Figure 12.3 Setup for determining Boyce-Clark radial shape index.

Finally, each of the lag measures is squared, and these values are summed to create lag-1^2, lag-2^2, lag-3^2 measures, and so on.

The value of Bunge's measure is that the set of all of the values derived for each shape is unique for that shape. This means that the shapes can be grouped on the basis of similar sets of values, as Bunge did with the Mexican communities. There are, however, a number of problems with the technique. The results obtained, for example, depend on the number of vertices chosen. This is an arbitrary number, governed by the investigator's judgment regarding how many line segments are needed to adequately describe a given group of shapes. Similarly, the locations of the vertices are arbitrary. They depend on the starting location selected by the investigator, as well as the standard distance between vertices. Finally, significant changes of direction in the outline may or may not be identified as the location of vertices. Again, this depends on the selected location of the starting vertex and the chosen standard distance between vertices. The existence of all these variables means that there is likely to be a lack of consistency from study to study in some of the choices made. As a result, different values may be obtained for the same shape. Also, there is no way that the values obtained can be used to regenerate the original shape.

Boyce-Clark Radial Shape Index

The **Boyce-Clark radial shape index (BCRSI)** is based on the lengths of radials extending outward from a node at the center of the shape.[4] The center node is located first, usually at the shape's center of gravity. A set of equally spaced radials is then drawn outward from the center node to the perimeter of the shape (Figure 12.3). The length of each of these radials is measured in any convenient units. Then, these lengths are added to obtain the total length of all of the radials. The percentage of the total that each radial represents is then computed, and the absolute value of the difference between that value and the percentage that each radial would represent if the shape were a circle is obtained. If there are eight radials, for example, each radial in the circle represents 12.5 percent of the total length. The index value for the shape is obtained by totaling the absolute differences between the actual and the expected percentage values. This procedure is summarized in the formula shown in Figure 12.3.

The index values that result from using this method range from 0 to 175, with 0 representing a perfect circle and 175 the value that would presumably be associated with a straight line (Figure 12.4). Other shapes have values that lie somewhere between. A square has an index of 12, a star 25, a specific rectangle 28, and so on.

The BCRSI suffers from several shortcomings. It does not, for example, have a one-to-one correspon-

[4]Boyce and Clark, "The Concept of Shape in Geography," 561–72.

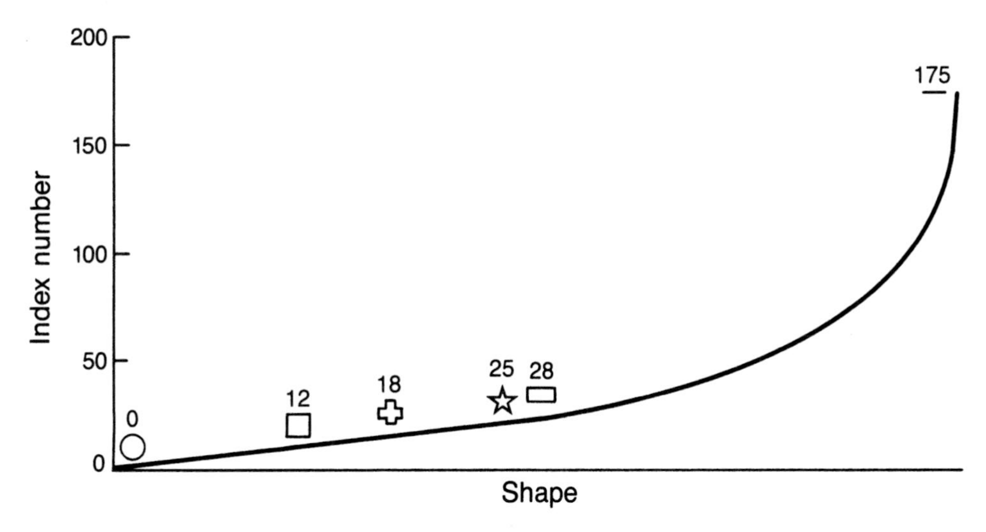

Figure 12.4 Index values for Boyce-Clark radial shape index range from 0 to 175.

Source: Modified from Ronald R. Boyce and W. A. V. Clark, "The Concept of Shape in Geography," *The Geographical Review* 54 (1964): 561–72. Reprinted by permission of The American Geographical Society.

dence between its value and the shape of the figure (Figure 12.5). This means that different shapes can have the same index value. Also, changes in the orientation of the radii in relation to the orientation of the figure result in different index values for the same shape. Changes in the number of radii used also result in different index values for the same figure, as do changes in the location of the center node. Finally, the index value does not permit regeneration of the original shape.

Advanced Techniques

Other approaches to the measurement of shape have been developed, including a technique called *Fourier analysis*. The mathematics involved in the use of such methods goes considerably beyond the scope of this book. The complex measure resulting from Fourier analysis permits regeneration of the original shape, but even it does not have a one-to-one correspondence between the shape and the index values.

Boyce-Clark radial shape index (BCRSI)

BCRSI = 24.24

BCRSI = 1.75

BCRSI = 24.24

BCRSI = 1.75

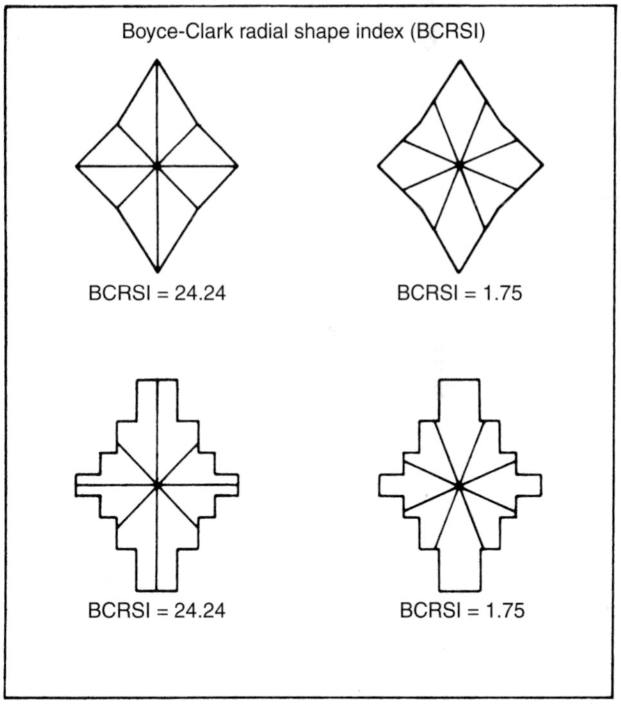

Figure 12.5 Lack of one-to-one correspondence between the value of the Boyce-Clark radial shape index and the shape of the figure. Different shapes can have the same index value, and the same shape can have different index values, depending upon the method of measurement.

Source: From Daniel A. Griffith, Michael P. O'Neill, Wende A. O'Neill, Lloyd A. Leifer, and Rodric B. Mooney, "Shape Indices: Useful Measures or Red Herrings?" *The Professional Geographer* 38, no. 3 (August 1986): 263–70. Reprinted by permission.

Comment on Shape Indices

It has been suggested that "the growing number of shape indices reflects the inability of any one measure to capture all elements of a given characteristic surface configuration. If one accepts this assessment, shape indices may represent a 'red herring' in geographic research."[5] If after reading the preceding discussion, you have the feeling that the shape measures described are not particularly satisfactory, you can at least be assured that your feeling is shared by others.

[5]Daniel A. Griffith, Michael P. O'Neill, Wende A. O'Neill, Lloyd A. Leifer, and Rodric G. Mooney, "Shape Indices: Useful Measures or Red Herrings?" *The Professional Geographer* 38, no. 3 (August 1986): 263–70.

POINT PATTERNS

Point symbols may be thought of as individual indications of the locations at which a particular type of feature is found. The features represented range from human-made phenomena, such as cities, towns, manufacturing plants, stores, or shopping centers, to natural phenomena, such as specific plant or animal species. At times, however, it is useful to consider a set of point symbols as a group, rather than as individual points, to see whether their distribution exhibits a regularity, or pattern. Although the eye provides a visual suggestion about whether or not a pattern seems to exist, analyzing the distribution of the symbols using a less subjective approach determines whether the apparent pattern is real.

A demonstrated order to the arrangement of points may signal that some process is at work to produce the arrangement. Further study may lead to an understanding of what this process is and how it operates. You may notice, for example, that the cities in a region seem to be rather uniformly distributed. Application of an analytic test may show that there is, indeed, a measurable order to their distribution, which may help in the development of a theory about why the pattern exists. One such theory, called *central place theory,* attempts to describe and explain just such an ordering process and is discussed in almost any economic geography textbook.

Types of Distributions

Theoretically, point distributions can assume three general types of patterns: random, uniform, and clustered. Many of the measures used to describe distributions are based on a comparison of the actual pattern of points to a pattern that would be developed randomly. A **random pattern** on a plane (such as a map surface) satisfies two conditions (Figure 12.6a). First, any point is equally likely to occur at any location and, second, the position of any point is not affected by the position of any other point. In such a pattern, then, there is no apparent ordering of the distribution. Instead, some points are clustered close to their neighbors, some are more remote from other points, and some are spaced at intermediate distances from one another. Completely **uniform patterns,** on the other hand, are distinguished by every point being as far from all of its neighbors as possible (Figure 12.6b). Finally, in a **clustered pattern,** many points are concentrated close together, and there are large areas that contain very few, if any, points (Figure 12.6c).

Point-distribution patterns are classified by comparing the specific pattern being tested to the basic patterns just discussed. Two techniques that take this approach are introduced here—quadrat analysis and nearest-neighbor analysis. Both of these approaches have shortcomings, especially in the simple form presented here, but they provide useful examples of analysis techniques. More advanced methods, or refinements of the methods discussed here, may be pursued through the study of some of the Suggested Readings at the end of the chapter.

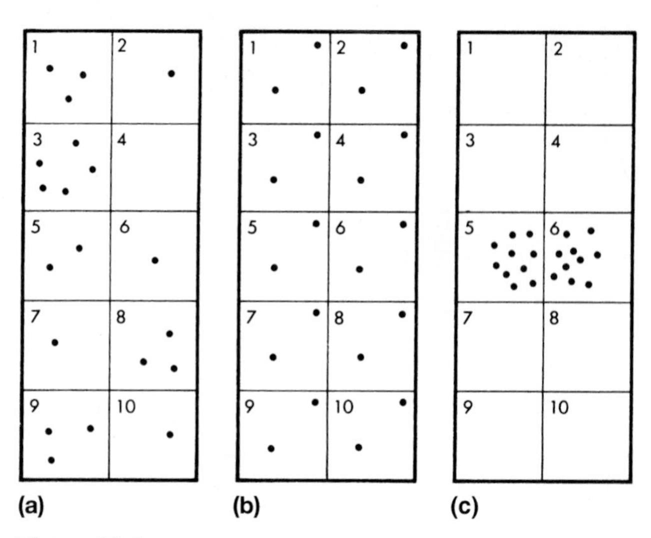

Figure 12.6 Standard types of point distributions. (*a*) Random. (*b*) Uniform. (*c*) Clustered.

Quadrat Analysis

Quadrat analysis is based on a measure derived from data obtained after a uniform grid network is drawn over a map of the distribution of interest. The openings in this grid, which are usually but not necessarily square, are called *quadrats.* The frequency count, which is the number of points occurring within each quadrat, is recorded first. These data are then used to compute a measure called the **variance.** The variance compares the number of points in *each* grid cell with the average number of points over *all* of the cells. The greater the difference between the value for a given cell and the average, the greater the effect on the variance.[6] Finally, the variance of the distribution being tested is compared to the characteristics of a distribution that consists of a random scatter of the same number of points over the same area. This comparison is done by using the variance-mean ratio, because research has demonstrated that the mean and variance of a random distribution are equal.

The simple example that follows shows the basic application of quadrat analysis, based on a test of the "random" map in Figure 12.6a.[7] To begin the process of computing the variance-mean ratio, an arbitrary

[6]Additional information about the variance may be found in any introductory statistics textbook, but, for our purposes, we will simply accept it as an appropriate measure.

[7]A random process was actually used to generate this map, but because of the small number of points involved, the results must be considered to only approximate a random distribution.

TABLE 12.1 Calculation of Variance-Mean Ratio for "Random" Distribution

(1) Quadrat Number	(2) x (Number of Points per Quadrat)	(3) x^2
1	3	9
2	1	1
3	5	25
4	0	0
5	2	4
6	1	1
7	1	1
8	3	9
9	3	9
10	1	1
	$\Sigma x = 20$	$\Sigma x^2 = 60$
	$(\Sigma x)^2 = 400$	

N = number of quadrats = 10

$$\text{Variance} = \frac{\Sigma x^2 - [(\Sigma x)^2/N]}{N-1} = \frac{60-(400/10)}{10-1} = \frac{20}{9} = 2.22$$

$$\text{Mean} = \frac{\Sigma x}{N} = \frac{20}{10} = 2$$

$$\text{Variance-mean ratio} = \frac{\text{variance}}{\text{mean}} = \frac{2.22}{2} = 1.11$$

TABLE 12.2 Calculation of Variance-Mean Ratio for "Uniform" Distribution

(1) Quadrat Number	(2) x (Number of Points per Quadrat)	(3) x^2
1	2	4
2	2	4
3	2	4
4	2	4
5	2	4
6	2	4
7	2	4
8	2	4
9	2	4
10	2	4
	$\Sigma x = 20$	$\Sigma x^2 = 40$
	$(\Sigma x)^2 = 400$	

N = number of quadrats = 10

$$\text{Variance} = \frac{\Sigma x^2 - [(\Sigma x)^2/N]}{N-1} = \frac{40-(400/10)}{10-1} = \frac{0}{9} = 0.0$$

$$\text{Mean} = \frac{\Sigma x}{N} = \frac{20}{10} = 2$$

$$\text{Variance-mean ratio} = \frac{\text{variance}}{\text{mean}} = \frac{0.0}{2} = 0.0$$

two-column by five-row grid is superimposed on the map (the importance of deciding on the appropriate number and size of quadrats is discussed later). The number of points contained in each of the resulting ten quadrats is determined and the values entered in column 2 of Table 12.1. The individual values in column 2 are totaled to give the sum of x, and this value is squared (sum of x)^2. Each of the x values is then squared, and the results are entered in column 3; the total of column 3 provides the (sum of x^2) value.

The formula shown in Table 12.1 allows the computation of the variance—in this case, 2.22. The next step is to interpret the meaning of this value. As mentioned previously, in a random distribution the variance and mean number of points per quadrat are the same, which means, of course, that the ratio between them is 1. In the example shown in Table 12.1, the variance-mean ratio is 1.11, which is very close to 1.00, so the distribution is, indeed, likely a random one. (If required, the statistical significance of the ratio may be determined with appropriate tests.)

Because the variance-mean ratio for a random distribution is 1, any other ratio indicates that the distribution being tested is *not* random. The calculation of the variance-mean ratio for two additional distributions—one uniform (Figure 12.6b and Table 12.2) and the other clustered (Figure 12.6c and Table 12.3)—shows

TABLE 12.3 Calculation of Variance-Mean Ratio for "Clustered" Distribution

(1) Quadrat Number	(2) x (Number of Points per Quadrat)	(3) x^2
1	0	0
2	0	0
3	0	0
4	0	0
5	10	100
6	10	100
7	0	0
8	0	0
9	0	0
10	0	0
	$\Sigma x = 20$	$\Sigma x^2 = 200$
	$(\Sigma x)^2 = 400$	

N = number of quadrats = 10

$$\text{Variance} = \frac{\Sigma x^2 - [(\Sigma x)^2/N]}{N-1} = \frac{200-(400/10)}{10-1} = \frac{160}{9} = 17.78$$

$$\text{Mean} = \frac{\Sigma x}{N} = \frac{20}{10} = 2$$

$$\text{Variance-mean ratio} = \frac{\text{variance}}{\text{mean}} = \frac{17.78}{2} = 8.89$$

that this is a correct assumption. Furthermore, the nature of the results obtained when the patterns exhibit uniformity or clustering helps to explain the manner in which different values are interpreted.

Quadrat Size and Orientation

The size of the quadrats, as well as their orientation, affects the outcome of the quadrat method. If the quadrats are too small, for example, they may each contain only one or two points. This means that any clusters that exist will escape detection because no quadrats will have large numbers of points. On the other hand, if the quadrats are too large, a single quadrat may contain all or almost all of the points, thus giving the appearance that a cluster exists even if it does not. One recommendation for deciding on an appropriate quadrat size—one that will not be likely to veer toward one or another of these extremes—is to make each quadrat approximately twice the size of the mean area per point.[8] An alternative is to test several quadrat sizes, observing the effect of each test on the results obtained. Similar tests using different quadrat orientations may also be useful because shifting the grid, or placing it at different angles to the distribution, introduces some variation into the results.[9]

Weaknesses of Quadrat Analysis

In addition to the quadrat-size and orientation problems just discussed, quadrat analysis has some other weaknesses. One weakness is actually a matter of terminology. Although the technique is considered a method for measuring pattern, it is actually more appropriately described as a method for measuring dispersion. This is because it is based on considerations of the density of points and not of their arrangement in relation to one another. Another problem is that it has been shown that different probability models can yield the same values, casting some doubt on the interpretation of the results. Finally, quadrat analysis results in a single measure for the entire distribution, so any variations *within* the region are not recognized. This problem can only be dealt with by subdividing the region, which would probably be done on the basis of a visual inspection indicating that variation in pattern seems to exist.

Nearest-Neighbor Analysis

A second method of analyzing distributions of point symbols is **nearest-neighbor analysis.** This method, unlike quadrat analysis, uses distances between points as its basis. Specifically, the mean of the distance observed between each point and its nearest neighbor is compared with the expected mean distance that would occur if the distribution were random. The underlying theme of testing against a random distribution occurs here, just as in quadrat analysis, because a random pattern is the outcome that would normally be expected if no ordering process were at work. In other words, something has likely caused the distribution pattern to deviate from a random arrangement if either a clustered or a uniform arrangement is found, and further investigation may reveal the cause.

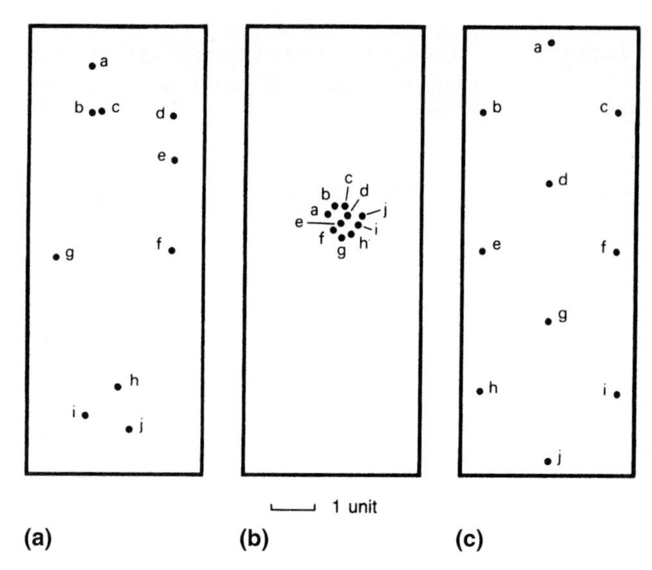

Figure 12.7 Source maps for calculation of the nearest-neighbor statistic. (*a*) Random. (*b*) Clustered. (*c*) Uniform.

Once again, three distinctly different distribution patterns are analyzed here to demonstrate how the nearest-neighbor statistic is calculated and to show the interpretation of the range of values that can result (Figure 12.7). In the case of the random pattern, which is analyzed first, the nearest neighbor of each point in the distribution is determined, as shown in column 2 of Table 12.4. The distance (r) from each point to its nearest neighbor is then recorded in column 3. Distances may be measured in any convenient units, although kilometers, meters, or similar units are usually selected, and the graphic scale of the map is used. In any case, the same units must be used throughout the process, including the calculation of density described in Table 12.4. The average spacing ($\bar{r}$) between the nearest neighbors is obtained by adding all of the individual spacing values (sum of r) and dividing the total by the number of points (n) in the distribution.

The *expected average spacing* [$\bar{r}(e)$] is then calculated. This calculation is based on the spacing of points that would occur in a random distribution containing the same number of points scattered over the same area. In general, the value of $\bar{r}(e)$ is $0.5/\sqrt{d}$,

[8]Based on P. Greig-Smith, *Quantitative Plant Ecology,* 2d ed. (London: Butterworths, 1964).

[9]See Arthur Getis and Barry N. Boots, *Models of Spatial Processes: An Approach to the Study of Point, Line, and Area Patterns* (New York: Cambridge University Press, 1978), 24–25, for more discussion of these concerns.

TABLE 12.4 Calculation of Nearest-Neighbor Statistic for "Random" Distribution

(1) Point	(2) Nearest Neighbor	(3) Distance to Nearest Neighbor
a	b	1.0
b	c	0.1
c	b	0.1
d	e	1.0
e	d	1.0
f	e	2.0
g	f	2.7
h	j	1.0
i	j	1.0
j	i	1.0
		$\Sigma r = 10.9$

Number of points $= n = 10$

$\bar{r} = \frac{\Sigma r}{n} = \frac{10.9}{10} = 1.09$

Area of region $= 5 \times 10 = 50$ sq. units

Density $= d = \frac{n}{\text{area}} = \frac{10}{50} = 0.2$

Expected mean spacing $= \bar{r}(e) = \frac{0.5}{\sqrt{d}} = \frac{0.5}{\sqrt{0.2}} = \frac{0.5}{0.447} = 1.12$

Nearest-neighbor statistic $= R = \frac{r}{\bar{r}(e)} = \frac{1.09}{1.12} = 0.97$

TABLE 12.5 Calculation of Nearest-Neighbor Statistic for "Clustered" Distribution

(1) Point	(2) Nearest Neighbor	(3) Distance to Nearest Neighbor
a	b	0.1
b	c	0.1
c	d	0.1
d	e	0.1
e	f	0.1
f	g	0.1
g	h	0.1
h	i	0.1
i	j	0.1
j	i	0.1
		$\Sigma r = 1.0$

Number of points $= n = 10$

$\bar{r} = \frac{\Sigma r}{n} = \frac{1.0}{10} = 0.1$

Area of region $= 5 \times 10 = 50$ sq. units

Density $= d = \frac{n}{\text{area}} = \frac{10}{50} = 0.2$

Expected mean spacing $= \bar{r}(e) = \frac{0.5}{\sqrt{d}} = \frac{0.5}{\sqrt{0.2}} = \frac{0.5}{0.447} = 1.12$

Nearest-neighbor statistic $= R = \frac{r}{\bar{r}(e)} = \frac{0.1}{1.12} = 0.09$

where d is the density of points over the study area (in other words, the number of points per unit area). In this case, there are 10 points in an area of 50 square units, so $\bar{r}(e)$ is 1.12.

Finally, the nearest-neighbor statistic (R) is calculated by dividing the actual by the expected average distances [$\bar{r}/\bar{r}(e)$]. If the two values are identical, as they obviously are if the distribution being tested is also random, R has a value of 1. In this case, the value of R is 0.97, which is very near to 1.00, so the distribution very closely approximates a random arrangement.

Let us now consider what the R values are for distributions that are not random. If all of the points in the distribution are clustered close together, their $\bar{r}$-value approaches 0, and R also approaches 0, as is shown in Table 12.5. In contrast, if all of the points are dispersed so that they are as far from one another as possible, $\bar{r}$ approaches its maximum value. This tendency is illustrated in Table 12.6, which shows the calculation of R for the uniform distribution. R, in this case, is 1.96. It has been shown that the theoretical maximum value of R, which would occur in a completely uniform distribution, approaches 2.1491. In Table 12.6, then, the distribution shows a strong tendency toward a uniform pattern. That it is not completely uniform is related to the boundary problems discussed later in this section.

TABLE 12.6 Calculation of Nearest-Neighbor Statistic for "Uniform" Distribution

(1) Point	(2) Nearest Neighbor	(3) Distance to Nearest Neighbor (r)
a	c	2.2
b	d	2.2
c	d	2.2
d	e	2.2
e	g	2.2
f	g	2.2
g	h	2.2
h	j	2.2
i	j	2.2
j	i	2.2
		$\Sigma r = 22.0$

Number of points $= n = 10$

$\Sigma r = \frac{\Sigma r}{n} = \frac{22}{10} = 2.2$

Area of region $= 5 \times 10 = 50$ sq. units

Density $= d = \frac{n}{\text{area}} = \frac{10}{50} = 0.2$

Expected mean spacing $= \bar{r}(e) = \frac{0.5}{\sqrt{d}} = \frac{0.5}{\sqrt{0.2}} = \frac{0.5}{0.447} = 1.12$

Nearest-neighbor statistic $= R = \frac{\bar{r}}{\bar{r}(e)} = \frac{2.2}{1.12} = 1.96$

The nearest-neighbor method has some advantages over the quadrat method for analyzing point distributions. First, there is no quadrat-size problem to be concerned with in the nearest-neighbor method, and, second, it takes distances into account.

The principal difficulties with the nearest-neighbor method are related to determining the limits of the area to be analyzed. If the area in which the points are distributed has a logical or absolute boundary, such as the shoreline of an island, that boundary can be used. If no such logical boundary exists, the boundary must be defined in other ways. The reason for this concern with establishing boundaries is that the value of the measure varies, depending on the density of the distribution, and the density itself obviously varies with differences in the size of the study area.

The importance of boundaries is illustrated by two of the previous examples of uniform and clustered distributions. The distances between nearest neighbors in the clustered distribution (Table 12.5) are all equal to one another, just as in the uniform pattern (Table 12.6). The only difference between the two sets of data is in the absolute values of the distances.[10] This means that if you placed the clustered distribution from Table 12.5 inside a smaller region, the nearest-neighbor statistic would indicate a more uniform distribution. (You may wish to try doing this, if it is not obvious to you.) The values obtained from a nearest-neighbor analysis, then, are dependent on the specific area analyzed.

The necessity for setting appropriate study-area boundaries is also apparent in the case where more than one distribution occurs within the analysis area. In this case, a nearest-neighbor statistic can be developed for each distribution, and, in addition, one distribution can be said to be more or less uniform, for example, than the other.

If the appropriate boundaries for the study area are not clear from the context of the study, some method is needed to determine them. One method for doing this is to select the extreme points in each direction of the distribution and to draw the limiting lines through them. In this approach, all of the points are contained within the limits.

Another boundary problem is that the measure may be biased if the nearest-neighbor distance for a point is greater than the distance from that point to the study-area boundary. The reason for this is that there may be a *nearer* neighbor on just the other side of the boundary. This is not encountered if it has been possible to establish a boundary that encloses all the points, as already suggested. In other cases, this may not be possible because the limits of the particular phenomenon being studied extend far beyond the area to be analyzed. Consider, for example, the study of the distribution of grass clumps in a specific field that lies in the middle of a grassland hundreds of hectares in extent. In cases of this type, measurements can be taken to nearest neighbors *whether or not* they are in the study area. When this is done, measurements *from* points outside the study area are not taken. Other methods of dealing with boundary problems have been developed, but this example at least points out the nature of the problem.

Finally, an R value of 1 does not necessarily indicate that the distribution is purely the result of chance, because other factors may result in a distribution that has such a characteristic. It is usually safer to say that values toward the extremes (0 and 2.1491) indicate tendencies toward, respectively, uniformity and clustering and, therefore, require some explanation.

Spatial Correspondence of Areal Distributions

Quadrat and nearest-neighbor analyses deal with the analysis of a single distribution. On occasion, however, two (or more) distributions must be compared. Various techniques are designed for that purpose. The coefficient of areal correspondence and the chi-square statistic are discussed here.

Coefficient of Areal Correspondence

The **coefficient of areal correspondence** is a simple measure of the extent to which two distributions correspond to one another. It might be used, for example, to compare the extent of wheat farming in a region with the area that receives some minimum level of rainfall. The approach is based on a technique called **overlay analysis.** In overlay analysis, the two distributions of interest are mapped at the same scale, and the outline of one is overlaid (superimposed) on the outline of the other (Figure 12.8). The regions formed on the resulting composite map fall into three different categories: (1) those that fall into only one category

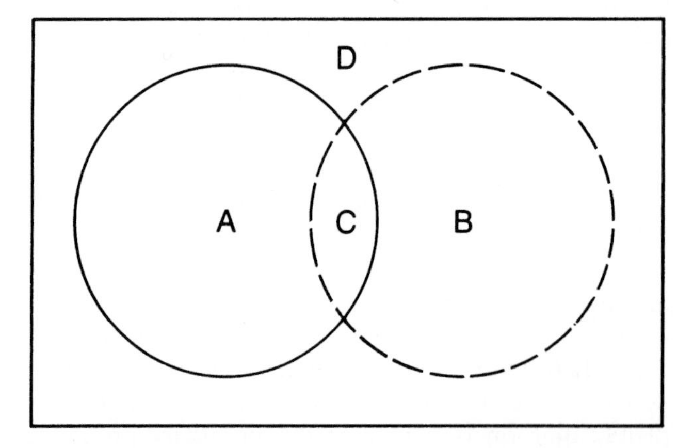

Figure 12.8 Composite map of two overlapping distributions.

[10] Such uniformity would seldom be encountered in a clustered distribution. The example was purposely designed to illustrate this problem.

(A and B in Figure 12.8); (2) those, if any, where the two categories overlap (C); and (3) those that fall outside either category (D). The area of each of these regions is measured, and the coefficient is computed on the basis of the measurements.

Specifically, the coefficient of areal correspondence is the ratio between the area of the region where the two distributions overlap (C) and the total area of the regions covered by the individual distributions and the region of overlap ($A + B + C$). Region D, which falls outside either category, is ignored. The computation takes the following form:

$$\frac{\text{Area of } C}{\text{Area of } A + B + C}$$

The three examples in Figure 12.9 show the range of values of this coefficient. Where there is no overlap, the value is 0. At the other extreme, where the two categories completely overlap, the value is 1. In situations where the amount of overlap lies somewhere between the two extremes, the value of the coefficient lies between 0 and 1. Where the three regions are of equal size, for example, the value is 0.33. The value of the coefficient of areal correspondence, then, provides a simple measure of the extent of spatial association between two distributions, but it cannot provide any information about the statistical significance of the relationship.

Chi-Square Statistic

If additional data are available, it may be possible to use other measures to indicate the strength of the association between two distributions. For example, to examine further the relationship between wheat yield and precipitation in a region, assume that data are available in the form of two simple maps (both drawn to the same scale). One map shows regions of high and low precipitation, and the other shows regions of high and low wheat yield (Figures 12.10a and 12.10b). On visual inspection, it seems that there is some level of correspondence between the two maps, with wheat yield generally higher where there is more precipitation. We can, therefore, reasonably compute a chi-square statistic to tell us more about the relationships between the two distributions, which is made clearer by plotting the two distributions on a combined map (Figure 12.10c). For simplicity's sake, this example is limited to two characteristics of two distributions, but the same technique can be applied to more complex situations.

The analysis process begins with the drawing of a square grid over the map of the combined distributions. The finer the grid, the more precise the measurements (and the greater the amount of work required for the analysis). Two values are then recorded for each quadrat of the grid: one for the precipitation class and one for the wheat production class (Table 12.7). One problem with this counting process is that some cells fall on the dividing line between the categories. The usual way of dealing with this is to count the entire cell in the category that covers 50% or more of the cell area. The inaccuracy introduced by this procedure is reduced by using a finer-mesh grid.

In this case, four data combinations are possible: low rainfall, low yield; low rainfall, high yield; high rainfall, low yield; and high rainfall, high yield. The total number of occurrences of each of these combinations is recorded in a **table of observed frequencies** (Table 12.8). In this example, we have a simple two-column by two-row table. Its four cells represent the four possible data combinations just mentioned. The column and row totals in the table are obtained by summing across each row and down each column, after the cells have been filled in. The value in the lower-right corner is the sum of either the rows or the

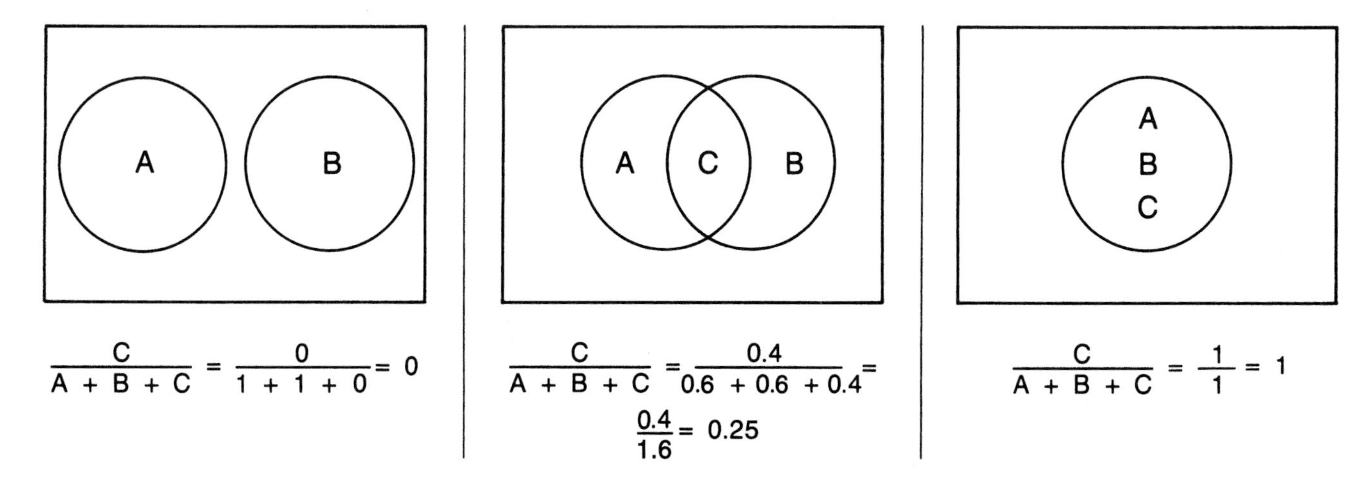

Figure 12.9 Range of values of the coefficient of areal correspondence. (*a*) With no overlap. (*b*) With partial overlap. (*c*) With complete overlap.

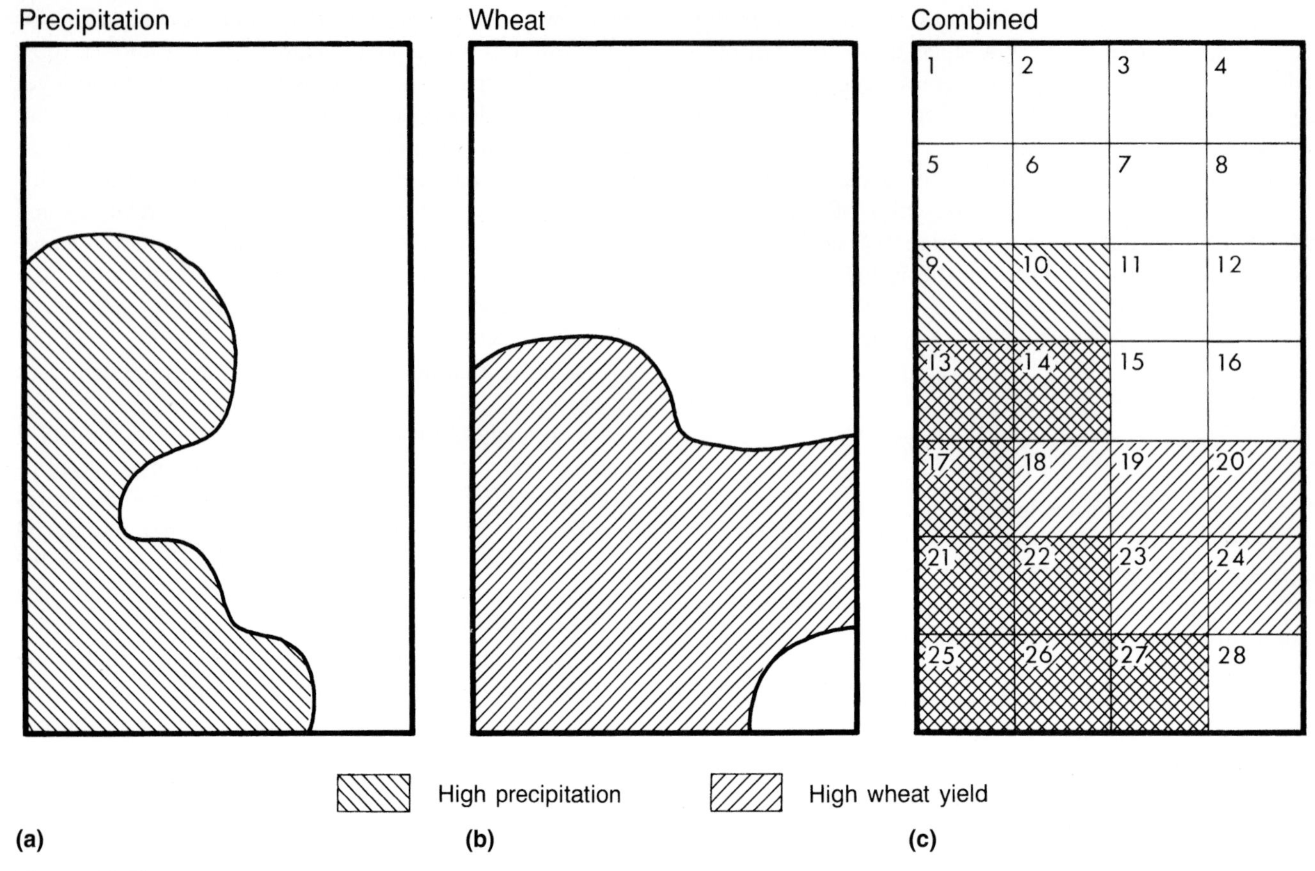

Figure 12.10 Relationships between precipitation and wheat yield. (*a*) Regions of high and low precipitation. (*b*) Regions of high and low wheat yield. (*c*) Combined map of the two distributions.

columns and equals the total number of cells in the study area.

Comparison of the observed frequencies with a similar **table of expected frequencies** (Table 12.9) provides the basis for computing a chi-square statistic, which gives a measure of the relationship between the two distributions. Determining these expected frequencies involves a **null hypothesis,** which states that there is *no* relationship between the two variables. This means that if the null hypothesis is correct, wheat yield will not vary in relation to rainfall. Of course, the null hypothesis can be rejected by demonstrating that rainfall (which is the **independent variable** in this example) has a strong effect on wheat yield (the **dependent variable**).

Each of the four entries in the interior of the table of expected frequencies (see Table 12.9) is calculated on the basis of the null hypothesis. The same row and column totals that occurred in the table of observed frequencies (see Table 12.8) are used as the starting point (the calculation of the table may be based on either the column or the row totals). In this case, for example, the "high rainfall" category represents 35.7% of the total number of cells (10/28 = 35.7%). This percentage is applied to the 13 cells that have "high yield" to determine that five cells would be expected to fall in the "high rainfall, high yield" category (13 × 35.7% = 5).

The remaining entries in the expected-frequencies table could be calculated in the same way, using either column or row data. In the case of Table 12.9, however, the table has limited degrees of freedom, which means that once the value of a single cell in the two-column by two-row matrix has been calculated, the values of the other three cells are determined. The entry for "high rainfall, low yield" (5) is determined by subtracting the "high rainfall, high yield" value (5) already calculated from the total number of high-rainfall cells (10), and so on for the remaining entries. With larger matrices, of course, many more entries must be calculated before the degrees of freedom are exhausted.

Next, the observed-frequency table (see Table 12.8) is compared with the expected-frequency table (see Table 12.9). The measure used for this purpose is called the **chi-square statistic,** which is given by the following:

χ^2 (chi-square) = $\Sigma[(O - E)^2/E]$ where

O = observed number of occurrences and

E = expected number of occurrences.

TABLE 12.7 Tabulation of Quadrat Characteristics

(1) Quadrat	(2) Rainfall Category	(3) Wheat Yield Category
1	L	L
2	L	L
3	L	L
4	L	L
5	L	L
6	L	L
7	L	L
8	L	L
9	H	L
10	H	L
11	L	L
12	L	L
13	H	H
14	H	H
15	L	L
16	L	L
17	H	H
18	L	H
19	L	H
20	L	H
21	H	H
22	H	H
23	L	H
24	L	H
25	H	H
26	H	H
27	H	H
28	L	L

In our example, the statistic is calculated as follows:

$$\frac{(8-5)^2}{5} + \frac{(2-5)^2}{5} + \frac{(5-8)^2}{8} + \frac{(13-10)^2}{10} = 5.625$$

The interpretation of the chi-square statistic is based on its value. Zero indicates no relationship between the variables, whereas a larger number indicates a stronger relationship. A table of significance can be consulted to determine whether the specific value is statistically significant—that is, is not likely to occur by chance.[11] Unfortunately, the significance of the chi-square value varies with the size of the table itself, as well as with the actual number of cases involved. For this reason, a correction factor that eliminates the influence of those two variables is calculated. Of the several correction procedures available, one known as

[11]See any introductory statistics textbook.

TABLE 12.8 Observed Frequencies

		WHEAT YIELD High	WHEAT YIELD Low	Total
RAINFALL	High	8	2	10
RAINFALL	Low	5	13	18
	Total	13	15	28

TABLE 12.9 Expected Frequencies

		WHEAT YIELD High	WHEAT YIELD Low	Total
RAINFALL	High	5	5	10 (35.7%)
RAINFALL	Low	8	10	18 (64.3%)
	Total	13	15	28

Yule's Q can only be used on two-column by two-row tables.[12]

Yule's Q is given by the following:

$$\frac{ad - bc}{ad + bc}$$

where a, b, c, and d are the cell values in the table of observed frequencies (the order of the cells is shown in Table 12.8). In this case, the value would be as follows:

$$\frac{(8 \times 13) - (2 \times 5)}{(8 \times 13) + (2 \times 5)} = \frac{104 - 10}{104 + 10} = \frac{94}{114} = 0.82$$

The value of Yule's Q always lies between $^+1$ and $^-1$, which makes its interpretation much more straightforward than that of the chi-square statistic. A Q value of 0, for example, indicates that there is no relationship between the two variables. Such a value would uphold the null hypothesis stated, and we would have to look elsewhere for explanations of the variations in wheat yields. A value of $^+1$, on the other hand, would indicate a perfect positive relationship between the two

[12]See Hubert M. Blalock, Jr., *Social Statistics,* 2d ed. (New York: McGraw-Hill Book Company, 1972), 298–99, for a discussion of the available correction procedures, including those for larger tables.

variables. Because the rainfall is the independent variable and the wheat yield is the dependent variable, a perfect positive relationship would mean that a high level of rainfall is associated with a high yield. In this case, 0.82 is quite close to 1.00, and rejection of the null hypothesis is justified.

If the value of Q were negative, it would mean that higher rainfall values are associated with lower wheat yields. In that case, we would want to rethink the problem and consider whether, for example, high rainfall is causing conditions that are too wet for the wheat to grow. For our purposes, it is sufficient to assume that, if the value of Q lies closer to 1 (either positive or negative), the relationship between the two variables is relatively strong, whereas if it lies closer to 0, the relationship is relatively weak. More specific interpretations lie beyond what can be explained here and can be pursued further in a standard statistics textbook.

Other Methods

The comparison of distribution patterns can be carried much further than discussed here. The correspondence between isoline maps (see Chapter 11) of two distributions can be measured, for example. This is shown in a study by Arthur H. Robinson, in which the relationship between farm population density and average annual precipitation is examined.[13] Studying this paper and other examples listed in the Suggested Readings at the end of the chapter will provide the basis for expanding your concepts with regard to pattern comparisons on maps.

[13]Arthur H. Robinson, "Mapping the Correspondence of Isarithmic Maps," *Annals of the Association of American Geographers* 52 (1962): 414–25.

SUMMARY

The description of shape is often subjective, and there has been a search for numerical indices that express shape objectively. Ideally, such an index always uses the same value to describe a given shape. Further, it provides enough information to permit the regeneration of the original shape. Unfortunately, the shape measures available are not totally satisfactory.

In Miller's measure, the shape index is the ratio between the area of a region and the area of a circle with the same perimeter. It ranges from almost 0, in the case of a very elongated shape, to 1, in the case of a perfectly circular shape. Unfortunately, different shapes can yield the same index value, and the original shape cannot be regenerated from the index value.

Bunge's measure is based on distances between vertices on the perimeter of the region. The same number of vertices must be used for all the shapes in a particular study, but the shapes may be of different sizes. Shapes can be grouped on the basis of similar index values. Among other problems, the results depend on the number and location of the vertices chosen, and, therefore, different values may be obtained for the same shape. Also, the index value cannot be used to regenerate the original shape.

The Boyce-Clark radial shape index is based on the lengths of equally spaced radials extended from a node at the center of the shape. The characteristics of these radials are compared with those of a similar set of radials superimposed on a circular region with the same area. The index values range from 0 for a perfect circle to 175 for a straight line. Unfortunately, changes in the orientation or number of radii or in the location of the center node result in different index values for the same shape. Also, the index value does not permit the regeneration of the original shape.

The three general types of point patterns are random, uniform, and clustered. Many of the measures for analyzing the distribution of point symbols are based on a comparison of the actual pattern with a random pattern to determine if there is an order to the arrangement of points. If the arrangement is other than random, investigation may lead to an understanding of the underlying processes leading to the establishment of uniformity or clustering.

One measure of point-distribution patterns is based on quadrat analysis. In this approach, a count is made of the number of points within each cell (quadrat) of a uniform grid drawn over a dot-distribution map. The variance is computed by comparing the number of points in each quadrat with the average number of points over all of the quadrats. This value is then compared to the variance of a random distribution of the same number of points over the same area to produce the variance-mean ratio. Because the variance-mean ratio for a random distribution is 1, any other ratio indicates that the distribution being tested is not random. In this measure, the size and orientation of the quadrats influence the outcome. Although used as a method for measuring pattern, the quadrat method is more appropriately described as measuring dispersion.

A second method of analyzing distributions of point symbols is the nearest-neighbor technique. In this method, the mean of the distances observed between each point and its nearest neighbor is compared with the mean distance that would occur if the distribution were random. If the tested pattern is also random, the resulting value is 1. If the pattern is dispersed, the value approaches its maximum (2.1491), and if the pattern is clustered, the value approaches 0. This method has the advantage that there is no quadrat-size problem, and it takes distances into account. On the other hand, the value of the measure varies with differences in the limits of the study area.

Measures of the spatial correspondence of areal distributions are used to compare two (or more) distributions. The coefficient of areal correspondence, a simple measure of this type, is the ratio between the area of overlap and the total area of the regions covered by the distributions, including the overlap. When there is no overlap, the value of the coefficient is 0. When the two regions completely overlap, the coefficient's value is 1. When the amount of overlap lies between these two extremes, the coefficient's value lies between 0 and 1. A chi-square statistic is sometimes used to tell more about the relationships between the two distributions. Because the significance of the chi-square value varies with the size of the table itself, as well as with the actual number of cases involved, a correction factor (such as Yule's *Q*) is needed.

SUGGESTED READINGS

Austin, R. F. "Measuring and Comparing Two-Dimensional Shapes." In *Spatial Statistics and Models,* edited by G. L. Gaile and C. J. Willmott. Boston: D. Reidel, 1984, 293–312.

Boyce, Ronald R., and W. A. V. Clark. "The Concept of Shape in Geography." *The Geographical Review* 54 (1964): 561–72.

Bunge, William. "Theoretical Geography." *Lund Studies in Geography,* Series C, General and Mathematical Geography, No. 1. Lund, Sweden: The Royal University of Lund, Department of Geography, C. W. K. Gleerup, 1962, 72–88.

Getis, Arthur. "Temporal Land-Use Pattern Analysis with the Use of Nearest-Neighbor and Quadrat Methods." *The Annals of the Association of American Geographers* 54 (September 1964): 391–99.

Griffith, Daniel A., Michael P. O'Neill, Wende A. O'Neill, Lloyd A. Leifer, and Rodric G. Mooney. "Shape Indices: Useful Measures or Red Herrings?" *The Professional Geographer* 38, no. 3 (August 1986): 263–70.

Lee, David R., and G. Thomas Sallee. "A Method of Measuring Shape." *The Geographical Review* 60, no. 4 (1970): 555–63.

McConnell, H. "Quadrat Methods in Map Analysis." Discussion Paper No. 3. Iowa City, Iowa: Department of Geography, University of Iowa, 1966.

Miller, Victor C. "A Quantitative Geomorphic Study of Drainage Basin Characteristics in the Clinch Mountain Area, Virginia, and Tennessee." *Department of Geology, Technical Report No. 3.* New York: Columbia University, Department of Geology, 1953.

Moellering, H., and J. N. Rayner. "The Dual Axis Fourier Shape Analysis of Closed Cartographic Forms." *The Cartographic Journal* 19 (1982): 53–59.

———. "The Harmonic Analysis of Spatial Shapes Using Dual Axis Fourier Shape Analysis (DAFSA)." *Geographical Analysis* 13 (1981): 64–77.

Rogers, A. *Statistical Analysis of Spatial Dispersion: The Quadrat Method.* London: Pion, 1974.

CHAPTER

13 Characteristics of Map Features: Networks and Trees

This chapter continues the discussion begun in Chapter 12 of the characteristics of map features. Many types of linear features are found on maps, including roads, railroads, rivers, and boundaries. Analysis of such features requires additional techniques.

Networks

In the field of geography, a **network** is generally defined as "a set of geographic locations interconnected in a system by a number of routes."[1] Many phenomena shown on maps, such as roads, railroads, airlines, pipelines, and river systems, fit this general definition. Describing and understanding the characteristics of these networks, then, is a legitimate part of map analysis. The field of mathematics known as *graph theory* provides a number of concepts useful for the analysis of networks that appear on maps. This type of analysis can serve a variety of purposes, some examples of which are presented in this section.

Graph theory introduces an element of abstraction into the world of reality shown on the map. The points and lines of the graph show **topological positions,** not real-world, geographical locations. What is important, in topological terms, is the location of an object, relative to other objects, and the connections between locations. Consider, for example, the case of two cities linked by a highway. Topological descriptions deal with matters such as whether one can reach city *a* from city *b*, and what political units, such as counties, are located on either side of the road that joins them. The topological descriptions do not, however, consider the geographical coordinates of the cities, the distances between them, or the direction of the road. Thus, a topological approach introduces a level of abstraction into the description of relationships between entities. This abstraction is often beneficial, because it results in a focus on the structure of the relationships, rather than on the more usual concerns of measuring distances and directions. As a result, the structure can be analyzed in an objective manner that allows a methodical interpretation of relationships. The relationships found at various locations can then be objectively compared.

Terminology

The terms used in network analysis show considerable diversity because workers from a variety of disciplines have contributed their unique terminology. The terms used most in geographic applications are defined here.

A **graph,** or **network,** is simply a figure composed of points and lines (Figure 13.1). The points are usually referred to as **places,** because that is what they most frequently represent on maps. On occasion, the more general term **node** is substituted. The connecting line between two places is called a **route,** and a series

[1]K. J. Kansky, "Structure of Transport Networks: Relationships Between Network Geometry and Regional Characteristics," University of Chicago, Department of Geography, *Research Papers* 84 (1963): 1.

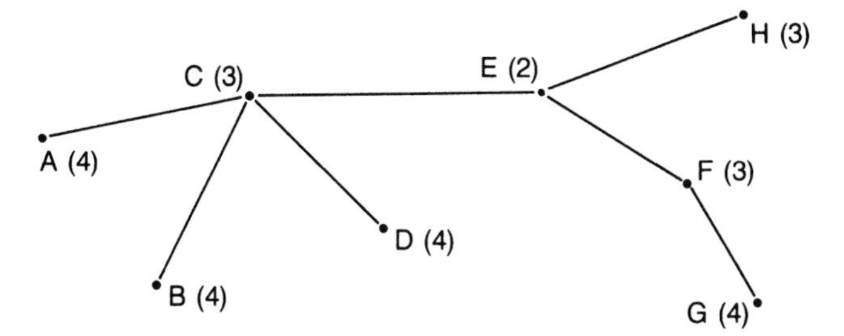

Figure 13.1 A graph, or network, with associated numbers.

of routes linking a series of places is called a **path.**[2] Again, the general term **link** may be substituted for route. The **length of a path** is given by the number of routes it contains. Also, the **topological distance** between two points is the length of the shortest path between them in terms of the number of routes.

Accessibility

Measures based on paths and routes are used to determine how well places are connected to other places. In other words, they show relative **accessibility.**[3] The **associated number** of a place, for example, is the topological distance from that place to the most remote place in the network. Associated numbers are indicated for each place in Figure 13.1. Notice that places *A, B, D,* and *G,* which are on the fringes of the network, have the greatest topological distances. The most central place in the network, then, is the one with the lowest associated number. In Figure 13.1, this is place *E.*

The highest associated number is called the network's **diameter.** The centrality or remoteness of any place in the network is evaluated in terms of its associated number relative to the associated numbers of other places and to the diameter of the network. Again, in Figure 13.1, places *C, F,* and *H* fall between the distant and central places already mentioned.

A frequently encountered real-world problem is the evaluation of possible locations for a warehouse or similar facility that will serve a distribution network. Preliminary ideas about the suitability of various potential locations can be obtained by as simple a process as comparing the associated numbers of the possible sites with one another. Locations with lower numbers usually merit more consideration than those with high numbers.

[2]Points are also called *junctions, intersections, terminals,* or *vertices,* and routes are called *links, arcs, sides, segments, branches,* or *edges.*

[3]Although it is not discussed here, for the sake of simplicity, matrix algebra is frequently used in the analysis of graphs. See, for example, Ronald Abler, John S. Adams, and Peter Gould, *Spatial Organization: The Geographer's View of the World* (Englewood Cliffs, N.J.: Prentice-Hall, 1971), 258–72.

Connectivity

A different type of network analysis is the comparison of the level of development of the transportation or communication systems in two regions. A study of this type might proceed by comparing the level of completeness, or **connectivity,** of the systems.[4] In this context, a highly connected network is presumably a sign of a greater level of development than is represented by a poorly connected one. An industrialized, economically developed country, for example, is likely to have much more fully developed road, railroad, and power networks than is an economically underdeveloped nation. Instead of being content with a subjective statement about the completeness of such networks, one can derive a numerical value that defines the level of development in the two situations, in network terms, and allows for a meaningful comparison.

The completeness of a network is measured by the **connectivity index.** The connectivity index provides a comparison between the actual connections, or routes, that exist in a given network, as compared to the total number of routes possible. A simple connectivity index is given by the formula $100\ (A/P)$, which converts the ratio between the actual number of routes (A) and the possible number of routes (P) to a figure that represents the percentage of completeness. To calculate this index, you must, of course, first determine the values of A and P. A is obtained by simply counting the number of connections that actually exist. P, however, must be computed. This is somewhat more complex than it might seem at first glance, because different types of graphs have different maximum numbers of possible routes.

Types

Assume, for the moment, that we are dealing with highway patterns. Considering the different highway configurations that we might encounter illustrates the types of networks that can be defined. First, a normal, two-way street pattern is called a **symmetric, planar network.** Such networks have two distinguishing characteristics: First, you can, for example, travel from point A to point B on a two-way street, and you can also travel the reverse, from point B to point A, which is the symmetry characteristic. Second, every place where two or more roads meet is a node, and you can enter or leave that intersection along any of the roads, which is the planar characteristic. The maximum number of possible links in a symmetric, planar network is $3\ (n - 2)$, where n equals the number of nodes.

[4]See, for example, E. J. Taaffe, R. L. Morrill, and P. R. Gould, "Transport Expansion in Underdeveloped Countries," *Geographical Review* 53 (1963): 503–29.

If the highway network is modified by the installation of overpasses and underpasses, the planar portion of the definition in the previous paragraph no longer holds, because not every crossing point is necessarily a connecting point. You can drive across an overpass over another road and not be able to get from one road to the other. A network with this limitation is called a **symmetric, nonplanar network.** An airline system is another example of a symmetric, nonplanar network. You cannot branch off to another destination every time the airplane you are on crosses another air route! The number of possible connections in a symmetric, nonplanar network is given by $[n(n - 1)]/2$.

If the highway system not only includes overpasses and underpasses but has a one-way-travel restriction on it, it becomes a **nonsymmetric, nonplanar network.** That is, you can travel from *A* to *B* but not from *B* to *A*, over any given route. The number of possible connections in a nonsymmetric, nonplanar network is given by $n(n - 1)$. In contrast, a **nonsymmetric, planar network** has one-way connections, but the paths are not allowed to cross without joining at a node. This is the equivalent of a one-way street system without overpasses or underpasses. The number of routes in such a network is given by $6(n - 2)$.

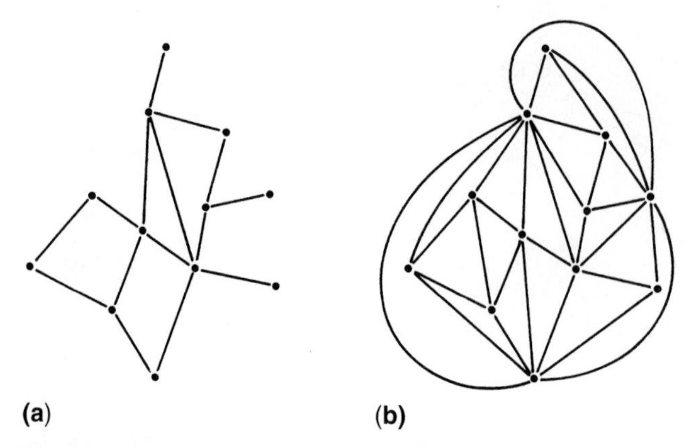

Figure 13.2 (*a*) Symmetric, planar highway network. (*b*) Example of a fully connected, symmetric, planar network—nonoptimum.

Measurement

Returning to the measurement of connectivity, let us assume that the map in Figure 13.2a represents a symmetric, planar highway network. On this map, there are 15 actual connections between cities. Because there are 12 cities in the network, the total possible number of connections is $3(12 - 2) = 30$, and the connectivity index is $100(15/30) = 50$. This indicates that only 50% of the potential direct, city-to-city linkages are in place. The interpretation of the connectivity index is somewhat ambiguous, however, because a fully connected symmetric planar network would not be particularly desirable. Figure 13.2b shows, for example, one way in which the 30 possible connections could be achieved. Clearly, this arrangement is not an optimum one for a highway network because of the circuitous nature of some routes. In addition, not every pair of places is directly connected—such complete connections would require a nonplanar network. Also, other considerations, such as cost, relative accessibility, importance of the various places involved, and so on, need to be taken into account. The connectivity index is useful, nevertheless, as a measure of network completeness.

Applications

Many other measures are used for the analysis of networks, but space does not permit discussing more than the few preliminary ideas presented here. Such studies can be applied to concerns such as the analysis of the hierarchy of cities, the location of a railroad route to maximize net revenues, the characteristics of the interstate highway system, and the shortest route for a traveling salesperson to follow to serve a series of customers. References in the Suggested Readings at the end of the chapter provide a variety of examples.

TREES

Up to now, we have implicitly assumed that networks include loops, so that there is more than one possible path between some pairs of places. Networks that permit such loops are called **circuit networks.** In contrast to circuit networks are **branching networks,** or **trees,** that do not contain circuits. In geographic terms, a river system is a particularly apt analogy for a graph-theoretic tree. Some transportation and distribution systems, including electrical transmission networks, are other examples of trees likely to be found on maps.

A tree with a given number of vertices always has a predetermined number of links. Specifically, because it consists simply of lines successively added to existing lines, a tree with n vertices has $n - 1$ edges, or links. For this reason, the concept of completeness, or connectivity, does not apply to branching networks as it does to circuit networks. The analysis of trees does, however, introduce additional graph-theoretic concepts.

Hierarchic Order

When a stream network, for example, is treated as an abstract tree, each stream segment assumes a specific level of importance or **hierarchic order.** Such order numbers assist in the description and analysis of the

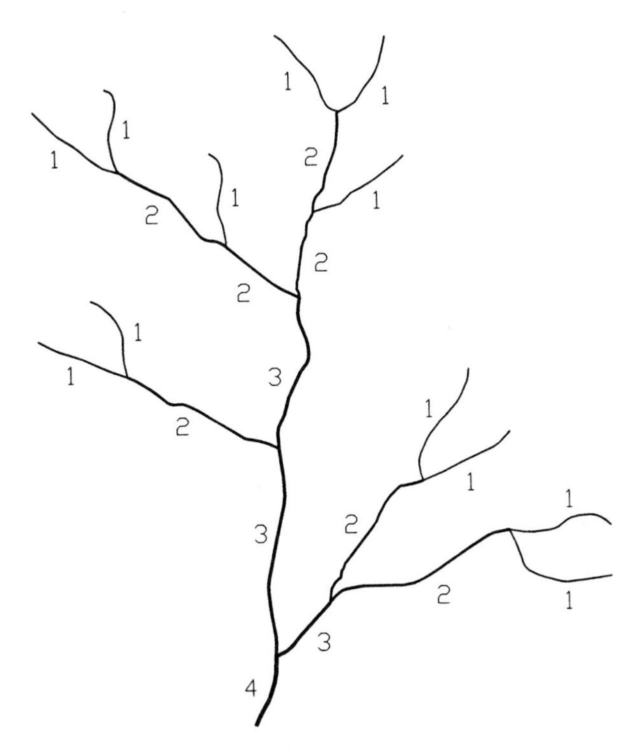

Figure 13.3 Strahler method of stream ordering.

characteristics of different types of stream-development patterns. The approach applies equally well to other treelike networks.

Different schemes are used to assign order numbers to trees, each with a somewhat different outcome. The **Strahler method** of stream ordering is a suitable example of such schemes. In this method, each headwater link in the network (in other words, the smallest tributary streams) is identified as being of order 1 (Figure 13.3). When two headwater (order-1) links join, the downstream link formed is classified as order 2. This process continues, and whenever two links of the same order join, the order assigned to the downstream link is that of the tributaries *plus one*. When two links of different order join, however, the downstream link is assigned an order number the *same* as that of its higher-order tributary. Counting and recording the number of links of each order provide information useful for analyzing the characteristics of the stream network.

Bifurcation Ratio

One use of the stream-order data obtained by the Strahler method is the calculation of **bifurcation ratios.** A bifurcation ratio is the ratio between the number of links of one order and the number of links of the next higher order:

$$\text{Bifurcation ratio} = R_b = \frac{N_{s1}}{N_{s2}}$$

where N_{s1} is the number of links of a given order, and N_{s2} is the number of links of the next higher order.

A single bifurcation ratio taken by itself has little meaning. Computing the mean bifurcation ratio for an entire drainage basin and matching the result with the same figure for other drainage basins, however, provides a method of comparing the characteristics of the various basins.[5] Differences in the patterns found from one basin to another, when compared with the geological and climatological characteristics of the basins, provide initial insights into differences in stream-development patterns under different conditions.

[5]The mean bifurcation ratio for basin 1, for example, equals $R_{bs1} = [(N_{s1}/N_{s2}) + (N_{s2}/N_{s3}) + \ldots + (N_{sn-1}/N_{sn})]/K$, where K is the number of such ratios for the whole basin.

SUMMARY

Graph theory provides methods for the objective analysis of networks on maps. Map networks consist of places (nodes), joined by routes (links), and paths, which are a series of routes linking a series of places.

Accessibility measures based on paths and routes are used to determine how well places are connected to other places. The associated number of a place, for example, is the topological distance (number of links) from that place to the most remote place in the network. This provides a measure of the centrality or remoteness of any place in the network. The highest associated number, which is the maximum topological distance in the network, is called the *diameter*.

Connectivity is a measure of the level of completeness of a network. The connectivity index provides a comparison between the actual connections, or routes, that exist in a given network and the total number of routes possible.

Networks are described in terms of their symmetric and planar characteristics. A pattern of two-way connections is symmetric, whereas a one-way pattern is nonsymmetric. If every place where two or more routes cross is a node, the network is planar, whereas if every crossing point is not a connecting point, the network is nonplanar.

Networks that include loops, so that there is more than one possible path between some pairs of places, are called *circuit networks*. In contrast, branching networks (trees), such as river systems, do not contain circuits. A tree always has a predetermined number of links, and each link has a specific hierarchic order.

The Strahler method is an example of hierarchic ordering of trees, applied to streams. In this approach, each headwater link is of order 1. When two order-1 links join, the downstream link formed is classified as order 2, and so on. A bifurcation ratio may be computed using the Strahler order. This is the ratio between the number of links of one order and the number of the next higher order. The mean bifurcation ratio for a drainage basin may be compared with the same measure for other basins. Differences among the ratios of different basins may be compared with the geological and climatological characteristics of the basins to provide initial insights into differences in stream-development patterns under different conditions.

Suggested Readings

Abler, Ronald, John S. Adams, and Peter Gould. *Spatial Organization: The Geographer's View of the World.* Englewood Cliffs, N.J.: Prentice-Hall, 1971, 255–78.

Garrison, W. L. "Connectivity of the Interstate Highway System." *Papers and Proceedings of the Regional Science Association* VI (1960): 121–37.

Haggett, Peter. "Network Models in Geography." In *Models in Geography,* edited by R. J. Chorley and Peter Haggett. London: Methuen, 1967, 609–68.

Haggett, Peter, and Richard J. Chorley. *Network Analysis in Geography.* London: Edward Arnold. New York: St. Martin's Press, 1969, chapter 15.

Kansky, K. J. "Structure of Transport Networks: Relationships Between Network Geometry and Regional Characteristics." University of Chicago, Department of Geography, *Research Papers* 84, 1963.

Nysteun, J. D., and M. Dacey. "A Graph Theory Interpretation of Nodal Regions." *Papers and Proceedings of the Regional Science Association* VII (1961): 29–42.

Ore, Oystein. *Graphs and Their Uses.* New York: Random House, New Mathematical Library, 1963.

C H A P T E R

Cartograms and Special-Purpose Maps

Maps are designed to serve a great variety of purposes. This chapter discusses the advantages and disadvantages of some maps designed to serve special purposes, including the maplike diagrams called *cartograms.* The proper use of cartograms and special-purpose maps is enhanced by an understanding of their special properties. Some presentations designed to serve a special purpose possess shortcomings for other applications, so they must be used cautiously.

Cartograms

Cartograms are unusual because they are modifications of the relationships that we are used to seeing on maps. They are produced by deliberately enlarging or reducing the size (area or length) of the data-collection units in proportion to the data value attached to each unit. These modifications reveal information about the data that would otherwise be difficult to observe. At the same time, on cartograms, locations, distances, and areas are not the same as they are on the globe. The nature of the differences between maps and cartograms must be kept in mind when interpreting cartograms. Two general types of cartograms are described here: linear cartograms and area cartograms.

Linear Cartograms

Linear cartograms, which are also called *distance-transformation maps* or *distance cartograms,* are designed to show relationships between locations but not in the usual distance terms. Instead, they show *relative* distances based on various measures, such as the cost of travel between points or the time it takes to travel between points.

Routed Linear Cartograms

Routed linear cartograms show certain characteristics of specific paths from one point to another. A typical routed linear cartogram shows the average travel times between various points (Figure 14.1). This is accomplished by drawing the length of each connecting segment so that it is proportional to the time required to negotiate it. If the scale is 1 centimeter per 10 minutes of travel, for example, a segment that takes an hour to traverse is drawn 6 centimeters long.

One complication with routed linear cartograms is that the time it takes to travel between two locations may vary, depending on the direction one is moving. For example, if there is an uphill gradient from point *A* to point *B,* it probably takes longer to negotiate the route from *A* to *B* than it does from *B* to *A,* especially if

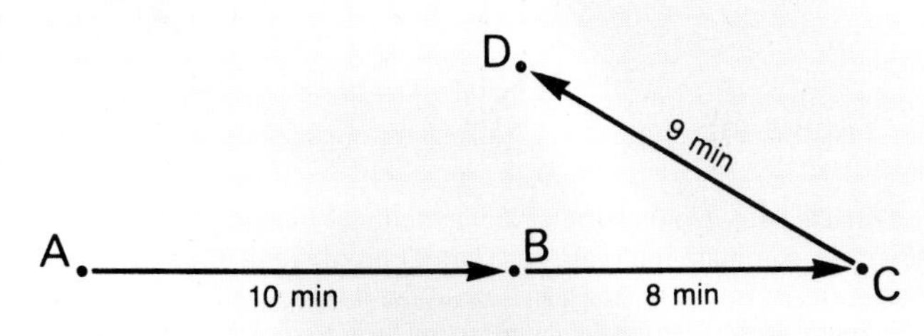

Figure 14.1 Routed linear cartogram.

one is riding a bicycle or walking. Similarly, if the cartogram is designed to show cost of travel, the freight rates may be less in one direction than in the other because the transport company wishes to attract traffic in a particular direction (perhaps to reduce the need for shipping empty cars on that leg of the network).

Such differences in the cost or time connected with traveling in opposite directions along the same route create difficulties in cartogram construction. The problem is that the connecting line between the two points *should* be drawn to two different lengths, to reflect the differing values. One solution is to use two lines, one for each direction, each drawn to its appropriate length and with the direction of movement specified by arrowheads. Obviously, when this is done, the lines cannot follow the exact paths followed in the real world, so distortions are introduced.

Even if the cost or time of travel between two points is the same in either direction, it is likely to be different for different segments of the cartogram. The airfare between two large cities may be less, for example, than the fare between two smaller cities located the same distance apart. If cost is being used to scale distances, the line connecting the smaller centers will be longer than the line connecting the larger ones. The result is that the points representing the cities have to be shifted to accommodate the different line lengths. As the number of connecting lines increases, accommodating their lengths through the shifting of locations may become impossible. When this is the case, a central-point cartogram may be used.

Central-Point Cartograms

A **central-point cartogram** is based on a single starting point and shows equal units of time or cost for reaching other locations from that central point (Figure 14.2). (Alternatively, it can show the values associated with reaching the central point from the other locations.) On cartograms of this type, the relative direction from the center to the other points is often retained, but directions and distances between other points on the cartogram are not correct.

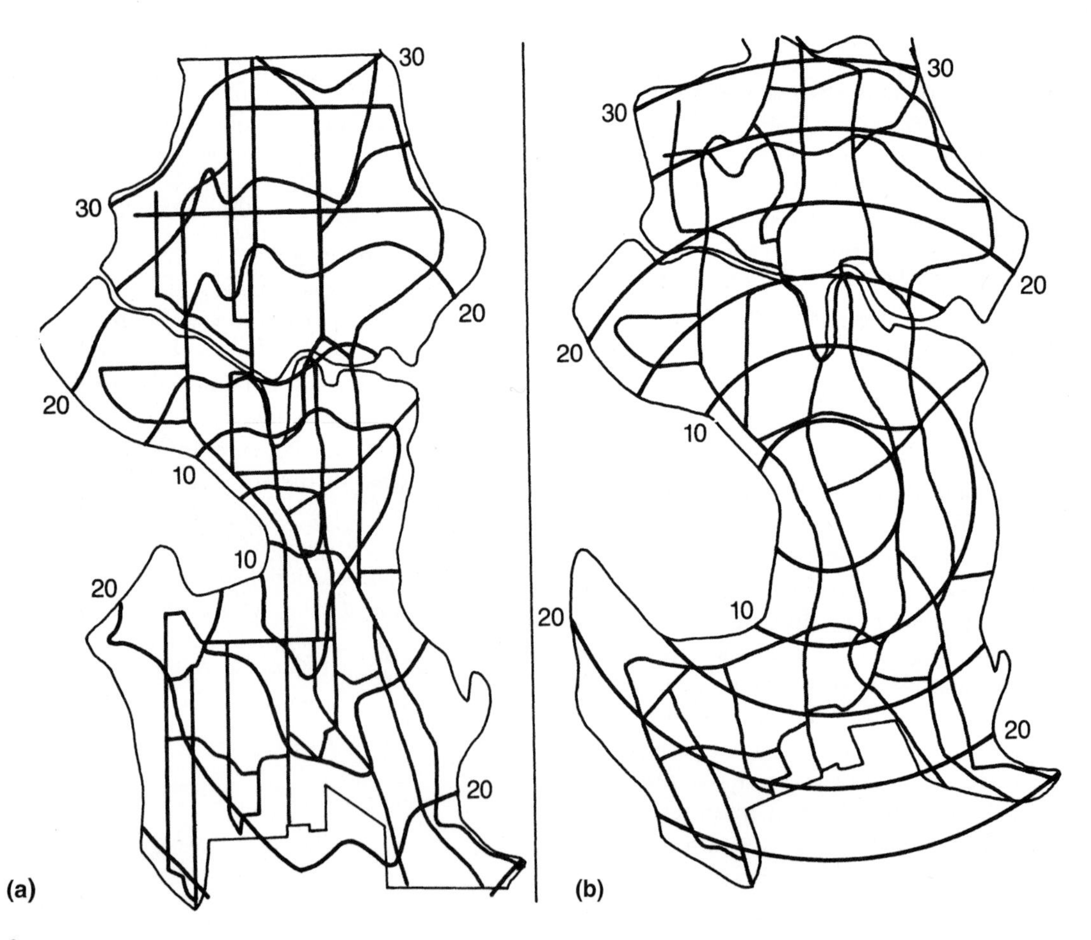

Figure 14.2 Conversion of a conventional map to a central-point cartogram. (*a*) Conventional map of peak-hour travel times from central Seattle, 5-minute intervals. (*b*) Cartogram of "real" time distances from downtown Seattle, 5-minute intervals.

Source: Redrawn from William Bunge, *Theoretical Geography:* Lund Studies in Geography, Series C, General and Mathematical Geography, no. 1 (Lund, Sweden: C. W. K. Gleerup, Publishers, for The Royal University of Lund, Sweden, Department of Geography, 1962), 55. Used by permission.

Area Cartograms

Cartograms in which areas are distorted are called **area cartograms** or, sometimes, **value-by-area cartograms.** On such cartograms, the areas of the various regions are not constructed in proportion to their actual areas on the earth's surface, as they are on conventional equal-area maps.[1] Instead, each region is drawn to a size that is proportional to some *other* value associated with it, such as income, population, or crop yield.

The rescaling involved in the construction of area cartograms introduces graphic difficulties. What happens is that the shapes of the regions and their locations in relation to one another cannot both be maintained when their sizes are varied in different proportions. One possible solution to this problem is to draw a noncontiguous cartogram, which retains shapes by separating the regions from one another. Another solution is to produce a contiguous cartogram, which retains the relationships between the regions, but varies the regions' shapes.

[1]Many map projections also introduce areal distortions. These distortions are derived in a systematic way from actual earth locations and dimensions and are, therefore, quite different from the scaling variations used in cartograms. It has been suggested, however, that cartograms "can be regarded as maps based on some unknown projection," so the distinction may be moot. See Waldo R. Tobler, "Geographic Area and Map Projections," *The Geographical Review* 53 (1963): 59–78.

Noncontiguous Cartograms

In a **noncontiguous cartogram,** the regions are easy to recognize because their shapes are retained, even though they are usually shown in a simplified outline form (Figure 14.3). Retaining the shapes, however, requires shifting the locations of the regions. This results in open areas of various sizes that reflect the discrepancy between the density in the most-crowded unit and the density in other units.[2] What is missing, of course, is the sense of the relationships between regions—for example, of where they meet.

Contiguous Cartograms

Some cartograms retain the desirable trait of contiguity and thus show the spatial relationships between regions. In such **contiguous cartograms,** the shapes of the regions must be changed, often drastically, to accommodate their differently proportioned areas and, at the same time, keep neighbors adjacent to one another (Figure 14.4). Not only are the shapes distorted, but the details of their outlines are also usually simplified. In some extreme cases, the end result is a diagram made up of a series of areas that look nothing like the regions represented. Such a cartogram is often somewhat confusing and difficult to read, because shape is

[2]Judy M. Olson, "Noncontinuous Area Cartograms," *The Professional Geographer* 28, no. 4 (1976): 377.

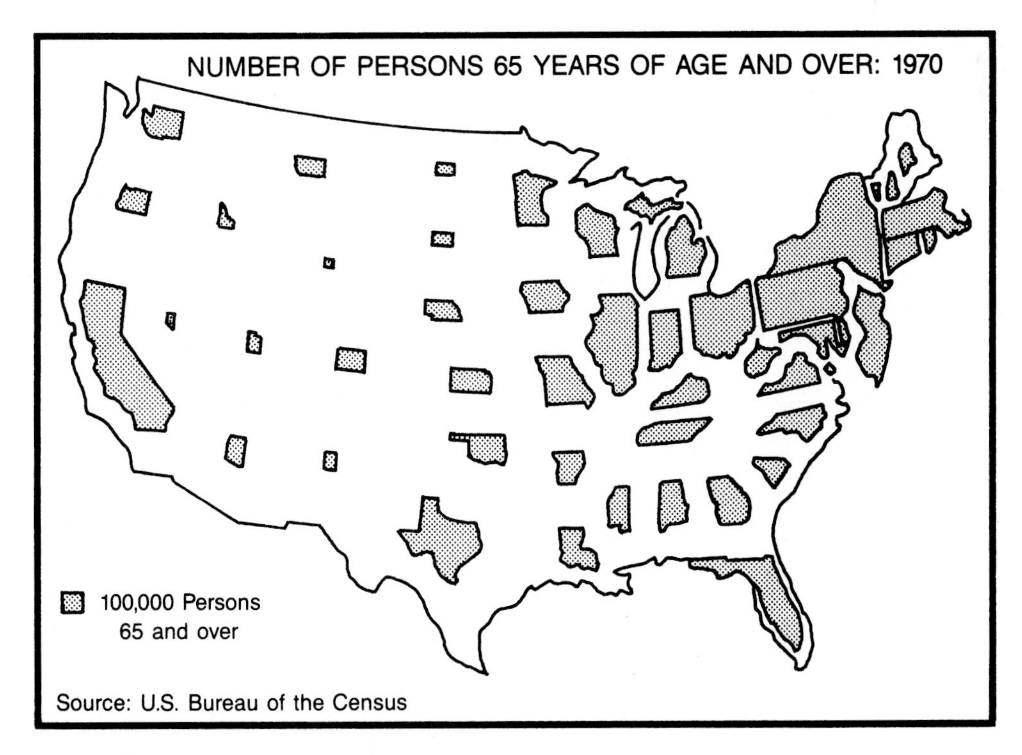

Figure 14.3 Noncontiguous, value-by-area cartogram. The area of each state represents the number of persons 65 and older in that state.
Source: From Judy M. Olson, "Noncontiguous Area Cartograms," *The Professional Geographer* 28, no. 4 (November 1976). Reprinted by permission.

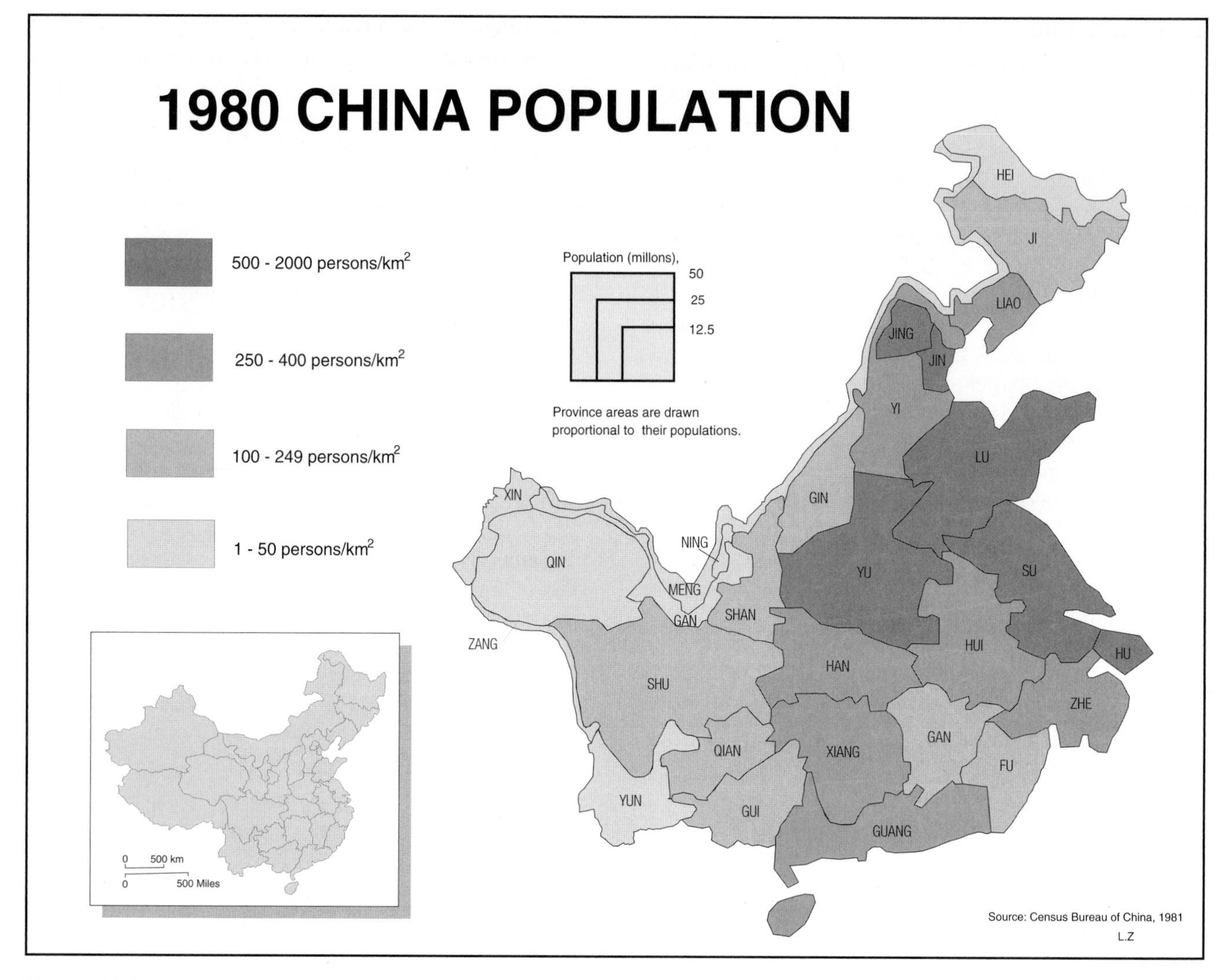

Figure 14.4 Contiguous cartogram showing provinces of China drawn in proportion to their populations in 1980.
Source: Courtesy of Liu Zhu.

one of the major clues used to identify regions.[3] An additional difficulty with some contiguous cartograms is that, because they are rather difficult to draw, the relationships between regions are not always completely retained. Regions may touch along boundaries that differ somewhat from their real boundaries, for example.

Interpreting Area Cartograms

Area cartograms are drawn as though the distribution of the phenomenon being represented is uniform within each region. They do not take into account that the density of occurrence may be very great in one area within a region and very sparse in another. Thus, on a population cartogram of the United States, the size of California is greatly affected by the major population concentration in the Los Angeles area. If the Los Angeles area were shown separately, the remainder of the state would shrink in relative importance. In the interpretation of an area cartogram, no assumptions can be made about the internal distribution of the phenomenon shown within the various regions.

The main point to keep in mind regarding either noncontiguous or contiguous cartograms is that the *sizes* of the regions have been drawn in proportion to their relative importance in terms of whatever value is being represented. One aid to the interpretation of an area cartogram is to compare it to an equal-area map of the same area. Indeed, cartographers sometimes include such a map as an adjunct to the cartogram. The map reference assists in identifying the internal divisions of the cartogram, the geographical relationships between them, and the size distortions introduced by

[3]Borden D. Dent, "Communication Aspects of Value-by-Area Cartograms," *The American Cartographer* 2, no. 2 (October 1975): 154–68.

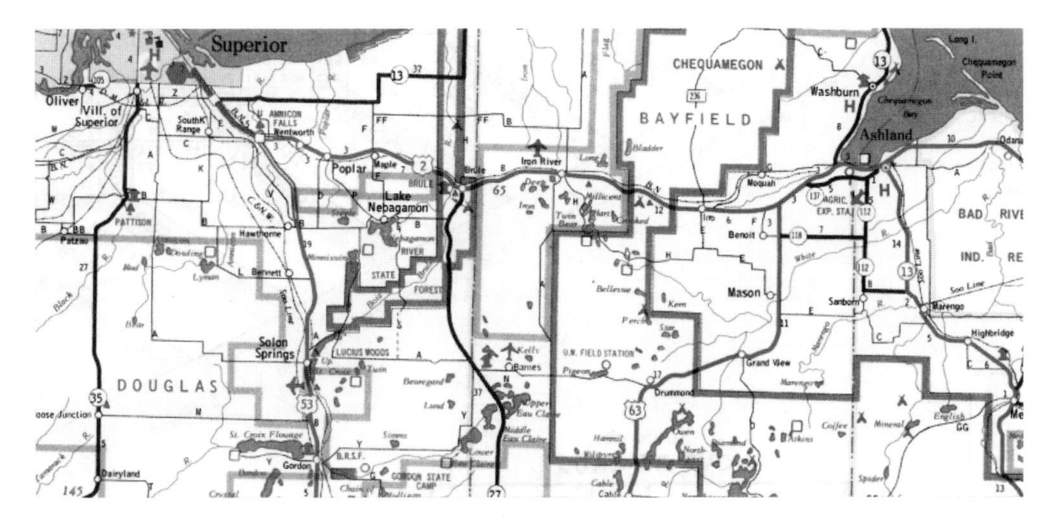

Figure 14.5 Portion of a typical highway map.
Source: Wisconsin State Department of Transportation, Wisconsin Highway Map, 1985–1986.

the cartogram. If no map is included with the cartogram, an atlas or other source can be used as a reference. Comparing the cartogram to an equal-area map is important, because the cartogram's message is contained in the contrast between the true size of the geographic areas, shown on the map, and the quantity of the variable in each area, shown on the cartogram.

Cartograms usually include a small, rectangular symbol with a value attached. This key shows the specific number of units associated with a given area on the cartogram and allows the determination of the approximate (but not exact) value associated with each region. Exact information is more readily and accurately obtained from census data or other sources. Thus, the key gives a sense of the general magnitudes involved, and the body of the cartogram gives an impression of the *importance* of the various regions relative to the variable mapped.

TRANSPORTATION MAPS

Maps are ideal aids for traveling, both when you are planning the trip and while you are en route. The type of map that is particularly useful for a specific journey depends, in large part, upon your mode of travel. If you are using an automobile, bicycle, or motorcycle or walking along roads or highways, for example, a street or highway map is usually preferable. If, on the other hand, you are traveling by public transportation, a route map is indispensable. In this section, we discuss each of these types of maps. The use of nautical and aeronautical navigation charts is covered in Chapter 7. Also, off-road travel on foot generally requires techniques described in sections on navigation and compass use (Chapter 7 and Appendix D).

Highway and Street Maps

For convenience, maps used for intercity travel are referred to here as **highway maps** (Figure 14.5), while maps used to show the road layout within cities are called **street maps** (Figure 14.6). **Road atlases** are bound volumes that usually contain maps of both types, so they are not dealt with separately.

Although the details of particular highway or street maps vary widely, these two types of maps share a number of characteristics. It is feasible, therefore, to consider some generally useful approaches to their use.

The amount of information shown on a highway or street map can be extensive. The map legend indicates the types of features shown on the map and the symbols used to represent the features. There is, of course, an emphasis on roads. On highway maps, national, state, county, and local roads are identified with distinctive route symbols. Types of road surface may be shown, and information such as whether the highway is divided, how many lanes it has, and whether access to it is limited is often included. In addition, towns, cities, parks, and other points of interest are usually shown, as are the more important rivers, lakes, and other water bodies. On street maps, major arterials and one-way streets are usually distinguished from secondary arterials and residential streets. Major buildings and water bodies typically are shown.

Street and highway maps do not usually contain much information about terrain, although hill shading or spot heights of major features may be included. Topographic maps are needed for detailed terrain information.

Scale considerations enter into the use of street and highway maps, just as with other classes of maps. This means that one map cannot serve all of your needs. If you are planning a trip between two major

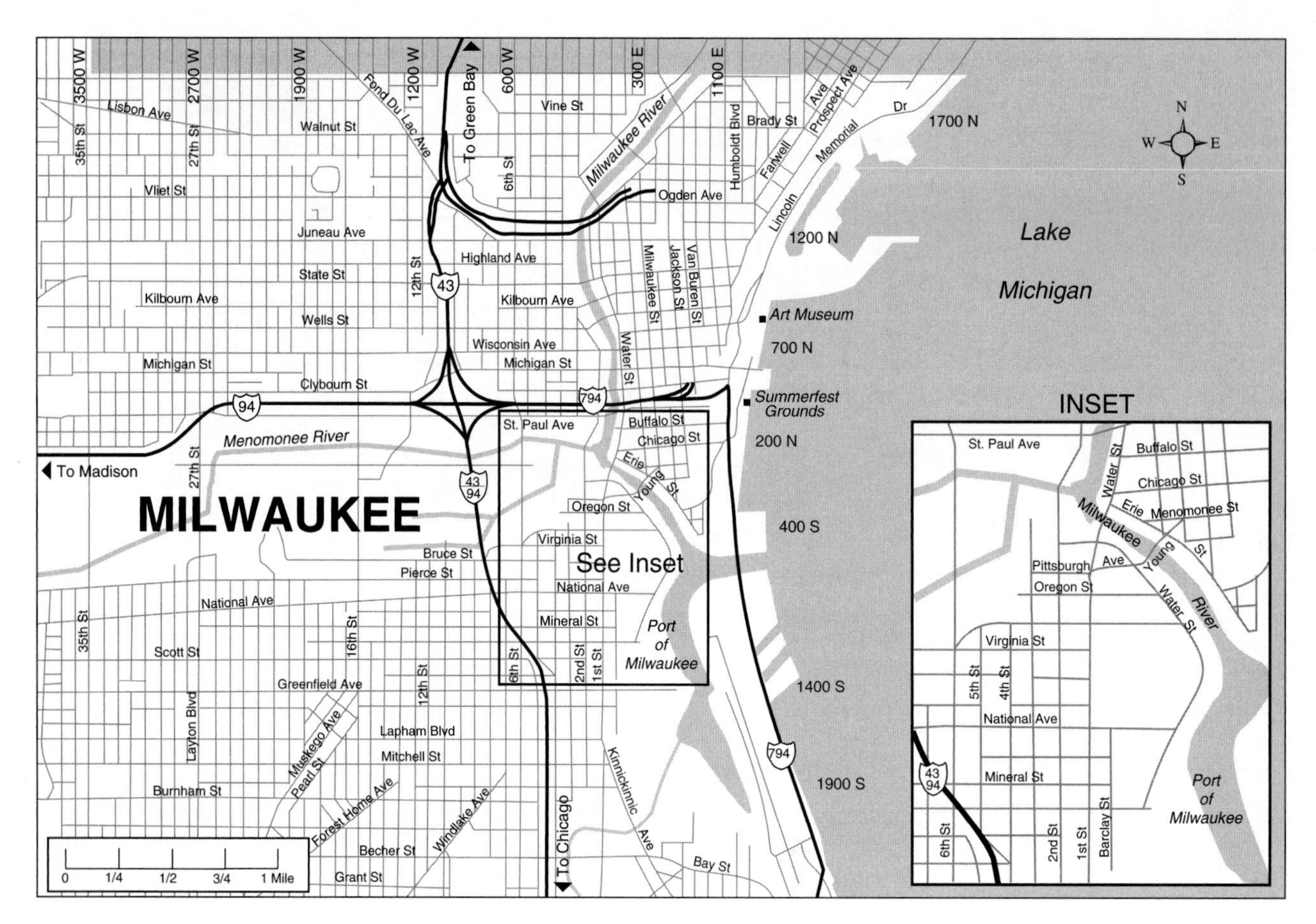

Figure 14.6 Typical street map, in this case derived from TIGER/Line files.
Source: Courtesy of *In Step Magazine* and University of Wisconsin-Milwaukee Cartographic Services.

cities, for example, you may select your general route by using a map that shows the major highways of all, or a large region of, the country. Such a small-scale map does not show local roads, however, so you will need larger-scale state or regional maps to pick the best route into or out of cities, parks, and other destinations. Also, you will need street maps to find your way to specific addresses or points of interest within each city you visit.

The electronic versions of highway and street maps are discussed in Chapter 20.

Highway Maps

Generally speaking, the relationships between features on highway maps are reasonably accurate, but exact locations of many features cannot be maintained. Because road widths on highway maps must be exaggerated to maintain legibility, city symbols and highway alignments are often shifted and generalized.

A general strategy for trip planning using a highway map involves first finding the starting point and destination. You may have to refer to an index if you are not familiar with the specific locations. You can then select the most convenient route by scanning the available roads. At this stage, you can also include any desired intermediate destinations. Planning should take into account the class of roads you wish to use. You may want to take a scenic route offered by secondary roads, for example, or you may prefer to use the faster, more direct route usually provided by major highways. The map key will contain a detailed indication of the various road types. These are typically symbolized by different-colored lines and line symbols. You can write down the route numbers or street names in the order that they will be traversed, as well as other information, such as the direction of turns that you will have to make and freeway exit and entrance numbers. You may also wish to note features along the way that may be important landmarks. A written travel plan is a useful supplement to your map and will help you to keep track of your travel progress and to anticipate the necessary turns.

Highway maps often include some indication of distances and travel times between major locations. The specific form of table or diagram used varies from map to map. A **distance matrix** is one common form. A typical distance matrix requires locating the names of your origin and destination. The names may be on

Appleton	249	130	186	196	176	38	31	259	241	213	135	118	138	166	102	45	83	103	100	149	20	175	201	126	134	148	257	58	68	73	315	91	101	79
Ashland		339	429	339	164	280	245	57	176	38	167	326	381	248	292	283	234	154	339	324	262	75	307	369	118	265	194	307	189	288	65	158	239	187
Beloit			97	94	223	103	161	323	289	309	265	13	67	177	50	152	216	185	73	33	110	264	136	69	244	109	304	125	154	203	372	186	99	154
Chicago, IL				184	314	148	204	414	380	398	308	109	57	273	146	172	259	276	93	130	166	355	233	68	313	205	395	145	243	233	463	275	190	246
Dubuque, IA					198	166	227	298	259	341	331	98	162	121	95	219	279	186	168	61	177	267	62	164	272	79	274	203	188	269	346	215	113	169
Eau Claire						192	193	111	66	182	250	211	269	87	176	221	214	80	241	208	177	106	146	264	149	141	82	231	110	236	152	100	124	97
Fond du Lac							61	283	257	245	165	91	98	155	71	54	114	127	62	118	19	207	171	87	166	119	276	39	92	104	341	124	76	97
Green Bay								261	252	207	104	148	147	196	132	38	55	124	114	179	50	176	232	141	127	180	268	62	89	43	310	93	132	100
Hayward									119	86	205	311	369	191	277	299	255	157	341	308	265	84	250	364	139	241	137	318	192	304	70	164	220	185
Hudson										205	299	276	337	137	242	287	271	145	306	274	242	156	196	329	198	205	18	297	176	295	145	159	190	162
Hurley											129	296	345	254	263	244	195	155	305	310	227	76	309	333	80	266	223	271	155	249	104	124	225	169
Iron Mtn., MI												252	251	302	236	142	72	199	218	283	154	149	336	245	101	283	316	166	190	147	232	159	236	204
Janesville													66	168	41	145	203	172	71	37	98	252	137	68	231	100	292	120	141	191	359	173	87	142
Kenosha														231	104	116	202	226	37	101	118	307	198	11	265	163	352	89	191	178	418	223	142	195
La Crosse															129	205	246	103	204	150	149	178	59	225	201	68	153	194	119	239	234	144	87	96
Madison																125	185	138	77	47	82	217	102	98	197	61	258	109	107	175	325	140	53	108
Manitowoc																	93	149	79	171	59	214	225	105	165	172	302	27	113	62	348	130	129	125
Marinette																		151	169	232	103	174	285	196	116	233	286	117	127	98	299	114	185	148
Marshfield																			184	170	108	79	153	214	98	111	161	161	35	167	219	45	86	33
Milwaukee																				107	80	265	179	26	227	138	322	52	149	141	389	181	116	158
Monroe																					129	249	102	103	244	82	289	156	154	222	357	186	85	141
Oshkosh																						187	181	106	147	129	258	58	73	93	327	105	82	80
Phillips																							233	295	59	190	173	233	114	219	137	84	165	112
Prairie du Chien																								192	251	53	212	211	167	275	295	193	102	147
Racine																									253	157	345	78	179	167	413	211	140	186
Rhinelander																										206	215	191	89	170	183	58	159	103
Richland Center																											221	158	124	223	290	149	49	103
St. Paul, MN																												312	192	311	162	175	206	178
Sheboygan																													126	89	372	149	115	137
Stevens Point																														131	254	32	76	21
Sturgeon Bay																															353	136	175	143
Superior																																223	273	252
Wausau																																	103	46
Wisconsin Dells																																		56
Wisconsin Rapids																																		

Figure 14.7 Typical distance matrix.
Source: Wisconsin State Department of Transportation, Wisconsin Highway Map, 1986–1987.

MILEAGE TABLE

This table shows distances between cities along state or federal routes. To find a distance, locate the two cities in the table at left. Find the smaller of the two index numbers in red in the table below. Read down the blue column to find the second index number. Mileage between the cities is in black to the right of this number.

1. Appleton
2. Ashland
3. Beloit
4. Chicago,IL
5. Dubuque, IA
6. Eau Claire
7. Fond du Lac
8. Green Bay
9. Hayward
10. Hudson
11. Hurley
12. Iron Mountain,MI
13. Janesville
14. Kenosha
15. La Crosse
16. Madison
17. Manitowoc
18. Marinette
19. Marshfield
20. Milwaukee
21. Monroe
22. Oshkosh
23. Phillips
24. Prairie du Chien
25. Racine
26. Rhinelander
27. Richland Center
28. St. Paul,MN
29. Sheboygan
30. Stevens Point
31. Sturgeon Bay
32. Superior
33. Wausau
34. Wisconsin Dells
35. Wisconsin Rapids

Index	Second index and mileage
1	2 249, 3 130, 4 186, 5 196, 6 176, 7 38, 8 31, 9 259, 10 241, 11 213, 12 135, 13 118, 14 138, 15 166, 16 102, 17 45, 18 83, 19 103, 20 100, 21 149, 22 20, 23 175, 24 201, 25 126, 26 134, 27 148, 28 257, 29 58, 30 68, 31 73, 32 315, 33 91, 34 101, 35 79
2	3 339, 4 429, 5 339, 6 164, 7 280, 8 245, 9 57, 10 176, 11 38, 12 167, 13 326, 14 381, 15 248, 16 292, 17 283, 18 234, 19 154, 20 339, 21 324, 22 262, 23 75, 24 307, 25 369, 26 118, 27 265, 28 194, 29 307, 30 189, 31 288, 32 65, 33 158, 34 239, 35 187
3	4 97, 5 94, 6 223, 7 103, 8 161, 9 323, 10 289, 11 309, 12 265, 13 13, 14 67, 15 177, 16 50, 17 152, 18 216, 19 185, 20 73, 21 33, 22 110, 23 264, 24 136, 25 69, 26 244, 27 109, 28 304, 29 125, 30 154, 31 203, 32 372, 33 186, 34 99, 35 154
4	5 184, 6 314, 7 148, 8 204, 9 414, 10 380, 11 398, 12 308, 13 109, 14 57, 15 273, 16 146, 17 172, 18 259, 19 276, 20 93, 21 130, 22 166, 23 355, 24 233, 25 68, 26 313, 27 205, 28 395, 29 145, 30 243, 31 233, 32 463, 33 275, 34 190, 35 246
5	6 198, 7 166, 8 227, 9 298, 10 259, 11 341, 12 331, 13 98, 14 162, 15 121, 16 95, 17 219, 18 279, 19 186, 20 168, 21 61, 22 177, 23 267, 24 62, 25 164, 26 272, 27 79, 28 274, 29 203, 30 188, 31 269, 32 346, 33 215, 34 113, 35 169
6	7 192, 8 193, 9 111, 10 66, 11 182, 12 250, 13 211, 14 269, 15 87, 16 176, 17 221, 18 214, 19 80, 20 241, 21 208, 22 177, 23 106, 24 146, 25 264, 26 149, 27 141, 28 82, 29 231, 30 110, 31 236, 32 152, 33 100, 34 124, 35 97
7	8 61, 9 283, 10 257, 11 245, 12 165, 13 91, 14 98, 15 155, 16 71, 17 54, 18 114, 19 127, 20 62, 21 118, 22 19, 23 207, 24 171, 25 87, 26 166, 27 119, 28 276, 29 39, 30 92, 31 104, 32 341, 33 124, 34 76, 35 97
8	9 261, 10 252, 11 207, 12 104, 13 148, 14 147, 15 196, 16 132, 17 38, 18 55, 19 124, 20 114, 21 179, 22 50, 23 176, 24 232, 25 141, 26 127, 27 180, 28 268, 29 62, 30 89, 31 43, 32 310, 33 93, 34 132, 35 100
9	10 119, 11 86, 12 205, 13 311, 14 369, 15 191, 16 277, 17 299, 18 255, 19 157, 20 341, 21 308, 22 265, 23 84, 24 250, 25 364, 26 139, 27 241, 28 137, 29 318, 30 192, 31 304, 32 70, 33 164, 34 220, 35 185
10	11 205, 12 299, 13 276, 14 337, 15 137, 16 242, 17 287, 18 271, 19 145, 20 306, 21 274, 22 242, 23 156, 24 196, 25 329, 26 198, 27 205, 28 18, 29 297, 30 176, 31 295, 32 145, 33 159, 34 190, 35 162
11	12 129, 13 296, 14 345, 15 254, 16 263, 17 244, 18 195, 19 155, 20 305, 21 310, 22 227, 23 76, 24 309, 25 333, 26 80, 27 266, 28 223, 29 271, 30 155, 31 249, 32 104, 33 124, 34 225, 35 169
12	13 252, 14 251, 15 302, 16 236, 17 142, 18 72, 19 199, 20 218, 21 283, 22 154, 23 149, 24 336, 25 245, 26 101, 27 283, 28 316, 29 166, 30 190, 31 147, 32 232, 33 159, 34 236, 35 204
13	14 66, 15 168, 16 41, 17 145, 18 203, 19 172, 20 71, 21 37, 22 98, 23 252, 24 137, 25 68, 26 231, 27 100, 28 292, 29 120, 30 141, 31 191, 32 359, 33 173, 34 87, 35 142
14	15 231, 16 104, 17 116, 18 202, 19 226, 20 37, 21 101, 22 118, 23 307, 24 198, 25 11, 26 265, 27 163, 28 352, 29 89, 30 191, 31 178, 32 418, 33 223, 34 142, 35 195
15	16 129, 17 205, 18 246, 19 103, 20 204, 21 150, 22 149, 23 178, 24 59, 25 225, 26 201, 27 68, 28 153, 29 194, 30 119, 31 239, 32 234, 33 144, 34 87, 35 96
16	17 125, 18 185, 19 138, 20 77, 21 47, 22 82, 23 217, 24 102, 25 98, 26 197, 27 61, 28 258, 29 109, 30 107, 31 175, 32 325, 33 140, 34 53, 35 108
17	18 93, 19 149, 20 79, 21 171, 22 59, 23 214, 24 225, 25 105, 26 165, 27 172, 28 302, 29 27, 30 113, 31 62, 32 348, 33 130, 34 129, 35 125
18	19 151, 20 169, 21 232, 22 103, 23 174, 24 285, 25 196, 26 116, 27 233, 28 286, 29 117, 30 127, 31 98, 32 299, 33 114, 34 185, 35 148
19	20 184, 21 170, 22 108, 23 79, 24 153, 25 214, 26 98, 27 111, 28 161, 29 161, 30 35, 31 167, 32 219, 33 45, 34 86, 35 33
20	21 107, 22 80, 23 265, 24 179, 25 26, 26 227, 27 138, 28 322, 29 52, 30 149, 31 141, 32 389, 33 181, 34 116, 35 158
21	22 129, 23 249, 24 102, 25 103, 26 244, 27 82, 28 289, 29 156, 30 154, 31 222, 32 357, 33 186, 34 85, 35 141
22	23 187, 24 181, 25 106, 26 147, 27 129, 28 258, 29 58, 30 73, 31 93, 32 327, 33 105, 34 82, 35 80
23	24 233, 25 295, 26 59, 27 190, 28 173, 29 233, 30 114, 31 219, 32 137, 33 84, 34 165, 35 112
24	25 192, 26 251, 27 53, 28 212, 29 211, 30 167, 31 275, 32 295, 33 193, 34 102, 35 147
25	26 253, 27 157, 28 345, 29 78, 30 179, 31 167, 32 413, 33 211, 34 140, 35 186
26	27 206, 28 215, 29 191, 30 89, 31 170, 32 183, 33 58, 34 159, 35 103
27	28 221, 29 158, 30 124, 31 223, 32 290, 33 149, 34 49, 35 103
28	29 312, 30 192, 31 311, 32 162, 33 175, 34 206, 35 178
29	30 126, 31 89, 32 372, 33 149, 34 115, 35 137
30	31 131, 32 254, 33 32, 34 76, 35 21
31	32 353, 33 136, 34 175, 35 143
32	33 223, 34 273, 35 252
33	34 103, 35 46
34	35 56

Figure 14.8 Typical mileage table.
Source: Wisconsin State Department of Transportation, Wisconsin Highway Map, 1987–1988.

two outside edges, or they may be consolidated on the diagonal (Figure 14.7). The box at the intersection of the column and row associated with these locations contains the desired distance information in miles, kilometers, or both. Distances are usually specified from city center to city center and assume the use of the most direct major route.

A **mileage table,** or **distance log,** is another form of mileage indicator. A mileage table consists of columns of mileage figures. Each city incorporated in the table is given an identification number. You determine the numbers assigned to your origin and destination and then follow the instructions provided to find the mileage data in the table (Figure 14.8).

Another method of providing distance information is the use of a special inset map. A map of this type is usually a simple diagram with straight lines between locations. Distances are printed next to each route

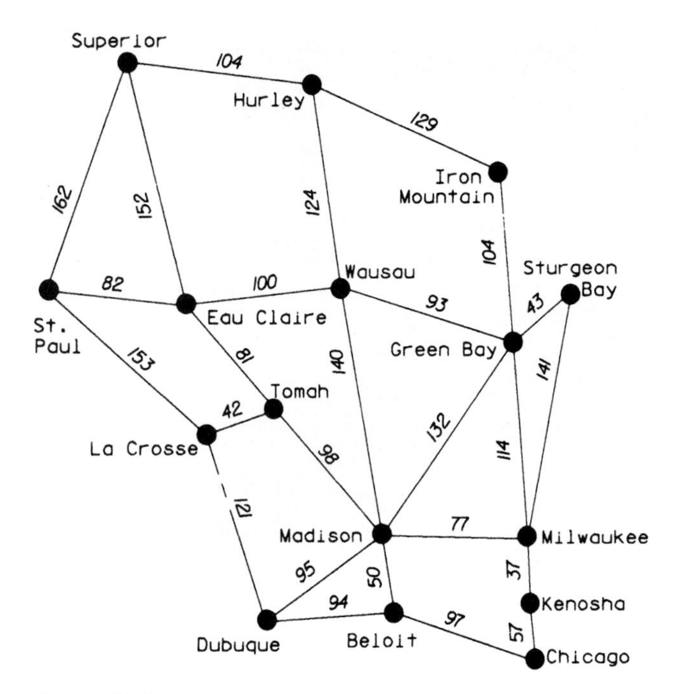

Figure 14.9 Simple distance diagram.
Source: Wisconsin State Department of Transportation, Wisconsin Highway Map, 1986–1987.

segment (Figure 14.9), and you simply total the distances for the appropriate segments between your origin and destination.

To determine distances between locations not shown on a distance table or diagram, you may use the map's graphic scale. This is rather awkward for any but the most direct route, so you may prefer to use the distance indications often placed along highway segments. Different symbols are used to indicate mileages, but those in Figure 14.10 are typical. Here, distances between intersections and towns are shown by a black number next to the highway segment. Longer distances, which may encompass several intersections, are indicated by a star at the start and end of the segment, with the distance printed in red (on the original map) adjacent to the middle of the segment. Simply adding the distances corresponding to each segment of your planned route gives you the total distance.

Information about travel times is sometimes indicated on highway maps, usually on inset maps or in tables similar to those used for distance information. In addition, travel-planning programs on personal computers or on the Internet provide a new source for travel distance and travel time information. These programs accumulate the lengths of each segment of a trip to obtain accurate total travel distances. Also, by taking into account average speeds for different types of roads, the programs produce estimates of travel times for trip segments and the full trip. Although distances can be calculated accurately, travel times are only estimates, regardless of the method used to calculate them, because speeds vary under different conditions.

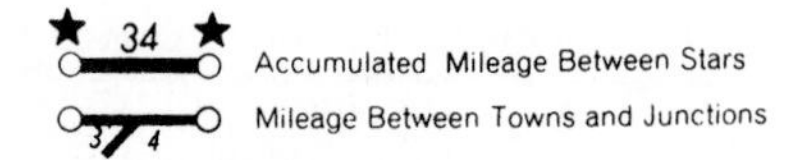

Figure 14.10 Distances printed next to route segments.
Source: Oregon Official Highway Map, December 1986.

Street Maps

On street maps, the roads within a city are usually identified by name and route number and are sometimes classified as major or minor arterials and neighborhood streets. Access roads and interchanges are usually shown, frequently with enough detail so that you can determine in advance whether you can transfer from one highway to another in the direction you wish to go. Address ranges may also be included so that you can determine the general location of a street address.

Street maps also include many other features, such as city and town names and boundaries, post offices, libraries, parks and zoos, cemeteries, museums, public and technical schools, colleges and universities, hospitals, major office buildings, major retail stores, shopping districts and centers, airports, bus and train stations, churches, convention centers, stadiums, golf courses and country clubs, yacht clubs and moorages, racetracks, and city, county, and other administration buildings. Suitable indexes are usually provided so that specific features can be located.

Route Maps

Route maps designed for bus, train, subway, and, to a lesser extent, airline passengers are similar in that their content is sharply limited. Usually, they show only the locations served by a particular carrier. They are often diagrammatic, rather than realistic. The simplified routes are usually just lines joining stations, and directions and distances are often not shown accurately (Plate 4). Different colors or line symbols may be used to distinguish routes. Route maps are often used in conjunction with timetable information.

Airline maps may be somewhat more colorful than other types of route maps because they are often printed on a background representation of the terrain (Plate 5). They usually show their routes as curved lines connecting the cities they serve. The paths of these lines seldom coincide with actual routes, which vary due to weather conditions and other factors.

WEATHER AND CLIMATE MAPS

Weather is the state of the atmosphere at a given moment in time, and a weather map presents a snapshot of these atmospheric conditions. Climate, on the other hand, is the average state of the weather over a long period of time. **Climate maps,** therefore, are classification maps that provide a summary of data gathered over time, rather than specific conditions at a given moment in time. If you are not familiar with the weather elements or climatic classifications, you may wish to consult a standard physical geography or weather and climate textbook for further information while reading this section.

Weather Maps

Information on a variety of elements is included on **weather maps.** Some weather maps are limited to individual topics, such as precipitation or temperature. A standard weather map, used for forecasting purposes, however, depicts a wide range of information. The information is gathered from many reporting locations and must be quickly read and understood by large numbers of people (weather forecasters, climatologists, and members of the public with an interest in the weather) located all around the world. For these reasons, weather maps use an internationally adopted set of standardized symbols designed to present information as clearly and succinctly as possible, without written language. Most weather maps are based on these symbols. Modified symbols often appear on newspaper and television weather maps, but even they usually bear a strong resemblance to the standard symbols. Familiarity with the standard symbols, then, allows understanding of the basic information shown on most weather maps.

As already suggested, the gathering of weather information involves a worldwide network of reporting stations. Observations are taken at all of these reporting stations at the **synoptic hours:** 0000, 0600, 1200, and 1800 Universal Coordinated Time (formerly Greenwich Mean Time). The observed information includes temperature, dew point, precipitation (type and quantity during the past six hours), visibility, clouds (types and percentage of sky covered), wind (direction and speed), sea-level barometric pressure (observed barometric pressure adjusted to take into account temperature and elevation of the station), and the change in barometric pressure during the past three hours. These data are plotted to create the basis for a weather map, using a prescribed format and special symbols.

A selection of the many symbols used on standard weather maps and a specimen plot for a single station are shown in Figure 14.11. The central point in the format is a circle that is blackened to indicate the amount of cloud coverage. Attached to this circle is a line that indicates the direction from which the wind is blowing. Tags of different types are added to the wind line to indicate the wind speed. If conditions are calm, there is no wind line, and the cloud-coverage symbol is enclosed by a second circle. The station symbols also indicate temperature, visibility, dew point, current barometric pressure, and a variety of other information.

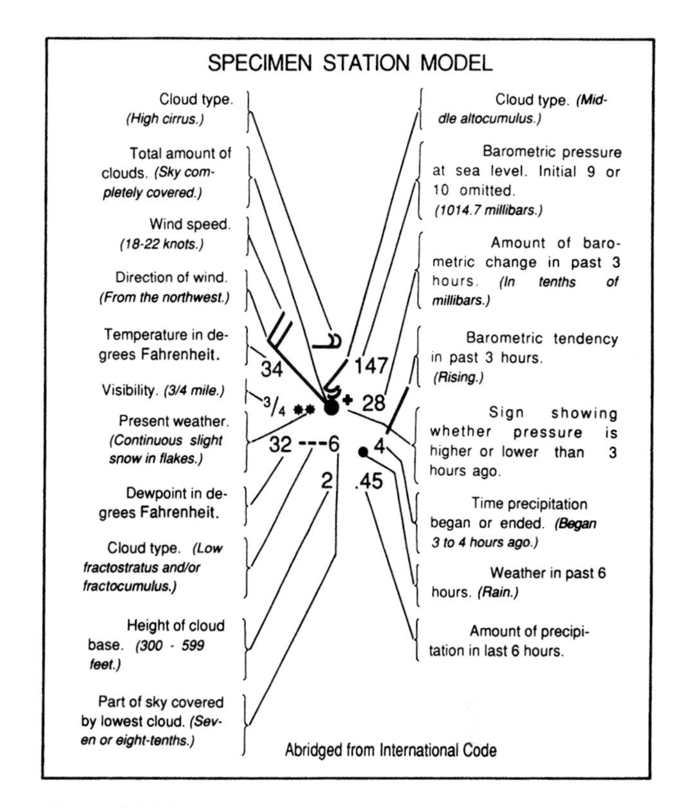

Figure 14.11 Specimen weather station model.

After the individual station observations have been plotted, two items are added to complete the map (Figure 14.12). The first of these are **isobars,** which are lines of equal barometric pressure. These indicate the locations of high- and low-pressure centers and pressure gradients. The locations of warm, cold, occluded, and stationary fronts are added next. These indicate the zones of contact between air masses with different characteristics, which are often the principal areas of cloudiness, precipitation, and the more violent forms of weather activity, such as thunderstorms and tornadoes.

Climate Maps

Climate is the summary of day-to-day weather conditions over a long time. Climatic types are based on the analysis of long-term patterns of such factors as temperature range, precipitation amounts (total and seasonal), and evaportranspiration. Locations with similar patterns of variation in the climatic factors are grouped to form climatic regions.[4] Maps of any or all of the climatic factors are produced (Figure 14.13), as are maps of the climatic regions (Figure 14.14).

Boundary lines on maps are usually shown as firm lines with a definite path that indicate that one region

[4]It is not surprising, given the large number of factors to consider and the variety of conditions found worldwide, that several climatic classification systems have been developed.

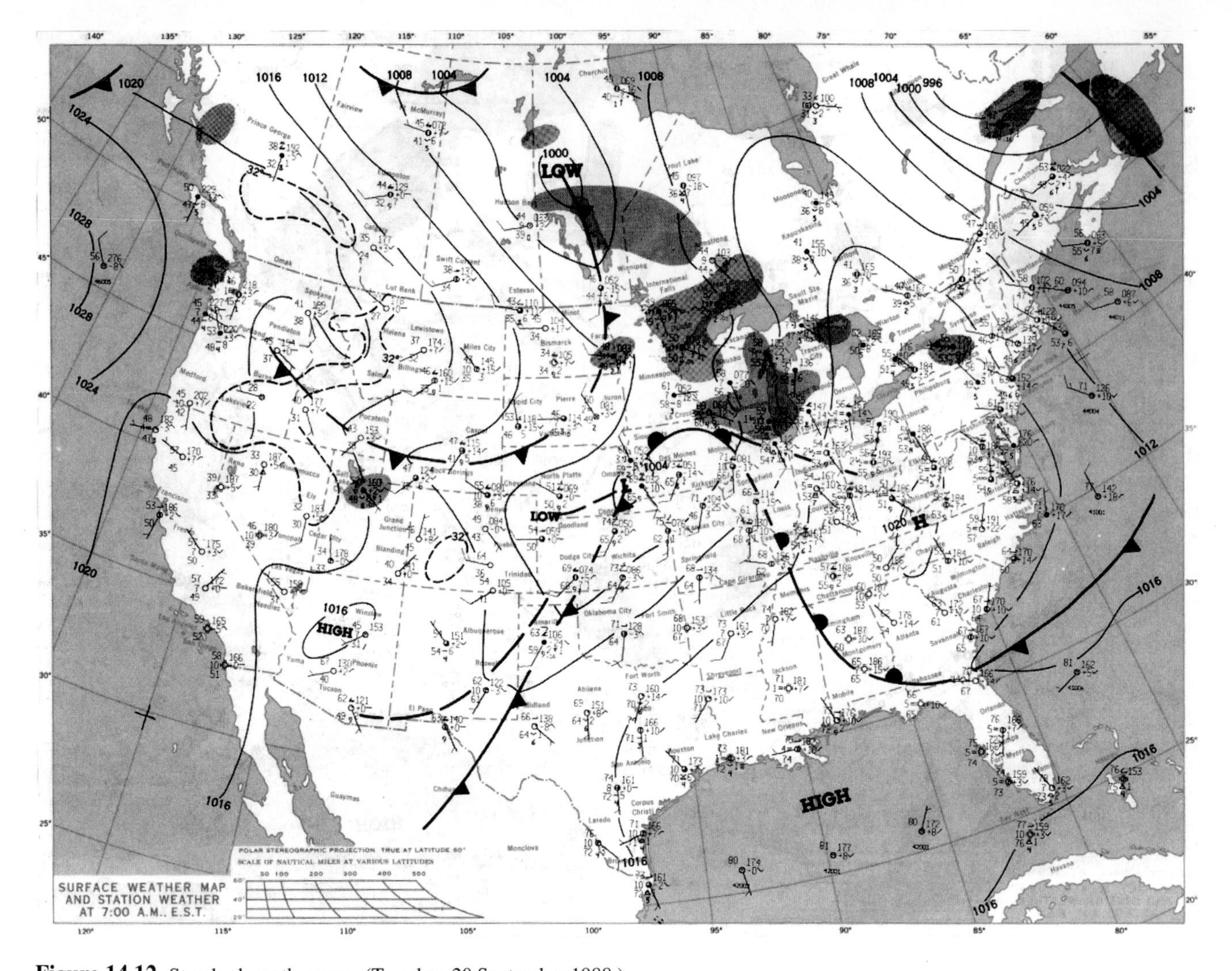

Figure 14.12 Standard weather map. (Tuesday, 20 September 1988.)

Source: *Daily Weather Maps: Weekly Series, September 19–25, 1988* (Washington, D.C.: Climate Analysis Center, National Oceanic and Atmospheric Administration).

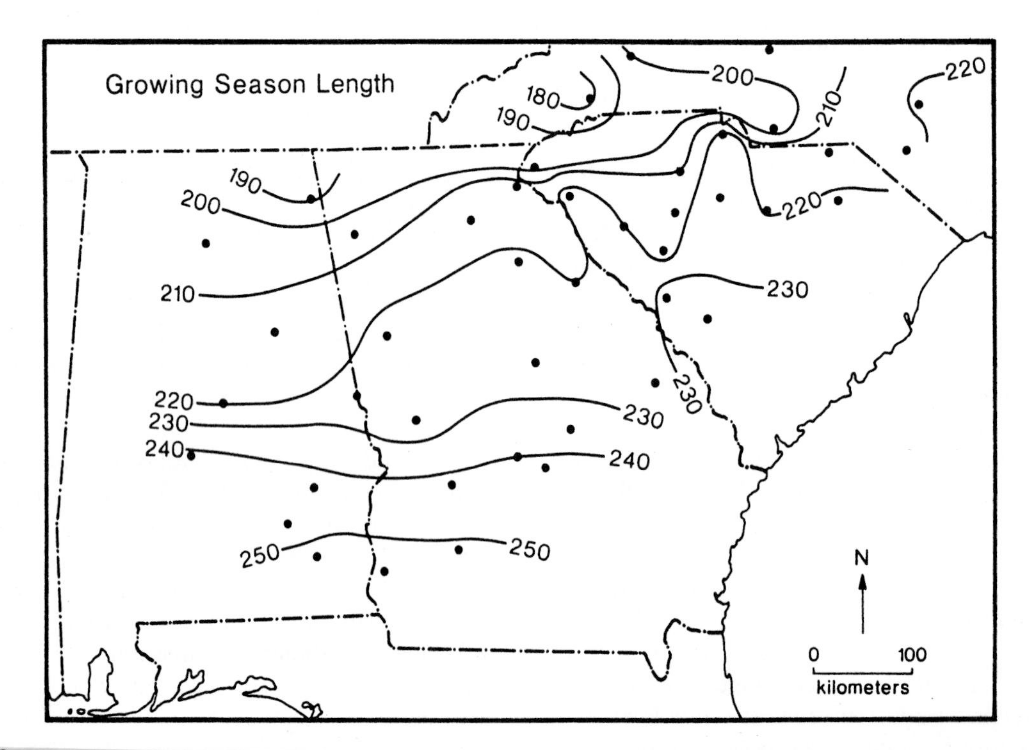

Figure 14.13 Map of an individual climatic element. Mean growing season length (in days) based on the 76-year period 1911–1986.

Source: From Philip W. Suckling, "Freeze Dates and Growing Season Length for the Southern Piedmont, 1911–1986," *Southeastern Geographer* 28, no. 1 (May 1988): 34–45. Reprinted by permission.

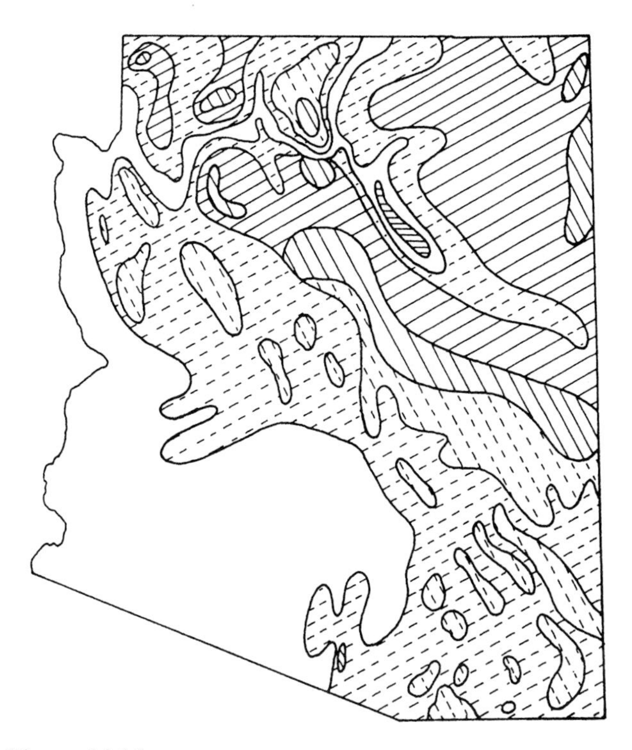

Figure 14.14 Map of climatic regions in Arizona.
Source: Stephen Bahre, ed., *Atlas of Arizona* (Yuma, Arizona: Arizona Information Press, 1976).

stops at the line and another begins. For example, political boundaries are definite dividing lines in that, on one side of the line, you are in Canada, for example, and, on the other, you are in the United States. Boundaries between climatic regions, however, are not exact, because there is almost always a transition zone between the core areas of adjoining climatic regions. This type of difficulty is associated with most maps that show boundary lines between regions with different characteristics.

Geologic Maps

Geologic maps provide information about the structure of the earth. Many types of maps fall within this general category. Some geologic maps show the types of rocks and other materials exposed at the earth's surface. Others use cross-sectional views to show the types of bedrock that underlie the surface materials. Still others illustrate the occurrence of petroleum or other mineral deposits, the arrangement of faults, the location of earthquake epicenters, the occurrence of fossils, the availability and quality of groundwater, aspects of the earth's thermal and magnetic characteristics, the location of different soil types, and many other topics.

Geologic maps are used for many purposes, including mineral exploration, searches for water sources, investigation of the suitability of different locations for dams or other structures, soil classification, and the selection of appropriate landfill garbage-disposal sites. They are also used as background for zoning studies related to groundwater percolation and to the ability of underlying structures to support major buildings.

The interpretation and use of most types of geologic maps require training in the field of geology. For this reason, the treatment of geologic maps in this section is limited to those maps similar to the geologic quadrangle type produced by the U.S. Geological Survey. Such maps, which show the nature of surface deposits, are produced on a standard, topographic-quadrangle base. Cultural features are eliminated on the topographic map, leaving only locational information and the contours. This base is overprinted, using a variety of colors, gray values, or patterns, to distinguish the various types of surface deposits (Plate 6). The colors give a good visual impression of the materials' distribution pattern. Because of the difficulty of exactly identifying the often somewhat subtle color differences, however, a supplementary letter code is often used to further identify the various materials.

The classification categories used on geologic maps first distinguish materials on the basis of their geologic age, such as Cretaceous or Carboniferous. Then, subcategories describe the composition of the material, such as gravel or sandstone. The result is a detailed description of what is often a large variety of materials (Figure 14.15).

Other information typically shown on geologic quadrangle maps includes the location of any fault, or fracture, lines. Also, information that helps to describe the attitudes of different beds or strata of rock is shown. One aspect that is indicated is the **strike** of the surface, which is the compass bearing of the intersection of the surface of the bed or strata with an imaginary horizontal plane (Figure 14.16). The second aspect, shown in conjunction with the strike, is the **dip,** which is the downward inclination of the bed relative to the same imaginary horizontal plane (also in Figure 14.16). The dip value ranges from 0 degrees for a horizontal bed to 90 degrees for a vertical bed.

Geologic maps sometimes indicate the surface configuration of any underlying structures. Such surfaces are shown by a separate set of contour lines, usually in a different color or with a distinctive line pattern. Finally, geologic cross sections frequently accompany the map, to reveal additional information about the structures found below the surface (Figure 14.17). The location and direction of each cross section are marked on the map.

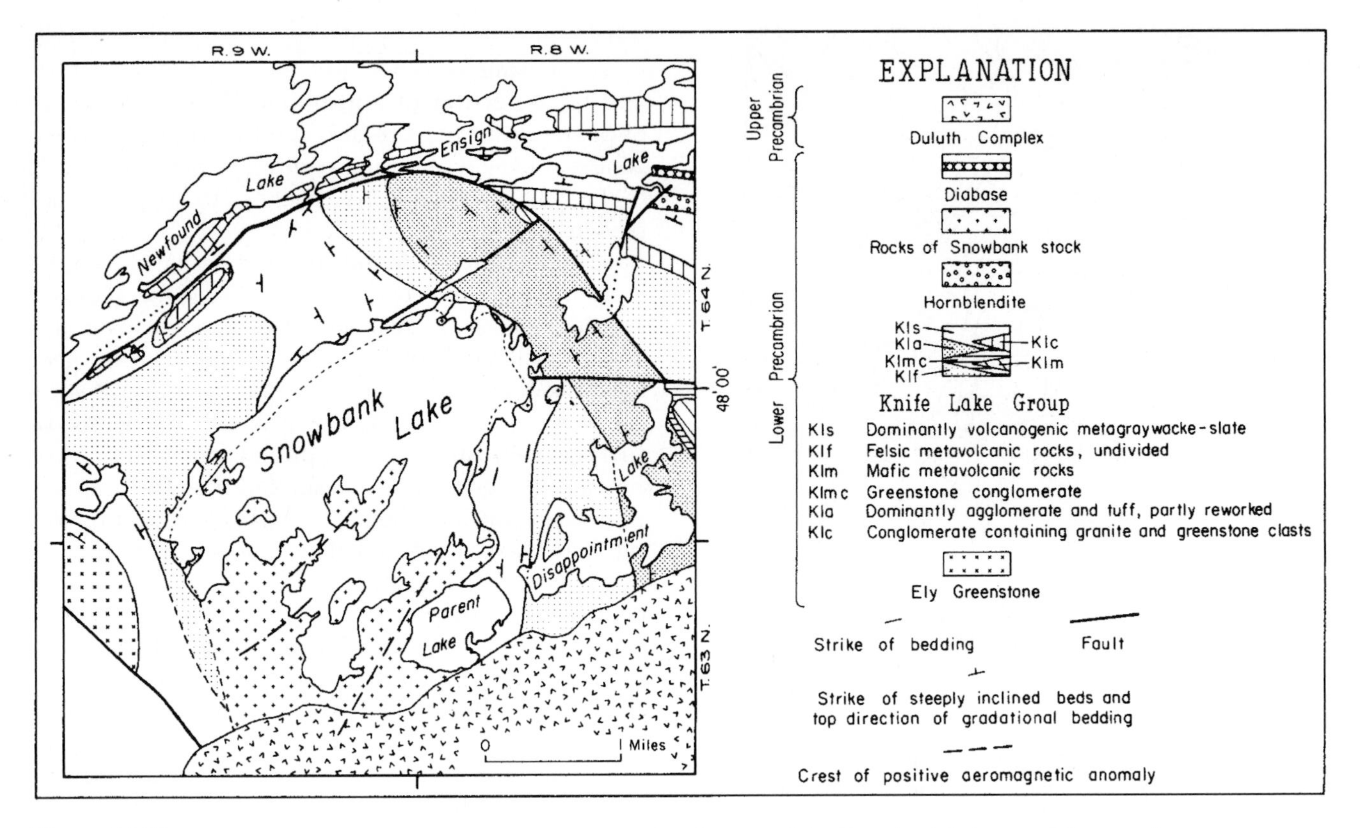

Figure 14.15 Generalized geologic map showing various types of surface materials. (Snowbank Lake area, Minnesota.)

Source: P. K. Sims and S. Viswanathan, "Giants Range Batholith," in *Geology of Minnesota: A Centennial Volume,* eds. P. K. Sims and G. B. Morey (St. Paul, Minn.: Minnesota Geological Survey, 1972), 120–96.

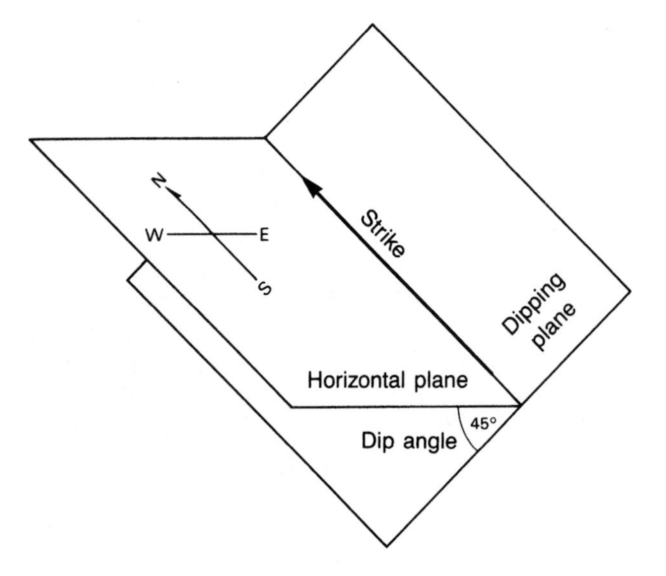

Figure 14.16 Interpretation of strike and dip.

Maps of the Past

Maps of the past fall into two categories. In the first category are maps that were produced at some time in the past and that have been preserved for examination today. These are called *historical maps.* In the second category are modern maps, produced to show some aspect of past events. These are called *history maps.*

Historical Maps

It has been suggested that, because of progress and change in every field of knowledge, every map is out-of-date from the moment it is produced. When a map is not only out-of-date but also "old," it may fall into the historical category. **Historical maps** are useful because they show things as they existed, or at least as they were represented, in the past. They were, as are modern maps, products of the knowledge, beliefs, and biases prevalent at the time of their production. A study of historical maps, therefore, is useful for understanding past knowledge and beliefs. Through the study of historical maps, for example, we can envision the extent of the knowledge that was available to Christopher Columbus regarding the shape, size, and location of the continents. Given that knowledge, we can better understand his belief that a route to Asia could be found by sailing west from Europe.

Careful analysis of historical maps also reveals some of the beliefs that people held when those maps were made. The **T-in-O maps** of the medieval period are an example (Figure 14.18). The name of these maps reflects their stylized appearance. The Mediterranean Sea formed the vertical portion of the "T," and the Don and Nile rivers constituted its crosspiece. The world ocean created the "O," which surrounded Asia, Africa, and Europe—the continents that were known to

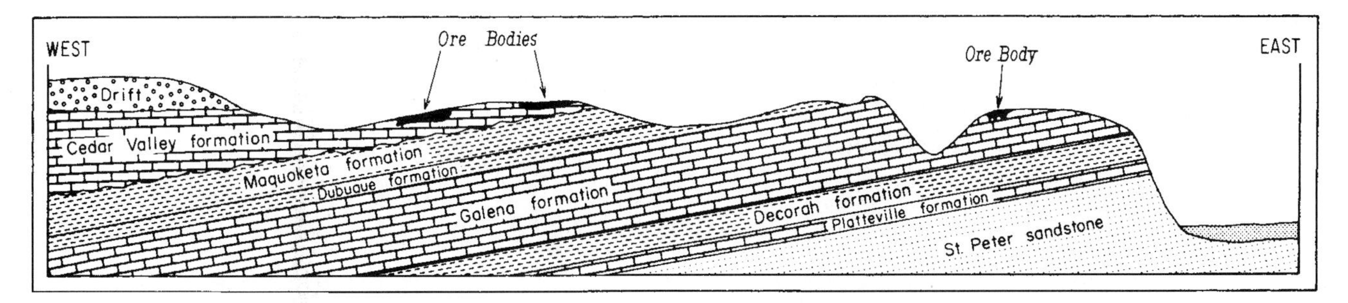

Figure 14.17 Generalized geologic cross section. (Fillmore County District, Minnesota.)
Source: Rodney L. Bleifuss, "The Iron Ores of Southeastern Minnesota," in *Geology of Minnesota: A Centennial Volume,* eds. P. K. Sims and G. B. Morey (St. Paul, Minnesota: Minnesota Geological Survey, 1972), 498–512.

Figure 14.18 T-in-O map by Isidore of Seville. (Diameter of the original: 64 cm.)
Source: Isidore of Seville, *Etymologiarum sive originum libri xx* (Augsburg: Günther Zainer, 1472). Courtesy of the Newberry Library.

the producers of the maps at the time. This arrangement probably did not represent the actual dominant view among educated persons of the true size and shape of the earth. Instead, it supported prevalent religious views. For example, the placement of Jerusalem at the center of the maps indicated its religious primacy. In addition, an imaginary Paradise was often shown at the top of the map, in the east, or *orient*. This pattern led to the use of the term **orientation** to describe the alignment of a map in its proper relationship to the earth's features. Today, north is usually at the top of the map, but the term persists.

Historical maps were subject to error and deliberate distortion, just as much as are contemporary maps. The careful interpretation of historical maps, therefore, requires equally careful investigation of other sources.

Many people enjoy collecting historical maps.[5] For some collectors, the map's information value is significant to the evolution of a branch of study in which they have an interest.[6] Frequently, however, maps are collected simply because old objects, such as antique furniture, dolls, cars, or trivets, are interesting and, sometimes, valuable as well. As with other types of antiques, reproductions of historical maps provide colorful and interesting collectibles and have the advantage of being much more affordable than the originals (Figure 14.19).

History Maps

History maps are different from antique, historical maps. They are modern products, produced to show what present-day researchers, in whatever field, believe happened in the past. Indeed, history maps are often based on antique maps produced at the time of the event. The topics included on history maps range from routes of exploration to summaries of the history of medicine, and from battle diagrams to maps of migration patterns. Whatever the topic, a history map shows, to the best of current knowledge, the status of the subject at some period in the past (Figure 14.20). A series of history maps on a topic can provide a very useful summary of the topic's historical evolution. They are tools for the study and explanation of past events and trends.

Fire Insurance Maps

Researchers in many fields, such as history, genealogy, urban planning, geography, and economics, are interested in understanding the history of urban development in the United States. In connection with their

[5]If you collect maps, you will find *Mercator's World* magazine interesting. For information, see the website for this chapter.
[6]If you are researching historical topics, the National Mapping Program brochure *Looking for An Old Map* provides useful source information. See also the sections on map collecting in Mark Monmonier and George A. Schell, *Map Appreciation* (Englewood Cliffs, N.J.: Prentice-Hall, 1988), 313–19, and Joel Makower (ed.), *The Map Catalog* (New York: Vintage Books, 1986).

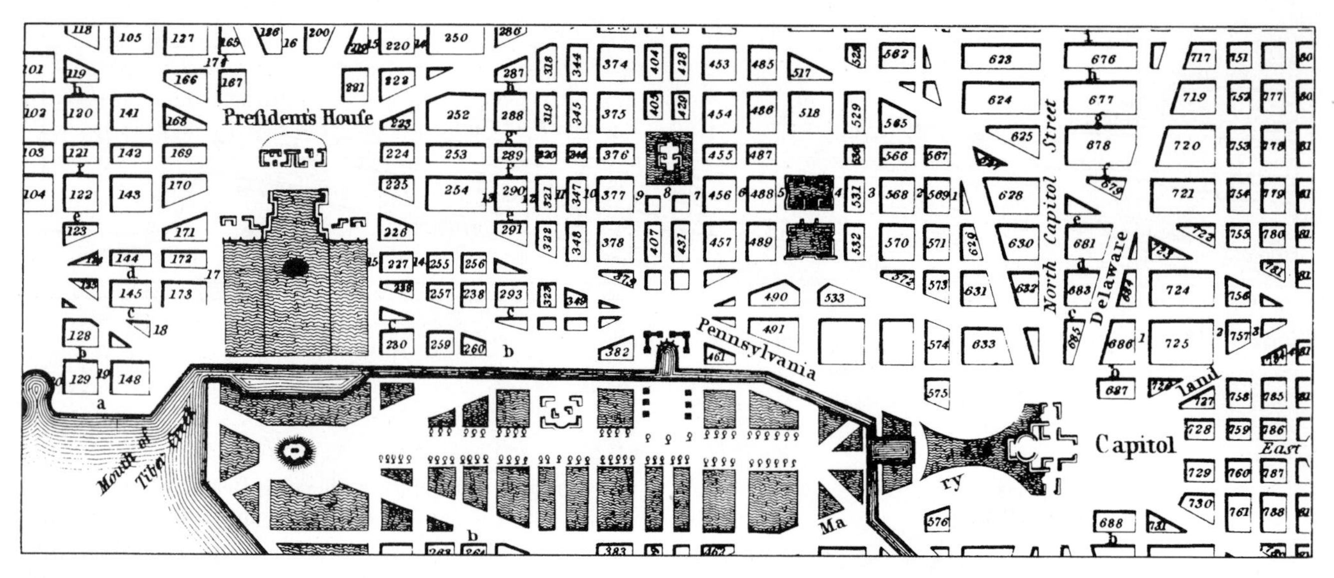

Figure 14.19 Historical map of the Washington Mall, 1792.
Source: *Looking for an Old Map* (U.S. Department of the Interior, Geological Survey, National Cartographic Information Center).

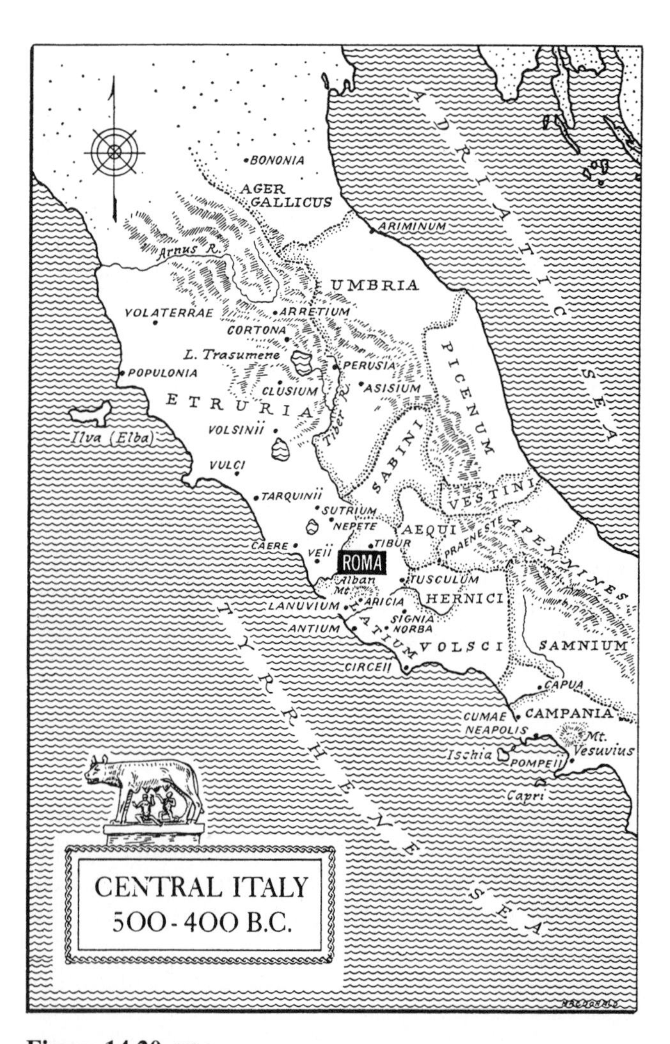

Figure 14.20 History map.
Source: From Richard Mansfield Haywood, *The Ancient World* (New York: David McKay Company Inc., 1971). Reprinted by permission.

studies, they often want to know when buildings were built and where they were located. They also may need to know the details of their size and shape, the uses to which they were put, and the materials with which they were built. It seems improbable that the demand for such detailed information about our cities could be retrieved but, fortunately, fire insurance maps often meet the need.

Fire insurance maps originated in London toward the end of the 18th century, when fire insurance companies needed up-to-date information about the buildings that they insured. By the 1850s, cities in the United States were built up to the extent that large fire losses were common. In response to the need for information that would assist in fire risk evaluation, maps showing such information for American cities soon became available. In 1867, a surveyor named D. A. Sanborn established the Sanborn Map Company, which soon became the predominant company in the field of fire insurance mapping. The company's domination was so complete that the name, **Sanborn Maps,** has become virtually synonymous with fire insurance maps in the United States. By 1961, when the publication of new fire insurance maps had virtually ceased, the company had published hundreds of thousands of such maps, including thousands of atlases, for 12,000 North American cities and towns. Although the Sanborn Map Company no longer produces its trademark fire insurance maps, it continues to produce maps used by municipalities in connection with planning, zoning, building inspection, and fire.

Specialized fire insurance maps, such as those produced by Sanborn, were copyrighted products prepared exclusively for the use of fire insurance specialists (Figure 14.21). Sanborn maps were printed in black on 21 ×

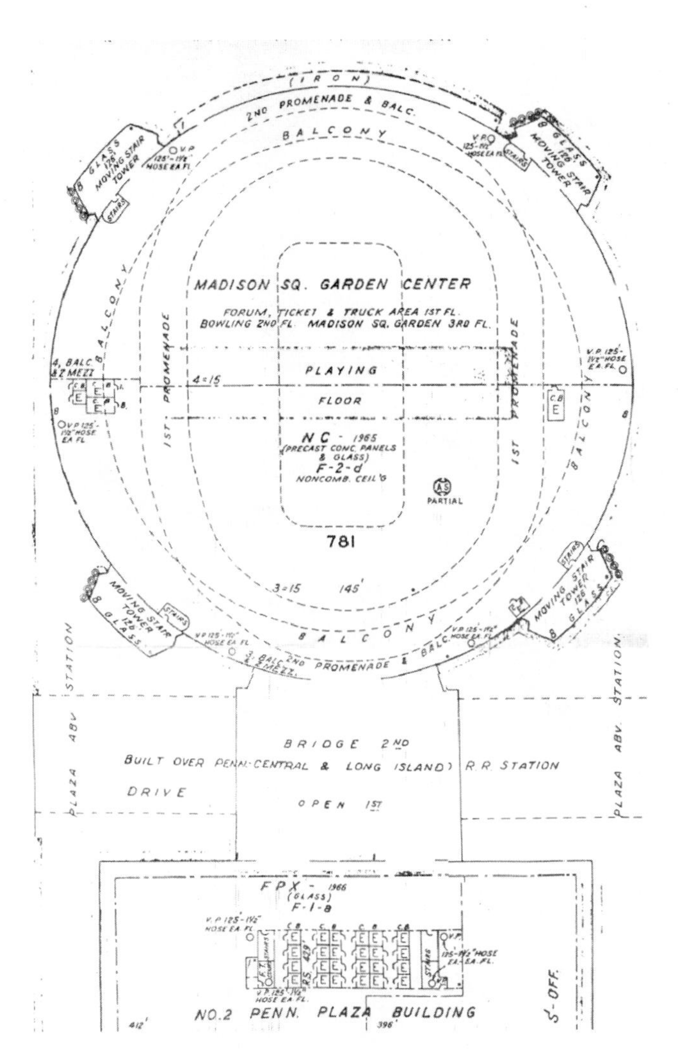

Figure 14.21 Portion of a 1993 Sanborn™ Map of New York, New York. (Scale reduced to approximately 1 inch to 175 feet.)

Source: Copyright 1993 Sanborn Map Company, The Sanborn Library, LLC. All rights reserved. This Sanborn™ Map has been reproduced with written permission from The Sanborn Library, LLC. All further reproductions are prohibited without prior written permission from the The Sanborn Library, LLC.

25-inch sheets. Hand coloring was added to show the type of materials used in the construction of the buildings. Many maps were issued as unbound sheets, but those for large cities were bound in volumes of approximately 100 plates (coverage of very large cities often required a number of such volumes). The maps were kept up to date by pasting correction stickers produced by the company at appropriate locations when minor changes were reported. When major changes had taken place, the maps were revised and reissued.

Most of the maps were drawn at the scale of 50 feet to an inch (1:600). This large scale allowed the inclusion of a great deal of accurate information. The maps incorporated the widths and names of streets, the location of property boundaries, house and block numbers, and the use to which the building was being put. Other information included the locations and sizes of water mains, and the location of fire alarm boxes and hydrants. In addition, structural information such as the location of firewalls, windows and doors, sprinkler systems, and types of roofs was recorded. The inclusion of such details allowed fire insurance specialists to determine the potential fire hazard associated with a particular property or district.

Because of changes in the operation of the fire insurance industry, the need for the types of maps offered by Sanborn declined by the 1960s. However, the historical paper maps, produced principally between 1867 and 1961, comprise a major resource for urban researchers. Collections of historical Sanborn maps and atlases are located at some local, state, or university libraries. The largest such collection, however, is housed in the Geography and Map Division of the Library of Congress. This collection consists of an estimated 700,000 Sanborn maps in bound and unbound editions. In addition, the Sanborn Map Company has recently donated more than 750,000 maps to the Library of Congress. Many of these maps are being scanned and will be available in digital form.

Maps of the Moon and Planets

Simple maps of the Moon have been drawn throughout history because the general arrangement of its surface features is visible from earth. Similarly, maps of Mars have been produced since the telescope brought its surface into view. Maps of the Moon and Mars were sometimes as much the result of fantasy as of reality but, as science progressed, their accuracy and usefulness improved. With the dawn of the age of space exploration, interest in producing detailed maps of the surfaces of the Moon, Mars, and other bodies in our solar system intensified.

The increased interest in nonterrestrial mapping stemmed, in part, from the need for detailed maps to assist in the selection of landing sites for both manned and unmanned exploration vehicles. Areas with suitable terrain for landing and with structures of interest to scientists could be more easily located with the help of accurate maps. In addition, the very technology that produced a need for better maps provided an increased ability to gather the necessary data. Early efforts at such mapping involved photogrammetric techniques and maps that were similar to topographic maps of the earth (Figure 14.22). Later, orbital and fly-by probes used remote-sensing systems to send back to earth detailed, digital information regarding detected features.

Since the 1960s, the U.S. Geological Survey has produced maps showing geologic information about the Moon and planets. Maps available in this series include geologic maps of Mars and the Moon and shaded-relief maps of Mars, Mercury, and Venus. More than 20 planets and satellites have been mapped by

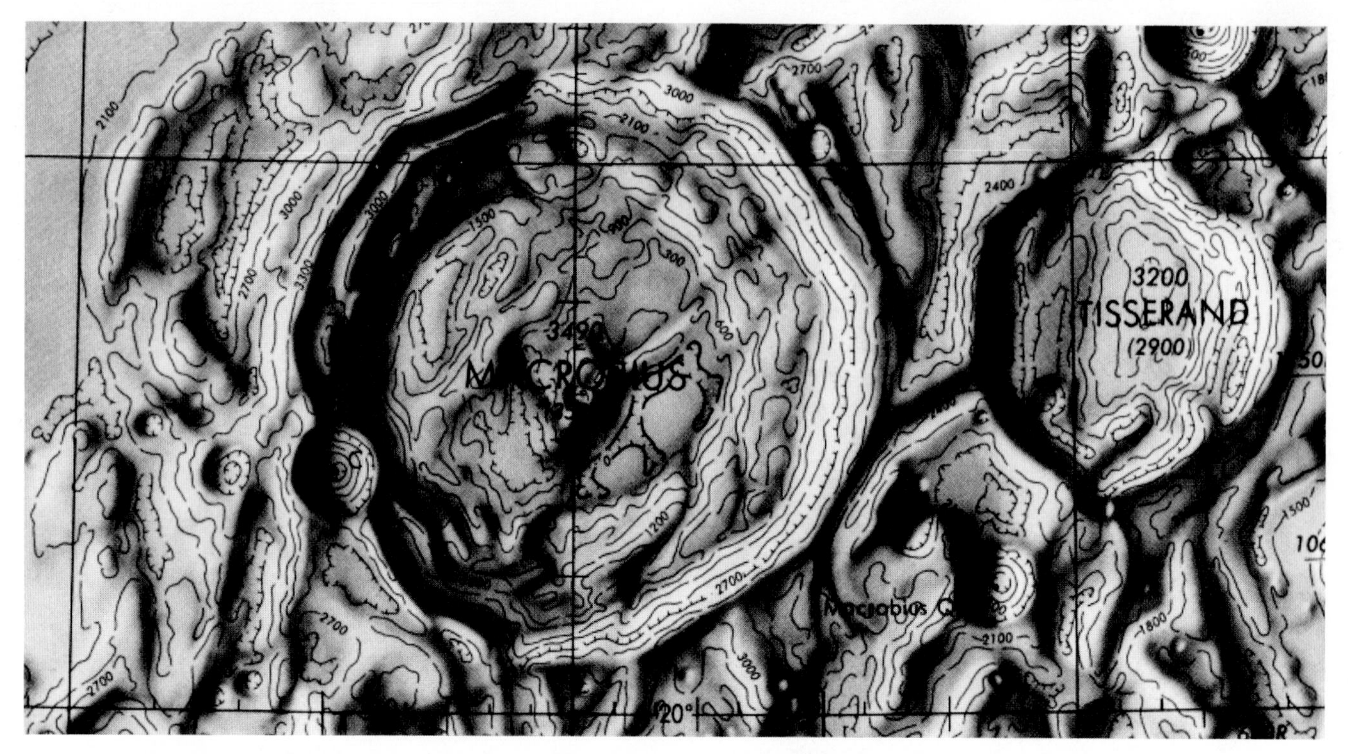

Figure 14.22 Map of a portion of the surface of the moon.
Source: Macrobius Chart (LAC 43), 1:1,000,000, Aeronautical Chart and Information Center, 1965.

Figure 14.23 Digital mosaic image of part of the south polar region of Mars.
Source: Courtesy of R. M. Batson, U.S. Department of the Interior, Geological Survey, Geologic Division, Branch of Astrogeology.

remote-sensing methods. Remotely sensed information is used to produce digital mosaic images whose appearance and accuracy rival similar images of the earth's surface (Figure 14.23). As space exploration continues, maps of other bodies in the solar system will be produced, with improving accuracy.

JOURNALISTIC CARTOGRAPHY

The term **journalistic cartography** refers to maps that appear in newspapers, weekly newsmagazines, and general and popular weekly or monthly magazines.[7] These maps are similar in some ways, but the publication circumstances faced by each of the media also result in some differences in how the maps are handled.

Daily newspapers typically lack the time and resources to develop detailed, fully researched map products. Newspaper maps are often limited to showing the locations of places in the news, whether proposed highway routes, accident sites, hijacking scenes, or battlefields. Maps that would effectively illustrate more complex situations are all too often missing. One informal study of maps that appeared in British newspapers revealed characteristics that seem to correspond to maps often encountered in their American counterparts.[8] This study pointed out the general lack of information about projection, scale, and orientation, in addition to frequent clutter and small size and, not infrequently, errors in boundaries and other information. These factors made it difficult to assess sizes and distances. Overall, the maps were less useful than they might have been.

Immediacy, especially when bad news is involved, is vital to the inclusion of maps in newspapers. A study of the use of newspaper maps during the Vietnam War, for example, found that the area devoted to maps and the number of maps appearing peaked sharply during the period that saw four coastal provinces fall under Communist control.[9] In addition, the "worse" or more dramatic the news, the larger the map's scale. On the other hand, maps are often used in newspapers to illustrate a feature article that is not faced with the same time constraints as the fast-breaking news of the day. In such cases, maps can present more complex subject matter, such as the evolution of a political campaign or the growth of world trade.

In addition to facing tight deadlines, newspaper maps face reproduction limitations that preclude the use of fine lines and delicate tones. These restraints, coupled with the desire to sharply focus the reader's attention on the topic at hand, result in maps with recognizable characteristics (Figure 14.24). They tend to be generalized and often do not have a specified scale or show the graticule. The symbolization tends to be bold and evocative, with elements such as graphic bomb bursts and tapered arrows indicating action and movement. In addition, explanatory labels are often incorporated to convey specific details of the event. Finally, color is sometimes used to provide even greater visual impact.

Weekly newsmagazine maps share some of the characteristics of newspaper maps. They, too, are usually generalized and use bold graphic techniques. On the other hand, they tend to be more sophisticated. This is partly because more time and resources are devoted to their production and partly because they can use finer line work and, frequently, take advantage of excellent color presentations.

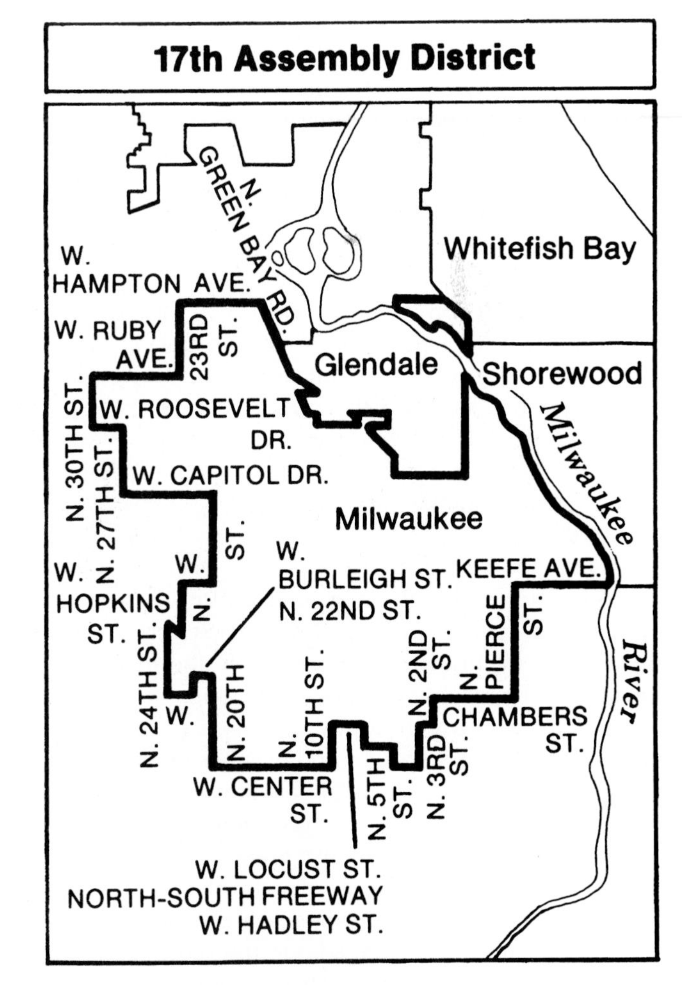

Figure 14.24 Typical newspaper map.
Source: Courtesy of the *Milwaukee Journal.*

[7] Walter W. Ristow, "Journalistic Cartography," *Surveying and Mapping* 17 (1957): 369–90.
[8] W.G.V. Balchin, "Media Map Watch," *The Geographical Magazine* 57 (August 1985): 408–9.
[9] Philip W. Porter, University of Minnesota, "Ten Years Ago Today—Vietnam and the Uses of Cartography" (Poster Session at Annual Meeting of Association of American Geographers, Detroit, 24 April 1985). This display presented an analysis of maps of the Vietnam War that appeared in the *Minneapolis Tribune,* January through August 1975.

Maps in other types of magazines run a wide gamut in terms of subject matter and style, from simple locational diagrams to maps that present statistical information, show climatic distributions, and represent trade flows. Overall, however, journalistic cartography tends to retain the predominant characteristics of simplicity, direct presentation, and dynamic style.

TELEVISION MAPS

Maps are often used in conjunction with television broadcasts. They seem to show up most frequently on news and weather reports, although they are also used on travelogues and other types of programs. Most **television maps** are similar to the maps that often appear in newspapers and newsmagazines. That is, they are usually highly generalized, with simplified (often oversimplified) coastlines, boundaries, and other features. They usually lack scales, and even when a latitude and longitude graticule is shown, the viewer often has no idea what map projection is being used.

Given the circumstances under which television maps are produced and viewed, their style characteristics are not surprising. For one thing, the maps are likely to be on the screen for only a very brief time, making it difficult for a viewer to absorb a great deal of map detail, so generalization is necessary. Similarly, because the map is on a screen and not in the viewer's hands, it is impossible to effectively measure distances, except by visual comparisons, so a scale is not likely to be very useful.

The frequent misuse of projections is probably the single most serious concern with television maps. The Mercator projection is frequently used, for example, even though its characteristics make it unsuitable for most circumstances. This type of misuse, however, is not surprising, because the characteristics of projections are obscure to the viewing public (and, in some cases, to the person responsible for creating the map). This does not make the problem any less serious, however.

One unique aspect of television is its ability to animate images, which is often used effectively to show motion and change on maps. News maps, for example, frequently use moving arrows to show the maneuvers of military forces or trade flow over time. On weather maps, the same techniques are used to show the movement of storm systems and weather fronts. In particular, the use of sequential satellite images gives a dramatic impression of the movement of cloud patterns, even over continental-size areas.

Map viewers must be as alert to misleading maps on television as they are when using other forms of maps—perhaps even more so. On the other hand, the television medium provides unique opportunities for developing distinctive map forms such as animations, and these opportunities probably will be used more extensively in the future.

TACTUAL MAPS

Most of this book deals with maps and map products designed for use by people with relatively unimpaired vision. Visually impaired people, however, have special map needs that require consideration. As has been noted, "we acquire and develop the concept of space, learn about our world and gather geographical information through the eye, with the help of maps and graphics. Most of these types of data are not available for the visually impaired and they have to be translated to a tactual format. Tactile maps and graphics are relevant because they help to overcome informational barriers for those who cannot see, facilitating their way in school, work, and everyday life."[10] The purpose of this section is to create an awareness of the existence and usefulness of **tactual maps**.

An important use of tactual maps is as a supplementary aid in travel and navigation. Such mobility maps help in making travel decisions and in organizing an individual's internal spatial images (mental maps; see Chapter 1). Thematic maps are another type of tactual map. As with traditional thematic maps, the information they provide supplements the user's knowledge of a great variety of topics.

Meeting the mapping needs of visually impaired people involves two major tasks. The first is learning what forms of information can be distinguished and understood by the map user. The second is to devise economically feasible and technically successful methods of producing the maps that record the information. Over the past 20 years, investigators have devoted a great deal of research to both tasks. The results of this research have contributed to the development of tactual maps that function successfully.

To convey spatial information, tactual maps depend on raised textures and surfaces that users can distinguish by touch. These symbols need to be distinct enough so that the user can readily distinguish between them but not abrasive or uncomfortable to the touch. Also, the materials involved have to be durable enough to withstand the abrasion and chemical exposure resulting from handling. The need to create distinguishable symbols often leads to large size, which reduces the amount of information that can be included within the available space, as compared to a conventional map (Figure 14.25). Labeling is also a problem. This is often added in braille but, again, the symbols take considerable space on a map, which reduces the

[10] "ICA Newsletter 22 (October 1993)," *Cartography and Geographic Information Systems* 21, no. 2 (April 1994): 121.

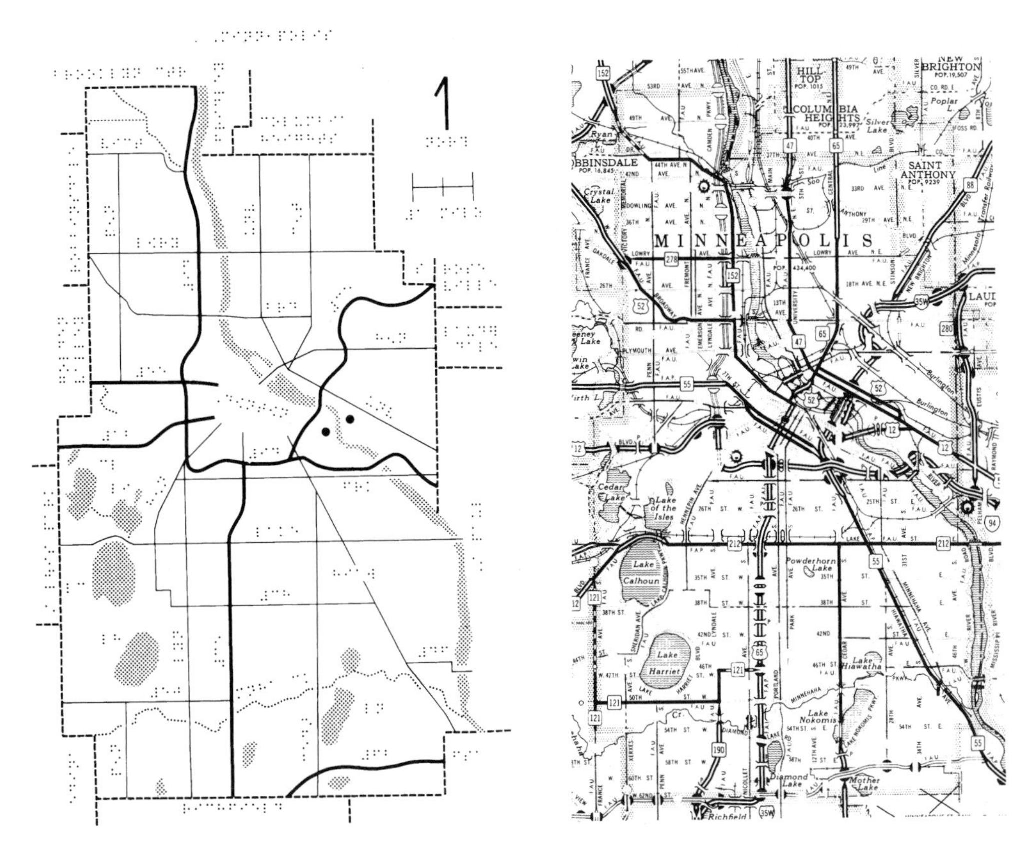

Figure 14.25 Comparison of a general reference tactual map (left) and a visual map (right). The tactual map in this example cannot contain the same level of detail displayed on the visual map.

Source: From Sona Karentz Andrews, "Applications of a Cartographic Communication Mode to Tactual Map Design," *The American Cartographer* 15, no. 2 (April 1988): 183–95. Copyright 1988 American Congress on Surveying and Mapping. Used with permission.

amount of information that can be effectively included. There is also the difficulty that many visually impaired persons do not know how to read braille. This has led to experimentation with other means of providing explanations, such as using audiotape to provide a supplementary verbal explanation of the mapped features.

Overall, a continuing challenge concerning tactual maps is to find an effective means of economic production. Because of the high cost of available production methods, tactual maps are not produced in quantities sufficient to meet the needs of all potential users.

SUMMARY

Maps are designed for many special applications. The types described in this chapter are often encountered but are only a sample of the vast variety produced.

Transportation maps include highway, street, and route maps. Because highway maps are designed to assist in intercity travel, they emphasize highways and roads. They tend to be highly generalized, and the locations of other features are often not accurate. Highway maps usually include distance and travel-time information, as well as insets of special features. Street maps concentrate on roads and routes within urban areas but also include supplementary information about scenic attractions and urban facilities. Route maps focus on the routes and facilities of public transportation systems, usually to the exclusion of other information. They are often diagrammatic and may include schedule information.

Weather maps show the state of the atmosphere at a given time, using a standard format to present detailed information gathered from many reporting locations. Included on these maps are the location of various types of fronts, temperature, wind speed and direction, cloudiness, isobars showing barometric pressure, and other factors useful for weather forecasting. Climate maps, in contrast, show the average distribution of various climatic factors or the extent of climatic regions.

Geologic maps show surface materials, bedrock types, or other information, ranging from the location of earthquake epicenters to groundwater quality. U.S. Geological Survey geologic quadrangle maps show surface deposits on a standard, topographic quadrangle base. The maps record the locations of any fault lines, as well as other information, such as the strike and dip of rock strata.

Historical maps were produced in the past and reflect the state of knowledge, beliefs, and biases prevalent at the time of their production. They provide useful insights into the past. History maps, on the other hand, show what present-day researchers understand about events or conditions in the past.

Fire insurance maps provide detailed information about urban development for researchers interested in history, genealogy, urban planning, geography, and economics. Maps produced by the Sanborn Map Company between 1867 and 1961 are a major source of such information. Sanborn maps, at the scale of 50 feet to an inch (1:600), can be consulted at many local, state, and university libraries, or the Geography and Map Division of the Library of Congress.

Maps of the Moon and planets have been produced in greater quantity, and with greater accuracy, since the advent of manned and unmanned space exploration. This partially reflects the greater need for detailed maps to assist in space flights and landings but also is a result of the great increase in the amount of information available.

Journalistic cartography refers to the maps that appear in newspapers and magazines. As a rule, these maps are highly generalized, often without a scale or graticule. They also tend to use bold and dramatic symbols and, unfortunately, often include errors. Typical television maps share the characteristic shortcomings of journalistic maps. They frequently use animation to effectively show motion and change.

The map needs of visually impaired people can often be met by the use of tactual maps as a supplementary aid in travel and navigation. Other tactual maps can provide thematic information on a great variety of topics to supplement the user's knowledge. To convey spatial information, tactual maps depend on raised textures and surfaces that users can distinguish by touch. This often leads to large size, which reduces the amount of information that can be included within the available space, as compared to a conventional map. Labeling is also a problem. Braille, for example, takes considerable space and many visually impaired persons do not know how to read it. Overall, the problem of providing mapped information for the visually impaired requires further research and development.

SUGGESTED READINGS

Bagrow, Leo. *History of Cartography*. 2d ed. Revised and enlarged by R. A. Skelton. Chicago: Precedent Publishing, 1985.

Balchin, W. G. V. "Media Map Watch." *The Geographical Magazine* 57 (August 1985): 408–9.

______. "Media Map Watch: A Report." *Geography* 70 (October 1985): 339–43.

Batson, R. M. "Digital Cartography of the Planets: New Methods, Its Status, and Its Future." *Photogrammetric Engineering and Remote Sensing* 53, no. 9 (September 1987): 1211–18.

Brown, Lloyd A. *The Story of Maps*. Boston: Little, Brown, 1949. New York: Dover, 1979.

Claussen, Hinrich. "Vehicle Navigation Systems." In *Geographic Information Systems: The Microcomputer and Modern Geography*, edited by D. R. Fraser Taylor. Vol. 1, *Modern Cartography*. Oxford: Pergamon Press Inc., 1991.

Crone, G. R. *Maps and Their Makers*. 5th ed. Folkestone, England: William Dawson and Sons. Hamden, Conn.: Archon Books, 1978.

Cykle, F. K., and J. M. Dixon. *International Directory of Tactile Map Collections*. Washington, D.C.: National Library Service for the Blind and Physically Handicapped, 1985.

Dent, Borden D. "Communication Aspects of Value-by-Area Cartograms." *The American Cartographer* 2, no. 2 (October 1975): 154–68.

Dougenik, James A., Nicholas R. Chrisman, and Duane R. Niemeyer. "An Algorithm to Construct Continuous Area Cartograms." *The Professional Geographer* 37, no. 1 (1985): 75.

Gauthier, Majella-J., ed. *Cartographie Dans les Medias* (Cartography in the media). Québec: Presses de l'Universite du Québec, 1988.

Gilmartin, Patricia. "The Design of Journalistic Maps: Purposes, Parameters and Prospects." *Cartographica* 22, no. 4 (1985): 1–18.

Griffin, T. L. C. "Recognition of Areal Units on Topological Cartograms." *The American Cartographer* 10, no. 1 (April 1983): 17.

Harley, J. B., and David Woodard. *History of Cartography. Vol. 1, Cartography in Prehistoric, Ancient, and Medieval Europe and the Mediterranean*. Chicago: University of Chicago Press, 1987.

Hodgkiss, Alan G. *Understanding Maps: A Systematic History of Their Use and Development*. Folkestone, England: William Dawson and Sons, 1981.

Holmes, Nigel. *Pictorial Maps*. New York: Watson-Guptill Publications, 1991.

Library of Congress, Geography and Map Division, Reference and Bibliography Section. *Fire Insurance Maps in the Library of Congress: Plan of North American Cities and Towns Produced by the Sanborn Map Company: A Checklist.* Washington, D.C.:Library of Congress, 1981.

Liner, D. S. *Tactile Maps: A Listing of Maps in the National Library Service for the Blind and Physically Handicapped Collection.* Washington, D.C.: National Library Service for the Blind and Physically Handicapped, 1987.

MacEachren, Alan M., and David DiBiase. "Animated Maps of Aggregate Data: Conceptual and Practical Problems." *Cartography and Geographic Information Systems* 18, no. 4 (October 1991): 221–29.

McGranaghan, Matthew, David M. Mark, and Michael D. Gould. "Automated Provision of Navigation Assistance to Drivers." *The American Cartographer* 14, no. 2 (April 1987): 121–38.

Monmonier, Mark S. *Maps with the News: The Development of American Journalistic Cartography.* Chicago: The University of Chicago Press, 1989.

______, and George A. Schnell. *Map Appreciation.* Englewood Cliffs, N.J.: Prentice-Hall, 1988, Chapters 8 and 9.

Moreland, Carl, and David Bannister. *Antique Maps: A Collector's Handbook.* London and New York: Longman, 1983.

Olson, Judy M. "Noncontiguous Area Cartograms." *The Professional Geographer* 28, no. 4 (1976): 371–80.

Oswald, Diane L. *Fire Insurance Maps: Their History and Applications.* College Station, Tex. Lacewing Press, 1997.

Ristow, Walter W. "Journalistic Cartography." *Surveying and Mapping* 17 (1957): 369–90.

Schiff, W., and E. Foulke, eds. *Tactual Perception: A Sourcebook.* Cambridge: Cambridge University Press, 1982.

Skelton, R. A. *Maps: A Historical Survey of Their Study and Collecting.* Chicago: University of Chicago Press, 1972.

Thrower, Norman J. W. *Maps and Man: An Examination of Cartography in Relation to Culture and Civilization.* Englewood Cliffs, N.J.: Prentice-Hall, 1972.

Tooley, R. V. *Maps and Map Makers.* 6th ed. London: Holland Press. New York: Richard B. Arkway, 1980.

Turner, Eugene, and John C. Sherman. "The Construction of Tactual Maps." *The American Cartographer* 13, no. 3 (1986): 199–218.

Wiedel, J. W., ed. *Proceedings of the First International Symposium on Maps and Graphics for the Visually Handicapped.* Washington, D.C.: Association of American Geographers, 1983.

CHAPTER

15 MAPS AND GRAPHS

Graphs are often associated with maps, either as map symbols or as supporting illustrations (Figure 15.1). The ability to read graphs properly, therefore, is an important map-use skill. In this chapter, we will discuss types of graphs commonly used in conjunction with maps, the interpretation of these graphs, and common interpretation difficulties. Some of these difficulties are due to deficiencies in graph design and others to misinterpretation on the part of the user. A benefit of this chapter is that its information is useful whenever graphs are encountered, whether or not they are associated with maps.

LINE GRAPHS

Line graphs are one of the most straightforward and commonly encountered types of graphs. Although their general appearance is likely to be familiar, a review of their characteristics will help avoid misinterpretations.

Simply stated, a line graph consists of a line joining points plotted on a grid. The path of the line shows the value of one variable relative to the value of another. The horizontal axis is called the x axis, and the vertical axis is called the y axis.

Types of Variables

Different types of variables are plotted on the axes of line graphs. Often, time periods are placed on the x axis, and the variable being considered in relation to time, such as trade volume, birthrate, automobile production, wholesale price index, output per worker hour, or total employment, is placed on the y axis (Figure 15.2a). However, time is not always one of the variables. Pairs of variables can show, for example, the number of students with different intelligence quotients or the number of towns of different size classes. In such cases, the categories themselves are usually shown on the x axis and the number of occurrences in each category on the y axis (Figure 15.2b).

Sometimes, one of the variables shown on a line graph is believed to be independent and to influence the value of the other (dependent) variable. In these situations, the independent variable is usually plotted on the x axis and the dependent variable on the y axis. The following list of paired variables illustrates the concept (the independent variables are listed first in each pair): temperature and disease rate; unemployment rate and frequency of strikes; alcohol consumption and accident rates; the thickness of glacial ice and its rate of flow.

Maps as Graphs

An even closer link exists between graphs and maps than has been suggested so far. For example, a line graph in which the longitude of an occurrence is plotted on the x axis and the latitude on the y axis could also be called a map. Another similar example is a profile. In this case, the graph shows elevations at given distances along a line.

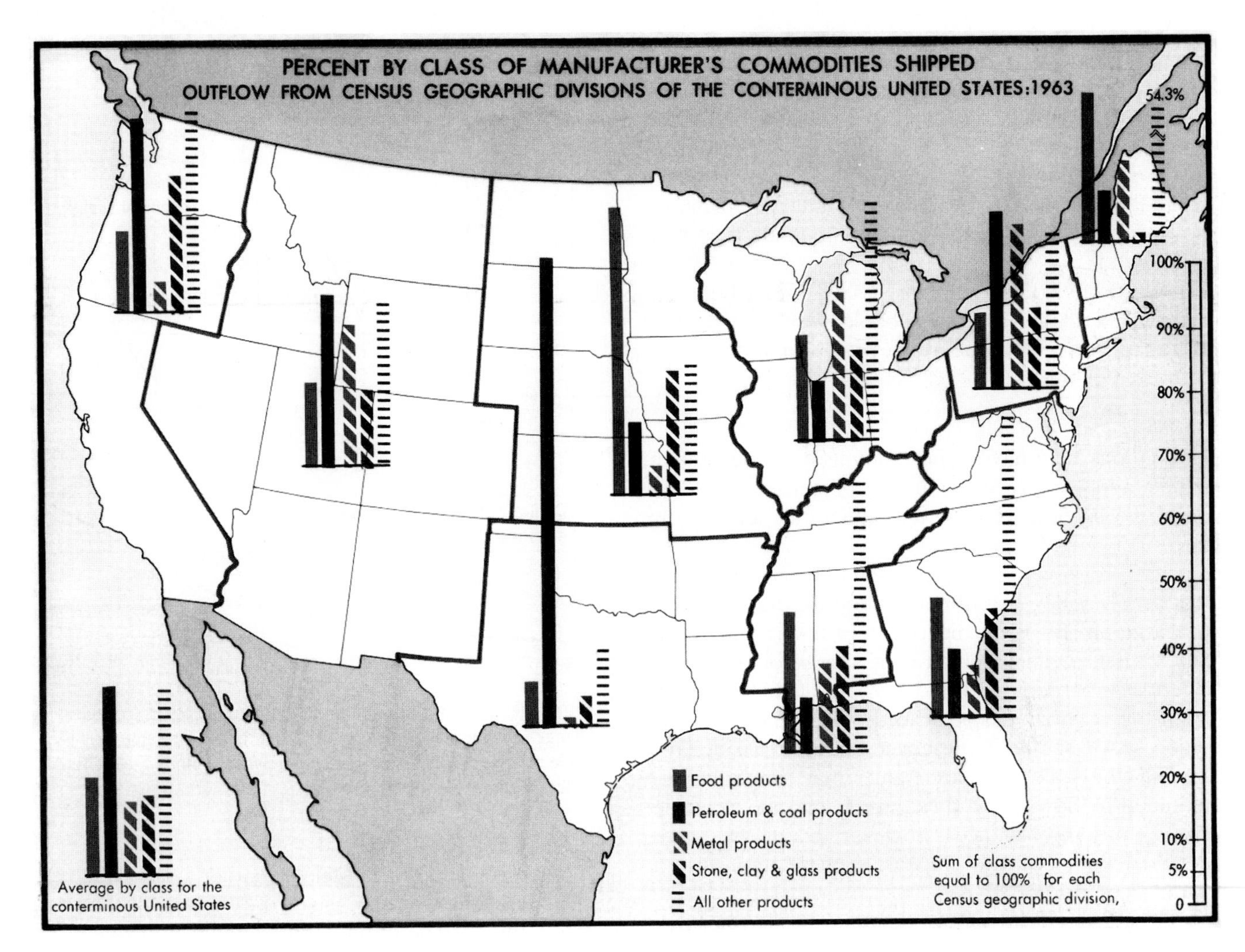

Figure 15.1 Graphs used as map symbols.

Source: Map sheet entitled Shipments of Commodities by Manufacturers in the Conterminous United States. Outflow from Census Geographic Divisions: 1963, Department of Commerce. Bureau of the Census (Washington, D.C.: U.S. Government Printing Office).

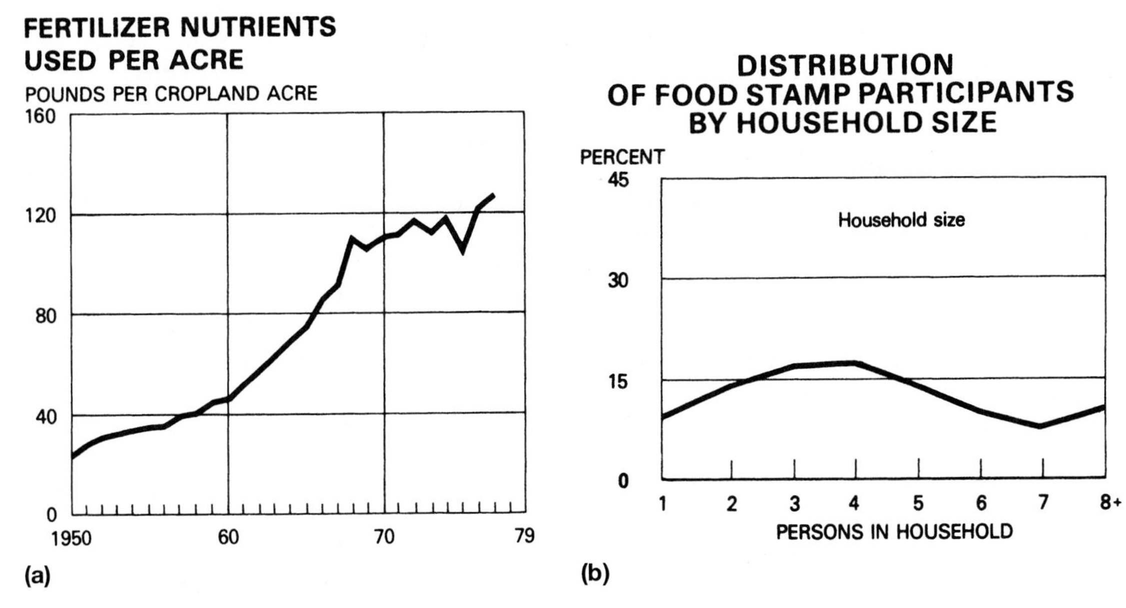

Figure 15.2 Typical simple line graphs. Such graphs usually show (*a*) time periods or (*b*) categories on the *x* axis and the number or rate of occurrences on the *y* axis.

Source: *1978 Handbook of Agricultural Charts,* Agriculture Handbook No. 551 (Washington, D.C.: U.S. Department of Agriculture, November 1978).

One fundamental difference between maps and graphs involves the notion of interdependence. On many graphs, one value (for example, age) may be related to another (for example, mobility). On maps, however, there is no implication that latitude is related to longitude.

We do not pursue this discussion further here because, after all, the rest of the book is about maps, but you may wish to give the matter some further thought.

Types of Scales

Line graphs may have either arithmetic or logarithmic scales, and each type has unique characteristics. The next two sections discuss the similarities and differences between the two types.

First, however, the general point needs to be made that graph scales are arbitrary. Suppose, for example, that a graph shows fluctuations in prices, ranging from 100 to 1000 dollars, spread over a span of ten years. There is no predetermined way to decide how far apart the years should be spaced on the graph or how much space should represent a value of 100 dollars. Instead, the scales for these values may simply be determined by how many units must fit into the available space. In some instances, however, graph scales are deliberately selected to emphasize or de-emphasize a particular aspect of the data.

Whatever considerations determine scale selection, the results are crucial to the graph's interpretation. If one axis is long relative to the other, for example, a specific change in value will look very different than if the same axis is made relatively short (Figure 15.3). The user must determine whether the scale of the graph is reasonable in relation to the range of values in the data or if some distortion is being introduced by the graph's dimensions.

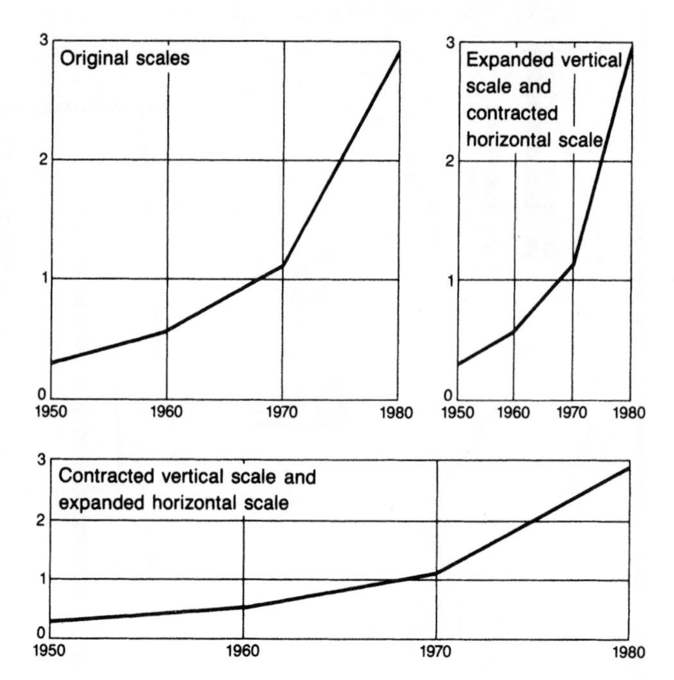

Figure 15.3 If one axis is long relative to the other, a specific change in value will look very different than if the same axis is made relatively short. Here, the same data are plotted using three different scale relationships.

Arithmetic Line Graphs

On an **arithmetic line graph,** the spacing between values is constant. That is, equal vertical (or horizontal) distances indicate equal changes in value, anywhere on the graph.

Some arithmetic line graphs are drawn using **absolute scales,** which means that each unit of distance on the graph represents the same *quantity* of a given variable. In Figure 15.4a, for example, a given vertical distance indicates a change of a given quantity, regardless of where on the graph it occurs. Other arithmetic line graphs are drawn using **percentage scales.** In this type, equal spacings indicate equal percentage ranges. For example, in Figure 15.4b, a given vertical distance indicates a certain percent difference, regardless of where on the graph it is measured. The comments on interpretation that follow apply equally to these two types of scales.

The spacing of the scale on either axis of an arithmetic line graph should usually be uniform. If the scale shows the passage of time, for example, a five-year period should cover five times as much distance as a one-year period. Similarly, if the scale shows values in dollars, every unit of distance should represent the same number of dollars. A nonuniform spacing on an arithmetic line graph often indicates that the graph is poorly drawn or purposely exaggerated and, therefore, is misleading. Logarithmic scales, which are discussed later, are a special exception to this.

A common fault in line-graph design that affects interpretation is the failure to start the vertical scale at zero. This often occurs when high total values are involved and the differences in those values are relatively small. In such cases, the omission of the zero baseline decreases the range of values that must be accommodated within the height of the graph. This allows variations in the values to be shown more clearly but also makes it difficult to judge the relative importance of the value changes (Figure 15.5).

An alternative way to deal with the problem of small changes in large total values is to show the zero baseline but break the vertical scale to indicate missing values (Figure 15.6). This is less likely to be misinterpreted because it calls attention to what has been done.

In either type of presentation, however, graph interpretation must be done cautiously, because the importance of the changes in value is exaggerated. The best thing the user can do is to visualize (or measure)

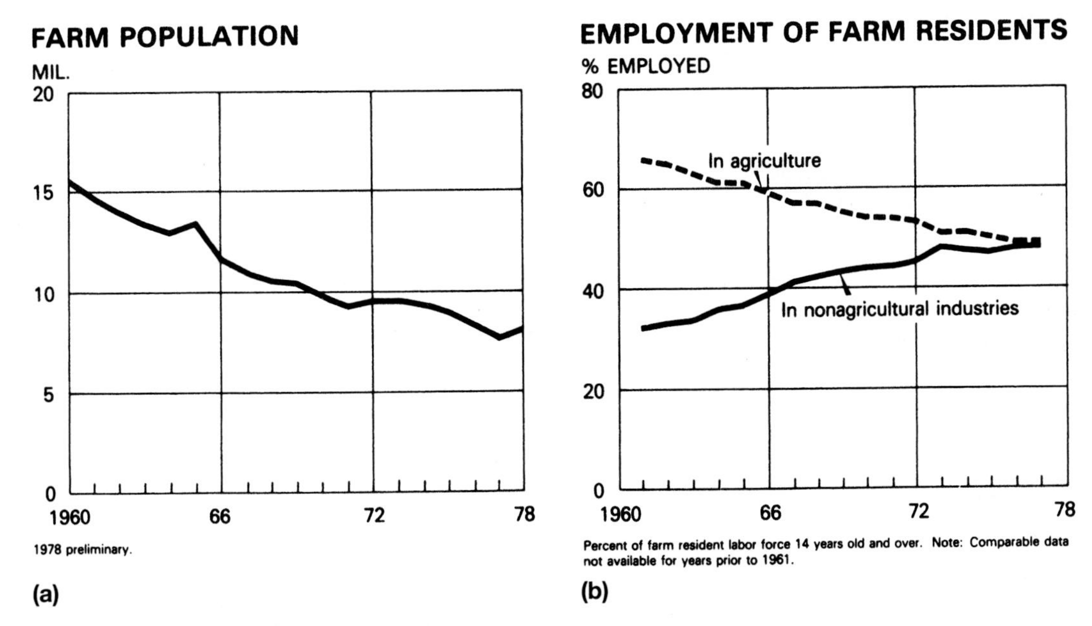

Figure 15.4 Line graphs using (*a*) absolute and (*b*) percentage scales.

Source: *1978 Handbook of Agricultural Charts,* Agriculture Handbook No. 551 (Washington, D.C.: U.S. Department of Agriculture, November 1978).

WORLD PRODUCTION AND CONSUMPTION OF RAW WOOL

BIL. LB.

4.0
3.5
3.0
2.5
2.0

Production

Consumption

1965 67 69 71 73 75 77 79

Clean-content weight. Production on marketing-year basis.

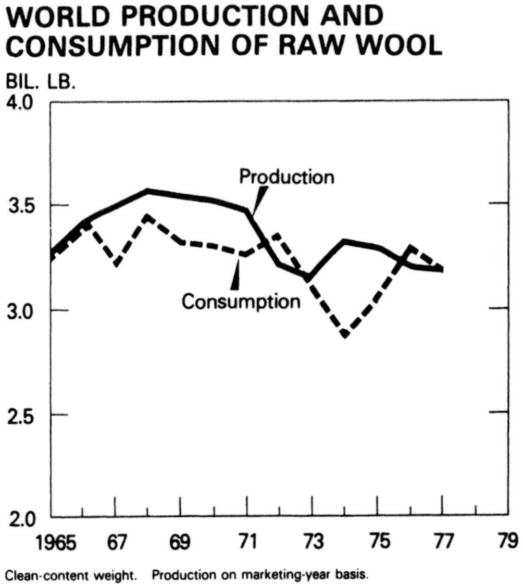

Figure 15.5 Omission of the zero baseline allows variations in the values to be shown more clearly but makes it difficult to judge the relative importance of the value changes.

Source: *1978 Handbook of Agricultural Charts,* Agriculture Handbook No. 551 (Washington, D.C.: U.S. Department of Agriculture, November 1978).

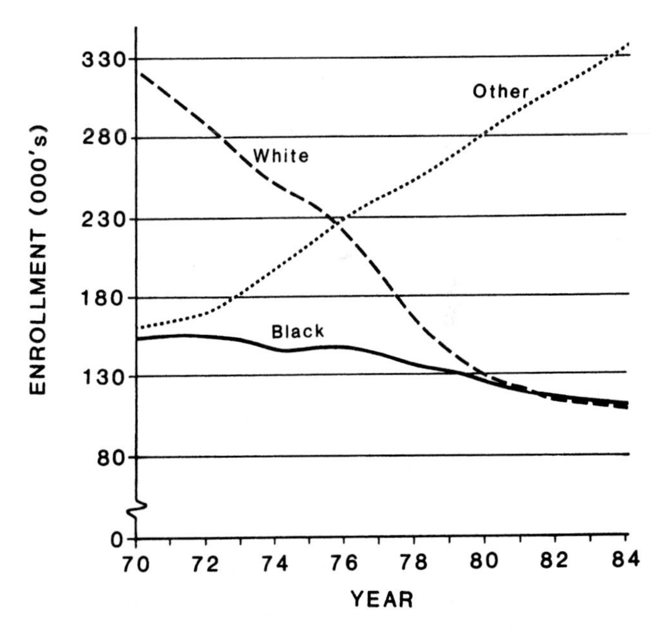

Figure 15.6 Zero baseline, with a break in the vertical scale to indicate missing values. (Total pupil enrollment in the Los Angeles school system, 1970–1984.)

Source: From W. A. V. Clark, "The Roepke Lecture in Economic Geography: Urban Restructuring from a Demographic Perspective," *Economic Geography* 63, no. 2 (April 1987): 103–25. Reprinted by permission.

where the baseline *should* be or what the total length of the scale should be. This gives a perspective on the amount of exaggeration present and on the real importance of the changes. Unfortunately, there seems to be a trend away from showing the zero baseline if the result would be "inartistic" and if the user can be assumed to be knowledgeable enough to realize that a zero origin does exist.[1]

[1]Cecil H. Meyers, *Handbook of Basic Graphics: A Modern Approach* (Belmont, Calif.: Dickenson, 1970), 39.

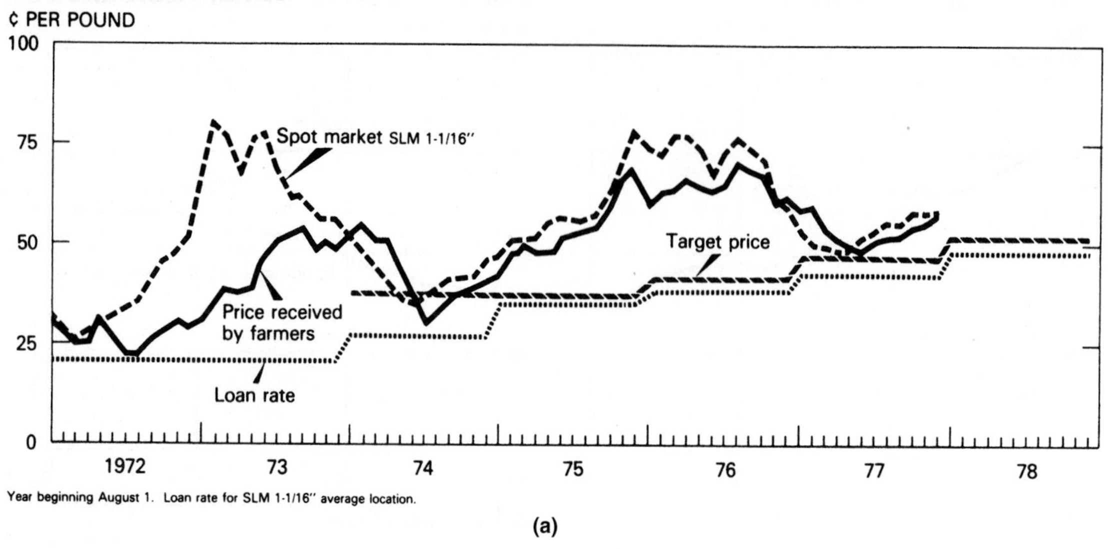

(a)

SHEEP NUMBERS AND LAMB AND MUTTON PRODUCTION

MIL. HEAD

BIL. LB.

Production, including farm

On farms (Jan. 1)

1978 production forecast.

(b)

Figure 15.7 Multiple-line graphs. (*a*) Several different lines plotted, all using the same scale. (*b*) Two lines plotted, each line showing a different type of quantity and each with its own scale. In this case, one line shows numbers of sheep on farms and the other shows lamb and mutton production in billions of pounds.

Source: *1978 Handbook of Agricultural Charts,* Agriculture Handbook No. 551 (Washington, D.C.: U.S. Department of Agriculture, November 1978).

Another interpretation problem is introduced when several different lines are plotted on a single graph. Normally, the main difficulty in such **multiple-line graphs** is simply keeping track of the trends of the separate lines as they intermingle with one another (Figure 15.7a). The problem is more complicated when the different lines are measured in different types of quantities. Such a graph might include population, measured in numbers of people; gross national product, petroleum prices, and stock-market averages, all measured in dollars; and gasoline and crude oil production, both measured in barrels. The interpretation of the multiple scales on such a multiple-scale graph requires that the user must, first, be alert enough to realize that different scales are being used and, second, be careful to use the proper scale for each variable (Figure 15.7b).

With a multiple-scale line graph, the relationships between the scales are arbitrary and would apparently change if the scale values were shifted. This means that little can be inferred from a comparison of the different lines, other than whether a trend relationship exists between them. A trend relationship, however, does not necessarily indicate a causal relationship between any of the variables. For example, if stock market prices and gasoline production seem to follow a similar pattern of ups and downs, is it because one is affecting the other? If so, which one is the cause and which the effect? Or is it possible that some *other* variable, perhaps one not shown on the graph, is affecting

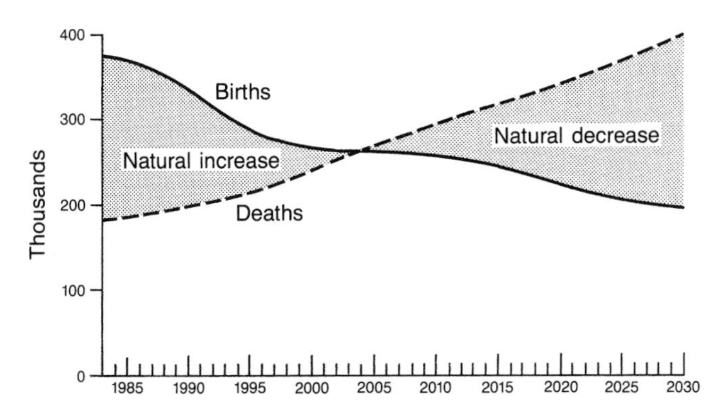

Figure 15.8 Bilateral graph. (Population forecast for Canada.)
Source: Redrawn from "Speaking Graphically," *Population Today* 13 no. 12 (December 1985): 5 (Population Reference Bureau Inc: Washington, D.C., 1985).

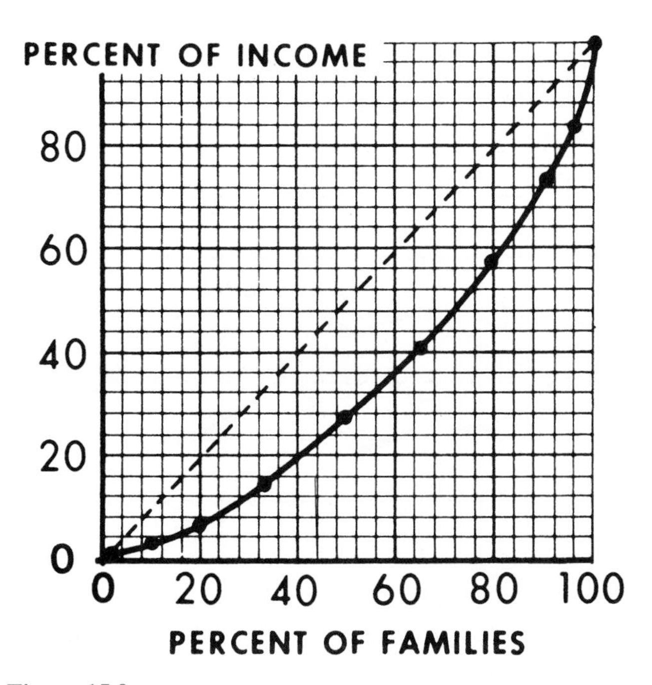

Figure 15.9 Accumulative line graph (Lorenz curve), showing families ranked by personal income, United States, 1954.
Source: Frederick V. Waugh, *Graphic Analysis in Agricultural Economics* (Washington, D.C.: U.S. Department of Agriculture, Agricultural Marketing Service, 1957), 9.

both stock-market values and gasoline production? In other words, various explanations for the patterns seen on a graph should be carefully considered. A graph, of whatever type, may suggest a variety of relationships, but the interpretation of those relationships cannot stand on its own without adequate theory.

One special type of single-scale line graph is a **bilateral graph.** A bilateral graph uses a single arithmetic scale to portray the values of two related variables. The overlap of the lines that represent these variables is then interpreted in terms of either positive or negative values. An excess of births over deaths yields a natural increase in population, for example, whereas an excess of deaths over births indicates a natural decrease (Figure 15.8).

Another special type of single-scale graph is a **silhouette graph.** This is simply a standard line graph in which the area below the curve is shaded or colored to give a silhouette effect. The shading highlights the trend of the line on the graph but has no other significance. Silhouette graphs should not be confused with the similar-appearing surface graphs discussed later in this chapter, because the two types are interpreted quite differently.

Accumulative line graphs, often called **Lorenz curves,** are special arithmetic line graphs used to determine the degree of concentration in one variable, as compared to another. Lorenz curves are frequently used on maps in connection with interregional comparisons of a variety of population, economic, or other information.

A Lorenz curve is constructed with equal percentage units on each axis. In the example shown in Figure 15.9, if income were equally distributed among all families, 20% of the families would have 20% of the income, 50% would have 50% of the income, and so on. The graph of these values would follow the diagonal straight line. In actuality, however, in 1954, the lower 20% of families received only about 6.5% of the income, 50% received about 27%, and 80% received about 57%. The plot of these values is shown by the solid, curved line. The greater the disparity in a distribution, the greater the departure of the Lorenz curve from the diagonal line of equality.

Some single-scale graphs are plotted on the basis of **index values** instead of absolute numbers. Some appropriate base period is selected, and index values for each category are then computed by dividing the individual data values by the base-period value and multiplying the results by 100. An increase from a base-period quantity of gasoline production of 85 million barrels to 170 million barrels in a later period, for example, represents a doubling of production. The index values for the two periods would be 100 and 200, respectively. The same index values might express a doubling in average wages from a base-period value of $19,500 to $39,000 in the later period. Index values, therefore, express the ratio between base-period absolute numbers and absolute numbers for other time periods. Index values make it possible to compare relative changes in the values of different variables by using the same base period for all of the categories, even when the units in which the variables are reported are different from one another.

When index values are plotted, the baseline of the graph is set at 100—the base-period value—not at 0. The index values of the different variables for other

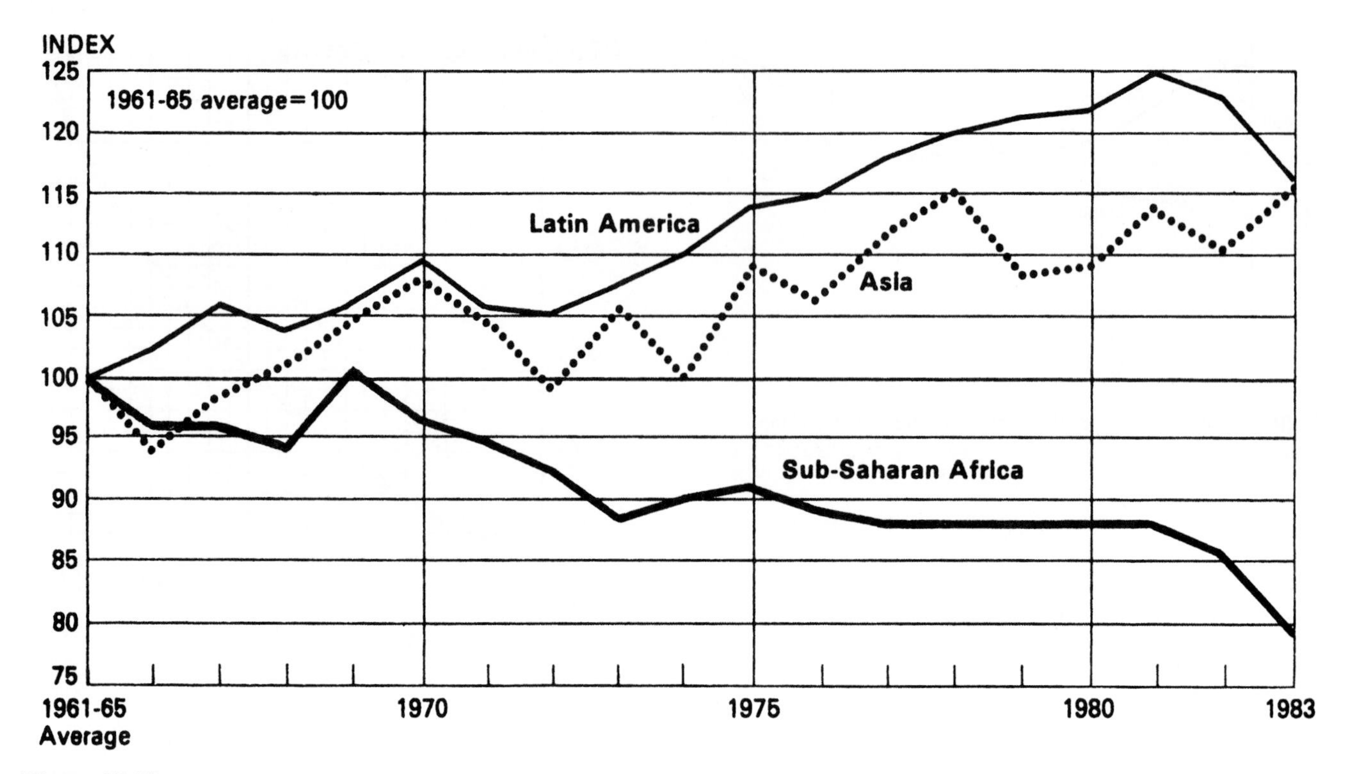

Figure 15.10 Line graph using index values. (Index of per capita food production, 1961–1965 to 1983: Sub-Saharan Africa, Latin America, and Asia.)

Source: Thomas J. Goliber, "Sub-Saharan Africa: Population Pressures on Development," *Population Bulletin* 40, no. 1 (1985): 16 (Washington, D.C.: Population Reference Bureau).

periods, which can be more or less than 100, are plotted above or below the baseline, as appropriate (Figure 15.10).

Different types of indexes are used in line graphs. Price indexes, for example, are based on the cost of goods purchased, in dollars. Quantity indexes measure the volume of variables, such as coal production in tons, number of workers employed in construction, or consumption of goods in units. Finally, value indexes are based on the total dollar volume of variables, such as income, payroll, or sales.

Logarithmic Line Graphs

The second major type of line graph uses a logarithmic, or ratio, scale. On a **logarithmic scale,** there is no zero.[2] Instead, the value range starts at some suitable value, such as 1 and increases until it reaches ten times the starting value—in this case, 10. A range of values with a multiple of ten is called a **cycle.** If the values shown on a graph cover more than a multiple of ten, additional cycles are required (such as 100 and 1000).

Because of their variable scale spacing, equal distances on logarithmic scales obviously do not represent equal absolute-value differences. Instead, equal distances indicate the same *percentage change* in value. In Figure 15.11, for example, a given distance represents a population range of 3000 persons between points *A* and *B,* but represents a range of 30,000 persons between points *C* and *D*. Notice, however, that *B* is two and one-half times the value of *A,* and *D* is two and one-half times the value of *C*. In other words, the ratio between the values is the same.

When a logarithmic scale is used on one axis of a graph (usually the *y* axis) and an arithmetic scale is used on the other, the graph is called a **semilogarithmic,** or **semilog, graph.** The most useful characteristic of a semilog graph is that the slope, shape, and direction of the lines plotted on the graph reveal trends in the changes of the values (Figure 15.12). A downward-sloping straight line on a semilog graph represents a constant rate of decrease, for example, whereas an upward-sloping straight line represents a constant rate of increase. Also, curved lines on a semilog graph indicate increasing or decreasing rates of change.

A semilog graph shows the *relative* rate and direction of change, and parallel lines anywhere on the graph indicate equal rates of change. The absolute

[2]See any standard mathematics textbook for a definition of *logarithms*.

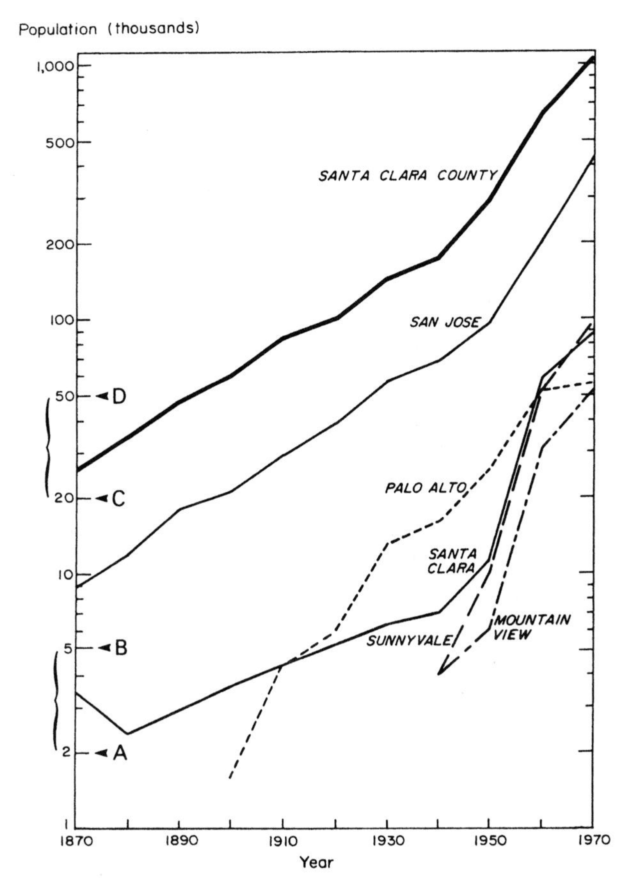

Figure 15.11 Graph using logarithmic cycles for the vertical scale, showing population growth in Santa Clara County, 1870–1970. On a logarithmic scale, equal distances represent equal percentage changes.

Source: Richard A. Walker and Matthew J. Williams, "Water from Power: Water Supply and Regional Growth in the Santa Clara Valley," *Economic Geography* 58, no. 2 (April 1982): 95–119. Reprinted by permission.

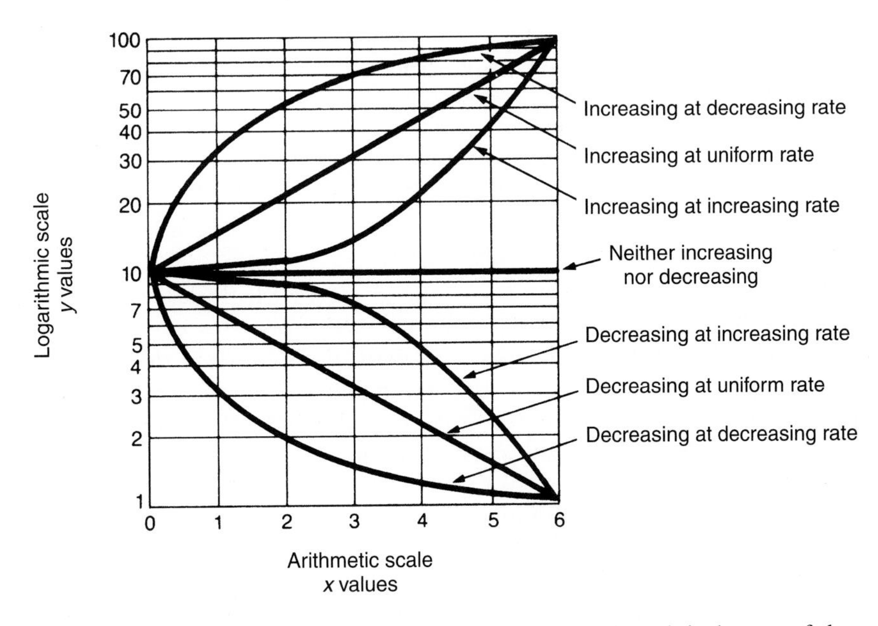

Figure 15.12 Slope, shape, and direction of lines on a semilogarithmic graph reveal trends in the rates of change of the values.

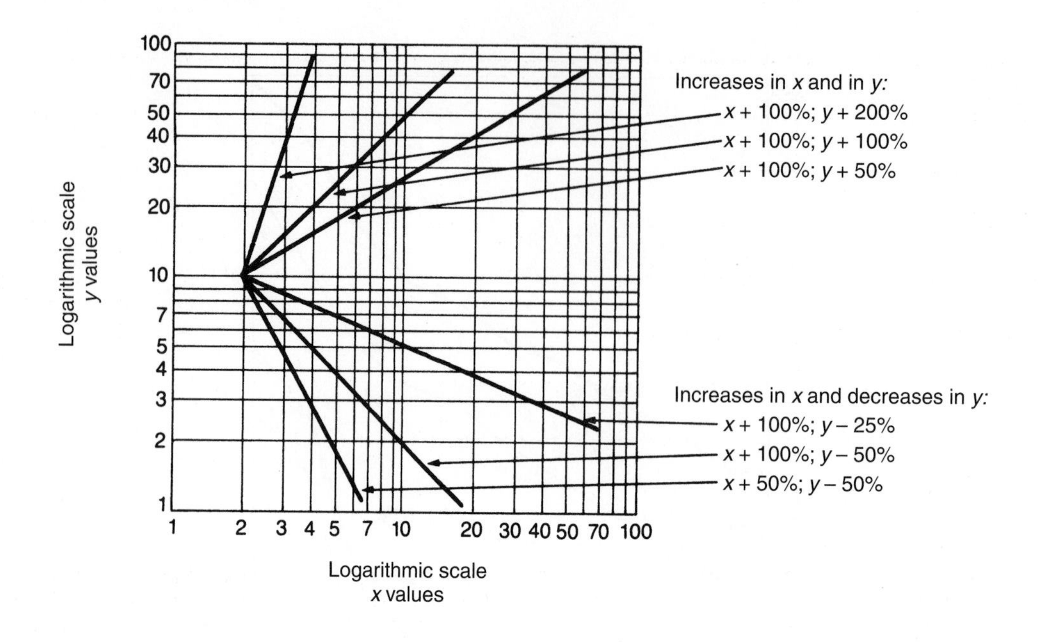

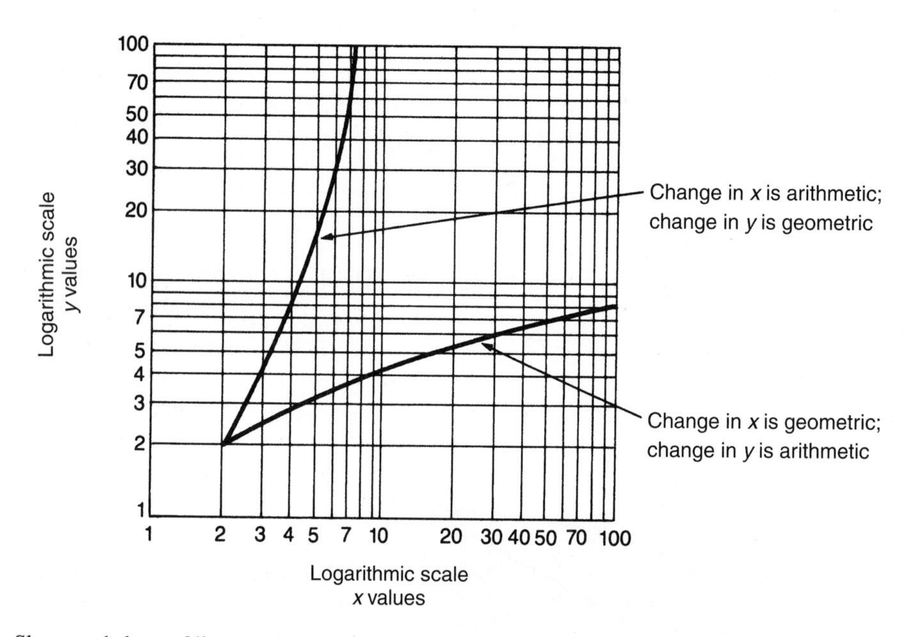

Figure 15.13 Shape and slope of lines on log-log graphs reflect percentage rates of change in both variables.

magnitudes of value changes are difficult to discern on a semilog graph. Arithmetic graphs, on the other hand, are appropriate for the interpretation of absolute magnitudes but not of rates of change.

Log-log graphs use logarithmic scales on both axes and show percentage changes in both variables simultaneously (Figure 15.13). A straight-line plot sloping upward at a 45-degree angle on a log-log graph indicates that the percentage rate of increase in both variables is the same. If the line slopes upward at greater than 45 degrees, the percentage increase in the y variable is greater than the percentage increase in the x variable. If the upward slope is less than 45 degrees, the opposite relationship is true. Downward-sloping lines indicate decreases in the value of the y variable. Curved lines on log-log graphs indicate that one of the rates of change is geometric and the other is arithmetic.

Surface Graphs

The discussion of the types of values and scales used with arithmetic-scale line graphs applies equally to **surface graphs.** Because line graphs and surface

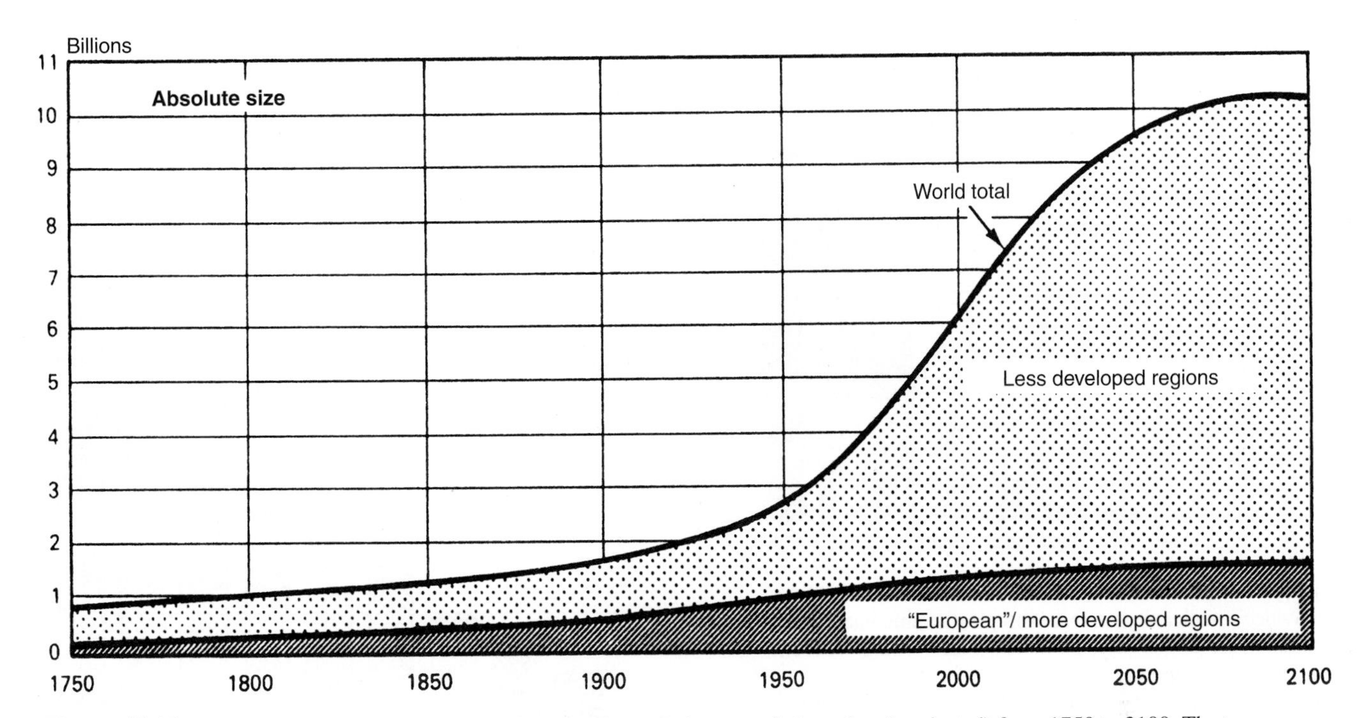

Figure 15.14 Absolute-scale surface graph showing world population growth (actual and projected) from 1750 to 2100. The two subcategories that are labeled "Less Developed" and "More Developed" yield total world population (labeled line).

Source: Thomas W. Merrick, with Population Research Bureau staff, "World Population in Transition," *Population Bulletin* 41, no. 2 (1986): 4 (Washington, D.C.: Population Reference Bureau).

graphs look very similar, they are sometimes confused. But how the two types are plotted and interpreted is significantly different. The values displayed on surface graphs are expressed as spaces, or bands, *between* the lines, whereas on line graphs, the values are measured from the baseline to each data line.

In contrast to line graphs, most surface graphs have a distinctive color, pattern, or shade of gray applied over each data band to call attention to the bands and emphasize the nature of the graph. If the bands are not marked in this way, the location of the graph labels should help to determine whether the graph is a line graph or a surface graph. In a line graph, the lines are labeled, as in the preceding illustrations. In contrast, on most surface graphs, the areas are labeled. (One exception is that the top line on a surface graph, which indicates the total value of the variable being graphed, is often labeled.)

Figure 15.14 shows how surface graphs are set up. The distance between the baseline and the top of the first band is based on the "European"/more developed regions' population for each year of the period. Similarly, the distance between the top of the first band and the top of the second band is based on the population of less developed regions. The line that defines the top of each band, then, is measured from the top of the band just below it, *not* from the baseline, as would be the case on a line graph. This method of construction results in the graph's top line reflecting a total value that is an accumulation of all of the values below it—World total, in Figure 15.14. When interpreting a surface graph, users should be aware that sharp changes in the value of a given band make it difficult to judge the widths of the bands above it.

Surface graphs are sometimes presented in percentage form. In these situations, the total distance between the baseline and the straight, horizontal line across the top of the graph represents 100%. Each band within the graph, then, represents a proportional share of the 100% total and not an absolute data value.

BAR CHARTS

A **bar chart** consists of bands or bars, each one representing a particular data range, such as an age grouping or a time period. The bars may be arranged either horizontally or vertically and are interpreted in the same way. Because the bars are kept a constant width, the area of each one, and hence its visual impact, is proportional to the value represented.

Types of Scales

Bar charts may show either absolute or percentage values, and either type may be subdivided. On maps,

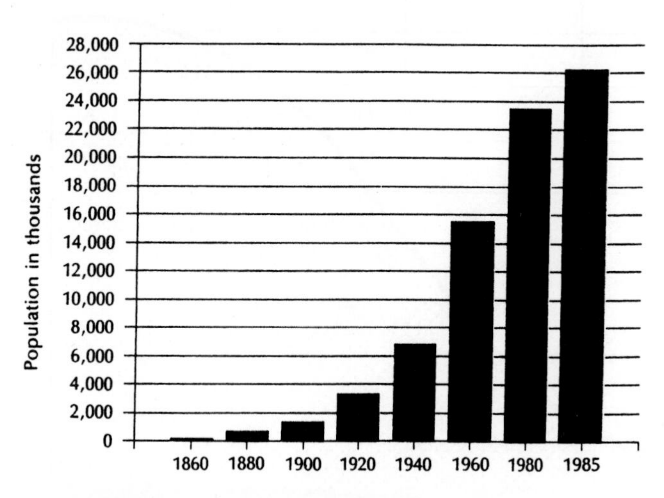

Figure 15.15 Bar chart with an absolute-value scale (undivided). Notice that, in this example, the time intervals between observations vary, and, ideally, the spacing between bars should vary in proportion to the actual intervals.

Source: "Speaking Graphically," *Population Today* (1984): 15 (Washington, D.C.: Population Reference Bureau).

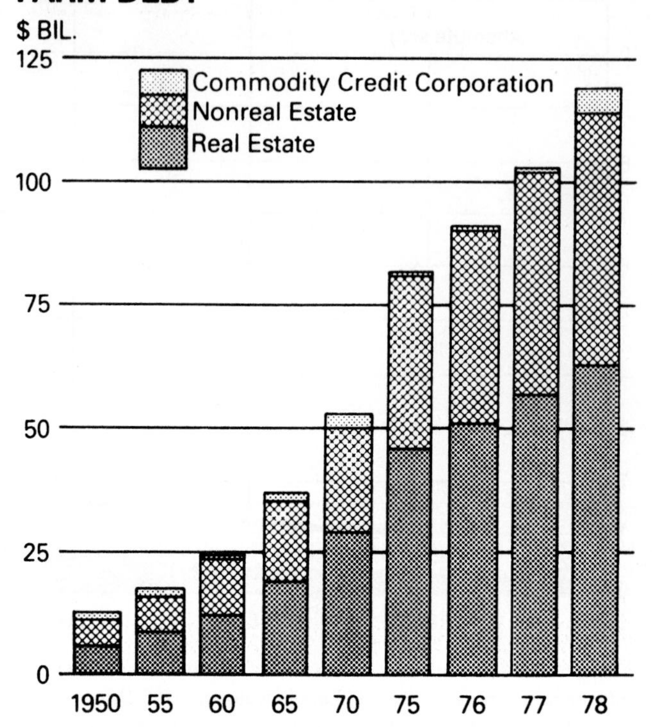

Figure 15.16 Bar chart with an absolute-value scale (subdivided).

Source: *1978 Handbook of Agricultural Charts,* Agriculture Handbook No. 551 (Washington, D.C.: U.S. Department of Agriculture, November 1978).

either a single bar or multiple bars may be shown for each region. A single bar represents the value of a certain category in a given time period. Multiple bars, on the other hand, show either the values of a single category for different time periods or the values of different categories during a single time period.

When an absolute-value chart is not subdivided, each bar is drawn to a length proportional to its value (Figure 15.15). When such a chart is subdivided, the lengths of the bars, and of each segment of each bar, are proportional to the value each represents (Figure 15.16).

Bar charts are also drawn with percentage scales. In one common type of percentage-value chart, each bar shows the subdivision of a single variable (such as a land-use type in Figure 15.17). All of the bars are the same length, which represents 100%, and are subdivided in proportion to the percentage of the total value represented by each subcategory (such as type of ownership).

Use of Symbols

Symbols, such as oil barrels or silhouettes of people, are often used in place of simple bars to introduce an element of interest into the bar graph. The proper use of such symbols should not interfere with the graph's interpretation. Each symbol is assigned a value, and the whole or partial symbols required to equal the total are lined up in a row (Figure 15.18). The result is that the row of symbols is the equivalent of a bar.

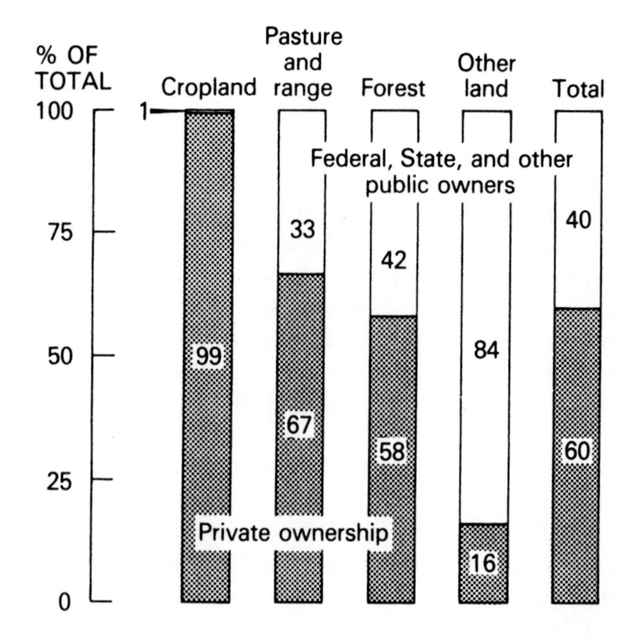

Figure 15.17 Bar chart with percentage-value scales (single variable).

Source: *1978 Handbook of Agricultural Charts,* Agriculture Handbook No. 551 (Washington, D.C.: U.S. Department of Agriculture, November 1978).

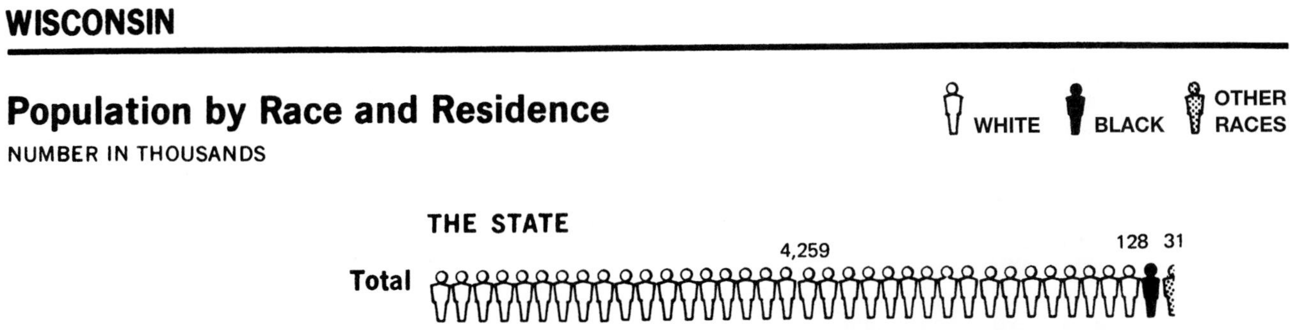

WISCONSIN

Population by Race and Residence

NUMBER IN THOUSANDS

WHITE BLACK OTHER RACES

THE STATE

Total 4,259 128 31

Figure 15.18 Chart with symbols used to create the equivalent of a bar.
Source: *General Population Characteristics: Wisconsin* (Washington, D.C.: U.S. Department of Commerce, Bureau of the Census, 1970 U.S. Census of Population).

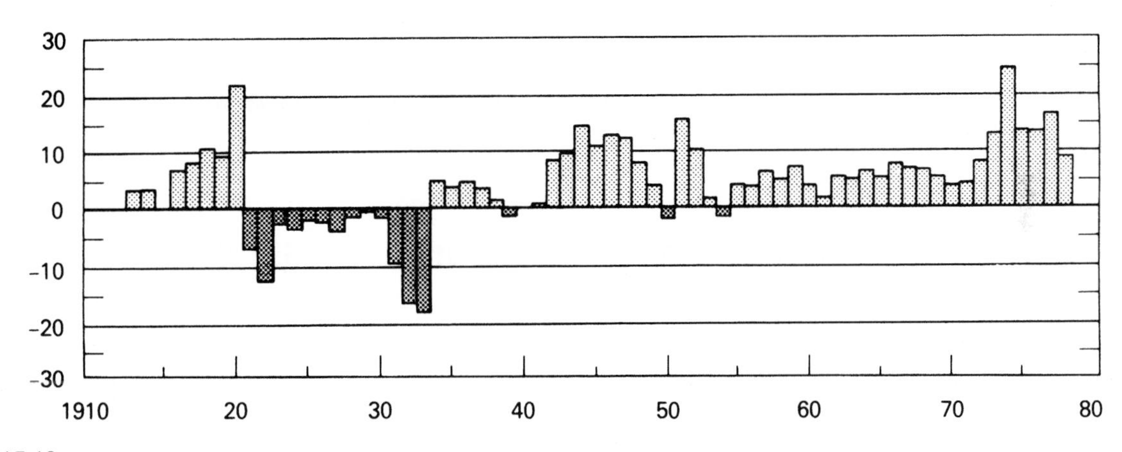

Figure 15.19 Loss-and-gain bar chart.
Source: *1978 Handbook of Agricultural Charts,* Agriculture Handbook No. 551 (Washington, D.C.: U.S. Department of Agriculture, November 1978).

Other Types of Bar Charts

Loss-and-Gain Bar Charts

Loss-and-gain bar charts differ from regular bar charts in that they have a central axis that is established as a zero line (Figure 15.19). The bars on the chart are drawn above and below a horizontal axis (or to the right and left of a vertical axis). Negative values (losses) are measured below the axis or to the left. Positive values (gains) are indicated by bars that lie above the axis or to the right.

High-Low Bar Charts

High-low bar charts usually are drawn as vertical bars representing a series of dates or times. Each bar shows the high and low values for each time period and may also include an average value.

Interpretation Problem

A zero baseline is always needed on a bar chart, because if it is missing, the relative importance of the values is very difficult to interpret. A scale break is generally unacceptable because it can be very misleading. Such breaks are sometimes introduced, however, if some of the values represented are significantly greater than the rest (Figure 15.20).

The introduction of a three-dimensional thickness to a bar chart presumably makes the bar look more interesting (Figure 15.21). Unfortunately, it also makes interpretation more difficult because of the visual impression of added height that the thickness gives to the portion of the bar closest to the top.

Each bar on a chart should represent an appropriate period of time or category of data. When time is involved, each bar must represent the same length of time. If lack of data causes the last bar to represent a shorter time period than the others, distortion is introduced (see Figure 15.15). In such a case, a projected figure should be used to create a comparable value for the full time period, or the spacing between the bars should be adjusted.

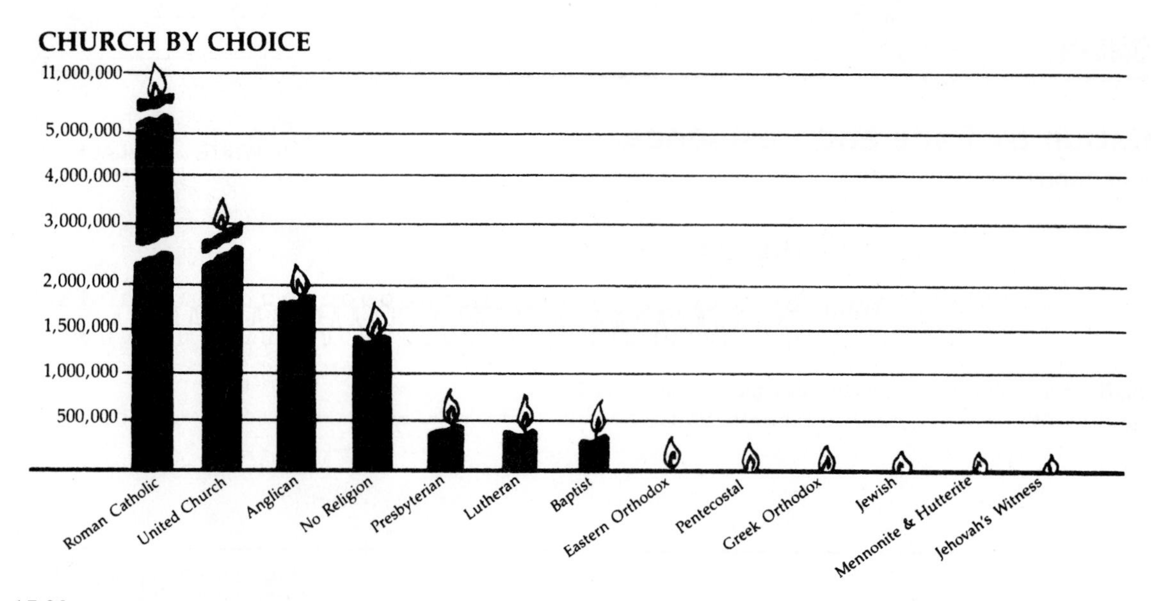

Figure 15.20 Scale breaks are sometimes used to make it easier to plot the differences in values. Doing so, however, de-emphasizes the range of values, so the technique is not generally acceptable.
Source: *Canada Today/d' aujourd' bui* 15, no. 2 (1984): 8.

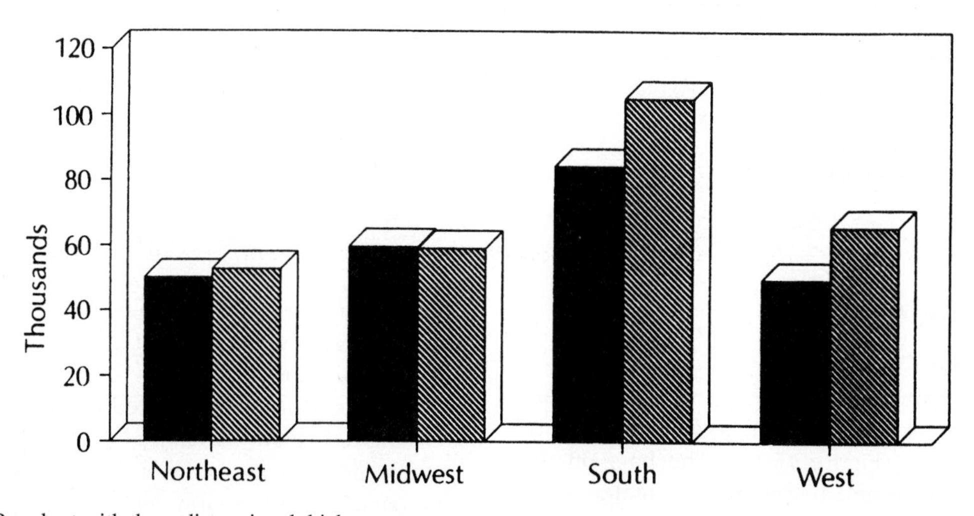

Figure 15.21 Bar chart with three-dimensional thickness.
Source: "Speaking Graphically," *Population Today* 16, no. 5 (May 1988): 2 (Washington, D.C.: Population Reference Bureau).

PIE CHARTS

A pie chart is simply a circle that is subdivided into proportional segments. For example, each subdivision of a particular chart could indicate the percentage of land ownership represented by different groups (Figure 15.22). On maps, pie charts are often used to represent values associated with different regions.[3]

Two common pieces of window dressing introduce difficulties into the interpretation of pie charts. One of these is the use of tilt, so that the pie is drawn as an ellipse, as though it were a circle being viewed from an elevated angle. Although this may look more "interesting," it is also more difficult to interpret. In Figure 15.23a, for example, each gray segment of the pie chart represents the same proportion of the total (20%). The use of an oblique viewpoint, however, makes it appear that the segment that lies at the top position represents a smaller value than the segment that lies at the side.

[3]When this is done, the circles are frequently drawn so that the area of each one is proportional to the value associated with its particular region. This use of size variation is discussed in the section "Proportional Symbols."

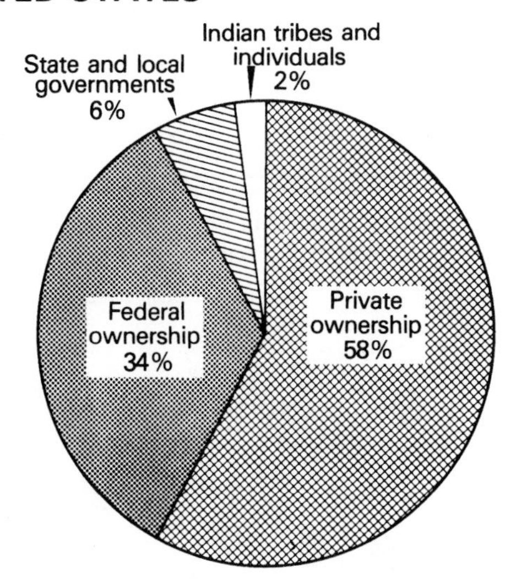

Figure 15.22 Pie chart.

Source: *1978 Handbook of Agricultural Charts,* Agriculture Handbook No. 551 (Washington, D.C.: U.S. Department of Agriculture, November 1978).

A second common visual enhancement is the addition of thickness to the pie chart (Figure 15.23b). Thickness increases the visual importance of the chart's bottom segment relative to the other segments, which is misleading.

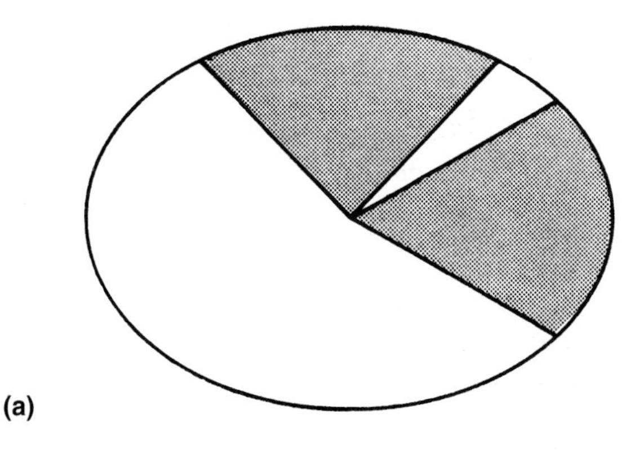

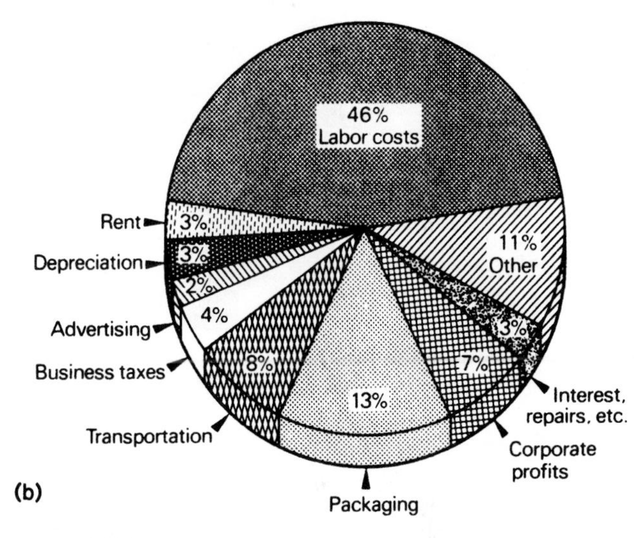

Figure 15.23 Enhanced pie charts. (*a*) Use of an oblique viewpoint. (*b*) Addition of thickness.

Source: *1978 Handbook of Agricultural Charts,* Agriculture Handbook No. 551 (Washington, D.C.: U.S. Department of Agriculture, November 1978).

AREA AND VOLUMETRIC DIAGRAMS

Proportional Symbols

Numerical data, such as the number of airplanes in an air force, the number of soldiers in an army, the amount of a region's agricultural output, or the quantity of energy resources available to a country, can be represented on a map in many different ways. One way is to use repeated symbols. In this approach, each symbol represents a given quantity, and the total number of symbols placed in each region represents the relevant total. This is essentially the same as the use of such symbols in a bar chart, except that the arrangement may be in a square or rectangular pattern instead of in a bar format.

Still another approach is to use proportional symbols, such as graduated circles, other geometric figures, or realistic figures.

Graduated Circles

The basic idea behind **graduated circles** is that the area of the symbol is proportional, more or less, to the value of the data that it represents. (The implication of the phrase *more or less* is discussed shortly.)

Following this rationale, a circle with one square unit of area may be used to signify 100 tractors. Then, if another circle is to represent 900 tractors, it should cover nine units of area. The *area* of the circle is important, not the circle's diameter.

When proportional circles are used, the cartographer should provide a legend. This allows the map user to compare the sizes of the symbols on the map with the samples in the legend to determine the value that each symbol represents.

There are two methods of sizing proportional circles. In one method, the sizes of the circles are continuously variable, and in the other, they are arranged in a graded series. When **continuously variable symbols** are used, each circle is mathematically scaled in proportion to the value associated with the particular point it represents (Figure 15.24a). This means that symbol sizes are extremely varied, with a separate size for each individual value. Because many cartographers feel that such continuously variable symbols are difficult to interpret, range-graded symbols are often used instead.

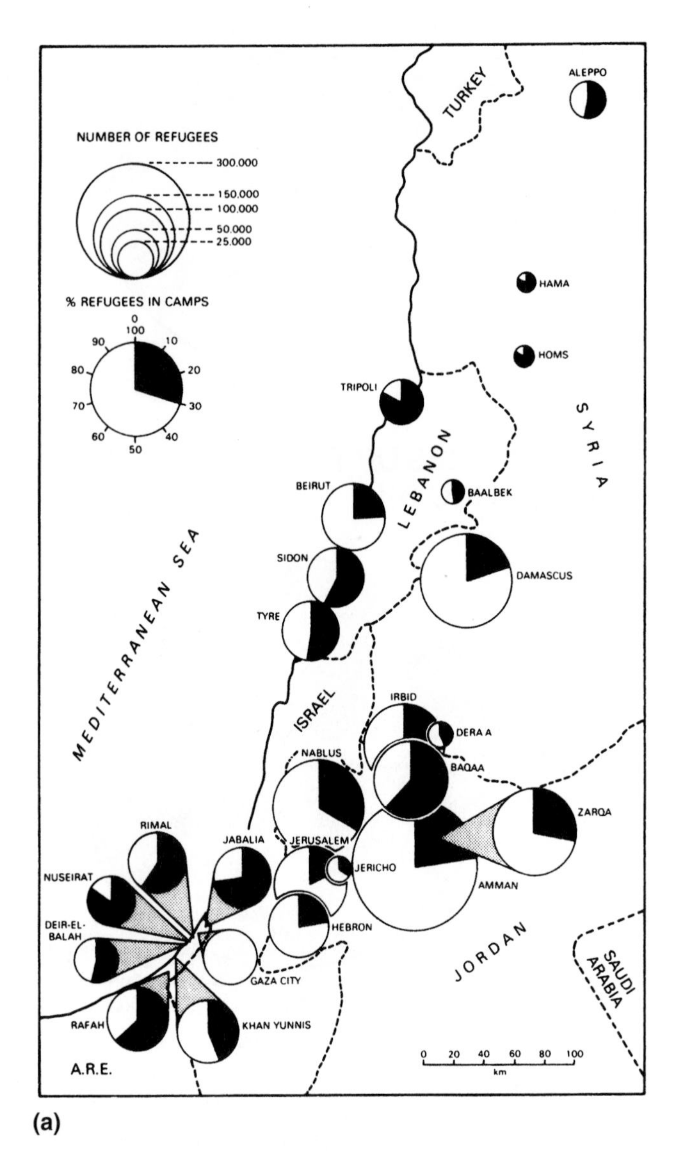

(a)

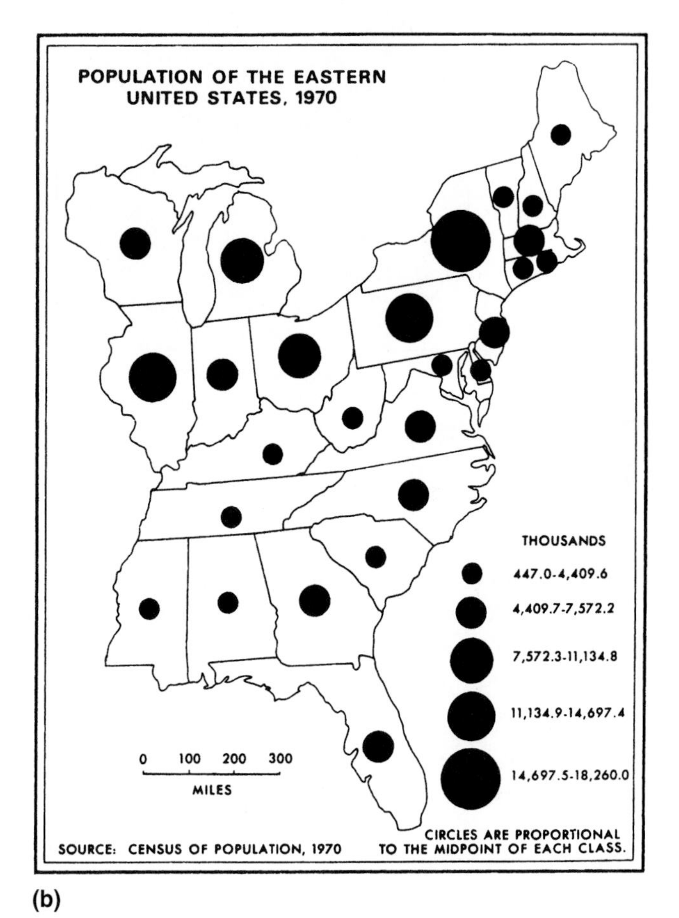

(b)

Figure 15.24 Proportional circle symbols. (*a*) Continuously variable symbols used to show Palestinian refugees registered with UNRWA (United Nations Relief and Works Agency), late 1987. (*b*) Range-graded symbols used to show the population of the eastern United States, 1970. Notice that each circle represents a range of values and is scaled in proportion to the midpoint of that class.

Source: (*a*) From *Focus* 37, no. 4. (Winter 1987). Copyright The American Geographical Society. Reprinted by permission. (*b*) From Borden D. Dent, "Communication Aspects of Value-by-Area Cartograms. *American Cartographer* 2, no. 2 (October 1975): 154–68. Reprinted by permission of the American Congress on Surveying and Mapping.

In the **range-graded symbol** approach, a number of specific symbol sizes are established, and each size represents a particular range of values (Figure 15.24b). Thus, on a given map, there may be only five circle sizes, even though a large number of different values are represented. The amount of information conveyed by such a range-graded symbol is less than that conveyed by a continuously variable symbol. This loss of information is, ideally, offset by the greater ease of interpreting the range-graded symbol. This view is reinforced by evidence that map users do not, in any case, accurately perceive the differences in value represented by continuously variable symbols.

Map users must examine the map legend to determine what type of graduated-circle symbol is being used and what level of detail can be ascribed to it. If range-graded symbols are used on the map, the information is generalized and, as the name implies, only a range of values can be applied to any given point and *not* an exact value. If continuously variable symbols are used, on the other hand, measuring the size of the symbols to obtain an exact data value is not recommended. It is more useful to interpret the symbols visually and to observe their spatial distribution but to refer to tabular information for exact values.

The preceding discussion implies that circle symbols are sized in direct proportion to data values. This is not always the case, however. It has been found that map users tend to underestimate the values represented by larger symbols, relative to smaller ones. If a

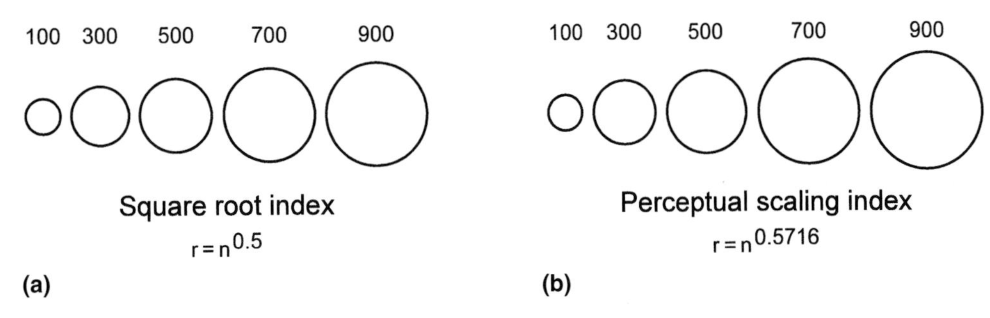

Figure 15.25 Principle of proportional circles using two different methods of scaling. (*a*) Square root index. (*b*) Perceptual scaling index.

given circle represents a value of 200, for example, a circle that occupies twice as much area is likely to be seen as representing *less than* a value of 400. In an attempt to offset this tendency toward underestimation, some maps are drawn using symbols that are not directly proportional in size to the values they represent. In this **perceptual scaling** approach, the symbols representing larger quantities are exaggerated in size rather than drawn in absolute proportion (see Figure 15.25). The result is that, if a symbol representing 100 tractors occupies one unit of area, one representing 900 tractors occupies *more* than nine units of area.[4]

The size of pie graphs often varies in proportion to the overall data value, in exactly the same manner as simple graduated circles. On a map showing timber harvested by region, for example, the overall size of each pie graph included on the map might be proportional to the value or volume of the total harvest for a given region. The types of trees contributing to each region's harvest can then be shown by the subdivisions of the pie graphs. The same interpretation problems apply to proportionally sized pie graphs as to simple proportional circles.

Other Geometric Symbols

Other proportional symbols, such as squares, rectangles, hexagons, and octagons, are sometimes used on maps. The areas of such geometric symbols are usually proportional to the values they represent, just as with proportional circles. The interpretation of such symbols is quite straightforward if the symbols are properly used. If not correctly scaled, however, their appearance may be very misleading. In either case, only careful comparison of the symbols on the map with the legend will result in accurate judgment of their size and value.

Realistic Figures

Instead of geometric figures, many maps use **realistic figures** to represent data values. These two-dimensional figures are chosen on the basis of the map topic and are meant to increase the map's graphic appeal. Airplanes are used to represent air power, soldiers for military forces, shocks of wheat for agricultural output, and oil tanks for energy resources, to cite a few typical examples. Unfortunately, such symbols often result in inadvertent errors in the presentation and may even introduce deliberate distortions.

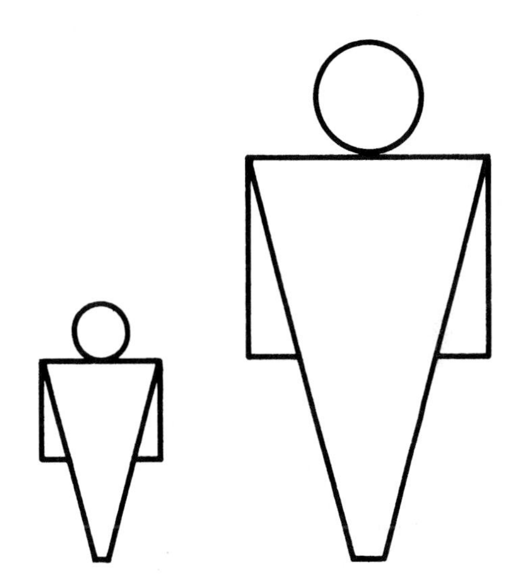

Figure 15.26 Interpretation problem with proportional realistic figures. (See text for explanation).

The concept involved in the use of realistic figures is precisely the same as for geometric figures. That is, the size of each symbol varies in relation to the value of the data associated with it. The use of irregular figures suffers from the same problems as the use of geometric figures, but it is even more difficult for the map user to compare the areas of the irregular symbols.

A major problem with realistic figures occurs when their size is determined incorrectly. This happens when the heights of the symbols are changed to represent different values and, at the same time, the symbols' widths are changed proportionally. Thus, for example, a figure that is made twice as tall also becomes twice as wide, and its area is four times as great (Figure 15.26). The person reading the map responds to the area of the

[4]The exact ratio used varies. For information on this aspect, see John Campbell, *Introductory Cartography,* 2d ed., (Dubuque, Iowa: Wm. C. Brown Communications, Inc., 1991).

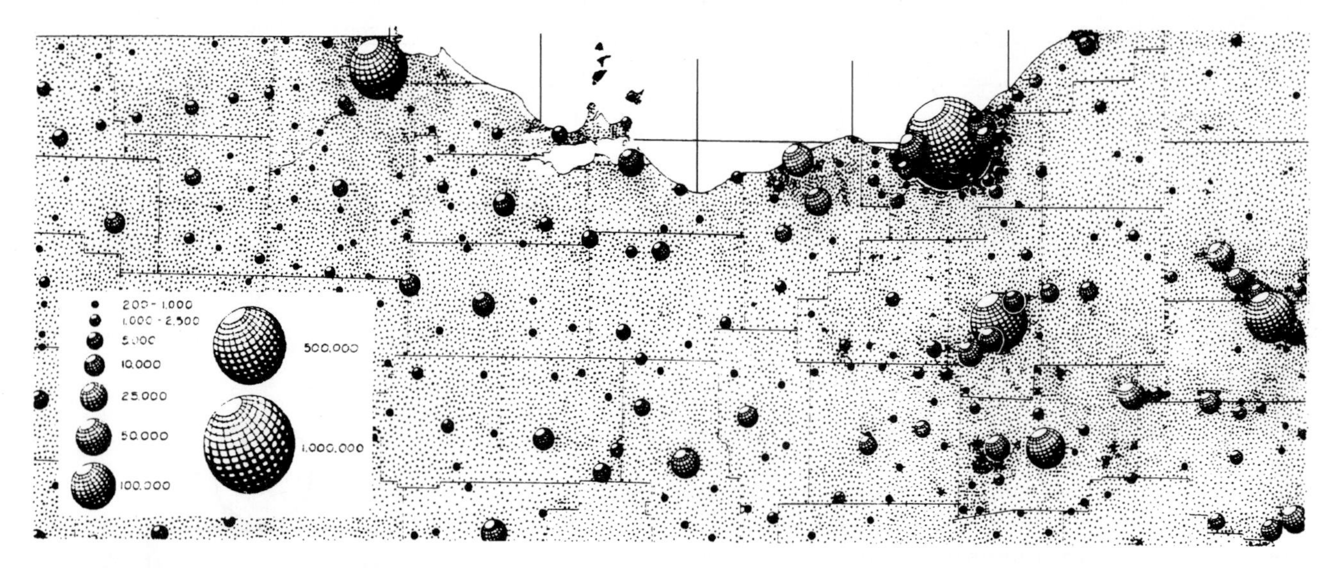

Figure 15.27 Portion of a classic example of the use of volumetric symbols for cities, combined with a dot map of rural population. Legend is at lower left.

Source: From Guy-Harold Smith, "A Population Map of Ohio for 1920," *Geographical Review* 18, no. 3 (July 1928): 422–27. Reprinted by permission of The American Geographical Society.

symbol rather than to just its height. The impression is given, then, that the second figure in this example represents four times the value of the first, instead of only two times.

Because this problem of improperly scaled symbols occurs frequently, it would be wise, for example, to measure some of the symbols on such a map and to compare the measurements with the values represented. If it is apparent that the areal scale is not consistent, the map is best disregarded.

Volumetric Diagrams

Sometimes, "solid" figures are used to represent values. These are drawings designed to look as much as possible like a sphere, a cube, or some other three-dimensional figure (Figure 15.27). Such diagrams are drawn so that the volume of each figure is proportional to the value that it represents; hence, the term **volumetric diagrams.**

Map users usually find it very difficult to correctly estimate the volumes represented by such diagrams. In particular, larger values are likely to be seriously underestimated. At a minimum, users should carefully compare each figure to the key provided to estimate the quantities represented. In addition, any numerical values that accompany the figures should be studied because they give exact information. Indeed, it might be better to ignore volumetric diagrams entirely because they are likely to be *less* informative than a table of data or some other form of graph, such as a bar chart.

SPECIAL DIAGRAMS

Scatter Diagrams

A special type of graph, called a **scatter diagram,** is often used to display the relationship between two variables, such as the use of multiple cropping techniques in agriculture and grain yields (Figure 15.28). Other examples from a practically endless set of possibilities include labor costs and product prices; percent of the labor force employed in primary activities and per capita gross national product; ground slope and percentage of woodland; distance from the capital city and population density; and amount of fuel used and amount of power output. A scatter diagram consists of a set of points plotted on the basis of paired observations of the variables. The value of the independent variable is measured on the x axis and the value of the dependent variable on the y axis.

A **trend line** is often superimposed over the cloud of dots on a scatter diagram, as in Figure 15.28. The line represents the relationship between the variables (Figure 15.29).[5] If the trend line slopes upward, from lower left to upper right, the relationship is positive. This means that, as the value of the independent variable increases, the value of the dependent variable also increases. A trend line that slopes downward, from upper left to lower right, indicates a negative relationship. A negative relationship is one in which the

[5]This trend line should be determined by a statistical technique, such as regression. Consult any introductory statistics textbook for information about regression computations and their interpretation.

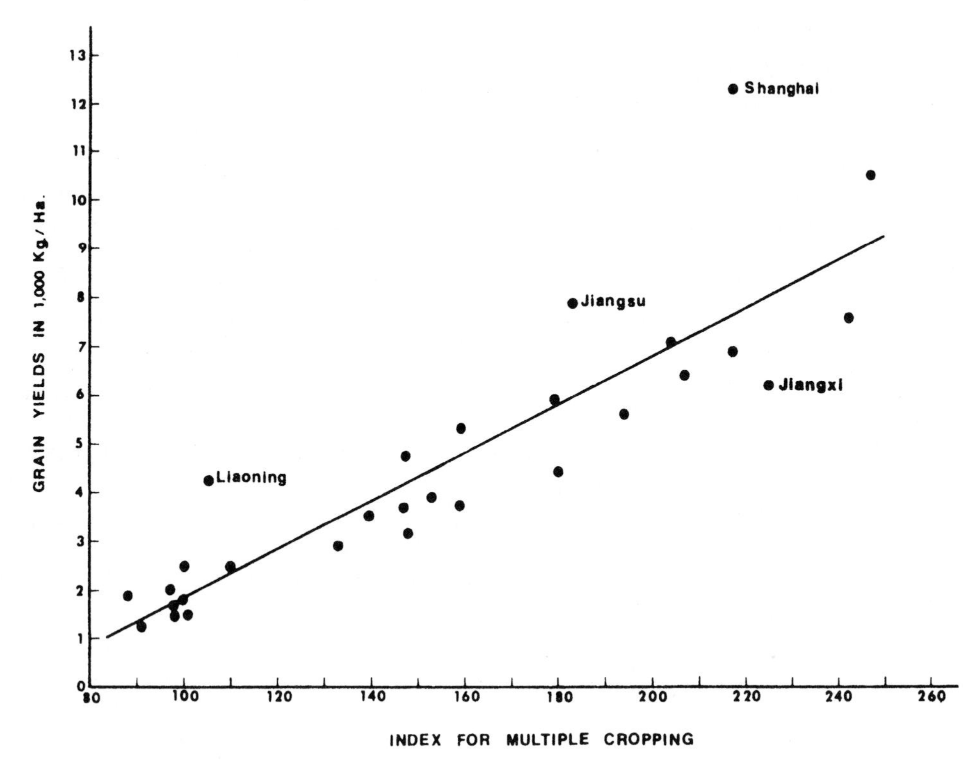

Figure 15.28 Scatter diagram with a trend line, showing the relationship between multiple cropping (independent variable) and grain yields (dependent variable) in China, 1979.

Source: From Clifton W. Pannell, "Recent Chinese Agriculture," *The Geographical Review* 75, no. 2 (April 1985): 170–85. Reprinted by permission of The American Geographical Society.

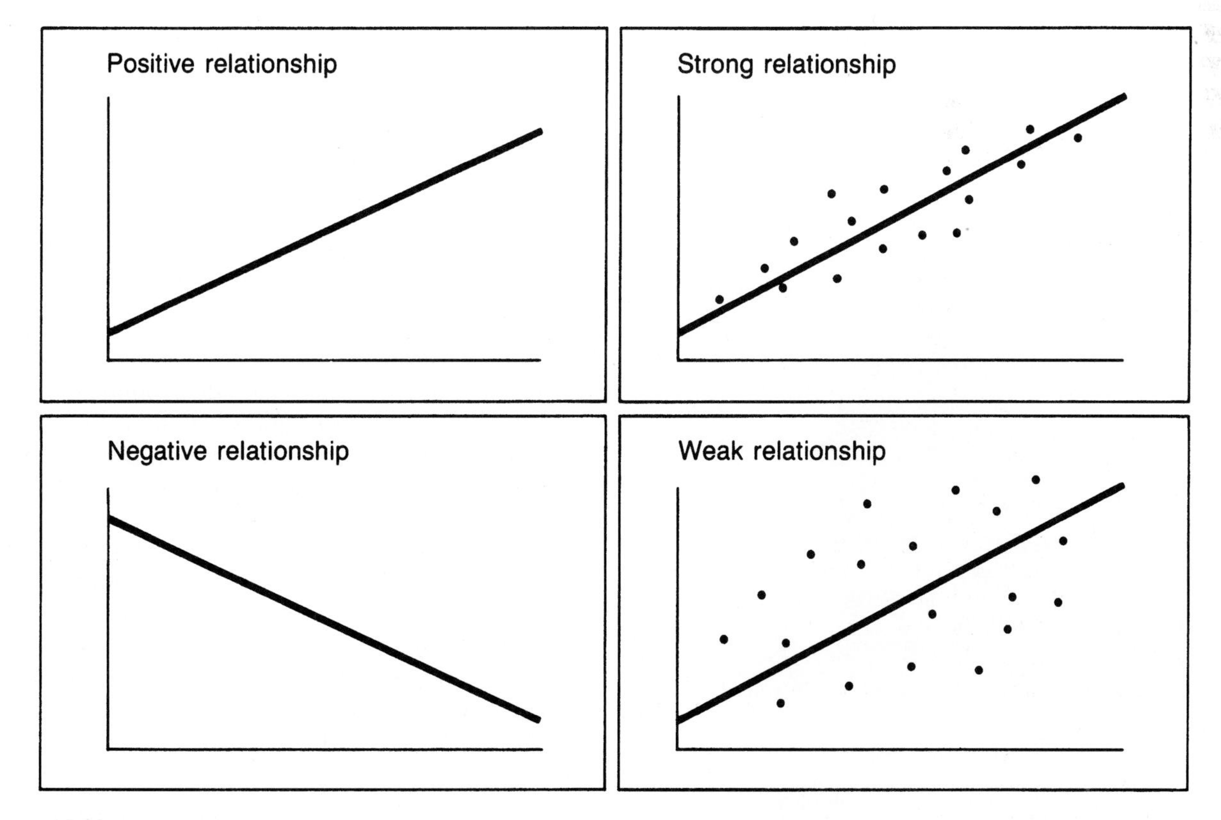

Figure 15.29 The slope and direction of a trend line and the grouping of a cloud of dots summarize the direction and strength of the relationship between variables.

value of the dependent variable decreases as the value of the independent variable increases. The strength of the relationship between the variables is indicated by the grouping of the cloud of dots. If they are closely grouped around the trend line, the relationship between the variables is strong. If the dots are widely scattered, however, the relationship between the variables is weak.

Population Profiles

A **population profile** (sometimes called a **population pyramid**) is a graphic display of the age-sex distribution of a given region's population. Population profiles are commonly placed on maps to facilitate the comparison of the population characteristics of different countries or regions.

The basic structure of a population profile consists of a horizontal baseline divided by a vertical centerline. Horizontal bars project from either side of the centerline, with the length of each bar representing the population in a specific age group. The bars on one side indicate the female segment of the population, and those on the other side represent males. The length of the bars may be based on either absolute numbers or on percentages, and the two types are interpreted differently.

When absolute numbers are used, the length of each bar is proportional to the population of that age-sex group (Figure 15.30). The result is that the overall size of each profile varies in proportion to the size of the population represented. If there is a great disparity in the population of the regions involved, the resulting variation in the size of the profiles makes it difficult to compare the population structures. When percentage values are used, on the other hand, the length of each bar is proportional to the percentage of the population represented (Figure 15.31). The overall size of each profile, however, is the same, regardless of the total population involved. Thus, a bar that represents 10% of a population of 1000 is the same length as the bar that represents 10% of a population of 10,000. This means that the proportion of the population in each age-sex group is emphasized, and the population structures of regions with different population sizes are more readily compared.

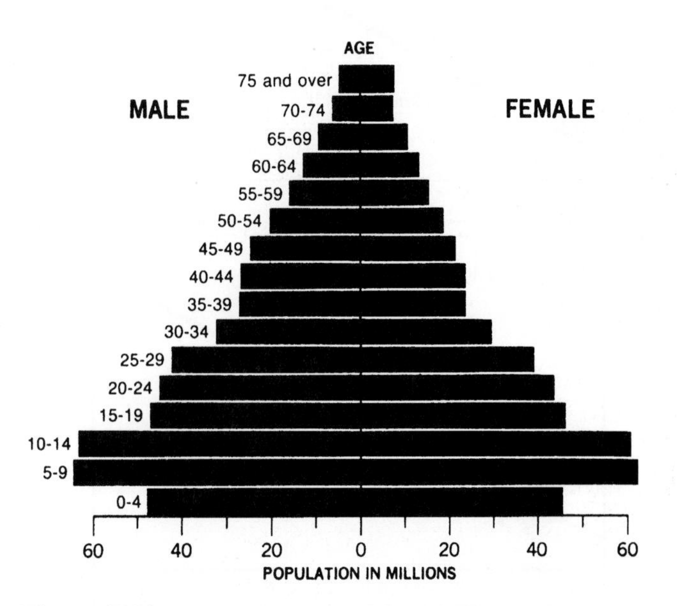

Figure 15.30 Population profile, using an absolute scale, for China, 1978.

Source: H. Yuan Tien, "China: Demographic Billionaire," *Population Bulletin* 38, no. 2 (1983): 18 (Washington, D.C.: Population Reference Bureau).

A population profile is an important tool for understanding the social and economic position of a region. A profile that is broad at its base and tapers rapidly to a peak, for example, is typical of the less-developed countries and regions. The high birth rates of such regions produce the profile's broad base, and the high death rates produce its rapid taper (Figure 15.32a). Regions with this type of population structure have a heavy social and economic burden because of their large, unproductive, young population. In contrast, the population profiles of economically advanced countries and regions display a relatively narrow base because of their lower birth rates. In addition, their relatively low death rates are reflected in the slower taper of the profile (Figures 15.32b and 15.32c). In these regions, the economic and social problems tend to focus more on the needs of the large older population.

Marked variations in the shape of a profile at certain age levels result from population changes imposed by shifts in birth rates or death rates, reflecting changing socioeconomic conditions or war or civil disturbances (see, for example, West Germany in Figure 15.32c).

Circular Graphs

Information about events that occur during a specific time period, such as the number of accidents per day, is often displayed on a **circular graph.** Averaged information about events of a recurring nature, such as weather and climate data, is also conveniently presented on circular graphs. Two examples of circular graphs used for the presentation of weather and climate information are described in this section because they are frequently plotted on maps to show the variations in conditions from place to place.[6] Other types of circular graphs have similar characteristics.

[6]Regular weather and climate maps are described in Chapter 14.

Climatograph

A **climatograph** is a circular graph that shows climatic information such as temperature and rainfall. The data, which often consist of averages taken over a long time period, are plotted along axes radiating outward from a central origin. Mean monthly temperatures, for example, are plotted from the center of the circle, which represents the minimum expected temperature (Figure 15.33). The outer circle then represents the maximum expected temperature. The plotted values are joined to form closed curves, and regions within the temperature circle are often shaded or colored to emphasize seasonal regimes. A climatograph is particularly useful for the depiction of seasonal changes because the graph is continuous throughout the year.

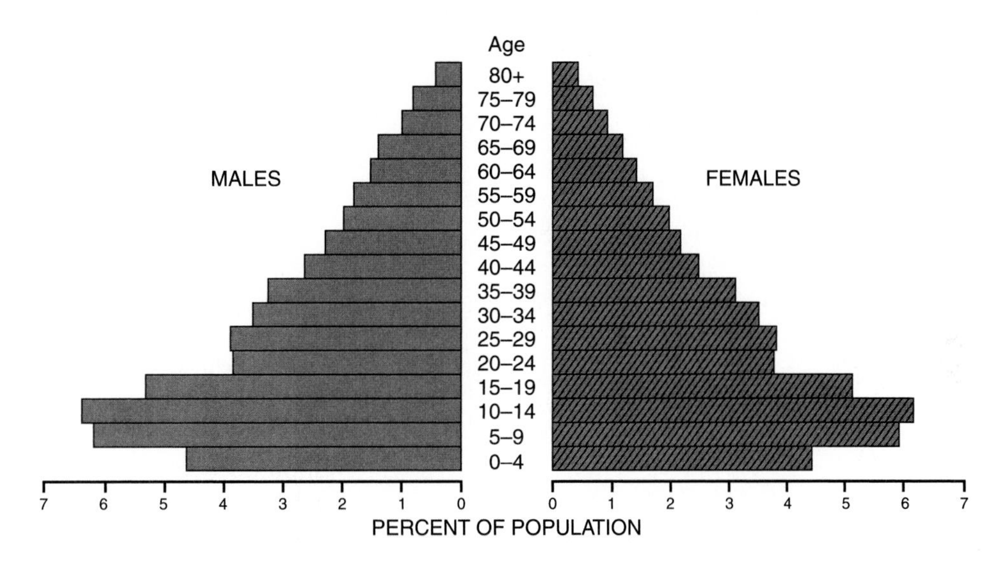

Figure 15.31 Population profile, using a percentage scale, for Cuba, 1979.

Source: Sergio Diaz-Briquets and Lisandro Perez, "Cuba: The Demography of Revolution," *Population Bulletin* 36, no. 1 (1981): 23 (Washington, D.C.: Population Reference Bureau).

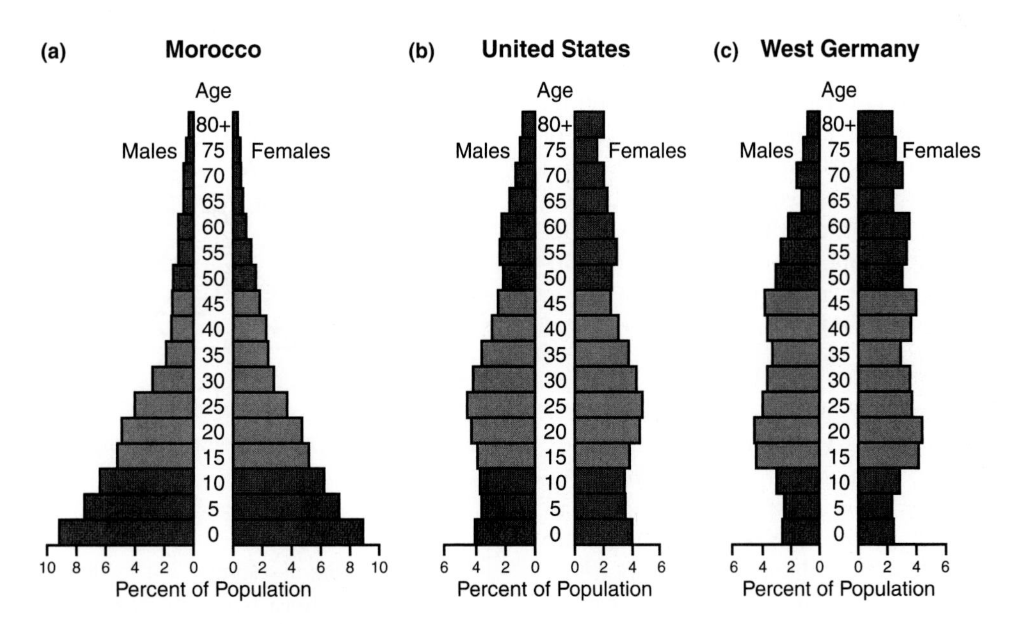

Figure 15.32 Comparative population profiles. (*a*) Morocco, a lesser-developed country with high fertility and death rates. (*b*) United States, an economically advanced country with relatively low fertility and death rates. (*c*) West Germany, an economically advanced country with the world's lowest fertility rate and a low death rate.

Source: From Carl Haub, "Understanding Population Projections," *Population Bulletin* 42, no. 4 (December 1987) (Washington, D.C.: Population Reference Bureau). Reprinted by permission.

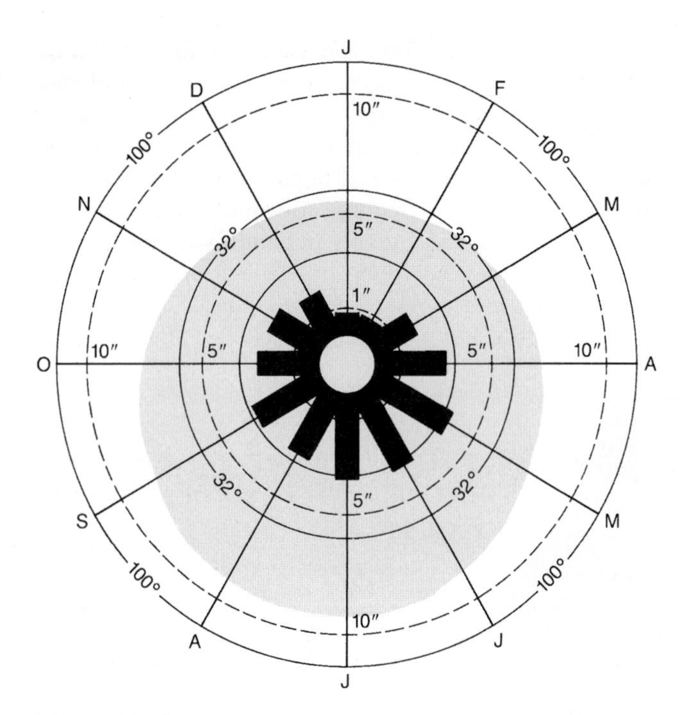

Figure 15.33 Circular climatograph for a hypothetical station. Bars represent total monthly rainfall. The edge of the shaded area represents average daily temperatures.

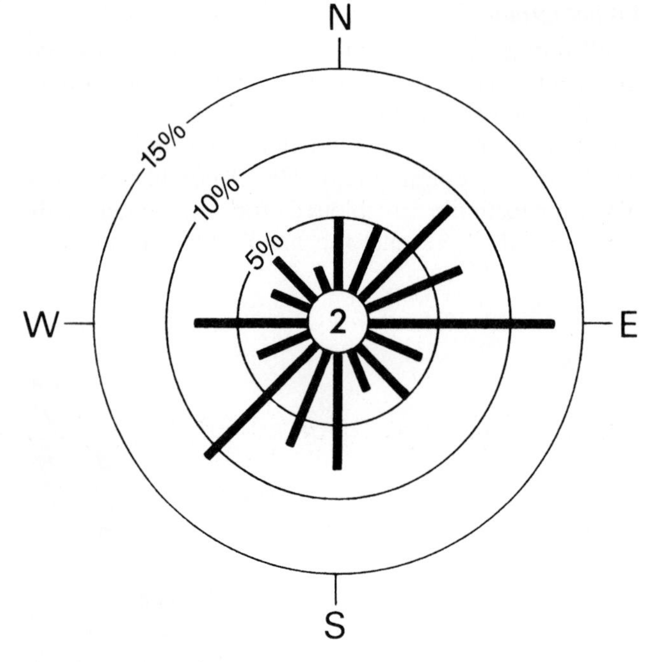

Figure 15.34 Wind rose for a hypothetical station. Lengths of bars denote the percentage of wind from the indicated direction. Figure in the center circle denotes the percentage of calm periods.

Wind Rose

A **wind rose** is a circular graph used to show the average frequency and direction of wind at one location over a given period of time. The top of the wind rose is oriented to the north, and bars are extended outward from the center (Figure 15.34). Many symbol systems are used on wind roses, but it is typical for the length of each bar to be proportional to the frequency of winds from the direction in which it is pointing. Barbs added to the bars, or variations in the widths of the bars, are often used to indicate average wind speed in each direction. Some indication of the frequency of calms is also commonly incorporated. Because of the variation in symbols and in the specific information shown, the key that accompanies any given wind rose must be examined to interpret it correctly.

Triangular Graphs

Triangular graphs, plotted on equilateral triangles, are used to show three components of some total (Figure 15.35). Each location on the graph indicates a particular mixture of the three components that make up the total. Topics that involve three main components, such as the composition of soils in different locations (silt, clay, or sand) or the composition of the work force in different countries (primary, secondary, or tertiary employment), are ideally suited for presentation on a triangular graph.

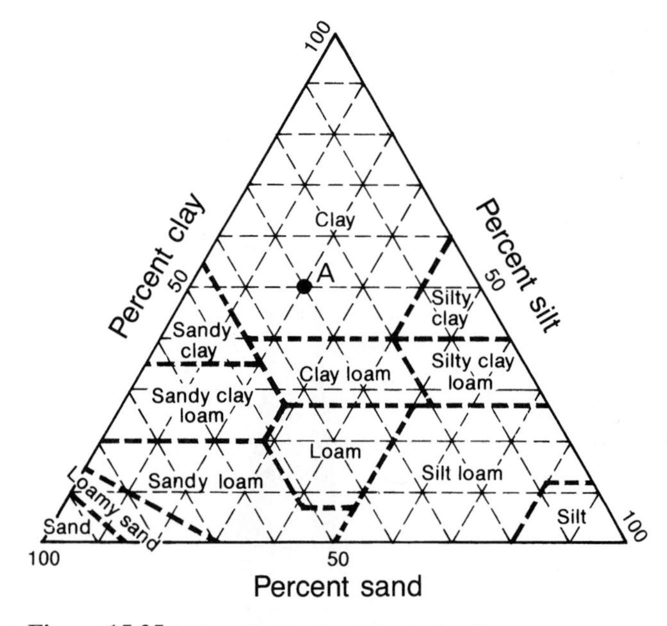

Figure 15.35 Triangular graph of standard soil types.

The values associated with a location on a triangular graph are read from the scales located along each side of the triangle. Point *A* in Figure 15.35, for example, has a value of 50% on the clay scale, 20% on the silt scale, and 30% on the sand scale. It is only necessary to read two values from the graph, because the third value always brings the total to 100%. Reading all three values and determining their sum, however, ensures accuracy.

Triangular graphs are often used to determine whether groupings of observations with similar structures exist. In Figure 15.35, for example, the soil groupings within the area identified as sand are similar because they have small amounts of silt (0 to 15%) and clay (0 to 10%) and large amounts of sand (85 to 100%). Silt soils, on the other hand, have large amounts of silt (80 to 100%) and small amounts of clay (0 to 12%) and sand (0 to 20%). Groupings in other portions of the graph have similarly distinctive characteristics.

The same information shown in Figure 15.35 could be displayed in other ways, such as in pie charts or bar graphs. The triangular graph, however, has distinct advantages. First, the other forms require a separate graph for each observation, which is much more unwieldy. Also, groupings of similar observations are easier to see on a triangular graph than on other types of graphs.

SUMMARY

Graphs are often associated with maps, either as map symbols or as supporting illustrations.

A line graph consists of a line, or lines, showing variations in the value of one or more variables (independent variables on the x axis and dependent on the y axis). Arithmetic or logarithmic scales are used. On arithmetic scales, equal distances indicate equal absolute (or percentage) changes in value. On logarithmic scales, equal distances represent equal percentage changes in values.

Semilog graphs use a logarithmic scale on one axis and an arithmetic scale on the other. On these graphs, the shape and slope of the lines reveal trends in the changes of the values. Sloping straight lines, for example, represent constant rates of change—the steeper the slope, the more rapid the rate of change. Curved lines, on the other hand, indicate increasing or decreasing rates of change.

Graphs that use logarithmic scales on both axes are called *log-log graphs*. The interpretation of these graphs is more complex, but a straight line sloping upward at a 45-degree angle, for example, indicates that the percentage rate of increase in both variables is the same, whereas a curved line indicates that one of the rates of change is geometric and the other is arithmetic.

There are several specialized types of line graphs. A bilateral graph, for example, portrays the values of two related variables. The relationship between the lines that represent these variables indicates either positive or negative net values. A silhouette graph is simply a standard line graph, with the area below the curve shaded to give a silhouette effect. Accumulative line graphs (Lorenz curves) are used to determine the degree of concentration in one variable, as compared to another. Surface graphs differ from other line graphs in that the values are expressed as bands, between the lines on the graph. Measurements are taken from one edge to the other of each band, not from the baseline. Thus, the top line in the graph reflects the accumulation of all of the values below it.

Bar charts consist of bands or bars, each representing a particular data range. On an absolute-value chart, the length of each bar, and of each segment of each bar, is proportional to the value it represents. Bar charts are also drawn with percentage scales. In one common type of percentage-value chart, all of the bars are the same length, which represents 100%. The bars are subdivided in proportion to the percentage of the total value represented by each subcategory.

Loss-and-gain bar charts have a central zero line. The lengths of the bars, measured from the zero line, indicate negative values in one direction and positive values in the other. On high-low bar charts, in contrast, each bar shows high and low values for one time period and may include an average value.

A pie chart is a circle that is subdivided into proportional segments. The circle may or may not be proportional in size to the data value.

When proportional symbols such as graduated circles are used, the area of each symbol is proportional to the value it represents. The sizes may be continuously variable or arranged in a graded series and may be determined on the basis of true proportions or on a perceptual-scaling approach.

Realistic, two-dimensional figures, such as airplanes or soldiers, may be used to represent data values. When the symbols are properly drawn, their areas are proportional to the quantities represented.

Volumetric diagrams are designed to look like three-dimensional figures. In these diagrams, the volume of each figure is proportional to the value it represents. The values represented by volumetric diagrams are difficult to estimate correctly.

A scatter diagram consists of a set of points, each of which represents a paired observation of two variables. The independent variable is measured on the x axis and the dependent variable on the y axis. A trend line is often superimposed on a scatter diagram to represent the relationship between the variables.

A population profile, or pyramid, displays the age-sex distribution of the population of a given region. The length of each bar in the profile is proportional to the population of the age-sex group it represents or to the percentage the age-sex group is of the total population.

Circular graphs can be used to present information related to time periods. A climatograph, for example, is a circular graph that shows temperature or rainfall variations. A wind rose is a circular graph that shows the average frequency and direction of wind over a given period of time.

Finally, triangular graphs show the relationships between three components of some total value.

SUGGESTED READINGS

Holmes, Nigel. *Designer's Guide to Creating Charts and Diagrams.* New York: Watson-Guptill, 1984.

Meyers, Cecil H. *Handbook of Basic Graphs: A Modern Approach.* Belmont, Calif.: Dickenson, 1970.

Rogers, Anna C. *Graphic Charts Handbook.* Washington, D.C.: Public Affairs Press, 1961.

Schmid, Calvin F. *Statistical Graphics: Design Principles and Practices.* New York: John Wiley & Sons, Inc., 1983.

Schmid, Calvin F., and Stanton E. Schmid. *Handbook of Graphic Presentation.* 2d ed. New York: A Ronald Press Publication, 1979.

CHAPTER

16

MAP MISUSE

Maps have credibility. Someone consulting a map expects an objective representation of the region it covers and the topic it presents. This expectation is especially strong today, given the high degree of technical capability involved in modern mapmaking. Nevertheless, maps can be complex, and the public generally lacks knowledge about their characteristics. Furthermore, no map is totally objective. Instead, map content and presentation are selective—sometimes deliberately, sometimes accidentally. The purpose of this chapter is to point out some implications of these factors.

THE POWER OF MAPS

Maps are powerful tools. This power may be hidden but is always present. Historically, it is exemplified by maps that were produced during expeditions of "discovery" by European explorers. Such expeditions were typically sponsored by a royal house, a nation-state, a religious order, a trading company, or some combination of such powers. Therefore, when the explorer came upon territory not previously known in Europe, his likely first act was to claim the land in the name of the sponsor. His likely second act was to draw the territory on a map, showing the newly claimed ownership. This mapping served to consolidate the power of the sponsor over the land. As has been said, "To map the land was to own it and make that ownership legitimate."[1]

Although maps of exploration are an extreme example of the power of maps, maps continue to reflect the interests of their producers. Recent studies of highway maps, for example, show this. Subtly, or not so subtly, highway maps and the illustrations and texts that typically are incorporated with them convey many messages. These messages reflect, for example, economic concerns as diverse as industrial development and tourism promotion, and they extend to cultural aspects such as patriotism, sexism, racism, and political partisanship.[2]

Part of the power of maps comes from their selectivity: Some features are included on a map and some are excluded. To some extent, selectivity stems from the need for generalization. As discussed in Chapter 5, showing every feature in the region being mapped is impossible for any map. If, somehow, "everything" were shown, the map would be totally cluttered and the map user probably could not use it effectively. However,

[1]J. B. Harley, *Maps and the Columbian encounter. An Interpretive Guide to the Travelling Exhibition* (Milwaukee: The Golda Meir Library, University of Wisconsin-Milwaukee, 1990), 99. The idea of discovery and the claims of ownership ignored the rights and interests of any groups already living in the territory. The negative consequences of these conceits of European culture are well-known.

[2]See, for example, Mark H. Bockenhauer, "Culture of the Wisconsin Official State Highway Map." *Cartographic Perspectives,* no. 18 (Spring 1994): 17–27.

decisions about what to show and what to exclude are not value free. It has been pointed out, for example, that USGS topographic maps show many visible aspects of the landscape but exclude many others that are equally visible. For example, gold mines are shown, but garbage dumps are considered unsuitable and are excluded. Thus, maps exert as much power by what they do not show as by what they do show, to the extent that, "if authoritative, official maps and atlases . . . fail to map places, the impression is given that those places do not exist."[3] On the other hand, some maps show places that do not really exist, sometimes in error, sometimes in the service of imagination or mythology, sometimes to mislead deliberately. In sum, no map is neutral in terms of content. The map user, therefore, must keep the biases of the map producer in mind, because the resulting map may be misleading.

However, misleading maps often go undetected, producing misunderstandings and misuse of the mapped data. Some distortions are accidental and stem from poor cartographic techniques. On occasion, however, misrepresentations are deliberately introduced by a propagandist. These may be based on such techniques as size exaggeration, the use of misleading symbols, or the use of colors with strong psychological impact. Or they may involve the selective omission of data, which can present an equally distorted view. Whatever the technique used, the resulting map is given a meaning that supports the aims of the propagandist. Because of the dangers of either accidental garbling or deliberate falsification, this chapter examines some warning signs of map misuse. The complexities of the power of maps, however, cannot be treated fully here. Several of the Suggested Readings provide an introduction to further considerations, especially the writings of Stephen Hall, Brian Harley, and Denis Wood.

MAP INACCURACY

As suggested, things are not always as they appear on maps. An interesting example of the truth of this statement is a dot map published in 1977, showing the distribution of histosols in the United States (Figure 16.1).[4] Histosols are a category of "soils with a surface layer of accumulated organic matter that is usually acid and wet."[5] The general category includes peat but also encompasses many other types with characteristics that differ significantly from those of peat. The original map of histosol distribution was copied for a series of other

[3]P. J. Stickler, "Invisible Towns: A Case Study in the Cartography of South Africa," *GeoJournal* 22, no. 3 (1990): 329.

[4]Philip J. Gersmehl, "The Data, the Reader, and the Innocent Bystander—A Parable for Map Users," *The Professional Geographer* 37, no. 3 (1985): 329–34.

[5]Gersmehl, "The Data, the Reader, and the Innocent Bystander," 331.

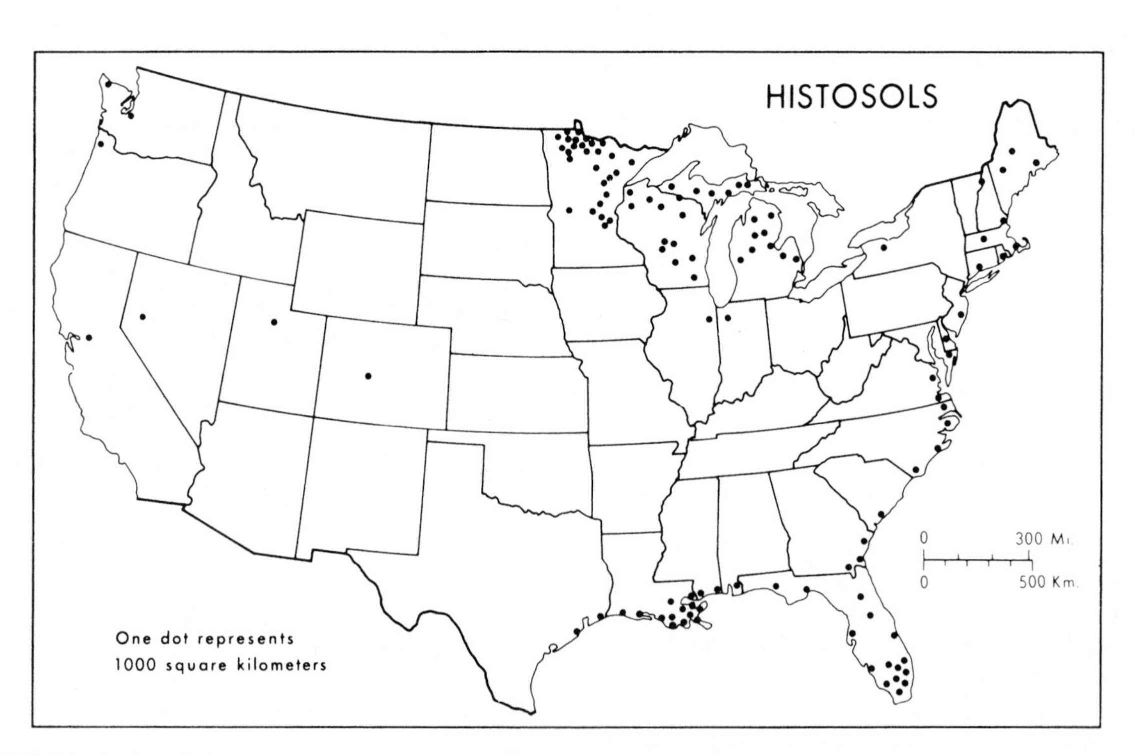

Figure 16.1 Distribution of histosols (soils with a surface layer of accumulated organic matter that is usually acid and wet) in the conterminous United States.

Source: From Philip J. Gersmehl, "Soil Taxonomy and Mapping," *Annals of the Association of American Geographers* 67, no. 3 (September 1977): 419–28. Reprinted by permission of the Association of American Geographers.

maps. At the end of this process, a modified version of the original map eventually appeared in a 1981 planning document as a shaded-area map with the title "Principal Peatlands of the United States" (Figure 16.2). Unfortunately, many misinterpretations, distortions, and errors had been introduced into the final map. Examples of these problems included the loss of southern New England's peat resources and the creation of an important peatland in the Nevada desert.

Cartographers can certainly learn some revealing lessons about their responsibilities in map compilation from this example. For a map user, the message is also clear, although what to do about it is less obvious. Basically, map users need to be aware that maps are *not,* unfortunately, always as accurate as they should be, even when produced by well-intentioned cartographers.

Frankly, it is difficult to know what advice to give a map user in this regard. The most useful approach may be simply to not depend entirely on one source when seeking information. At the very least, the map user should consider the sources listed as the basis of each map and, if possible, compare the copied map with them. Any major departures of the copy from the source should be detected by this kind of check. If no sources are given, there is double reason to be wary, because there is no direct way to check the map's accuracy. In such a situation, the map user should seek additional information.

PROPAGANDA

Maps are excellent vehicles for **propaganda,** because the public puts great faith in map information. They are, therefore, often used to convey impressions that are helpful to a given cause, regardless of the truth of the information shown. A number of devices are common to propagandistic presentations.[6]

General Characteristics

The major general characteristic of propaganda maps is the use of simple design, omitting background and other detail that is extraneous, from the standpoint of the propagandist. This often includes scale, latitude and longitude lines, roads, and other items that would provide an accurate frame of reference. Map projections, which are carefully selected by the propagandist for their effect on the presentation of the data, are usually unnamed. Projections that are not equal-area or conformal are particularly common. The exaggeration of the polar areas on the Mercator projection may be employed, for example, to show the threat (or dominance, depending upon the point of view to be expressed) of countries located in the higher latitudes. In

[6]Louis O. Quam, "The Use of Maps in Propaganda." *Journal of Geography* 42 (January 1943): 21–32; Judith A. Tyner, "Persuasive Cartography," *Journal of Geography* 81, no. 4 (July/August 1982): 14–44.

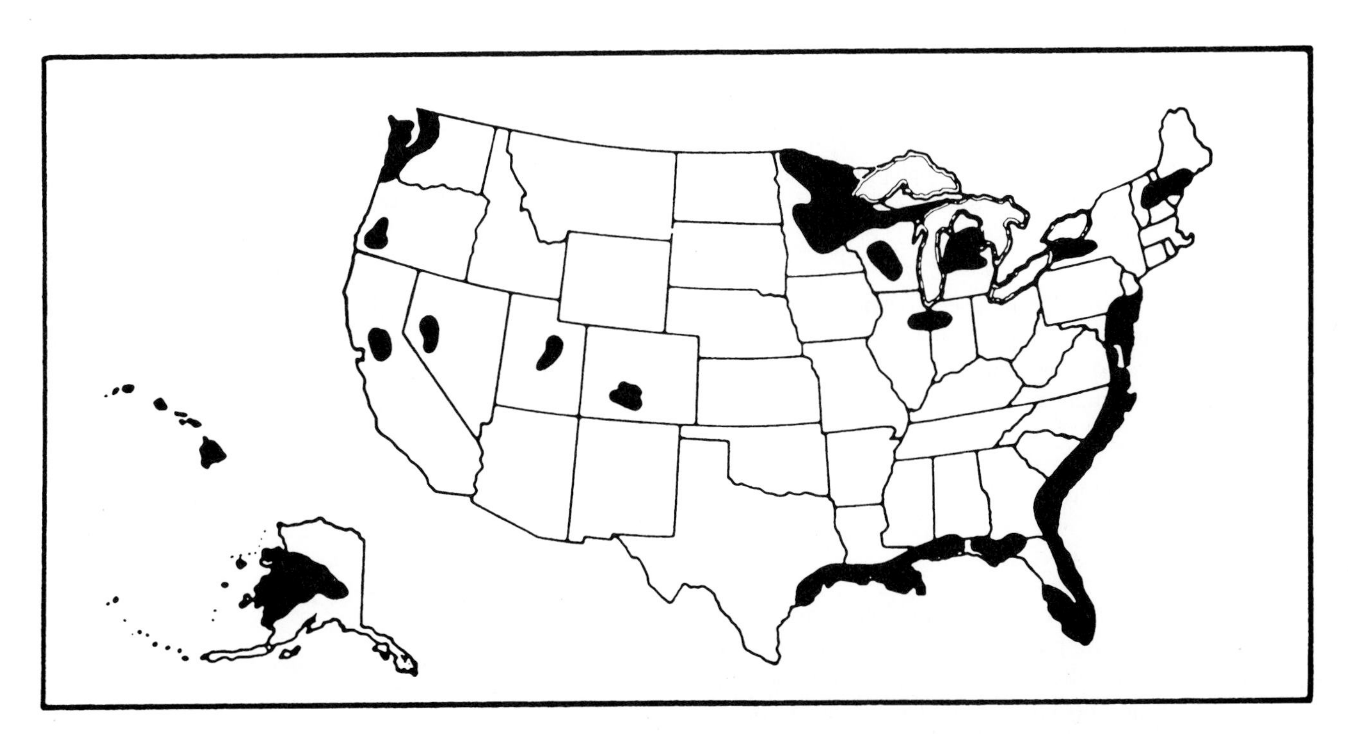

Figure 16.2 "Principal Peatlands of the United States."
Source: U.S. Department of Energy, Division of Fossil Fuels Processing, *Peat Prospectus* (Washington, D.C.: U.S. Department of Energy, 1979).

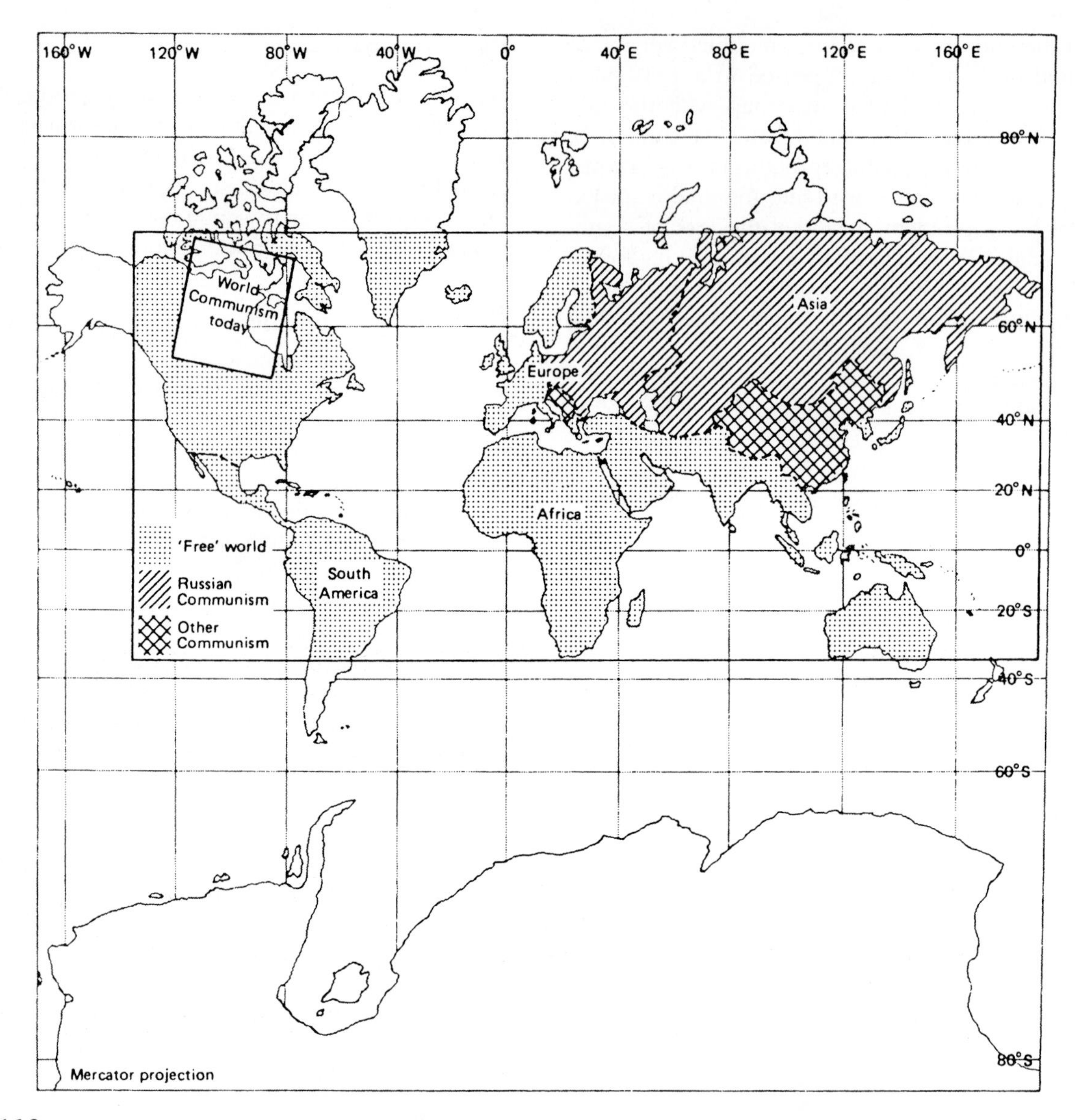

Figure 16.3 Mercator projection used for propaganda purposes.

Source: From Roger Prestwich, "Maps and the Perception of Space," in *An Invitation to Geography,* eds. David A. Lanegran and Risa Palm (New York: McGraw-Hill, 1973), 35–60. Reprinted by permission.

Figure 16.3, the projection is used to show the threat of "World Communism," as it was perceived at the time. The position of the map neatline excludes non-Communist areas, such as Antarctica, Greenland, and Alaska, and much of Canada is covered by the label. On the other hand, virtually all of the exaggerated area of the former Soviet Union is included.

Symbolization

Propaganda maps often use strong colors and bold symbols that make the map appear to represent a reality that does not actually exist. Only those few viewers who take the time and effort to check the accuracy of the data discover whether the information presented is correct. In one atlas, the maps were drawn using this approach. In this case, however, the authors included, in an explanatory section at the end of the atlas, such statements as, "Some of the allegations recorded may be mere figments of propaganda," and "The information falls far short of being complete."[7] The producers of propaganda maps are not likely to provide such candid statements to explain their distorted maps.

Action symbols are used extensively on propaganda maps (Figure 16.4). Arrows, for example, exemplify speed, emphasizing the rapid movements of modern war, with its use of tanks, motorized infantry, and aircraft. Pincers, concentric rings, or pictorial symbols

[7]Michael Kidron and Dan Smith, *The War Atlas: Armed Conflict—Armed Peace* (New York: Simon and Schuster, Pluto Press, 1983).

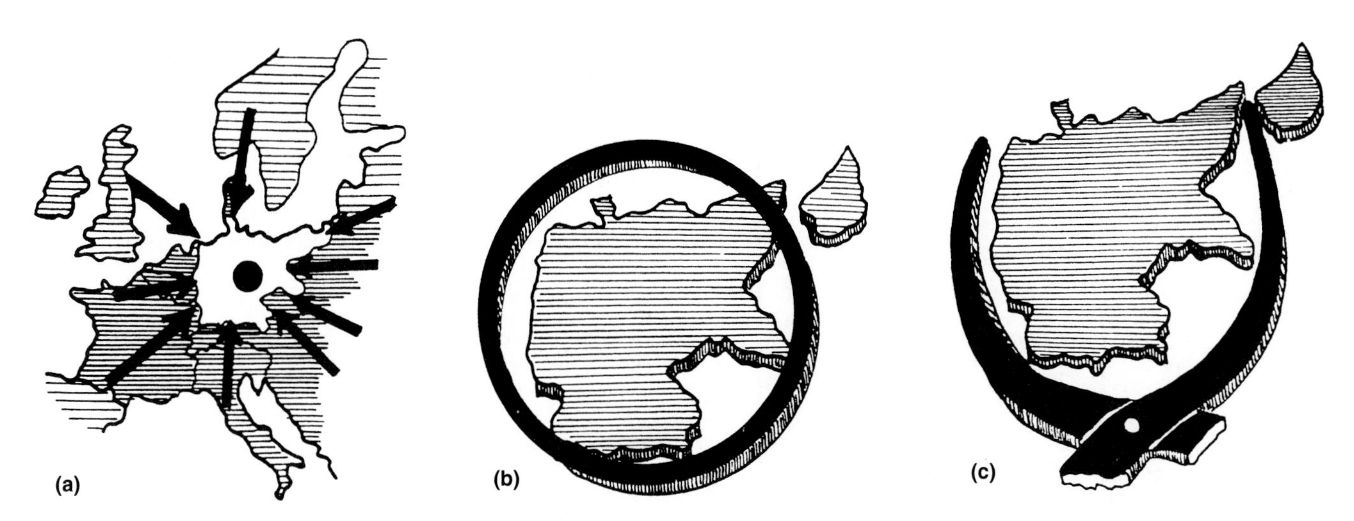

Figure 16.4 Typical propaganda map symbols. (*a*) Arrows, representing pressure on Germany from all sides. (*b*) Circle, showing the encirclement of Germany before and after the First World War. (*c*) Pincers, signifying pressure against Germany from France and Poland.
Source: Karl Springenschmid, *Die Staaten als Lebewesen: Geopolitisches Skizzenbuch* (Leipzig: Verlag Ernst Wunderlich, 1934).

of chains, shown around areas threatened by encirclement, imply the certainty of defeat and the uselessness of resistance.

In addition, symbols are often designed to confuse the relative importance of topics. Oversized symbols, for example, can make it appear that an area is dominated by a certain ethnic group. In one case, thirteen large symbols were imposed on a German map of Latvia, giving the impression that the country was dominated by a German population (Figure 16.5). In reality, Germans comprised only 3.7% of the total population[8]; other nationalities were simply omitted from the map. A similar approach can suggest that weapons cover every square foot of land, thus emphasizing (or, perhaps, exaggerating) the threat that the weapons represent.

When graphs are drawn, either separately or as map symbols, a common type of distortion is to use inconsistent scales to represent statistical data. The result is that one cannot obtain a correct visual impression of the relative numbers involved.

Figure 16.5 Oversized symbols in this map make it incorrectly show that Latvia is dominated by a German population.
Source: *Facts in Review* 1, no. 16 (30 November 1939): 3.

Use of Imaginary Events

Frequently, propaganda maps are simply the products of creative imaginations. An example of the "what might be" approach is provided by the "secret map" that showed purported plans for the reapportionment of South America into five republics (Figure 16.6). The aim of this map was to elicit support for the Nazi regime from the major South American countries that would, presumably, benefit from the addition of territory previously under other jurisdictions. There was a twist to this particular case because British intelligence apparently modified a copy of the Nazi map to show a more drastic change in national boundaries than the Nazis had actually proposed. U.S. President Franklin Delano Roosevelt then used *this* version "to generate support for the de facto war being waged in

[8]John Ager, "Maps and Propaganda," *Bulletin of the Society of University Cartographers* 11, no. 1 (1977): 1–15, quoting Hans Spier, "Magic Geography," *Social Researcher* 8 (1941): 310–30.

Figure 16.6 "Secret map" showing the purported plan for the reapportionment of South America into five republics.
Source: Courtesy of Franklin D. Roosevelt Library.

the Atlantic by the United States Navy against German submarines."[9]

Propaganda maps are also frequently used to support national territorial claims. The appearance of such claims presumably adds to their validity or, at least, reflects a certain element of national pride. Even tiny postage-stamp maps have served this purpose. For example, Argentina advanced its claim to sovereignty over the Falkland Islands (Islas Malvinas) through the publication of a postage stamp with a map showing the islands as Argentine territory.[10]

Misrepresentation

In 1904, Professor Halford J. Mackinder published a scholarly paper, "The Geographical Pivot of History."[11] The paper included a map that showed a pivot area, or "heartland," in a protected position in north-central Asia (Figure 16.7). Mackinder presented the concept that this heartland had been the source of dominant forces in world history and, furthermore, would continue its importance. For our present purposes, the most interesting aspect of this map was its manner of construction, which was designed to support Mackinder's concept.

[9]John F. Bratzel and Leslie B. Rout Jr., "FDR and the 'Secret Map,'" *The Wilson Quarterly* 9, no. 1 (New Year's 1985): 167–73.
[10]Bruce Davis, "Maps on Postage Stamps as Propaganda," *The Cartographic Journal* 22, no. 2 (December 1985): 125–30.
[11]Halford J. Mackinder, "The Geographical Pivot of History," *Geographical Journal* 23, no. 4 (April 1904): 421–37.

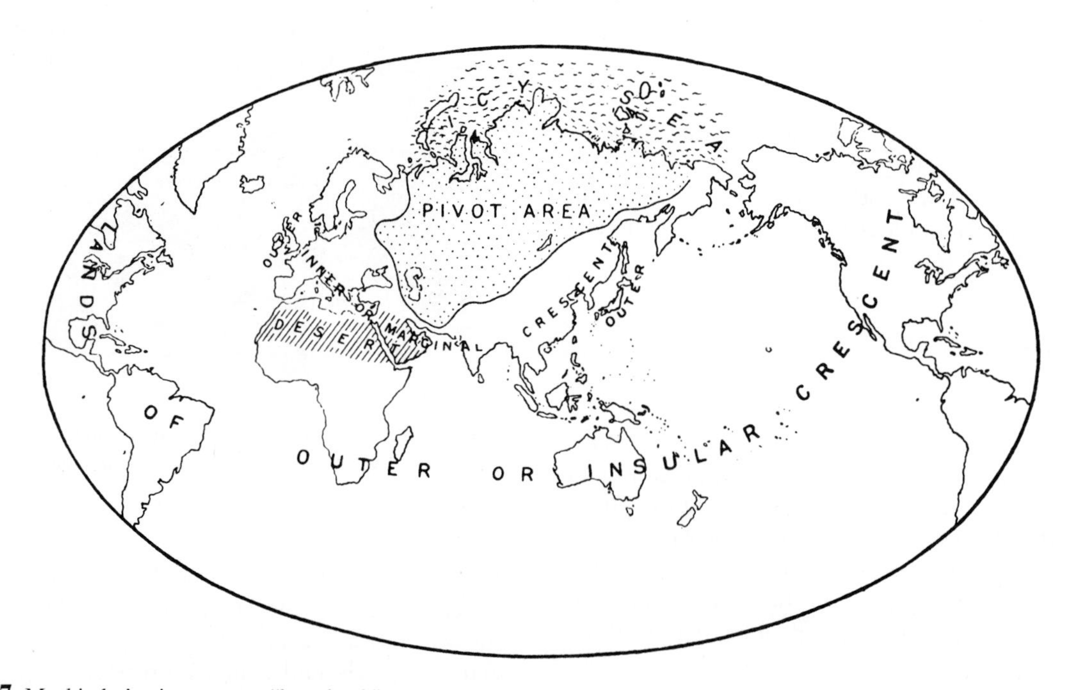

Figure 16.7 Mackinder's pivot area or "heartland."
Source: From Halford J. Mackinder, "The Geographical Pivot of History," *Geographical Journal* 23, no. 4 (April 1904): 435. © Royal Geographical Society. Reprinted by permission.

The base of the heartland map was a Mercator projection, which exaggerated the size and importance of northern Asia. In addition, the map was enclosed within an unusual elliptical border, which bears no regular relationship to the projection itself. Because of the placement of this border, portions of North and South America appeared *twice* on the map, whereas the polar regions were not shown at all.[12] Mackinder elaborated on the relationships based on this unusual map design. In addition to identifying the heartland, for example, he identified an "outer crescent," consisting of North and South America, Australia, and southern Africa. This crescent was illusory because, on the globe, the region is anything but crescent-shaped and, furthermore, does not seem to be related to the heartland in at all the manner of Mackinder's interpretation.

Mackinder's map, and his heartland concept, attracted considerable attention, despite what, in retrospect, seem to be rather strained interpretations. The map, in particular, "was reproduced at least four times in the Nazi literature . . . , with perversions of the author's original intent which were designed to serve malevolent purposes in propaganda."[13]

Other distortions of concepts are also utilized on propaganda maps, but the previous examples illustrate the major techniques. The overall result is often very similar to the style of the journalistic maps discussed in Chapter 14.

Map users need to be wary of accepting every map as correct. It was recently acknowledged, for example, "that, for the last fifty years, the Soviet Union had falsified virtually all public maps of the country, misplacing rivers and streets, distorting boundaries, and omitting geographical features. . . ."[14] Map users must always be careful to evaluate the maps they use, either on the basis of information provided by the producer or, preferably, by checking independent information sources. Indeed, map users must decide whether each particular presentation is a valid picture of reality or simply a propaganda trick.

The Peters Projection

It may be somewhat harsh to cite the **Peters** projection as an example of a propaganda map. The atmosphere in which the use of the projection has been advocated, however, does fit the definition rather well.

The Peters projection is promoted as a substitute for the Mercator projection, for purposes of thematic mapping. The Mercator projection has, over the years, been consistently used as a general-purpose map. This practice has continued, despite the fact that, because of its scale exaggeration, it is *not* suitable for use as a general reference map or for presentation of most thematic information (see Chapter 3). With the scale exaggeration of the Mercator in mind, German historian Arno Peters promoted the substitution of a particular equal-area projection for the Mercator. The main advantage claimed for the Peters projection, as compared to the Mercator, is that it represents the continents and countries at their correct relative size. In addition, Peters claimed that the Mercator is based on "a colonialist and racist mentality" and that its continued use fostered those abhorrent views. This statement was, at best, extremely doubtful. The popular use of the Mercator is more likely the simple result of uncritical acceptance based on familiarity and, possibly, the fact that its rectangular shape easily fits a wall-map format.

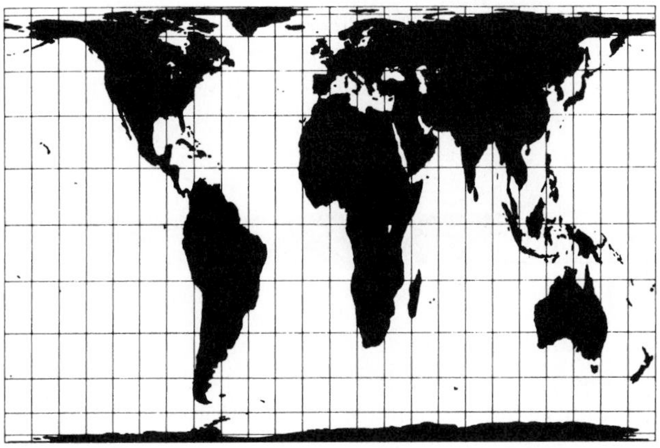

Figure 16.8 Peters (Gall) projection. Areas toward the poles are stretched east-west and compressed north-south, whereas equatorial areas are stretched north-south and compressed east-west.
Source: Courtesy of Cartographic Laboratory, University of Wisconsin-Madison.

In recent years, the shortcomings of the Mercator have resulted in the increasing use of a variety of other projections for wall maps and for illustrations in textbooks and atlases. The truth of the matter is that Peters raised a rather false argument by consistently comparing the Peters to the Mercator, and only the Mercator, as though it were the predominant (or, indeed, the only) alternative projection.

On the Peters projection, areas toward the poles are stretched east-west and compressed north-south, whereas equatorial areas are stretched north-south and compressed east-west (Figure 16.8). Only in the middle latitudes are the general shapes of the landmasses relatively undistorted. This is because the Peters projection is actually a cylindrical equal-area projection with standard parallels at 45°.[15] A large number of

[12]S. W. Boggs, "Cartohypnosis," *The Scientific Monthly* 64, no. 6 (June 1974): 471.
[13]Boggs, "Cartohypnosis," 471.
[14]Which Way to Lenin's Tomb? "*Time* 132, no. 11 (12 September 1988): 49.

[15]The Peters projection is "one of a family of cylindrical equal-area projections initially developed by Johann Lambert in 1772," and is also "indistinguishable" from a cylindrical equal-area projection exhibited by Reverend James Gall in 1855. See Phil Porter and Phil Voxland, "Distortion in Maps: The Peters Projection and Other Devilments," *Focus* (Summer 1986): 27.

other equal-area projections are available, all of which provide the proper size relationships, and many of them have much less-distorted shapes than those on the Peters projection.

Above all else, map users should remember that *no* single projection provides the best answer for all purposes, regardless of what boasts may be generated. This is as true of the Peters projection as of any other.

SUMMARY

Map misuse may be accidental, or it may involve systematic propaganda. The existence of accidental inaccuracy points out that care should be taken when deciding whether a map presentation should be believed, even when the map is not a propagandistic product.

Propaganda maps share some general, recognizable characteristics, such as simple design and lack of a scale, a graticule, or other frame-of-reference elements. They may use inappropriate projections, especially those that exaggerate size or distance. Strong colors and bold symbols are used to imply speed of movement, threat of encirclement, or dominance. Purely imaginary events or features may be mapped as well. Given all of the possible ploys, map users must be wary of propagandistic products.

SUGGESTED READINGS

Ager, John. "Maps and Propaganda." *Bulletin of the Society of University Cartographers* 11, no. 1 (1977): 1–15.

Boggs, S. W. "Cartohypnosis." *The Scientific Monthly* 64, no. 6 (June 1974): 469–76.

Fleming, Douglas K. "Cartographic Strategies for Airline Advertising." *Geographical Review* 74, no. 1 (January 1984): 76–93.

Hall, Stephen S. *Mapping the Next Millennium.* New York: Random House, 1992, concluding chapter.

Harley, J. B. "Deconstructing the Map." *Cartographica* 26, no. 2 (1989): 1–20.

Haushofer, Hans. "Magic Geography." *Social Research* 8 (1941): 310–30.

Loxton, John. "The Peters Phenomenon." *The Cartographic Journal* 22, no. 2 (December 1985): 106–8.

Monmonier, Mark. *How to Lie with Maps.* Chicago: University of Chicago Press, 1991.

———. "Maps, Distortion, and Meaning." *Association of American Geographers Resource Paper 75–4.* Washington, D.C.: Association of American Geographers, 1977.

Pickles, J. "Bibliography on Propaganda Maps." Special Libraries Association, Geography and Map Division, *Bulletin* no. 146 (December 1986): 2–7.

Porter, Phil, and Phil Voxland. "Distortion in Maps: The Peters Projection and Other Devilments." *Focus* (Summer 1986): 22–30.

Quam, Louis O. "The Use of Maps in Propaganda." *The Journal of Geography* 42 (January 1943): 21–32.

Robinson, Arthur H. "Arno Peters and His New Cartography." *The American Cartographer* 12, no. 2 (October 1985): 103–11.

Robinson, Arthur H., Joel L. Morrison, Phillip C. Muehrcke, A. Jon Kimerling, and Stephen C. Guptill. *Elements of Cartography.* 6th ed. New York: John Wiley & Sons, Inc., 1995, 20–36.

Thomas, Louis B. "Maps as Instruments of Propaganda." *Surveying and Mapping* 9, no. 2 (April, May, June 1949): 75–81.

Tufte, Edward R. *The Visual Display of Quantitative Information.* Cheshire, Conn.: Graphics Press, 1983.

Turnbull, D. *Maps Are Territories; Science Is an Atlas: A Portfolio of Exhibits.* Geelong, Victoria, Australia: Deakin University Press, 1989. Chicago: University of Chicago Press, 1993.

Tyner, Judith A. "Persuasive Cartography." *Journal of Geography* 81, no. 4 (July/August 1982): 140–44.

Wood, Denis. "How Maps Work." *Cartographica* 29, nos. 3 & 4 (Autumn/Winter 1992): 66–74.

———. "Designs on Signs/Myth and Meaning in Maps." *Cartographica* 23, no. 3 (1986): 54–103.

———, with John Fels. *The Power of Maps.* New York: The Guilford Press, 1992.

CHAPTER

17

Remote Sensing from Airborne Platforms

The field of cartography is undergoing rapid changes due to the introduction of advanced technology. These changes involve the adoption of new techniques for acquiring and analyzing data, as well as of new methods of producing maps. This chapter and the three following describe how the use of high technology affects map users and map-related data. The emphasis in these chapters, therefore, is more on description and explanation and less on direct, "hands-on" methods than has been the case in other chapters. This chapter and Chapter 18 discuss remote sensing, whereas Chapters 19, 20, and 21 deal with computer-assisted cartography, digital map applications, and geographic information systems.

Methods and Images

Remote sensing involves gathering information by means of a sensor that is not in contact with the object being detected. Remote sensing is not new; for millennia, humans have relied on their eyes, a type of remote sensor, for gathering information about their environment. Remarkable as they are, however, visual images have two specific limitations in terms of mapping applications. First, they reveal only a very small portion of a wide range of potentially useful information. Second, they are not recorded in a form that is easily and accurately measured, stored, or analyzed.

Since 1850, when the first experimental aerial photographs were taken, and especially within the last 50 years, technology has greatly expanded capabilities for detecting and using remotely sensed information for mapping purposes. The discussion here focuses on the characteristics and uses of remotely sensed data about the earth, with emphasis on remotely sensed images.

Electromagnetic Spectrum

The **electromagnetic spectrum** consists of wavelengths of electromagnetic radiation. Everything on earth, living or inert, reflects electromagnetic radiation and, in addition, some objects emit radiation. Each source object has a characteristic radiation pattern associated with it. These patterns of energy reflection and emission are called **signatures.** Signatures are recorded by remote-sensing devices, and the information thus obtained is stored and analyzed to obtain knowledge about the properties of objects and to distinguish between objects of different types. The information is also used to produce images.

The radiation commonly used for remote-sensing applications ranges from relatively short (ultraviolet) to relatively long (microwave) wavelengths, which include the narrow range of wavelengths of the **visible spectrum** (Figure 17.1).

Types of Remote-Sensing Devices

Remote-sensing devices are generally categorized by two sets of characteristics. The first category is based on the range of wavelengths the device detects. The second is based on the source of the signal detected.

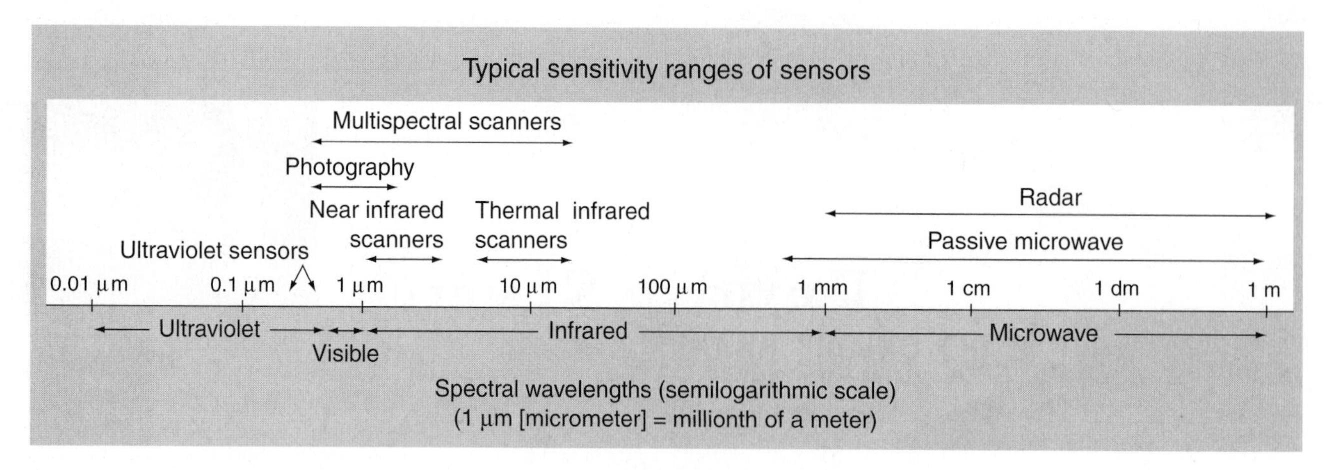

Figure 17.1 The portion of the electromagnetic spectrum commonly used for remote-sensing applications.

Spectral Range

The **spectral range** of wavelengths used by remote-sensing devices may be broad, narrow, or multispectral. **Broad-band devices** use a sensor that receives energy from many wavelengths and integrates the information into a composite signal. A common broad-band sensor is the panchromatic film used in aerial photography. **Panchromatic film** forms an image through the action of a range of wavelengths that generally duplicates those involved in human vision. A **narrow-band device** transmits and receives a limited range of frequencies, as is the case with radar. Finally, a **multispectral device** simultaneously records radiation in two or more separate wavelength bands, as is the case with Landsat satellite images.

Signal Source

The **signal source** of a detected signal determines whether a sensor system is classified as active or passive. An **active remote-sensing system,** such as radar, includes special equipment that transmits signals of the required wavelength. When the transmitted signal strikes the surfaces of objects, a portion of the energy is reflected back to the receiving station, where it is detected and converted into an image. **Passive remote-sensing systems,** in contrast, rely on the detection of naturally generated energy that is either reflected or emitted by an object. Aerial photography is an example of a passive system that uses reflected energy. Some sensors detect the varying amount of electromagnetic energy radiated by all objects that have a temperature above absolute zero. Such systems operate without either visible light or a special source of transmitted energy.

Detector Location

Remote sensors operate in a variety of locations, on, under, or above the earth's surface. The types that are mounted in aircraft or satellite vehicles above the earth's surface are emphasized here because they are most commonly used for mapping applications.

AERIAL PHOTOGRAPHY

In Chapter 2, aerial photography's prominent role in the creation of maps was discussed. **Aerial photographs** are equally important in their function as "real" maps. In this section, we will discuss some of the types, characteristics, and uses of aerial photographic products.

Types of Aerial Photographs

The two major types of aerial photographs are vertical and oblique.

Vertical Photographs

Vertical aerial photographs, as the term implies, are taken with the camera pointed as nearly perpendicular to the earth's surface as possible (Figure 17.2). The use of stereo pairs of vertical photos for mapping purposes was discussed in Chapter 2, but stereo pairs are also of major importance as maps. For this reason, most of the discussion that follows deals with vertical photos.

Oblique Photographs

Oblique aerial photographs are taken with the camera lens deliberately pointed at a nonperpendicular angle to the earth's surface. **High-oblique aerial photographs** include the horizon, whereas **low-oblique aerial photographs** do not (Figure 17.3). Both high and low obliques are used for their visual effect, in advertising or publicity photographs, for example, but they are also useful for feature-recognition purposes. We are accustomed to viewing our surroundings from near ground level and, therefore, usually see the sides, rather than the tops, of most objects. Oblique aerial photos, for this reason, provide a more familiar perspective than do vertical photos.

Photographic Materials

The relationships among the various components involved in the creation of photographic images strongly

Figure 17.2 Vertical aerial photograph. Original scale 1:20,000. Reduced to approximately 1:26,300. (Little Gypsum Valley, Colorado.)
Source: U.S. Geological Survey photo.

influence the final product. Photographic materials, including transparent films and opaque materials like photographic paper, are coated with light-sensitive emulsions. Various emulsions are available, each sensitive to specific electromagnetic wavelengths. The images obtained from the different types differ greatly from one another. Most of this discussion of aerial photography deals with the effects of wavelengths within the visible spectrum on panchromatic film. Some aspects of the use of film that is sensitive to shorter or longer wavelengths are also mentioned, as are color films.

Negatives and Positives

When the film in a camera is exposed to light and later developed, the usual result is a photographic **negative,** made on a transparent medium. In the scene being photographed, objects vary in the amount and wavelength of light that they reflect. Objects that have a high level of energy reflection are recorded on the

(a)

(b)

Figure 17.3 Oblique aerial photographs. (*a*) High oblique of downtown Chicago. (*b*) Low oblique of downtown Sheboygan.
Source: Courtesy of Aero-metric Engineering Inc.

(a)

(b)

Figure 17.4 The tones in a positive (*a*) are the reverse of the tones in a negative (*b*). (Mount St. Helens during eruption, May 18, 1980.)
Source: U.S. Geological Survey photo.

negative film as dark images, whereas those that reflect very little energy appear as transparent areas.

For most mapping purposes, a **positive** print or **transparency** is made from the photographic negative. In a positive print, the areas in the original scene that emit high levels of energy are light, whereas those that emit little energy are dark. In other words, the tones in a positive are the reverse of the tones in a negative (Figure 17.4). The descriptions that follow are based on the characteristics of positive prints made from photographic negatives of different types.

(a)

(b)

Figure 17.5 (*a*) Panchromatic and (*b*) black-and-white infrared photographs of the same area. (Golden Gate Bridge, San Francisco.)
Source: National High Altitude Program photos.

Emulsion Types

Many aerial photographs are made on film coated with a black-and-white **panchromatic emulsion** that is sensitive to approximately the same range of wavelengths as the human eye. Although the images obtained with this emulsion lack color, they are useful for many purposes. In prints made from panchromatic negatives, the visual brightness of objects in the original scene is closely duplicated. The resulting image is familiar and rather easily interpreted. Black-and-white images from panchromatic film, therefore, are in common use. In fact, for a long time, panchromatic images were the predominant type produced. Other types of emulsions are often encountered, however, because their characteristics make them useful for particular purposes.

A second type of black-and-white film emulsion, **infrared emulsion,** is sensitive to the **infrared wavelengths.** These wavelengths, which fall outside the range of human vision, are particularly useful in the analysis of vegetation conditions and the identification of water bodies. Broadleaf vegetation is highly reflective in the infrared range, whereas coniferous vegetation is less so. On infrared print images, then, the two types of vegetation can be differentiated because broad-leaves show up in light tones and evergreens in darker tones. Also, water bodies are usually very dark on infrared photos, which assists in determining the locations of shorelines. This is especially helpful in swampy or marshy areas that have indistinct margins on panchromatic film (Figure 17.5). Infrared film also has some haze-penetrating ability that is useful when conditions are not suitable for regular panchromatic film.

Features that have distinctive colors may appear in similar shades of gray on black-and-white film, which makes them difficult to identify. Because color film is particularly helpful in identifying soil types, rock outcrops, water bodies, and other features that exhibit distinctive coloration, **color emulsion** has supplanted panchromatic black-and-white film emulsion for many purposes (Plate 7). This is economically feasible because, although color film is more costly to purchase and process than black-and-white film, the differences are often negligible when the total cost of an aerial photography project is considered.

Color film is also available with an infrared-sensitive emulsion (Plate 8). The image obtained from **color-infrared emulsion** has colors unlike those we are used to seeing. The colors are also different from those that appear on regular color film. As with black-and-white-infrared, the sensitivity of color infrared beyond the visible spectrum makes it useful for vegetation studies. Healthy deciduous plants show up in red tones on color-infrared images, whereas evergreens range into the bluish-purple colors. The result is that different plant types can be differentiated quite easily. Also, plants that are under stress due to disease or insect infestation appear as bright green on color-infrared film. This allows detection of the stress, which may not even be visible in the environment or show up on other types of film.

Photo Characteristics

Aerial photographs to be used for mapping purposes are taken using a flight plan and equipment that provide for a high-quality product with certain desired characteristics. The **flight plan** is designed so that each photograph in a sequence overlaps the next by approximately 60 to 65% (Figure 17.6) and so that the photos are taken in a zigzag sequence of **legs.** The photos in one leg overlap those in the adjacent legs by approximately 20 to 35%. These conditions assure that the area of interest is completely covered and that the photos may be viewed stereoscopically, as is discussed later in the chapter (see also Chapter 2).

Although it may not be apparent to a casual observer, aerial photographs invariably contain distortions. Some of these distortions are due to technical irregularities in the camera system, but these are likely to be very slight. Conditions encountered at the time the photos are actually taken, however, may result in photographs that are less than perfect. The aircraft may deviate from the desired course, for example, or the timing between photos may not be accurate. Either of these conditions may result in inadequate coverage between photos. Other conditions may cause variations in the elevation and altitude of the airplane at the moment the picture is taken. The aircraft may have to crab into the wind, for example, resulting in photos that do not overlap properly (Figure 17.7). Similarly, the aircraft may tilt or change altitude unexpectedly, causing scale variations.

Most aerial photographs are produced under government or commercial contracts. Because of the high standards required, such photos are not likely to suffer significantly from the distortions just described. For

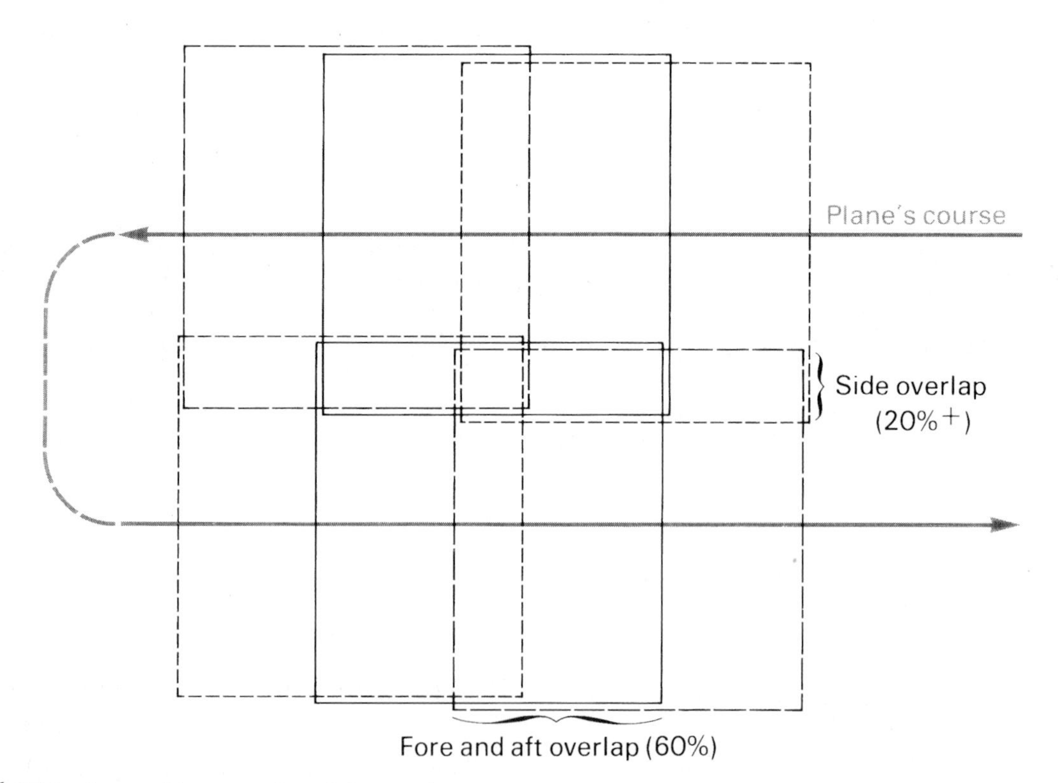

Figure 17.6 Flight plan provides an overlap of photographs.

purposes of this discussion, aerial photographs are assumed to be free from significant technical flaws. They are also assumed to be close to absolutely vertical and to have been taken from the correct elevation (or to have been corrected to remove flaws from either of these sources). In particular, rectified prints have been specially processed to correct such irregularities.

Nevertheless, distortions are introduced into aerial photos through variations in topography and are particularly important when aerial photos are used as maps. An understanding of these distortions requires consideration of those aspects of the geometry of aerial photography that relate to scale determination and variation.

Scale

The scale of an aerial photo depends on two factors: (1) the focal length of the lens used to take the photo and (2) the elevation of the camera above the surface of the earth when the photo is taken. The scale results from the following simple geometric relationships, which are shown in Figure 17.8.

1. The angles a and a' are right angles because the photograph was taken vertically.
2. The film in the camera is parallel with the surface of the earth. The lines in Figure 17.8 that represent these features are, therefore, parallel to one another.
3. Given items 1 and 2, angles b and b' are equivalent angles.
4. Given items 1, 2, and 3, triangles ABL and $A'B'L$ are similar triangles.
5. Because of their location in the similar triangles, the ratio between the lengths of sides AL and $A'L$ is the same as the ratio between the lengths of sides AB and $A'B'$. This is equivalent to saying that the scale of the photo is given by the ratio between the altitude and the focal length of the lens ($AL{:}A'L$), provided that both are in the same measurement units:

$$\text{Scale} = \frac{f\,\text{(focal length of lens)}}{H\,\text{(altitude above ground)}}$$

Once the scale of a photo is known, the distance between two points on the photo (such as A' and B') can be measured. The scale ratio is then applied to this measurement, and the distance between the equivalent points on the ground (A and B) can be calculated. The reverse is also true. This means that the scale of an aerial photo can be determined by taking the ratio of a known distance between two objects on the ground to the distance between the same objects on the photo. The known distances for such a calculation can be determined either by direct measurement in the field or by scaling from a map. They can also be derived from other knowledge, such as knowing that the football field on a photo is 100 yards long or that the gauge of the railway tracks is 4 feet, 8 1/2 inches.

Unfortunately, scale varies throughout an aerial photograph (the reasons for this are discussed next). Because of these scale variations, several known distance relationships are usually measured, and the resulting scales are averaged to produce the photograph's average scale. The scale obtained by use of the focal length and altitude formula is also an average scale.

Planimetric Shift

Why do variations of scale occur in aerial photographs, and what effect do these variations have on the use of the photos as maps? The answers to these questions can be obtained by looking at some simple geometric relationships. The key factor is that an aerial photograph is a **perspective view,** which is a view from a single viewpoint (Figure 17.9). A map, on the other hand, is an **orthographic view,** which is a view

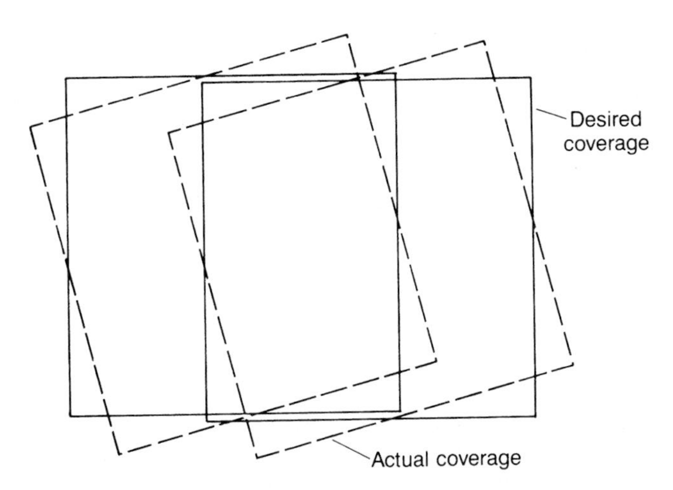

Figure 17.7 Effect of crab on aerial photo coverage (exaggerated).

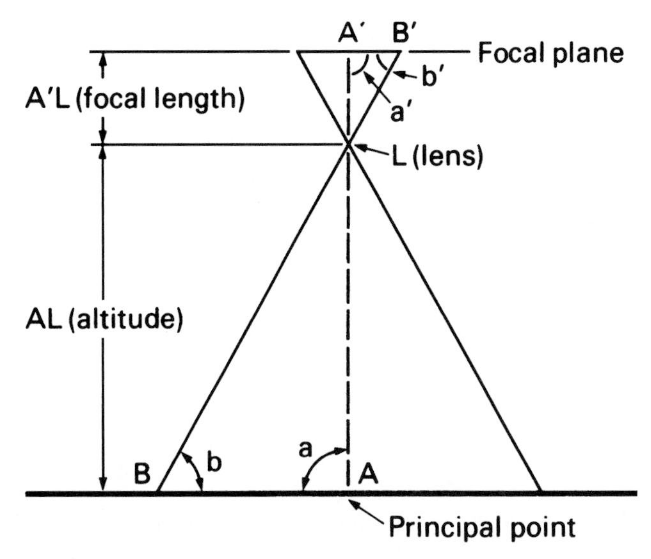

Figure 17.8 Geometric relationships of scale in an aerial photograph.

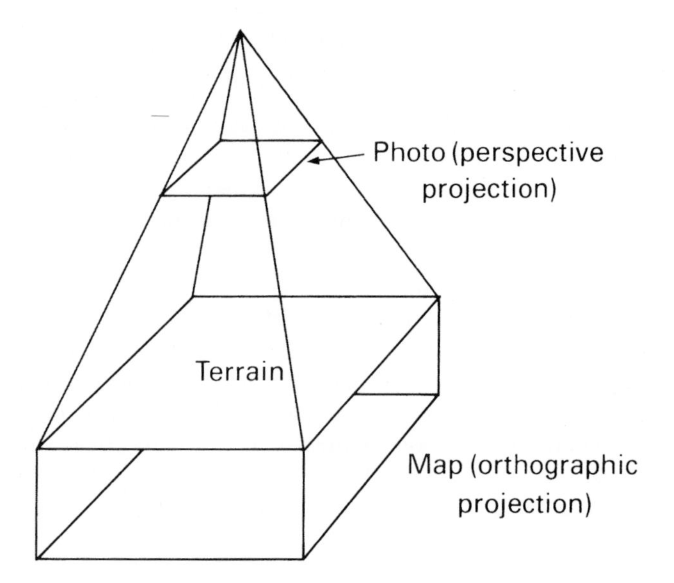

Figure 17.9 Relationship between the terrain, an orthographic map projection, and the perspective projection of an aerial photograph.

drawn as though every point were seen from directly overhead. The perspective nature of photographs would pose few problems if the terrain were always perfectly flat. When it is not, which is most of the time, the perspective view results in the **planimetric shift** of features that are above or below the base level of the photograph.

The amount of planimetric shift of a feature is related to its height above the datum level of the photograph and its distance from the center (called the **principal point**) of the photograph. In Figure 17.10a, the top of the tower (*T*) appears, on the photograph, to be located at T′. This point is shifted away from the location of the bottom of the tower (*B*). On a map, in contrast, both the top and the bottom of the tower would be located at *B*. The farther the feature is from the principal point and the greater the height of the feature above the datum, the greater the amount of shift that occurs (Figure 17.10a–c). On the other hand, the amount of shift from either source decreases as the altitude of the camera above the earth's surface increases (Figure 17.10d). Planimetric shift also increases with increasing distance from the principal point and is dis-

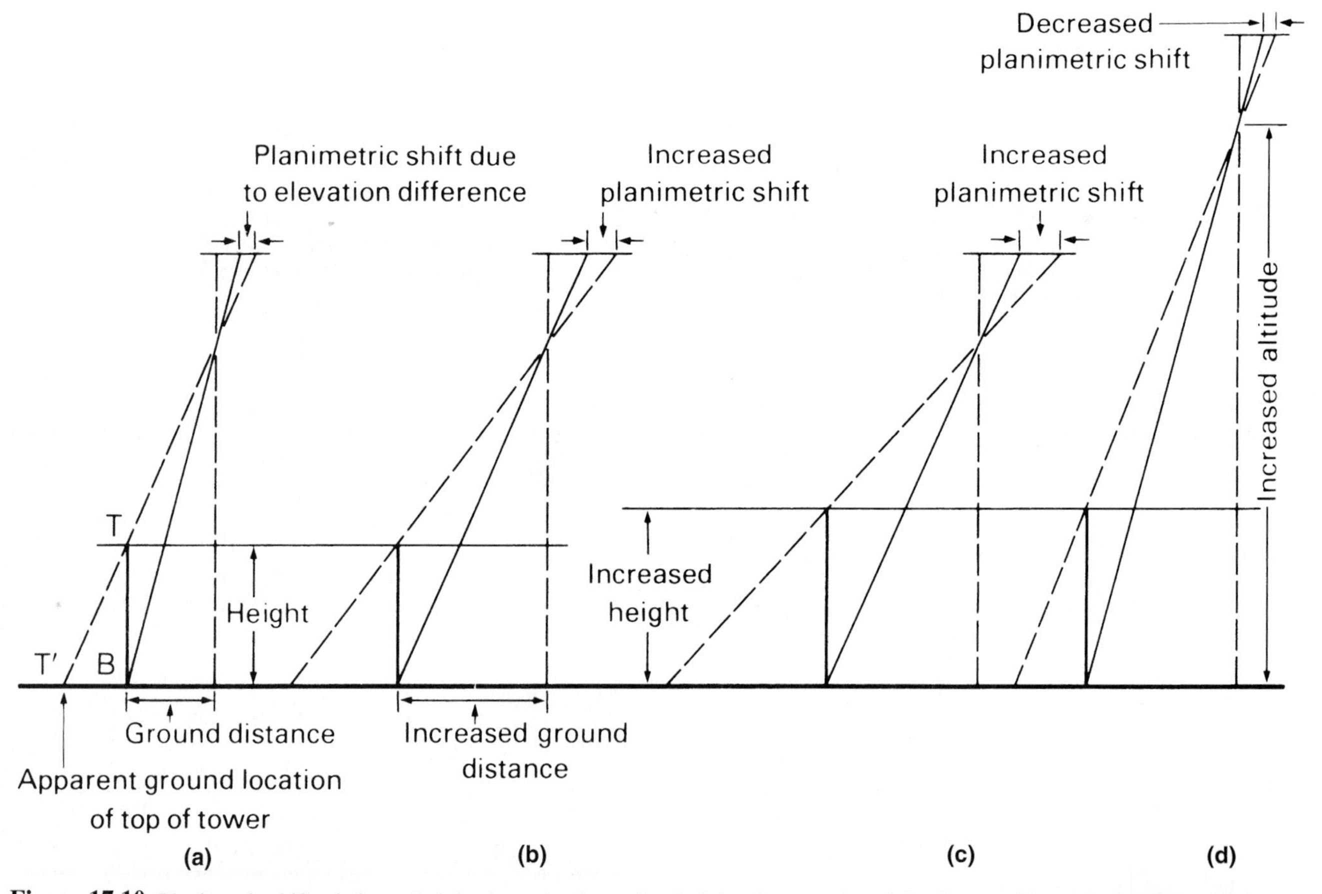

Figure 17.10 Planimetric shift relative to height above the datum level of the photograph and the distance from the principal point. (*a*) Planimetric shift due to elevation difference between the top and bottom of the tower. (*b*) Increased planimetric shift due to increased distance of the tower from the principal point. (*c*) Increased planimetric shift due to the increased height of the tower. (*d*) Decreased planimetric shift due to the increased altitude of the camera.

tributed in a radial pattern around the principal point. Figure 17.11 shows the direction and amount of displacement of the images of several objects of the same height that are located at varying distances and directions from the principal point.

Planimetric shift applies not only to tall towers and buildings that appear in an aerial photograph; it also applies to the earth's surface and the location of any objects on that surface. Thus, on an aerial photograph, trees, road intersections, buildings, rock outcrops, fence lines, and any other features may be shifted some distance from their correct map location.

Planimetric shift results in scale variations within an aerial photograph, because the ratio of the distance between objects on the photograph and the distance between the same objects on the ground varies from place to place. As a result, the representative fraction that expresses the scale of the photo also varies from place to place.

The errors introduced by planimetric shift cannot be corrected without specialized training and equipment. Map users, however, should at least be aware of planimetric shift and of the fact that features on an ordinary aerial photograph do not always line up properly with the same features on a map that is nominally of the same scale. For this reason, specially processed aerial photos, called *orthophotos,* are often useful. **Orthophotos** are aerial photographs that have been processed in a special way to remove the effect of planimetric shift. Figure 17.12 shows the type of locational shift that exists in a normal aerial photograph in comparison to an orthophoto.

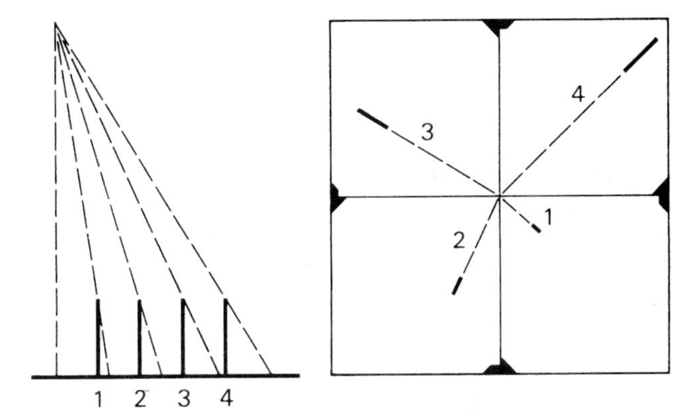

Figure 17.11 Planimetric shift increases with increasing distance from the principal point and is distributed in a radial pattern around the principal point.

(a)

(b)

Figure 17.12 Locational shift in a normal aerial photograph (*a*), in comparison to an orthophoto (*b*).
Source: Courtesy of Aero-metric Engineering Inc.

Aerial Photo Products

Single Prints

The basic product in any aerial photography project is a single print of the specific area photographed (or a set of single prints of a larger area). Most single prints measure 9 by 9 inches, although other sizes are sometimes encountered. As already indicated, single prints are often used as map substitutes, although the problems of scale variation and locational displacement must be dealt with if the photos are not orthophoto products. One thing to keep in mind when using such prints is that what appears to be the top, based on the orientation of numbering and lettering on the print, is not necessarily north.

Photomosaics

When a single aerial photo does not cover all of the area of interest, adjoining photos can be pieced together to cover the larger area with a composite image called a **photomosaic.** The photographs used to construct a photomosaic must overlap one another. Usually, this is not a problem because standard procedures for most aerial photography flown for mapping purposes provide such overlap.

Photomosaics may be either uncontrolled or controlled. The assembly of an **uncontrolled photomosaic** is begun by fastening the print of the center of the area in the middle of a flat board. An adjoining print is then overlapped with the first one, so that features common to both are aligned with one another as closely as possible. The excess portions of the overlapping photo are trimmed away, and it is temporarily fastened in place. The same procedure is followed with additional photos, with each one matched as closely as possible with those previously put in place. Aligning the additional photos usually requires some adjustment in the locations of the photos already positioned. After the necessary adjustments are made, everything is glued in place. The result resembles a large aerial photograph of the area (Figure 17.13).

Assembling a photomosaic in the manner just described quickly results in difficulties due to the scale variations in the photos. Assume, for example, that a road intersection that shows in the first two photos is selected as one matching point. The scale variations within the two photos may make it impossible to line up the images of other features when the road intersections are aligned. An arrangement that comes as close as possible to aligning the desired points may be a necessary compromise. The alignment problems become more and more difficult to resolve as additional photos are matched to one another, progressing farther and farther from the starting point. In the end, therefore, very few points are likely to match precisely.

Orthophotos, if they are available, produce a better end product. Nonetheless, some means of structuring the mosaic's assembly is desirable, even when orthophotos are used.

Structure is provided for a **controlled photomosaic** by introducing a base map, drawn at the average scale of the aerial photos. This need only be a simple,

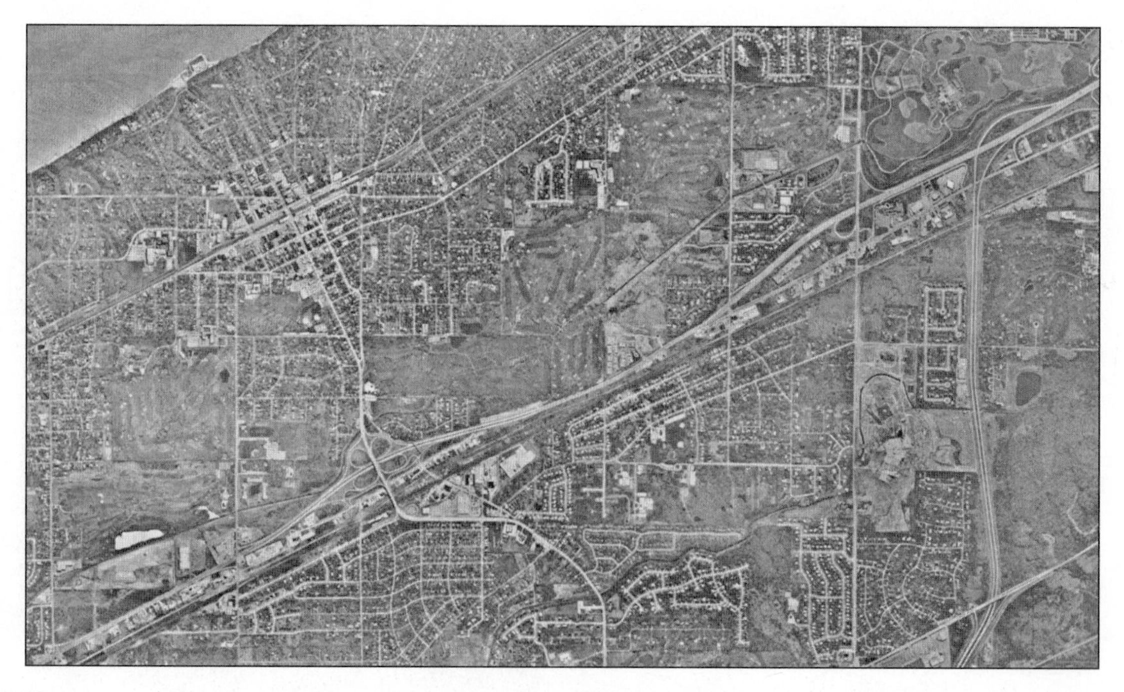

Figure 17.13 Mosaic of aerial photographs, covering a large area in northeastern Illinois.
Source: Courtesy of Aero-metric Engineering Inc.

skeletal map, either an existing one or one drawn for the purpose. The primary requirement is that it show the correct planimetric location of major features that are recognizable on the aerial photographs. These include, for example, road and railroad alignments and intersections, fence lines, property corners, and major buildings.

A controlled mosaic is assembled by securing the base map to the mounting board and fastening the individual photos over it, as with an uncontrolled mosaic. The difference is that the features on the photos are matched to their plotted map locations as closely as possible before they are fastened in place. Individual features may not line up exactly when this is done, but any errors do not accumulate, as they do in an uncontrolled mosaic. As before, overlapping areas are removed so that adjoining photos blend together as well as possible before the photos are glued down.

Completed photomosaics are frequently photographed and reproduced and, with information such as place names and boundary lines added, made into a form of photomap.

Orthophoto Products

Because their locational characteristics are equivalent to those of maps, orthophoto products have been incorporated into the map series published by the U.S. Geological Survey. One such product is an orthophotoquad, and another is an orthophotomap. These products are similar to regular quadrangle maps in that they are produced at a scale of 1:24,000 and cover 7.5 minutes of latitude and longitude. **Orthophotoquads** are orthophoto images, printed in black and white (Figure 17.14). They show photographic detail, but very little cartographic treatment is added, and contours are not shown. **Orthophotomaps,** in contrast, include contour lines and feature names and are color-enhanced to improve readability (Plate 9). Orthophoto products often show environmental information, such as the locations of trees and wetlands. As a result, they may be more effective than maps in some applications.

Photo Interpretation

The image captured in an aerial photograph can be subjected to detailed interpretation and analysis. The field of **photo interpretation** involves the recognition and identification of the great variety of natural and human-made features on the earth's surface. Feature identification is achieved by considering such attributes as the shape, size, tone, color, texture, pattern, relative locations, and shadows visible in the photo. Photo interpretation supports a great variety of applications, including archaeology, waterway charting, crop-yield predictions, fire protection, glacial studies, land-use analyses, mapping, range management, real-estate zoning, transportation studies, and wildlife management, to name only a few examples. In addition, in any field of application, photographs taken at different times can be compared to determine the nature and extent of changes in either the natural or human-made environment.

Effective photo interpretation requires a considerable amount of training and experience. The sections that follow examine some of the basic aspects of the subject to give some idea of what can be accomplished. Some of the sources in the Suggested Readings at the end of the chapter provide information about more advanced applications.

Visual Clues

Usually, in aerial photographs, certain features can be positively identified, even if the interpreter does not have any previous experience or training. Other features that may not have been immediately recognizable

Figure 17.14 Portion of a U.S. Geological Survey orthophotoquad. (Frederic SE quadrangle, Wisconsin, 1974, 1:24,000.)

can be identified as the interpreter gains experience. Various clues in aerial photographs assist this type of interpretation (Figure 17.15). Aerial photos that are of as large a scale as possible are preferable. Large-scale photos, supplemented by the use of a magnifying lens, allow the interpretation of many details that might otherwise be overlooked.

Even in the absence of other clues, some features are immediately recognizable because of their distinctive shape. A baseball park is an obvious example (Figure 17.15a), but other, more subtle features are also often

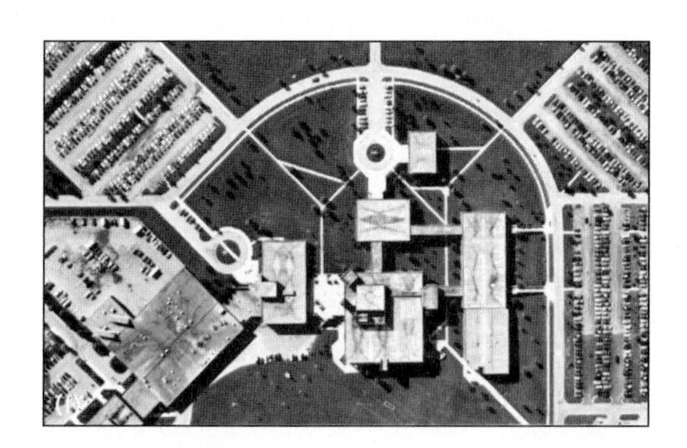

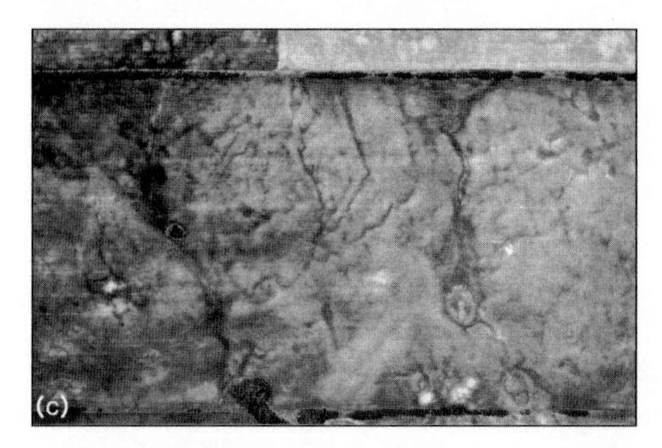

Figure 17.15 Examples of aerial photo interpretation clues. (*a*) Major-league baseball park. (*b*) Farmstead with a house, barn, silo, and outbuildings. (*c*) Agricultural land with moist and dry areas. (*d*) Different crops resulting in a variety of textures. (*e*) Regular pattern of orchard, contrasting with natural woodlots. (*f*) Office complex. (*g*) Distinctive shadow of the Statue of Liberty.

Source: (a), (c), (d), (f), and (g) courtesy of Aero-metric Engineering Inc. (b) and (e) courtesy of Racine County Planning and Development.

revealed by their shapes. Different features may have similar shapes, however, so that an assessment of the size of the feature may be necessary. Size can be described in two ways: relative and absolute. Relative size is the size of one object as compared with another, and it helps to distinguish between what might otherwise appear to be similar features. Two buildings, for example, may have similar rectangular outlines, but the house on a farmstead is readily distinguishable from a barn, simply on the basis of relative size (Figure 17.15b). Absolute size, on the other hand, refers to the actual measurement of features, based on knowledge of the photo's scale. A railroad car, for example, may look like a simple rectangle, but measurement of its size will indicate its probable capacity and may give a clue to its use.

Other, slightly less obvious, clues assist the photo interpreter. One of these is that the features on the earth are of different colors and appear in black-and-white photographs as different tones, or values, of gray. The darker areas in Figure 17.15c are areas of wet soil, for example, whereas the lighter areas are dry. When a color photo is available, of course, differences in color are more immediately obvious and are usually easier to interpret than are differences in tone (see Plate 8). The shade of green associated with evergreen trees is usually different from that of deciduous trees, for example. Another useful identifying characteristic is texture, which is the relative coarseness or smoothness of areas in the image. Fields planted to different crops display a variety of textures (Figure 17.15d). The disparate textural characteristics help the photo interpreter to measure and tabulate the fields planted to each crop or left fallow throughout the area.

The arrangement of features can create patterns that are clues to their identification. The orderly pattern of the orchards in Figure 17.15e differentiates them clearly from the random-appearing pattern of a natural woodlot. Features with regular patterns and straight edges are almost certain to represent human intervention in the environment. Also, the relative location of certain features gives a clue to their identity, even though their shapes or other characteristics may not. An example of this is the large building in Figure 17.15f. The extensive parking lots around this building complex mean that it is more likely to be an office structure than a school, despite the adjoining softball diamond and football field.

As was previously mentioned, vertical views may not reveal the distinctive features of some objects that would be obvious in a ground-level view. In such cases, a shadow cast on the surface provides the equivalent of a side view. The shadow of the Statue of Liberty is certainly distinctive (Figure 17.15g), and the shadow of a smokestack, for example, looks distinctly different from that of a church steeple.

Interpretive Keys

Extensive **interpretive keys** have been developed that list logically organized clues that assist an interpreter in the analysis of particular environmental features. Some keys are useful for understanding agricultural environments, some are designed for use in industrial settings, and so on. The detail provided by keys often allows quite specific interpretations that would be impossible if the interpreter relied strictly on personal knowledge of particular feature types.

Stereoscopic Pairs

The use of stereoscopic pairs of aerial photographs assists in the interpretation of scenes, as well as in the measurement of elevations. A **stereoscopic pair** consists of two sequential photographs that normally overlap one another by approximately 60%.

Stereoscopic Viewing

We see in three dimensions. Individual aerial photographic prints, in contrast, show only two dimensions. Special techniques for viewing aerial photos, however, provide apparent three-dimensional views. Such **stereoscopic viewing** is vital to the production of topographic maps and is also a valuable adjunct to the interpretation of aerial photographs.

The three prerequisites for stereoscopic viewing of aerial photographs are (1) a stereo pair of prints, (2) some type of stereoscope, and (3) normal vision. A **stereo pair** consists of two overlapping photographs of the same scene, usually consecutive prints from a single flight line. In the area of overlap, the features in the scene are imaged from a different viewpoint on each photograph, just as each of our eyes registers a slightly different view of the scene before us.

A **stereoscope** is a simple instrument that permits the simultaneous viewing of a stereo pair so that the desired three-dimensional effect is achieved. Stereoscopes are arranged so that one eye focuses on objects in one photograph and the other eye focuses on the same objects in the other photograph (see Figure 2.18). Individuals with normal binocular vision should experience no difficulty in viewing a stereo pair through a stereoscope.

When you wish to prepare photographs for stereoscopic viewing, first draw light lines joining the **fiducial marks,** which are located in the margins of each photo (Figure 17.16). The crossing of these lines identifies the **principal point** of the photograph, which is the point on the ground that was directly below the vertical camera. Put a pinprick at the principal point, and, for clarity, draw a small, inked circle (about 5 millimeters in diameter) around the point.

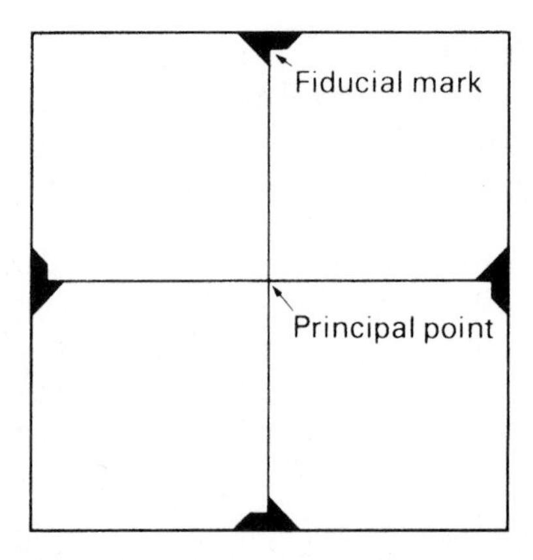

Figure 17.16 Relationship between fiducial marks and the principal point.

After the principal point has been identified, locate and mark the equivalent point, which is called the **conjugate principal point,** on the adjoining photo.[1] Then draw a straight line between the principal point and the conjugate principal point on each photograph. This line is called the **flight line,** because it joins consecutive points over which the airplane flew while taking the photographs.

Place the prepared photographs side by side on a flat surface, with the shadows falling toward you. The shadows must fall toward you to avoid a **pseudoscopic effect,** which causes hills to look like valleys and valleys to look like hills. (You may wish to test this effect by viewing a pair of photographs with the shadows falling away from you.) Move the photographs so that the flight line is aligned horizontally. Overlap the photographs so that the principal point on one and the conjugate principal point on the other are the correct distance apart for the type of stereoscope being used. This distance varies but is usually about 6.0 centimeters (2.36 inches) for pocket stereoscopes and about 25.4 centimeters (10 inches) for the mirror type. Then place the stereoscope over the prints so that the left-hand lens is over the principal point of the left-hand photograph and the right-hand lens is over its conjugate principal point on the right-hand photograph. Looking through the lenses of the stereoscope then produces a three-dimensional view of the terrain in the area of overlapping coverage. Small adjustments of the prints usually remove any difficulty with focus.

Stereo pairs viewed through a stereoscope appear dramatically three-dimensional because the photos were taken many meters apart, as the airplane moved between exposures. Our brains, in contrast, are accustomed to interpreting scenes viewed through our eyes, which are only about 65 millimeters (2.5 inches) apart. The result is a much-exaggerated three-dimensional effect in the stereoscopic view.

[1]This is best done using stereoscopic viewing. It can be done with reasonable accuracy, however, simply by visual inspection.

Interpretation

The addition of the third dimension to aerial photographic images is an important adjunct to their interpretation. The relative elevations of points on the surface, as well as the heights of objects, become distinctive features. Also, relatively slight slopes stand out clearly. An obvious disadvantage is that the actual slope may be overestimated. Experience, however, assists in making the correct interpretation. In addition, drainage patterns, fence lines, ditches, changes in slope, types of trees, relative heights of buildings, and similar details become much more apparent.

The availability of stereoscopic images also allows the mapping of relief and the measurement of elevations. These useful techniques fall outside the scope of coverage provided here. Consult the references listed in the Suggested Readings at the end of the chapter for information on measurement and mapping techniques, as well as for further discussion of the interpretative uses of aerial photographs.

OTHER AIRBORNE IMAGERY

Radar

Radar is an active remote-sensing technique involving the transmission of microwave signals that are reflected back to a receiver from objects in the viewing area. Aircraft-mounted radar systems provide information that can be mapped to indicate the location of objects, the texture of surfaces, the presence of subsurface moisture, surface temperatures, and so on. Radar signals penetrate clouds effectively, a characteristic that allows the acquisition of images under conditions that would prevent the use of aerial photography. Synthetic aperture radar (SAR) is particularly useful for mapping purposes, as the image in Figure 17.17 suggests. Because of the technical nature of radar images, we do not explore their use further. Consult the *Manual of Remote Sensing* listed in the Suggested Readings at the end of the chapter for more information.

Scanners

Scanners are passive devices that detect and record energy reflected or emitted from objects within their range. They can be mounted on aircraft or space platforms, and the information they generate can be converted into an image form.

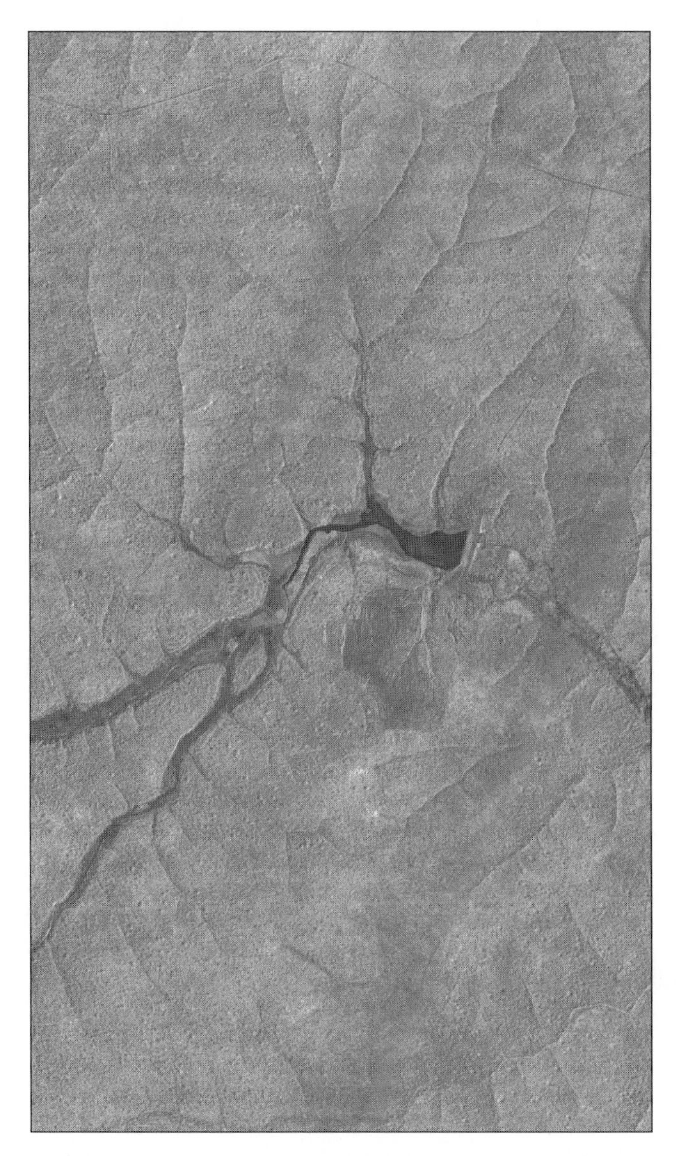

Figure 17.17 Synthetic aperture radar (SAR) image, near Camalaú, northeastern Brazil. This is a dry area (Caatinga), with small and low trees, often without leaves. Water reservoirs for irrigation and small structures are also visible.

Source: Courtesy of Radarsysteme GmbH, Wessling, Germany.

Figure 17.18 Thermal infrared scanner image of a thermal plume. (Savannah River, South Carolina, 31 March 1981.)

Source: From John R. Jensen, *Introductory Digital Image Processing: A Remote Sensing Perspective* (Englewood Cliffs, N.J.: Prentice-Hall, 1986). Reprinted by permission.

Infrared scanners depend on the detection of infrared energy emitted by any object, whether animal, vegetable, or mineral. The amount of radiation emitted at the longer, invisible wavelengths is related to the temperature and surface characteristics of the object. Cooler surfaces within the target area appear as dark areas on the image, and warmer surfaces are lighter (Figure 17.18). Infrared images, therefore, are useful for applications in which temperature differences are important, such as detecting forest fires, monitoring volcanic activity, determining the status of irrigation moisture, monitoring the decay of sea ice, counting animal populations, and checking the thermal characteristics of buildings.

SUMMARY

Remote sensing consists of gathering information by means of a sensor that is not in contact with the object being detected. The radiation commonly used for remote sensing ranges from relatively short (ultraviolet) to relatively long (microwave) wavelengths, including the visible spectrum.

Remote-sensing devices are categorized according to the range of wavelengths they detect (broad, narrow, or multispectral) and the source of the signal detected (passive or active). The devices operate in a variety of locations, but those mounted in aircraft or satellites are most important for mapping applications.

Aerial photographs, especially vertical ones, are important for mapping purposes and as map substitutes. Various film emulsions are available, each sensitive to specific electromagnetic wavelengths. The images obtained from the various types differ greatly. Black-and-white panchromatic emulsion is sensitive to approximately the same range of wavelengths as the human eye. Prints made from it closely duplicate the visual appearance of the original scene, which aids in photo interpretation. Black-and-white infrared emulsion is useful in the analysis of vegetation conditions and the identification of water bodies. Normal color film is helpful in identifying features that exhibit distinctive coloration, and color-infrared is useful for vegetation studies.

Aerial photographs for mapping are taken so that each photograph in a sequence overlaps the next by approximately 60 to 65% and photos in one leg overlap those in adjacent legs by approximately 20 to 35%. These requirements are particularly important in relation to stereoscopic viewing.

The scale of an aerial photo depends on the focal length of the lens used to take the photo and the elevation of the camera above the surface of the earth. Distortions in scale and planimetric locations of features are introduced through variations in topography. This means that regular aerial photos are not identical with maps of the same area.

The amount of planimetric shift of a feature is related to the feature's height above the photo's datum and its distance from the photo's principal point. The amount of shift increases as the altitude of the camera above the earth's surface decreases. It also increases with increasing distance from the principal point and is distributed in a radial pattern around that point. Planimetric shift results in scale variations within an aerial photograph because the ratio of the distance between objects on the photograph and the distance between the same objects on the ground varies from place to place.

A single aerial photo may not cover all of an area of interest, so uncontrolled or controlled photomosaics may be assembled. Uncontrolled mosaics are subject to difficulties, but controlled mosaics, especially when made from orthophotos, are effective map substitutes. Orthophoto products published by the U.S. Geological Survey include orthophotoquads and orthophotomaps.

Photo interpretation involves the recognition and identification of natural and human-made features on the earth's surface. The process uses shape, size, tone, color, texture, pattern, relative location, and shadow.

Stereoscopes permit the simultaneous viewing of a stereo pair so that a three-dimensional effect is achieved. This is an important adjunct to photo interpretation. It also allows the mapping of relief and the measurement of elevations.

Radar is an active remote-sensing technique involving transmitted microwave signals that are reflected back to a receiver from objects in the viewing area. Aircraft-mounted radar systems provide a variety of information that can be mapped. Scanners, in contrast, are passive devices that detect and record energy reflected or emitted from objects within their range.

SUGGESTED READINGS

Avery, Thomas Eugene, and Graydon Lennis Berlin. *Fundamentals of Remote Sensing and Airphoto Interpretation*. 5th ed. New York: Macmillan Publishing Company, 1992.

Burnside, Clifford D. *Mapping from Aerial Photographs*. New York: John Wiley & Sons, Inc., A Halsted Press Book, 1979.

Dickinson, G. C. *Maps and Air Photographs*. 2d ed. New York: John Wiley & Sons, Inc., A Halsted Press Book, 1979.

Moffitt, Francis H., and Edward M. Mikhail. *Photogrammetry*. 3d ed. New York: Harper and Row, 1980.

Reeves, Robert G., ed. *Manual of Remote Sensing*. Washington, D.C.: American Society of Photogrammetry, 1975.

Slama, Chester C., ed. *Manual of Photogrammetry*. 4th ed. Falls Church, Va.: American Society of Photogrammetry, 1980.

Thrower, Norman J. W., and John R. Jensen, "The Orthophoto and Orthophotomap: Characteristics, Development, and Applications." *The American Cartographer* 3, no. 1 (April 1976): 39–52.

CHAPTER 18

REMOTE SENSING FROM SPACE

Since the early 1970s, a number of space programs have produced imagery that has been made available to the public. Many such images are products of the U.S. Landsat program or, more recently, the French SPOT program. Some characteristics of these two programs are briefly described here because they are good examples of the methods by which space images are obtained and of the types of products available (see Appendix A and this book's website for information regarding the purchase of products).

Astronauts have brought many photographs back to earth, and some photographic film has been parachuted from satellites. Most information collected from space, however, is encoded onboard the spacecraft and then transmitted by radio to earth stations for conversion into the form required for interpretation. Knowledge of the basic characteristics of the processes involved in obtaining information from instruments in space gives some idea of what can and cannot be expected from the information.

Imagery from space is usually acquired and interpreted with the help of computers. If images are created from the information (which is not necessary for many applications), they are frequently viewed on a computer screen, and hard-copy images are produced only occasionally. This text provides a broad overview of the process of obtaining and processing remotely sensed information, and the discussion concentrates on visual images. Although remotely sensed images are often spectacular in appearance, the information contained in them is often more useful in digital, rather than visual, form. The Suggested Readings at the end of the chapter provide some guidance for further study in this complex and important field.

This chapter follows the customary practice of stating wavelengths and pixel sizes in metric units, so U.S. Customary equivalents are not included.

SPECTRAL CHARACTERISTICS

Only a portion of the tremendously wide range of electromagnetic radiation wavelengths is used for remote-sensing purposes (see Figure 17.1). The most frequently used wavelengths fall between about 0.4 micrometers and 22.0 centimeters, because these wavelengths are more useful than others. Some objects are most easily detected by the use of radiation in certain ranges. For example, healthy vegetation is a strong reflector of energy in the infrared wavelengths, whereas vegetation under stress is not. For this reason, a project involving the detection of areas of unhealthy vegetation is likely to use imagery based on the infrared wavelengths. In addition, the atmosphere scatters and absorbs some wavelengths of energy, whereas other wavelengths pass easily through so-called atmospheric windows. Appropriate wavelengths that match these **atmospheric windows** are used to obtain as much useful information as possible.

Collection Systems

Systems for collecting remotely sensed information involve the use of sensors that respond to electromagnetic radiation. The systems are arranged so that, at any given moment, each detector receives the radiation gathered from a specific area on the ground. The amount of radiation received affects detector voltage. This voltage is sampled frequently so that a series of values is obtained. Subsequently, each of these values is converted to a numerical value, which is assigned to the appropriate image area, called a **pixel.** Thus, each pixel value reflects the strength of the signal received for a particular ground location, within a specific band of the electromagnetic spectrum. A single value is recorded for each pixel, for each band.

The pixel values frequently range from 0 to 255, with 0 indicating no signal in the particular band and 255 indicating a very strong signal. The frequent occurrence of this range in numerical values is explained by the fact that the strength of the remote-sensing values is recorded in a computer. In computers, binary codes, called **bits** (0 or 1), are used to designate numbers, letters, and symbols. This is done by combining several bits into a string, called a **byte.** The number of different codes that can be recorded in a single byte depends on the number of bits it contains. Because there are two binary states, a byte can represent 2^n values (where n is the number of bits in a byte). In eight-bit system, which has commonly been used in remote-sensing satellites, 256 (2^8) values can be recorded. These are binary values ranging from 00000000 through 11111111, which represent the integers 0 through 255.

When the pixel values have been encoded, they are transmitted to a ground station. At the ground station, the information is stored on a computer-compatible tape, which is subsequently read into a computer for processing.

Ground Resolution

Ground resolution is the size of the area on the ground that is imaged at any one time. It determines the size of objects or areas that can be detected.

A remotely sensed image is composed of myriad individual pixels, which are minuscule rectangular units, each representing a specific area on the ground. Within each band, each pixel is imaged as a single shade of gray or a single color, determined by the numerical value attached to it. The size of objects that can be detected by an examination of such an image is a function of the system's ground resolution.

Although this relationship exists, it is sometimes assumed to mean, for example, that 30-meter-square objects can be reliably detected on a 30-meter resolution system. Unfortunately, this is not usually true. Instead, the rule of thumb is that the smallest dimension of an object must be at least twice as great as the system's ground resolution before there is a good likelihood of detection. This means that a football stadium that surrounds a 50-by-100-yard (approximately 46-by-92 meter) playing field is likely to be detectable in a 30-meter resolution system, whereas a large house that measures 10 by 15 meters (33 by 49 feet) is not. It should be noted that long, narrow objects, such as rivers or highways, are frequently detectable, although their width is much less than the system's resolution.

The resolution capability of commercial systems is improving, however. Another, newer system has a maximum resolution of 10 by 10 meters. Military systems have greater resolutions than those available to the public, thus allowing the detection of smaller objects. The specific capabilities of military systems are subject to speculation and are undoubtedly often exaggerated. However, high-resolution civilian imagery is under development. The firms planning to produce this type of imagery expect to provide images with between 1- and 4-meter resolution.

Image Creation

One use of remotely sensed information that is particularly important in the context of this book is the creation of images for visual interpretation. This is done by using a computer-controlled exposure device. The gray tone assigned to each pixel in the image is determined by the numerical value associated with the corresponding pixel in the original scene. Pixels that have a value of 0, for example, appear black in the image, whereas those with a value of 255 appear white. Cells with values between 0 and 255 have intermediate gray values proportionate to their numerical value.

An image can be created for each band for which information is available. By assigning a different color to each band, a composite color image is produced on color film. The colors that result from this process are *not* necessarily the colors that would be visible to an observer of the original scene. Some portions of the image might even be made from data from the infrared, microwave, or other bands that are not visible to the human eye. Microwave detectors, for example, are useful for applications such as determining the relative roughness of the surface of the sea, monitoring the extent of sea ice, estimating the extent of snow cover, and locating oil films on water surfaces (Figure 18.1).

To take advantage of the information gathered from invisible wavelengths, remote-sensing images are typically reproduced in false color. In a **false-color image,** the colors assigned to each wavelength band are shifted. One typical false-color arrangement, for example, reproduces the green band as though it

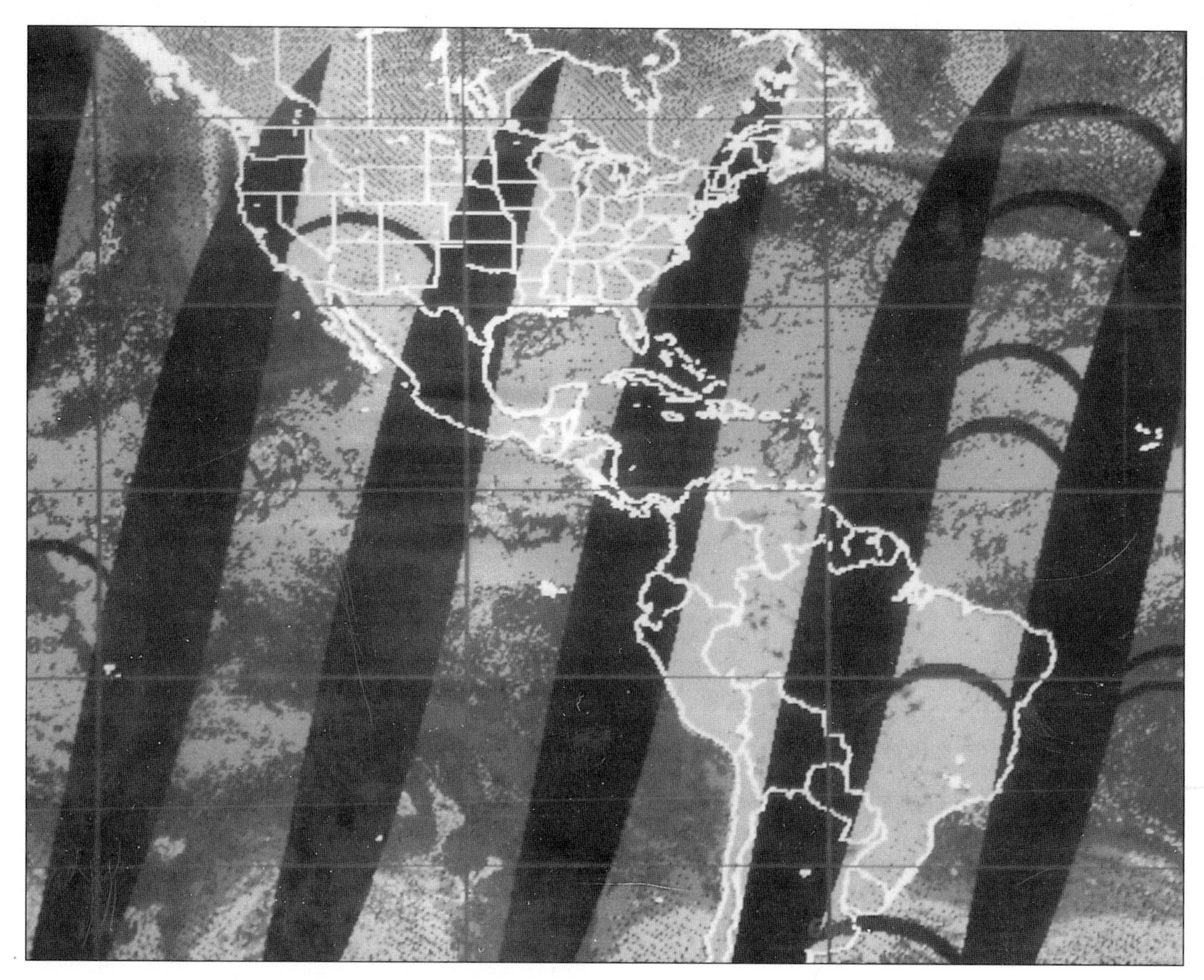

Figure 18.1 Microwave image, produced from Special Sensor Microwave Imager (SSM/I) data. The purpose of such low-resolution images (12-by-12 km resolution) is the remote sensing of atmospheric and surface parameters, not of terrain detail.
Source: Courtesy of Benjamin Watkins, National Environmental Satellite, Data, and Information Service (NESDIS).

were blue, the red band as though it were green, and the near-infrared band as though it were red. In this type of image, healthy vegetation appears as bright red, clear water is black, sediment-laden water is powder blue, and urban centers are blue or blue-gray (Plate 10). Other color assignments are used for special purposes.

Examples of Remote-Sensing Systems

Landsat

Since 1972, a series of **Landsat** satellites has been providing information about the earth's surface. Landsats 1 through 4 are no longer operational. Landsat 5 has been in operation since 1985 and continues to provide useful information. Landsat 6 was launched in 1995 but was lost before it became operational. Landsat 7 was launched in April 1999, and it is expected to provide useful information well into the next century. Landsats 5 and 7 orbit at an altitude of 705 kilometers (437 miles). Each of these systems provides repeat coverage every 16 days, but they have been positioned so that, together, coverage is available every 8 days. The major sensor systems used on Landsat 5 are the multispectral scanner and thematic mapper. Landsat 7 is equipped with a sensor system called Enhanced Thematic Mapper Plus (ETM+).

The **multispectral scanner (MSS) system** was used on the first five Landsat satellites and has collected a vast amount of information. The MSS multidetector array records data for four spectral bands, designated as bands 4, 5, 6, and 7 (Table 18.1). Because the four bands are scanned continuously and 6 lines of the image are recorded at a time, a total of 24 sensors is required.

TABLE 18.1 Landsat MSS Spectral Bands

Band*	Frequency
4	0.5–0.6 micrometers (green)
5	0.6–0.7 micrometers (red)
6	0.7–0.8 micrometers (reflective infrared)
7	0.8–1.1 micrometers (reflective infrared)

*Renumbered as bands 1 through 4 on Landsats 4 and 5.

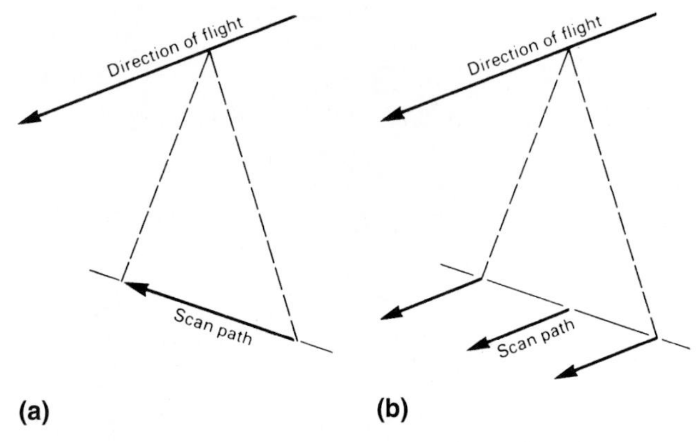

Figure 18.2 (*a*) Whisk-broom scanning. (*b*) Push-broom scanning.

The Landsat MSS scans a 185-kilometer-wide (115-mile-wide) field, using a **whisk-broom scanning** movement, in which an oscillating mirror scans the ground across the satellite's direction of movement (Figure 18.2a). The MSS scanner has a ground resolution of 79 by 79 meters, and sampling is conducted in such a manner that each of the final picture elements covers an area of 56 by 79 meters.

Portions of the same MSS scene, using bands 4, 5, 6, and 7, are shown in Figure 18.3. Comparison of these scenes gives a sense of the variations in the images, such as the relative sharpness of the shorelines in band 6. Detailed analysis of the data that lie behind the scenes is necessary, however, to gain more than a general impression of the possible uses of such images. Composite false-color images provide easier recognition of features such as urbanized areas, wetlands, forested areas, and geologic structures.[1]

The **thematic mapper (TM) system** was used on Landsat 4 and continues in use on Landsat 5. TM has a ground resolution of 30 by 30 meters (except the thermal-infrared band, band 6, which has a resolution of 120 by 120 meters). The TM detector arrays are arranged to scan the 185-kilometer-wide (115-mile) ground track in whisk-broom fashion. The frequencies of the TM spectral bands are shown in Table 18.2.

[1]See, for example, the many scenes included in the *Atlas of North America: Space Age Portrait of a Continent* (Washington, D.C.: National Geographic Society, 1985).

The **Enhanced Thematic Mapper Plus (ETM+)** installed on Landsat 7 replicates the capabilities of the thematic mapper instruments on Landsats 4 and 5. Like TM, the ETM+ instrument is a whisk-broom, multispectral scanner, which scans a 185-kilometer wide (115-mile) swath from a orbit altitude of 705 kilometers (437 miles). However, it also includes new features that make it a more versatile and efficient instrument for global change studies, land cover monitoring and assessment, and large area mapping. Six of the ETM+ spectral bands are virtually identical to those of TM. One significant improvement, however, is the increase in the spatial resolution of band 7, the thermal-infrared channel, from 120 by 120 meters to 60 by 60 meters. A second major improvement is the addition of an eighth band, which provides panchromatic coverage with 15-by-15-meter spatial resolution.

SPOT

A satellite observation system called Système Probatoire d'Observation de la Terre **(SPOT)** has been operating since February 1986. Two early satellites, SPOT 1 and SPOT 2, are still in operation, along with the newer SPOT 4. SPOT operations are currently funded primarily by France, with funding contributions from Belgium and Sweden and other contributions from Italy, Spain, and the European Union. The system is operated by the French space agency, Centre National d'Études Spatiales (CNES). The SPOT 4 satellite carries three sensors that operate in the visible spectrum, as well as a shortware infrared sensor (Table 18.3). The bands are selected to support applications related to land management, topographic and relief mapping, and ecosystem, crop, and plant health monitoring.

The satellites orbit at altitudes of 830 kilometers (515 miles). At this altitude the sensors, which are arranged in linear arrays, scan along a 60-kilometer-wide (37-mile) ground track of the satellite. In what is known as **push-broom scanning,** a complete line of the ground scene is imaged at one time (Figure 18.2b).

The sensors can operate in any of three modes: multispectral, monospectral, or combined. In multispectral (X) mode, all four bands are recorded simultaneously, with a ground resolution of 20 by 20 meters. In monospectral (M) mode, the sensors record only band B2, with a ground resolution of 10 by 10 meters. This produces extremely detailed images, with accuracy comparable to that of 1:24,000 topographic quadrangles. Finally, in combined mode, the X and M modes are combined.

A unique feature of the SPOT system is a ground-controlled mirror system that can be pointed away from the vertical. This capability makes it possible to view selected areas across a 950-kilometer-wide (589-mile) path. The result is the observation can be

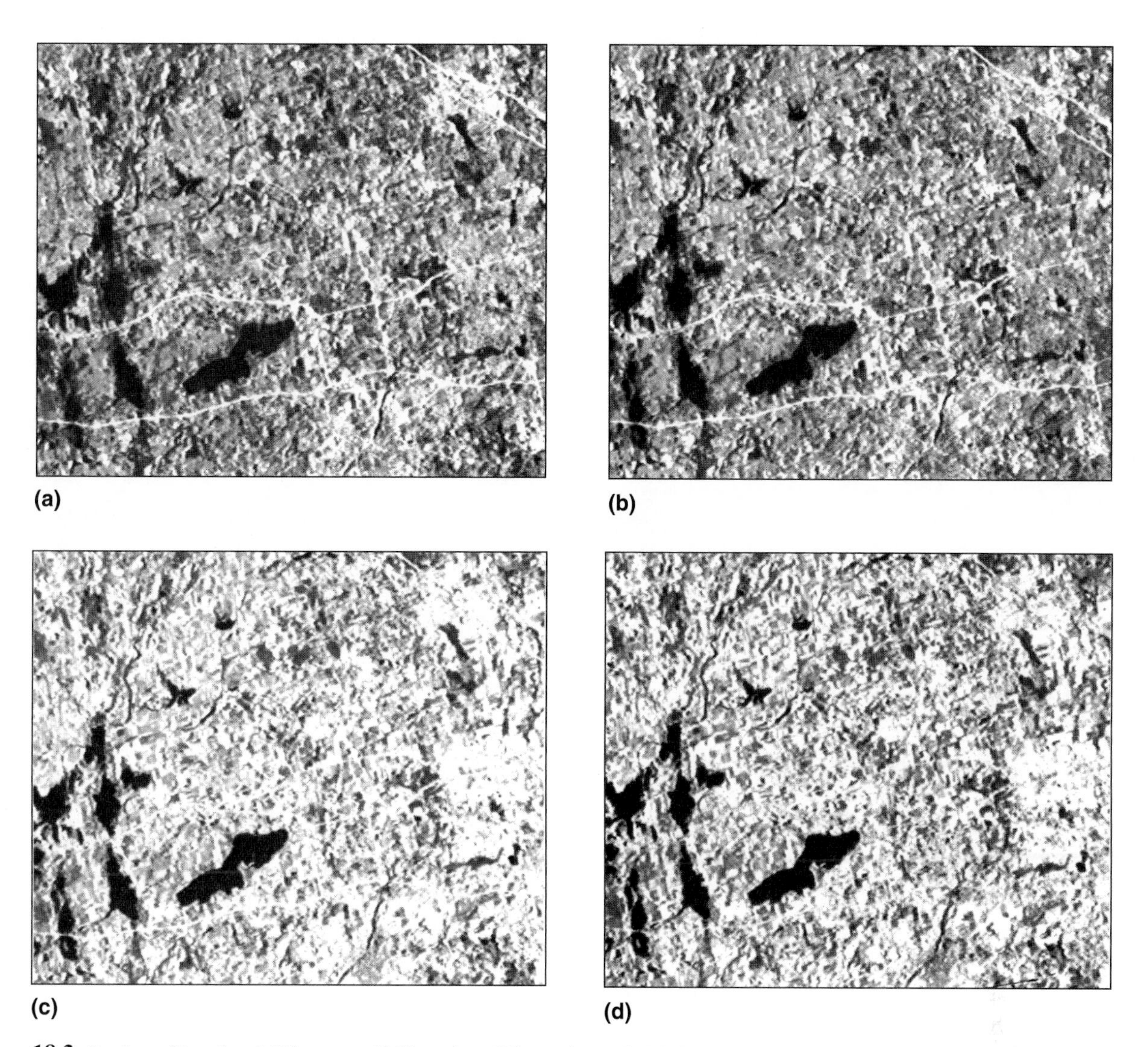

Figure 18.3 Portion of Landsat MSS scene of Milwaukee, Wisconsin, and vicinity (2 April 1976). (*a*) Band 4. (*b*) Band 5. (*c*) Band 6. (*d*) Band 7.
Source: U.S. Geological Survey EROS Data Center.

TABLE 18.2 TM Spectral Bands

Band	Frequency
1	0.45– 0.52 micrometers (blue)
2	0.52– 0.60 micrometers (green)
3	0.63– 0.69 micrometers (red)
4	0.76– 0.90 micrometers (reflective infrared)
5	1.55– 1.75 micrometers (mid-infrared)
6	10.40–12.50 micrometers (thermal infrared)
7	2.08– 2.35 micrometers (mid-infrared)

TABLE 18.3 SPOT Spectral Bands

Band	Frequency
Visible B1	0.50–0.59 micrometers
Visible B2	0.61–0.68 micrometers
Visible B3	0.78–0.89 micrometers
Shortwave infrared	1.58–1.75 micrometers

made more often than the 26-day interval between successive passes of the satellite over a given ground location. In addition, the three operating satellites are positioned to improve the frequency of imagery opportunities. Such opportunities currently occur at less than five-day intervals at the equator. The frequency of imaging opportunities increases toward the poles, sometimes occurring daily above 70° latitude. This more frequent viewing capability is useful when cloud cover interferes with normal vertical viewing or when other circumstances make more frequent viewing desirable. Changing the angle of view also allows the production of stereo pairs for photogrammetric applications.

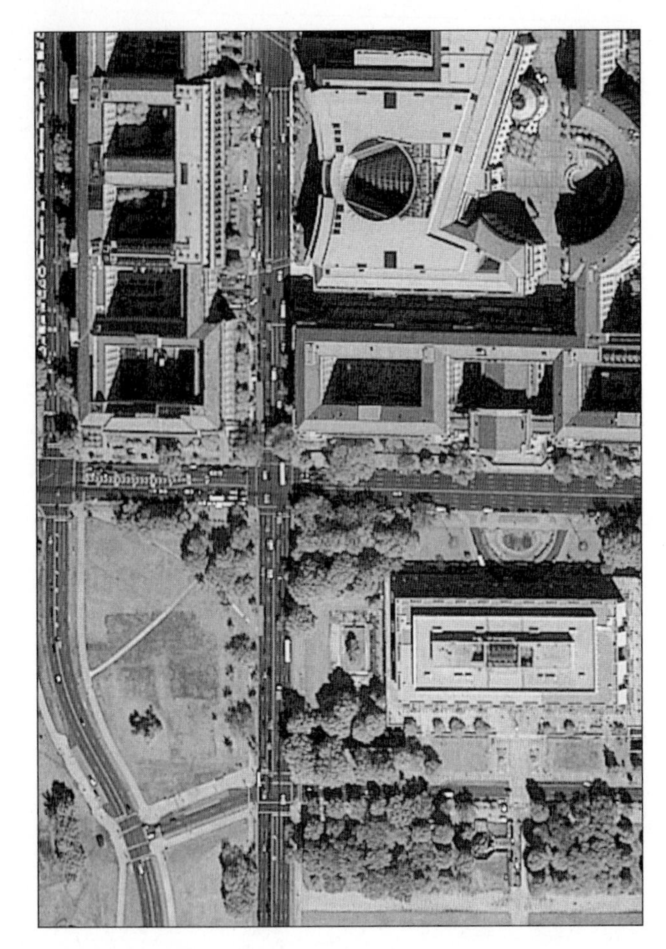

Figure 18.4 Portion of first 1-meter resolution *IKONOS* Satellite panchromatic image of Washington, D.C.

Source: Photo courtesy of spaceimaging.com.

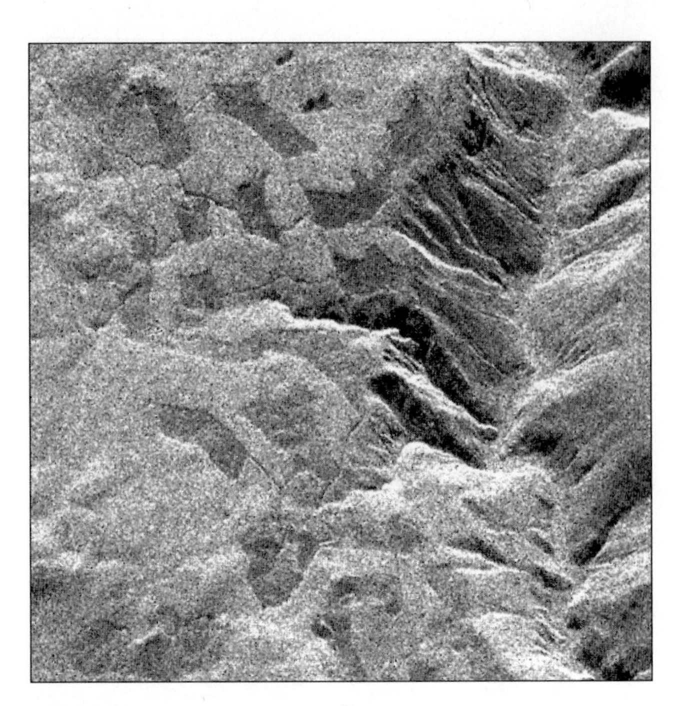

Figure 18.5 Satellite radar image of a site in the Okanagan Valley, British Columbia. Demonstrates RADARSAT's ability to detect clearcut areas in a boreal forest as shown by well-defined boundaries and land differences—forested regions are light gray and cut blocks are dark gray.

Source: © Canadian Space Agency/Agence Spatiale Canadienne, 1996. Data received by the Canada Centre for Remote Sensing. Data courtesy RADARSAT International.

Other Programs

As mentioned, several companies plan to produce high-resolution imagery. The expected resolutions of these images will be about 1 meter in panchromatic mode and 4 meters in multispectral mode. This means that the images will often be effective substitutes for conventional aerial photographs. In addition, it is anticipated that acquisition and processing costs may be lower than for comparable air photos and that delivery times will be very rapid. An example of the 1-meter resolution panchromatic imagery that is available from one of these systems is shown in Figure 18.4.

Another development is the introduction of satellite-based radar imagery (Figure 18.5). Radar is an active system, in which the satellite transmits pulses of microwave energy toward the earth. It then measures the energy reflected back to the satellite from the earth's surface. The data thus obtained allow the production of images based on variations in the strength of the reflected energy.

A major advantage of radar images is that they do not depend on daylight or on clear atmospheric conditions. In particular, microwave energy penetrates clouds, rain, dust, and haze. Therefore, obtaining an image in darkness or under any atmospheric conditions is possible. This means that radar images are advantageous in areas often subject to heavy cloud cover, such as the humid tropical or marine West Coast climates. They also are useful in polar regions, which are subject to long periods of darkness, and in areas hidden by dust from volcanic eruptions or dust storms. Also, radar effectively images surface roughness, topography, land/water boundaries, fabricated features such as buildings, and differences in moisture levels.

Image Processing

Simple conversion of remotely sensed pixel values into an image does not usually produce a particularly useful product. Computer-aided manipulations of the original data, however, improve the viewing quality of the final product. Some of these manipulations are typically done before the data are released to the prospective user. These **preprocessing techniques** are not discussed here because the user is not involved in them. Other manipulations are left to the user's discretion, because they must be modified to suit particular applications. These processes are referred to as **image**

enhancements. In the past, image enhancements required large-scale computer facilities. Recently, however, useful work has been performed on minicomputers and microcomputers.

Data adjustments often enhance the quality of the image for visual analysis purposes. Representative operations of this type include linear contrast stretching and edge enhancement. Other enhancements, such as density slicing and band ratioing, are sometimes used but are not discussed here.

Linear Contrast Stretching

Certain detector characteristics, or the presence of atmospheric haze or of certain types of materials on the surface, may result in a low-contrast image. In a **low-contrast image,** the values of the pixels in the image fall in a relatively narrow range, so that variations in the gray tones are very difficult to discern (Figure 18.6a). **Linear contrast stretching** expands the range of tones so that they cover the entire available range and are much easier to differentiate (Figure 18.6b).

Edge Enhancement

Edge enhancement increases the contrast between the values on either side of a dividing line, rendering shapes and details along the edges of areas more conspicuous (Figure 18.6c).

Thematic Information Extraction

The major value of remotely sensed data is that they provide information about conditions on the earth's surface. Interpretative work with the data, however, is necessary for **thematic information extraction.** This is often done without the aid of images, using computer-controlled statistical methods. Manipulation of the images themselves, however, is also an important aspect of analysis, because the images reveal the spatial arrangement of different land-use types, surface elevation, and biophysical information, such as vegetation types and conditions, soil moisture, temperature, and texture. The extraction of thematic information is sometimes difficult. For example, different surfaces may have signatures that are similar enough that they cannot be reliably separated. Or similar surfaces may not necessarily appear the same because of slightly varying conditions at different locations or at different times.

Classification

The data signature of a given image pixel is a composite mixture of the energy wavelengths and signal strengths of the ground surface within the area in the pixel. These data signatures range from uniform to highly mixed in any given region. In a nonurban region covered by a single crop type or by a uniform forest cover, surface signatures and image values may be quite uniform over many pixels.

Mixed uses, on the other hand, result in a greater diversity of signatures and a more complex range of values. In an urban area, for example, a great variety of cover types exists in close proximity. In such areas, asphalt, concrete, and other building materials are mixed in greater or lesser proportions with vegetation of various types, as well as with areas of water. Because each surface has its own signature, the end result is a great variety of signal combinations among the various pixels. An image produced from the original, unprocessed pixel data in such an area, therefore, is also extremely variable.

The goal of **data classification** is to group pixels so that each one is associated with others that have relatively similar characteristics and is separated from those with distinctly different characteristics. This classification is done by statistically analyzing the characteristics of the radiation collected for each pixel and assigning the pixel to the appropriate category. This information can be easily mapped (see Plate 11). How classification is accomplished is not of direct concern to us here. Regardless of the method used, however, pixels are not always placed in the appropriate category—that is, the category they would be placed in if the classification were checked on the ground. These **classification errors** obviously reduce to some extent the usefulness of the image for particular applications.

Change Detection

The concept of **change detection** is simple. Suppose that two remotely sensed images of a specific region exist. One of these images may have been obtained prior to the current planting season and the other during the season. Change detection involves the careful alignment of the two images, followed by a comparison of the two on a pixel-by-pixel or area-by-area basis. This comparison should make it possible to determine which areas were planted (were changed) and which areas were not (were unchanged).

Many problems are encountered with real-world applications of change detection. Changes in the environment other than those under study may well have occurred during the time period between the two images. Weeds may have grown, brushfires may have eliminated vegetation, heavy rain may have soaked the soil or drought may have dried it, and so on. Any such changes affect the data values recorded, and their effects must be separated from changes due to the process of interest.

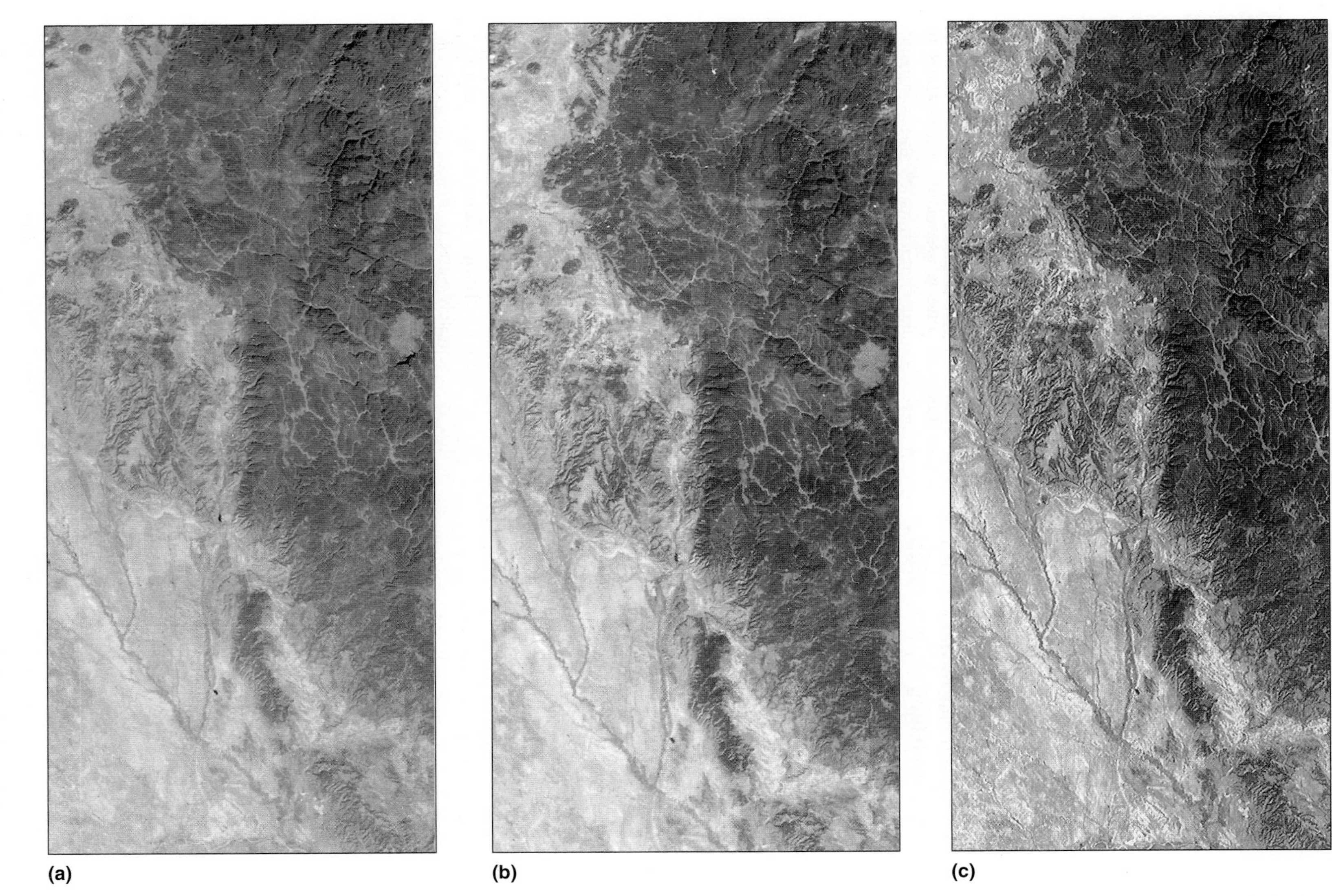

Figure 18.6 Enhancements of a remote-sensing image. (*a*) Unprocessed image. (*b*) Image after linear contrast stretching. (*c*) Image after linear contrast stretching and edge enhancement.

Source: Courtesy of U.S. Geological Survey EROS Data Center.

Other problems must also be considered, including equipment variability from one image to another and the fact that the pixels in one image are not aligned perfectly with the pixels in the other. Ideally, these factors are eliminated by appropriate processing so that *actual* changes, rather than *apparent* ones, are located. All in all, change detection is not an easy task.

INTERFACE WITH GEOGRAPHIC INFORMATION SYSTEMS

In recent years, geographic information systems (GISs) have come into use. The data used in such systems are drawn from many sources, including maps, aerial photography, field investigations, and statistical studies. Incorporation of remote-sensing data into a GIS adds a dimension to the system's capabilities. Remote-sensing systems collect data from the same area on multiple dates, for example. This allows the analyst to not only inventory surface characteristics on an up-to-date basis, but also to monitor changes in them. The characteristics of GISs, and the types of applications for which they are used, are discussed in Chapter 21.

SUMMARY

The electromagnetic radiation wavelengths in atmospheric windows within the range from about 0.4 micrometers to 22 centimeters are most frequently used for remote sensing from space. Detectors mounted in satellites receive signals gathered from a specific area on the ground, and the strength of radiation in specified bands is converted to digital values. Each value, which often lies between 0 and 255, is assigned to an image area, called a *pixel*. The encoded pixel values are transmitted to a ground station, where they are stored on tape.

Ground resolution determines the size of objects that can be detected. The smallest dimension of an object must be at least twice as great as the ground resolution of the system before it is likely to be detectable.

Images are produced by assigning a different color to each data band. In typical false-color images, the green band may be reproduced in yellow, the red band in green, and the near-infrared in red. Although images of this and other types are frequently created, the information from which they are constructed is often more useful in digital than in visual form.

Landsat satellites have been in operation since 1972. The major sensor systems used are the multispectral scanner (MSS) and, since 1982, the thematic mapper (TM). The MSS whisk-broom scanner records data for four spectral bands, from a 185-kilometer-wide field, with a ground resolution of 79 by 79 meters. Sampling is conducted in such a manner that the final picture elements cover 56 by 79 meters. TM also scans a 185-kilometer-wide ground track in whisk-broom fashion in seven bands. It has a ground resolution of 30 by 30 meters (and of 120 by 120 meters in the thermal-infrared band). The Enhanced Thematic Mapper Plus (ETM+) installed of Landsat 7 replicates the capabilities of the TM instruments on Landsats 4 and 5. It also increases the spatial resolution of the thermal-infrared channel to 60 by 60 meters and adds a panchromatic band with 15-by-15-meter resolution.

The French SPOT system carries sensors that operate in multispectral, monospectral, or combined modes. Its push-broom sensors scan a 60-kilometer-wide ground track, with ground resolution of 10 by 10 meters in monospectral mode and 20 by 20 meters in multispectral mode.

Data manipulations, such as linear contrast stretching and edge enhancement, modify images for visual analysis purposes. Linear contrast stretching expands the range of tones so that they are easier to differentiate, whereas edge enhancement increases the contrast between the values on either side of a dividing line.

Thematic information extraction may be done without the aid of images, using computer-controlled statistical methods. Image interpretation is also important, however, because a human observer can often detect spatial arrangements that a computer routine may not. On the other hand, data differences that are meaningful to the computer are often too subtle to be seen, even on enhanced images. Data classification is used to group pixels so that each one is associated with others with similar characteristics and is separated from those with different characteristics.

Change detection involves comparisons of remotely sensed images taken at different times, to determine changes in conditions related to some factor under study. Extraneous changes, however, may mask those due to the factor of interest, and their effects must be separated if the study is to yield useful results.

SUGGESTED READINGS

Barrett, E. C., and L. F. Curtis. *Introduction to Environmental Remote Sensing*. London: Chapman and Hall. New York: John Wiley & Sons, Inc., A Halsted Press Book, 1976.

Campbell, James B. *Introduction to Remote Sensing*. New York, London: Guilford Press, 1987.

Colwell, R. N., ed. *Manual of Remote Sensing*. 2d ed. Vols. 1 and 2. Falls Church, Va.: American Society of Photogrammetry, 1983.

Curran, Paul J. *Principles of Remote Sensing*. New York: Longman, 1985.

Jensen, John R. *Introductory Digital Image Processing: A Remote-Sensing Perspective*. 2d ed. Englewood Cliffs, N.J.: Prentice-Hall, 1996.

CHAPTER

19

Computer-Assisted Cartography

An important trend in the world of maps and map products is an increasing use of computers. The search for the realization of two different objectives has led to this trend. The first objective is simply to use computers directly in map production. This can involve making the same types of maps that have always been made or making new types of maps that were not possible without the computer's computational power. This approach, which is called **computer-assisted cartography,** is designed to improve the speed and accuracy of production. The second objective is the development and use of cartographic databases in geographic information systems. In this approach, computers are used to improve the analysis of geographic data. Some of the results obtained from this approach have the same appearance as traditional map products, but others can be quite different products or applications. Some of the results had probably never been thought of before or, if they had been visualized, could not be produced by conventional means.

Overview

The boundary between computer-assisted cartography and digital geographic information systems is indistinct. To the extent that a separation is possible, however, the discussion in this chapter focuses on maps and map-related products produced with computer-assisted techniques. The concern here, therefore, is mostly with the production of maps. Chapter 21 deals with the extension of those techniques into geographic information systems, which allow maps to be manipulated to help solve problems. The computer allows rapid data analysis and manipulation, often in ways that are beyond the capability of noncomputer approaches.

Advantages of Computer-Assisted Techniques to the Map Producer

Computer-assisted techniques provide many production advantages that have attracted the attention of cartographers and have absorbed considerable research effort and financial resources. While map-production considerations are not discussed in detail here, an awareness of certain aspects is helpful in understanding the impetus behind the trend toward computer assistance, as well as the nature of the map products that result.

Computer-assisted mapping methods are considered advantageous because the necessary cartographic database is in digital form. Computer techniques enable the cartographer to quickly and easily manipulate the digital data to produce the required **output,** including any necessary mathematical or statistical operations, scale or projection changes, and symbol selection. The ease of using such a system may even encourage the testing of alternative designs, designs that may turn out to produce better products. If a permanent copy of the map is desired, computer-driven devices produce it with accuracies that far exceed the accuracy of manual drafting. Another advantage of such a system, provided the database is kept current, is that each new map produced incorporates information that is correct and up-to-date. Finally, computers allow map data to be rapidly transmitted and exchanged between users. This ease of interchange assures the production of better, more complete, and up-to-date maps. The end result of all of this is the economical production of a greater variety of more accurate and better-quality maps.

Beyond map production, digital techniques allow textual or tabular information to be tied to cartographic features. This allows the system user to analyze selected features, recall information about them, examine the relationships between features, carry out measurements, and so on, by simply calling the appropriate functions from the computer terminal. This aspect of computer-assisted cartography blends into the digital geographic information systems discussed in Chapter 21.

The attractiveness of digital production techniques is easy to understand. Over the past decade or so, the problems that affected their earlier adoption have become less significant. This is partly due to the increased numbers of experienced users and partly due to advances in the computer hardware and software that provide the underpinning for digital mapping. As a result, an ever-increasing volume of computer-assisted map products has appeared and been adopted, and this trend is likely to accelerate in the years ahead.

Effects of Computer-Assisted Techniques on the Map User

One aspect of the application of computer techniques that may be unexpected is that many of the resulting products are simply replications of traditional map products. Such products not only look the same but are used in the same ways. Often, however, computer-assisted methods result in nontraditional products that provide opportunities for new applications.[1]

To the extent that they are developed, new forms of map products will force some changes in the way users approach maps. For one thing, they will need to become familiar with the use of computer screens for viewing both data and maps. Also, they may have to use systems that provide databases and mapping software, rather than printed maps. In such systems, users will have access to the desired information and will be able to format it as they wish, thus producing do-it-yourself interactive maps. This possibility provides further incentive for learning about those aspects of computer-assisted mapping techniques that are important to the effective use of interactive map products.

This chapter provides general background on the topic of computer-assisted cartography. Two special applications related to this discussion, electronic atlases and automated vehicle navigation systems, are presented in Chapter 20.

FORMAT AND DATA CAPTURE

Data input techniques and data storage structures differ from system to system, depending on the type of data needed, the form in which the data exist, and the system's operating characteristics. Most of these matters are beyond the control of the system user, and some make little practical difference in terms of system capabilities, so they are not discussed here. An understanding of the methods used to capture the data used in computer-assisted cartography, however, will help you to gain an appreciation for the differences between two distinctly different types of systems: vector (or polygon) and raster (or gridded). These systems are discussed after the data-capture methods are introduced.

Digitizing

Maps act as storehouses for spatial information about features on the earth's surface. This information consists of the locations of points, the paths of lines, the outlines of areas, and the complex interrelationships between all of the various types of features. The process of converting map data from their original, visual form to a digital format that can be handled by a computer is called **digitizing** or **conversion.** High-quality conversion work requires specialized equipment and trained personnel. In addition, the work is often sporadic but, when it is needed, a large volume may be required within a short time. For these reasons, conversion work is often done by specialized service bureaus, rather than by the data users.

Labeling

A detailed discussion of the digitizing process is not presented here, because digitizing is completed before map users' use of the system. However, one aspect of digitizing that is important to map users is that, as the various map elements are recorded, they must have an identification code attached to them so that they can be selectively retrieved from the data file. This is done by **labeling** each record with a standardized set of characters, such as "050 0412" for a stream or "170 0201" for a class-1 highway.[2] Some knowledge of these labels is helpful as you work with a geographic information system so that you understand the contents of its files.

[1]Some excellent examples of unusual computer-assisted maps are included in Mark Monmonier and Alan M. MacEachren, guest editors, "Geographic Visualization," *Cartography and Geographic Information Systems* 19, no. 4 (October 1992).

[2]One set of codes, the standard U.S. Geological Survey attribute codes for digital line graphs, is found in appendix D of U.S. GeoData, *Data Users Guide 1: Digital Line Graphs from 1:24,000-Scale Maps* (Reston, Va.: U.S. Geological Survey, 1986).

Vector and Raster Digitizing

Another aspect of digitizing that map users should understand is the difference between vector and raster methods. This is because the format of a map file controls the uses to which it can be put.

In **vector digitizing,** the records consist of the coordinates of specific points (or a series of points, in the case of linear features or area outlines; Figure 19.1a). Traditional maps are vector products, drawn using point symbols, along with lines connecting points or outlining areas.

An alternative to vector digitizing is **raster digitizing.** This means that a map is divided into myriad tiny cells. These cells, each of which may be as small as 0.001 square inch, contain the feature records (Figure 19.1b). A major difference between vector and raster digitization, then, is that a raster image is presented as a set of individual cells, and the appearance of such an image is quite different from that of a traditional line map. The smaller the cells in a raster image, however, the more its appearance resembles that of a vector image.

Other Considerations

Maps are not the only source of data used in computer-assisted cartography. Original survey data, aerial photos, and data obtained from remote sensing are also used. Aerial photos are typically digitized in much the same manner as conventional maps. Data obtained from original surveys and remote-sensing systems may already be in digital form. This means that they are especially suited to incorporation into a digital-mapping process.

Machine Coordinates

The types of coordinates used in digitizing need further clarification. The coordinates obtained directly from many digitizing methods are arbitrary **machine coordinates.** This means that they are recorded in inches or centimeters, measured from a control point, such as the corner of the digitizing board. Coordinates of this type have the shortcoming that they do not have any direct relationship to locations on the earth's surface. In practice, therefore, machine coordinates are usually converted to some more general coordinate system, such as latitude/longitude, Universal Transverse Mercator (UTM), or State Plane Coordinates (SPC), in a process called *geocoding*.

Geocoding

Geocoding is accomplished by first determining the actual coordinates of known points on the map using the desired coordinate system. The machine coordinates of these control points are then compared to their actual coordinates. The relationship between the two sets of coordinates is subsequently used to automatically record actual coordinates, as features on the map are digitized, instead of simply using machine coordinates.

A major advantage of geocoding is that data from a variety of sources are recorded in a single locational system. This allows materials mapped at different scales or on different projections to be brought to a common scale and projection so that they can be used in conjunction with one another.

Format and the User

Whether a GIS user would prefer a raster or vector format depends on both the application at hand and the scale at which the work is being done. On the other hand, the selection of the system format is beyond the user's control, because the system already exists and must be accepted as it stands, regardless of the data format. For this reason, and acknowledging the importance of differences between raster versus vector systems, we now move on to a consideration of other aspects of computer-assisted cartography.

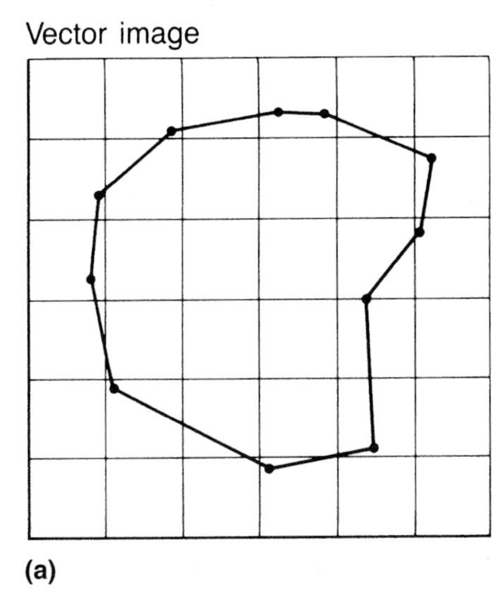

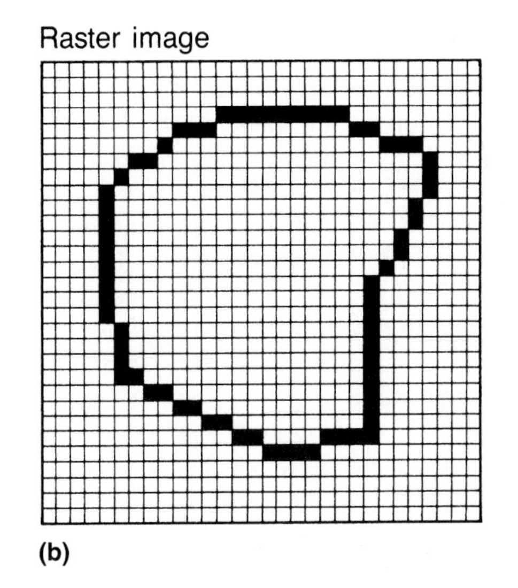

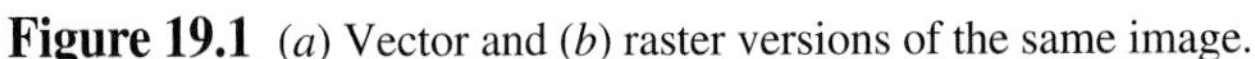

Figure 19.1 (*a*) Vector and (*b*) raster versions of the same image.

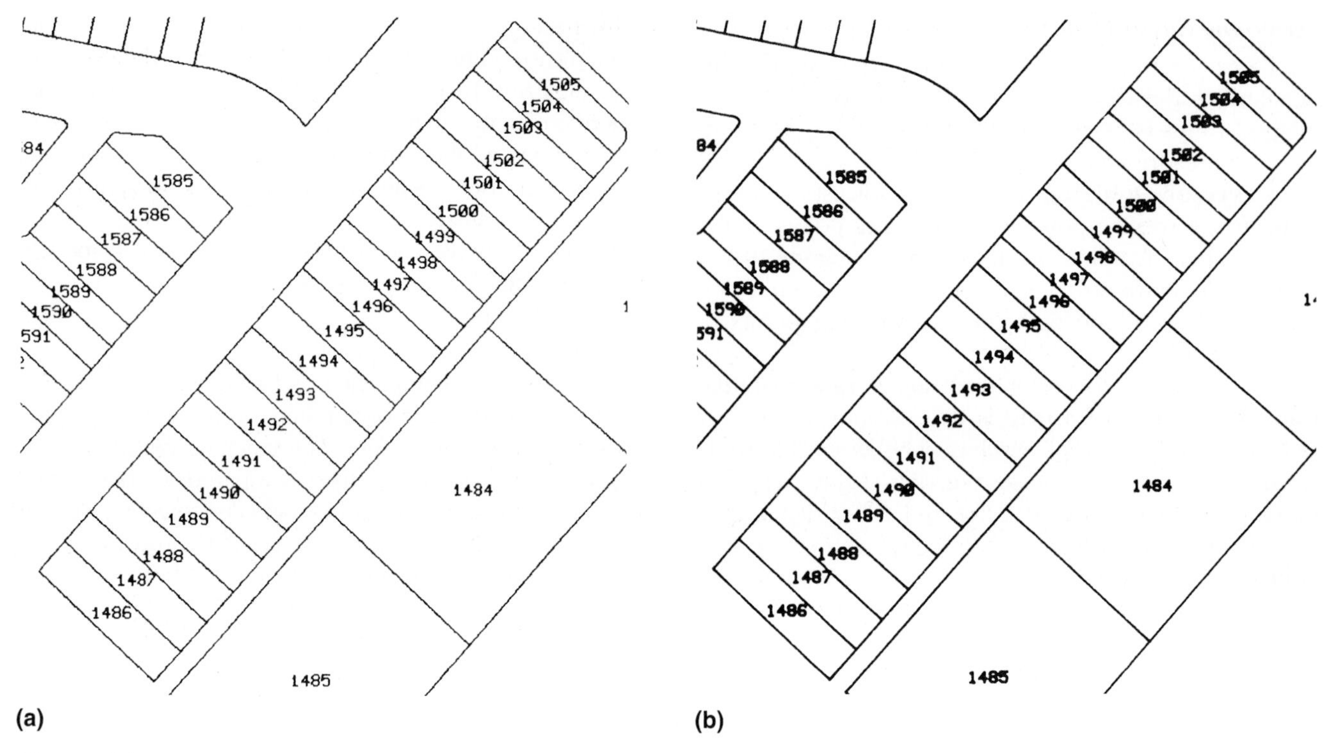

Figure 19.2 (*a*) Raster plotter image and (*b*) line plotter image of the same area. Each address contains a node that is linked to a database containing attributes associated with that parcel, such as acreage, ownership, and zoning.
Source: Courtesy of City of Boston and Donohue Intelligraphics.

OUTPUT

One simple and direct way of producing hard-copy maps in a computer-assisted cartography system is simply to photograph the image displayed on the computer screen. Direct photography is useful for either black-and-white or color images and has the advantage of being quick and easy.

Probably the most common method of producing hard-copy maps, however, is to print or plot them, using a peripheral device that is part of the computer system. Depending on the characteristics of the particular system, either raster or vector devices can be used.

Raster output consists of an image produced by creating a pattern of dots on the paper (Figure 19.2a). The **inkjet printers** that are often used with word processing and similar office systems can be used to produce graphic images, including maps. Special **raster plotters,** however, can produce larger images. Technological changes have led to substantial increases in the resolution of raster plotters and printers. Therefore, it is often difficult to distinguish raster images from line plotter output. **Line plotters** are vector output devices that draw an image using continuous lines, in the manner of traditional drafting (Figure 19.2b). Most users seem to prefer the quality of line-plotter output, but raster devices with fine resolutions are also capable of producing high-quality images. Raster plotters have the advantage of operating much more rapidly than vector plotters. Devices of both types are capable of producing color as well as black-and-white products.

Small-sized plots are commonly produced by the user. On the other hand, large-format output devices, especially color plotters, are expensive to own and maintain. Therefore, it is common for users to send digital files of their large map products to specialized service bureaus for plotting.

APPLICATIONS

Microcomputer Programs

A number of computer-mapping programs that run on microcomputers have been marketed recently. Many of these programs allow the user to create special-purpose maps without elaborate equipment (Figure 19.3). These programs typically limit the user to thematic databases, such as population and income data, supplied by the software manufacturer. In addition, the manufacturer usually supplies base-map data for a variety of areas, ranging from the entire country, through states, counties, Standard Metropolitan Statistical Areas (SMSAs, used in U.S. census), and zip-code areas. More flexible programs provide utilities that permit the user to create and use diverse data sets and map bases and to produce a variety of map types.

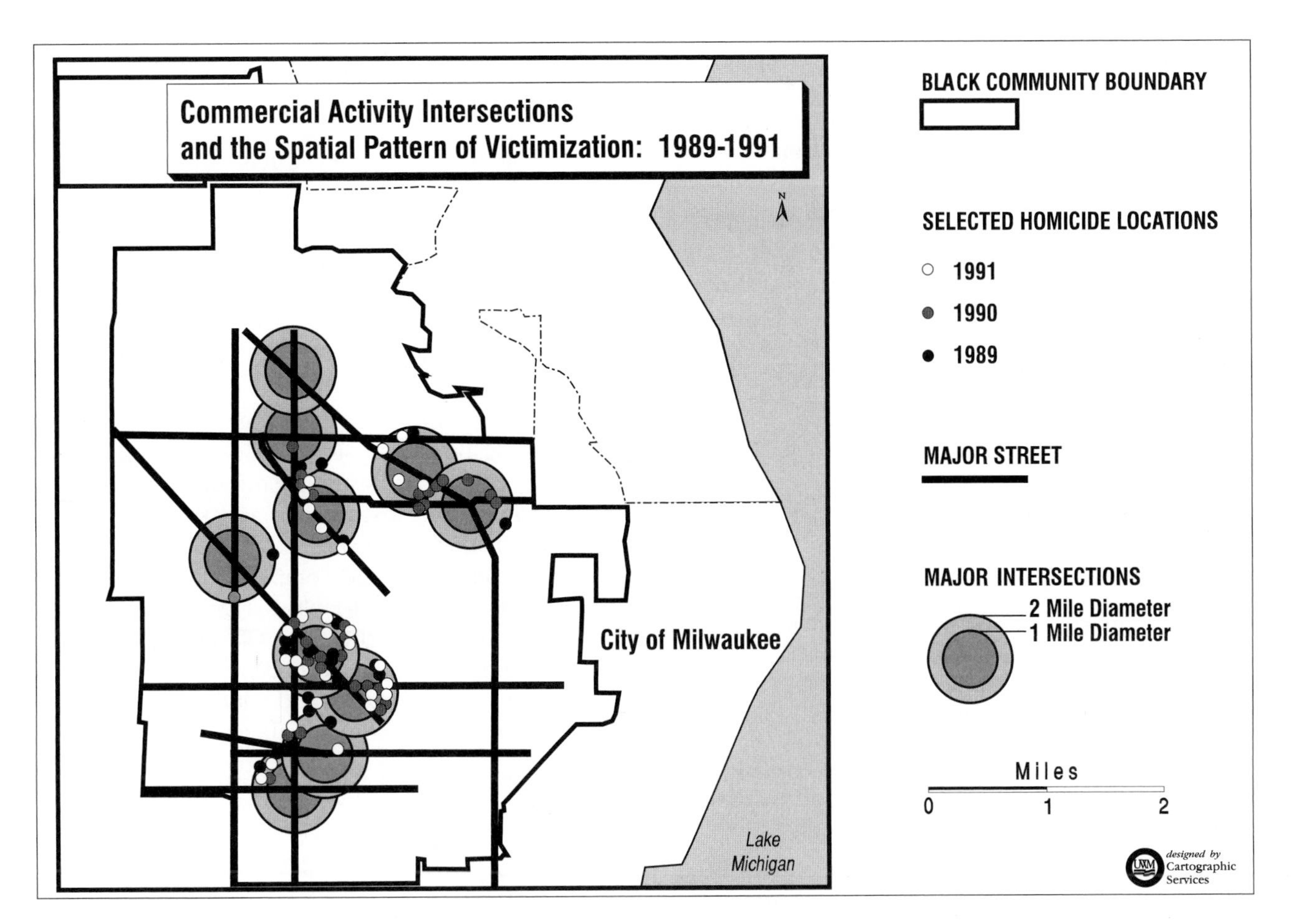

Figure 19.3 Map produced by a microcomputer.
Source: Courtesy of Professor Harold M. Rose and University of Wisconsin-Milwaukee Cartographic Services.

Because of the lack of standardization in the computer field, programs may only operate with particular computers, printers, plotters, and other hardware. If you are planning to implement a computer-assisted mapping project, the best approach is first to find the software that can produce the products you need. Then you will know what hardware to purchase to use the program effectively. Buying hardware first and then looking for software to do the job you require is likely to lead to extra expense and frustration.

Larger-Scale Applications

Many applications of digital cartographic information have been undertaken or are in the development stages. We look at a few examples of such applications here to provide some insight into the range of problems for which such techniques are helpful and to illustrate the concept of the "smart map."

Irrigation Potential

In a study of the Umatilla River basin in Oregon, a map showing the irrigation development potential of areas within the basin was produced by computer-assisted methods.[3] First, digital elevation models (DEMs) were used to compute surface slopes in the region (DEMs are discussed later in the chapter). This information was then combined with land-cover, soils, and land-ownership data, also in digital form. The results were used to produce a map of irrigation potential in the study area. The elevation and location data in the DEMs were also used to help determine the costs of irrigation. This calculation was based on the simple fact that higher elevations and greater distances from the water source require greater amounts of energy to deliver water to a site. This information, combined with pumping-plant data, determined the total energy requirements for bringing irrigation to the land. Then, information on the cost of energy was incorporated, and a map of the relative cost of irrigation was developed. The maps of irrigation potential and irrigation costs were used to produce the final map of irrigation development potential. This map, in turn, was used in making resource-efficient water-development decisions.

[3]Survey of Oregon Irrigation Development Potential," In T. R. Loveland and Ben Ramey, *Applications of U.S. Geological Survey Digital Cartographic Products, 1979–1983* (Washington, D.C.: U.S. Government Printing Office, U.S. Geological Survey Bulletin 1583, 1986), 29–33.

Terrain Analysis

In military applications, digital-analysis methods produce important tactical information used in making a variety of important decisions. Such information was previously obtained by traditional map-analysis methods but with less accuracy and with much less speed. The digital methods can provide information about sites that are screened from the enemy's electronic line of sight, that are away from enemy high-speed avenues of approach, or that have more than one access route.[4] Such information is vital to conducting an effective military operation.

Other Applications

A great variety of other applications of digital cartographic techniques has been developed. Examples that give an idea of the range of such applications are studies of intervisibility for forest-fire lookout sites, selection of the optimal areas for clear-cut logging; military observation, access, and concealment studies; selection of locations for microwave towers and of routes that would hide power lines; engineering applications, such as cut and fill, and overburden and reservoir volume calculations; forest-fire hazard evaluation; earthquake hazard prediction; rangeland management; and nuclear disaster planning.

Although these examples are classified as digital cartography applications, they cannot be clearly separated from applications of geographic information systems. The main distinction is that geographic information systems provide a higher level of data handling and analytic capabilities, as is discussed more fully in Chapter 21.

DATABASE AVAILABILITY

The creation of digital databases for use in computer-assisted cartography applications is time-consuming and expensive, especially because of the cost of the hardware required for some of the operations. In addition, because of the repetitive nature of many mapping projects, users may waste resources (both time and money) if they create databases that are essentially the same as those created by others. Indeed, the capability of sharing data is one of the advantages of computer-assisted cartography. For these reasons, already-prepared databases (of both base-map data and thematic information) are increasingly available for purchase. This development, especially given the often very reasonable prices of the available databases, frequently renders the production of new individual files not cost efficient. The availability of materials of this nature is constantly changing, so only some of the more important products are described here.

[4]Tom Hills, "Terrain Information Used by Field Artillery," *Tech-Tran* 11, no. 2 (Spring 1986): 2.

U.S. Geological Survey Products

The U.S. Geological Survey sells a number of digital cartographic products. The brief descriptions that follow provide a general idea of the characteristics of each.

Digital Chart of the World

The **Digital Chart of the World (DCW)** is a comprehensive base map of the world. It was digitized from the former Defense Mapping Agency (DMA) Operational Navigation Chart series, at a scale of 1:1,000,000. The vector database is stored on four CD-ROMs, along with attribute and textual data and software that permits information to be selected from the database and displayed on a personal computer (Figure 19.4). This high-quality, low-cost database is divided into four regions, arranged on the four CD-ROMs as follows: North America; Europe and northern Asia; South America, Africa, and Antarctica; and southern Asia and Australia. There are 17 layers of thematic information, in-

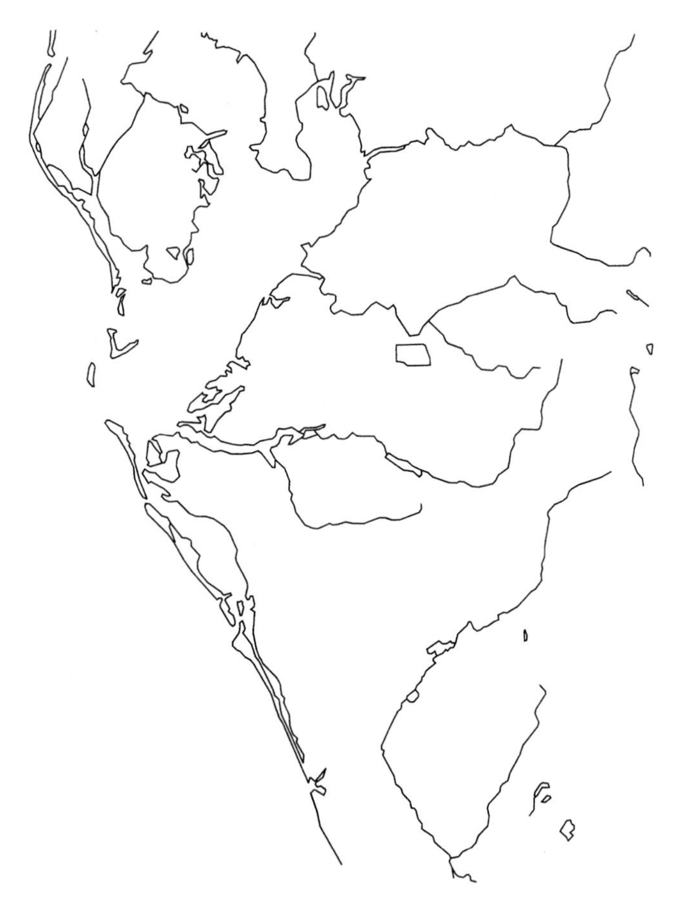

Figure 19.4 1:1,000,000-scale plot of coastline and hydrographic features from a 1-degree-by-1-degree section of the "Digital Chart of the World" (DCW) on an Albers equal-area projection. (Tampa Bay area, Florida, 27° N to 28° N, 82° W to 83° W.)

cluding ocean coastlines, drainage (rivers and lakes), political boundaries, cities, transportation networks, land cover, and elevation contours. It also contains a worldwide index containing more than 100,000 place names.

The original database is available from USGS Earth Science Information Centers (ESICs). However, private vendors also market databases derived from the DCW. The major selling point of these modified DCW files is a compressed format, with the worldwide coverage on one CD-ROM. Also, the formats are designed to provide greater ease of access for many commonly used mapping programs.

Digital Line Graphs

Digital line graphs (DLGs) contain digital map data in vector format (Figure 19.5 and Table 19.1). They are of three types: (1) large scale, primarily digitized from 1:24,000-scale quadrangle maps; (2) medium scale, from 1:100,000-scale quadrangle maps; and (3) small scale, from the 1:2,000,000-scale *National Atlas* sectional maps. The files contain point, line, and area data, with each category of information (boundaries, hydrography, Public Land Survey System, or transportation) stored as a separate file or subfile. The features are coded by size and type so that it is possible to group or select on the basis of those characteristics.

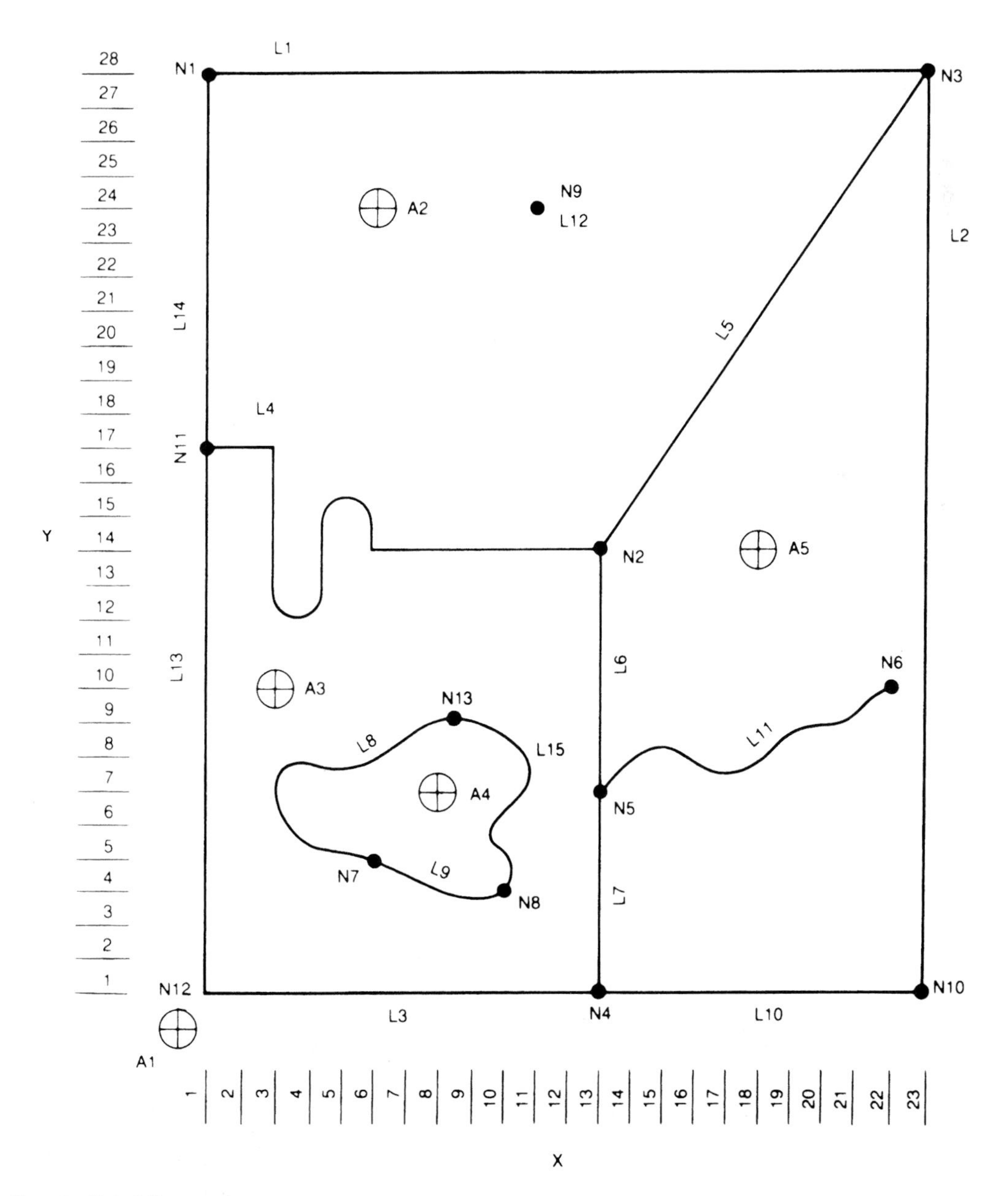

Figure 19.5 Sample digital line graph.
Source: U.S. GeoData. *Data Users Guide 1: Digital Line Graphs from 1:24,000-Scale Maps* (Reston, Va.: U.S. Geological Survey, 1986).

TABLE 19.1 Digital Description of Topological Elements and Relationships of a Digital Line Graph (See Figure 19.5)

Nodes and Areas

Internal ID Number	x Coordinate	y Coordinate	Internal ID Number	x Coordinate	y Coordinate
N1	1	28	N2	13	14
N3	23	28	N4	13	1
N5	13	7	N6	22	10
N7	6	5	N8	10	4
N9	11	24	N10	23	1
N11	1	17	N12	1	1
N13	9	9			
A1	0	0	A2	6	24
A3	3	10	A4	8	7
A5	18	14			

Lines

	Nodes		Area		Coordinates	
Number	Starting	Ending	Left	Right	(First x, y)	(Last x, y)
L1	1	3	1	2	1, 28	23, 28
L2	3	10	1	5	23, 28	23, 1
L3	4	12	1	3	13, 1	1, 1
L4	11	2	2	3	1, 17..........	13, 14
L5	2	3	2	5	13, 14	23, 28
L6	2	5	5	3	13, 14	13, 7
L7	5	4	5	3	13, 7	13, 1
L8	13	7	4	3	9, 9	6, 5
L9	7	8	4	3	6, 5	10, 4
L10	4	10	5	1	13, 1	23, 1
L11	5	6	5	5	13, 7..........	22, 10
L12	9	9	2	2	11, 24	11, 24
L13	12	11	1	3	1, 1	1, 17
L14	11	1	1	2	1, 17	1, 28
L15	8	13	4	3	10, 4..........	9, 9

Source: U.S. GeoData, *Data Users Guide 1: Digital Line Graphs from 1:24,000-Scale Maps* (Reston, Va.: U.S. Geological Survey, 1986).

The files currently being distributed have been spatially structured to define all of the topological relationships that exist between the component elements of the file and are designed for use in geographic information systems.

Small-scale DLGs are divided into twenty-one multistate-blocks. The files contain boundaries, hydrography, and transportation information.

Medium-scale DLGs are provided in groups that cover 30-by-30-minute areas, corresponding to the east or west half of the 1:100,000-scale source map. Each 30-minute area group usually consists of four 15-by-15-minute files, although as many as 16 files may be used for areas of high feature density, such as for roads and trails in large cities. Current medium-scale DLGs contain hydrography and transportation files.

Most large-scale DLG files cover the same area as a U.S. Geological Survey 7.5-minute quadrangle map. The geographical extent of large-scale coverage is presently limited, although it will eventually include the entire country. Most current, large-scale DLGs contain boundary, hydrography, U.S. Public Land Survey (USPLS), and transportation files.

The extensive, computer-compatible data contained in DLG files are a great boon to users. The files provide the raw material on which to base the production of customized maps, without the necessity of creating new base files. Once the DLGs are read into a computer-mapping system, they can be extensively modified. Information can be added or deleted, and entirely new overlay files can be created as well. Hard-copy maps can be produced from various combinations of these files, as desired.

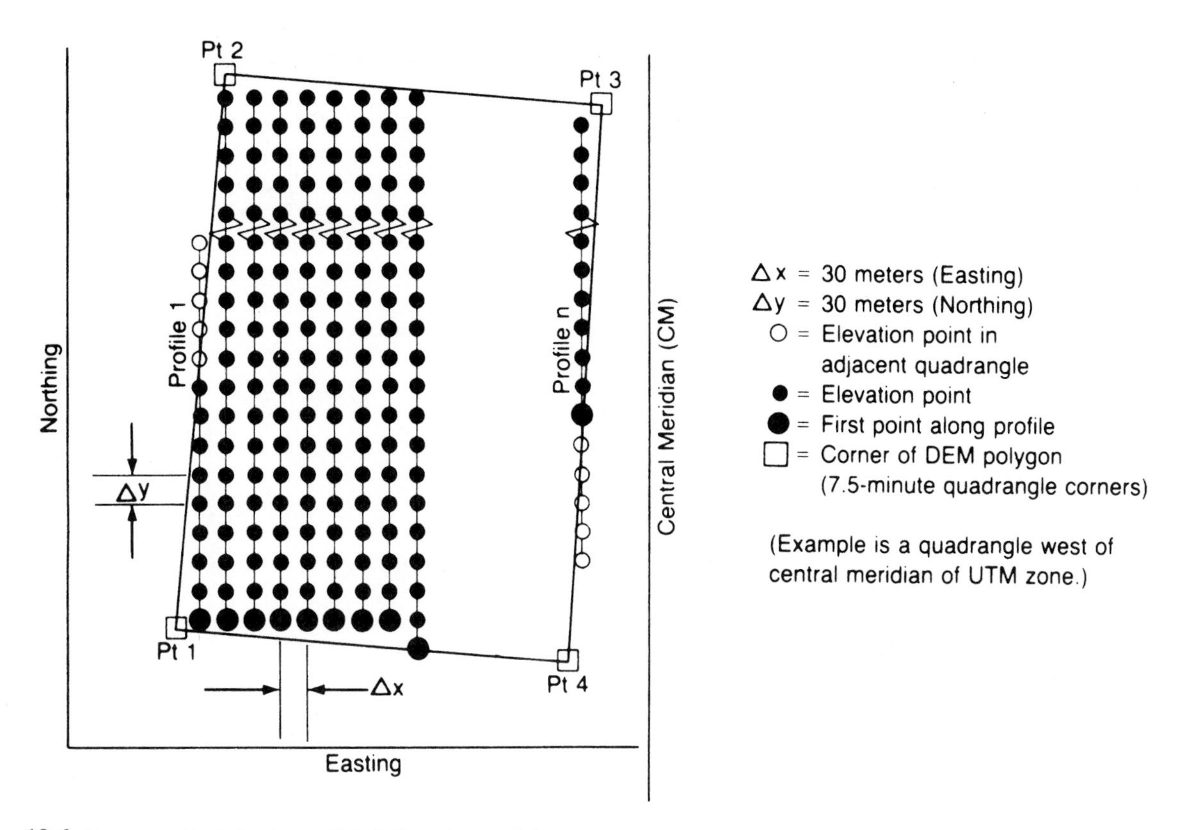

Figure 19.6 Structure of a 7.5-minute digital elevation model.
Source: U.S. GeoData. *Data Users Guide 5: Digital Elevation Models* (Reston, Va.: U.S. Geological Survey, 1987).

DLG files are now available in CD-ROM format, which is compatible with microcomputers.

Digital Elevation Models

Digital elevation models (DEMs) are digital records of terrain elevation values. They are produced in four different formats: 7.5-minute, 15-minute, 30-minute (2 arc second), and 1-degree.

Each 7.5-minute DEM corresponds to the equivalent U.S. Geological Survey quadrangle. The files contain elevation values, in meters, spaced at regular, 30-meter intervals (Figure 19.6). The location of each value is referenced to the Universal Transverse Mercator (UTM) coordinate system.

Fifteen-minute DEMs each correspond to a 15-minute topographic quadrangle map in Alaska. The files contain elevations spaced at 2 arc seconds of latitude by 3 arc seconds of longitude.

Thirty-minute DEMs correspond to the east half or west half of USGS 30- by 60-minute quadrangles for the conterminous United States and Hawaii. Each 30-minute unit is distributed as four 15-minute by 15-minute cells. The spacing of elevation values is 2 arc seconds of latitude by 2 arc seconds of longitude.

The 1-degree DEMs cover 1-degree-by-1-degree blocks, which are the east or west half of the 1:250,000-scale source maps from which they are digitized. The spacing between the data values is generally 3 arc-seconds, which means that the east-west ground-distance spacing varies with latitude. At the equator, the spacing is approximately 90 meters in each direction, for example. At 50° latitude, however, the spacing is approximately 90 meters north-south by 60 meters east-west. Elevation values are recorded to the nearest meter and are referenced by latitude and longitude coordinates.

DEMs provide data that can be used for the production of contour maps and slope maps. They also provide the raw material for intervisibility maps and profiles. As with DLGs, DEMs are now available in CD-ROM format.

Digital Orthophotos

A **digital orthophoto** is a digital image scanned from an aerial photograph. Displacements caused by the camera and by differences in terrain elevation have been removed from the scanned image, so it can be used directly as a map. A standard digital orthophoto produced by the USGS is called a **digital orthophoto quadrangle (DOQ).** Each image is scanned from a 1:12,000-scale black-and-white, color, or color-infrared orthophoto and covers 3.75 minutes of latitude and 3.75 minutes of longitude (a quarter of a standard 7.5-minute quadrangle). The product includes the coordinates of ground control positions for georeferencing purposes. Digital orthophotos are also produced by private contractors, and the specific characteristics of such products vary.

Because they are already in digital form, digital orthophotos are often used as background images in geographic information systems. In this type of application, other layers of information can be overlaid and standard GIS manipulations carried out. Typical applications include revision of existing maps, vegetation and timber management, habitat analysis, and flood analysis. Such applications are often assisted by the environmental information, such as the locations of trees and wetlands, that is visible on the orthophoto.

Digital Raster Graphics

A **digital raster graphic (DRG)** is a scanned image of a USGS topographic map (Figure 19.7). The entire map sheet is scanned, including the map collar information. (Map collar information consists of the title, scale, declination diagram, date of publication, list of data sources, and similar material printed outside the neatline of USGS topographic maps.) The image within the map neatlines is georeferenced to UTM coordinates at the 2.5-minute grid ticks. Horizontal accuracy matches that of the published source map. Production of this product has begun on the 1:24,000-scale topographic map series and will be extended to include other series.

DRGs are distributed on CD-ROMs as color images at 250 dots per inch (dpi) resolution in compressed TIFF format. File sizes range from 5 to 15 megabytes.

An entire DRG, or part of one, can be used as a substitute for a printed map in digital manuscripts. DRGs are also useful as backdrops on which to overlay other digital map data.

DRGs are available through the usual USGS map distribution channels. Information is available on the World Wide Web.

Digital Land-Use and Land-Cover Data

Digital land-use and land-cover data are produced by the U.S. Geological Survey from 1:250,000- and, occasionally, 1:100,000-scale U.S. Geological Survey land-use and land-cover maps. The files contain information on up to 37 categories of land-use and land-cover data. They have a minimum-size polygon of 16 hectares (40 acres) in rural areas and of 4 hectares (10 acres) in urban and built-up areas, water areas, and some areas of other types of land use. Separate files provide the outlines of state and county boundaries and of census county subdivisions. A hydrologic file contains watershed boundaries, and a file of federal land ownership delimits areas that contain 16 hectares or more. In some cases, a file of state land ownership is also produced.

Digital land-use and land-cover data are often used in connection with land-use planning studies, such as the irrigation study mentioned earlier. They are also frequently incorporated into digital geographic information systems.

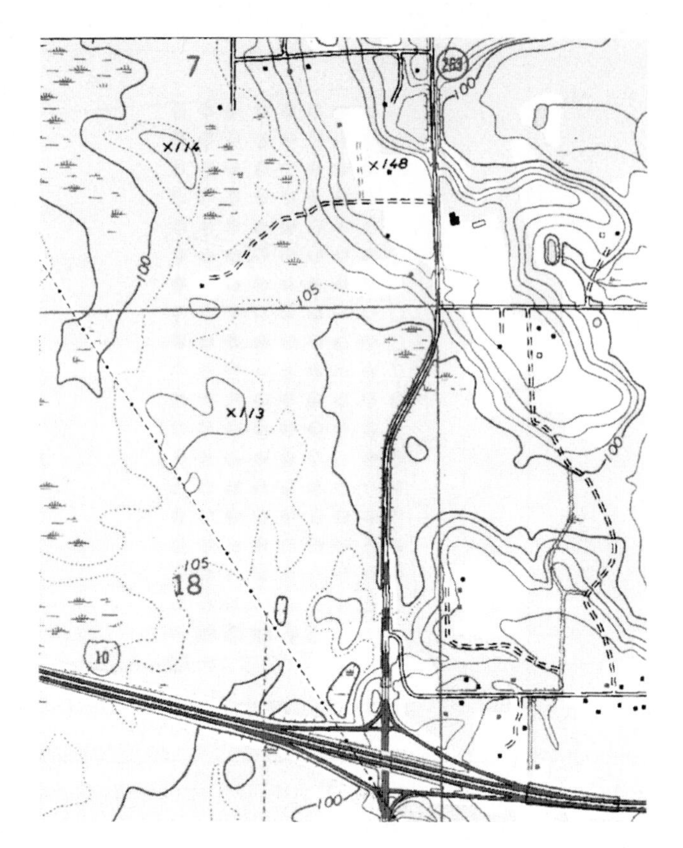

Figure 19.7 Portion of a digital raster graphic (DRG). (Tallahassee quadrangle, Florida, 1:24,000.)

Geographic Names Information System

An additional element of the U.S. Geological Survey digital cartographic data is the **Geographic Names Information System (GNIS).** This automated system is designed to provide an index of standardized geographic names in the United States, as well as other information about the names. GNIS contains records of almost 2 million geographic features in the United States, garnered from USGS 1:24,000-scale maps, as well as charts, maps, and databases of several federal agencies. The features include settlements, and the parks, schools, and other facilities associated with them, as well as natural features, such as mountains, valleys, streams, and lakes. Information available includes the official name of the feature, the state and county in which it is located, its geographic coordinates, and the name of the 1:24,000-scale USGS topographic map on which it appears.

The database is searchable online. In addition, standard reports and digital data sets are available, by state, and customized reports can be ordered. USGS Professional Paper 1200, *The National Gazetteer of the United States of America,* and a concise *National Gazetteer* are also available.

Bureau of the Census Products

In support of its Decennial Census of Population, the U.S. Bureau of the Census maintains a digital cartographic database for the United States. This database is called **TIGER,** an acronym for *T*opologically *I*ntegrated *G*eographic *E*ncoding and *R*eferencing system. The goal of TIGER is to provide a geographical database for the entire country.

At the Bureau of the Census, TIGER is used to assist in the collection, tabulation, and dissemination of census data. This includes the assignment of location codes to addresses and the provision of a geographic structure for the tabulation and publication of census data. Also, it involves the production of cartographic products to support both collection field work and publication efforts.

For users outside the Bureau of the Census, TIGER represents a valuable database for a wide range of applications. In particular, TIGER/Line files are records of topologically consistent networks of line segments, representing roads, boundaries, and other features. Each linear feature in the file is represented by one or more segments, identified by a unique record number. Each of these segments has a recorded beginning and ending point. If needed, shape points are added to the record to represent the path followed by the line segment. The coordinates of each of these points is recorded in degrees of latitude and longitude, expressed to six decimal places. Additional information is recorded for each line segment to describe its attributes. This includes the source of the information; the type of feature (census feature class code [CFCC]); topological information, including the state, county, and census divisions to the left and right of the line; and, in certain urban areas, address and zip code information.

In urban areas, each record in a TIGER/Line file describes a segment of a street (Figure 19.8). The record includes the name of the street, the range of addresses on both sides of the street segment, and the node numbers that occur at each end of the segment. Also included are the numbers that identify the block, census tract, county, or other areal units that adjoin either side of the segment.

One somewhat controversial aspect of TIGER/Line files is their level of accuracy. Especially in the case of early releases of the files, some features were not located accurately, did not have the proper shape, or were missing altogether. Often such problems occurred when information about new subdivisions or similar changes was entered from field notes without surveyed data. The presence of apparent inaccuracies does not necessarily indicate that the census is based on faulty maps. Instead, it suggests that the census requires a level of accuracy that may be too low for other purposes. For example, in the conduct of the census, the absence of a highway on-ramp or the exact location of a street in a subdivision is not particularly significant. To a company using the files as the basis for delivery routing, however, the absence of the on-ramp would obviously be troublesome, as it would be to many map users. As is the case with any database, TIGER/Line files should be used only if they meet the potential user's accuracy requirements.

Figure 19.8 Portion of a TIGER file for Fairfax City, Virginia.

It is important to note that the Bureau of the Census gives continuing attention to the improvement of the TIGER/Line files. Also, a number of private companies produce enhanced files based on TIGER. These files have been improved by additional checking and correction of locational data and updating of details on roads, highways, and other features. Often, the enhanced files are used in electronic street and highway maps or for automated vehicle navigation systems.

TIGER/Line files may be used to produce maps of roads and other features. In addition, they may be combined with statistical information about population, housing, or income, obtained from the census or from other sources. With proper software, various types of thematic maps may be produced for use in market research, routing problems, address-matching problems, redistricting plans, and many other situations. As a service to mapping and geographical information system practitioners, TIGER/Line files are periodically released to the public, although application software that permits such uses is not available from the Bureau of the Census. There are, however, several private companies that produce software capable of using the TIGER/Line files. These same companies usually provide data files that have been enhanced by additional editing.

World Data Banks

Two worldwide databases are **World Data Bank I** and **II.** World Data Bank I contains data about coastlines, national boundaries, and place names. It was digitized from 1:12,000,000-scale source maps. World Data Bank II contains data about coastlines, major rivers, national boundaries, state or province boundaries, and place names. It was digitized from 1:3,000,000-scale source maps, supplemented by some larger-scale coverage. These databases will likely see less use, now that the Digital Chart of the World is available. Both databases are available from the National Technical Information Service (NTIS):

National Technical Information Service
Technology Administration
U.S. Department of Commerce
Springfield, VA 22161
Phone: (703) 605-6000

Other Databases

The field of computer-assisted cartography is changing rapidly, and a large number of government agencies, university researchers, and private companies are involved in new developments. Indeed, new products are introduced (and some old ones disappear) with great frequency. For this reason, a book of this nature cannot provide the latest information about available materials. Current information on these topics is provided more efficiently through software reviews and through information about databases that appears in periodical publications. *Cartography and Geographic Information Systems, Cartographica,* and *The Professional Geographer,* for example, regularly publish information of this nature.[5]

Finally, mention must be made of the increasing access to data sources provided by the Internet. This source is potentially so important that anyone interested in obtaining a variety of mapping data should learn how to use it. The operation of the Internet lies well beyond the scope of this book, but useful manuals and tutorials are available on this topic. The range of resources on the Internet changes daily. Consequently, giving clear guidance about what is available there is impossible. This book's website provides a list of Internet addresses that are useful as starting points for the exploration of this interesting resource.

ACCURACY ISSUES

National Map Accuracy Standards were discussed in Chapter 6. As was noted, those standards do not directly apply to the digital databases discussed in this chapter. Nonetheless, accuracy and related issues are important in computer-assisted cartography and geographic information systems applications.

Scale

A crucial difference between a conventional, printed map and a digital database is that the scale of the printed map is explicit. This means that, when its scale is changed, it is immediately apparent and the map user is alerted to the fact that measurement accuracy is affected. With a digital database, however, scale is implicit—it is directly related to the method of digitization and the scale of the source map. However, a map can be produced from the data at any desired scale. The scale of the map is dependent on its size and the area covered, whether it is on a computer screen or a plotted hard copy. As a result, the map may be at a scale that implies far greater accuracy than is warranted. For example, if a database is digitized from a 1:100,000-scale source map, producing a map from that data at a scale of 1:24,000 is unwarranted.

Metadata

Lacking map accuracy standards for digital map databases, the approach taken toward their evaluation has been to provide metadata about them. **Metadata** are data about data. That is, a metadata file is provided to accompany a digital database. This file contains information that describes the characteristics of the database, so that a potential user can evaluate it. Thus, metadata help a potential user establish what data are available, whether the data meet specific needs, and where and how to obtain the data. Because data are provided by a broad range of producers, a National Geospatial Data Clearinghouse has been established to help users find suitable data. Information about metadata and the Clearinghouse can be obtained on the Internet and from

Federal Geographic Data Committee
U.S. Geological Survey
590 National Center
Reston, VA 20192
Phone: (703) 648-5514

Metadata also benefit the producing organization by documenting the origins and characteristics of each data set. Approaches to metadata differ, but the USGS standards are a typical example. Briefly stated, the contents of these standards are as follows:

Identification information. Covers basic information such as the title, geographic area covered, and date of preparation.

[5]One useful source is Dean T. Edson, "Data Bases," in Section 5, "Automation in Cartography," *U.S. National Report to ICA, 1984,* ed. Judy M. Olson (Falls Church, Va.: American Congress on Surveying and Mapping, 1984), 51–54.

Data quality information. Provides an assessment of the quality of the positional accuracy, completeness, and consistency of the data and the sources and methods used to produce it.

Spatial reference information. Describes the map projection or grid-coordinate system used, horizontal and vertical datums, and resolution.

Entity and attribute information. Provides information about types and attributes of features.

Distribution information. Includes information about the formats in which the data are available, how to contact the distributor, and fees charged.

Metadata reference information. Consists of information about the party responsible for preparing the metadata and how the report was prepared.

For a map user, the importance of the availability of metadata is twofold. First, metadata provide background regarding the concerns of map producers and how data quality can affect the quality of computer-assisted maps. Second, if one is seeking digital data for use in a computer program, metadata provide a framework for the evaluation of possible data sources.

SUMMARY

Computer-assisted cartography involves the direct use of computers for map production. The anticipated advantages of this approach include quick and easy map design, integrated mathematical or statistical operations (including scale or projection changes), and symbol selection. Also, the rapid exchange of data between users means that each new map can incorporate information that is correct and up-to-date. Beyond facilitating simple map production, digital techniques allow textual or tabular information to be tied to cartographic features. Most of these advantages stem from the use of digital cartographic databases. Implementation problems, however, have resulted in the advantages not always being achieved, although this situation is rapidly improving.

Digitizing is the process of converting map data from their original visual form to digital form. For maximum benefit, the various map elements must be labeled with identification codes during the digitizing process so that they can be selectively identified in the data file.

Digital records in vector form contain the coordinates of specific points. Linear features, including the outlines of regions, consist of linked coordinates of a series of points. In contrast, digital records in raster form consist of information about whether features fall within individual cells in a matrix. Arbitrary machine coordinates, recorded in inches or centimeters, are measured from a control point on the digitizing board. More usefully, however, machine coordinates are usually converted to latitude and longitude, Universal Transverse Mercator, or State Plane Coordinates, in a process called *geocoding*.

To generate output, the image may be photographed on the computer screen, but printing or plotting, using either raster or vector devices, are more often used. Raster output devices, either printers or plotters, produce an image by creating a pattern of dots. Line plotters are vector output devices that draw an image, using continuous lines, in the manner of traditional cartography.

Computer mapping programs that run on microcomputers allow the user to create special-purpose maps without elaborate equipment. Larger-scale computer systems are usually needed for major applications, mostly because of the greater database capacity and higher operating speeds that they offer.

Because the creation of digital databases is time-consuming and expensive, already-prepared databases of both base map data and thematic information are becoming increasingly available. U.S. Geological Survey products are among the more important of these digital cartographic products. The Digital Chart of the World (DCW) provides worldwide coverage at a scale of 1:1,000,000. Digital line graphs (DLGs) contain digital map data in vector format, and digital elevation models (DEMs) provide elevation values. Digital raster graphics (DRGs) are scanned images of USGS topographic maps, and digital orthophoto quads are scanned from 1:12,000-scale orthophotos. Digital land-use and land-cover data are also available, as are Geographic Names Information System (GNIS) data.

The U.S. Bureau of the Census has developed a digital cartographic database for the United States, called *TIGER* (*T*opologically *I*ntegrated *G*eographic *E*ncoding and *R*eferencing system). This database is used by the Bureau of the Census to assist in the collection, tabulation, and dissemination of census data. It also provides other users with a database for a wide range of applications. In general, TIGER/Line files include street names and address ranges, as well as block, census tract, county, and zip code data.

World Data Banks I and II are other examples of easily obtainable files that contain data about coastlines, national boundaries, and other features. The field of computer-assisted cartography is changing rapidly, and a large number of government agencies, university researchers, and private companies are involved in new developments. Therefore, additional map data and thematic-information files are continually becoming available. The Internet, in particular, may prove to be an important source of data.

National Map Accuracy Standards do not directly apply to digital databases, but accuracy and related issues are important in computer-assisted cartography and geographic information systems applications. One difference between a conventional, printed map and a digital database is that the scale of the printed map is specified explicitly. With a digital database, however, scale is related to the method of digitization and the scale of the source map. This means that a map can be produced from digital data at any desired scale, even if the accuracy of the database does not warrant it.

Metadata, which are data about data, are used to evaluate digital map databases. Metadata help a potential user establish what data are available, whether the data meet specific needs, and where and how to obtain the data. USGS metadata standards contain information identifying the data set, as well as information about data quality, the spatial reference system, types and attributes of features, how to obtain the data, and information about the metadata themselves.

SUGGESTED READINGS

Bauer, Michael. "Value-Added Data Bases Are Foundation for Diverse AM/FM/GIS Applications," *Geo Info Systems* 2, no. 5 (May 1992): 36–41.

Bickmore, D. P. "Objectives in Digital Mapping." In *Proceedings, International Seminar on Computer-Assisted Cartography*. Organized by the Survey of India in collaboration with the International Cartographic Association, New Delhi, India, November 22–29, 1983, 38–40.

Computer Mapping of Natural Resources and the Environment, Plus Satellite-Derived Data Applications. Vol. 15. Cambridge, Mass.: Harvard University, Laboratory for Computer Graphics and Spatial Analysis, 1981.

Douglas, David H., and A. Raymond Boyle, eds. *Computer-Assisted Cartography and Geographic Information Processing: Hope and Realism*. Ottawa, Canada: Canadian Cartographic Association, Department of Geography, University of Ottawa, 1982.

Goodchild, M., and S. Gopal. *The Accuracy of Spatial Databases*. New York: Taylor and Francis, 1989.

Guptill, Stephen C. "1:100,000-Scale Digital Cartographic Data Base for Federal Requirements." In *United States Geological Survey Yearbook,* Fiscal Year 1984. Washington, D.C.: U.S. Geological Survey, 1984.

McEwen, R. B., and H. W. Calkins. "Digital Cartography in the USGS National Mapping Division." *Cartographica* 19, no. 2 (1982): 11–26.

Monmonier, Mark S. *Computer-Assisted Cartography: Principles and Prospects*. Englewood Cliffs, N.J.: Prentice-Hall, 1982.

Olson, Judy M., ed. *U.S. National Report to ICA, 1984*. Falls Church, Va.: American Congress on Surveying and Mapping, 1984, Section 5, "Automation in Cartography."

Peucker, Thomas K. *Computer Cartography*. Washington, D.C.: Association of American Geographers, Commission on College Geography, Resource Paper No. 17, 1972.

Peuquet, Donna J., and A. Raymond Boyle. *Raster Scanning, Processing and Plotting of Cartographic Documents*. Williamsville, N.Y.: Spad Systems, 1984.

Slonecker, E. Terrence. "National Map Accuracy Standards: Out of Sync, Out of Time." *Geo Info Systems* 2, no. 1 (January 1992): 20, 24–26.

Taylor, D. R. Fraser, ed. *The Computer in Contemporary Cartography*. Vol. 1., *Progress in Contemporary Cartography*. New York: John Wiley & Sons, Inc., 1980.

CHAPTER

20

Digital Map Applications

Maps can now be provided on computers and can provide the basis for geographic information systems, as described in Chapter 21. In addition, this technological advance means map users are no longer restricted to using maps in printed form but have new and interesting possibilities for map access and use. This chapter describes two such applications: the direct use of electronic atlases, road maps, and trip planning software and the development of automated automobile navigation systems.

The range of software available for mapping and related applications on microcomputers changes rapidly, with new programs constantly appearing and others disappearing. The available programs display a wide range of capabilities and an equally wide range of prices. Before considering the purchase of any program, obtain as much information as possible about it. Developers and vendors will be happy, of course, to provide literature about their products. Such information should be supplemented, whenever possible, by reports from users of the programs (again, developers and vendors will often provide the names of users of their programs). In addition, reviews of some programs are available in journals such as *GeoInfo Systems, GEOWorld, GISWorld, Cartography and Geographic Information Systems, Mercator's World,* and *The Professional Geographer.* Reviews give a good idea of the hardware requirements as well as the capabilities and limitations of a program. Information about both of these aspects will help in assessing a program's suitability for a particular purpose.

Because of rapid changes in computer applications to cartography, it is impractical to provide a list of programs and databases in this book. Instead, information is provided at the book's website, at http://www.mhhe.com/earthsci/geography/campbell.

Electronic Atlases

Electronic atlases are designed to resemble traditional printed atlases. They provide access to general reference maps and specialized thematic maps of the world, countries, or regions, for example. The two types differ, however. With a traditional atlas, a map user typically consults an index or gazetteer and then turns to the appropriate page. The user then searches the page for the specific place of interest. With an electronic atlas, access is generally easier. Search criteria are typically entered by selection from menus and the software then displays the desired map, often with the place name of interest highlighted. This approach eliminates the flipping of pages of what is often a large and cumbersome book and usually saves time as well. This advantage is scarcely worthwhile by itself, however, and other advantages make electronic atlases more useful and interesting.

A major shortcoming of traditional atlases is that their content is fixed in time and scope. The main method of solving this problem is to publish periodic new editions, but this is costly, and developing adequate sales volume to justify frequent new editions is sometimes difficult. Some producers have tried to solve this problem by publishing atlases in loose-leaf format. Then, maps of additional topics or updates of existing topics are published to expand the coverage and keep it current. This is an awkward approach, however, and has never gained popularity. By contrast, in an electronic atlas, new maps, either updates or of new topics, can be added to the database anytime; the ease and cost of doing this vary, depending on the technology in use and the publishing and pricing practices of the producer, but it is generally effective. More important, however, the electronic form provides an unparalleled opportunity to create maps tailored to the specific needs of the map user. Joel L. Morrison outlines some possibilities in an article on which the following comments are based (see the Suggested Readings at the end of the chapter).

A national atlas is generally a large, expensive book. The 1970 U.S. Geological Survey publication, *The National Atlas of the United States of America,*™ was typical. The 400-page, 12-pound volume sold for $100. Therefore, most copies are found in libraries, and few people have copies of their own. In addition, when the maps became out of date, it was impractical to issue a new, printed atlas. Now, the new edition of this atlas, which is partially completed, includes both electronic and paper map products, with a concentration on electronic products. A major goal is to develop a comprehensive reference atlas that is easily and inexpensively available to home computer users on demand and which can be kept up to date.

A comprehensive national atlas is useful because it contains information that is not included in commercial atlases. This is because commercial producers face high production and distribution costs that must be offset by sales revenue. Therefore, the content of the work must appeal to the largest possible number of readers. A national atlas, on the other hand, "should contain information that is of interest and utility to every person in the country" (Morrison, *op. cit.*), including technical information that may lack general appeal.

Morrison suggests the following format for a "personalized national atlas," his term for a national atlas in electronic form. Because of the need for broad coverage, he visualizes production by a consortium of federal and state agencies and private companies. The atlas would contain base and thematic data, a set of standards, and optional additional features. Base data would include items that change infrequently, such as rivers and coastlines, state and county boundaries, watersheds, national parks, wildlife refuges, and national forests. Other important base data, such as zip code areas, telephone area code regions, census subdivisions, and congressional districts would also be included, though they might change more frequently. The atlas would also contain a listing of geographical names, with their locations, and a digital elevation model (DEM) depicting the terrain of the entire country. In addition, it would include the basic highway network and, perhaps, nationwide orthophoto coverage. Thematic data sets would include items such as weather data, socioeconomic and census statistics, and information provided by commercial companies. The system would provide some data free of charge, but accessing proprietary data might require the payment of a fee to the provider.

A personalized atlas is an attractive idea because it would allow individuals to produce maps to suit their own needs by selecting and combining base and thematic data. Thus, it would avoid being limited to existing maps. Also, because the system databases would be updated regularly, the atlas would not become obsolete with passing time. Electronic technology also makes it possible to add graphics (photos, existing maps, diagrams) and sound to produce maps that go far beyond what publishers can include on the pages of even the most imaginatively produced book.

Although a personalized national atlas may sound visionary, a prototype that incorporates many of its features already exists. This is the *National Atlas of Canada,* which is now accessible on the World Wide Web. Using the facilities provided, one can browse, select, and view existing maps. Also, new maps can be created, displayed, printed, or saved to disk. This option allows users to select standard layers (coastlines, political boundaries, roads, and geological, hydrological, and environmental zones) and to combine them with selected thematic information. The available information consists of the distribution ranges of endangered species of birds, mammals, and reptiles, but this is likely to be greatly expanded. Another option provides access to the Canadian Geographical Names Data Base. This is a useful utility for checking the location and spelling of toponyms.

This discussion of electronic atlases is couched in terms of using Internet access to databases and mapping utilities. In addition, however, many electronic atlases are now available for individual microcomputer users. Because of the rapidly changing array of such programs, with new titles appearing and old ones changing or disappearing from the market, providing individual descriptions of these programs is not feasible. Instead, we will mention some typical features. No program is likely to have all these features, but the discussion will help in the selection of programs with useful features to meet specific needs.

Typically, electronic atlases contain already prepared locational and statistical maps, text, and other graphic material. Electronic atlases are designed to make it easy to select specific topics and usually provide facilities for adapting and printing the text, tables, and maps for specialized needs. As mentioned, a potential advantage of electronic atlases is the ease of updating them with current information, although this may require the purchase of additional files. Also as already mentioned, these atlases free the user from turning pages. Another feature that differs from traditional atlases is the ability to zoom in to enlarge the features being viewed—the electronic equivalent of using a magnifying glass. Depending on the detail included in the database, zooming in may make additional detail available. Zooming out provides an overall view of a region. At least one program provides detail comparable to a 1:1,000,000 topographic map. This scale is more detailed than typical, traditional atlas maps and is likely to be quite common, now that the Digital Chart of the World (DCW) is available as a base.

Commonly, electronic atlases provide built-in thematic databases. These are usually derived from traditional sources, such as United Nations publications, Bureau of the Census tables, and informational almanacs of various types. Typical topics include population size and composition, educational characteristics, crime statistics, political inclinations, wildlife distribution, and travel information. The types of maps that can be produced include topographic maps, political maps, thematic maps (agriculture, history, demographics, environment, government, health, international organizations, language, religion, ethnic characteristics, economic activity, climate, time zones). Besides offering maps, graphs, and tables, many programs provide information in text form. This commonly includes coverage of the people, economy, history, and politics of individual countries and travel information. Some programs include descriptions of known regions of political unrest around the globe. The presentation varies, so that some programs are more suitable for young people and some for high school, college, or adult users.

Importantly, some electronic atlas programs provide a data entry option so that the user can enter data obtained from other sources. This greatly improves the flexibility and potential usefulness of the program, including the timeliness of the data. An additional option that is sometimes included is the display of the national flag and the playing of the national anthem when a national map is viewed. Although this feature may seem to have little more than curiosity value, it may be of some educational value and may spark interest, especially for younger users. Another frequently offered function involves the ability to find the distance between any two cities, by either selecting them on the screen or entering their names from the keyboard. At least one atlas program provides a unique view of the earth, including a spinning globe with color hill shading. In addition, it provides access to satellite images and time-lapse photos. Simulated flyover views of the terrain are included, along with a variety of photos of peoples and natural features of different regions.

User Concerns

Potential purchasers of electronic atlases need to consider several factors that affect the potential usefulness of a particular product. Usually, the first concern is compatibility of the product with the hardware and operating systems on which it is expected to run. If the program requires a different operating system or different or additional hardware, the expense of upgrading the system to meet those requirements must be considered.

The cost of the program itself is another obvious consideration. A costly program may be assumed to be better than a less expensive one. However, cost in relation to the program's capabilities is a more important consideration. A less expensive program, for example, may have the desired features, and some capabilities of a more expensive program may not be needed. In these circumstances, the cheaper program would represent a considerable value. On the other hand, if the less expensive program cannot meet all of the performance requirements, paying the additional money for a more capable program may be preferable. Therefore, before buying any product, carefully read the literature describing its capabilities for a basis for evaluation. Even more insight can be gained by reading published reviews of the programs under consideration. Reviews of specific programs are not necessarily easy to find, however (see the Suggested Readings at the end of the chapter). When reading reviews, also remember that each reviewer has a point of view regarding program features and that this point of view may not coincide with one's own needs. A further problem is that, because of the rapid changes in software, reviews often lag considerably behind the release of new versions of programs. Thus, problems with an early version of a program may be corrected in a later version. On the other hand, a new version of a program is not always a definite improvement over an earlier version. New products often have to undergo further revision to bring them up to their full potential—as with other products, it sometimes pays to be a later adopter of the latest version.

Two capabilities are worth special consideration. One of these is the possibility of importing databases (both geographic and thematic) from outside sources, as opposed to being restricted to the program's internal databases. This adds flexibility and can be a major cost saving because databases are often available from World Wide Web sources. Another useful capability is the ability to export the maps to a word processing program for inclusion in a report or article. Be careful, however, to consider copyright matters, because the rules for usage may differ among personal, student, and commercial uses. To avoid copyright violations, investigate and resolve these matters before putting the maps to use (see Appendix C).

Finally, although many electronic atlases produce cartographically sound maps, some suffer from poor use of cartographic symbolization. For example, the color sequences used for choropleth maps may not be suitable, the size of proportional symbols may not be calculated properly, or the classification methods provided may not produce optimal results. Furthermore, several on-line mapping programs were tested, and it was found that they could create *only* choropleth maps, even if the data were absolute values (see Chapter 11 for an explanation of why this is incorrect). These and other problems can be reduced by keeping in mind the characteristics of properly prepared maps described elsewhere in this book.

ELECTRONIC ROAD MAPS

The functions of traditional highway and road maps and atlases can also be fulfilled by electronic maps. Note that some programs provide only one type of function and others offer both.

Street and Highway Maps

A person might need a local street map showing the location of a friend's home, a store in a neighboring town, a theater or park, or another feature of interest, without necessarily needing the travel planning functions described in the next section. Programs are available to provide complete street and highway coverage (and, sometimes, railroad routes) for a country or region, such as North America.

Assume that the map user has the street address of a person or feature that is to be found. The first step in the search is to enter the name of the city or town, usually by simply typing it on the keyboard. If two locales have the same name, the correct one is selected, usually from a list posted on the screen. Then, a map is plotted, perhaps covering an area as large as the contiguous 48 states, highlighting the selected town or city. Only major highways are likely to be shown and identified at this level. The user would instruct the program to focus on the town or city. Then, a map of the major streets in the community is plotted, along with some street names. Next, the search is narrowed by entering the desired street name. At this point the address may have to be designated more fully. For example, is it North or South Main Street, College Drive or College Street, and so on? When the specification is narrowed to an unambiguous designation, the map is centered on the street and magnified to show its full length. Finally, after the address number of the feature is entered, the specific location is highlighted. It is usually possible, then, to zoom to higher levels of magnification, which typically show and label all of the streets in a small region around the target address. The user can pan to recenter the display and can also change zoom level to provide the map coverage wanted. Then, the map can be annotated with names, telephone numbers, and other information and, finally, can be printed so that it is available for use during a trip to the selected destination.

In many programs, finding the general location of a feature without knowing an address is also possible. In these cases, the telephone area code or postal zip code can be specified. The program then displays the place name associated with the area. It is also possible to run a search within a community, based on landmarks, such as parks, churches, cemeteries, schools, airports, water features, and even quarries.

Trip-Planning Aids

Another type of individual mapping program is geared to trip-planning activities—usually for trips between cities. This kind of program typically allows the user to specify an origin and a destination. It then automatically produces a route map and driving directions. Commonly, the user may request a route that meets specified preferences. For example, the user may specify the shortest route in either travel time or distance. Other constraints that may be specified include selection of the most scenic route or avoidance of certain road classes, such as interstate highways or dirt roads. Also, the user can usually alter portions of the trip as desired. Preplanned scenic tours may also be provided by the program.

Some programs permit placing annotations on maps or otherwise personalizing them. Also, most programs allow the map, along with the travel itineraries, distances, travel times, and other information, to be printed.

A variety of supporting information may be provided. A review of currently available programs shows that information may be available about the following features, although not all programs include all of the features: airports, national land, state parks, monuments, historic sites, mountains, lakes, ski resorts, border crossing points, toll roads, car rentals, and hotel, restaurant, and campground prices and ratings. Also included may be descriptions of local or regional history, geography, economy, recreational activities, and other points of interest, complete with pictures and sounds. Another feature that may help is information about typical rush-hour traffic times and routes to avoid at those times.

Trip-planning aids of the type described here are also available on the Internet.

AUTOMATED AUTOMOBILE NAVIGATION

Futurists have anticipated **automated automobile navigation** for years, and some systems have been in use since 1985. In recent years, such aids have become more common. The capabilities and operational characteristics of the available systems vary considerably, and no single system has emerged as the definitive automated navigation solution. However, based on the operation of existing systems and the clearly feasible plans for their improvement, it is now possible to predict with some confidence the general form future systems will take.

An ideal automated navigation system automatically determines and displays the location of the car in which it is mounted. In addition, it automatically selects the best route to follow from that location to a specified destination and provides directions to guide the driver along that route. The dominant feature of most useful existing systems is the capability of continually updating the vehicle's location for the benefit of the driver.

Position Display

Most automated navigation systems use graphic displays that show a digital map on a monitor (Figure 20.1). Manufacturers design these displays to provide essential information to the driver as succinctly as possible, to reduce distraction from driving tasks. The information about the road pattern and names of features included in the display is kept to a minimum.

Most systems provide the capability of zooming in or out on the map display. Thus, if the destination is some distance away, a system displays a small-scale map with an appropriately limited selection of roads. As the driver requires more detail, either at interchange points or near the destination, the system zooms the display to a larger scale, showing more street detail.

Usually, the system orients the display to conform with the direction of travel. That is, instead of always placing north at the top of the display, the direction toward which the vehicle is moving is placed at or near the top. The advantage of this approach is that features to the right of the driver's line of travel appear on the right side of the road shown on the screen, and so on for the other directions.

Even with simple and readable map displays, some experts are concerned about the safety hazards involved in looking at the display screen. As a result, some systems use voice instructions, instead of map displays. Other experimental systems use a "heads-up" display, in which the system projects the map image onto the windshield. This display may reduce distraction, because the view of the road ahead remains in

Figure 20.1 Components of an automated vehicle navigation system, including a CD-ROM with a map and supplementary information, a monitor, a GPS antenna and receiver with CD-ROM player, and a wireless remote control unit.

Source: Courtesy of Sony Corporation of America.

the driver's line of vision, even when the display is being examined. It is not clear, however, that visibility will be adequate under all lighting conditions or that there will not be confusion between the projected image and the driving environment.

Initial Vehicle Location

The most basic navigational systems require the user to provide the vehicle's initial position. One way for the driver to do this is to key in the starting location by pushing buttons on the system display. Another way is to use voice-recognition technology, which allows the driver to provide the initial location verbally.

Another approach to providing initial vehicle location information is to use externally derived data, without operator intervention. This type of system provides easier operation and can be used either on- or off-road. The system may obtain external location information from a special local radio beacon, the LORAN-C radio-location system or the satellite-based global positioning system (see Chapter 7). Because of their accuracy and worldwide availability under nearly all conditions, GPS-based systems seem likely to dominate in the years ahead.

Pathfinding

Besides simply displaying a road map, some guidance systems provide pathfinding capabilities. This means that the system automatically tests possible routes from the current position to a specified destination and selects the most suitable. This function relies on a database that includes street classifications, one-way streets, and turn restrictions. It may even take into account current traffic conditions and situations such as road closings due to adverse conditions.

In a pathfinding operation, after the initial point has been determined, the destination is specified, using a street address, a landmark, or a street intersection. After selection, the destination usually appears on the screen, along with a map of the recommended route. The driver can then use the map to navigate to the destination. In a more advanced installation, the system provides driving directions. For example, it might use arrows or other symbols to show the action required at each intersection along the route. In another approach, the selected path is highlighted on the display map. This requires close inspection of the map by the driver, however, so this type of display may be reinforced with directional arrows. Alternatively, because of concern that visual displays contribute to safety hazards, some systems use spoken instructions, instead of maps or arrows, to provide navigational information. When the system combines spoken instructions with a voice-recognition capability, the driver can also issue instructions to the system verbally.

As the vehicle moves, the system monitors its direction and distance of travel. One system relies on a dead-reckoning process (see Chapter 7). Sensors mounted on the vehicle's wheels detect movement, and many systems also involve solid-state compasses, gyroscopes, or other sensors. These sensors report the distance and direction of movement from the initial point, which allows the updating of the dead-reckoning position. This type of system has the problem that sensor limitations cause position errors that accumulate. Therefore, the system needs some type of periodic correction, such as matching the dead-reckoning position against a digital map. For example, when a turn is made, the location of the nearest mapped intersection is checked and the system assumes that the vehicle is at that location, even if the vehicle has entered a parking lot, for example. Because of this necessity for checking against the digital map, this approach is limited to on-road operation; the system can become "lost" if the route strays away from the roads included in the database.

Added Features

Many automated navigation systems include special guides and maps, digital photos, and sound, to provide supplementary information. This information often includes locations, facilities, and ratings of restaurants, hotels, shops, parks, arts and cultural facilities, sports and entertainment centers, and airports. Once the desired destination is selected from such information listings, the pathfinding capabilities of the system can select the best route to it.

Many experts anticipate that automated automobile navigation aids will eventually be integrated with the Intelligent Vehicle Highway Systems (IVHS) being tested by a variety of manufacturers. Possible benefits from this integration include informing drivers and onboard computers of traffic conditions. Based on this information, the onboard unit could reprogram a planned route, thus reducing traffic delays. Still more benefits could be gained if the information were integrated with information about other modes, such as rail, bus, or ferry. Some experts even anticipate the incorporation of radar into IVHS facilities to provide a collision avoidance capability.

Another related technology is automated vehicle location (AVL). AVL requires vehicles to have external location capability (such as GPS). In commercial applications, the system radios the vehicle's position to a dispatcher at a central facility. At the facility, the system displays the location of the vehicle on the screen of a computerized mapping system that incorporates the locations of streets, addresses, and other features. The dispatcher then radios routing orders to the vehicle as needed. In addition, the driver of an AVL-equipped vehicle can key in information, such as the existence of a downed utility pole, and the system automatically informs the dispatcher of the location of the problem.

Similar AVL capabilities provide emergency services in at least one system now available in limited areas. This system uses a simplified GPS receiver, linked to a radio or cellular phone. In an emergency, the system contacts a monitoring station, either under operator control or automatically when the vehicle's airbag deploys. Operators at the station then interact with the driver, or operate independently, to arrange the dispatch of required emergency assistance. Another application allows tracking of a stolen vehicle.

User Concerns

Prospective purchasers of vehicle guidance systems will want to know the answers to several questions before making a choice. The foremost question concerns purchase and installation costs. Most systems are currently in the $1500 to $2000 range but, as manufacturers develop the technology and sales volumes increase, experts expect that the typical price will be about $500 (in current dollars). The cost question involves not simply the dollars involved but also how helpful the prospective purchaser expects the system to be. If most trips involve driving short distances in familiar surroundings, a system may not be particularly useful. However, if frequent travel to obscure locales is involved, or if finding addresses in unfamiliar cities is often necessary, the investment may be very worthwhile.

If a system seems potentially useful, other questions come into play. These include the timeliness and accuracy of the road network data. In particular, does the system depend on stock databases, or is it extensively field checked? Also important are the frequency and cost of updates to the database. Obviously, a database starts to age when it is created and, over time, may become misleading or worse, as new roads are built, highways are renumbered, and other items of information change. Equally important is the quantity of information stored in the system. For example, is a single database sufficient for the entire country, or is a different one needed for a given region, state, or city? One advantage of CD-ROM-based systems is that one CD can hold all the street names and address ranges for the United States. This has obvious implications for the convenience and cost of operation of systems, as well as of updating the database.

Another question concerns what supplementary information is available through the system, such as locations of motels, gas stations, or restaurants. Again, the information stored, its accuracy, and the ease and cost of updates come into play.

Finally, as a system is used, its speed of operation assumes great importance. If the time required for the retrieval of information averages 1 second, for example, the system will probably be adequate. If, however, it averages 15 seconds or longer, the slow response is likely to become an irritant. Again, CD players with acceptably fast retrieval times are available and are likely to be the device of choice.

SUMMARY

The availability of maps on computers means map users are no longer restricted to using maps in printed form but have new and interesting possibilities for map access and use. These include electronic atlases, road maps, trip-planning software, and automated automobile navigation systems.

Electronic atlases provide access to general reference maps and specialized thematic maps of the world, countries, or regions, as do traditional atlases. The use of an electronic atlas, however, eliminates the flipping of pages. In addition, an electronic atlas can provide new maps or revisions simply through an update of its database. The electronic form can also provide for the creation of maps tailored to the specific needs of the map user, using built-in thematic databases. The *National Atlas of Canada* is a current example of this approach.

Some electronic atlases suffer from poor use of cartographic symbolization. The map user should keep in mind the characteristics of properly prepared maps when using these atlases.

For travel-planning purposes, programs are available to provide complete street and highway coverage for a country or region. These products provide search capabilities that allow the production of maps centered on a desired location. Such maps can be annotated with names, telephone numbers, and other information and printed for future reference.

A typical trip-planning program allows a user to specify an origin and a destination and to specify preferences such as selection of the shortest route, selection of the most scenic route, or avoidance of certain road classes. The program then produces a map with the route highlighted. This map, along with a travel itinerary, can often be annotated and printed. Such programs also provide a variety of supporting information, such as the locations of airports, monuments, historic sites, ski resorts, hotels, and restaurants.

Automated automobile navigation systems are now becoming generally available. An ideal system of this type automatically determines and displays the location of the car in which it is mounted. It also automatically selects the best route to follow from that location to a specified destination and provides directions to guide the driver along that route. Most systems use graphic displays showing a minimal road pattern and selected feature names. Because of safety concerns, however, some systems use voice instructions instead of map displays. Some systems require the user to provide the vehicle's initial position by pushing buttons on the system display, for example. Other systems obtain external location information, frequently from the satellite-based global positioning system (GPS). In addition, some guidance systems provide pathfinding capabilities, in which the system determines the preferred route between the vehicle's position and the destination. Driving directions may be provided by arrows on a visual display or by voice. More directions are issued as the vehicle moves along the route.

Automated automobile navigation aids may be integrated with Intelligent Vehicle Highway Systems (IVHS) and automated vehicle location (AVL). These systems provide for external information about traffic conditions and other concerns, as well as dispatcher control or the provision of emergency services.

The selection of an appropriate automated vehicle navigation system requires careful evaluation of such factors as initial cost, range of capabilities provided, currentness of the database, and ease of update.

SUGGESTED READINGS

McGranaghan, Matthew, David M. Mark, and Michael D. Gould. "Automated Provision of Navigation Assistance to Drivers." *The American Cartographer* 14, no. 2 (April 1987): 121–38.

Mersey, Jan. "Coming Soon to a Library Near You . . ." *Cartouche* (Special Issue, Autumn/Winter 1994): 12.

Morrison, Joel L. "A Personalized National Atlas of the United States." *Cartographic Perspectives* no. 20 (Winter 1995): 40–44.

Software reviews in *Cartography and Geographic Information Systems, The Electronic Atlas Newsletter* (P.O. Box 75394, Seattle, WA 98125-0394) and other journals.

"The State of IVHS in the United States: An Information Superhighway for Drivers." *Maps Alive* 4, no. 1 (January 1995): 8–11.

White, M. "Car Navigation Systems." In *Geographical Information Systems: Principles and Applications,* edited by D. J. Maguire, M. F. Goodchild, and D. W. Rhind. London: Longman, 1991, 115–25.

CHAPTER 21

Geographic Information Systems

Overview

Before discussing the applications of **geographic information systems (GISs),** we need the answer to a very basic question: Just what *is* a GIS?

Definition

A GIS is an information system, which means that it provides a specific set of facts (the information) arranged in an orderly manner (the system). It also provides facilities to help the user work with the facts it contains. Many types of information systems are used today. They deal with concerns such as payroll, financial information, and library holdings. A GIS, however, has another primary characteristic that distinguishes it from the bulk of such information systems: It deals with information that has a spatial, or geographic, component. This means that each piece of information in the system includes a reference to its location. This spatial component means that most GISs can produce maps from the data. It also means that variations from place to place in a particular variable or geographical interrelationships among different variables can be analyzed.

A GIS, or any other type of information system, need not be computer operated. Today, however, the use of digital data in computer-based systems is the rule, rather than the exception. In this chapter, assume, unless otherwise indicated, that the term implies that the systems under discussion use a computer to work with **digital data.** Digital information processing is of particular importance, because it provides increased speed and expanded capabilities in the analysis of spatial information.

In summary, we may accept a definition of a GIS used by the U.S. Geological Survey: "A GIS is a computer system capable of assembling, storing, manipulating, and displaying geographically referenced information."[1]

Antecedents of GISs

Relatively simple GIS procedures have been in use for a long time. Rudimentary elements of digital techniques for the recording and analysis of geographic information were in use as early as the 1800s. Probably the most notable early example was Hollerith's use of a punched-card system for the manipulation and analysis of data during the 1890 U.S. Census of Population.

Early GIS methods evolved from techniques that geographers, planners, geologists, and others have used over the years to analyze mapped information. One such method is **map-overlay analysis.** This technique involves tracing the outline of some variable of interest, such as soil types, from a source map onto a transparent overlay. The overlay is then placed over another map of the same area at the same scale, such as a vegetation map. The overlap of the soil and vegetation types shows the relationship between the two phenomena (Figure 21.1). The investigator might go further and apply statistical techniques to determine

[1]Brochure, "Geographic Information Systems" (U.S. Department of the Interior, U.S. Geological Survey, 1992).

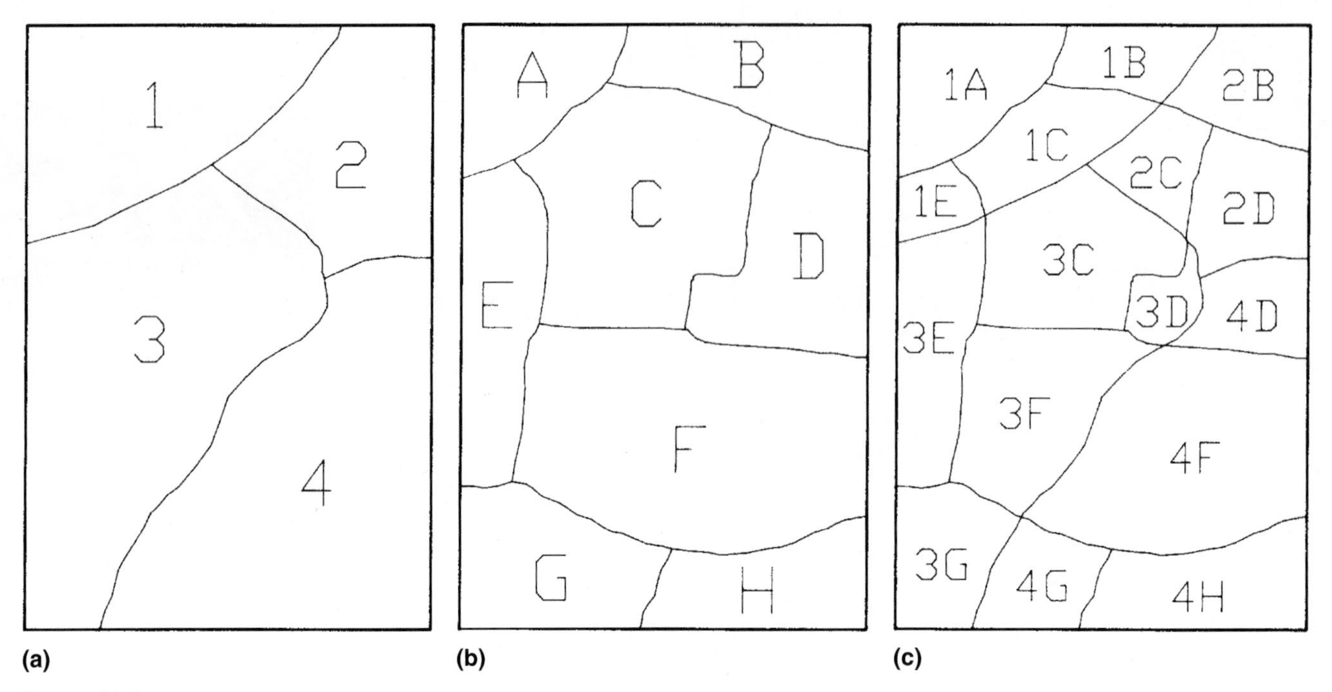

Figure 21.1 Relationship between sample soil and vegetation maps. (*a*) Soil map. (*b*) Vegetation map. (*c*) Overlap between maps.

the level of coincidence between the two distributions (for example, the coefficient of areal correspondence; see Chapter 11). Although such methods are useful, they are cumbersome and time-consuming and provide only a limited type of analysis.

A particularly well-developed extension of map-overlay analysis techniques was the **photographic-overlay analysis** method, used by planner Ian McHarg in the late 1960s.[2] This technique used maps as tools for analysis, taking into account the physical and social environments, and thus qualified as a GIS approach.

The photographic-overlay method was aimed at finding "the solution of maximum social utility" for the particular problem being studied. It involved transparent, photographic overlays on which information was represented in shades of gray. Physiographic factors were recorded so that higher dollar costs and higher social values were both represented by darker tones. When the various overlays were placed over one another, the resulting map showed a complex pattern of gray values, ranging from light to dark. The areas that could best be used for a particular project, such as a highway, were the areas with the "least social cost." These areas, which were generally zones of lower dollar cost and of lesser social value, were shown by the lightest tone, whereas those of higher dollar cost and/or greater social value were shown by the darkest tone.

This method was used in real-world decision making, such as in route selection for the Richmond Parkway in New York in the 1960s. A wide variety of factors was considered in this study, resulting in a very complex analysis. Factors considered included surface drainage, soil foundation and drainage, bedrock foundation, susceptibility to erosion, land values, tidal inundation, historic values, scenic values, recreation values, water values, forest values, wildlife values, residential values, and institutional values.[3] The Tri-State Transportation Commission ultimately accepted the alignment of least social cost developed in McHarg's study.

Useful as the photographic-overlay approach was, there were technical problems, such as determining the appropriate weighing of factors (a problem that continues, even with computer-assisted methods). Also, the method had reached "the limits to the photographic resolution of many factors." McHarg felt that "it may be that the computer will resolve this problem, although the state of the art is not yet at this level of competence."[4] As indicated here, the "level of competence" of computer techniques has increased significantly in the two decades that have passed since these words were written, providing greatly expanded capabilities.

[2]See Ian L. McHarg, *Design with Nature* (Garden City, N.Y.: Doubleday/Natural History Press, Doubleday and Company, 1969).

[3]The maps of these factors, which cannot be reproduced here, are shown on pages 36–41 of *Design with Nature* by McHarg.

[4]McHarg, *Design with Nature,* 115.

Formal GISs

Many early GISs were developed for land-use planning purposes. These typically involved mapping and recording various natural resources at the state, regional, or national level. The first full GIS of this nature, used on a production basis, was the Canada Geographic Information System (CGIS). The CGIS was authorized in 1964 and put into operation in 1969. The system involved a land-inventory database that was used to locate and identify marginal lands currently being used for agricultural purposes. This information was particularly useful for resource management projects, such as erosion control.

Throughout the 1960s, similar projects proliferated, including early projects in New York, Minnesota, and Kentucky. The complexity of these projects meant that for them to be implemented, large agencies were required. The number of such systems continues to grow, because a broad range of functions can now be carried out on microcomputers.

The comments in the previous chapter regarding rapid changes in computer applications to cartography apply equally to geographic information systems. Again, it is impractical to provide a list of programs and databases in this book. Instead, information is provided at the book's website, at http://www.mhhe.com/earthsci/geography/campbell.

GIS COMPONENTS

A GIS acts as a link between the real world and the user. The real world is the source of the information stored in the system. The user employs the GIS to analyze the information and applies the results to the solution of a problem. The essential considerations in any GIS, from the standpoint of such a user, are data input and storage structure, manipulation and analysis techniques, and output capabilities. We will now examine these components in somewhat more detail. Other aspects, such as input processing and data management, are extremely important but generally fall outside the focus of this discussion.

Data Input and Storage Structure

The data incorporated into GISs are obtained from sources that are either graphic and spatial or nongraphic and nonspatial in nature. Examples of graphic, spatial sources include existing maps, charts, graphs, and aerial photographs and other remotely sensed images. Nongraphic, nonspatial sources include tabular information, digital databases, and textual materials. Data from each type of source are converted into computer-compatible form, using techniques ranging from typing tabular data at a computer terminal to scanning source

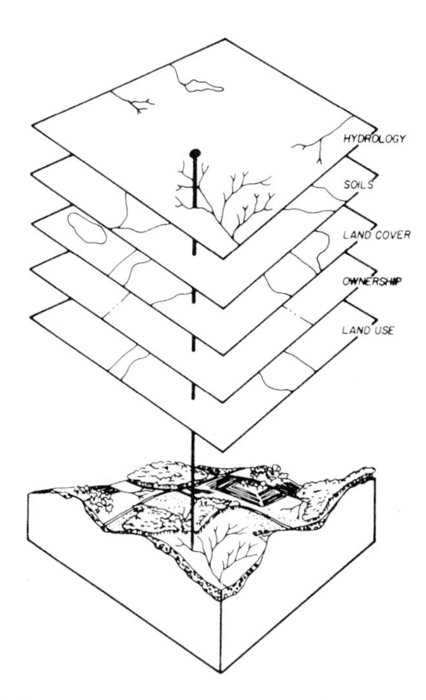

Figure 21.2 Example of multiple data planes in a database.
Source: Reproduced with permission of Joseph K. Berry.

maps on a digitizer. The details of these techniques lie beyond the scope of this discussion, although the Suggested Readings section at the end of the chapter provides sources of information about them.

In a GIS, data are divided into separate themes, chosen from the multitude of topics possible. Typical themes include physical phenomena, such as landforms, climate, soils, and vegetation. They also include social aspects, such as income levels, racial heritage, population density, and zoning designations. Information about the characteristics of each theme throughout the region covered by the system is stored in a separate file, called a **data plane.** Data planes are the computer equivalent of map overlays. The data planes do not exist in the system as physical entities, of course; instead, they are in digital form.

The number and specific contents of the data planes in a particular GIS depend partly on the applications anticipated and partly on the resources, including computer capacity, devoted to the system. In a simple situation, for example, separate data planes could record water features, vegetation, urbanized areas, roads, pipelines, railroads, and so on (Figure 21.2). It might be desirable to include a number of data planes within each of the broad categories. For example, road classes might include separate data planes for local, county, state, federal, and interstate highways. Water features might be similarly separated into intermittent and permanent streams, lakes, and marshes and other wetlands, and vegetation categories might include deciduous trees, conifers, brush, and grassland. In a real-world GIS, then, a multitude of data planes may break the information about the region into finely dissected categories.

Data Adequacy

The characteristics of the data in a GIS place the ultimate limits on what can and cannot be accomplished within the system. Even a system with excellent data storage, manipulation, and other characteristics cannot provide useful information if its data are inadequate. There are, however, practical limits to the level of detail contained in any given GIS. Greater detail means higher costs (in both time and money) of acquiring, storing, maintaining, and using the GIS. As a potential user of a GIS, therefore, you need to know the specific topics included in a particular system to see whether your data needs are met. In addition, you need to evaluate the accuracy and timeliness of the data content.

Data accuracy can be generally evaluated if enough is known of the manner in which the data were collected. If remotely sensed data are used, for example, resolution (that is, how small an area on the earth's surface can be discriminated) is specifically limited. If an application demands the detection of objects that measure 10 meters on a side, and the system contains image pixels that are 79 meters on a side, the resolution is obviously inadequate for that specific application. Evaluations of accuracy are *relative,* however. That is, data may be sufficiently accurate for one use even though inadequate for another.

If existing maps are used as a database, their scale, accuracy, and date of production are important quality-control factors. In the case of one county, for example, it is claimed that the base map currently used for recording property locations was created many years ago by gluing together the 1899 U.S. Geological Survey 1:62,500-scale topographic maps for the county. The resulting mosaic was then photographed onto a 2-by-2-inch slide, which was projected onto a sheet of paper tacked on the wall. The resulting outline was traced to create the base map! Digitizing a base map of this "quality" would not result in a very useful database, in terms of either locational accuracy or timely information.

The information contained in a GIS must be timely as well as sufficiently accurate. Information on most topics is subject to frequent change. The management capabilities of a GIS, therefore, must include procedures for incorporating new information, as well as for deleting obsolete information. Also, information is often published years after it is gathered. The date of publication must not be confused with the date of acquisition, which is the relevant one. Old data, however, are not necessarily bad data. If a study includes a historical dimension, the use of older material may be essential. Data showing the land-use patterns of a region at 10-year intervals, for example, would be invaluable for conducting a study of land-use evolution. For each application, it is necessary to decide which date or dates are relevant to the task at hand.

The discussion of accuracy issues in Chapter 19 applies equally here.

Data Forms

Each data plane contains information that occurs in some combination of four basic types: points, lines, polygons (areas), and surfaces. A few comments about these types are appropriate, in the context of GISs.

Points

Features that are small in areal extent are thought of as occurring at a point in space. The classification of a feature as a point is relative and depends on the scale of the data plane. A city, for example, may be considered to occur at a point on the earth's surface if the data plane is at a small scale. On the other hand, if the data plane is at a large scale, the same city would cover a relatively large area and could no longer be considered a point feature.

Lines

When features have length but lack significant width, they are treated as lines. Line features include roads, railroads, canals, pipelines, power lines, and national, state, and county boundaries.

Areas

Features that cover a region on the earth's surface are represented as areas. Such regions are defined on the basis of distinctive characteristics, such as specific crop types, dominant religious denominations, climatic type, income per capita, and population density.

Surfaces

Finally, features that cover an area but that also have a vertical dimension are called surfaces. The earth's surface is probably the easiest to visualize. A surface has an outline, whether it is the shoreline of an island or a political boundary. Within that outline, however, a surface also has a series of elevation values, one for each point defined on the surface. Other phenomena that vary in value from place to place, such as rainfall, temperature, or total income, are also represented as surfaces.

Georeferencing

Each data plane also includes locational information about the features recorded. Part of the input processing involves the **georeferencing** of the various data planes so that the features recorded are in correct alignment from one data plane to another. Locational information is specified either in terms of latitude and longitude or according to a locational grid, such as the Universal Transverse Mercator (UTM) or State Plane Coordinate (SPC) systems. Positions also can be recorded using a special-purpose, localized reference system. Such a system is a problem, however, when the data need to be related to other sources of information that are organized around one of the standard locational systems.

Labeling

In addition to having locational information, the various elements in a data plane each have an identification code so that they can be selectively retrieved from the data file. This is done, as in computer-assisted cartography, by **labeling** each record with a set of identifying characters called an **attribute code.**

Format Considerations

The discussion of vector and raster formats in Chapter 19 also applies to GISs, but some additional points need to be made. Some GISs are vector-based, some are raster-based, and mixed systems are becoming more common. Part of the reason for this mixture of types is that different sources of information are used as input into GISs. For example, many GISs are built on a base of remotely sensed information. Because such data are usually already in raster format, such systems often are built on raster methods

On the other hand, maps are also a major source of GIS data, and they may be digitized in either mode. The influences favoring one or the other format in systems based on map inputs are less obvious. One influence is that, in the early phases of the development of computer-assisted cartography, most digitizing was done using vector-based digitizers. This influence carried over to related systems in the GIS arena. As more emphasis was placed on automating and speeding up the digitizing process, however, raster-scanning devices were developed. As these devices became more important, the development of raster-based systems was emphasized.

Another important factor that influences the vector versus raster decision is that some operations are performed more efficiently in one mode than in the other. For example, a raster format is favored when overlaying operations are performed. Routing and network linkage problems, on the other hand, are handled more efficiently within a vector format. When geographic accuracy and the appearance of a traditional map are important, vector systems tend to be favored. Improvements in output devices and an increase in the accuracy levels of raster digitizing seem to be reducing the importance of this factor, however.

Finally, because many different types of operations are performed within a given GIS, it is desirable to have data in one format to facilitate some operations and in the other format for other operations. For this reason, some systems provide for raster-to-vector and vector-to-raster conversions. This flexibility allows the use of the format best suited to each particular operation.

Manipulation and Analysis Techniques

A GIS is a storehouse of information for use in the analysis and solution of a variety of problems. To provide the flexibility required for such applications, the system must include capabilities for manipulating, comparing, or combining the data in various ways. This section describes some techniques typically available for these purposes.

System Capabilities

GIS capabilities usually include the following:[5]

1. Comparing the distribution of two types of data within the study area
2. Searching for a selected set of characteristics that occur together, either with or without some other specific characteristic
3. Searching for the nearest neighbor of a specific feature, with added criteria regarding the presence or absence of other characteristics
4. Handling comparisons between data recorded at different scales or on different projections

The basic operations that support these capabilities include reclassification, overlaying, distance functions, and modeling.

Basic Operations

The user of a GIS calls selected data planes from the computer memory and carries out any operations desired. In the process, new data planes are created and, henceforth, are treated as separate entities. These data planes are subjected to further manipulation, discarded, or stored, as desired. However, the original data planes are almost invariably retained in the system, in unchanged form, so that they are available for future applications. This protection of the original data is usually assured by the use of a "read-only" arrangement, which allows the user to access and manipulate the original data but prevents any changes to the original files by anyone but system maintenance personnel.

[5]R. F. Tomlinson, "A Geographical Information System for Regional Planning," in *Land Evaluation,* edited by G. A. Stewart (New York: St. Martin's Press; South Melbourne: Macmillan of Australia, 1968), 200–10.

Reclassification

One method of altering an existing data plane to create a new one is **reclassification.** Reclassification is the replacement of the initial category designations by new values, based on specified criteria. Assume, for example, that one data plane includes soils in several categories, with designations such as 1 for gravel, 2 for clay loam, 3 for silt loam, and 4 for loam (Figure 21.3a). The areas suitable for agriculture could be delineated by arbitrarily giving them a value of 1 and giving all of the other areas a value of 0. The reclassified data would be stored as a new, separate overlay of areas of soil suitable for agriculture (Figure 21.3b).

Reclassification may also be done on the basis of the relative importance of a particular type of feature for a specific purpose. For example, conifers might be considered more suitable than deciduous trees for high-quality recreational facilities, and brushland might be altogether unsuitable. To represent this situation in the GIS, the vegetation data plane could be reclassified so that areas designated as conifers were shown with a value of 2, deciduous trees with a value of 1, and brushland with a value of 0. Although there would be no difference in the map's general appearance, the assigned numerical values would represent an ordinal classification of the vegetation types in terms of recreation potential.

Still another type of reclassification requires simple arithmetic or statistical operations. A data plane recording average temperatures in degrees Fahrenheit could be converted to one of temperatures in degrees Celsius by applying the appropriate conversion factor to the original values. Again, there would be no difference in the appearance of the two maps, but the difference in values would be important in relation to any subsequent calculations.

In another situation, reclassification might be done on the basis of feature size. Woodland areas, for example, could be reclassified on the basis of the number of acres contained in a contiguous forested area. The new data plane would have entries indicating feature size, rather than simply indicating wooded sites (Figure 21.4).

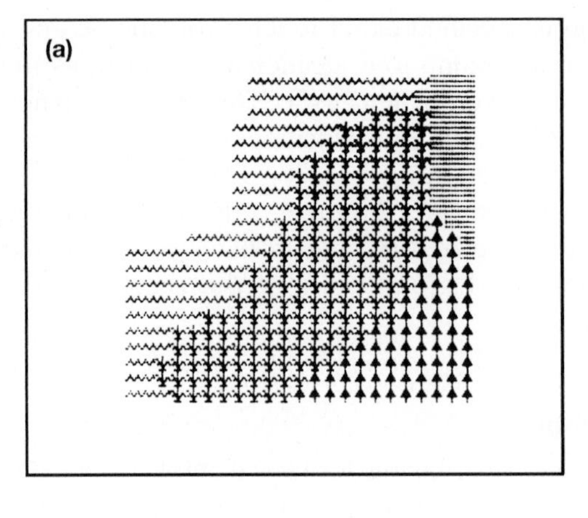

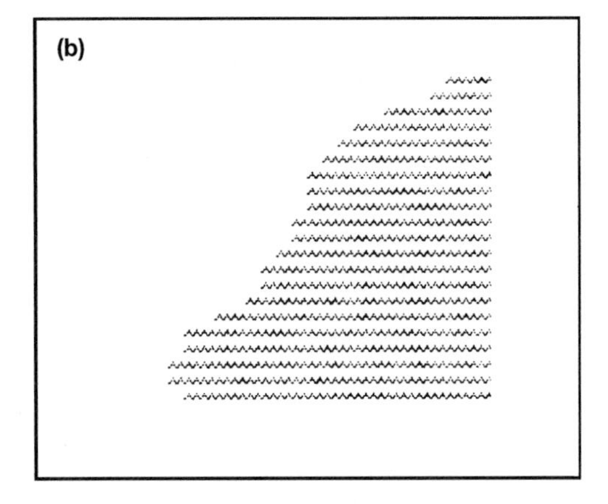

Figure 21.3 Reclassification by renumbering. (*a*) Original classification. (*b*) Reclassified data.
Source: Map by Dean Boyer.

Overlaying

Another way of generating a new data plane is **overlaying** two or more existing data planes on one another. This is the digital equivalent of carrying out map-overlay analysis by physical methods. In a GIS, however, the task is accomplished much more quickly and easily. In addition, other considerations, such as weighting of the various factors, can be accommodated.

Overlaying may be done in several ways, including establishing some minimum, maximum, or other specific value as a requirement. A certain minimum average rainfall might be required for a specific crop, for example. For planning purposes, a map could be generated to show only those areas that meet or exceed that minimum requirement. In another situation, several complementary characteristics might be desired for development purposes, such as selected soils, certain vegetation types, and specific zoning. Overlaying the data planes for each of these categories would give those areas that had all three required categories a value indicating that they were suitable (Figure 21.5). Those areas that did not contain all three categories would be given a value indicating that they were unsuitable.

Alternatively, ratings can be obtained by weighting the categories on the basis of their relative importance and adding the layers together. If slope is the most important limiting factor for a particular project, for example, slope ratings might be given a weight of 3, whereas vegetation ratings, which are slightly less im-

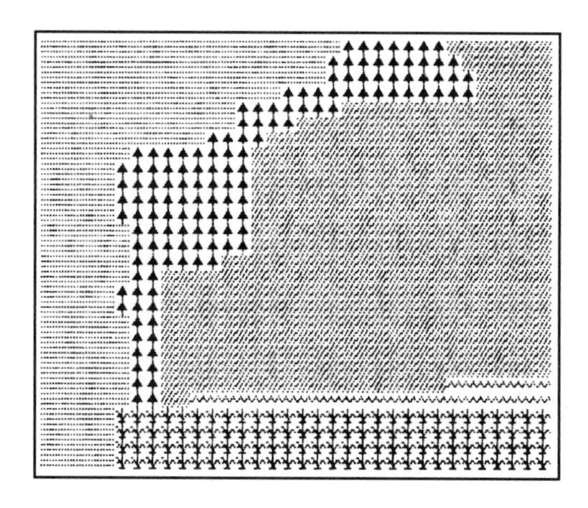

Figure 21.4 Reclassification on the basis of feature size.

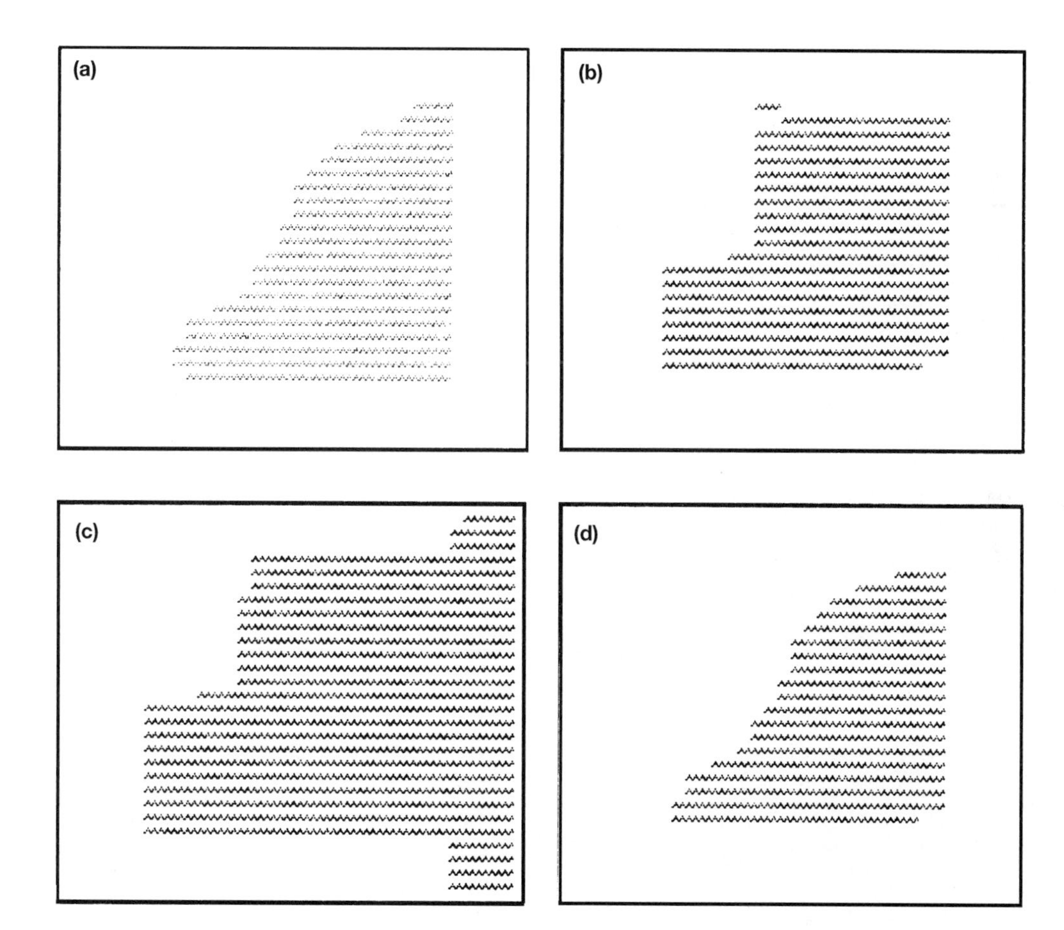

Figure 21.5 Overlaying to determine areas that meet or exceed a minimum requirement. (*a*) Suitable soils. (*b*) Suitable vegetation. (*c*) Suitable zoning. (*d*) Area suitable for development.
Source: Map by Dean Boyer.

portant, might have a weight of 2, and so on with other factors. In this type of application, a range of ratings for each location is possible. The weighted total rating values can then be used to select suitable sites.

In another application, industrial development activities might need to be excluded from wildlife habitat areas. In this case, areas for potential industrial development can be evaluated on a variety of characteristics. The information on wildlife habitat areas, however, will be used to exclude development from a given area, even if the other characteristics of that area indicate suitability for industrial use.

Distance Functions

Distance functions are also available and are used for a variety of purposes. Assume that a particular location is being considered for a shopping center, for example. With a distance function, each point in the region can be assigned a value based on its distance from the proposed site (Figure 21.6a). The resulting information can then be analyzed along with population and income information to evaluate the new shopping center's potential market zone. In situations in which travel time is more important than simple distance, the calculation of the zone of influence can be modified. By incorporating road network information (Figure 21.6b), in particular, distance can be expressed in terms of travel time, rather than as direct distance (Figure 21.6c).

Other types of analysis of the effects of distance are equally possible. A cost analysis can be used, for example, to compute the minimum cost path from a factory to a number of potential warehouse locations. Or, in another application, a distance function can determine the location and extent of desirable environmental buffer zones. Such zones are often used to protect important aspects of the environment, such as bald-eagle nests, scenic areas, or high-quality water sources (Figure 21.7). In a GIS, a buffer zone is established by appropriately renumbering locations within a specified distance of the protected feature. The designated buffer zones can then be excluded from any development or activity plan that might affect them.

Modeling

The production of new data planes and statistics is one of the basic tools of a GIS. The utility of such a system is based on its capability for handling a logical sequence of analytical operations, called **modeling.** Models use the database, in a series of steps, to produce information regarding the alternative outcomes that would result from different courses of action.

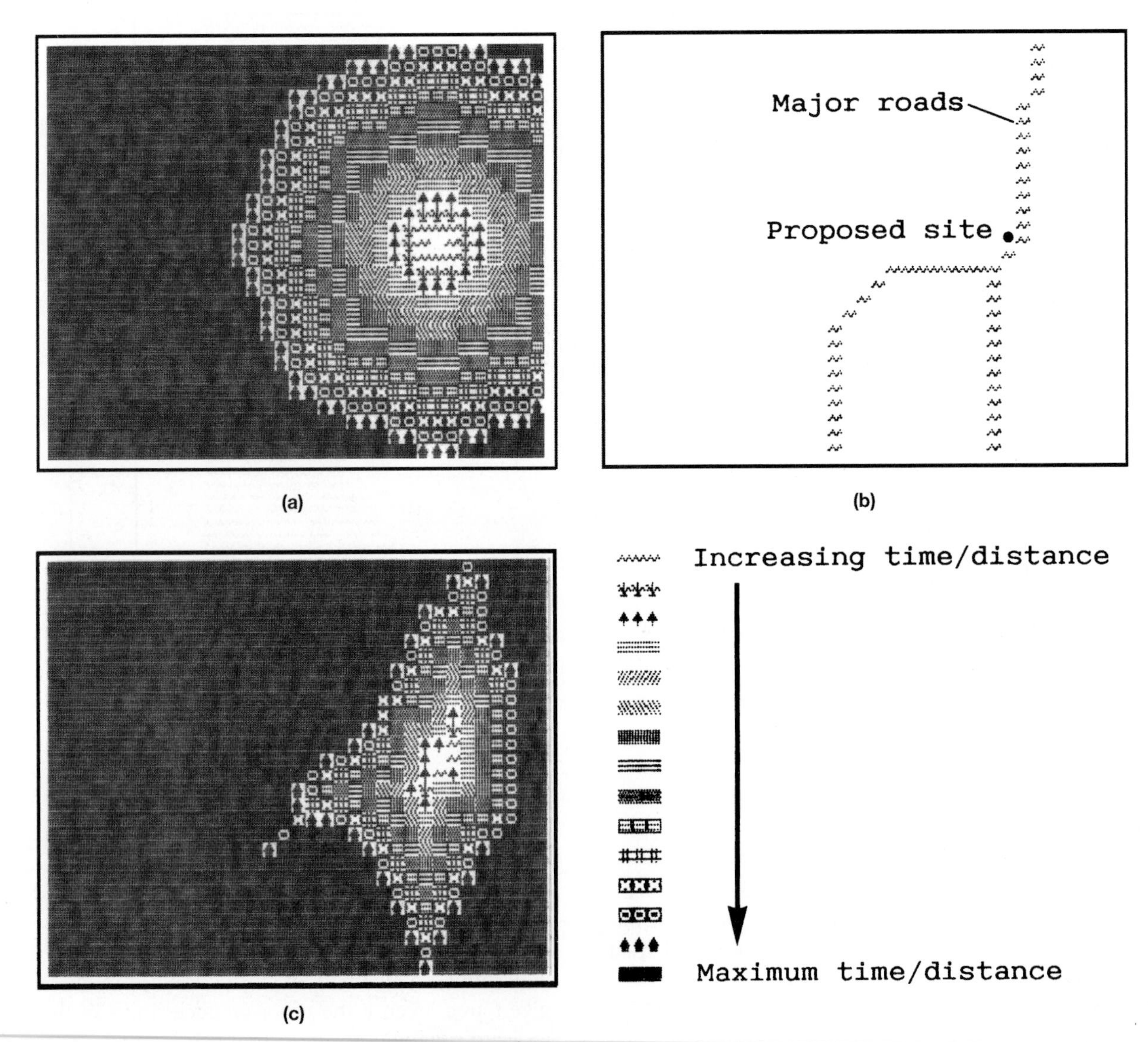

Figure 21.6 Time/distance relationships. (*a*) Simple distance. (*b*) Principal roads. (*c*) Weighted distance (travel time).

The sequence of operations in a model varies from case to case. The appropriate model is determined by the analyst, who takes into consideration the relationships among the variables of importance in a particular situation. A simple model for the selection of the optimal corridor for a highway is an example. In this case, it is assumed that it is preferable for the highway to avoid steep slopes and also to be as well hidden as possible.[6]

In this example, the model is designed to select a minimum-cost highway alignment between two towns. To accomplish this, the model uses information about topography and residential housing locations. A portion of the cost is measured in dollars, reflecting the higher construction costs encountered in areas with steeper slopes. Also involved, however, are the social costs associated with having a highway built within view of residential locations. Costs of the latter type cannot be converted directly into dollar terms. One portion of the model, therefore, uses a sequence of steps to develop an appropriate cost surface, based on exposure to view. Each location is then assigned a relative cost based on its particular combinations of steepness and exposure.

The model works as follows (Figure 21.8): First, it uses elevation information to develop data on steepness of slope. The same elevation information is used to produce an intervisibility map that shows areas within view of existing residential areas. Values on the cost surface are then determined. They are low in areas that are flat and not visible from houses and reach a maximum in visually exposed areas on steep slopes. Finally, the model identifies the minimum-cost route between the two towns, taking into account the values on the cost surface.

A particular advantage of computer-assisted modeling is its flexibility, which encourages the testing of a variety of possible approaches to a problem. The highway-siting model could be modified extensively. Several different relative weightings of the criteria could be tested, for example, including variations in the relative weighting of the terrain steepness and visual-exposure factors. In addition, new factors, such as the barrier nature of lakes, could be recognized. In each case, the effect of the changes on route selection would be determined. Thus, the response of the model to various

[6]Based on Joseph K. Berry, "Computer-Assisted Map Analysis: Fundamental Techniques," paper presented at National Computer Graphics Association, Computer Graphics 1985 Conference (Fairfax, Va.: 1985).

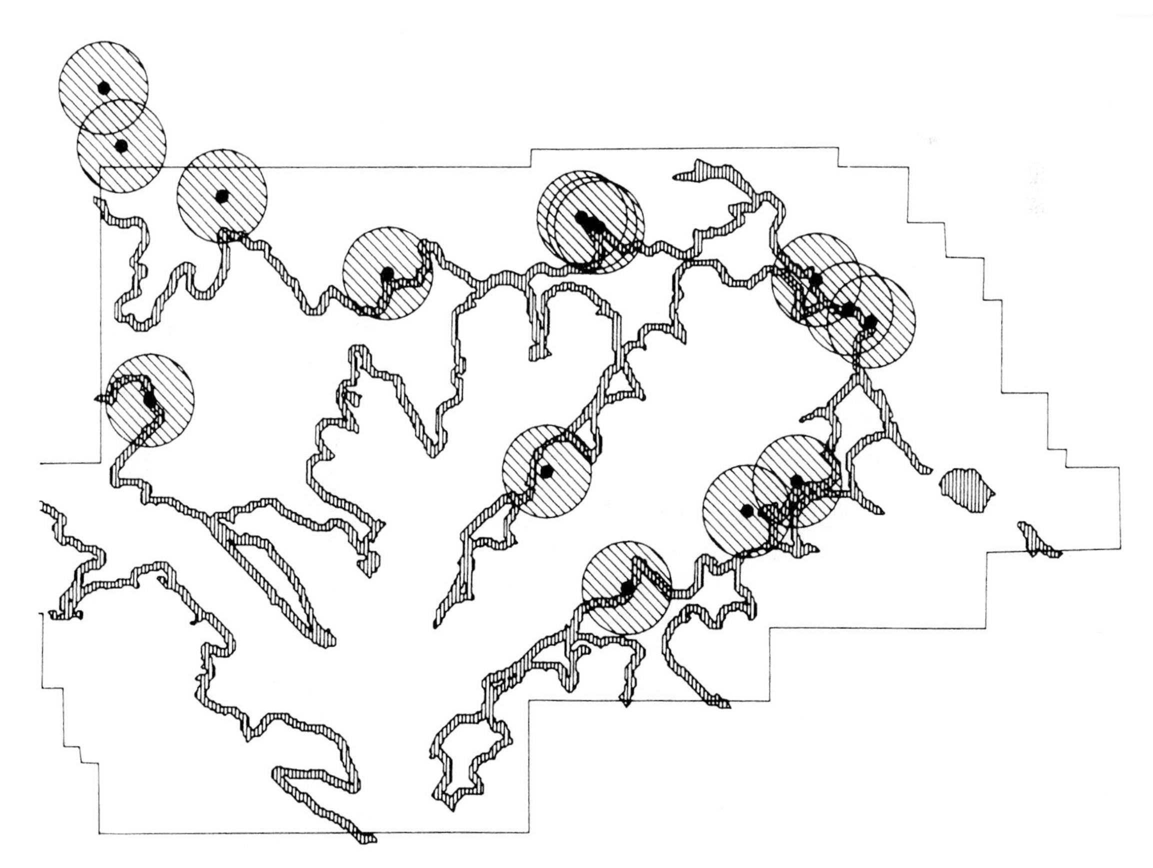

Figure 21.7 Sample final map showing buffers away from cliff edges (bands) and around raptor sites (circles) in the Tar Sand Lease Unit, Utah.

Source: *FDC Newsletter* 3 (Winter 1986): 7. Contributed by Harvey Fleet, National Park Service.

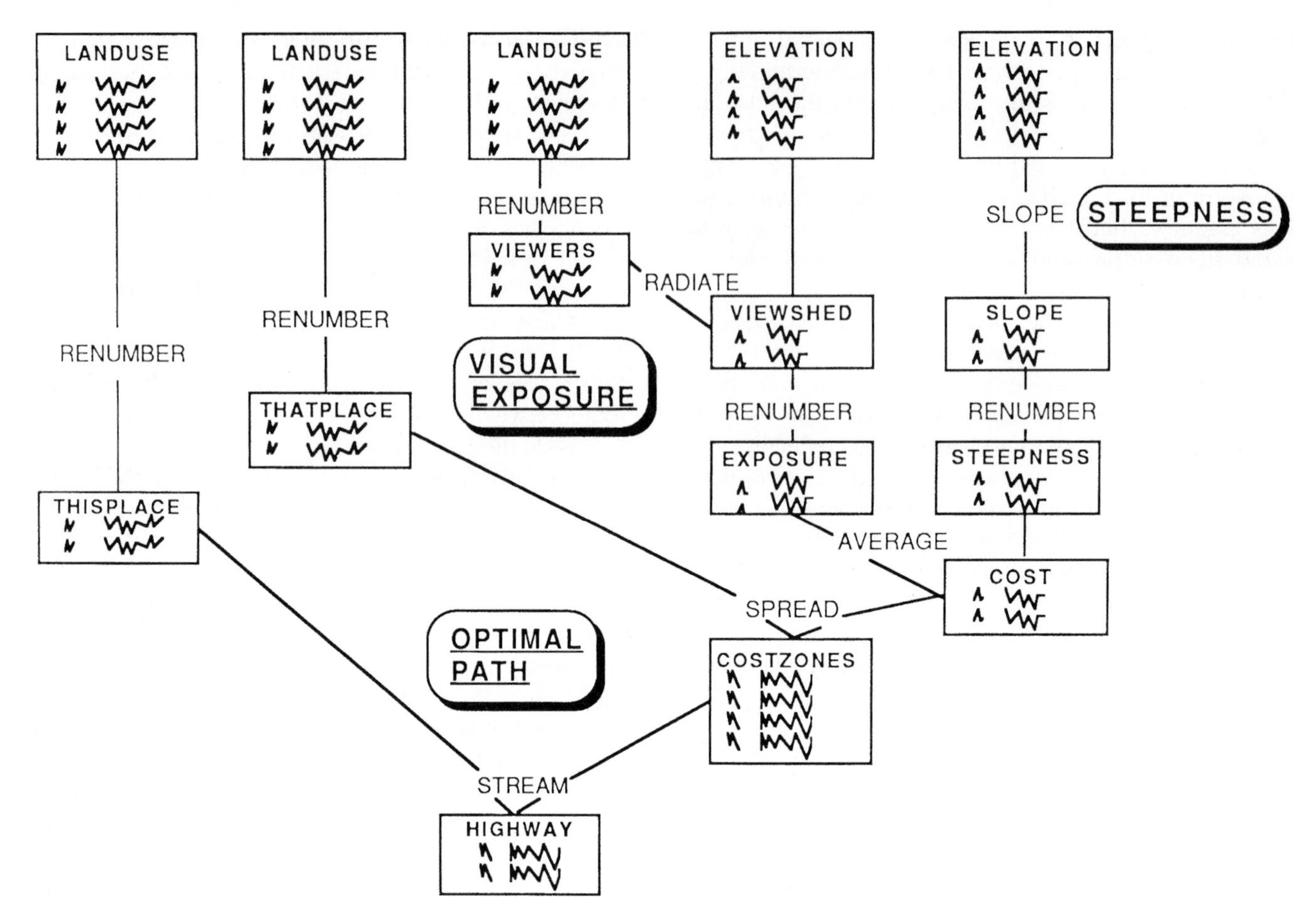

Figure 21.8 Example of simple modeling.
Source: Reproduced with permission of Joseph K. Berry.

possible factor relationships provides information useful for informed decision making. This is a more rational approach than simply providing a single "take-it-or-leave-it" outcome, with no alternatives.[7]

Determining the relative importance of the factors involved in any problem is the most difficult aspect of modeling techniques. This problem is just as severe as it was when McHarg used the photographic-overlay method. Although the GIS provides the capabilities of dealing with ranks and weights, the system itself cannot determine what those values should be. Decisions of that nature can only be made by persons who have the necessary technical expertise in the fields of study pertinent to a particular problem. Even then, the decisions are subjective and must be exposed to some type of societal evaluation before they are accepted.

Output Capabilities

In addition to offering their complex information processing and modeling capabilities, GISs incorporate large amounts of data. This fact places a premium on an adequate means for getting the information and analyses to the user. Typically, the data planes can be viewed, one at a time, directly on the computer screen (CRT). At times, however, some type of relatively permanent (**hard-copy**) output is needed for distribution purposes, and most systems provide methods for producing such copy.

Statistical Reports

A GIS database is often used for the production of **statistical reports,** as well as for the creation of maps. Depending on the data available in the system, such statistics might include calculations of average income level, average elevation, maximum elevation, and similar information. In addition, total land values within a specific distance of a given origin, total hectares of agricultural land, total area in a forest with percentages of different tree types, or any of myriad other possible pieces of statistical information could be generated. Such statistical information is typically printed out in the form of tables, using a printer attached to the computer system.

[7]Berry, "Computer-Assisted Map Analysis."

Maps and Graphs

The methods available for the production of hard-copy maps in a GIS are identical to those mentioned in Chapter 19 in connection with computer-assisted cartography. These methods include photographs of the CRT image, raster prints, or vector plots (see Figures 19.1 and 19.2).

GISs commonly have the capability of converting the statistical information into, in addition to statistical reports and maps, various graphs, such as pie, bar, and line graphs. Hard-copy graphs are produced by the same methods as hard-copy maps.

GIS APPLICATIONS

To conclude this discussion, some examples of real-world applications will illustrate the usefulness of GIS techniques. The range and scope of applications are increasing rapidly, so those mentioned here only suggest the extent of the possibilities.[8]

Urban Planning

GIS applications related to urban planning are numerous and include the development of master plans. Information about vegetation classes, water quality zones, slope categories, unique landforms, locations of endangered or valuable species, and so on can be put into a planning database. By integrating this information, one can produce a composite suitability map for development. Such a map shows the relative suitability of areas for development. Areas of special water quality, critical environmental features, steep slopes, endangered species habitats, and so on would be given low rankings, whereas areas lacking these characteristics would likely be ranked as more suitable for development.

Often, the first step in an urban GIS is the production of improved property descriptions. The resulting high-quality database is often immediately useful for many tasks. Applications range from generating address lists for notification of public hearings to finding variances from zoning controls. Over a longer time, the same database may be used for reevaluating a zoning plan; planning sewerage routes by overlaying soils, roads, housing, and contours; and many other tasks.

A GIS can also be used to monitor patterns of urban growth. This may be done, for example, by overlaying urbanized area maps for two different times to produce a change map showing urban growth. This type of information is useful in planning schools, utilities, and other services.

GIS data can also be used in neighborhood planning applications. For example, in one city, it was used to evaluate neighborhoods requesting funding to help increase owner occupancy of housing. Another useful application is the development of crime prevention programs targeted at areas of crime concentration revealed by GIS analysis of crime statistics.

[8]New examples are published monthly in periodicals such as *Business Geographics, Geo Info Systems,* and *GIS World.*

Traffic Control

Many communities use GIS to monitor and control traffic flows on streets and highways. Information is fed into a typical system from video monitors, vehicle detectors, and other sources. When congestion develops, this information is used as background for decisions about changes in message signs, lane signals, and other controls. If accidents occur, motorist assistance is quickly dispatched to the scene. Other elements of some systems monitor public transportation, using GPS to record bus locations, for example. This allows immediate distribution to customers of timely information about route operations, available stops, and schedules.

Emergency Management

Emergency management teams often use a variety of GIS and GPS techniques before, during, and after disasters of natural or human origin. Earthquakes, floods, fires, hurricanes, tornados, and other emergencies generate a variety of tasks that public and private agencies must accomplish rapidly and accurately. Often, background information about utilities, highways, population, terrain, and myriad other elements of the local environment is stored in a GIS system. This information can be rapidly accessed to provide a framework for emergency management.

During an emergency, GPS can be used to gather information to be added to the GIS. Then, current maps of damage to roads, utilities, and buildings can be generated. This information is useful for making road closure decisions, planning alternate routes for emergency vehicles, posting the necessary signs, and many related tasks. During the cleanup after the emergency, the same information can be used for many purposes. Depending upon the situation, these include the refinement of maps useful for determining how many homes and businesses were involved and assessing the type of damage and the value of losses.

After an emergency, the GIS is useful for recording, monitoring, and coordinating rebuilding efforts, setting up service centers, and other activities. Also, information collected during the event can be combined with information about high-risk parameters collected earlier. This information includes factors such as steep slope, vegetation cover, building construction types and materials, and building density. This type of information is useful for evacuation planning, planning for improved response time in future emergencies, and general reduction of risks.

Airport Noise Management

Data regarding airport configurations, combined with aircraft noise generation monitoring, have been used to map noise levels around airports. Such programs seek to reduce noise impacts near the airport by following up on noise incidents and by planning future land uses and airport designs to reduce exposure levels.

Education

GISs are used in schools ranging from elementary schools through universities. One elementary school, for example, developed a project in which pupils updated a regional GIS database. They used the results to show state legislators that a proposed road-building project could adversely affect natural and historical points of interest in their state. The quality of their report may have influenced the outcome of the project. A university designed a campus-wide GIS to fit research and curricular developments in business, social science, and environmental science departments. Education programs such as these are frequently involved in training future GIS professionals. At all levels, such programs are important for developing student interest in geography and other resource-oriented studies.

Scientific Research

A GIS developed for an underwater scientific research application combines information about a variety of factors, such as deep-sea hydrothermal vents, lava flows, fissures and faults, rock-core samples, sediment samples, and earthquake locations. The availability of this diverse information in a single system allows biologists, chemists, and geologists to view the area in map form. The system then provides a background against which to interpret interrelationships among variables for development and testing of hypotheses and planning for further research.

Social Programs

GIS methods are used in a variety of social programs. One jurisdiction, for example, uses GIS to manage a program in which probationer addresses are geocoded to the street network. Probation officers employ the system to monitor clients, to plan home visitation routes, and to report the locations of clients in risk areas, such as school yards or high-crime areas. In addition, maps are produced comparing addresses to census tract statistics regarding employment status, length of residence, and average rents. These maps are used for risk assessment of residence locations. Finally, a variety of records from the system have proven valuable in crime prevention education.

In another community, GIS has been used to study factors related to the need for homeless shelters. Socioeconomic variables, such as income, education, marital status, race, and disability, were studied in relation to triggering factors such as utility shutoff, eviction, job loss, and family conflict. Tracking of experience with these relationships provided guidelines for developing programs aimed at stabilizing people who are at risk of becoming homeless.

Public Health

Public health applications of GIS continue to be varied. In one case, for example, GIS was used to set priorities for the replacement of lead water pipe. This application generated maps showing the statistical relationship between lead water pipes and the number of children younger than 5 years old. Areas where many children were exposed to lead water pipes were given a high priority for pipe replacement, thus reducing the exposure of young children to the risk of lead poisoning.

In one rural setting, the distribution of different vegetation types was determined by remote-sensing analysis. This information was combined with elevation data, so that potential mosquito habitats within regions of favorable vegetation and elevation were identified. Because the flight range of malaria-carrying mosquitos was estimated to be about 1 mile, potential habitats in a 1-mile buffer zone around villages in the region were identified as areas of maximum risk for exposure to malaria. This information aided in the planning of cost-effective control strategies.

Business Planning

Business planning applications of GIS are numerous, ranging from generating data about potential markets to creating targeted mailing lists. Such systems are also used in the analysis of potential store sites. One aspect of such analysis involves modeling predicted consumer spending and comparing expected profitability at various proposed locations. Such studies can also incorporate relationships with existing stores, proximity to housing developments, population numbers and income characteristics, and other factors. In addition, transportation information can be used to show the direction and volume of traffic at potential sites. GIS techniques also make it possible to highlight building sites that avoid wetlands, ravines, rights-of-way, and other problem areas.

Utilities Planning

Manually maintained utility records often involve duplication of and conflict between records gathered at different times, and, may also be incomplete. With GIS applications, utility providers find it much more efficient and accurate to maintain information such as locations and types of equipment and facilities and maintenance records. These applications allow better service planning, provide accurate information to personnel answering service calls, reduce maintenance costs, and improve system reliability.

Resource Management

Resource management problems occur in many regions, some large and some small, on land and under the sea.

In one coastal harbor region, GIS was used to map, analyze, and manage a complex ecosystem upon which the local economy relied. This ecosystem involved timber production, oyster farming, maintenance of fisheries, farming, harvesting of cranberry bogs, and other activities. It also involved the often disparate interests of large timber companies, local Native American tribes, cranberry growers, commercial fishers, oyster harvesters, local residents, and visitors. Information in the GIS database included a variety of topics, ranging from slope instability and erosion potential to pesticide- and herbicide-use levels. This information, plus the locations of roadways, wells, aquifers, prime farmland, and other data, was used to predict and illustrate the potential outcomes of alternative watershed management decisions.

Another system uses remote-sensing techniques to monitor forest change, including the detection of forest damage caused by pests or disease outbreaks. Such information influences many aspects of resource management, including harvest scheduling, wildlife management, and the evaluation of the effects of development on scenery.

Wildlife Management

GIS is often used in wildlife management applications, as is illustrated by two examples. One case involved the identification of favorable wolf habitats in the upper midwest. The approach involved a study to find the territories of existing wolf packs by radio tracking. Characteristics of these territories were analyzed in a GIS. The information gained was then used to search for similar habitats throughout the region. These potential habitats were then analyzed to evaluate their potential for the introduction and protection of new wolf packs.

A second program involved finding alternative habitats for desert tortoises, whose population is declining because of diminishing creosote shrub habitats. First, a classification of existing cover types was developed from remote-sensing data. The cover information was combined with land-ownership information, data about typical habitats for terrestrial animal species, and records of actual sightings. Identifying potential sites for desert tortoise protection was then possible. Information about these sites was considered when planning biodiversity protection for these areas, including the possible establishment of wilderness area status.

Military Base Management

Increasingly, the military has recognized an obligation to protect base residents and elements of the environment, such as plant and animal habitats, during weapons and missile testing, training maneuvers, and other activities. GIS systems have been developed for many military facilities to help in honoring this obligation. The data layers in such systems typically include traditional topics, such as buildings, streets, bridges, railroads, airport facilities, communications antennas, topography, training areas, and buried pipes and cables. However, to help meet environmental obligations, information about factors such as wetlands, flood zones, archeological sites, recreation areas, trails, and plant and animal habitats is often added. GIS capabilities are then used for a variety of purposes. For example, levels of constraint are established on training maneuvers, according to the intrusiveness of the planned activity. Also, emergency response plans may be developed for dealing with chemical plumes, wildfires, or other problems. In addition, the system may be used to help in the selection of safe areas suitable for storage of explosives needed for various types of weaponry and a variety of other applications.

Agriculture

A combination of GIS and GPS technologies, called *precision farming,* promises to increase agricultural productivity and, simultaneously, reduce farming costs. The approach is based on a GIS map of the soil types found on a farm. This map is created by a field crew that samples soil types at regular intervals in each field. Each sample location is stored in a GPS recorder. When the soil samples are analyzed, the information obtained is tied to the appropriate sample locations and a soil map of the farm is generated in a GIS system. Based on knowledge of crop requirements, decisions are made regarding the appropriate amount and type of fertilizer and herbicide for each soil type. These requirements are mapped and saved in another GIS file. This product application map is stored in a computer mounted in the tractor that applies the materials (Figure 21.9). As the tractor is driven across the fields in a regular pattern, its position and speed are continuously monitored by an onboard GPS. The onboard computer compares the location of the tractor with the product application map and continuously adjusts the mix and application rate of the fertilizer and herbicide.

In the past, fertilizers and herbicides were mixed and applied at an average rate over entire fields. This

(a)

(b)

Figure 21.9 (*a*) In-cab view of the FALCON™ computer. The FALCON control system monitors ground speed and field position and adjusts rates and/or blends of crop inputs such as fertilizer and herbicides based on digital product application maps. (*b*) Ag-Chem Terra-Gator®, equipped with the SOILECTION® system.

Source: Courtesy of Ag-Chem Equipment Co. Inc.

meant that some areas received appropriate applications, whereas others received too much or too little. With precision farming, all areas receive appropriate applications. This reduces costs while also benefitting the environment by reducing runoff.

Other Examples

Other examples abound, including military mission planning, provision of tourism information, managing services and facilities for parades, siting landfills, planning efficient street repair management, and monitoring duck habitat. The principles involved in these applications are similar to those mentioned. They involve mapping variables important to a particular activity, combined with modeling of the interactions between the variables. These capabilities produce information about outcomes that is helpful to the solution of some important problem.

SUMMARY

Geographic information systems (GISs) are defined as computer systems capable of assembling, storing, manipulating, and displaying geographically referenced information. This definition implies an automated, digital system that provides an orderly set of facts, as well as facilities for file-handling and data-processing operations. Each piece of data is tied to a locational reference system so that the system can produce maps from the data.

Information in a GIS is classified according to physical or social themes and stored on separate data planes on the basis of points, lines, polygons (areas), and surfaces. Each data plane is rectified so that it is in correct alignment with the other data planes. In addition, the various elements are assigned attribute codes so that they can be identified and selectively retrieved.

GIS capabilities usually include the comparison of distributions, searching for features with selected characteristics, finding nearest neighbors, and comparing data recorded at different scales or on different projections. The methods used include reclassification, overlaying, distance functions, and modeling.

Reclassification is the replacement of the initial categories by new values, based on such criteria as relative importance, arithmetic or statistical specifications, or feature size. Overlaying is the digital equivalent of carrying out map-overlay analysis by physical methods and is based on some minimum, maximum, or other specific value requirement. Distance functions assign values based on simple distance, travel time, or minimum cost paths. Modeling pulls the capabilities of the system together by establishing a logical sequence of analytical operations and using the database to produce information about the alternative outcomes that would result from different courses of action. The output from this analysis is made available by direct viewing on the CRT or in the form of hard-copy maps and statistical reports.

Geographic information systems entail, in one way or another, virtually all of the topics discussed earlier in this book. The techniques involved in the use of conventional maps, of remote-sensing products, and of computer-assisted cartographic output are all used. These range from the determination of scale and the interpretation of symbols to the measurement of distances, areas, and other characteristics of features on the earth's surface. In addition, GISs bring greater flexibility and accuracy to the analysis of the relationships between such features. Indeed, when one considers the capabilities of geographic information systems, it is apparent that they constitute the "state of the art" in the field of map use and analysis. It is not surprising, therefore, that GISs are currently being applied to a wide variety of "real-world" projects. In addition, the hardware and software that support such systems are becoming more refined and less expensive. For these reasons, GISs are becoming more widely available. These developments virtually assure the increased use of geographic information system technology and concepts in a more varied range of applications.

SUGGESTED READINGS

Antennucci, John C., Kay Brown, Peter L. Croswell, Michael J. Kevany, with Hugh Archer. *Geographic Information Systems: A Guide to the Technology*. New York: Chapman & Hall, 1991.

Aronoff, Stan. *Geographic Information Systems: A Management Perspective*. Ottawa, Canada: WDL Publications, 1989.

Berry, J. K. "Computer-Assisted Map Analysis: Extending the Utility of the GIS Technology." *Proceedings,* Tenth Canadian Symposium on Remote Sensing, Ottawa, Ontario, 1986, sponsored by the Canada Centre for Remote Sensing.

Douglas, David H., and A. Raymond Boyle, eds. *Computer-Assisted Cartography and Geographic Information Processing: Hope and Realism*. Ottawa: Canadian Cartographic Association, Department of Geography, University of Ottawa, 1982.

Huxhold, William E. *An Introduction to Urban Geographic Information Systems*. New York: Oxford University Press, 1991.

Maguire, David J., Michael Goodchild, and David W. Rhind, eds. *Geographical Information Systems: Principles and Applications*. Vols. 1 & 2. Essex, England: Longman Scientific & Technical. New York: John Wiley & Sons, Inc., 1991.

McHarg, Ian L. *Design with Nature*. Garden City, N.Y.: Doubleday/Natural History Press, Doubleday and Company, 1969.

Puequet, Donna J., and Duane F. Marble, eds. *Introductory Readings in Geographic Information Systems*. New York: Taylor & Francis, 1990.

Star, Jeffrey, and John Estes. *Geographic Information Systems: An Introduction*. Englewood Cliffs, N.J.: Prentice Hall, 1990.

Taylor, D. R. Fraser, ed. *Geographic Information Systems: The Microcomputer and Modern Cartography*. Vol. 1, *Modern Cartography*. Oxford: Pergamon Press, 1991.

Appendix A

U.S. and Canadian Map Producers and Information Sources

A tremendous variety of mapping programs, both public and private, exists in the United States and Canada. This chapter provides some basic background data about the major map producers and the types of maps they publish. In addition, some of the libraries and depositories where maps can be consulted are discussed, as are sources from which maps can be purchased. Additional information about these and other mapping programs can be obtained by consulting the sources listed in the Suggested Readings at the end of the chapter. Telephone numbers are provided, when available, to help in contacting the agency or company.

Many of the companies, educational institutions, libraries, and agencies mentioned in this chapter maintain sites on the World Wide Web. These sites, although they vary considerably in content, often offer significant resources that can supplement and support the information in this book. For example, some sites provide product listings and ordering instructions. Others describe the sponsoring organization, sometimes providing information about the availability of data, programs, or services. Still other sites are instructional in nature. The addresses of these sites are not listed in this book, because they are subject to frequent change. Instead, a website has been developed to provide up-to-date links to sites selected on the basis of their relevance to topics in the book. If you have access to the World Wide Web, you may find it helpful to visit this site, at http://www.mhhe.com/earthsci/geography/campbell/.

U.S. Federal Map Producers

Major Agencies

U.S. Geological Survey

The U.S. Geological Survey (USGS) is an agency of the Department of the Interior. Its role is to gather, analyze, and make available information about the composition and structure of rocks, data about surface water and groundwater, knowledge of earth history and natural processes, and appraisals of potential energy and mineral resources. In addition, USGS classifies federal lands for mineral and water-power purposes and supervises oil, gas, and mineral lease operations on federal and Native American lands and the outer continental shelf.

In connection with its assigned tasks, USGS publishes a wide variety of reports and maps and is probably the preeminent U.S. map producer. A listing of the types of products the agency publishes includes topographic maps; floodplain maps; geodetic control data; geologic maps (general geology, minerals, coal, oil, gas); land-use, land-cover, and river basin/watershed studies; the *National Atlas of the United States;* aerial photographs; orthophotomaps; seismicity maps; clinometric (slope) maps; gravity survey charts; isogonic charts; water-availability maps; hydrologic maps; Landsat scenes and mosaics; and digital data. Examples of these products appear frequently throughout this book, especially in Chapter 9.

A monthly catalog of USGS publications, entitled *New Publications of the U.S. Geological Survey,* includes information about the various books, professional papers, reports, and special maps published each month, as well as about new and revised topographic maps produced by the agency. The catalog is also available only on the World Wide Web.

Distribution services for all USGS maps and text products (books and open-file reports) are handled by

U.S. Geological Survey, Information Services
P.O. Box 25286
Denver, CO 80225
Phone: 1-888-ASK-USGS (1-888-275-8747)

Many USGS products, especially topographic maps, are also available from certain U.S. Geological Survey Earth Science Information Centers (ESICs), discussed later in this chapter, and from private map dealers throughout the nation.

Bureau of Land Management

Another Department of the Interior agency with extensive mapping responsibilities is the U.S. Bureau of Land Management (BLM). The BLM is the final authority on all matters pertaining to the survey of public lands.

Maps and related products produced by the BLM include maps of lands administered by the BLM; wildlife and scenic-river jurisdictions; and recreation maps (Figure A.1). Information about these products is available from ESICs or from

U.S. Department of the Interior
Customer Service, Room 1000 LS
Bureau of Land Management
1849 C Street NW
Washington, DC 20240
Phone: (202) 452-5171

National Ocean Service

The U.S. National Ocean Service (NOS) is a component of the National Oceanic and Atmospheric Administration (NOAA), in the U.S. Department of Commerce. The NOS produces aeronautical charts, hydrographic surveys, nautical charts, geodetic control data, aerial photographs, orthophotomaps, gravity survey charts, and storm evacuation maps. Examples of aeronautical and nautical charts are shown in Chapter 7. Information can be obtained from

NOAA Distribution Division N/ACC3
National Ocean Service
6501 Lafayette Avenue
Riverdale, MD 20737-1199
Phone: (301) 436-8301 or (800) 638-8972

National Imagery and Mapping Agency

The U.S. National Imagery and Mapping Agency (NIMA) provides imagery and geospatial information to national policy makers and military forces. The agency incorporates the former U.S. Defense Mapping Agency (Figure A.2) and other imagery and mapping agencies. Further information about the agency and its products is available on the World Wide Web.

Information regarding the purchase of NIMA nautical and aeronautical products can be obtained from the National Ocean Service Distribution Branch. Information about topographic maps and gazetteers can be obtained from the USGS Distribution Branch. Both of these sources are listed on previous pages.

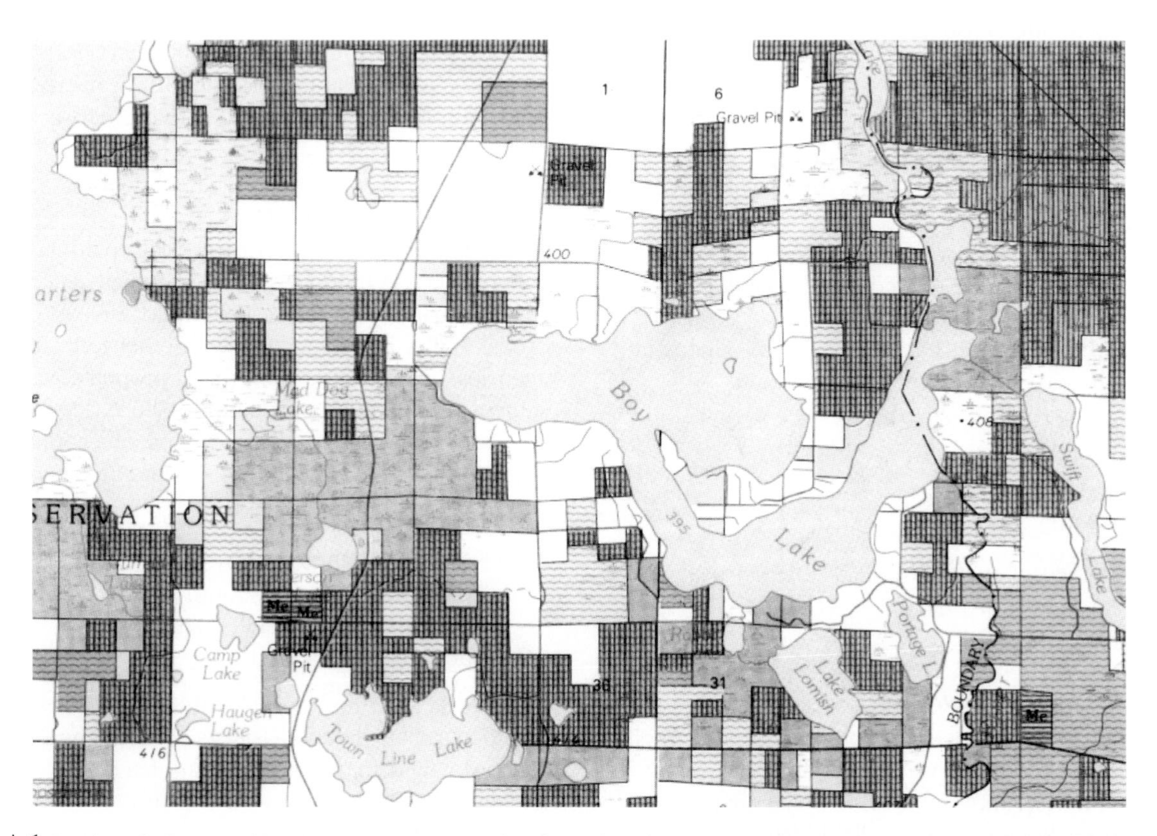

Figure A.1 Portion of a Bureau of Land Management map of surface-minerals management status. Tones (colors on original) indicate public land-management status, and patterns indicate federal mineral rights. (Cass Lake quadrangle [planimetric], Minnesota, 1:100,000, 1978.)

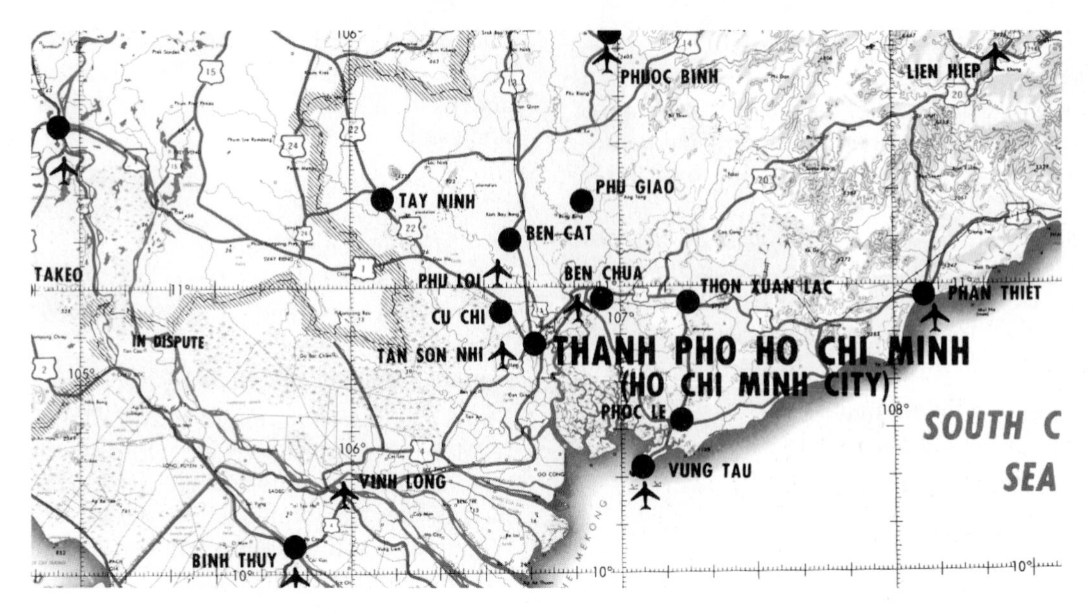

Figure A.2 Portion of former Defense Mapping Agency Southeast Asia Briefing Map. (Series 5213, Edition 2, original scale 1:2,000,000. Reduced.)

Other Agencies

Bureau of the Census

The U.S. Bureau of the Census is a branch of the Department of Commerce. It produces a wide variety of maps, ranging from large-scale coverage of census subdivisions, including census tracts, blocks, and enumeration districts, to maps of incorporated places and other densely settled areas, including Standard Metropolitan Statistical Areas (SMSA). The bureau also produces a map of U.S. counties and a variety of color-shaded and dot-distribution maps of U.S. census social and economic data (Figure A.3).

Computerized geographic coding of block-by-block census records is related to the TIGER system, which is described in Chapter 19.

Detailed information regarding available maps and other census materials can be obtained from

Census Customer Services
Bureau of the Census
Washington, DC 20233
Phone: (301) 457-4100

A great deal of material is available on the World Wide Web.

Central Intelligence Agency

Most of the maps produced by the U.S. Central Intelligence Agency (CIA) are not available to the public. A number of high-quality, multicolor, relief maps of foreign countries, however, are available from

National Technical Information Service
U.S. Department of Commerce
5285 Port Royal Road
Springfield, VA 22161
Phone: (703) 605-6000

These maps sometimes include thematic insets that provide summary information about economic and social topics and features of the physical environment. An example of an economic activity map is shown in Figure A.4.

Department of Agriculture

Some organizations within the U.S. Department of Agriculture (USDA) produce maps and other materials of interest to map users. The results of soil surveys, conducted by the USDA in conjunction with other federal and state agencies, are available from the appropriate National Resources Conservation Service State Office (Figure A.5).

Aerial photos and photomaps of cropland areas (Figure A.6), as well as some satellite images, are available from

USDA Aerial Photography Field Office
2222 West 2300 Street South
Salt Lake City, UT 84119-2020
Phone: (801) 975-3500

The U.S. Forest Service publishes maps showing the location of national forests. The agency also publishes detailed maps of each national forest, usually at a scale of 1:24,000, as well as various visitors' maps and guides (Figure A.7).

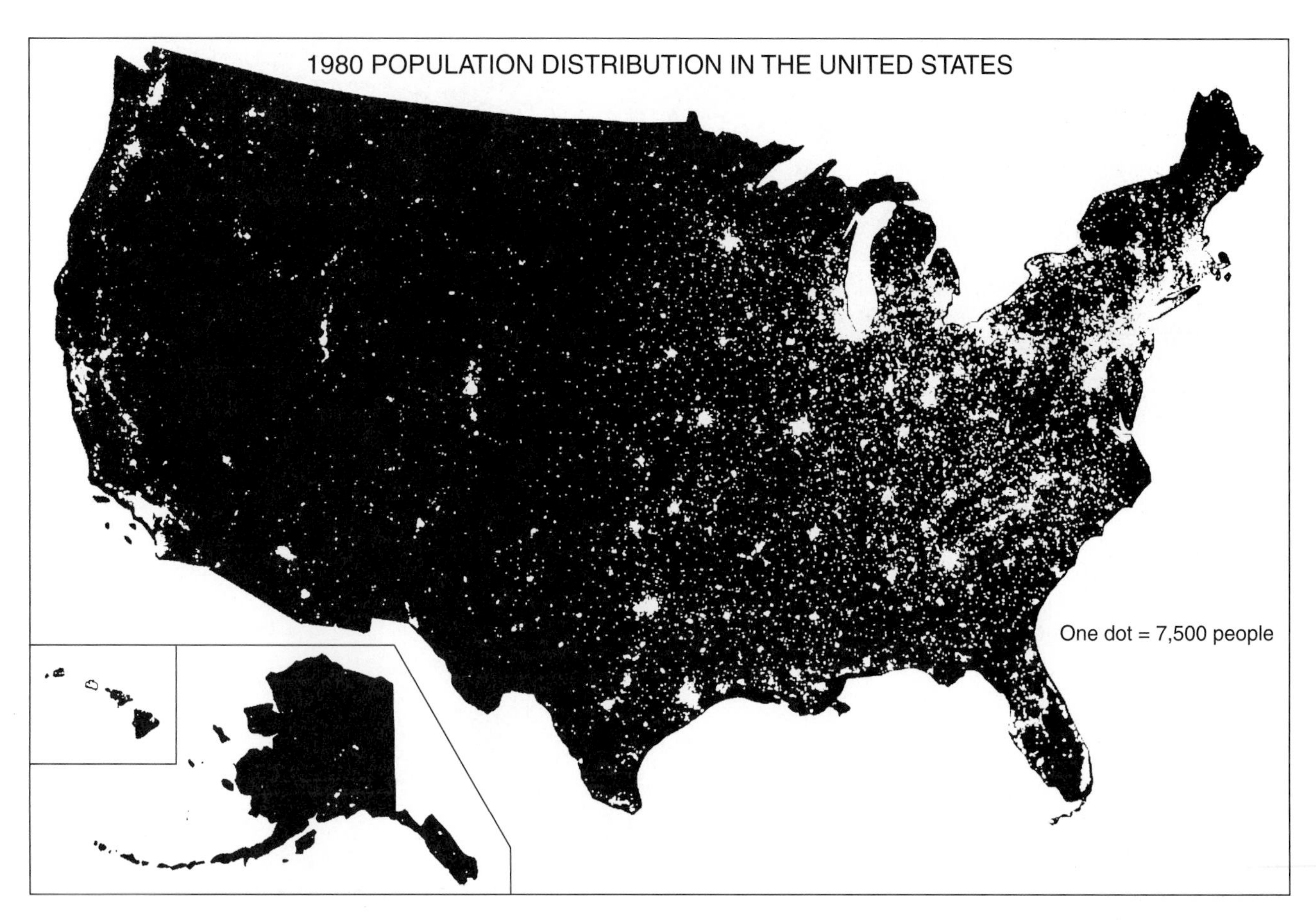

Figure A.3 1980 population distribution in the United States.
Source: Courtesy of U.S. Department of Commerce, Bureau of the Census.

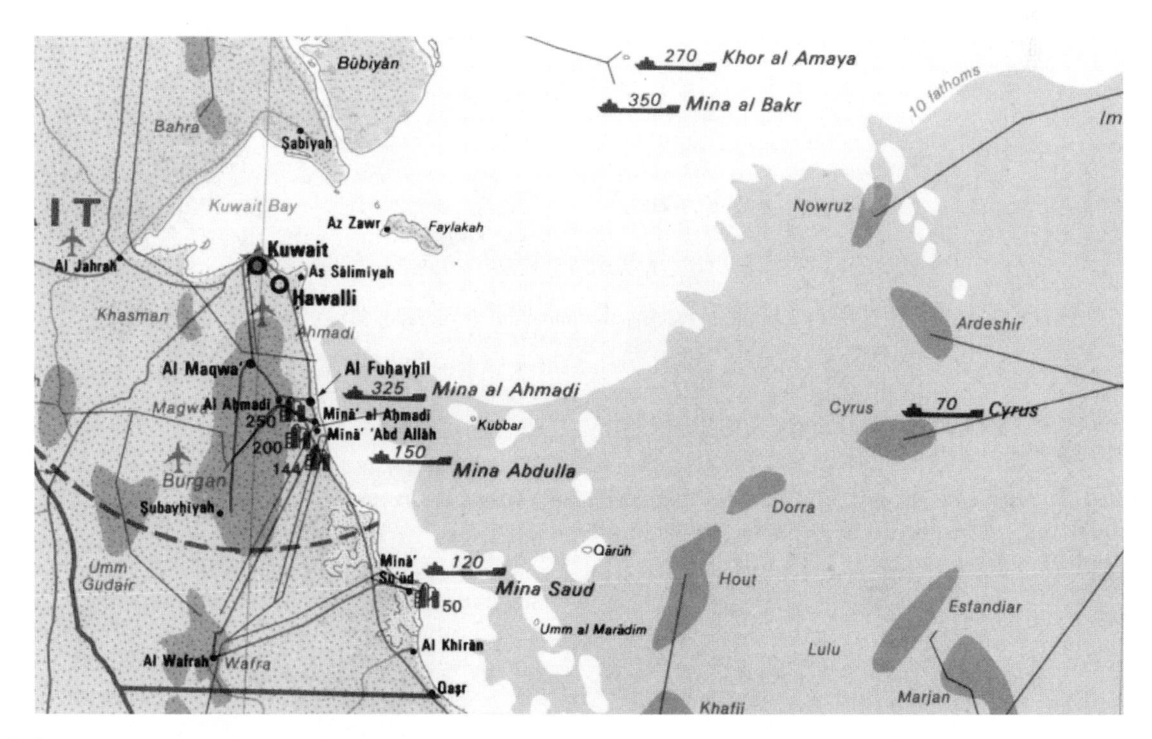

Figure A.4 Portion of Central Intelligence Agency map, "The Persian Gulf" (1:1,600,000, 1980).

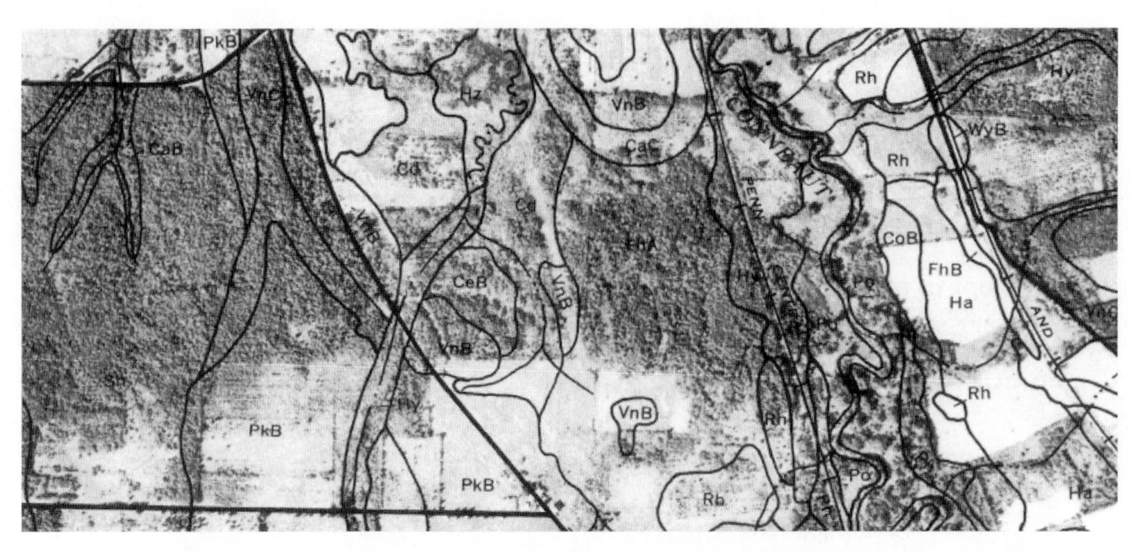

Figure A.5 Portion of Natural Resources Conservation Service soil survey map. (Crawford County, Pennsylvania, 1:20,000.)

Figure A.6 Portion of National High Altitude Program black-and-white photo (Colorado, 1983).
Source: Courtesy of U.S. Department of Agriculture, FSA Aerial Photography Field Office.

Information regarding these maps can be obtained by writing to the following address or by contacting any U.S. Forest Service regional office, or from

U.S. Forest Service, Public Affairs Office
Department of Agriculture
P.O. Box 96090
Washington, DC 20090-6090
Phone: (202) 205-1760

National Aeronautics and Space Administration

A variety of aerial photographs, Landsat imagery, manned spacecraft imagery, Skylab imagery, and other materials resulting from programs of the U.S. National Aeronautics and Space Administration (NASA) are available from the EROS Data Center (described later in the chapter). An example of one of these products is shown in Figure 2.1.

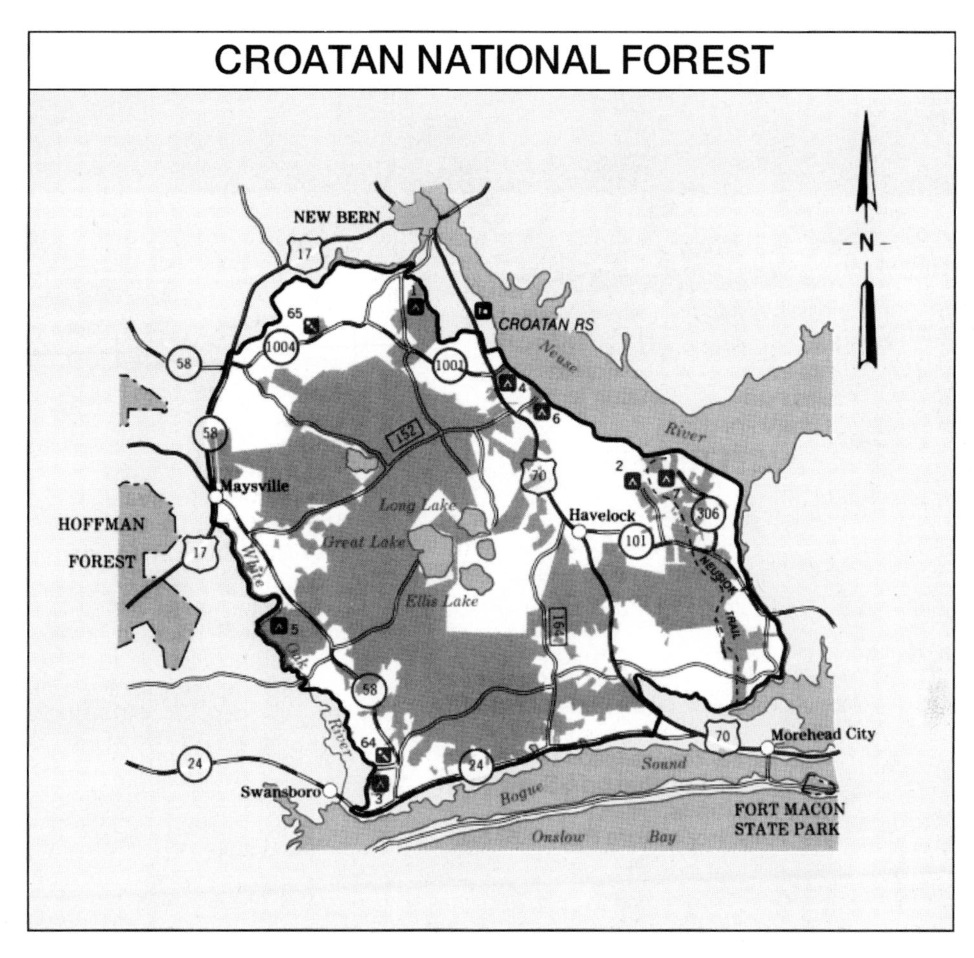

Figure A.7 Map of Croatan National Forest, North Carolina.
Source: Courtesy of U.S. Department of Agriculture, Forest Service.

National Park Service

The U.S. National Park Service (NPS) produces topographic maps, as well as visitors' maps and guides, for the national parks and monuments (Figure A.8). Information about these materials can be obtained from

National Park Service
1849 C Street NW
Washington, DC 20240
Phone: (202) 208-6843

Tennessee Valley Authority

The Tennessee Valley Authority (TVA) produces various topographic, land-use, navigation, and recreation maps of the area within its jurisdiction. For a free catalog of these materials, examples of which are shown in Figure A.9 contact

TVA Map Store
1101 Market Street
Chattanooga, TN 37402-2801
Phone: (423) 751-6277

MAJOR MAP SERIES

U.S. Geological Survey Topographic Maps

The USGS produces topographic maps in a wide variety of scales. Table A.1 indicates the customary (English) system scales, as well as the metric system scales that are available.

Quadrangles are individual map sheets that are laid out as four-sided figures, each of which is bounded by lines of latitude and longitude.

Various map scales make it necessary to vary the extent of coverage of quadrangle maps. For example, a 1-degree quadrangle covers an area bounded by a line of latitude on the southern boundary and a second line of latitude, 1 degree greater in value, on the northern boundary. Similarly, the area is bounded by a line of longitude on the eastern boundary and a second line of longitude, 1 degree greater in value, on the western boundary. In the same fashion, a 30-minute quadrangle has boundaries that are 30 minutes apart on all four sides, and similar relationships hold for 7.5-minute quadrangles.

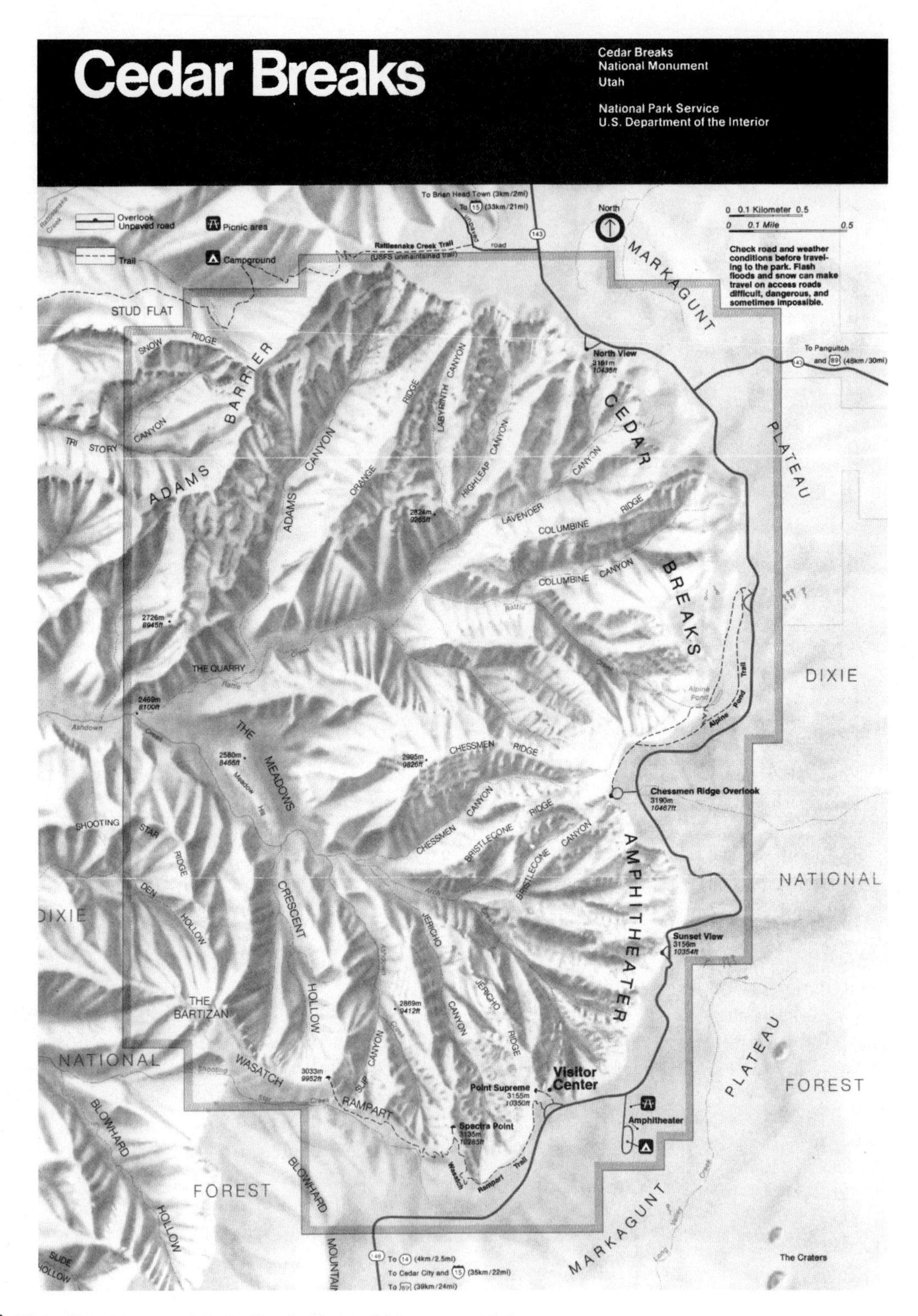

Figure A.8 Visitor's guide map of Cedar Breaks National Monument, Utah.
Source: Courtesy of U.S. Department of the Interior, National Park Service.

A 1-degree quadrangle covers the same area as is covered by four 30-minute quadrangles and a 30-minute quadrangle the same as sixteen 7.5-minute quadrangles.

As discussed in Chapter 2, lines of longitude converge toward the poles. In the Northern Hemisphere areas covered by the USGS, the shape of a quadrangle, which is bounded by two lines of longitude, is closer to a trapezoid because the northern boundary is somewhat narrower than the southern. The difference between this trapezoid and a true rectangle is extremely small, in the lower and midlatitudes, but becomes greater at higher latitudes.

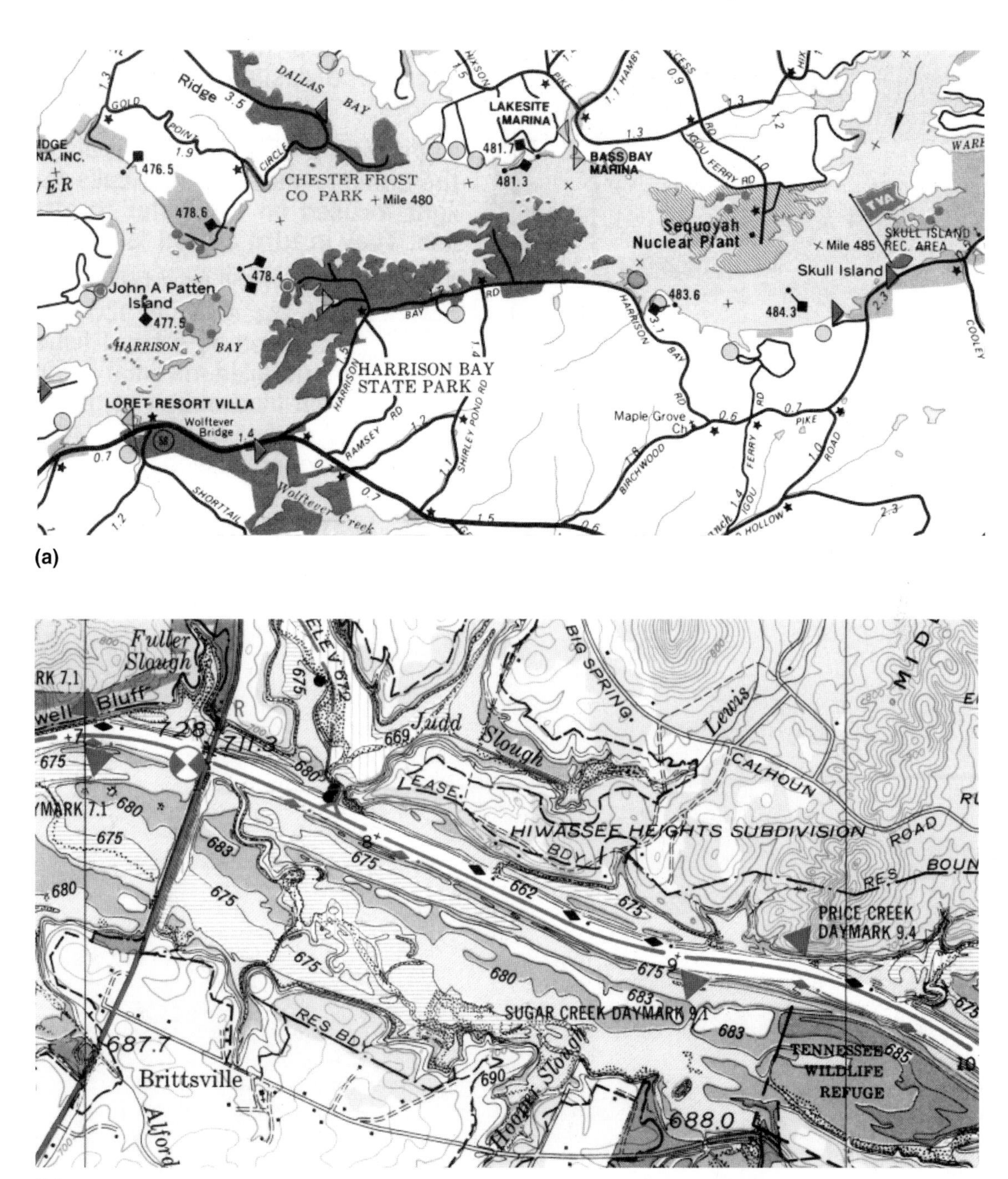

Figure A.9 Sample Tennessee Valley Authority products. (*a*) Portion of Chickamauga Lake Recreation Map (1982). (*b*) Portion of Navigation Chart 703, Tennessee River, Chickamauga Lake (1968). (Not for navigational purposes.)

Another variation stems from meridian convergence in the high latitudes. If all quadrangles covered the same number of degrees and minutes in the east-west direction, those in Alaska would be significantly narrower than those in Texas. For this reason, and to keep sheet sizes relatively constant for printing and filing convenience, the east-west extent of quadrangles is adjusted in Alaska.

National Ocean Service Nautical Charts

The U.S. National Ocean Service (NOS) is a component of the National Oceanic and Atmospheric Administration (NOAA). Along with its other responsibilities, the Office of Charting and Geodetic Services (C&GS) within NOS produces nautical charts for U.S. oceanic coastal waters and the Great Lakes and connecting waterways. These charts are produced in different scales, ranging

TABLE A.1 Publication Scales of U.S. Geological Survey Topographic Maps

Customary System

Series	Scale	1 Inch Represents	Standard Quadrangle Size (Latitude × Longitude)
7.5 minutes	1:24,000	2000 feet	7.5 minutes × 7.5 minutes
15 minutes (no longer published)	1:62,500	1 mile (approx.)	15 minutes × 15 minutes
1:63,360 Alaska	1:63,360	1 mile	15 minutes × 20–36 minutes
United States*	1:250,000	4 miles (approx.)	1 degree × 2 or 3 degrees†
State maps*	1:500,000	8 miles (approx.)	Not applicable‡
United States	1:1,000,000	16 miles (approx.)	4 degrees × 6 degrees

Metric System

Series	Scale	1 Centimeter Represents	Standard Quadrangle Size (Latitude × Longitude)
7.5 minutes Puerto Rico	1:20,000	200 meters	7.5 minutes × 7.5 minutes
7.5 minutes × 15 minutes	1:25,000	250 meters	7.5 minutes × 15 minutes
Intermediate	1:50,000	500 meters	
Intermediate	1:100,000	1 kilometer	30 minutes × 60 minutes

Source: Adapted from National Cartographic Information Center pamphlet *National Mapping Program, Map Scales.*
*Coverage of Antarctica is also provided at this scale.
†Antarctic quadrangles are 1 degree × 3–15 degrees.
‡Antarctic quadrangles are 2 degrees × 7.5 degrees.

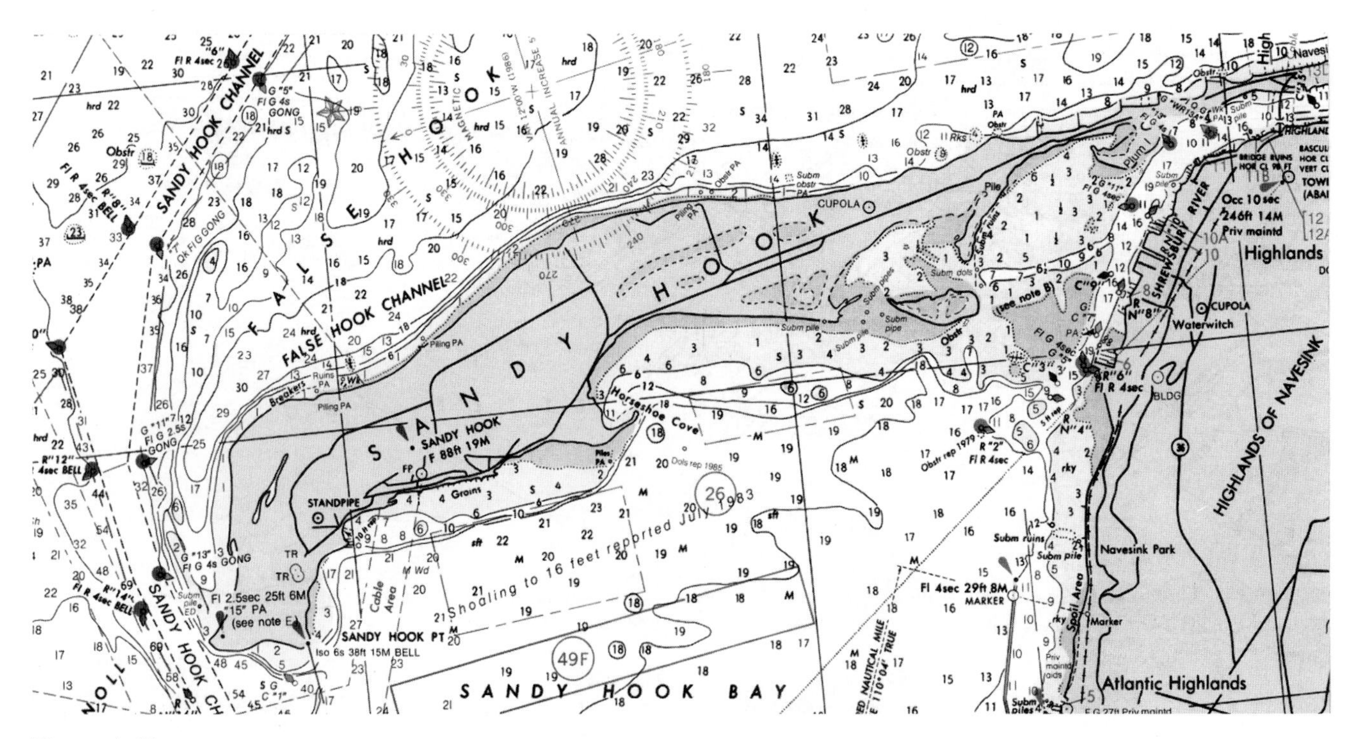

Figure A.10 Portion of National Ocean Service small-craft navigational chart. (Chart 12324. original scale 1:40,000. Reduced. November 1986. Not for navigational purposes.)

from small scales adequate for sailing between coastal ports and in the open ocean to large scales that are needed to navigate in harbors and restricted channels. Chapter 6 provides information about the use of nautical charts and more details regarding their contents.

The two general series of charts are (1) small-craft charts (Figure A.10) and (2) conventional charts (see Figure 7.9). The small-craft charts are specially designed to meet the needs of small-craft operators. The conventional charts fall into the following categories:

1. **Harbor charts,** at scales of 1:50,000 and larger, are intended for navigation and anchorage in harbors and small waterways.
2. **Coast charts,** at scales from 1:50,000 to 1:150,000, are for coastwise navigation inside offshore reefs and shoals, entrance into bays and harbors of considerable size, and navigation of certain inland waterways.
3. **General charts,** at scales from 1:150,000 to 1:600,000, are for vessels whose courses are well offshore but whose position can be fixed by landmarks, lights, buoys, and characteristic soundings.
4. **Sailing charts,** at scales smaller than 1:600,000, are for offshore sailing between distant coastal ports and for approaching the coast from the open ocean.
5. Special charts, such as **intracoastal waterway charts,** at a scale of 1:40,000, provide coverage of the inside route from Miami to Key West, Florida, and from Tampa to Anclote Anchorage, Florida. **Canoe charts** of the Minnesota-Ontario border lakes are designed to suit the needs of small, shallow-draft vessels.

Information regarding these products is included in a series of *Nautical Chart Catalogs*. Each catalog lists and describes the charts and other products, such as tide tables, coast pilots, current tables, tidal current charts, and Great Lakes water-level information, available for a particular region.

NOS charts and related publications may be ordered from

NOAA Distribution Division (N/ACC3)
National Ocean Service
Riverdale, MD 20737-1199
Phone: (301) 436-8301 or (800) 638-8972

or from many authorized sales agents located in cities around the country.

NONFEDERAL MAP PRODUCERS

State Agencies[1]

Government agencies in the 50 states are involved in many mapping activities. The products that these agencies produce, which are generally available to the public, vary considerably in type and quality from state to state. They have some common characteristics, however, because the needs of the various state governments tend to be very similar. All of the states, for example, are involved in site-specific functions, such as highway planning and construction. This type of work requires topographic coverage that is more detailed and larger scale than that provided by USGS quadrangle maps. On an entirely different front, the individual states vigorously promote tourism. One effective method of doing this is distributing promotional pieces, such as state highway maps or maps of state park facilities.

Products likely to be produced by every state include (1) planimetric base maps of the counties, typically at a scale of 1 inch to 1 mile, with larger-scale coverage of metropolitan areas; (2) multicolor, folded highway maps, which usually contain illustrations and promotional copy; (3) large-scale, photogrammetric engineering maps for highway planning and construction; (4) maps showing the state's geologic characteristics; (5) outline maps of civil division boundaries and names; and (6) land-use maps.

Obtaining state-produced maps is not always straightforward, because production and distribution responsibilities are often spread throughout several agencies. Helpful information in this regard is provided by the list of addresses and phone numbers, listed by state, contained in the brochure *State Information Sources*. This brochure can be obtained from the ESICs, discussed later in this appendix.

Major Private Firms

Private companies produce many of the map products used in the United States. Some of the major private cartographic firms are listed in this section to provide a general overview of the types of maps and related products available from each producer. The list also provides a basis for searching out maps to serve special interests and needs. No evaluations, either pro or con, are implied by the inclusion or exclusion of a particular company from this listing or by the necessarily brief descriptions.

AAA Cartographic Services

The American Automobile Association (AAA) supplies to its members a variety of travel-related publications, including road atlases, tour books, camping guides, city maps, and regional and national road maps. AAA Cartographic Services produces all of these publications, as well as a road atlas available to the public. Information can be obtained from

American Automobile Association
1000 AAA Drive
Heathrow, FL 32746-5063
Phone: (407) 444-7000

[1]This section is based on Paul McElligott, "Mapping by State Governments," in *U.S. National Report to ICA, 1984,* edited by Judy M. Olson (Falls Church, Va.: American Congress on Surveying and Mapping, 1984), Section 1, "Government Mapping Activities," 27–28.

Hammond Incorporated

Hammond Incorporated produces a comprehensive line of maps, world and road atlases, reference books, and wall charts. In addition, the company has a line of educational products for classroom use. Hammond also produces maps for the advertising and publishing markets. Information can be obtained from

Hammond World Atlas Incorporated
95 Progress Street
Union, NJ 07083
Phone: (908) 206-1300

National Geographic Society

The Cartographic Division of the National Geographic Society produces supplement maps that accompany the *National Geographic* magazine (60 million map copies per year), as well as smaller maps that illustrate the magazine's stories. The Cartographic Division also produces atlases, physical and political globes, and mural maps. The maps cover a wide range of topics, from general maps that show cities and countries, to detailed maps of historical events. National Geographic Society maps are noted for their accuracy and attractive rendering. Information is available from

National Geographic Society
1145 17th Street N.W.
Washington, DC 20036-4688
Phone: (800) 647-5463

Nystrom

Nystrom produces physical and political maps (Figure A.11) and globes in various series for different school-grade levels. The company's Sculptural Relief Wall Maps feature a physical-features base map without names, mounted on the same roller with a political-information transparent overlay. The company also produces laminated maps for individual student use; raised-relief maps, both wall-size and desk-size; maps for U.S. and world history; outline maps; and supplementary teaching programs. Nystrom's vacuum-formed vinyl globes are coordinated with the map series. Information can be obtained from

Nystrom
3333 Elston Avenue
Chicago, IL 60618
Phone: (800) 621-8086

Rand McNally and Company

Rand McNally and Company produces and markets a wide range of map, atlas, directory/guide, and related cartographic products and services for the consumer, business, and education markets. Included in these products are over 130 state and city road maps and 50 educational wall maps. More than 30 atlases are produced, ranging from classroom atlases to commercial and marketing guides and from a variety of road atlases to fishing guides and a zip code atlas. Information is available from

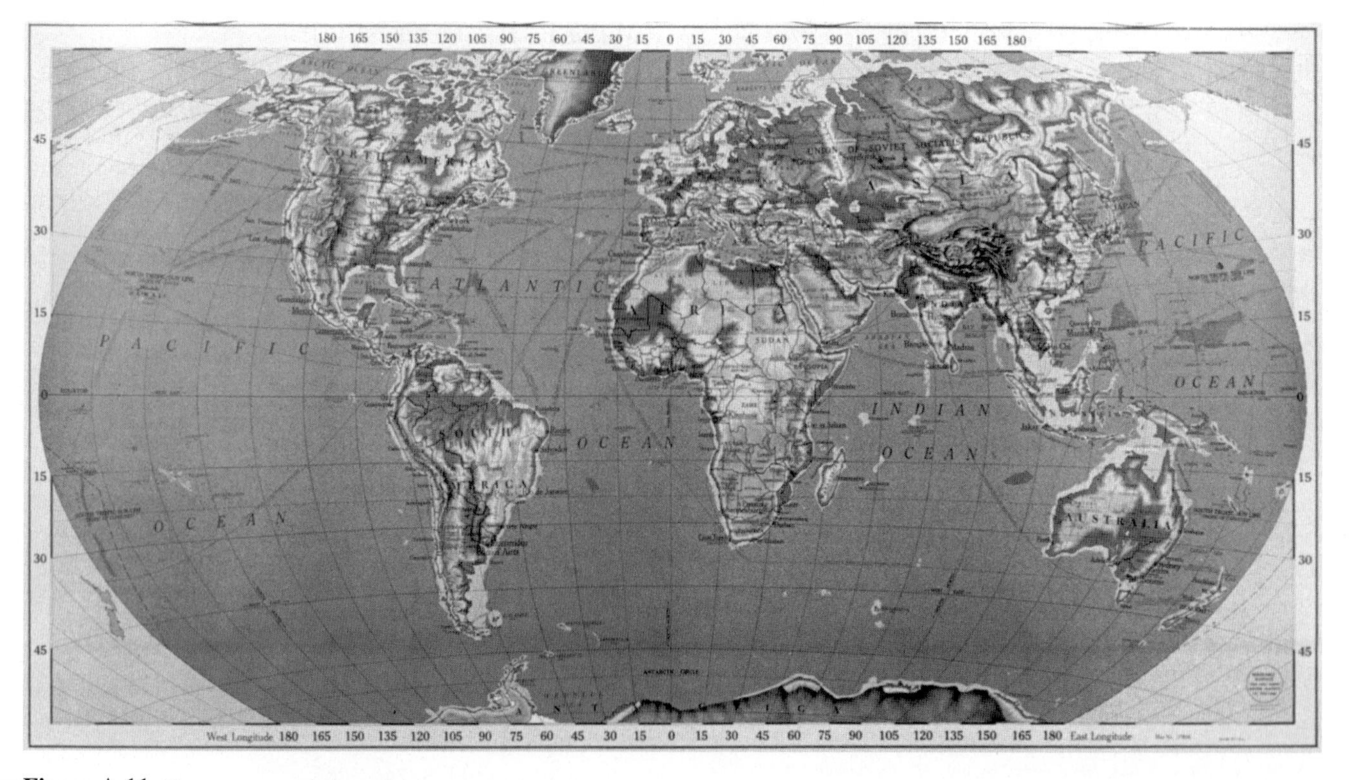

Figure A.11 Nystrom map of the world.
Source: Courtesy of Nystrom Division of Herff Jones Inc.

Rand McNally and Company
Customer Service Department/Publishing Group
P.O. Box 7600
Chicago, IL 60680
Phone: (800) 333-0136, ext. 6171

INFORMATION SOURCES FOR MAPS AND RELATED MATERIALS

The greatest challenge facing the potential map user sometimes seems to be simply determining what maps and related materials are available. Almost equally difficult, however, is the problem of how to obtain the items when their existence is known or suspected. This section outlines some of the more likely sources of both types of information.

Federal Sources of Information

Earth Science Information Center Network

The Earth Science Information Center Network (ESIC) is part of the U.S. Department of the Interior, Geological Survey (USGS). ESIC's mission is to help users gain better, faster, low-cost access to the cartographic holdings of federal, state, and private agencies. Several ESIC offices provide information about the availability, cost, and method of ordering almost any cartographic product.

A free *MiniCatalog of Map Data* can be obtained by writing or calling ESIC, and a more complete *Catalog of Map Data* is for sale by the Superintendent of Documents. These catalogs describe and give ordering information for such products as advance prints; color separates; feature separates; out-of-print maps; land-use, land-cover, and associated maps; slope maps; digital terrain tapes; maps on microfilm; orthophotoquads; aircraft photos of various types; manned spacecraft photos; Landsat imagery; geodetic control data; and a number of other materials.

For up-to-date information about ESIC addresses or for a state affiliate near you, call 1-888-ASK-USGS. A link to ESIC information is also provided at this book's website.

National Geodetic Survey

The National Geodetic Survey provides information regarding the horizontal and vertical control networks established by the National Geodetic Survey and other federal and state agencies. To order NGS products, contact

NOAA, National Geodetic Survey, N/NGS12
1315 East-West Highway, Station 9202
Silver Spring, MD 20910-3282
Phone: (301) 713-3242

EROS Data Center

The EROS Data Center, in Sioux Falls, South Dakota, is part of the Earth Resources Observation Systems (EROS) program of the U.S. Geological Survey. It is the national center for the processing and dissemination of photographic images and electronic data obtained from aerial photographic and space missions. Data from a number of programs, including U.S. Department of the Interior aerial photography and National Aeronautics and Space Administration (NASA) photography and other remotely sensed data from research aircraft and from Skylab, Apollo, and Gemini spacecraft, are maintained at the center. The center, in conjunction with Space Imaging, maintains and sells information from the Landsat program. Space Imaging is described later in this appendix.

The entire United States is covered by aerial photography produced by the National Aerial Photography Program (NAPP). Under NAPP, photography is color-infrared or black-and-white, with photos centered on one-quarter USGS quadrangles. In addition, there is a wide range of choice in terms of enlargements and the season of the photography.

Inquiries about the availability of various types of coverage for a specific location and orders for photos and images may be directed to the center at

Customer Services
U.S. Geological Survey
EROS Data Center
Sioux Falls, SD 57198-0001
Phone: (605) 594-6151 or (800) 252-4547

Nonfederal Sources of Information

Publications

Information about current maps is available in at least two journals. One source is the "New Maps" and "New Atlases" sections of *Cartographic Perspectives,* the journal of the North American Cartographic Information Society. Another source that has published similar information is the "Distinctive Recent Maps" section of some issues of *Cartography and Geographic Information Systems.* Both of these journals are available in many libraries.

Space Imaging

Space Imaging is a world leader in digital earth information. The company owns or has marketing rights to a constellation of remote-sensing satellites already in orbit. These include the U.S. Landsat satellites, the Indian Remote Sensing (IRS) satellites, Canada's RADARSAT, and the European Space Agency's Radar Satellite. In addition, the company has its own IKONOS satellite, launched in 1999, which provides 1-meter resolution imaging. Space Imaging also sells aerial imagery in the sub-meter resolution scales.

Inquiries should be directed to

Space Imaging
12076 Grant Street
Thornton, CO 80241
Phone: (303) 254-2000
Customer Service: (301) 552-0537 or (800) 232-9037

SPOT Image Corporation

The French, Belgian, and Swedish governments have undertaken the development of a satellite observation system called Système Probatoire d'Observation de la Terre (SPOT).[2] The system became operational in 1986, and images are commercially available in the United States from the SPOT Image Corporation.

Black-and-white images achieve a 10-meter ground resolution, whereas color images in three spectral bands have a 20-meter resolution. The full-scene format covers a 60-by-60-kilometer area on the ground, and images are available with differing degrees of correction and in the form of computer tapes or film or paper prints. Stereoscopic images are also available through the use of "off-nadir" stereoscopic pairs, which are two images of the same area recorded from different angles.

Inquiries should be directed to

SPOT Image Corporation
1897 Preston White Drive
Reston, VA 20191-4368
Phone: (703) 715-3100

Map Collections

Some of the largest map collections in the United States are described here, as are several others of special interest (academic and state libraries are generally not discussed). More complete information is provided by specialized publications, such as the *Guide to U.S. Map Resources* (2d ed.) or *Map Collections in the United States and Canada* (4th ed.), which are listed in the Suggested Readings at the end of this appendix. A short discussion of the characteristics of the map storage and cataloging systems in map libraries is provided in Appendix E.

Library of Congress

The Geography and Map Division of the Library of Congress maintains "the largest and most comprehensive cartographic collection in the world."[3] Included in the collection are more than 4.5 million print and manuscript maps, 60,000 atlases, and 6000 reference works. Over 300 globes, 2000 three-dimensional plastic relief models, and a large number of other cartographic materials are also part of the collection. This collection may be consulted in the division's reading room. Specialists of the division are available to assist in the use of the collection, which emphasizes the United States and the other areas of the Western Hemisphere. The collection also includes historical holdings, such as Ptolemy's *Geography* (dated 1482) and some 700,000 fire-insurance maps.

The Library of Congress map collection is a treasure trove for anyone with an interest in maps. Persons interested in genealogical research, for example, often find that the collection of county and state maps and atlases provides valuable documentary information. Similarly, persons interested in colonial America, the Revolutionary War, the War of 1812, and the Civil War find manuscript and printed maps that provide useful information.

The National Digital Library Program for Cartographic Materials is beginning to convert all the maps in the collection to digital format. This project has a goal of 5 million digitized maps by the year 2000. As the project continues, the assets of the library will become accessible through the Internet.

The mailing address is

Library of Congress
Geography and Map Division
Washington, DC 20540

The location is

James Madison Memorial Building
Room LM B01
101 Independence Avenue SE
Washington, DC 20540-4650
Phone: (202) 707-6277

National Archives

The U.S. National Archives and Records Administration maintains a collection that includes two million maps and 100,000 aerial photographs produced by federal agencies. These materials include information pertaining to Native American treaties and related matters, surveys and settlement records compiled by the Bureau of Land Management, immigration and population growth and enumeration district maps from the records of the Bureau of the Census and the Bureau of Agricultural Economics, and other records related to urban development. The address for cartography-related materials is

National Archives II
8601 Adelphi Road
College Park, MD 20740-6001
Phone: (301) 713-6780

[2]For a detailed description of the system, see Michael Courtois and Gilbert Weill, "The SPOT Satellite System," in *Monitoring Earth's Ocean, Land, and Atmosphere from Space—Sensors, Systems, and Applications,* Vol. 97 of Progress in Astronautics and Aeronautics series, edited by Abraham Schnapf (New York: American Institute of Aeronautics and Astronautics, 1985).
[3]From a description brochure published by the Geography and Map Division.

The American Geographical Society Collection

Located in the Golda Meir Library of the University of Wisconsin-Milwaukee, the American Geographical Society Collection is a major geographical reference and research collection. "The scope of the collection is broad, encompassing all aspects of geography and selected facets of related disciplines, such as economics, demography, regional science, urban studies, history, anthropology, archaeology, sociology, geology, earth science, oceanography, and meteorology. The materials collected range from the early travel narratives to the scientific results of polar and oceanographic expeditions to the technical reports and discussion papers on quantitative techniques that characterize much of contemporary research."[4]

Further information can be obtained from

The American Geographical Society Collection
Golda Meir Library
University of Wisconsin-Milwaukee
P.O. Box 399
Milwaukee, WI 53201
Phone: (414) 229-6282 or (800) 558-8993

Specialized Examples

Although they do not have the largest collections, the libraries discussed here exemplify the rich variety of specialized holdings that exist. A search of the various sources in the Suggested Readings at the end of the chapter may suggest map libraries with collections of particular personal interest.

The Newberry Library specializes in materials in the humanities, including a research collection in the history of cartography. Admission to the library's reading rooms is granted to researchers whose research projects require extensive use of the collections, and application for admission must be based on defined research needs.

The Hermon Dunlap Smith Center for the History of Cartography has been established at the library. The center offers courses and lectures, publishes manuscripts, and participates in special projects related to the expansion of knowledge regarding the history of cartography.

Information can be obtained from

Newberry Library
Map Section
60 West Walton Street
Chicago, IL 60610-3305
Phone: (312) 255-3509

[4]From a descriptive brochure, *The American Geographical Society Collection, The University of Wisconsin-Milwaukee Library,* Roman Drazniowsky, Curator.

The John Carter Brown Library contains a collection of primary map materials. These materials relate to virtually all aspects of the discovery, exploration, settlement, and development of the New World. The collection includes printed books, major collections of maps and prints, and a large number of textual manuscripts. The address is

Brown University
John Carter Brown Library
P.O. Box 1894
Providence, RI 02912
Phone: (401) 863-2725

CANADIAN GOVERNMENT MAPS

Maps produced by Canadian government agencies are generally available only from authorized dealers in Canada and the United States. Agencies that sell maps directly are noted in the text.

Topographic Mapping

The Centre for Topographic Information of Natural Resources Canada is Canada's topographic mapping agency. Among its products are maps at scales ranging from 1:50,000 to 1:250,000.

Maps of the National Topographic System of Canada are produced at scales of 1:250,000 and 1:50,000. The 1:250,000-scale maps provide the only complete topographic coverage of the country. These maps include detailed representation of relief, drainage, transportation, forest cover, and centers of population. Coverage of the more detailed 1:50,000-scale series is concentrated in the southerly, most densely populated portions of the country. Additional coverage at this scale is available for selected areas of particular interest. The system used to identify the separate sheets within these series is described in Appendix E.

Aeronautical Charts

Aeronautical charts are also produced by the Aeronautical Charts Service. Charts are available in several formats suited for particular needs.

Visual flight rules aviation is served by three sets of charts. The first of these is the World Aeronautical Charts, at a scale of 1:1,000,000, on the Lambert conformal conic projection. These serve the needs of medium speed, medium range operations. The second set is the VFR Navigation Charts, at a scale of 1:500,000. They are on the Lambert conformal conic projection to 80° North and on the polar stereographic from 80° North to the North Pole. These serve the needs of low- and medium-speed aircraft at low and intermediate altitudes. In addition, the third type, which are VFR Terminal Area charts, provide greater detail for several terminal areas. They are at a scale of 1:250,000, on the transverse Mercator projection.

Charts for instrument operations, as well as a variety of supplementary data, are also produced.

Defense Mapping

Topographic surveying and mapping for defense purposes is carried out by Defence Geomatics of the Department of National Defence. A series of Military City Maps is available for civilian use. These color street maps of selected urban areas are at a 1:25,000 scale.

National Atlas of Canada

The *National Atlas of Canada* is now in its sixth edition. The atlas portrays information about the natural environment, people's use of resources, demography, and the historical development of the country.

Although conventional map products are available as part of the program, the main activities related to the atlas are conducted on-line. Base maps and thematic maps may be purchased in printed form or as digital files. In addition, on-line facilities are provided so the user can construct and store personalized maps from available base maps and data sets. Printed maps from the fifth edition of the atlas may be purchased as long as copies last. After printed copies are exhausted, the maps will be available as digital files.

Toponyms

Toponyms (geographical names) in Canada are coordinated by the Canadian Permanent Committee on Geographical Names (CPCGN) (see the box feature "Names on Maps"). The National Atlas Information Service listings of toponyms, as well as National Atlas information and connections to other Canadian geographical material, are available on the World Wide Web. General information about the place-name program is available from

CPCGN Secretariat
615 Booth Street, Room 634
Ottawa, Ontario K1A 0E9
Phone: (613) 992-3892

Aerial Photography

Extensive aerial photographic coverage is available in Canada. A large part of this coverage, especially for recent years, consists of standard 9-inch-by-9-inch black-and-white photographs at a nominal scale of 1:50,000. Some color coverage is available, mostly in major metropolitan areas.

Aerial photographs are ordered by specifying geographic location. This may be done by listing the latitude and longitude or by outlining the area desired on a topographic map. In addition, representative photographs of typical topographic features (eskers, for example) are available. Requests for aerial photographs are handled by

National Air Photo Library
615 Booth Street
Ottawa, Ontario K1A 0E9
Phone: (613) 995-4560 or (800) 230-6275

Hydrographic Charts

Hydrographic charts for Canadian waters are published by the Canadian Hydrographic Service of the Department of Fisheries and Oceans. There are two major chart series: standard nautical charts and small-craft charts.

Standard nautical charts are of three types. Relatively small-scale general and offshore charts are the least detailed and may not show all navigational aids. As the name suggests, they are not suitable for navigation close to shore. Larger scale, more detailed coastal and approach charts provide sufficient information for navigation close to shore. Finally, harbor charts are used for navigation in confined waters.

Small-craft charts are designed to serve the operators of recreational small craft. They generally cover rivers, lakes, and portions of other waterways heavily used for recreational purposes.

Indexes, charts, related publications, and ordering information are available from authorized dealers in the United States and Canada, or

Canadian Hydrographic Service
615 Booth Street
Ottawa, Ontario K1A 0E6

Box A.1 Names on Maps

Occasionally, different maps attach different names (**toponyms**) to the same town, city, river, or other feature. In the case of maps produced in foreign countries, such differences may occur because of language differences. However, a feature may be given quite different names by different countries, and each country's maps are likely to follow the national preference, leading to conflicting toponyms.

Place-name difficulties that occur on maps of the United States produced within the country are usually a result of name changes. Old maps, of course, continue to show the earlier names. Even maps published after a change is made may mistakenly continue to use the earlier name. In extreme cases, a series of name changes may be involved, and considerable confusion may occur, as in the case of Cape Canaveral, Florida. The cape was renamed Cape Kennedy in 1963 but, after considerable controversy, the Cape Canaveral name was reinstated some 10 years later. In this case, publishers who were slow to make the change to Cape Kennedy benefited because the former name was restored! Another occasional source of name confusion occurs when local usage differs from the officially recognized version and a map publisher prefers one version to the other.

The names used on maps of other parts of the world produced in the United States sometimes vary considerably. Map producers may pick up different names from maps produced in other countries thus spreading confusion. Name confusion first came about because the English rendering of non-English place names by early explorers and travelers was inaccurate or inconsistent. It is not unusual, therefore, to encounter an English language name for many cities, such as Vienna or Rome, even though the inhabitants of those cities know them as Wien and Roma. In fact, the local name was often simply ignored and an unrelated English name substituted, such as in the case of Bangkok being substituted for Krung Thep in Thailand. In addition, because of language differences, foreign names may be incorrectly rendered or may be rendered differently under different systems of transliteration. An example of this type of problem is the name of the city of Guangzhou, China. Guangzhou is the currently accepted name of the city, which is better known in the United States as Canton. Other versions of the name of this city are Kwangchow and Kuang-Chow. The tendency, in recent years, has been to give preference to the local version of such names but to include the English version when it is well-known.

The Board on Geographic Names (BGN) is the authority charged with establishing the official names of settlements and natural features in the United States. The names adopted by the BGN, both foreign and domestic become the official names for use by U.S. government agencies. Because of the difficulties involved in connection with foreign place names, BGN decisions take into account the recommendations of the United Nations Group of Experts on Geographical Names. Although there is no legal requirement that private map producers follow BGN recommendations, many if not all, do so. Because of this trend, maps produced by reputable concerns probably will contain few name conflicts in the future. When such conflicts are encountered, however, library-owned BGN gazetteers often can be consulted. Further information about available lists of geographic names can be obtained from Earth Science Information Centers.

In Canada, place name decisions are under the guidance and coordination of the Canadian Permanent committee on Geographical Names (CPCGN). This committee consists of provincial and federal representatives. The actual place-name decision process if decentralized. Each provincial or territorial government is responsible for geographic names decisions within its jurisdiction. Responsibility is shared with the appropriate federal government department in the case of federal lands. In reaching their decisions, the governments follow principles established by CPCGN. Decisions of the provincial and territorial authorities are provided to the CPCGN Secretariat. The Secretariat maintains the national toponymic data bank, which records the authorized names (see the list of Canadian maps and related products in this chapter).

Suggested Readings and Reference Materials

American Geographical Society. *Current Geographical Publications, Additions to the Research Catalogue of the American Geographical Society.* Vol. 1 (1938–). New York: 1938–February 1978. Milwaukee: American Geographical Society Collection, March 1978- .

Carrington, David K., and Richard W. Stephenson, eds. *Map Collections in the United States and Canada: A Directory.* 4th ed. New York: Special Libraries Association, 1985. Provides information about 804 map collections. Includes the name of the head of the collection; address and telephone number; hours of operation; size, area, and subject specialization; and so on.

Cobb, David A., comp. *Guide to U.S. Map Resources.* 2d ed. Chicago: American Library Association, 1990. A comprehensive guide to the holdings of U.S. federal, state, university, historical society, public library, and private map collections. Provides names, addresses, telephone numbers, size and areas of strength of the collection, access information, listings of available equipment, and other characteristics.

Dubriel, Lorraine. *Directory of Canadian Map Collections.* 5th ed. Ottawa: Association of Canadian Map Libraries, 1986.

Ehrenberg, Ralph E. *Scholars' Guide to Washington, D.C. for Cartography and Remote-Sensing Imagery: Maps, Charts, Aerial Photographs, Satellite Images, Cartographic Literature, and Geographic Information Systems.* Woodrow Wilson International Center for Scholars, *Scholars' Guide to Washington, D.C.,* no. 12. Washington, D.C.: Smithsonian Institution Press, 1987. Provides information about the cartography-related holdings of over 200 collections and organizations in Washington, D.C., including libraries; archives; galleries; museums; embassies; U.S. and international organizations; federal, state, and local government agencies; data banks; and academic programs and departments.

Feild, Lance. *Map User's Source Book.* New York: Oceana Publications, 1982.

Hodgkiss, A. G., and A. F. Tatham. *Keyguide to Information Sources in Cartography.* New York: Facts on File Publications, 1986. Part IV of this guide provides a directory of map-making agencies, as well as of agencies that distribute or conserve maps, from which one may obtain information about products or services. There is also an index of names, titles, organizations, topics, and geographical locations. Other portions of the book are also likely to be of interest: Part I includes a brief historical review of cartography in the Western world, lists the major map-producing agencies and map collections (primarily in Europe and the United States but with some information about other regions as well), and discusses some of the major cartographic literature, including books and journals. It also provides information about map collecting, including cataloging, storage, and other related topics. Part II is an annotated bibliography of reference sources on the history of cartography, and Part III provides a similar bibliography on contemporary cartographic research.

Kister, Kenneth F. *Kister's Atlas Buying Guide: General English-Language World Atlases Available in North America.* Phoenix, Ariz.: Oryx Press, 1983.

Makower, John, ed. *The Map Catalog.* New York: Vintage Books, 1986.

National Cartographic Information Center, National Mapping Program. *Map Data Catalog.* Washington, D.C.: U.S. Government Printing Office, 1981.

Olson, Judy M., ed. *U.S. National Report to ICA, 1984.* Falls Church, Va.: American Congress on Surveying and Mapping, 1984. Published as a special issue of *The American Cartographer,* supplement to vol. 11 (Summer 1984). The report to ICA (International Cartographic Association) includes information regarding government and nongovernment mapping activities; cartographic education; new products and automation; cartographic literature; societies, sources, and personnel; remote sensing; and other topics. Although it focuses on the period from January 1980 through December 1983, a considerable portion of the contents is useful for general reference purposes.

Perkins, C. R., and R. B. Parry, eds. *Information Sources in Cartography.* New Providence, N.J.: K. G. Saur, 1991.

Rabenhorst, Thomas D., and Paul D. McDermott. *Applied Cartography: Source Materials for Mapmaking.* Columbus, Ohio: Merrill Publishing Company, 1989.

Shupe, Barbara, and Colette O'Connell. *Mapping Your Business.* New York: Special Libraries Association, 1983. Despite the narrow focus suggested by its title, this is an excellent reference source for all map users. It includes lists of map collections in the United States, sources of remote-sensing imagery and federal, state, and foreign maps, and information regarding commercial publishers and firms and agencies involved in computerized cartography.

U.S. Geological Survey. *New Publications of the U.S. Geological Survey.* Reston, Va.: U.S. Geological Survey, monthly. A monthly publication listing books, maps, professional papers, reports, and other publications of the U.S. Geological Survey.

U.S. Geological Survey, Kurt Dodd, H. Kit Fuller, and Paul F. Clarke, comps. *Guide to Obtaining USGS Information.* U.S. Geological Survey Circular 900. Washington, D.C.: U.S. Government Printing Office, 1986.

U.S. Geological Survey and Morris M. Thompson. *Maps for America: Cartographic Products of the U.S. Geological Survey and Others.* 3d ed. Washington, D.C.: U.S. Government Printing Office, 1987.

Wolter, John A., Ronald E. Grim, and David K. Carrington, eds. *World Directory of Map Collections.* International Federation of Library Associations (IFLA) Publications 31. New York: K. G. Saur, 1986.

Appendix B

Foreign Maps

Using a map produced in a foreign country involves the same general techniques as using a map produced in one's own country. Special problems sometimes are encountered, however, especially when the map is printed in an unfamiliar language. Map content is more difficult to comprehend without the clues usually provided by an understanding of the printed language. Even when language problems are absent, there are other obstacles to map use, including the use of unusual signs and symbols, alternative locational systems, and measurements expressed in uncommon units. When these difficulties are encountered, map interpretation requires additional effort.

Language

Map use and interpretation are obviously difficult when the place names, legends, and other map information are printed in an unfamiliar language. The degree of difficulty depends on the characteristics of the language, as well as on the map's subject matter. If the map is a standard topographic sheet, for example, much of the content is likely to be self-explanatory. This is because the symbols used for coastlines, contours, and other features are frequently universal. Additional references for translating some of the words used on such a map, therefore, are often unnecessary because the symbols themselves are easily recognized.

An example would be a Hungarian map with many blue lines that obviously represent the drainage pattern. Each blue line is labeled with two words. One of the words changes every time and is apparently the name of the water course, but the other word is always *folyam*. A Hungarian-English dictionary is not necessary to determine that *folyam* is the Hungarian word for some type of stream, although the specific type remains uncertain. The uncertainty is greater if there are two types of blue lines, with one set labeled *folyam* and the second set labeled *patak*. The comparative size and apparent importance of the streams might suggest that those labeled *folyam* are more important than those labeled *patak*. Only a dictionary, however, would reveal that *folyam* specifically means "river" and *patak* means "brook."

Thus, in a simple case, basic understanding of a foreign map is often possible without resorting to reference materials. When a more advanced level of understanding is required, however, additional reference materials are often needed. The need is especially acute when the map content is more unusual. In the case of a thematic map showing economic activity, for example, the symbols used to represent different types of factories, production levels, or other information may be anything but self-evident. In this type of situation, the user must expend additional effort if map content is to be adequately understood.

If the language on a map uses the Roman alphabet, the user can simply translate the terms of interest into English, using a foreign language/English dictionary. An even more useful reference is a **glossary** that concentrates on the definitions of geographical or cartographic terms that frequently occur on maps. Such glossaries are often included in the reference sections of atlases.[1]

If the language in question does not use the Roman alphabet, translation is likely to be more difficult. The first step in the process is parallel to the previous example and requires looking for the repeated use of a certain character, or group of characters, in conjunction with a specific class of map features. The unfamiliar nature of the characters used often limits this approach, however, and a dictionary or glossary is usually necessary in all but the simplest cases.

[1]Useful glossaries are included in E. C. Olson and A. Whitmarsh, *Foreign Maps* (New York: Harper and Brothers, 1944) and U.S. Department of the Army, *Foreign Maps,* U.S. Department of the Army Technical Manual TM 5-248 (Washington, D.C.: Headquarters, Department of the Army, June 1956).

Depending on the nature of the reference materials themselves, the use of dictionaries and glossaries may be either a one- or a two-step process. If the reference materials are presented in the characters of the original language, the only step needed is to look up the terms in the dictionary. Even this process may be somewhat difficult, however, because the structure of the characters needs to be understood before the method of listing them in the dictionary makes sense. In some cases, the dictionary contains a guide that assists in locating the characters and the needed translations.

Other dictionaries require an intermediate step known as **transliteration,** which means that each character used in the source language is converted into one or more Roman letters. The transliteration of some languages, such as Greek and Russian, involves the substitution, virtually letter for letter, of letters of the Roman alphabet for the original characters. After the transliteration is completed, the Romanized word and its translation are located in the dictionary, using standard alphabetization procedures. Thus, even though an extra step is involved, the transliteration of words prior to determining their meaning is often a fairly easy process, even for the user who is unfamiliar with the language.

The transliteration of some languages, such as Chinese and Japanese, is more difficult because the characters stand for whole syllables, words, or phrases. Locating these characters in a dictionary requires some study, because their complex structure must be understood first.

Specialized glossaries are sometimes available, arranged so that common English terms for map features are listed alphabetically. Each of the English terms is followed by the equivalent foreign-language translation. In this case, the translation process consists of simply scanning the most likely English terms that describe the feature of interest until the foreign-language version that appears on the map is located. This process is somewhat akin to searching for a needle in a haystack, but it at least has the advantage that some clues are provided regarding the likely location of the needle.

Signs and Symbols

One of the difficulties of interpreting signs and symbols is that they are often culturally related. This problem is similar to the difficulties encountered in the translation of culturally related words in a language. Many languages have an array of specialized terms that do not necessarily have direct equivalents in other languages. This is because each language has developed to meet the needs of the particular physical and social environment of its people.

People native to the Arctic, for example, use a large number of specialized terms to distinguish a great variety of snow and ice conditions. This level of detail has evolved because differentiating between these conditions may be vital to survival. Similarly, the Arabic language includes a range of terms that describe particular types of desert-surface conditions.

In the same way, cultural factors often influence the types of features included on maps, as well as the amount of detail provided about them. Interest in certain features of the landscape may result in the development of specialized symbols to differentiate among them. Different types of windmills shown on maps of Holland are an example. Maps designed for people in other locations, and with other needs, might not show these specialized features at all or, at most, might present them much more simply. When specialized symbols appear on a map, some means of translating the legend description into a familiar language is mandatory.

Locational Systems

There are two basic systems for designating locations on the earth's surface. One system involves the use of the latitude and longitude graticule, and the other involves the use of plane rectangular grids. Both of these methods were discussed in earlier chapters, but a few points should be added regarding their use on foreign maps.

Graticule

Using the latitude and longitude graticule is not always straightforward. In the United States and most other countries, maps are commonly based on the familiar 360-degree system of coordinates, called the **sexagesimal system.** This system is not universal, however, and some maps use a 400-unit system, called the **centesimal system.** In addition, no matter which system is used, the starting point for longitude designations is not always the same, so that the same place may have different longitude designations on different maps. A foreign map, then, may possibly use either the centesimal coordinate system, an unusual origin for longitude designations, or both.

Centesimal System

The centesimal coordinate system is sometimes encountered on maps produced in Europe, especially in France and Belgium. This system involves dividing the circle of the globe into four quadrants, with each quadrant further subdivided into one hundred units, called **grads.** In addition, in the centesimal system, there are 100 minutes in each grad and 100 seconds in each minute. Grads are sometimes indicated by a superscript *g*(^{g}), in place of the degree symbol of the sexagesimal system.

Prime Meridian

There is no obvious meridian to use as a starting point for measuring longitude. In general, most maps produced today adhere to the convention of measuring longitude from Greenwich. Some countries, however, still base their system of longitude on a prime meridian that passes through their capital or some other important city. This practice was even more common on older maps. The prime meridians that have sometimes been used by a number of countries are listed in Table B.1.

In general, on any map produced outside the United States, the user should check the prime meridian employed before using the longitude designations. (Even this rule is not entirely safe, however, because some older U.S. maps used a prime meridian based on Washington, D.C.) The origin point for the prime meridian is sometimes stated in the map's marginal information. If it is not, or if the information is in an inaccessible foreign language, the source map will need to be compared to other maps to determine the origin's location. The difference between the local meridian and the Greenwich meridian is then used to calculate the Greenwich longitude of the places of interest.

TABLE B.1 Foreign Prime Meridians

Location	Greenwich Longitude			
	°	′	″	
Amsterdam, Netherlands	4	53	01	E
Athens, Greece	23	42	59	E
Bern, Switzerland	7	26	22	E
Bruxelles (Brussels), Belgium	4	22	06	E
Helsinki, Finland	24	57	17	E
Hierro Island (Ferro Island), Canary Islands*	17	39	46	W
Istanbul, Turkey	28	58	50	E
Jakarta, Indonesia	106	48	28	E
København (Copenhagen), Denmark	12	34	40	E
Lisboa (Lisbon), Portugal	9	07	55	W
Madrid, Spain	3	41	15	W
Oslo, Norway	10	43	23	E
Paris, France	2	20	14	E
Pulkovo, former U.S.S.R.	30	19	39	E
Roma (Rome), Italy	12	27	08	E
Stockholm, Sweden	18	03	30	E
Tiranë, Albania	19	46	5	E

Source: Modified from U.S. Department of the Army, *Map Reading,* U.S. Department of the Army Field Manual FM 21–26 (Washington, D.C.: Headquarters, Department of the Army, January 1969), 3–6.

*Often suggested as the datum for the prime meridian before the general acceptance of the Greenwich meridian in 1884. Adopted for French maps, for example, by Louis XIII, in 1634.

Plane Rectangular Grids

Topographic maps, in particular, often incorporate plane rectangular grids because they are relatively easy to use. In the past, there was considerable variation from country to country in the characteristics of these grids, but there is a tendency toward more uniformity in recent years. The Universal Transverse Mercator (UTM) system is often encountered, for example. On older maps, or in other rare cases, however, an unusual grid system is still possible. For this reason, a few comments regarding the characteristics of plane rectangular grid systems are in order.

Plane rectangular grid systems are universally based on the Cartesian coordinate concept, with measurements taken in the x and y directions from an origin point. The northeast (first) quadrant is usually used for map purposes so that all coordinates are positive values, but this is not always the case. A brief examination of the grid designations on a map quickly reveals the use of any other quadrant, because numbers in such a situation do not increase to the right and upward, and negative numbers may be encountered.

Grid origins are of two types: false origins and geographic origins. False origins are determined by measurement from some arbitrary starting point. This point might be a certain distance south of the region being mapped and, perhaps, a specified distance west of the central meridian of the map projection on which the grid has been superimposed. Geographic origins are based on a specific latitude and longitude location, usually outside the bounds of the area being mapped. The difference between the two is not likely to be of importance to a map user, because the designation of grid locations is handled in the same manner under either system. One exception occurs when the origin is inside the region being mapped. This is rarely done, however, because it results in some positive and some negative coordinates.

Although numbers are almost always used to designate the grid lines, letters are sometimes used. In some cases, numbers may be used on one axis and letters on the other, or, rarely, letters may be used on both. The intervals between the grid lines are usually uniform so that the grid is square, but rectangular grids are sometimes encountered. Direct measurement of the spacing between the lines will quickly reveal the distances involved. Taking such measurements is especially important if letter designations are used.

Grids are used to designate either specific point locations or map areas. In the first case, the coordinates of the location are measured and specified, usually east and then north of the origin. If a location within a grid square is sufficiently accurate, the coordinates of the corner of the square nearest to the origin are given.

Finally, the method of stating a plane rectangular grid location may vary from system to system. These variations include the directional sequence—whether eastings are given first and then northings or whether some other sequence is called for. Also, conventions vary with regard to the abbreviations of coordinates. Some systems routinely leave off initial digits, for example, and some omit trailing zeros.

If specifying a location in terms of an unfamiliar system, one need simply follow a method that is consistent and logical. All directions should be indicated, and abbreviations of numbers should be avoided. In particular, dividing coordinate pairs, possibly with a semicolon, helps to indicate the two elements of the coordinate (easting and northing, for example).

In summary, a bit of care and a few minutes spent examining the coordinates of the system shown around the map's neatline will pay off. Usually, understanding and using virtually any grid coordinate system with reasonable accuracy is not difficult. Information regarding the British National Grid, which is often encountered, is provided in Appendix F.

UNITS OF MEASUREMENT

Units of measurement are fundamental to the mapping of features on the earth's surface. In the past, many different countries used their own specialized units, such as the Chinese *tang-su* (4.41 kilometers) or the Swedish *geografisk mil* (approximately 7.5 kilometers). In recent years, however, most countries have either adopted the metric system as a standard or continue to use units of the English system. For this reason, uncommon units are only likely to be encountered on older maps.

If a map with an unusual unit of measurement is encountered, one needs to determine the unfamiliar unit's equivalency in familiar units.[2] If this effort is unsuccessful, the problem can still be solved if the scale is stated as a representative fraction, which allows drawing of the user's own appropriately labeled graphic scale. For example, a scale of 1:10,000 means that 1 unit on the map represents 10,000 units on the earth. A 10-centimeter line, therefore, represents 1 kilometer (10 × 10,000 centimeters = 100,000 centimeters = 1 kilometer).

[2]See, for example, the table in Olson and Whitmarsh, *Foreign Maps,* 169, or a table of measures in an encyclopedia.

FOREIGN MAP PRODUCTS

An incredible variety of maps is produced worldwide. Virtually every country has at least one government agency that produces official maps (and many countries have literally dozens of such agencies). In addition, private companies throughout the world produce a multitude of maps. Simply finding out about the maps that these agencies and companies produce is a formidable task (and obtaining the maps is even more daunting). These aspects of worldwide mapping cannot be discussed in any systematic way in the space available here. Instead, the interested map reader should consult the specialized sources of information available.

An outstanding source of information about map coverage throughout the world is *World Mapping Today*. As the authors of this book indicate, they have attempted "to put together within one cover as much as possible of the various kinds of information needed for finding out about and acquiring modern topographic and thematic maps." The emphasis is on government mapping, but some information about commercial mapping is also included. A particularly helpful aspect is that many of the ins and outs of map acquisition are explained in Chapter 3 of the book. *World Mapping Today* is listed in the Suggested Readings, along with a number of other useful information sources.

Map dealers can be of great assistance in locating maps produced by major foreign agencies and companies. Often, they have a selection of such maps in stock. Also, they have catalogs that list maps that can be ordered. In addition, this book's World Wide Web site provides links to sources of foreign maps.

SUGGESTED READINGS

Allin, J. *Map Sources Directory*. Downsview, Ontario: York University Libraries, 1982– (periodically updated).

Bibliographia Cartographica. München, W. Ger.: K. G. Saur, 1974– (annual).

Cartinform. Budapest, Hungary: Cartographia, 1971– (supplied with periodical *Cartactual*).

GeogKatalog. Stuttgart, W. Ger.: Geocenter ILH, 1977– (systematically updated).

Hodgkiss, A. G., and A. F. Tatham. *Keyguide to Information Sources in Cartography*. London: Mansell, 1983.

Kister, K. F. *Kister's Atlas Buying Guide*. Phoenix, Ariz.: Oryx Press, 1984 (first of a series).

Lawrence, G. P. "Maps, Atlases and Gazetteers." In *A Guide to Geographical Information Sources,* edited by S. Goddard. Beckenham, England: Croom Helm, 1985, 189–210.

Lock, C. B. M. *Modern Maps and Atlases.* London: Clive Bingley, 1969.

Olson, Everett C., and Agnes Whitmarsh. *Foreign Maps.* New York: Harper and Brothers, 1944. Includes glossaries of foreign-language terms commonly encountered on maps.

Parry, R. B., and C. R. Perkins. *World Mapping Today.* London: Butterworths, 1987.

Speak, P., and A. H. C. Carter. *Map Reading and Interpretation: New Edition with Metric Examples.* London: Longman Group Limited, 1970. Provides sample color reproductions of official maps published in the United Kingdom.

U.S. Department of the Army. *Foreign Maps.* U.S. Department of the Army Technical Manual TM 5–248. Washington, D.C.: Headquarters, U.S. Department of the Army, June 1956. Includes translations and transliterations of many commonly encountered mapping terms in a variety of languages.

Winch, K. L. *International Maps and Atlases in Print.* 2d ed. London and New York: Bowker, 1976.

Appendix C Copyright[1]

Maps are protected under copyright laws, which means that it is illegal for anyone to reproduce copyrighted maps, in whole or in part, without permission from the copyright owner. This is true regardless of the final purpose of the reproduction.

Essentially, all maps are copyrighted, including those published by foreign governments. The only notable exception to this rule are U.S. government publications, which are not protected by copyright.

Penalties for copyright violations can be severe and may include the payment of royalties, damages, costs, and fees. Avoiding such penalties requires scrupulously avoiding making unauthorized copies. This even extends to a single photocopy. In addition, a copy made from a previous copy is in violation, even if the copyright identification is not visible.

Requests for permission to reproduce a copyrighted map should be sent to the copyright holder. If permission is granted, a royalty fee may be required, depending upon the quantity of maps involved and the purpose for which they will be used.

The era of computerization has, if anything, exacerbated the problem of copyright protection. It has become all too easy to copy digital files or to scan printed maps and to use the results with apparent impunity. A useful discussion of these and related issues is provided by Patrick McHaffie, Sona Karentz Andrews, Michael Dobson, and two anonymous employees of a federal mapping agency in "Ethical Problems in Cartography: A Roundtable Commentary," *Cartographic Perspectives* no. 7 (fall 1990): 3–13.

[1]From "Questions and Answers about Map Copyrights," with permission from International Map Dealers Association, P.O. Box 1789, Kankakee, IL 60901 (Copyright 1984 Wide World of Maps, Inc.).

Magnetic Compass Use

Appendix D

Magnetic compasses are simple but useful devices. A magnetic compass contains a needle or dial that aligns with the magnetic force field of the earth and, hence, points to magnetic north. Although this basic concept is constant, compasses are produced in a variety of forms, each designed to be especially suitable for a particular task. In the end, however, *any* type of compass can be used for *any* of the tasks; you simply follow the procedures appropriate for the type available. This appendix is limited to providing basic information about the use of hand-held magnetic compasses in conjunction with maps.

Hand-Held Magnetic Compass Types

The two primary types of hand-held magnetic compasses are the floating-dial and floating-needle. A **floating-dial compass,** as the name suggests, contains a dial that is free to rotate. This dial has a north arrow printed on it, and degrees are printed clockwise around its periphery. **A lensatic compass** is of this type (Figure D.1). A **floating-needle compass,** on the other hand, has a needle that is free to rotate within its housing. In some floating-needle compasses, degrees are printed in a clockwise direction around the periphery of a movable housing. An **orienteering compass** is of this type (Figure D.2). Another type of floating-needle compass is called a **pocket transit** (Figure D.3). In this type, degrees are printed in a counterclockwise direction around a fixed housing.

Directions are usually marked on modern compasses in degrees, although the cardinal directions (N, S, E, W) are often also indicated. Less expensive compasses are often calibrated in 2-degree increments. More expensive models have finer calibrations and often have sighting devices built in so that directions can be determined more accurately. As with many other products, much better accuracy is obtained with a compass of good quality.

Compass Tasks

Two fundamental tasks—map orientation and azimuth determination—are performed with the help of compasses. When these tasks are mastered, they contribute to laying out and following specified routes, to locating the position of objects by intersection, and to determining one's position by resection. For simplicity's sake, the discussion that follows assumes the use of the commonly encountered lensatic compass for these purposes. Differences involved in the use of other types will be apparent when you have them in hand. The instructions that accompany a newly purchased compass provide additional guidance.

Map Orientation

A map is oriented most simply by relating it to visible landmarks (see Chapter 7). Often, however, there are no suitable landmarks within view, or more accuracy is required. In either case, map orientation is accomplished with the help of a compass.

Declination must be taken into account when orienting a map. The declination diagram on the map provides information regarding the amount and direction of local declination at the time the map was published. The direction and amount of annual change may also be included.

Before attempting to use a map in the field, do some preparatory work. First, calculate the current declination and note it on the map. Second, add a magnetic-north reference line to the map. To do this, align a protractor with one of the meridians, measure the declination angle, and draw a pencil line across the face of the map. You may wish to add a series of parallel pencil lines, spaced about an inch apart, so that there will be a magnetic-north line handy to any map location. These steps will preclude having to make calculations in the field, where working conditions are often not ideal and confusion is more likely.

Figure D.1 Lensatic compass.
Source: John Kehoe photo.

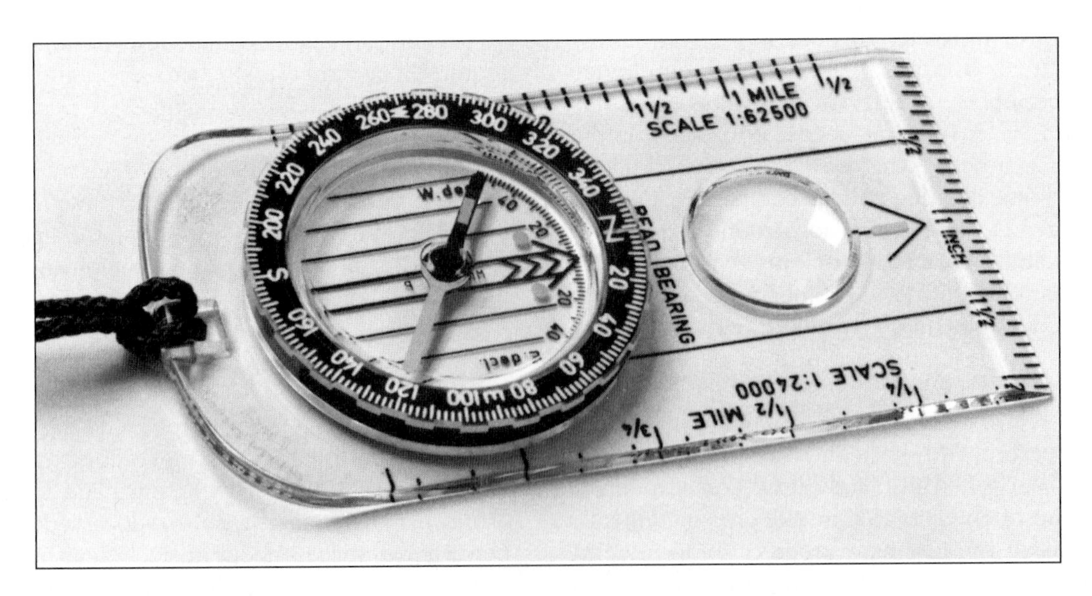

Figure D.2 Orienteering compass.
Source: Courtesy of Silva Division, Johnson Camping Inc.

Figure D.3 Pocket transit.
Source: Courtesy of Brunton Company.

A tempting alternative method of obtaining a magnetic-north reference line is to simply extend the north magnetic line in the declination diagram. The preceding procedure is preferable for three reasons. First, the declination diagram does not take into account changes in declination since the map's publication date. Second, the declination diagram is usually quite small, and a slight drawing misalignment can result in considerable error. Third, especially in the case of small angles, the declination diagram may not be accurately drawn in the first place.

When the prepared map is taken into the field, orientation is easily accomplished. Select a location that provides a good view of the surrounding terrain and that is well away from automobiles, wire fences, power lines, or other metal objects, including your belt buckle. Place the map on a flat surface, open the compass out flat, and lay it on the map. Then, turn the map so that a magnetic-north reference line is aligned with the north arrow on the compass. (The compass sights can be used to provide a longer baseline and, therefore, greater accuracy of alignment.)

If you have not had the opportunity to draw magnetic-north lines on the map, simply lay the compass on the map so that its body is in alignment with a meridian. Then rotate the map until the north arrow matches the required declination.

Magnetic Azimuth Determination

Determining a magnetic azimuth from your location to a landmark is extremely simple. With the front and rear sights set vertically, hold the compass to your eye and line up the landmark with the rear peep sight and the vertical wire in the front sight. While maintaining this alignment, read the azimuth from the dial through the lens on the rear sight.

Route Following

You may wish to follow a particular azimuth that you have drawn on your map. Simply measure this azimuth with a protractor, using the magnetic-north reference lines for alignment. If you do not have a protractor, lay the compass flat on the oriented map. Align the compass body with the desired line and read its azimuth at the 0 mark on the case.

When you have determined the desired azimuth, hold the compass to your eye and read the azimuth angle through the lens. Sighting through the front and rear sights, pick out a landmark along the indicated alignment and walk to it. Repeat this procedure as necessary until you reach your destination or until a change in direction is required.

For quicker compass reference, set the compass before you start your route. First, align the 0 mark on the compass housing with the north arrow. Then, holding the compass to this alignment, turn the bezel until the course marker on the cover (a short, radial line painted on the glass) is set at the desired magnetic azimuth. With this arrangement, whenever you need to take a sighting, simply hold the compass in front of you so that the 0 mark is aligned with the north arrow and sight along the course marker.

When following a route, you may also need to keep track of the distance you have traveled. Prepare for this ahead of time by pacing a measured distance, such as a football field, and determining the average length of your pace. Then, in the field, maintain a normal pace and count your steps to determine the distance you have traveled.

Intersection and Resection

The methods of determining the map position of an object by intersection and of finding your own position by resection are discussed in Chapter 7. All that needs to be noted here is that you can use your compass to determine the necessary azimuths for these procedures.

When locating an object by intersection, obtain magnetic azimuths from two known locations toward the object whose location you wish to map. Plot these azimuths on the map, using the magnetic-north reference lines to align the protractor. The intersection of the lines gives the map location of the object.

When you wish to determine your position by resection, determine azimuths from your location to two or three identifiable landmarks. Then, compute the back-azimuths (add 180 degrees to azimuths of less than 180 degrees; subtract 180 degrees from azimuths of 180 degrees or more). Identify the landmarks on the map and draw the back-azimuths from each landmark, in turn. The intersection of these back-azimuths identifies your location.

SUGGESTED READINGS

Andresen, Steve. *The Orienteering Book.* Mountain View, Calif.: Anderson World, 1977.

Kjellstrom, Bjorn. *Be Expert with Map and Compass.* New York: Charles Scribner's Sons, 1976.

Muehrcke, Phillip C. *Map Use: Reading, Analysis, and Interpretation.* 3d ed. Madison, Wis.: JP Publications, 1992, 584–90.

Map Storage and Cataloging Systems

Appendix E

Map sheets are difficult to store and retrieve. They are also fragile and must be protected from damage. A typical storage system involves large cabinets containing horizontal file drawers (Figure E.1). Individual maps, or groups of maps, are placed in large file folders. The file folders are then stacked in the appropriate drawers. Tubes or vertical files are sometimes used as alternative storage methods. Although the physical storage arrangements are important, the vital consideration, from the standpoint of the map user, is the *organization* of the system.

Assume, for example, that you are looking for a map showing the political subdivisions of Africa in 1970. In a small collection, you might simply go to a drawer labeled "Africa" and rummage through the maps to see if any of them suit your needs. If the collection is large, however, you will be involved in a time-consuming, physically tiring, and possibly frustrating search. Indeed, you might even overlook the map that you want because you fail to guess its file location. Some type of *cataloging system* is of obvious assistance in locating and retrieving specific maps.

The catalog is an important component of the organization of any but the smallest map libraries, because it provides a framework that facilitates the search for and retrieval of particular maps. A catalog eliminates the time-consuming and inefficient process of rummaging through random stacks of maps to locate the one needed.

Figure E.1 Typical map storage drawers.
Source: Courtesy of Hamilton Industries Inc.

The cataloging systems used in map collections vary in complexity, depending upon the anticipated needs of the users and the size and characteristics of the collection. This appendix, therefore, does not describe the details of such systems but simply indicates the general approach typically involved.

One of the most commonly used map cataloging systems is the **Library of Congress system.** Maps are classified under *Class G* in this system, which is described in detail in the Library of Congress manual listed in the Suggested Readings at the end of this appendix. The other sources indicated in the Suggested Readings provide the detail necessary to fully understand some of the other commonly used systems.

CATALOGING

Map cataloging systems typically begin with the physical and political subdivisions of the world as an overall framework.[1] Each continent, for example, may be identified by a specific letter or number. Each country within the continent is then indicated by an additional letter or number code, with internal political subdivisions indicated by an extension of the call number. In a particular system, then, a map covering Racine County, Wisconsin, might be indexed as BA25.10. In this case, the *B* indicates that the area is located in North America, the *A* that it is in the United States, the 25 that it is in Wisconsin, and the 10 that it is Racine County.

The Library of Congress Class G schedule operates on a hierarchical numerical sequence. Under this system, world maps are designated G3200, and maps of the hemispheres are subdivisions (such as G3210 for the Northern Hemisphere). Maps of continents are given individual numbers (G3300 for North America, for example), and countries within each continent are given subnumbers (such as G3700 for the United States). Within a country, each state or province has its own designation (such as G4120 for Wisconsin). A helpful map index of this schedule is available.[2]

In addition to indicating the areal reference, the catalog usually indicates the map's subject matter. Typically, this indication involves a numerical system that identifies economic, physical, political, agricultural, and other topics or combinations of topics. The topics that appear on a specific map are indicated by the appropriate code numbers. Each map is then recorded in the catalog, using a cross-referencing system. One entry, under the regional designation, would list the characteristics of the map and the topics that appear on it. Then, within each subject heading category, there would be a cross-reference to the map. Each map may be referenced by several entries, depending on the range of topics that it includes. Each reference specifies the number of the map to which it refers. In addition, the call number indicates the location of the map within the collection.

[1]It has been suggested that the main entry for cataloging maps should be the geographical area, followed by the subject. Other elements, such as date, scale, publisher, and so on, are likely to occur in a variety of sequences. See Arch C. Gerlach, "Geography and Map Cataloging and Classification in the Libraries," 317–22, and Joan Winearls, "Some Problems in Classifying and Cataloging Maps," 352–58, in *Map Librarianship: Readings,* compiled by Roman Drazniowsky (Metuchen, N.J.: Scarecrow Press, 1975).

[2]Janice C. Lorrain and James A. Coombs, comps., *A Map Index to the LC "G" Schedule,* Open File Reports 88–3, Map and Geography Round Table, American Library Association (Springfield, Mo.: Map Library, Southwest Missouri State University, 1988).

STORAGE

Once the index has identified the map of interest, the next problem is to physically locate the map. The arrangement of the map drawers and cabinets in the library is likely to be along the same basic lines as the index system. That is, areas may be set aside for maps of each continent, with cabinets or drawers assigned for maps of each country and then of regions within the country.

Nonseries Maps

Some maps are produced as individual sheets, complete in and of themselves, and are not part of a series. This is particularly true of thematic maps, which often cover some political entity, such as a city, county, state, or country, or an area such as a continent or a hemisphere. Each individual thematic map is usually filed according to the physical/political arrangement mentioned previously. Then, if there are a number of maps of the same area, they are placed in order, according to topic, within the area file.

Series Maps

Many maps, especially topographic maps, are produced as part of a series. Such series are often produced by government agencies and extend over an entire political entity. In addition, map series frequently cover areas outside of the region of origin, extending over continental areas or, even, the world.

The sheet outlines of series maps are usually systematically arranged, often along lines of latitude and longitude. The area covered by individual sheets in such a system is approximately rectangular, although the exact shape depends on latitude and the map projection used. This means that individual map sheets may cross political boundaries. When this occurs, it is not always clear where the map should be filed. For this reason, series maps are often filed together on the basis of the total extent of the series. A series that covers the entire world, for example, may be housed in one area of the library. An alternative is to break the series into parts, with each part filed with the maps of the appropriate region.

The main goal in filing the sheets of a series is to follow a procedure that allows individual sheets to be quickly identified, located, and retrieved. The producers of the series usually provide an identification and indexing system that allows users to organize their holdings systematically and to decide where each sheet belongs.

One common practice is to give each sheet the name of the most prominent feature located within its boundaries, whether it is a cultural feature, such as a city, dam, or town, or a natural feature, such as a river, lake, or mountain. The logical method of filing sheets identified by name is to simply put them in alphabetical order.

Obviously, the sequence of maps arranged in alphabetical order is not likely to bear any relationship to the geographical arrangement of the areas covered by the individual sheets. In the U.S. Geological Survey 1:250,000-scale National Topographic Map Series, for example, the sheet named "Beaumont" is located just east of the one named "Austin" (Figure E.2). On the other hand, the "Waco" sheet is located north of the "Austin" sheet, and the "Seguin" sheet is located south of it. When the Texas sheets are placed in alphabetical order, then, sheets from one area of the state are intermixed with sheets from other areas of the state and are not located next to one another. Furthermore, if the file contains sheets for the entire United States, the Texas sheets are further intermixed with sheets from all of the other states. Despite this apparent difficulty, the use of sheet names is a very common indexing method.

USGS Map Reference Coding

A useful method for filing USGS series maps is to use the **Map Reference Code** system as a guide. In this system, each map is assigned a unique reference code, based on its location (Figure E.3). This code can be used to establish a filing hierarchy. In this approach, each element in the code is considered in sequence.

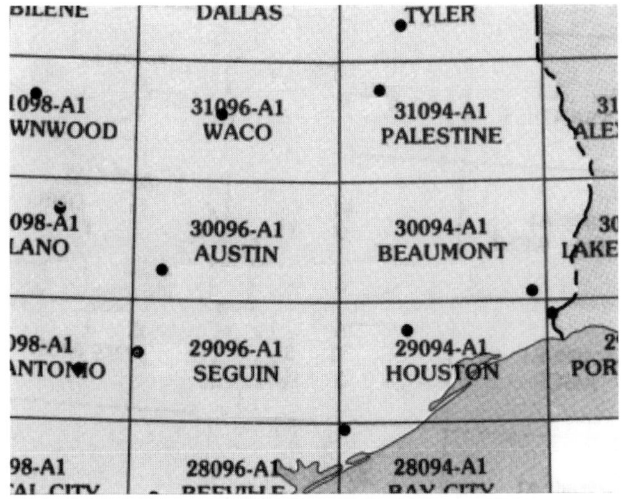

Figure E.2 Arrangement of U.S. Geological Survey 1:250,000-scale topographic sheets for a portion of Texas.

Source: U.S Geological Survey, "Texas Index to Topographic and Other Map Coverage."

Consider, for example, how maps that are part of a 7.5-minute topographic quadrangle series (scale 1:24,000) could be filed. The first two elements of the code for each map represent the latitude and longitude of the southeast corner of the 1-degree square quadrilateral in which it falls. Thus, quadrangles between 30° and 31° North are filed before those between 31° and 32° North. Then, within each latitude group, the maps are arranged so that those between 97° and 98° East are filed before those between 98° and 99° East. As is apparent from Figure E.3, sixty-four 7.5-minute quadrangles fall in each 1-degree square region. The sequence of filing these 64 maps is based on the next element of the map reference code. This element consists of a letter/number pair, arranged in order from A1 to H8. It is worth mentioning that DLG data (see Chapter 19) are often filed on the basis of this map reference code.

National Topographic System of Canada Coding

The coding used to identify quadrangles in the National Topographic System of Canada provides an excellent framework for map filing. In this system, primary quadrangles are identified by a number based on columns and rows that generally cover 4 degrees north to south and 8 degrees east to west (Figure E.4). (Because of the convergence of the meridians, the columns are widened to 16 degrees east to west, north of 80° North.) These quadrangles correspond to the coverage of the sheets of the 1:1,000,000-scale topographic series.[3]

Larger-scale topographic maps are identified by subdivision of the primary quadrangles (Figure E.5). Within each scale group, the identification numbers provide a logical filing sequence. Thus, 1:500,000-scale sheets cover an area that is 2 degrees north-south by 4 degrees east-west. These sheets are identified by the primary quadrangle number, followed by the designation NE, NW, SE, or SW. 1:250,000-scale sheets cover sixteen 1-degree north-south by 2-degree east-west quadrangles within each primary quadrangle. North of 64° North there are eight subdivisions, which cover 1 degree north-south and 2 degrees east-west. North of 80° North there are four subdivisions, each covering 1 degree north-south and 4 degrees east-west. These sheets are identified by the primary quadrangle number, followed by the letters *A* through *P*, in the case of sheets south of 68° North, and *A* through *H* north of 68° North. Sheets at larger scales are identified by further subdivision, as indicated in Figure E.5.

[3]Note that these quadrangles do not correspond to the sheets of the 1:1,000,000-scale International Map of the World.

MAP REFERENCE CODE

U.S. Geological Survey maps are identified by name and series. In addition, a system has been initiated whereby each map is assigned a unique reference code which indicates the geographic coordinates, type, unit of measurement, and scale of map.

Various examples are illustrated at the right and a key to abbreviations is shown below.

MAP TYPES

AB - Alaska Boundary Series Index
AF - Alaska Boundary Series-National Forest System
AL - Alaska Boundary Series-Bureau of Land Management System
AP - Alaska Boundary Series-National Park System
AW- Alaska Boundary Series-National Wildlife Refuge System
CF - County (feet)
CM - County (metric)
EI - Ecological Inventory
LB - Land Use and Land Cover (base)
LM - Land Use and Land Cover (multicolor)
MM- Surface-Minerals Management Status
NS - National Atlas (separate sales)
OM - Orthophotomap
OQ - Orthophotoquad
PF - National Park, Monument (feet)
PL - Planimetric
PM - National Park, Monument (metric)
PR - National Park, Monument (shaded relief)
RA - Radar (airborne)
RF - Regional (feet)
RM - Regional (metric)
RP - Regional (planimetric)
RS - Radar (satellite)
SI - Satellite Imagery
SL - Slope Maps
SM - Surface Management Status
SP - State Base (planimetric)
SR - State Base (shaded relief)
ST - State Base (topographic)
TB - Topographic Bathymetric
TF - Topographic (feet)
TM - Topographic (metric)
UB - United States (base)
UG - United States (general)
UM - United States (magnetic)
UO - United States (outline)
UT - United States (topographic)
WB - World (base)
WG - World (general)
WO - World (outline)
WP - World (political)
WT - World (topographic)

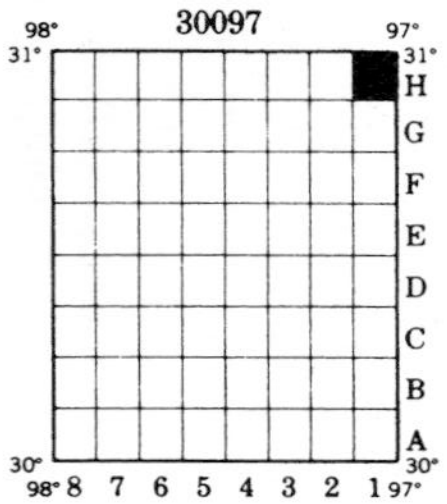

7.5 MINUTE SERIES

Reference Code 30097-H1-T F-024
Latitude (30°)
Longitude (97°)
Index Number (H1)
Map Type (Topographic)
Unit of Measurement (Feet)
Map Scale (1:24 000)

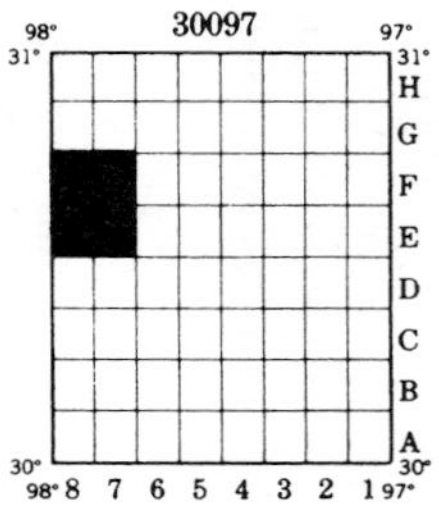

15 MINUTE SERIES

Reference Code 30097-E7-T F-062
Latitude (30°)
Longitude (97°)
Index Number (E7)
Map Type (Topographic)
Unit of Measurement (Feet)
Map Scale (1:62 500)

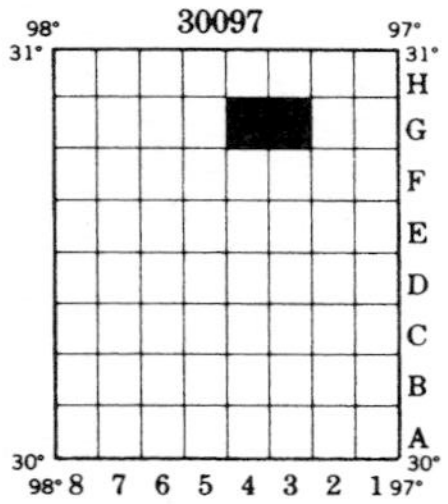

7.5 x 15 MINUTE SERIES

Reference Code 30097-G3-T M-025
Latitude (30°)
Longitude (97°)
Index Number (G3)
Map Type (Topographic)
Unit of Measurement (Meters)
Map Scale (1:25 000)

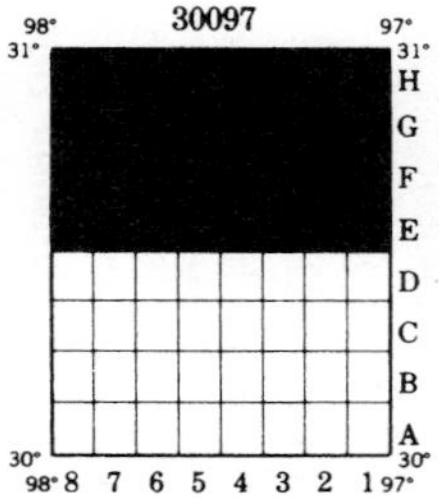

30 x 60 MINUTE SERIES

Reference Code 30097-E1-T M-100
Latitude (30°)
Longitude (97°)
Index Number (E1)
Map Type (Topographic)
Unit of Measurement (Meters)
Map Scale (1:100 000)

MAP SCALES

020-1:20 000	35M-1:3 500 000
024-1:24 000	05M-1:5 000 000
025-1:25 000	06M-1:6 000 000
050-1:50 000	07M-1:7 000 000
062-1:62 500	75M-1:7 500 000
063-1:63 360	08M-1:8 000 000
100-1:100 000	10M-1:10 000 000
250-1:250 000	11M-1:11 875 000
500-1:500 000	14M-1:14 000 000
01M-1:1 000 000	16M-1:16 500 000
02M-1:2 000 000	22M-1:22 000 000
25M-1:2 500 000	30M-1:30 000 000
31M-1:3 168 000	

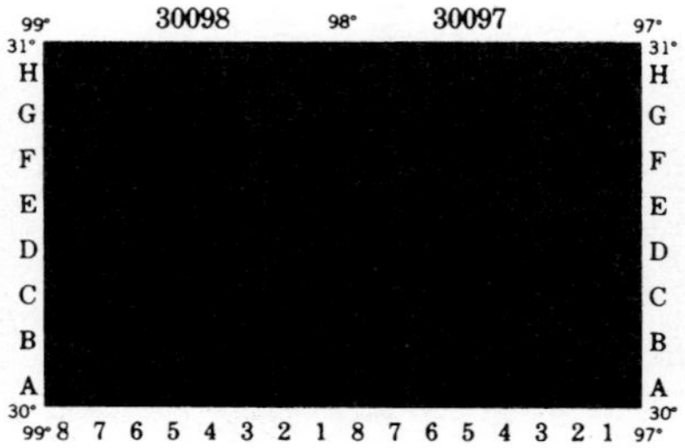

1 x 2 DEGREE SERIES

Reference Code 30097-A1-T F-250
Latitude (30°)
Longitude (97°)
Index Number (A1)
Map Type (Topographic)
Unit of Measurement (Feet)
Map Scale (1:250 000)

NOTE: A file number, also unique to each published map, is assigned for internal USGS order filling purposes.

Figure E.3 USGS Map Reference Code System.

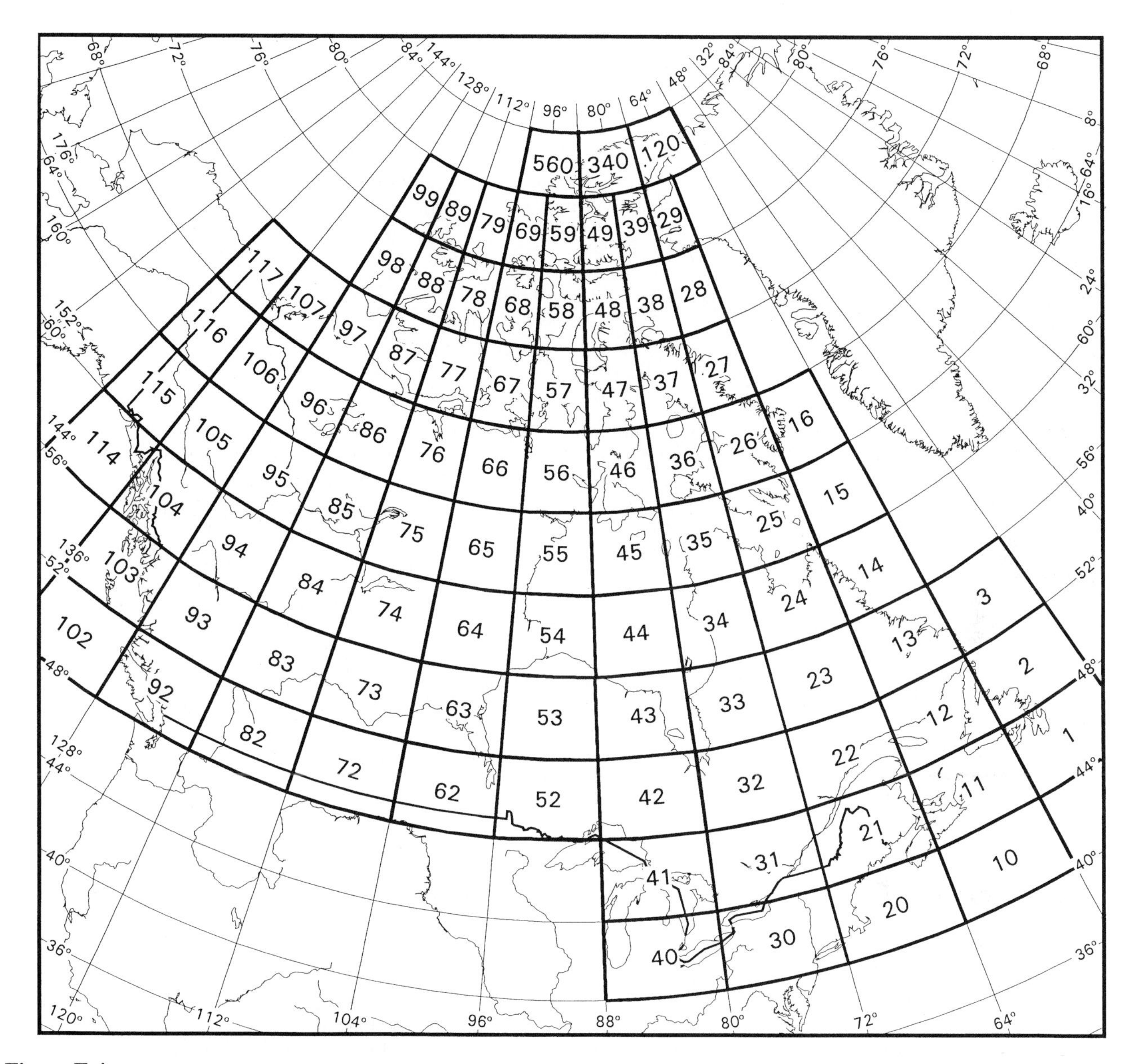

Figure E.4 Primary quadrangles of the National Topographic System of Canada.

SERIES INDEXES

A quick summary of the holdings in a map series is often useful and is provided by a series index. Such an index may be in the form of a table or a map. Each form has advantages and disadvantages.

A **tabular index** is simply an alphabetically or numerically ordered list of map names or identification numbers (Figure E.6). Such a list is useful for inventory or ordering purposes. If an inventory is needed, a quantity indication is added next to the identifier for each map of the series in the collection. Those sheets that are not in the collection are then easily identified, and, if desired, an order list can be created directly from the inventory sheet.

An **index map** is most satisfactory from the standpoint of determining the geographic coverage of the maps from a given series that are included in a collection. It consists of a base map of the area of interest, whether it is a county, state, country, continent, or the entire world. The outlines of the sheets that are published, or that are planned for publication, are drawn on this base map. The sheet name or number of each map is also indicated. An index map of this type is often available from the producer of the map series.

The map librarian places appropriate marks on the index map to indicate the status of the series. The markings may simply indicate which sheets are in the collection. Often, however, they are more extensive and may indicate, for example, the edition or editions that are in the collection, whether the hill-shaded version is included, the sheets that are on order, and so on (Figure E.7).

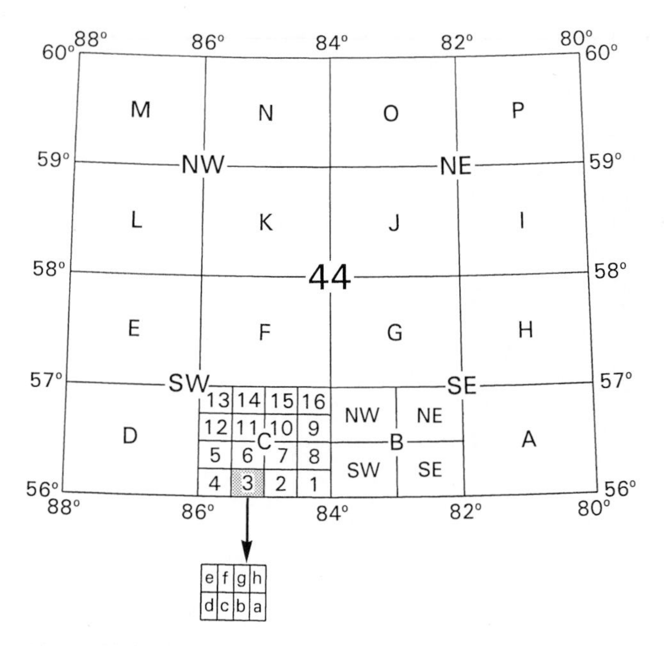

Figure E.5 Subdivision of a primary quadrangle of the National Topographic System of Canada.

When an index map is provided, it is a simple matter to locate the city, region, or other feature of interest and to determine the specific map that provides the desired coverage. Also, the index map shows at a glance the geographical extent of the coverage available for a general area, as well as what areas are absent from the collection. This is, therefore, often the most convenient form of index, because typical map users almost always know what *area* or *feature* is of interest to them but are less likely to know the name or number of the map sheet(s) that provides the required coverage.

7.5 MINUTE QUADRANGLES

43088–F3	Eden	43088–G4	Fond du Lac	43088–G8	Green Lake
44089–H8	Edgar	42089–F2	Footville	44088–C1	Greenleaf
42089–G1	Edgerton	45092–B3	Forest	44088–C5	Greenville
45091–F4	Edgewater	44087–F4	Forestville	44090–G5	Greenwood
45087–A3	Egg Harbor	42088–H7	Fort Atkinson	44088–G7	Gresham
45089–D2	Elcho	46088–A4	Fortune Lakes	46090–D5	Gurney
43088–G5	Eldorado	44091–B6	Fountain City	42091–G1	Guttenberg
44091–D4	Elk Creek	44091–A8	Four Corners		
44091–G6	Elk Creek Lake	46092–D3	Foxboro	46088–A7	Hagerman Lake
43088–G1	Elkhart Lake	42088–D2	Fox Lake	44089–G7	Halder
42088–F5	Elkhorn	43088–E8	Fox Lake	42088–H1	Hales Corners
44092–E1	Ella	43087–G8	Franklin	45089–A8	Hamburg
42090–G5	Ellenboro	42087–G8	Franksville	44089–B5	Hancock
45087–C1	Ellison Bay	45088–A5	Fredenburg Lake	45087–A7	Harmony
46091–C5	Ellison Lake	45092–F4	Frederic	43091–B2	Harpers Ferry
44092–F4	Ellsworth	44088–D3	Freedom	46089–C7	Harris Lake
44092–G3	El Paso	46092–A1	Frog Lake	45089–D5	Harrison
44088–F6	Embarrass	46092–E3	Frogner	45089–F6	Harshaw
45092–A3	Emerald	45089–B8	Fromm Lookout Tower	43088–C3	Hartford East

Figure E.6 Example of a tabular index of map series holdings.

SUGGESTED READINGS

Akers, B. "Care and Handling of a Map Collection." *The Map Collector* 4 (1978): 2–5.

American Library Association. *Anglo-American Cataloging Rules*. 2d ed. Chicago: American Library Association, 1978.

Cartographic Materials: A Manual of Interpretation for AACR2. Chicago: American Library Association, 1982.

Larsgaard, Mary Lynette. *Map Librarianship: An Introduction*. 2d ed. Littleton, Colo.: Libraries Unlimited, 1987.

Neddermeyer, G. "Map Cataloging—An Introduction." *Drexel Library Quarterly* 9, no. 4 (1973): 18–25.

Nichols, H. *Map Librarianship*. Hamden, Conn.: Linnet Books, 1976.

United States, Library of Congress, Subject Cataloging Division. *Classification, Class G*. 4th ed. Washington, D.C.: Library of Congress, 1976.

30 × 60 MINUTE SERIES
1:100 000 SCALE

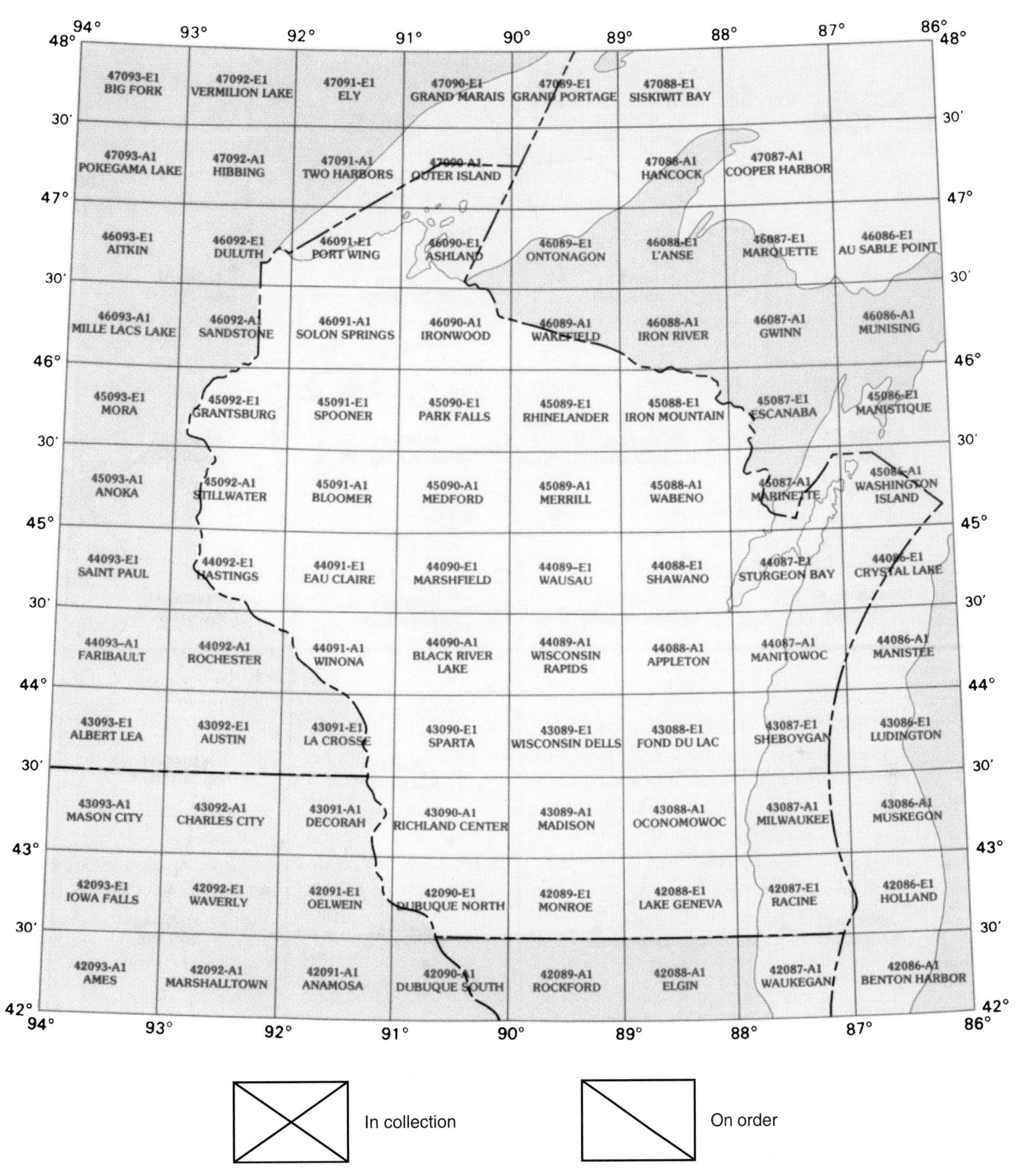

Figure E.7 Series index map modified for use as an index of map series holdings. (U.S. Geological Survey 1:100,000-scale maps of Wisconsin and vicinity.)

Source: U.S. Geological Survey. "Wisconsin Index to Topographic and Other Map Coverage."

APPENDIX F British National Grid

A specialized grid reference system, called the **British National Grid,** is used by the British Ordnance Survey for maps of Britain. Similar systems are used in other parts of the world where British influence is or was strong.

The system involves major and minor grid squares (Figure F.1). The major squares measure 500 kilometers on a side and are identified by a letter of the alphabet. The first major square is located north and west of Scotland and is lettered *H;* the last major square is located south and east of London and is lettered *T.*

Within each 500-kilometer square on the grid are 25 grid squares that are each 100 kilometers on a side. These 100-kilometer squares are lettered *A* to *Z* (with the letter *I* omitted), in a row-by-row pattern beginning in the northwest corner of the 500-kilometer square and ending in the southeast corner. The result is that each 100-kilometer grid square is uniquely identified by a pair of letters, with the letter that identifies the 500-kilometer square given first and the identifier for the 100-kilometer square given second. The Isle of Man, for example, is located principally in grid square SC.

Grid references for specific points within each 100-kilometer grid square are written as either four- or six-digit numbers. If a location is to be specified with an accuracy of 1 kilometer, the four-digit reference is used; this is called the *Normal Kilometre Grid Reference.* In essence, the Normal Kilometre Grid Reference identifies a square, 1 kilometer on a side, within which the point of interest is located. For example, the Normal Kilometre Grid Reference for a point in Figure F.2 is 4972. If the point is to be located to the nearest 100 meters, the Normal National Grid Reference is given as a six-digit figure—in this case, 498727. Because the numerical references could identify a point within *any* 100-kilometer grid square, the identifying letters for the appropriate square must be given if there is any possibility of confusion on the part of the map user. In the example given in Figure F.2, since the point is within grid square SC, then the Full Kilometre Grid Reference is SC/4972, and the Full National Grid Reference is SC/498727.

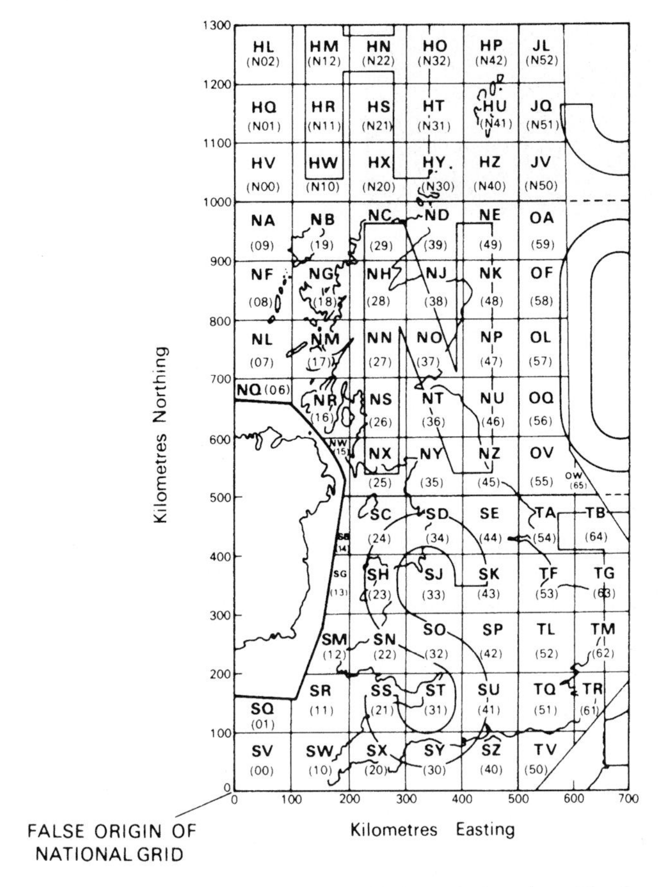

Figure F.1 Major (100-km) grid squares of the British National Grid, with letter designations. Numbers formerly used to designate each square are shown in parentheses.

Source: Reproduced from "The Projection for Ordnance Survey Maps and Plans and the National Reference System," with permission of the Controller of Her Majesty's Stationary Office, © Crown copyright reserved.

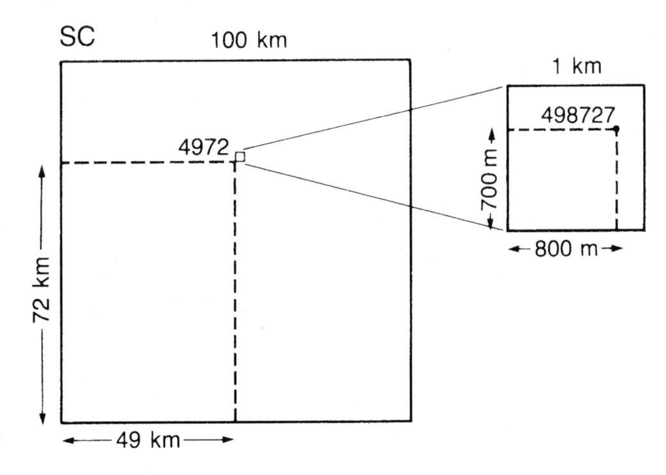

Figure F.2 Grid references for a specific point.

Glossary

Absolute altitude Altitude above the earth's surface, not above sea level.

Absolute scale Scale, on a graph, in which each unit of distance represents the same quantity of a given variable.

Accessibility In graph theory, how well-connected places (nodes) are to other places.

Accumulative line graph (Lorenz curve) Arithmetic line graph used to illustrate the degree of concentration in one variable as compared to in another.

Acre Unit of areal measurement in the U.S. Customary system. Contains 43,560 square feet.

Active remote-sensing system Remote-sensing system that includes special equipment that transmits signals of the required wavelength. Radar is an example.

Admiralty nautical mile *See* **Nautical mile.**

Aerial photograph Remote-sensing product obtained by photography from an airborne platform.

Aerial survey camera Large-format camera used to take vertical photographs for mapping purposes.

Aeronautical chart Special map used for aerial navigation.

Agonic line Line on an isogonic chart that joins points of zero declination.

Aliquot parts Division of a U.S. Public Land Survey section into various combinations of quarters (160 acres), quarter-quarters (40 acres), and halves (80 acres).

Altitude of the sun *See* **Declination of the sun.**

Analemma Figure-eight-shaped diagram that gives information about the latitudes at which the sun is directly overhead throughout the year and about the equation of time.

Aneroid altimeter Altimeter that, when properly calibrated, converts current barometric pressure into a direct altitude reading.

Apparent solar day Period between apparent solar noon on two consecutive days.

Apparent solar noon Time at which the sun passes over the meridian on which an observer is located.

Apparent solar time Time based on the passage of the sun.

Approximate contour Dashed or dotted line that joins points of approximately equal elevation.

Area cartogram (value-by-area cartogram) Maplike diagram on which the areas of the various regions are drawn in proportion to some value associated with them, such as income or population, instead of their actual area.

Areal averaging Situation in which the information presented on a map is the average value for each areal unit.

Areal scale Ratio between the area of a feature plotted on a map and the area of the same feature on the earth's surface.

Areal symbol Map symbol that represents a feature that is spread out over the earth's surface.

Arithmetic line graph Line graph in which equal distances (either vertical or horizontal) indicate equal changes in value, anywhere on the graph.

Associated number In graph theory, topological distance from a given place (node) to the most remote place in the network.

Astronomic latitude Latitude based on the observation of astronomic bodies, such as Polaris.

Atmospheric window Wavelength region in which electromagnetic energy passes through the atmosphere with relatively little scatter or absorption.

Attribute code In computer-assisted cartography and geographic information systems, set of characters by which a feature is identified.

Automated automobile navigation Driving aids that automatically determine and display the current location of a vehicle, determine the best route to follow between selected locations, and/or provide en route guidance.

Azimuth (forward azimuth) Direction of line, defined at its starting point by its angle in relation to a baseline. Usually measured in a clockwise direction, starting from north.

Azimuthal equidistant projection Map projection on a plane, in which all points are plotted at their true distance from the center of the projection and are in their true global direction (azimuth) from the center.

Azimuthal projection Map projection made on a plane surface.

Back-azimuth Direction that is the exact reverse of an azimuth; that is, 180 degrees from an azimuth.

Back-bearing The direction that is the exact reverse of a bearing.

Bar chart Diagram on which bars of various lengths are used to express the values of the variables graphed. May use either absolute or percentage values.

Bar scale *See* **Graphic Scale.**

Baseline Line that serves as the basis for a survey, with endpoints, direction, and length that are accurately known. Or east-west reference line used in conjunction with a U.S. Public Land Survey principal meridian.

Base map Background framework on which a thematic map is drawn.

Bathymetric map Map that shows water depths and the configuration of the underwater topography.

Bearing Direction, not exceeding 90 degrees, measured from either a north or south baseline, whichever is nearer to the direction being designated. May be expressed as a true, magnetic, or grid direction. A bearing in marine navigation is the same as an azimuth in land navigation.

Bench mark Monument used to mark a point whose elevation above or below a specified datum has been established.

Bifurcation ratio In graph theory, ratio between the number of segments (links) of one order, in a tree, and the number of segments of the next higher order.

Bilateral graph Line graph that uses a single arithmetic scale to portray the values of two related variables so that the overlap of the lines indicates either positive or negative net values. For example, an excess of births over deaths yields a natural increase in population, whereas an excess of deaths over births indicates a natural decrease.

Bit A binary digit (0 or 1).

Bivariate map Map that shows the relationships between the values of two variables at once.

Block diagram Pictorial representation of a portion of the earth's surface from an elevated, oblique point of view.

Boyce-Clark radial shape index (BCRSI) Measure of shape based on the lengths of radials extending outward from a node placed at the center of the shape.

Branching network (tree) In graph theory, network that does not contain circuits (loops).

British National Grid Specialized grid reference system used by the British Ordnance Survey for maps of Britain.

Broad-band device Remote-sensing device that uses a sensor that receives electromagnetic energy of many wavelengths and integrates the information into a composite signal. An example is panchromatic film used in aerial photography.

Bunge's measure Measure of shape based on a set of distance measurements taken between systematically placed vertices on the perimeters of the region.

Byte A string of binary digits (bits).

Cadastral map Map that includes the location of property-ownership lines, along with their bearings and lengths, the ownership and size of land parcels, and similar information.

Cadastral survey Survey used to determine land-ownership boundaries and property descriptions. In the United States, preferably reserved for land-ownership surveys of the public lands.

Canada Land Survey System (CLSS) Survey system established in 1870 to provide an orderly land-ownership framework in the prairies and mountains of western Canada. Uses principal meridians and baselines, townships, ranges, sections, and subdivisions of sections.

Canoe chart Nautical chart of the Minnesota-Ontario border lakes produced by NOS. Designed to suit the needs of operators of small, shallow-draft vessels.

Carrying contour Single contour drawn to represent several contours that fall extremely close together in a particularly steep area.

Cartesian coordinate system Four-quadrant system of locational coordinates.

Cartogram Special type of map in which a different standard of measurement, such as time or cost, is substituted for distance measurements or in which the area of regions is made proportional to some other measure, such as population or income.

Cartographic map *See* **Real map.**

Cartography Art, science, and technology of making maps, together with their study as scientific documents and works of art. (Definition from International Cartographic Association.)

Centesimal system Coordinate system sometimes encountered on maps, especially French and Belgian. The circle on the globe is divided into four quadrants, with each quadrant subdivided into 100 units, called grads. There are 100 minutes in each grad and 100 seconds in each minute.

Central-point cartogram Maplike diagram that shows travel costs or travel times from a single starting point to other locations in the region.

Change detection In remote-sensing analysis, detection of changed conditions by comparison of images obtained at different time periods.

Chart Special map used for navigational purposes.

Chi-square statistic Measure used to indicate the strength of association between two distributions.

Choropleth map Quantitative areal map in which the average magnitude of a phenomenon is indicated by a distinctive tone, pattern, or color applied over each unit area. Unit areas are delimited by state or county boundaries or other arbitrary boundary lines.

Chronometer Extremely accurate clock. Essential for making longitude determinations before the development of radio time signals.

Circuit network In graph theory, network that includes loops so that there is more than one possible path between some pairs of places (nodes).

Circular graph Graphic display often used to record information about events that occur during a specific time period, such as 12 hours or a year.

Classification Division of a range of data into groups. *See also* **Data classification.**

Classification error In remote-sensing analysis, classification of pixels into inappropriate categories.

Climate map Map of any climatic factor or of climatic regions.

Climatograph Circular graph that shows climatic information, such as temperature or rainfall.

CLSS *See* **Canada Land Survey System (CLSS).**

Clustered pattern Pattern in which many points are concentrated close together and outlying areas contain very few, if any, points.

Coast chart Nautical chart for coastwise navigation inside offshore reefs and shoals, entrance into bays and harbors of considerable size, and navigation of certain inland waterways.

Coefficient of areal correspondence Measure of the extent to which two distributions correspond to one another. Compares areas that fall into only one category, areas where the two categories overlap, and areas that fall outside either category.

Color emulsion Photographic emulsion that reproduces the range of colors seen by the human eye.

Color-infrared emulsion *See* **Infrared emulsion.**

Combination Means of map generalization in which two or more similar features are combined into one.

Compass deviation Difference between the direction to the north magnetic pole and the direction in which the compass needle actually points.

Compromise projection Map projection that does not preserve any of the globe properties but also does not result in extreme distortion of any property.

Computer-assisted cartography Use of the computational power of the computer to improve the speed and accuracy of map production.

Cone Developable surface used for conic projections.

Conformality Retention of correct angles on a map. Requires that the lines of latitude and longitude cross one another at right angles and that the map scale be the same in all directions at any given point.

Conjugate principal point On one photograph of a stereo pair, location of the principal point of an adjacent photograph.

Connectivity In graph theory, measure of the level of completeness of the graph, given by the connectivity index.

Connectivity index In graph theory, comparison between the actual connections (routes) that exist in a given network, as compared to the total number possible.

Constant azimuth (rhumb line) Directional line that crosses each succeeding meridian at a constant angle.

Contiguous cartogram Cartogram in which the relationships among regions are retained but, of necessity, into which variations in the regions' shapes are introduced.

Continuously variable symbol Symbol, such as a circle, whose size is mathematically scaled in proportion to the value it represents.

Contour Imaginary line that joins points of equal elevation above or below some datum.

Contour interval Vertical distance between contours.

Control diagram Map based on a survey used to establish ground locations of control points for aerial photography.

Controlled photomosaic Mosaic of aerial photos assembled with reference to ground control.

Control network *See* **Geodetic control network.**

Conversion *See* **Digitizing.**

Conversion diagram Diagram used to assist in the change of azimuths from one reference direction to another.

Coordinated Universal Time (UTC) Basis of the times observed in each time zone. Formerly known as *Greenwich Mean Time (GMT).*

Cut and fill Engineering process of balancing the amount of material removed from the high areas (cut) with that required to fill low areas.

Cut-line Path along which a profile is drawn.

Cycle On a graph with a logarithmic scale, range of values with a multiple of ten (such as 1 to 10 or 10 to 100).

Cylinder Developable surface used for cylindrical projections.

Dasymetric map Quantitative areal map in which the variation in the magnitude of a phenomenon that occurs in a region is indicated by different tones, patterns, or colors applied over appropriate subdivisions of the region.

Data classification Process of grouping values so that each one is associated with others that have relatively similar characteristics and is separated from those with distinctly different characteristics. Applies to pixels in remote-sensing analysis or to data values in thematic mapping.

Data plane Computer equivalent of a map overlay. Each data plane contains information about the spatial arrangement of a specific variable.

Datum Fixed elevation, such as mean sea level, used as the starting point for a vertical survey.

Dead reckoning Method used by navigators for route planning or for keeping track of progress toward a destination. Depends on knowing (or predicting) the starting point and the direction, speed, and time of travel away from that point.

Declination diagram Diagram that shows the relationships between true, magnetic, and grid north.

Declination of the sun Latitude that the sun is directly over on a given day.

Defilade Location that is hidden from view.

Deflection of the vertical Difference, at a given location, between a line drawn to the center of the earth and a line drawn to the earth's center of gravity.

Degree (of a slope) Angle of slope, measured in degrees.

DEM *See* **Digital elevation model (DEM).**

Dependent variable Value predicted on the basis of the value of an independent variable.

Depression contour Contour line drawn with short ticks at right angles to the contour line. The ticks point downslope, toward the bottom of the depression.

Depth curve (isobath) Underwater contour that shows the depth of water below a designated datum, such as mean low water.

Developable surface Projection surface, especially a cone or a cylinder, that is not flat at the time the projection is created but that can be flattened later.

Diameter In graph theory, highest associated number in a network.

Differential leveling Method of establishing elevations relative to a starting level, or datum.

Digital Chart of the World (DCW) Comprehensive digital base map of the world, digitized at a scale of 1:1,000,000.

Digital data Tabular or locational information converted to digital form, especially for use with computer-assisted methods.

Digital elevation model (DEM) U.S. Geological Survey database that contains elevation values in digital form.

Digital land-use and land-cover data U.S. Geological Survey digital database that contains information on the location of up to 37 categories of land-use and land-cover data.

Digital line graph (DLG) U.S. Geological Survey database that contains digital map data in vector format.

Digital orthophoto Digital image scanned from an aerial photograph from which displacements caused by the camera and by differences in terrain elevation have been removed.

Digital orthophoto quadrangle (DOQ) Standard digital orthophoto produced by the U.S. Geological Survey. Scanned from 1:12,000-scale black-and-white, color, or color-infrared orthophotos, covering 3.75 minutes of latitude and 3.75 minutes of longitude.

Digital raster graphic (DRG) Scanned image of a U.S. Geological Survey topographic map.

Digitizing In computer-assisted cartography, process of converting map data from their original visual form to a digital format that can be handled by a computer.

Dip In geology, downward inclination of a bed or stratum relative to an imaginary horizontal plane.

Direction The path along which something is pointing or moving.

Distance function In a digital geographic information system, function that produces information regarding, for example, the distance or the cost of access from a given point to every other point in the region.

Distance log *See* **Mileage table (distance log).**

Distance matrix Diagram that provides information about distances and/or travel times between major locations.

DLG *See* **Digital line graph (DLG).**

Dominion Land Survey System *See* **Canada Land Survey System (CLSS).**

Dot-distribution map Thematic map in which each dot represents a specified quantity of a particular variable.

Dot-planimeter method Method of areal measurement using regularly spaced pattern of dots.

DRG *See* **Digital raster graphic (DRG).**

Ecliptic, plane of the Imaginary plane that passes through the sun and the earth at all positions of the earth's orbit around the sun.

Edge enhancement Image enhancement of remotely-sensed data that increases the contrast between the values on either side of a dividing line, rendering shapes and details along the edges more conspicuous.

Electromagnetic spectrum Wavelengths of electromagnetic radiation.

Ellipsoid Regular solid formed by the rotation of an ellipse on its minor axis. Has a greater radius at right angles to the axis of rotation and a smaller radius along the axis. Better approximation of the shape of the earth than a sphere.

Engineering map Detailed map (plan) for guiding engineering projects.

Enhanced Thematic Mapper Plus (ETM+) Eight-band, whisk-broom, remote-sensing system installed on Landsat 7. Bands 1 through 6 provide 30 by 30 meters resolution, band 7 provides thermal-infrared resolution of 60 by 60 meters, and the panchromatic band provides 15 by 15 meters resolution.

Ephemeris Set of tables that gives information about the movements of the sun, moon, stars, and planets.

Equal-area (equivalent) projection Projection on which equal areas are retained, so that a unit area drawn anywhere on a map always represents the same area on the earth's surface.

Equation of time Measure obtained by subtracting mean solar time from apparent solar time.

Equator Great circle located midway between the poles of rotation.

Equidistance Retention of correct distance on a map.

Equidistant map Map with the characteristic of correct distance relationships.

Equivalence *See* **Equal-area (equivalent) projection.**

Eratosthenes Greek mathematician who, in 250 B.C., estimated the diameter of the earth using simple observations and geometric principles.

Estimated location Dead-reckoning location adjusted for the effects of winds and tides. Used if a locational fix cannot be obtained.

ETM+ *See* **Enhanced Thematic Mapper Plus (ETM+)**

Exaggeration *See* **Locational shift and size exaggeration.**

Expected frequencies, table of Table that shows expected data combinations in the study area. Used in chi-square analysis.

False-color image In remote sensing, image in which color bands are shifted away from the colors in the actual scene. For example, the green band may be reproduced in yellow, the red band in green, and the near-infrared band in red.

Fathom Nautical measurement that equals 6 feet (approximately 1.83 meters).

Fiducial mark Mark on an aerial photograph used to locate its principal point.

Fire insurance map Specialized, large-scale map used to assist in determining the potential fire hazard associated with a particular property or district.

Fix Point of crossing of lines of position when plotted on a navigation chart, establishing the location of the vessel. *See also* **Line of position.**

Flight line Straight line between the principal point and the conjugate principal points on an aerial photograph. Consecutive points over which the airplane flew while the series of photographs was taken.

Flight plan Plan based on dead-reckoning procedures filed by aircraft crews prior to departure. Includes courses that will be flown, important landmarks, distances, estimated flying time, and altitude for each leg. In aerial photography, flight path designed to provide the desired photographic coverage.

Floating-dial compass One of two primary types of hand-held, magnetic compasses that has a dial that is free to rotate within its housing.

Floating mark In a stereoplotter, point of light that appears to be suspended over the stereomodel. Used to transfer locational information for both elevation (vertical) and planimetric (horizontal) features from the stereomodel to the manuscript map.

Floating-needle compass One of two primary types of hand-held, magnetic compasses that has a needle that is free to rotate within its housing.

Flood-prone area map Map that provides information about areas subject to flooding.

Flood-zone map Map showing zones subject to flooding under various conditions.

Flow map Map that uses quantitative linear symbols to represent connections or flows that exist between locations on the earth's surface.

Foot Unit of measurement in the U.S. Customary system, defined as 1:0.30480061 meter. Since 1959, also known as **U.S. survey foot.** *See also* **International foot.**

Form line Generalized contour frequently used on small-scale, less-detailed maps.

Forward azimuth (azimuth) Directional designation. Usually measured in a clockwise direction from north.

GBF/DIME *See* **Geographic base file/dual independent map encoding (GBF/DIME) files.**

General chart Nautical chart for use when a vessel's course is well offshore but its position can be fixed by landmarks, lights, buoys, and characteristic soundings.

Generalization Process of adjusting the content of a map to provide as useful and recognizable a representation of the real world as is feasible within the map's space limits and scale.

Geocoding In computer-assisted cartography, assigning locational coordinates to map features based on an earth-based coordinate system, such as latitude/longitude, Universal Transverse Mercator, or State Plane Coordinates.

Geodetic control network Geodetic survey used to provide the overall control, or framework, for mapping.

Geodetic control survey Detailed survey that takes into account the curvature of the earth and that determines locations in terms of latitude and longitude.

Geographical mile *See* **Nautical mile.**

Geographical north *See* **True north.**

Geographic base file/dual independent map encoding (GBF/DIME) files U.S. Bureau of the Census computerized records regarding addresses within urban areas. Supplanted by the TIGER system. *See also* **Topologically Integrated Geographic Encoding and Reference System TIGER.**

Geographic information system (GIS) Spatial information system designed for data management, mapping, and analysis.

Geographic Names Information System (GNIS) U.S. Geological Survey digital index of standardized geographic names, as well as other information about the names.

Geoid Irregular shape of the surface of equigravitational potential of the earth.

Geologic map Map that provides information about the structure of the earth, such as types of rocks exposed at the earth's surface, types of bedrock that underlie the surface materials, aspects of the earth's thermal and magnetic characteristics, and many other characteristics.

GEOREF *See* **World Geographic Reference System (GEOREF).**

Georeferencing In computer-assisted cartography and geographic information systems, the process of aligning the various data planes with one another.

GIS *See* **Geographic information system (GIS).**

Globe Spherical model of the earth.

Global Positioning System (GPS) Worldwide positioning system based on a constellation of 24 satellites operated by the U.S. Department of Defense. Makes accurate position determination possible 24 hours per day, in any weather.

Glossary Table of definitions of specialized terms. Especially useful glossaries are those showing foreign-language designations for frequently encountered map features.

GNIS *See* **Geographic Names Information System (GNIS).**

Gnomonic position Location of the projection light source at the center of the globe.

Gnomonic projection Map projection on a plane, with the light source at the center of the earth. All straight lines drawn on a gnomonic projection represent great circles.

Gore Tapered strip map used in globe construction.

GPS *See* **Global Positioning System (GPS).**

Grad *See* **Centesimal system.**

Gradient *See* **Slope.** *Also,* rate of change of slope at a given point.

Gradient path Steepest path down a hill.

Graduated circle Circular symbol whose area is proportional to the value of the data that it represents.

Graduated point symbol Map symbol used to indicate the value of the phenomenon (quantitative information) that occurs at a given location or within a given region.

Graph (network) Figure composed of points and lines.

Graphic scale Line drawn on a map whose length (and subdivisions) represents a specified distance on the earth.

Graph theory Method of analyzing the topology of a graph.

Graticule Pattern of meridians and parallels on the earth.

Great circle Line that is defined by the intersection of the earth's surface and a plane passing through the center of the earth.

Great-circle distance Shortest distance between two points on the earth's surface, represented by the path of a great circle.

Grid north The northerly direction of the lines of a locational grid, such as the Universal Transverse Mercator or State Plane Coordinate systems.

Grid-squares method Method of areal measurement using a superimposed, regular grid.

Grid tick Short line indicating the intersection of a grid or graticule line with the map neatline.

Ground control point Easily identifiable point on the ground established to tie remote-sensing images to a ground survey.

Ground resolution In remote sensing, size of the area on the ground that is imaged at any one time. Determines the size of objects or areas that can be detected. *See* **Pixel.**

Guide meridian In the U.S. Public Land Survey, north-south reference line established at 24-mile spacing.

Gyrocompass Nonmagnetic compass that indicates true north.

Harbor chart Nautical chart intended for navigation and anchorage in harbors and small waterways.

Hard copy Computer jargon meaning a tangible copy of computer information, for example, a map.

Hierarchic order In a branching network (tree), level of importance of a segment (link).

High-low bar chart Diagram on which each bar represents a date or time and shows the high, low, and, sometimes, average values for the time period.

High-oblique aerial photograph Aerial photograph taken with the lens pointed at an angle away from the vertical so that the horizon is included in the image.

Highway map Map used for intercity road travel.

Hill shading Gray values added to a map, as though illuminated by a northwest light source, so that the modeling of the surface is apparent.

Historical map Map produced in the past. Often provides useful insights into past knowledge and beliefs.

History map Modern map produced to show a situation from the past, including routes of exploration, battle diagrams, migration patterns, and many other topics.

Horizontal accuracy In National Map Accuracy Standards, accuracy of horizontal placement of well-defined points on the map.

Horizontal datum Starting elevation to which elevations on a map are referred.

Hypsometric tint *See* **Layer tint.**

Illuminated contour Contour drawn as though illuminated by a northwest light source.

Image enhancement Computer-aided manipulations of remotely sensed data to improve the interpretation and classification of the image.

Independent variable Variable whose value is used to predict the value of a dependent variable.

Index contour Contour line that is accentuated and has elevation values inserted at intervals along its length.

Index map Base map showing outlines, and sheet numbers or names, of the sheets of a map series that are published or planned for publication. May be annotated to indicate the sheets that are in a map collection.

Index value Computed by dividing individual data values by an appropriate base-period value and multiplying the results by 100.

Infrared emulsion In aerial photography, film emulsion that is sensitive to the infrared wavelengths. May be black and white or color.

Infrared wavelengths Wavelengths that fall outside the range of human vision. Particularly useful in remote sensing for the analysis of vegetation conditions and the identification of water bodies.

Inkjet printer Computer output device that prints graphic images, including maps.

Inset Small map included within a main map.

Inspection Process of orienting a map in the field by aligning recognizable features in the landscape with the same features on the map.

Instrument flight rules Rules that govern flights under conditions that prevent reliable visual contact and, therefore, that require instruments to determine location, direction, and altitude.

Intermediate contour Contour line drawn in lighter line weight, usually without printed elevation values.

International Date Line Internationally accepted line, located near the 180th meridian, where one day ends and another day begins. Deviates from the meridian to avoid the occurrence of two different days of the week in a single nation.

International foot Defined as 1:0.3048 meter. Compare with **U.S. survey foot.**

International nautical mile *See* **Nautical mile.**

Interpolation Method of logically determining the elevation of a point that is not on a contour or a spot height. Alternatively, method of determining the coordinates of a point not located on a grid or graticule line.

Interpretive key Logically organized listing of clues to assist a photo interpreter in the identification of features on an aerial photograph.

Interruption (of a map) Separation of a map projection along several dividing lines to show areas of particular interest (such as the continents or the oceans) with less distortion.

Interval scale Quantitative information that provides the exact value of some phenomenon on a standard scale with uniform intervals but an arbitrary zero value. An example is Celsius degrees.

Intervisible Characteristic ascribed to two points that are mutually observable. Determined by the use of a profile.

Intracoastal waterway chart Nautical chart that provides coverage of the inside route from Miami to Key West, Florida, and from Tampa to Anclote Anchorage, Florida.

Isobar Line that joins points of equal barometric pressure.

Isobath *See* **Depth curve (isobath).**

Isogonic chart Map that shows magnetic declination.

Isogonic line Line on an isogonic chart that joins points of equal declination.

Isoline Line that joins points of equal value. *See also* **Isometric line** and **Isopleth.**

Isoline map Map that represents the values that occur on a surface by a series of lines that join points of equal value.

Isometric line Isoline based on values actually observed at control points located on a continuous surface.

Isopleth Isoline based on areal averages assigned to arbitrary control points that are not directly observable and are not located on a continuous surface.

Journalistic cartography Production of maps for use in newspapers, weekly newsmagazines, and general and popular weekly or monthly magazines.

Knot Speed of 1 nautical mile per hour.

Labeling In computer-assisted cartography, process of attaching identification codes to the map elements.

Lambert conformal conic Map projection frequently used for midlatitude air charts because straight lines drawn on it closely approximate great-circle routes.

Landsat Series of U.S. satellites in operation since 1972 that carry the multispectral scanner (MSS) and thematic mapper (TM) remote-sensing systems.

Landscape map Map that provides detailed site information and planting plans for gardens and parks.

Land survey Cadastral survey. In the United States, preferably used to describe such surveys outside the public lands.

Large scale Comparative term for map scale in which a relatively small area of the earth's surface is represented in considerable detail.

Latitude Angular measurement of north-south location, relative to the equator.

Latitude, line of *See* **Parallel.**

Layer tint (hypsometric tint) Tone or color used to fill in between contour lines or isobaths to indicate a zone of general elevation or water depth.

Leg In dead reckoning, portion of a route traveled in a specific direction for a known length of time at a known speed. In aerial photography, portion of the flight plan.

Legal subdivision In the Canada Land Survey System, division of a section into sixteen 40-acre subdivisions, which are the equivalent of quarter-quarter sections in the U.S. Public Land Survey.

Legend Explanation of map symbols.

Length of a path In graph theory, number of routes or links contained in a path between two nodes.

Lensatic compass Type of hand-held, floating-dial, magnetic compass.

Leveling *See* **Differential leveling.**

Library of Congress system Commonly used map cataloging system. Maps are classified under class G.

Linear cartogram Maplike diagram (also called a *distance-transformation map* or *distance cartogram*) designed to show relative distances, based on such measures as cost or time.

Linear contrast stretching Image enhancement of remotely sensed data that expands the range of tones so that they are easier to differentiate.

Linear interpolation On a contour map, logical method of estimating the unknown elevation of a point that falls between contours. Based on the assumption that the slope between contours is constant.

Linear symbol Map symbol used to represent a feature that has a linear path on the earth's surface.

Line graph Diagram consisting of a line or lines joining points plotted on a grid. Shows the value of one variable relative to the value of another.

Line of latitude *See* **Parallel.**

Line of longitude *See* **Meridian.**

Line of position Observation of known locations or alignments (ranges) by navigators operating within sight of land to monitor the location of their vessels. Application of the method of resection.

Line plotter Computer output device that draws an image using continuous lines in the manner of traditional drafting.

Line-route map Specialized cadastral map used by utility companies to show transmission line or pipeline routes and rights-of-way.

Link (route) In graph theory, connecting line between two places, or nodes.

Locational shift and size exaggeration Means of map generalization in which the size of features is exaggerated to assure their visibility, and their location is shifted to accommodate the larger-than-scale size.

Logarithmic scale On a line graph, scale in which equal distances indicate the same percentage change in value.

Log-log graph Line graph that uses logarithmic scales on both axes. The shapes of lines plotted on the graph reflect percentage rates of change in both variables.

Longitude Angular measurement of east-west location, relative to the prime meridian.

Longitude, line of *See* **Meridian.**

Loran Navigational, position-determination system based on radio transmission travel times from pairs of master and slave broadcast locations.

Lorenz curve *See* **Accumulative line graph (Lorenz curve).**

Loss-and-gain bar chart Bar chart with a central axis that serves as a zero line. Bars drawn on one side of the axis represent positive values (gains), and those on the other side represent negative values (losses).

Lot In U.S. Public Land Survey, subdivisions used when the normal system of fractional sections cannot be applied, such as along the northern and western edges of some townships or along the shores of navigable streams or lakes 25 acres or more in extent.

Low-contrast image Remotely sensed image in which the values of the pixels fall in a relatively narrow range so that variations in the image's gray tones are difficult to discern.

Low-oblique aerial photograph Aerial photograph taken with the lens pointed at an angle away from the vertical but not including the horizon.

Machine coordinates In computer-assisted cartography, locational coordinates of map features based on measurements (in inches or centimeters) from a control point, such as the corner of the digitizing board.

Magnetic declination Difference between true north and magnetic north at any given location.

Magnetic north Direction to the north magnetic pole.

Map Any concrete or abstract representation of the distributions and features that occur on or near the surface of the earth or other celestial bodies.

Map measurer Instrument with a tracing wheel. Used for measuring the length of curved or irregular lines. An opisometer.

Map orientation Alignment of a map so that the mapped features are in the proper relationship to the landscape.

Map-overlay analysis Technique that involves tracing and overlaying the outlines of variables of interest. The overlap of the areas shows the relationship between the phenomena mapped.

Map projection Systematic rendering on a flat sheet of paper of the earth's graticule (lines of latitude and longitude).

Map Reference Code System by which a USGS quadrangle is identified based on the latitude and longitude of the southeast corner of the 1-degree square quadrilateral in which it falls and its location within that quadrilateral.

Maximum slope limit Maximum permissible slope for a route. Established because of the limited capabilities of the equipment to be used.

Mean datum Vertical datum that averages long-term tidal variations.

Meander line In the U.S. Public Land Survey, line surveyed along the mean high-water elevation of a navigable water body.

Mean solar time Average length of the day throughout the year. Equal to 24 hours. Required because the earth's speed on its elliptically shaped orbit varies, resulting in apparent solar days of different length from one part of the year to another.

Mental map Generalized mental image of the physical world.

Mercator Conformal map projection that shows the world (less the polar regions) in a rectangular format. Useful for navigational purposes because all straight lines drawn on it are lines of constant compass direction.

Meridian Portion of a meridian circle located between the North and South Poles. Used to specify east-west location. Line of longitude. Meridians are spaced farthest apart on the equator and converge to a single point at the poles. Meridians and parallels are at right angles to one another.

Metadata Evaluative description that allows a potential user to assess the quality of a digital map database.

Meter Basic unit of measurement of the metric system. Originally defined as one ten-millionth of the great-circle distance from the earth's equator to the earth's pole. Since 1983, defined as the distance that light travels in 1/299,792,458 of a second.

Metes-and-bounds surveying Method of land survey used to describe unsystematic land ownerships. Records the direction and length of each portion of the boundary line of the property in sequence, often using distinctive markers, such as trees of particular species.

Metonic cycle Period of 18.6 years during which the complete cycle of relationships among the sun, moon, and earth evolves. Time period observed for the establishment of an accurate mean elevation datum.

Metric system System of measurement originally defined by the French Academy of Sciences in 1791.

Mile *See* **Statute mile** and **Nautical mile.**

Mileage table (distance log) Table consisting of columns of mileage between major locations.

Miller cylindrical Map projection drawn on a cylinder showing the world in a rectangular format. Distorts the size of areas less than does the Mercator.

Miller's measure Measure of shape based on the ratio between the area of a region and the area of a circle with the same perimeter as the region.

Modeling In digital geographic information systems, logical sequence of analytical operations used to produce information, for example, regarding the alternative outcomes that would result from different courses of action.

Monument (survey) Plaque placed at survey control point to designate location and/or elevation.

MSS *See* **Multispectral scanner (MSS) system.**

Multiple line graph Diagram on which separate lines are used to represent changes in each of two or more variables, sometimes with separate scales.

Multispectral device Remote-sensing device that simultaneously records radiation in two or more separate wavelength bands. An example is Landsat.

Multispectral scanner (MSS) system Four-band, whisk-broom, remote-sensing system, with ground resolution of 79 by 79 meters, carried by Landsat.

Multivariate map Map that shows the relationships among the values of three or more variables at once.

Narrow-band device Remote-sensing device that uses a sensor that receives electromagnetic energy within a limited range of frequencies. An example is radar.

National Map Accuracy Standards U.S. standards for evaluation of accuracy of vertical and horizontal locations on large-scale maps.

Nautical chart Special map used for navigation at sea.

Nautical mile Unit of measurement used for measuring distances at sea. Former U. S. nautical mile (or geographical mile) measured 6080.2 feet (1853.24 meters) and British admiralty nautical mile equaled 6080 feet (1853.18 meters). International nautical mile, established in 1959, equals 1852 meters (6076.12 feet).

NAVSTAR Global Positioning System *See* **Global Positioning System (GPS).**

Nearest-neighbor analysis Point pattern measure based on comparison of the mean of the distance observed between each point and its nearest neighbor with the expected mean distance that would occur if the distribution were random.

Neatline Outline at which map detail terminates.

Negative In photography, image on transparent film in which objects with a high level of energy reflection are recorded as dark images and those that reflect little energy appear as transparent areas.

Network In general, set of geographic locations interconnected in a system by a number of routes. In graph theory, figure composed of points and lines.

Node (place) In graph theory, location.

Nominal areal symbol Map symbol that represents only the type or category and the areal extent of some phenomenon.

Nominal linear symbol Map symbol used to indicate only the type or category and the alignment of some phenomenon.

Nominal measurement Qualitative information that divides a group into subgroups, with no implication that one subgroup is more or less important than another.

Noncontiguous cartogram Area cartogram that retains shapes by separating regions from one another.

Nonproportional point symbol Map symbol used to indicate only the type or category and the point location of some phenomenon.

Nonspatial information system Information system that deals with data that have no locational component.

Nonsymmetric, nonplanar network Network with one-way connections in which a place where two or more routes (links) meet is not necessarily a connecting point (node).

Nonsymmetric, planar network Network with one-way connections in which every place where two or more routes (links) meet is a connecting point (node).

Normal orientation (of a projection surface) Commonly accepted standard arrangement of the projection surface relative to the globe. Normal orientations include the following: in cylindrical projections, the cylinder is tangent along the equator; in conic projections, the cone is tangent along a parallel, with its apex over the pole; and in azimuthal projections, the plane is tangent at the pole.

North Pole One of the earth's poles of rotation.

NOTAM *See Notice to Airmen (NOTAM).*

Notice to Airmen (NOTAM) Publication of the Federal Aviation Administration (FAA) that provides up-to-date information about changed aeronautical navigational aids.

Notice to Mariners (NTM) Weekly publication of the Marine Navigation Department of the National Imagery and Mapping Agency (NIMA) that provides information needed to update nautical charts. Local notices are also published by U.S. Coast Guard District Offices.

Null hypothesis Statement that there is no relationship between variables. If the null hypothesis can be rejected, some relationship exists.

Oblique aerial photograph Aerial photograph taken with the lens pointed at an angle away from the vertical. High obliques include the horizon; low obliques do not.

Oblique orientation (of a projection surface) Alignment of a projection surface at an angle somewhere between the normal and the transverse positions.

Observed frequencies, table of Used in chi-square analysis, shows actual relationships between two variables in the study area.

100-year flood Level of flooding that has one chance in 100 of being equaled or exceeded in any one-year period.

Opisometer *See* **Map measurer.**

Ordinal linear symbol Map symbol used to differentiate the relative importance of what they represent.

Ordinal measurement Quantitative information classified on the basis of comparative values only (greater than, less than).

Orientation (of a map) Alignment of the map relative to features on the earth's surface.

Orientation (of a projection surface) Arrangement of the projection surface relative to the globe. *See also* **Normal, Transverse,** and **Oblique orientations.**

Orienteering compass Type of hand-held, floating-needle, magnetic compass.

Orthodrome Direction of a great circle. Also known as the true azimuth.

Orthographic position Location of the projection light source at infinity.

Orthographic projection Map projection made with the light source positioned at infinity.

Orthographic view View drawn as though every point were seen from directly overhead, such as in a map.

Orthophoto Aerial photograph processed to remove the effect of planimetric shift. Has correct scale throughout.

Orthophotomap U.S. Geological Survey product, with the same scale (1:24,000), coverage (7.5 minutes of latitude and longitude), and locational grids (Universal Transverse Mercator and State Plane Coordinates) as a regular quadrangle map. Color-enhanced orthophoto image with added contour lines and feature names.

Orthophotoquad U.S. Geological Survey product, with the same scale (1:24,000), coverage (7.5 minutes of latitude and longitude), and locational grids (Universal Transverse Mercator and State Plane Coordinates) as a regular quadrangle map. Orthophoto image, printed in black and white, without contours.

Outline map Map that provides a framework for plotting information.

Output In computer-assisted cartography, production of data.

Overlay analysis As used in the computation of the coefficient of areal correspondence, mapping of two distributions at the same scale and placing of the outline of one over the outline of the other.

Overlaying In digital geographic information systems, digital equivalent of map-overlay analysis. Used for determining areas of overlap among categories on various data planes.

Panchromatic emulsion Black-and-white photographic emulsion that is sensitive to a range of wavelengths within the electromagnetic spectrum similar to that seen by the human eye.

Panchromatic film Film coated with an emulsion sensitive to visible wavelengths within the electromagnetic spectrum.

Parallax Difference in viewpoint of one eye versus the other, which results in a three-dimensional, stereoscopic image.

Parallel Set of locational lines used to specify location north or south of the equator. Parallels are true east-west lines, equally spaced between the equator and the poles and always parallel to one another, so any two parallels are always the same distance apart. Parallels and meridians are at right angles to one another.

Passive remote-sensing system Remote-sensing system that relies on the detection of naturally generated energy that is either reflected or emitted by an object. An example is aerial photography.

Path In graph theory, series of routes linking a series of places, or nodes.

Percent (of a slope) Vertical distance (rise) per 100 units of horizontal distance (run) on a slope.

Percentage scale In a graph, scale in which each unit of distance represents an equal percentage range.

Perceptual scaling Method of sizing proportional circles so that the symbols representing larger quantities are exaggerated in size, rather than drawn in absolute proportion. Designed to offset a common tendency to underestimate the values represented by larger circles.

Perspective view View of a scene from a single viewpoint, such as an aerial photograph.

Peters Equal-area projection promoted for use in thematic mapping, despite its shortcomings in terms of shapes.

Photographic-overlay analysis Method of map-overlay analysis, developed by Ian McHarg in the late 1960s, aimed at finding "the solution of maximum social utility" for land-use applications.

Photo interpretation Recognition and identification, using aerial photographs, of the natural and human-made features on the earth's surface.

Photomosaic Composite image formed by piecing adjoining aerial photographs together to cover a larger area.

Physical model Three-dimensional model of the earth's surface.

Physiographic diagram Pictorial representation of the earth's surface, from an elevated, oblique point of view, using stylized representations of the various types of relief features.

Pie chart Circular diagram that is subdivided into proportional segments.

Pixel In remote sensing, small portion of an image that contains information about a specified area on the ground.

Place (node) In graph theory, location.

Plan Detailed, large-scale map of a limited area.

Plane Flat surface used for azimuthal projections.

Plane of the ecliptic *See* **Ecliptic, plane of the.**

Plane rectangular grid (reference grid) System of lines superimposed on a map for locational purposes that does not take the curvature of the earth's surface into account.

Plane survey Survey of limited extent conducted without taking the earth's curvature into account. Contrast with **Geodetic control survey.**

Planimetric map Map that does not show relief features in measurable form.

Planimetric shift Variation in the planimetric location of objects in an aerial photograph due to differences in elevation. Results in scale variations within the photograph.

Plat Detailed map of a land subdivision.

Pocket transit Type of hand-held, floating-needle, magnetic compass with degrees printed in a counterclockwise direction around a fixed housing.

Point symbol Map symbol used to represent a feature that occurs at a single location (point) on the earth's surface.

Polar flattening Result of the polar diameter of the earth being less than the equatorial diameter, due to the rotation of the earth on its axis. Amounts to about 1 part in 298, which is very close to a sphere.

Polaris Northern Hemisphere star, almost exactly overhead at the North Pole, that provides an excellent point of reference for determining latitude.

Polar planimeter Instrument that measures areas on a map.

Poles of rotation Location of points at which the earth's axis meets the surface (North Pole and South Pole). Poles provide the starting point for the system of latitude and longitude.

Polygon method Method of areal measurement based on subdividing the area to be measured into regular geometric figures.

Population profile Graphic display of the age-sex distribution of a given region's population.

Population pyramid *See* **Population profile.**

Positive In photography, print or transparency, usually made from a photographic negative, in which the areas in the original scene that reflect high levels of energy are light in tone and those that reflect little energy are dark in tone.

Preprocessing techniques Computer-aided manipulations of original remotely sensed data. Typically done prior to release of the data to the user.

Prime meridian Starting line for measurements of east and west longitude. On most modern maps, the meridian of Greenwich, England.

Principal meridian North-south reference line used in a U.S. Public Land Survey zone.

Principal point Location on the ground at which an aerial camera is pointed. In a vertical photograph, coincides with the geometric center of the photograph. Located by joining the fiducial marks on the photograph.

Prism map Map drawn from an elevated oblique viewpoint that shows each region raised above the base level an amount proportional to the value, in that region, of the variable being mapped.

Profile Cross-sectional view through the terrain.

Propaganda Deliberate introduction of misleading information into map products.

Pseudoscopic effect In stereoscopic viewing of aerial photographs, effect that causes hills to look like valleys and valleys to look like hills. Created when the photographs are oriented with shadows falling away from the viewer.

Public Land Survey, U.S. (USPLS, systematic subdivision) Method of land survey, utilizing townships and ranges. Started in Ohio and ultimately extended throughout most of the western states and parts of the south. Originally known as the rectangular survey system.

Push-broom scanning In remote-sensing system, scanning along the satellite's direction of movement.

Quadrangle Individual map sheet laid out as a four-sided figure bounded by lines of latitude and longitude.

Quadrat analysis Point pattern measure based on a frequency count of points occurring within a superimposed grid (quadrats).

Qualitative information Nominal measurement that carries no implication of importance.

Quantitative areal symbol Map symbol used to indicate the magnitude of the phenomenon (quantitative information) occurring in a given region.

Quantitative linear symbol Map symbol used to indicate the alignment and magnitude (quantitative information) of the connection or flow that exists between locations on the earth's surface.

Radar Active remote-sensing technique involving the transmission, reflection, and detection of radio waves in the microwave wavelengths.

Radar image Remotely sensed image from a radar unit.

Radio altimeter Altimeter that provides direct information about the absolute altitude of an aircraft above the earth's surface, instead of above sea level.

Radio direction finding (RDF) Navigational, position-determination system that uses directional information obtained from radio signals.

Random pattern Pattern in which a phenomenon is equally likely to occur at any location and in which the position of any point is not affected by the position of any other point.

Range In U.S. Public Land Survey and Canada Land Survey System, 6-mile-wide, north-south zone. In navigation, alignment of two known objects. *See also* **Line of position.**

Range-graded symbol Set of symbols, such as circles, in which each size represents a particular range of values.

Raster digitizing In computer-assisted cartography, digitizing map features on the basis of their presence or absence in myriad tiny cells (rasters).

Raster output In computer-assisted cartography, image produced by creating a pattern of dots.

Raster plotter Computer output device that prints an image in the form of a pattern of dots.

Ratio measurement Quantitative information that provides the exact value of some phenomenon on a standard scale that has uniform intervals and a real (nonarbitrary) zero value. An example is distance in meters.

RDF *See* **Radio direction finding (RDF).**

Realistic figures Symbols drawn in recognizable form, used instead of geometric figures to represent data values.

Real map Any tangible map product that has a permanent form and can be directly viewed (also called *hard copy*).

Reclassification In digital geographic information systems, replacement of the initial category designations on a data plane with new values based on specified criteria.

Rectangular survey system *See* **Public Land Survey, U.S. (systematic subdivision).**

Reference grid *See* **Plane rectangular grid (reference grid).**

Remote sensing Gathering data by means of a sensor that is not in contact with the objects in the scene being observed. Includes aerial photographs, images from radar and other airborne sensors, and satellite images.

Remote-sensing system Active or passive system that gathers information by means of sensors that are not in contact with the object being detected.

Representative fraction (RF) scale Expression of map scale as a ratio or fraction.

Resection Method of determining one's location by reference to three known features on the ground.
RF *See* **Representative fraction (RF).**
Rhumb line (constant azimuth) Directional line that crosses each succeeding meridian at a constant angle.
Rise Vertical distance between two points on a slope.
Road atlas Bound volume containing highway and/or street maps.
Robinson Compromise world map projection designed to minimize visually disturbing distortions of many world projections.
Roture Lot allocated to an individual farmer in the seigneurial land ownership system of French Canada.
Route (link) In graph theory, connecting line between two places, or nodes.
Routed linear cartogram Maplike diagram that shows characteristics of specific paths from one point to another, such as average travel time.
Route map Map that provides information regarding bus, train, subway, or airline routes.
Run Horizontal distance between two points on a slope.
S/A *See* **Selective availability (S/A).**
Sailing chart Nautical chart for offshore sailing between distant coastal ports and for approaching the coast from the open ocean.
Sanborn Map Fire insurance map produced by the Sanborn Map Company.
Satellite image Remotely sensed image captured by instruments taken aloft in a spacecraft, with or without a human crew.
Scale Ratio of the distance between two points on a map and the earth distance between the same two points.
Scale conversion Transforming a map scale from one form to anther, such as from graphic to representative fraction.
Scanner Passive remote-sensing device that detects and records energy reflected or emitted from objects within its range. May be mounted on aircraft or space platforms.
Scatter diagram Graph used to display the relationship between two variables. Consists of a set of points plotted on the basis of paired, x and y values.
Secant projection Projection in which the projection surface intersects the globe.
Section In U.S. Public Land Survey and Canada Land Survey System, 1-square-mile (640-acre) subdivision of a township.
Sectional charts Aeronautical chart series, based on the Lambert conformal conic projection, with a scale of 1:500,000.
Seigneur Nobleman, yeoman, or farmer who received a seigneurial land grant in exchange for promoting settlement and development.
Seigneury Seventeenth century French land grant in the region that is now Quebec and the Maritime provinces except Newfoundland. Granted for the purpose of promoting settlement and development.
Selection Means of map generalization in which the more important features in an area are retained and the less important ones are eliminated.
Selective Availability (S/A) Deliberate degradation by the Department of Defense of the accuracy of GPS positions.
Semilogarithmic graph Line graph on which a logarithmic scale is used on one axis (usually the y axis) and an arithmetic scale is used on the other. The shapes of lines plotted on the graph reflect rates of change in the variable plotted.
Semilog graph *See* **Semilogarithmic graph.**
Sexagesimal system Coordinate system in which the circle of the globe is divided into 360 degrees. Each degree is divided into 60 minutes and each minute into 60 seconds.
Shift *See* **Locational shift and size exaggeration.**
Signal source Origin of a signal that is detected by a remote-sensing system.
Signature Characteristic electromagnetic radiation pattern associated with a source object.
Silhouette graph Line graph on which the area below the curve is shaded or colored to give a silhouette effect.
Simplification Means of map generalization in which the shapes of the features retained on the map are presented in simplified form.
Sinusoidal projection An equal-area (equivalent) projection.
Slope A measure of the vertical difference in the elevation of a surface at two different points, in relation to the horizontal distance between the same points.
Slope error Variation in horizontal distance introduced by surface undulations. (The surface distance between two objects is longer than the map distance, except where the ground surface is perfectly flat.)
Slope fraction Ratio between rise and run, written as a fraction.
Slope ratio Ratio between rise and run, stated in the same units and with the rise reduced to 1.
Slope-zone map Map showing, by means of coloring or shading, areas of relatively similar slopes.
Small circle Circle created when a plane passing through the earth divides the earth into two unequal portions. Called *line of latitude,* or *parallel,* when parallel to the equator.
Small scale Comparative term for a map in which a generalized representation of a relatively large area of the earth's surface is shown in a relatively small space.
Smoothing *See* **Simplification.**
Sounding Water depth at a given point, printed on a hydrographic chart.
Spacing dividers Eleven-point dividers used for measuring interpolated locations.
SPC *See* **State Plane Coordinate system.**
Spectral range Range of wavelengths used by a remote-sensing device (broad-band, narrow-band, or multispectral).
Sphere Regular solid formed by the rotation of a circle on one of its diameters. Less-satisfactory approximation of the shape of the earth than an ellipsoid.
SPOT Push-broom remote-sensing system that carries sensors that operate in multispectral, monospectral, or combined modes. Ground resolution is 10 by 10 meters in monospectral mode and 20 by 20 meters in multispectral mode.
Spot height Point symbol used to indicate the location to which an elevation value refers.
Standard line The line of tangency between a projection surface and the surface of the globe. Along the standard line, the map has no distortion.

Standard parallel In the U.S. Public Land Survey, east-west reference line established at 24-mile spacing. In normally oriented conic or cylindrical map projections, parallel of intersection of the projection surface with the globe.

State Plane Coordinate (SPC) system Plane rectangular coordinate system that is individually applied to zones in each state in the United States.

Statistical report In digital geographic information systems, production of statistical information from the database.

Statistical surface Imaginary surface representing differences in the value of a nonphysical variable over a region.

Statute mile Unit of measurement in the U.S. Customary system. Defined as 5280 feet (1069.35 meters).

Stereographic position Location of the projection light source at the antipode.

Stereographic projection Map projection made with the light source positioned at the antipode (the point exactly opposite the point of tangency of the projection surface).

Stereomodel Apparent three-dimensional view of the terrain, as viewed in a stereoplotter.

Stereo pair *See* **Stereoscopic pair (stereo pair).**

Stereoscope Simple instrument that permits the simultaneous viewing of a stereo pair of aerial photographs so that the desired three-dimensional effect is achieved.

Stereoscopic image Three-dimensional image (especially of aerial photographs) seen through a viewer that allows one eye to view one photograph and the other eye to see another, taken from a slightly different location.

Stereoscopic pair (stereo pair) Two adjacent aerial photographs showing the same scene from slightly different points of view. Used for stereoscopic (three-dimensional) viewing.

Stereoscopic viewing Method of looking at a stereo pair of aerial photographs through a stereoscope. Results in an apparently three-dimensional view of the scene.

Stilling basin Basin cut off from the direct influence of wind and waves; used in the establishment of mean sea level.

Strahler method Method of assigning hierarchic order to the segments (links) in a branching network (tree).

Street map Map showing the road layout within a city.

Strike In geology, compass bearing of the intersection of the surface of a bed or stratum with an imaginary horizontal plane.

Strip method Method of areal measurement using a superimposed series of parallel lines.

Summer solstice Moment when the noon sun is directly overhead on the Tropic of Cancer.

Supplementary contour Additional contour, usually dashed or dotted, drawn between intermediate and index contours at intervals that are some regular fraction of the basic contour interval. Used to provide additional information in areas of flat terrain.

Surface Variation in elevation representing differences in the value of a (physical or statistical) variable over a region.

Surface graph Diagram on which values are expressed as spaces, or bands, between the lines on the graph.

Symmetric, nonplanar network Network with two-way connections, in which a place where two or more routes (links) meet is not necessarily a connecting point (node).

Symmetric, planar network Network with two-way connections, in which every place where two or more routes (links) meet is a connecting point (node).

Synoptic hours Times at which weather information is recorded by the worldwide network of reporting stations: 0000, 0600, 1200, and 1800 Universal Coordinated Time (formerly Greenwich Mean Time).

Systematic subdivision *See* **Public Land Survey, U.S. (USPLS systematic subdivision).**

Table of expected frequencies *See* **Expected frequencies, table of.**

Table of observed frequencies *See* **Observed frequencies, table of.**

Tabular index Alphabetically or numerically ordered list of the sheet numbers or names of the sheets of a map series that are published or are planned for publication. May be annotated to indicate the sheets that are in a map collection.

Tactual maps Maps designed for use by visually impaired persons, using the sense of touch.

Tangency (of a projection surface) Point or line at which the projection surface and the globe touch.

Television map Map used in conjunction with television broadcasts. Often highly generalized.

Thematic information extraction Image enhancement of remotely sensed data that provides information about conditions on the earth's surface.

Thematic map Map that shows information about a special topic, often statistical in nature.

Thematic mapper (TM) system Seven-band, whisk-broom, remote-sensing system, with a resolution of 30 by 30 meters, carried by Landsats 4 and 5.

Tick *See* **Grid tick.**

TIGER *See* **Topologically Integrated Geographic Encoding and Reference System (TIGER).**

Time zones Twenty-four zones, each spanning 15 degrees of longitude, into which the earth is divided. All localities within a zone keep the same time as the zone's central (time) meridian. Zonal boundaries are adjusted to follow political boundaries or to fall in less-populated areas.

T-in-O map Medieval map that was not a literal representation of the true size and shape of the earth but a stylized presentation used to support prevalent religious views.

Title Explanation of a map's subject, time period, and other aspects of its contents.

TM *See* **Thematic Mapper (TM) system.**

Topographic map Map that shows the shape and elevation of the terrain.

Topography Configuration of the surface of the earth.

Topological distance In graph theory, length of the shortest path (number of routes or links) between two places, or nodes.

Topologically Integrated Geographic Encoding and Reference System (TIGER) U.S. Bureau of the Census digital geographic and cartographic data file being developed for the 1990 Census of Population and future census operations, supplanting the GBF/DIME system.

Topological positions Abstract relationships between the elements of a graph.

Toponym Geographical name.

Township Six-mile-wide, east-west zone in the U.S. Public Land Survey and Canada Land Survey System. Also, 36-square-mile area formed by the intersection of an east-west township and a north-south range.

Township plat Detailed map of the land-ownership boundaries within a U.S. Public Land Survey township.

Trafficability Ability of an area to support vehicles of a specific type.

Transliteration Conversion of characters in a source language into one or more Roman letters.

Transparency *See* **Positive.**

Transverse orientation (of a projection surface) Arrangement of the projection surface (relative to the globe) so that it is turned 90 degrees from normal.

Tree (branching network) In graph theory, network that does not contain circuits (loops).

Trend line Straight line superimposed over the cloud of dots on a scatter diagram representing the direction and strength of the relationship between the variables.

Triangle of error Triangle formed by failure of back-azimuth lines to intersect exactly during determination of location by resection.

Triangular graph Graphic display plotted on an equilateral triangle to show the mixture of three components of some total quantity.

Triangulation network Surveyed network established by measuring the angles between the baseline and additional observation points.

Trilateration network Surveyed network that serves the same purpose as a triangulation network but based on measurement of distances, instead of angles.

True azimuth Direction of a great circle. Also known as an *orthodrome.*

True north (geographical north) Direction to the geographical north pole, along a line of longitude.

Typographical information Titles, legends, names, and notes printed on maps.

Unclassed choropleth map Type of choropleth map that utilizes a separate symbol for each value represented (that is, the values are not classified or grouped).

Uncontrolled photomosaic Mosaic of aerial photos assembled without reference to ground control.

Underwater contour Contour that represents the land surface that was exposed before an area was flooded. Differs from **Depth curve.**

Uniform pattern Pattern in which every point is as far from all of its neighbors as possible.

Univariate map Map that shows the distribution of the values of a single variable.

Universal Polar Stereographic (UPS) system Plane rectangular grid system that covers the polar caps, north of 84° North and south of 80° South. Used in conjunction with the Universal Transverse Mercator (UTM) system, which covers the rest of the world. Civilian system that uses only numbers, while the U.S. Military Grid Reference System uses number/letter combinations to designate grid zones.

Universal Transverse Mercator (UTM) system Plane rectangular grid system that covers the earth's surface between 80° South and 84° North. Used in conjunction with the Universal Polar Stereographic (UPS) system, which covers the polar caps. Civilian system that uses only numbers, while the U.S. Military Grid Reference System uses number/letter combinations to designate grid zones.

Unsystematic subdivision Land-ownership pattern developed in an unplanned manner, using, for example, property boundaries established on the basis of settlers' claims.

UPS *See* **Universal Polar Stereographic (UPS) system.**

U.S. Customary Units Units of measurement, such as feet, inches, and miles, based on the units of the older English system.

U.S. National Map Accuracy Standards *See* **National Map Accuracy Standards.**

U.S. nautical mile *See* **Nautical mile.**

USPLS *See* **Public Land Survey, U.S. (USPLS, systematic subdivision).**

U.S. survey foot Defined as 1:0.30480061 meter.

UTM *See* **Universal Transverse Mercator (UTM) system.**

Value-by-area cartogram *See* **Area cartogram (value-by-area cartogram).**

Variance As used in quadrat analysis, compares the number of points in each grid cell with the average number of points over all of the cells.

Vector digitizing In computer-assisted cartography, digitizing map features on the basis of the coordinates of points or, in the case of linear features or the outlines of areas, a series of points.

Vertical accuracy In National Map Accuracy Standards, accuracy of the designated elevation of well-defined points on the map.

Vertical aerial photograph Aerial photograph taken with the camera pointed as nearly perpendicular to the earth's surface as possible.

Vertical exaggeration Ratio between the vertical earth distance represented by a unit of map distance and the horizontal earth distance represented by the same measurement.

Virtual map Image that can be directly viewed but is not permanent, such as a map image on a computer monitor, a mental map image, or information gathered by researchers in the field or by remote-sensing methods.

Visible spectrum Portion of the electromagnetic spectrum detected by the human eye.

Visual flight rules Simplest type of flight rules, governing flights under favorable weather conditions, in which visual references to landmarks can be used to reliably control the operation of the plane.

Volumetric diagram Apparently solid, three-dimensional figure, drawn so that the volume of each figure is proportional to the value it represents.

WAC *See* **World Aeronautical Chart (WAC) series.**

Weather map Standard weather map used for forecasting purposes. Depicts a wide range of information regarding the status of the weather at a specific time.

Whisk-broom scanning In the remote-sensing system, oscillating mirror scanning that sweeps across the satellite's direction of movement.

Wind rose Circular graph that shows the average frequency and direction of wind at one location over a given period of time.

Winkel Tripel Compromise world map projection that minimize visually disturbing distortions.

Word statement (of map scale) Statement of scale ratio in words.

World Aeronautical Chart (WAC) series Aeronautical chart series, with a 1:1,000,000 scale, based on the Lambert conformal conic projection.

World Data Bank I *See* **World Data Banks.**

World Data Banks Two frequently used worldwide databases that contain data about coastlines, national boundaries, and place names. World Data Bank II, which is more detailed, also contains data about major rivers and national, state, and province boundaries.

World Data Bank II *See* **World Data Banks.**

World Geographic Reference System (GEOREF) System, primarily used for military air operations, that designates latitude and longitude by a simple set of letters and numbers.

Yule's *Q* Correction factor that can be applied to two-column by two-column chi-square table to eliminate the influence of the number of cases involved.

Index

D

E

F

G

H

Q

R

S

T

U

V

W

Y